OF THE ELEMENTS

Atomic weights are based on carbon-12. Atomic weights in parentheses indicate the most stable or best-known isotope. Slight disagreement exists as to the exact electronic configuration of several of the high-atomic-number elements. Names and symbols for elements 104, 105, and 106 are unofficial.

								2 Helium **He** 4.00260
			IIIA	IVA	VA	VIA	VIIA	
			5 Boron **B** 10.81	6 Carbon **C** 12.011	7 Nitrogen **N** 14.0067	8 Oxygen **O** 15.9994	9 Fluorine **F** 18.99840	10 Neon **Ne** 20.179
	IB	IIB	13 Aluminum **Al** 26.98154	14 Silicon **Si** 28.086	15 Phosphorus **P** 30.97376	16 Sulfur **S** 32.06	17 Chlorine **Cl** 35.453	18 Argon **Ar** 39.948
28 Nickel **Ni** 58.71	29 Copper **Cu** 63.546	30 Zinc **Zn** 65.38	31 Gallium **Ga** 69.72	32 Germanium **Ge** 72.59	33 Arsenic **As** 74.9216	34 Selenium **Se** 78.96	35 Bromine **Br** 79.904	36 Krypton **Kr** 83.80
46 Palladium **Pd** 106.4	47 Silver **Ag** 107.868	48 Cadmium **Cd** 112.40	49 Indium **In** 114.82	50 Tin **Sn** 118.69	51 Antimony **Sb** 121.75	52 Tellurium **Te** 127.60	53 Iodine **I** 126.9045	54 Xenon **Xe** 131.30
78 Platinum **Pt** 195.09	79 Gold **Au** 196.9665	80 Mercury **Hg** 200.59	81 Thallium **Tl** 204.37	82 Lead **Pb** 207.2	83 Bismuth **Bi** 208.9804	84 Polonium **Po** (210)[a]	85 Astatine **At** (210)[a]	86 Radon **Rn** (222)[a]

Inner transition elements

63 Europium **Eu** 151.96	64 Gadolinium **Gd** 157.25	65 Terbium **Tb** 158.9254	66 Dysprosium **Dy** 162.50	67 Holmium **Ho** 164.9304	68 Erbium **Er** 167.26	69 Thulium **Tm** 168.9342	70 Ytterbium **Yb** 173.04	71 Lutetium **Lu** 174.97
95 Americium **Am** (243)[a]	96 Curium **Cm** (247)[a]	97 Berkelium **Bk** (249)[a]	98 Californium **Cf** (251)[a]	99 Einsteinium **Es** (254)[a]	100 Fermium **Fm** (253)[a]	101 Mendelevium **Md** (256)[a]	102 Nobelium **No** (254)[a]	103 Lawrencium **Lr** (257)[a]

FOUNDATIONS OF COLLEGE CHEMISTRY
5th EDITION

THE BROOKS/COLE SERIES IN CHEMISTRY

Foundations of College Chemistry, FIFTH EDITION, *by Morris Hein*

Foundations of College Chemistry in the Laboratory, FIFTH EDITION, *by Morris Hein, Leo R. Best, and Robert L. Miner*

Study Guide for Foundations of College Chemistry, FIFTH EDITION, *by Peter Scott*

College Chemistry: An Introduction to Inorganic, Organic, and Biochemistry, SECOND EDITION, *by Morris Hein and Leo R. Best*

College Chemistry in the Laboratory, SECOND EDITION, *by Morris Hein, Leo R. Best, Robert L. Miner, and Jim Ritchey*

Study Guide for College Chemistry: An Introduction to Inorganic, Organic, and Biochemistry, SECOND EDITION, *by Peter Scott*

Foundations of College Chemistry: The Alternate Edition, *by Morris Hein*

Study Guide for Foundations of College Chemistry: The Alternate Edition, *by Peter Scott*

A Laboratory Project in Modern Coordination Chemistry, *by Lawrence C. Nathan*

*NOTE: Foundations of College Chemistry in the Laboratory, FIFTH EDITION, *is also compatible with* Foundations of College Chemistry: The Alternate Edition.

FOUNDATIONS OF COLLEGE CHEMISTRY

MORRIS HEIN
MOUNT SAN ANTONIO COLLEGE

5th EDITION

Brooks/Cole Publishing Company

MONTEREY, CALIFORNIA

Printed in the United States of America

10 9 8 7 6 5 4 3 2 1

Library of Congress Cataloging in Publication Data

Hein, Morris.
   Foundations of college chemistry.

   (The Brooks/Cole series in chemistry)
   Includes index.
   1. Chemistry.   I. Title.   II. Series.
QD33.H45   1982   540   81-18003
ISBN 0-8185-0476-5   AACR2

Acquisition Editor: *Michael V. Needham*
Project Development Editor: *James F. Leisy, Jr.*
Manuscript Editor: *Gloria Joyce*
Production Staff: *Phyllis Niklas, Joan Marsh*
Interior and Cover Design: *Stan Rice*
Illustrations: *Cyndie Jo Clark, Felix Cooper*
Typesetting: *Typothetae*

*To my colleagues at Mt. San Antonio College
for the many wonderful years of our association:*

*George Andreville, Leo Best, Grant Cooper,
Robert Miner, Lynn Pendleton, Patricia Perez,
Harley Reifsnyder*

Preface

The original purpose of *Foundations of College Chemistry* was to instruct students in the basic concepts of chemistry so that they would be qualified to enter courses in general college chemistry. This is still the primary purpose of the fifth edition. The text remains designed for a one-semester or a two-quarter course in beginning or preparatory chemistry. The overall sequence of topics and the level of the material remain approximately the same as in the previous edition.

However, in planning for this edition a national survey was conducted to obtain suggestions from the many instructors using the text. The changes made in accordance with these suggestions have resulted in a text that is easier to use than any of the previous editions. For example:

- A list of achievement goals is given at the beginning of each chapter to serve as a guide for students.
- Since many students using this text need to develop computational skills early in the course, Chapter 2 gives a detailed discussion of significant figures, rounding off numbers, scientific notation, and determining significant figures in calculations.
- Discussion of problem solving has been expanded.
- All chapter-end problems are new or have been revised. More drill problems are included. A few more difficult and challenging problems (marked with asterisks) have been added. Answers to all mathematical problems are given in Appendix VI.
- Five review exercise sections, following Chapters 4, 8, 12, 16, and 20, have been added to aid student self-evaluation.
- A comprehensive glossary has been added (Appendix V).
- More emphasis has been given to using the mole in calculations.
- An expanded treatment of how to write Lewis-dot structures has been added.
- Quantum mechanics terminology has been deleted in discussion of atomic structure. However, the use of energy levels and energy sublevels of electrons has been retained.

- A two-color format is used to emphasize noteworthy aspects of the contents and to highlight illustrations.
- Each new term is identified by boldface type where it is defined and is also printed in color in the margin.

In addition to these changes, considerable refinements have been made throughout the book to further improve the readability and the clarity of the material. Because *Foundations of College Chemistry* is written for students who have a limited science background, the material is developed on the assumption that the student has not had a previous course in chemistry. Accordingly, one of the major goals is to present the subject matter in a well-organized and easily understandable fashion, making the book one that students can read, understand, and study by themselves.

Although this edition contains more pages than the previous one, the additional pages are due primarily to expanded discussions, more sample problems, more detail in example problems, and new features such as the lists of achievement goals, glossary, and review exercises. No additional topics have been included.

Lastly, it is not usually possible to cover all this material in one semester. I have provided some chapters in order to accommodate the variety of courses this text is used in. The abundance of material should provide some flexibility for the individual teacher and student.

Many people have been involved in this revision of the *Foundations of College Chemistry* teaching package. I would like to acknowledge the suggestions given by the reviewers of this edition: Alice S. Corey of Pasadena City College, Elisheva Goldstein of California State Polytechnic University, Pomona, Arthur H. Hayes of Santa Ana College, and Paul Lauren of Suffolk County College. I want to give special thanks to Robert L. Miner of Mt. San Antonio College for his assistance with the end-of-chapter questions and problems; to Lynn Pendleton of Mt. San Antonio College for preparing the solutions manual; and to Peter C. Scott of Linn-Benton Community College, who prepared the student study guide. I also want to thank Leo R. Best, my coauthor of *College Chemistry,* and all my colleagues at Mt. San Antonio College for their helpful suggestions.

No textbook can be completed without the untiring work of many professionals. I owe a great deal to James F. Leisy, Jr., and to Michael V. Needham, Chemistry Editors at Brooks/Cole; to Phyllis Niklas, an exceptionally capable production manager; to Gloria Joyce, a superb manuscript editor; and to the entire staff at Brooks/Cole Publishing Company.

Last, but certainly not least, I am forever grateful to my dear wife, Edna, for her patience with me throughout the project.

M. H.
Walnut, California

Contents

4 Elements and Compounds   61

Review Exercises for Chapters 1–4   83

5 Atomic Theory and Structure   87

6 The Periodic Arrangement of the Elements   121

7 Chemical Bonds—The Formation of Compounds from Atoms   137

8 Nomenclature of Inorganic Compounds   173

Review Exercises for Chapters 5–8   189

16 Chemical Equilibrium   383

Review Exercises for Chapters 13–16   414

17 Oxidation–Reduction   423

18 Radioactivity and Nuclear Chemistry   447

19 Organic Chemistry 475

20 Introduction to Biochemistry 537

Review Exercises for Chapters 17–20 570

FOUNDATIONS OF COLLEGE CHEMISTRY
5th EDITION

1

Introduction

1.1 The Nature of Chemistry

What is chemistry? A popular dictionary gives this definition: Chemistry is the science of the composition, structure, properties, and reactions of matter, especially of atomic and molecular systems. Another and somewhat simpler dictionary definition is: Chemistry is the science dealing with the composition of substances and the transformations that they undergo. Neither of these definitions is entirely adequate. Chemistry, along with the closely related science of physics, is a fundamental branch of science. Chemistry is also closely related to biology—not only because living organisms are made of material substances, but because life itself is essentially a very complicated system of interrelated chemical processes.

The scope of chemistry, as implied by the definitions just quoted, is extremely broad—it includes the whole universe and everything, animate and inanimate, to be found in the universe. Chemistry is concerned not only with the composition and changes of composition of matter, but also with the energy and energy changes associated with matter. Through chemistry we seek to learn and to understand the general principles that govern the behavior of all matter.

The chemist, like other scientists, observes nature at work and attempts to unlock its secrets: What makes a rose red? Why is sugar sweet? Why is water wet? Why is carbon monoxide poisonous? Why do people wither with age? Problems such as these—some of which have been solved, some of which are still to be solved—are part of what we call chemistry.

A chemist may interpret natural phenomena, devise experiments that will reveal the composition and structure of complex substances, study methods for improving natural processes, or, sometimes, synthesize substances unknown in nature. Ultimately, the efforts of successful chemists advance the frontiers of knowledge and at the same time contribute to the well-being of humanity. Chemistry helps us to understand nature; however, one need not be a professional chemist or scientist to enjoy natural phenomena. Nature and its beauty, its simplicity within complexity, are for all to appreciate.

The body of chemical knowledge is so vast that no one can hope to master it all even in a lifetime of study. However, many of the basic concepts can be learned in a relatively short period of time. These basic concepts have become part of the education required for many professionals, including agriculturists, biologists, dental hygienists, dentists, medical technologists, microbiologists, nurses, nutritionists, pharmacists, physicians, and veterinarians, to name a few.

1.2 History of Chemistry

From the earliest times, people have practiced empirical chemistry. Ancient civilizations were practicing the art of chemistry in such processes as wine-making, glass-making, pottery, dyeing, and elementary metallurgy. The early Egyptians, for example, had considerable knowledge of certain chemical processes. Excavations into ancient tombs dated about 3000 B.C. have uncovered workings of gold, silver, copper, and iron, pottery from clay, glass beads, beautiful dyes and paints, as well as bodies of Egyptian rulers in unbelievably well-preserved states. Many other cultures made significant developments in chemistry. However, all these developments were empirical; that is, they were achieved by trial and error and did not rest upon any valid theory of matter.

Philosophical ideas relating to the properties of matter (chemistry) did not develop as early as those relating to astronomy and mathematics. The ancient Greek philosophers made great strides in philosophical speculation concerning materialistic ideas about chemistry. They led the way to placing chemistry on a highly intellectual, scientific basis. They first introduced the concepts of elements, atoms, shapes of atoms, chemical combinations, and so on. The Greeks believed that there were four elements—earth, air, fire, and water—and that all matter was derived from these elements. The Greek philosophers had keen minds and perhaps came very close to establishing chemistry on a sound basis similar to the one that was to develop about 2000 years later. The main shortcoming of the Greek approach to scientific work was a failure to carry out systematic experimentation.

The Greek civilization declined and was succeeded by the Roman civilization. The Romans were outstanding in military, political, and economic affairs. They continued to practice empirical chemical arts such as metallurgy, enameling, glass-making, and pottery-making, but they did little to advance new and theoretical knowledge. Eventually, the Roman civilization declined and was succeeded in Europe by the Dark Ages. During this period, European civilization and learning were at a very low ebb.

In the Middle East and in North Africa knowledge did not decline during the Dark Ages as it did in Western Europe. During this period Arabic cultures made contributions that were of great value to the later development of modern chemistry. In particular, the Arabic number system, including the use of zero, gained acceptance; the branch of mathematics known as *algebra* was developed; and alchemy, a sort of pseudochemistry, was practiced extensively.

One of the more interesting periods in the history of chemistry was that of the alchemists (A.D. 500–1600). People have long had a lust for gold, and in those days gold was considered the ultimate, most perfect metal formed in nature. The principal goals of the alchemists were to find a method of prolonging human life indefinitely and to change the base metals—such as iron, zinc, and copper—into gold. They searched for a universal solvent to transmute base metals into gold and for the "philosopher's stone" to rid the body of all diseases and to renew life. In the course of their labors, they learned a great deal of chemistry. Unfortunately, much of their work was done secretly because of the mysticism that shrouded their activity, and very few records remain.

Although the alchemists were not guided by sound theoretical reasoning and were clearly not in the intellectual class of the Greek philosophers, they did something that the philosophers had not considered worthwhile. They subjected various materials to prescribed treatments under what might be loosely described as laboratory methods. These manipulations, carried out in alchemical laboratories, not only uncovered many facts of nature but paved the way for the systematic experimentation that is characteristic of modern science.

Alchemy began to decline in the 16th century when Paracelsus (1493–1541), a Swiss physician and outspoken revolutionary leader in chemistry, strongly advocated that the objectives of chemistry be directed toward the needs of medicine and the curing of human ailments. He openly condemned the mercenary efforts of alchemists to convert cheaper metals to gold.

But the real beginning of modern science can be traced to astronomy during the Renaissance. Nicolaus Copernicus (1473–1543), a Polish astronomer, succeeded in upsetting the generally accepted belief in a geocentric universe. Although not all the Greek philosophers had believed that the sun and the stars revolved about the earth, the geocentric concept had come to be accepted without question. The heliocentric (sun-centered) universe concept of Copernicus was based on direct astronomical observation and represented a radical departure from the concepts handed down from Greek and Roman times. The ideas of Copernicus and the invention of the telescope stimulated additional work in astronomy. This work, especially that of Galileo Galilei (1564–1642) and Johannes Kepler (1571–1630), led directly to a rational explanation of the general laws of motion by Sir Isaac Newton (1642–1727) from about 1665 to 1685.

Modern chemistry was slower to develop than astronomy and physics; it began in the 17th and 18th centuries when Joseph Priestley (1733–1804), who discovered oxygen in 1774, and Robert Boyle (1627–1691) began to record and publish the results of their experiments and to discuss their theories openly. Boyle, who has been called the father of modern chemistry, was one of the first to practice chemistry as a true science. He believed in the experimental method. In his most important book, *The Sceptical Chymist,* he clearly distinguished between an element and a compound or mixture. Boyle is best known today for the gas law that bears his name. A French chemist, Antoine Lavoisier (1743–1794), placed the science on a firm foundation with experiments in which he used a chemical balance to make quantitative measurements of the weights of substances involved in chemical reactions.

The use of the chemical balance by Lavoisier and others later in the 18th century was almost as revolutionary in chemistry as the use of the telescope had been in astronomy. Thereafter, chemistry was a highly quantitative experimental science. Lavoisier also contributed greatly to the organization of chemical data, to chemical nomenclature, and to the establishment of the Law of Conservation of Mass in chemical changes. During the period from 1803 to 1810, John Dalton (1766–1844), an English schoolteacher, advanced his atomic theory. This theory (see Section 5.2) placed the atomistic concept of matter on a valid rational basis. It remains today as a tremendously important general concept of modern science.

Since the time of Dalton, knowledge of chemistry has advanced in great strides, with the most rapid advancement occurring at the end of the 19th century and during the 20th century. Especially outstanding achievements have been made in determining the structure of the atom, understanding the biochemical fundamentals of life, developing chemical technology, and mass-producing chemicals and related products.

1.3 The Branches of Chemistry

Chemistry may be broadly classified into two main branches: *organic* chemistry and *inorganic* chemistry. Organic chemistry is concerned with compounds containing the element carbon. The term *organic* was originally derived from the chemistry of living organisms—plants and animals. Inorganic chemistry deals with all the other elements as well as with some carbon compounds.

Other subdivisions of chemistry, such as analytical chemistry, physical chemistry, biochemistry, electrochemistry, geochemistry, and radiochemistry, may be considered specialized fields of, or auxiliary fields to, the two main branches.

Chemical engineering is the branch of engineering that deals with the development, design, and operation of chemical processes. A chemical engineer generally begins with a chemist's laboratory-scale process and develops it into an industrial-scale operation.

1.4 Relationship of Chemistry to Other Sciences and Industry

Besides being a science in its own right, chemistry is the servant of other sciences and industry. Chemical principles contribute to the study of physics, biology, agriculture, engineering, medicine, space research, oceanography, and many other sciences. Chemistry and physics are overlapping sciences, since both are based on the properties and behavior of matter. Biological processes are chemical in nature. The metabolism of food to provide energy to living organisms is a chemical process. Knowledge of the molecular structure of proteins, hormones, enzymes, and nucleic acids is assisting biologists in their investigation of the composition, development, and reproduction of living cells.

Chemistry is playing an important role in alleviating the growing shortage of food in the world. Agricultural production has been increased with the use of chemical fertilizers, pesticides, and improved varieties of seeds. Chemical refrigerants make possible the frozen food industry, which preserves large amounts of food that might otherwise spoil. Chemistry is also producing synthetic nutrients, but much remains to be done as the world population multiplies with respect to the land available for cultivation. Expanding energy needs have brought about difficult environmental problems in the form of air and water

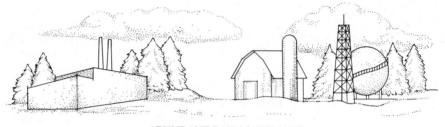

ABUNDANT RAW MATERIALS
from mine, forest, sea, air, farm, oil, brine, and gas wells

THE CHEMICAL INDUSTRY
in more than 13,500 plants in the United States
converts these raw materials into more than 50,000
CHEMICALS
such as acids and alkalies, salts, organic compounds,
solvents, compressed gases, pigments, and dyes,
which are used

BY THE CHEMICAL INDUSTRY ITSELF
To Produce

Cosmetics
Detergents and soap
Drugs and medicines
Dyes and inks
Explosives
Fertilizers
Paints
Pesticides
Plastic materials
Sanitizing chemicals
Synthetic fibers
Synthetic rubber
And many others

BY OTHER INDUSTRIES

In the Production of

Durable Goods

Aircraft and equipment
Building materials
Electrical equipment
Hardware
Machinery
Metal products
Motor vehicles and equipment
Other products of
metal, glass, paper, and wood

Nondurable Goods

Beverages
Food products
Leather and leather products
Packaging
Paper and paper products
Petroleum and coal products
Rubber products
Textiles

THE ULTIMATE MARKET
(Fundamental human needs)

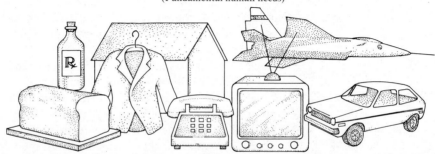

Health, food, clothing, shelter, transportation, communication, defense, and other needs

Figure 1.1 Broad scope of the chemical industry today. *(Courtesy Chemical Manufacturers Association.)*

pollution. Chemists as well as other scientists are working diligently to alleviate these problems.

Advances in medicine and chemotherapy, through the development of new drugs, have contributed to prolonged life and the relief of human suffering. More than 90% of the drugs and pharmaceuticals being used in the United States today have been developed commercially within the past 45 years. The entire plastics and polymer industry, unknown 55 years ago, has revolutionized the packaging and textile industries and is producing more durable and useful construction materials. Energy derived from chemical processes is used universally for heating, lighting, and transportation. There is virtually no industry that is not dependent on chemicals—for example, the petroleum, steel, rubber, pharmaceutical, electronic, transportation, cosmetic, garment, aircraft, and television industries (the list could go on and on). Figure 1.1 illustrates the conversion of natural resources by the chemical industry into useful products for commerce, industry, and human needs.

1.5 Scientific Method

Chemistry as a science or field of knowledge is concerned with ideas and concepts relating to the behavior of matter. Although these ideas and concepts are abstract, their application has had an extraordinarily concrete impact on human culture. This impact is due to modern technology, which may be said to have begun about 200 years ago and has continued to grow at an accelerating rate.

There is a very important difference between the science of chemistry and technology. The science represents a basic body of knowledge; technology represents the physical application of this knowledge to the real world in which we live.

Why has the science of chemistry and its associated technology flourished so abundantly in the last two centuries? Is it because we are growing more intelligent? No, there is absolutely no reason to believe that the general level of human intelligence is any higher today than it was 1000 years ago in the Dark Ages. The use of the scientific method is usually credited with being the most important single factor in the amazing development of chemistry and technology. Although complete agreement is lacking on exactly what is meant by "using the scientific method," the general approach is as follows:

1. Collect facts or data that are relevant to the problem or question at hand. This is usually done by planned experimentation.
2. Analyze the data to find trends (regularities) that are pertinent to the problem. Formulate a hypothesis that will account for the data accumulated and that can be tested by further experimentation.
3. Plan and do additional experiments to test the hypothesis. Such experiments extend beyond the range that is covered in Step 1.
4. Modify the hypothesis as necessary so that it is compatible with all the pertinent experimental data.

Confusion sometimes arises regarding the exact meanings of the words *hypothesis, theory,* and *law.* A well-established hypothesis is often called a theory. Hypotheses and theories explain natural phenomena, whereas scientific laws are simple statements of natural phenomena to which no exceptions are known.

While the four steps listed above are a broad outline of the general procedure that is followed in much scientific work, they do not provide a recipe for doing chemistry or any other science. But chemistry is an experimental science, and much of its progress has been due to application of the scientific method through systematic research. Occasionally, a great discovery is made by accident, but the majority of scientific achievements are accomplished by well-planned experiments.

Many theories and laws are studied in chemistry. They make the study of chemistry or any science easier, because they summarize a particular aspect of the science. Although the student will see that theories advanced by great thinkers have been subject to change, this change does not mean that their contributions are of lesser significance than the discoveries of today. Change is the natural evolution of scientific knowledge.

1.6 How to Study Chemistry

How do you as a student approach a subject such as chemistry, with its unfamiliar terminology, symbols, formulas, theories, and laws? All the normally accepted habits of good study are applicable to the study of chemistry. Budget your study time and spend it wisely. In particular, you can spend your study time more profitably in regular, relatively short periods rather than in one prolonged cram session.

Chemistry has its own language, and learning this language is of prime importance to the successful study of chemistry. Chemistry is a subject of many facts. At first you will simply have to memorize some of them. However, you will also learn these facts by referring to them frequently in your studies and by repetitive use. For example, you must learn the symbols of 30 or 40 common elements in order to be able to write chemical formulas and equations. As with the alphabet, repetitive use of these symbols will soon make them part of your vocabulary.

Careful reading of assigned material cannot be overemphasized. You should read each chapter at least twice. The first time, read the chapter rapidly, noting especially topic headings, diagrams, and other outstanding features. Then read more thoroughly and deliberately for better understanding. It may be profitable to underline and abstract material during the second reading. Isolated reading may be sufficient for learning some subjects, but it is not sufficient for learning chemistry. During the lectures, become an active mental participant and try to think along with your instructor—do not just occupy a seat. Lecture and laboratory sessions will be much more meaningful if you have already read the assigned material.

Your studies must include a good deal of written chemistry. Chemical symbolism, equations, problem solving, and so on, require much written practice for proficiency. One does not become an accomplished pianist by merely reading or listening to music—it takes practice. One does not become a good baseball player by reading the rules and watching baseball games—it takes practice. So it is with chemistry. One does not become proficient in chemistry by only reading about it—it takes practice.

You will encounter many mathematical problems as you progress through this text. In solving a numerical problem, you should read the problem carefully to determine what is being asked. Then develop a plan for solving the problem. It is a good idea to start by writing down something—a formula, a diagram, an equation, the data given in the problem. This will give you something to work with, to think about, to modify—and finally to expand into an answer. When you have arrived at an answer, consider it carefully to make sure that it is a reasonable one. The solutions to problems should be recorded in a neat, orderly, stepwise fashion. Fewer errors and time saved are the rewards of a neat and orderly approach to problem solving. If you need to read and study still further for complete understanding, do it!

QUESTIONS   **A. Review Questions**

1. Classify the following statements as observation, law, hypothesis, or theory:
   (a) The volume of a gas is directly proportional to the absolute temperature, when the pressure remains constant.
   (b) The water in a closed test tube boiled at 83°C.
   (c) Iron gets heavier when it rusts because it attracts particles of rust from the air.
   (d) All matter is composed of tiny particles called atoms.
   (e) Glass turns a flame yellow as it approaches its melting point.
   (f) Molecules in a gas are always moving.
   (g) When wood burns it decomposes to its elements, which all escape as gas.
2. Name at least three products of chemical research you use regularly that probably did not exist 35 years ago.
3. Which of the following statements are correct?
   (a) Chemistry is the science that deals with the composition of substances and the transformations they undergo.
   (b) Robert Boyle, in the 17th century, clearly distinguished between an element and a compound or mixture.
   (c) The microscope was the key instrument in early modern chemistry, as the telescope had been in astronomy.
   (d) From 1803 to 1810, John Dalton advanced his atomic theory.
   (e) Most of the drugs and pharmaceuticals used in the United States today have been available for at least a century.
   (f) A key feature of the scientific method is to plan and do additional experiments to test a hypothesis.
   (g) Scientific laws are simple statements of natural phenomena to which no exceptions are known.

2

Standards for Measurement

After studying Chapter 2 you should be able to:

1. Understand the terms listed in Question A at the end of the chapter.
2. Differentiate clearly between mass and weight.
3. Know the basic metric units of mass, length, and volume.
4. Give the numerical equivalents of the metric prefixes *deci, centi, milli, micro, deka, hecto, kilo,* and *mega.*
5. Express any number in scientific notation form.
6. Express and round off the results of arithmetic operations to the proper number of significant figures.
7. Set up and solve problems by the dimensional analysis, or factor-label, method.
8. Convert any measurement of mass, length, or volume in American units to metric units and vice versa.
9. Make conversions between Fahrenheit, Celsius, and Kelvin temperatures.
10. Differentiate clearly between temperature and heat.
11. Make calculations using the equation

    calories = (Grams of substance) × (Heat capacity of substance) × (Δ*t*)

12. Calculate density, mass, or volume of an object (or substance) from appropriate data.
13. Calculate the specific gravity when given the density of a substance and vice versa.
14. Recognize the common laboratory measuring instruments illustrated in this chapter.

2.1 Mass and Weight

Chemistry is an experimental science. The results of experiments are usually determined by making measurements. In elementary experiments, the quantities that are commonly measured are mass (weight), length, volume, temperature, and time. Measurements of electrical and optical quantities may also be needed in more sophisticated experimental work.

mass

Although *mass* and *weight* are often used interchangeably, the two words have quite different meanings. The **mass** of a body is defined as the amount of matter in that body. The amount of mass in an object is a fixed and unvarying quantity that is independent of the object's location.

weight

The **weight** of a body is the measure of the earth's gravitational attraction for that body. Unlike its mass, the weight of an object varies in relation to (1) its position on or its distance from the earth and (2) whether the rate of motion of the object is changing with respect to the motion of the earth. Consider an astronaut of mass 70.0 kilograms (154 pounds) who is being shot into a space orbit. At the instant before blastoff the weight of the astronaut is also 70.0 kilograms. As the distance from the earth increases and the rocket turns into an orbiting course, the gravitational pull on the astronaut's body decreases until a state of "weightlessness" is attained. However, the mass of the astronaut's body has remained constant at 70.0 kilograms during the entire process of lift-off and going into orbit.

The mass of an object may be measured on a chemical balance by comparing it with other known masses. Two objects of equal mass will also have equal weights if they are measured in the same place. Thus, under these conditions the terms *mass* and *weight* are used interchangeably. Although the chemical balance is used to determine mass, it is said to *weigh* objects, and we often speak of the *weight* of an object when we really mean its mass.

2.2 Measurement, Significant Figures, and Calculations

To understand certain phases of chemistry it is necessary to set up and solve problems. Problem solving requires an understanding of the elementary mathematical operations used to manipulate numbers. Numerical values or data are obtained from measurements made in an experiment. A chemist may use these data to calculate the extent of the physical and chemical changes occurring in the substances that are being studied. By appropriate calculations, the results of an experiment may be compared with those of other experiments and summarized in ways that are meaningful.

When expressing a quantity of something, one must state both the numerical value and the units in which the quantity is expressed. For example, in the statement "1 kilometre contains 1000 metres," kilometre and metre are the units in which this length is expressed. In every experimental measurement, there is some degree of uncertainty due to both the limitations of the measuring instrument and the skill of the person who uses it. The value recorded for a measurement should give an indication of its reliability (precision). For maximum precision, this value should contain all the digits that are known plus one digit that is estimated. Because of the uncertainty expressed in the estimated digit, every number that expresses a measurement can have only a limited number of digits. These digits, used to express a measured quantity, are known as **significant figures** or **significant digits.**

significant
figures

Suppose we weigh an object on a balance that is calibrated in grams, and we observe that the balance scale stops between 15 and 16 (see Figure 2.1). We know for sure that the weight of the object is at least 15 grams but less than 16 grams. To express the weight with greater precision, we estimate that the indicator on the balance is about four-tenths the distance between 15 and 16. The weight of the object is, therefore, recorded as 15.4 grams. The last digit

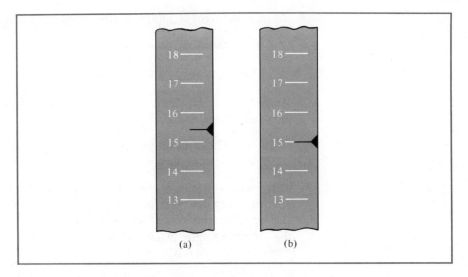

Figure 2.1 (a) Scale on a balance indicating a weight of 15.4 grams. (b) Scale on a balance indicating a weight of 15.0 grams.

(4) has some uncertainty since it is an estimated value. The recorded weight, 15.4 grams, is said to have three significant figures. Now suppose the balance scale shown in Figure 2.1 stops right on 15. The weight of the object should be recorded as 15.0 grams. It is necessary to use the zero to indicate that the object was weighed to a precision of one-tenth gram. The weight of the object is 15.0 grams, not 14.9 grams, or 15.1 grams, or 15 grams.

Some numbers are exact and have an infinite number of significant figures. Exact numbers occur in simple counting operations; when you count 25 dollars, you have exactly 25 dollars. Defined numbers, such as 12 inches in 1 foot, 60 minutes in 1 hour, and 100 centimetres in 1 metre, are also considered to be exact numbers.

Evaluating zeros.   In any measurement all nonzero numbers are significant. However, zeros may or may not be significant, depending on their position in the number. Rules for determining when zero is significant in a measurement follow:

1. All zeros between other digits are significant:

>   205 has three significant figures.
>   2.05 has three significant figures.
>   61.09 has four significant figures.

2. Zeros to the left of the first nonzero digit are not significant. These zeros are used to locate a decimal point:

>   0.0025 has two significant figures (2, 5).
>   0.0108 has three significant figures (1, 0, 8).

3. Zeros to the right of a number after a decimal point are significant:

0.500 has three significant figures (5, 0, 0).
25.160 has five significant figures.
3.00 has three significant figures.

4. Zeros to the right of nonzero digits but before the decimal point may or may not be significant:

1000 (The zeros may or may not be significant.)
590 (The zero may or may not be significant.)

One way of indicating whether these zeros are significant is to write the number using a decimal point and a power of 10. Thus, if the value 1000 has been determined to four significant figures, it is written as 1.000×10^3. If 590 has only two significant figures, it is written as 5.9×10^2 (in this case the zero is not significant).

PROBLEM 2.1   How many significant figures are in each of these numbers?
(a) 4.5 inches   (e) 25.0 grams
(b) 3.025 feet   (f) 12.20 litres
(c) 125.0 metres   (g) 100,000 people
(d) 0.001 mile   (h) 205 birds

Answers: (a) 2   (b) 4   (c) 4   (d) 1   (e) 3   (f) 4
(g) unknown but probably 1   (h) 3

2.3 Rounding Off Numbers

In calculations we often obtain answers that have more digits than we are justified in using. It is necessary, therefore, to drop the nonsignificant digits in order to express the answer with the proper number of significant figures. When digits are dropped from a number, the value of the last digit retained is determined by a process known as **rounding off numbers.** There are three rules for rounding off numbers:

rounding off numbers

Rule 1. When the first digit after those you want to retain is 4 or less, that digit and all others beyond it are dropped. The last digit retained is not changed.

Examples rounded off to four digits:

74.693 = 74.69
This digit is dropped.

1.00629 = 1.006
These two digits are dropped.

Rule 2. When the first digit after those you want to retain is 6 or greater, increase the last digit retained by 1 and drop all digits to the right of the last digit retained.

Examples rounded off to four digits:

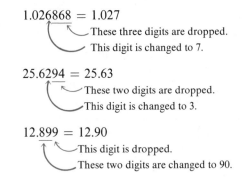

1.026868 = 1.027

       These three digits are dropped.

       This digit is changed to 7.

25.6294 = 25.63

       These two digits are dropped.

       This digit is changed to 3.

12.899 = 12.90

       This digit is dropped.

       These two digits are changed to 90.

Rule 3. When the first digit after those you want to retain is 5 and all others beyond are zeros, increase the last digit retained by 1 if it is an odd number; leave it unchanged if it is an even number. Drop all digits to the right of the last digit retained. When any digit beyond the 5 is other than zero, increase the last digit retained by 1.

Examples rounded to four digits:

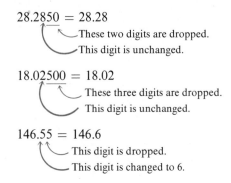

28.2850 = 28.28

       These two digits are dropped.

       This digit is unchanged.

18.02500 = 18.02

       These three digits are dropped.

       This digit is unchanged.

146.55 = 146.6

       This digit is dropped.

       This digit is changed to 6.

PROBLEM 2.2   Round off these numbers to the number of digits indicated:
(a) 42.246 (four digits)   (d) 0.08965 (two digits)
(b) 88.015 (three digits)   (e) 225.3 (three digits)
(c) 0.08965 (three digits)   (f) 14.150 (three digits)

Answers: (a) 42.25 (Rule 2)   (b) 88.0 (Rule 1)   (c) 0.0896 (Rule 3)
(d) 0.090 (Rule 2)   (e) 225 (Rule 1)   (f) 14.2 (Rule 3)

2.4 Scientific Notation of Numbers

The age of the earth has been estimated as about 4,500,000,000 (4.5 billion) years. Since this is an estimated value, let us say to the nearest 0.1 billion years, we are justified in using only two significant figures to express it. To express this number numerically with two significant figures we write it using a power of 10 as 4.5×10^9 years.

Very large and very small numbers are often used in chemistry. These numbers can be simplified and conveniently written using a power of 10. Writing a number as a power of 10 is called **scientific notation.**

scientific notation

To write a number in scientific notation, move the decimal point in the original number so that it is located after the first nonzero digit. This new number is multiplied by 10 raised to the proper power (exponent). If the decimal was moved to the left, the power of 10 will be a positive number. If the decimal was moved to the right, the power of 10 will be a negative number. Study the examples that follow.

Example 1: Write 5283 in scientific notation.

5283. Place the decimal between the 5 and the 2. Since the decimal was moved
321   three places to the left, the power of 10 will be 3 and the number 5.283 is multiplied by 10^3.

5.283×10^3 (Correct scientific notation)

Example 2: Write 4,500,000,000 in scientific notation (two significant figures).

4,500,000,000. Place the decimal between the 4 and the 5. Since the decimal was
9   1   moved nine places to the left, the power of 10 will be 9 and the number 4.5 is multiplied by 10^9.

4.5×10^9 (Correct scientific notation)

Example 3: Write 0.000123 in scientific notation.

0.000123 Place the decimal between the 1 and the 2. Since the decimal was moved
4   four places to the right, the power of 10 will be -4 and the number 1.23 is multiplied by 10^{-4}.

1.23×10^{-4} (Correct scientific notation)

PROBLEM 2.3 Write the following numbers in scientific notation:
(a) 1200 (four digits)   (c) 0.0468
(b) 6,600,000 (two digits)   (d) 0.00003

Answers: (a) 1200 $= 1.200 \times 10^3$ (b) 6,600,000 $= 6.6 \times 10^6$
           3   6

(c) 0.0468 $= 4.68 \times 10^{-2}$ (d) 0.00003 $= 3 \times 10^{-5}$
           2   5

■

2.5 Significant Figures in Calculations

The results of a calculation based on measurements cannot be more precise than the measurement that has the greatest uncertainty (that is, the least precise measurement).

Multiplication or division.   In calculations involving multiplication or division the answer must contain the same number of significant figures as are contained in the measurement that has the least number of significant figures. Consider the following calculations:

(a) 134 in. $\times$ 25 in. = 3350 in.^{2}

The answer should contain two significant figures because the factor 25 has two significant figures. The answer, therefore, is rounded off and expressed in scientific notation:

$$134 \text{ in.} \times 25 \text{ in.} = 3.4 \times 10^3 \text{ in.}^2$$

(b) $\dfrac{213 \text{ miles}}{4.20 \text{ hours}}$ = 50.714285 miles/hour

A hand calculator was used to obtain the eight-digit answer given. Since each factor in the problem has three digits, only three digits can be used in the answer:

$$\frac{213 \text{ miles}}{4.20 \text{ hours}} = 50.7 \text{ miles/hour}\quad \text{(Answer)}$$

(c) $\dfrac{2.2 \times 273}{760}$ = 0.79026315

The answer should contain two significant figures (2.2 is the limiting factor). The correct, rounded-off answer is 0.79.

Addition or subtraction.   The results of an addition or a subtraction must contain the same number of digits to the right of the decimal point as is contained in that term with the least number of digits to the right of the decimal point. Consider the following calculations:

(a) Add 125.17, 129.2, and 52.24.

$$
\begin{array}{r}
125.17 \\
129.2 \\
\underline{52.24} \\
306.61
\end{array}
$$

The quantity 129.2 is the least precise of these figures. Therefore, the answer is rounded off to a value having one decimal place.

306.6   (Answer)

(b) Subtract 14.1 from 132.56.

$$\begin{array}{r} 132.56 \\ -\ \ 14.1 \\ \hline 118.46 \end{array}$$

14.1 contains the least number of digits to the right of the decimal point. Therefore, the answer is rounded off to a value having one decimal place.

118.5 (Answer)

PROBLEM 2.4   How many significant figures should the answer contain in each of these calculations?

(a) $14.0 \times 5.2 =$ (d) $8.2 + 0.125 =$

(b) $0.1682 \times 8.2 =$ (e) $119.1 - 3.44 =$

(c) $\dfrac{160 \times 33}{4} =$ (f) $\dfrac{94.5}{1.2} =$

Answers: (a) 2   (b) 2   (c) 1   (d) 2 (one decimal place)
(e) 4 (one decimal place)   (f) 2

■

PROBLEM 2.5   Do the following calculations and round off your answers to the proper number of significant figures.

(a) $190.6 \times 2.3 =$

The answer obtained with a hand calculator is 438.38. The answer should have two significant figures, so it must be expressed in scientific notation:

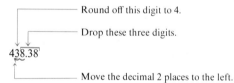

Round off this digit to 4.

Drop these three digits.

Move the decimal 2 places to the left.

The correct answer is 4.4×10^2.

(b) $\dfrac{13.59 \times 6.3}{12} =$

The answer obtained with a hand calculator is 7.13475. The answer should contain two significant figures:

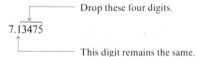

Drop these four digits.

This digit remains the same.

The correct answer is 7.1.

(c) $\dfrac{1.039 - 1.020}{1.039} =$

The answer obtained with a hand calculator is 0.018286814. When the subtraction in the numerator is done,

1.039 − 1.020 = 0.019

the number of significant figures changes from four to two. Therefore, the answer should contain two significant figures after the division is carried out:

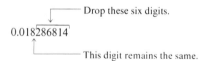

The correct answer is 0.018 or 1.8 × 10⁻².

■

Additional material on mathematical operations is given in the Mathematical Review in Appendix I. You are urged to review Appendix I and to study carefully any portions that are not familiar to you. This study may be done at various times during the course as the need for additional knowledge of mathematical operations arises.

2.6 The Metric System

metric system
or SI

The **metric system,** or **International System (SI),** is a decimal system of units for measurements of mass, length, time, and other physical constants. It is built around a set of basic units and uses factors of 10 to express larger or smaller quantities of these units. To express larger and smaller quantities, prefixes are added to the names of the units. These prefixes represent multiples of 10, making the metric system a total decimal system of measurement. Table 2.1

TABLE 2.1 **Prefixes used in the metric system and their numerical values.**

Prefix	Symbol	Numerical value	
tera	T	1,000,000,000,000	or 10^{12}
giga	G	1,000,000,000	or 10^{9}
mega	M	1,000,000	or 10^{6}
kilo	k	1,000	or 10^{3}
hecto	h	100	or 10^{2}
deka	da	10	or 10^{1}
deci	d	0.1	or 10^{-1}
centi	c	0.01	or 10^{-2}
milli	m	0.001	or 10^{-3}
micro	μ	0.000001	or 10^{-6}
nano	n	0.000000001	or 10^{-9}
pico	p	0.000000000001	or 10^{-12}
femto	f	0.000000000000001	or 10^{-15}
atto	a	0.000000000000000001	or 10^{-18}

shows the names, symbols, and numerical values of the prefixes. These are also shown in Appendix III. Some of the more commonly used prefixes are

kilo   One thousand (1000) times the unit expressed
deci   One-tenth (0.1) of the unit expressed
centi   One-hundredth (0.01) of the unit expressed
milli   One-thousandth (0.001) of the unit expressed
micro   One-millionth (0.000001) of the unit expressed

Examples are

1 kilometre = 1000 metres
1 kilogram = 1000 grams
1 microsecond = 0.000001 second

The seven base units of measurement in the International System are given in the following table. Other units are derived from these base units.

Quantity	Name of unit	Symbol
Length	Metre	m
Mass	Kilogram	kg
Time	Second	s
Electric current	Ampere	A
Temperature	Kelvin	K
Luminous intensity	Candela	cd
Amount of substance	Mole	mol

The metric system, or International System, is currently used by most of the countries in the world, not only for scientific and technical work, but also in commerce and industry. The United States is currently in the process of changing to the metric system of weights and measurements.

2.7 Measurement of Length

Standards for the measurement of length have had an interesting historical development. The Old Testament mentions such units as the cubit (the distance from a man's elbow to the tip of his outstretched hand). In ancient Scotland the inch was once defined as a distance equal to the width of a man's thumb.

Reference standards of measurements have undergone continuous improvement in precision. The standard unit of length in the metric system is the **metre.** When the metric system was introduced in the 1790s, the metre was defined as one ten-millionth of the distance from the equator to the North Pole measured along the meridian passing through Dunkirk, France. In 1889, the metre was redefined as the distance between two engraved lines on a platinum–iridium alloy bar maintained at 0° Celsius. This international metre bar is stored

metre

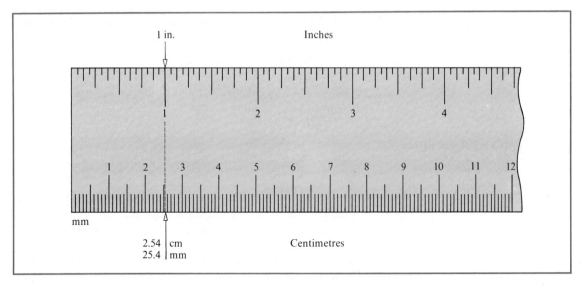

Figure 2.2   Comparison of the metric and American systems of length measurement: 2.54 cm = 1 in.

in a vault at Sèvres near Paris. Duplicate metre bars have been made and used as standards by many nations.

By the 1950s, length could be measured with such precision that a new standard was needed. Accordingly, in 1960, by international agreement the metre was again redefined, this time as 1,650,763.73 wavelengths of a particular spectral emission line of krypton-86. This is the reference standard for length presently in use.

A metre is 39.37 inches, a little longer than 1 yard. One metre contains 10 decimetres, 100 centimetres, or 1000 millimetres. A kilometre contains 1000 metres. Table 2.2 shows the relationships of these units. A comparison of centimetres, millimetres, and inches is shown in Figure 2.2.

TABLE 2.2 Metric units of length.

Unit	Symbol	Metre equivalent	Exponential equivalent
Kilometre	km	1000 m	10^3 m
Metre	m	1 m	10^0 m
Decimetre	dm	0.1 m	10^{-1} m
Centimetre	cm	0.01 m	10^{-2} m
Millimetre	mm	0.001 m	10^{-3} m
Micrometre	μm	0.000001 m	10^{-6} m
Nanometre	nm	0.000000001 m	10^{-9} m
Angstrom	Å	0.0000000001 m	10^{-10} m

The angstrom unit (10^{-8} cm) is used extensively in expressing the wavelength of light and atomic dimensions. Other important relationships are

$$1 \text{ m} = 100 \text{ cm} = 1000 \text{ mm} = 10^6 \text{ } \mu\text{m} = 10^{10} \text{ Å}$$
$$1 \text{ cm} = 10 \text{ mm} = 0.01 \text{ m}$$
$$1 \text{ in.} = 2.54 \text{ cm}$$
$$1 \text{ mile} = 1.61 \text{ km}$$

2.8 Problem Solving

One of the most consistently troublesome areas in chemistry involves the solving of mathematical problems. Since many chemical principles are illustrated by mathematical concepts, it is necessary to learn to solve problems dealing with these concepts.

There are usually several methods by which a problem can be solved. But in all methods it is best, especially for beginners, to use a systematic, orderly approach. The dimensional analysis, or factor-label, method (Appendix I, Section 11) is stressed in this book for the following reasons:

1. It provides a systematic, straightforward way to set up problems.
2. It gives a clear understanding of the principles involved.
3. It helps in learning to organize and evaluate data.
4. It helps to identify errors because unwanted units are not eliminated if the setup of the problem is incorrect.

The basic steps for solving problems are

1. Read the problem very carefully to determine what is to be solved for. Write down what you are solving for.
2. Tabulate the data given in the problem. Even in tabulating data, it is important to label all factors and measurements with the proper units.
3. Determine which principles are involved and which unit relationships are needed to solve the problem. It may be necessary to refer to certain tables to obtain other data needed. Use sample problems in the text to help set up and solve the problem.
4. Proceed with the necessary mathematical operations. Make certain that the answer contains the proper number of significant figures.
5. Check the answer to see if it is reasonable.
6. Do your work in a neat and organized form.

Label all factors with the proper units.

Just a few more words about problem solving. Don't allow any formal method of problem solving to limit your use of common sense and intuition. If the solution to a problem is obvious and seems simpler to you by another method, by all means use it. But in the long run you should be able to solve many otherwise difficult problems by using the dimensional analysis method.

The dimensional analysis method of problem solving converts one unit to another unit by the use of conversion factors.

$$\text{Unit}_1 \times \text{Conversion factor} = \text{Unit}_2$$

Suppose you want to know how many millimetres there are in 2.5 metres. As far as units are concerned you need to convert metres (m) to millimetres (mm). Therefore, you can start by writing down

$$\text{m} \times \text{Conversion factor} = \text{mm}$$

This conversion factor must accomplish two things. It must cancel or eliminate metres and it must introduce millimetres, the unit wanted in the answer. Such a conversion factor will be in fractional form and have metres in the denominator and millimetres in the numerator:

$$\cancel{\text{m}} \times \frac{\text{mm}}{\cancel{\text{m}}} = \text{mm}$$

We know that 1 m = 1000 mm. From this relationship we can write two factors: 1 m per 1000 mm, and 1000 mm per 1 m:

$$\frac{1 \text{ m}}{1000 \text{ mm}} \quad \text{and} \quad \frac{1000 \text{ mm}}{1 \text{ m}}$$

Using the factor 1000 mm/1 m we can set up the calculation for the conversion of 2.5 m to millimetres:

$$2.5 \cancel{\text{m}} \times \frac{1000 \text{ mm}}{1 \cancel{\text{m}}} = 2500 \text{ mm} \quad \text{or} \quad 2.5 \times 10^3 \text{ mm}$$
$$\text{(two significant figures)}$$

Note that in making this calculation, units are treated the same as numbers; metres in the numerator are canceled by metres in the denominator.

Now suppose you need to change 215 centimetres to metres. First you must determine that you need to convert centimetres to metres. We can start with

$$\text{cm} \times \text{Conversion factor} = \text{m}$$

The conversion factor must have centimetres in the denominator and metres in the numerator:

$$\cancel{\text{cm}} \times \frac{\text{m}}{\cancel{\text{cm}}} = \text{m}$$

From the relationship 100 cm = 1 m we can write a factor that will accomplish this conversion:

$$\frac{1 \text{ m}}{100 \text{ cm}}$$

Now set up the calculation using all the data given:

$$215 \text{ cm} \times \frac{1 \text{ m}}{100 \text{ cm}} = \frac{215 \text{ m}}{100} = 2.15 \text{ m}$$

Some problems may require a series of conversions to reach the correct units in the answer. For example, suppose we want to know the number of seconds in 1 day. We need to go from the unit of days to seconds and can proceed in this manner:

$$\text{Day} \longrightarrow \text{Hours} \longrightarrow \text{Minutes} \longrightarrow \text{Seconds}$$

This series requires three conversion factors, one for each step. We convert days to hours (hr), hours to minutes (min), and minutes to seconds (s). The conversions can be done individually or in a continuous sequence:

$$\text{day} \times \frac{\text{hr}}{\text{day}} \qquad \text{hr} \times \frac{\text{min}}{\text{hr}} \qquad \text{min} \times \frac{\text{s}}{\text{min}}$$

$$\text{day} \times \frac{\text{hr}}{\text{day}} \times \frac{\text{min}}{\text{hr}} \times \frac{\text{s}}{\text{min}} = \text{s}$$

Inserting the proper factors, the calculation for the number of seconds in 1 day is

$$1 \text{ day} \times \frac{24 \text{ hr}}{1 \text{ day}} \times \frac{60 \text{ min}}{1 \text{ hr}} \times \frac{60 \text{ s}}{1 \text{ min}} = 86,400 \text{ s}$$

All five digits in 86,400 are significant, since all the factors in the calculation are exact numbers.

The dimensional analysis, or factor-label, method used in the preceding work shows how unit conversion factors are derived and used in calculations. After you become more proficient with the terms, you can save steps by writing the factors directly in the calculation. The problems that follow give examples of the conversion from American to metric units.

PROBLEM 2.6   How many centimetres are in 2.00 ft?

The stepwise conversion of units from feet to centimetres may be done in this manner: Convert feet to inches; then convert inches to centimetres.

$$\text{ft} \longrightarrow \text{in.} \longrightarrow \text{cm}$$

The needed conversion factors are

$$\frac{12 \text{ in.}}{1 \text{ ft}} \qquad \text{and} \qquad \frac{2.54 \text{ cm}}{1 \text{ in.}}$$

$$2.00 \text{ ft} \times \frac{12 \text{ in.}}{\text{ft}} = 24.0 \text{ in.}$$

$$24.0 \text{ in.} \times \frac{2.54 \text{ cm}}{\text{in.}} = 61.0 \text{ cm} \quad \text{(Answer)}$$

Since 12 in./ft and 2.54 cm/in. are considered to be exact numbers, the number of significant figures allowed in the answer is three, based on the number 2.00.

■

PROBLEM 2.7   How many metres are there in a 100 yd football field?

The stepwise conversion of units from yards to metres may be done in this manner, using the proper conversion factors:

$$yd \longrightarrow ft \longrightarrow in. \longrightarrow cm \longrightarrow m$$

$$100 \, \cancel{yd} \times \frac{3 \, ft}{\cancel{yd}} = 300 \, ft \qquad\qquad (3 \, ft/yd)$$

$$300 \, \cancel{ft} \times \frac{12 \, in.}{\cancel{ft}} = 3600 \, in. \qquad\qquad (12 \, in./ft)$$

$$3600 \, \cancel{in.} \times \frac{2.54 \, cm}{\cancel{in.}} = 9144 \, cm \qquad\qquad (2.54 \, cm/in.)$$

$$9144 \, \cancel{cm} \times \frac{1 \, m}{100 \, \cancel{cm}} = 91.4 \, m \qquad (1 \, m/100 \, cm) \qquad (three \, significant \, figures)$$

■

Problems 2.6 and 2.7 may be solved using a running linear expression, writing down each conversion factor in succession. Very often this saves one or two calculation steps, and numerical values may be reduced to simpler terms leading to simpler calculations. The single linear expressions for Problems 2.6 and 2.7 are

$$2.00 \, \cancel{ft} \times \frac{12 \, \cancel{in.}}{\cancel{ft}} \times \frac{2.54 \, cm}{\cancel{in.}} = 61.0 \, cm$$

$$100 \, \cancel{yd} \times \frac{3 \, \cancel{ft}}{\cancel{yd}} \times \frac{12 \, \cancel{in.}}{\cancel{ft}} \times \frac{2.54 \, \cancel{cm}}{\cancel{in.}} \times \frac{1 \, m}{100 \, \cancel{cm}} = 91.4 \, m$$

2.9 Measurement of Mass

kilogram

The standard unit of mass in the metric system is the **kilogram.** This amount of mass is defined by international agreement as exactly equal to the mass of a platinum–iridium weight (*Kilogramme de Archive*) kept in a vault at Sèvres. A kilogram contains 1000 grams. Comparing this unit of mass to 1 pound (16 ounces), we find that a kilogram is equal to 2.2 pounds. A pound is equal to 454 grams (0.454 kilogram). The same prefixes used in length measurement are used to indicate larger and smaller gram units (see Table 2.3, page 24).

It is convenient to remember that

$$1 \, g = 1000 \, mg \qquad\qquad 1 \, kg = 2.2 \, lb$$

$$1 \, kg = 1000 \, g \qquad\qquad 1 \, lb = 454 \, g$$

TABLE 2.3 **Metric units of mass.**

Unit	Symbol	Gram equivalent	Exponential equivalent
Kilogram	kg	1000 g	10^3 g
Gram	g	1 g	10^0 g
Decigram	dg	0.1 g	10^{-1} g
Centigram	cg	0.01 g	10^{-2} g
Milligram	mg	0.001 g	10^{-3} g
Microgram	μg	0.000001 g	10^{-6} g

To change grams to milligrams, multiply grams by the conversion factor 1000 mg/g. The setup for converting 2.5 grams to milligrams is

$$2.5 \, g \times \frac{1000 \, mg}{1 \, g} = 2500 \, mg \quad \text{or} \quad 2.5 \times 10^3 \, mg \quad \text{(Answer)}$$

Note that multiplying a number by 1000 is the same as multiplying the number by 10^3 and can be simply done by moving the decimal point three places to the right.

$$6.428 \times 1000 = 6428 \quad (6.428)$$
$$123$$

To change milligrams to grams, multiply grams by the conversion factor 1 g/1000 mg. For example, to convert 150 mg to grams:

$$150 \, mg \times \frac{1 \, g}{1000 \, mg} = 0.150 \, g \quad \text{(Answer)}$$

PROBLEM 2.8   A 1.50 lb package of sodium bicarbonate costs 80 cents. How many grams of this substance are in this package?

We are solving for the number of grams equivalent to 1.50 lb. Since 1 lb = 454 g, the factor to convert pounds to grams is 454 g/lb.

$$1.50 \, lb \times \frac{454 \, g}{lb} = 681 \, g \quad \text{(Answer)}$$

[*Note:* The cost of the sodium bicarbonate has no bearing on the question asked in this problem.]

■

PROBLEM 2.9   How many milligrams does 1.00 kg of water weigh?

The conversion from kilograms to milligrams is done in two steps:

$$kg \longrightarrow g \longrightarrow mg$$

Two conversion factors are needed. They are

$$\frac{1000\ \text{g}}{1\ \text{kg}} \quad \text{and} \quad \frac{1000\ \text{mg}}{1\ \text{g}}$$

$$1.00\ \cancel{\text{kg}} \times \frac{1000\ \cancel{\text{g}}}{1\ \cancel{\text{kg}}} \times \frac{1000\ \text{mg}}{1\ \cancel{\text{g}}} = 1.00 \times 10^6\ \text{mg} \quad \text{(Answer)}$$

The answer is 1 million milligrams, but it must be expressed in scientific notation with only three significant figures, since 1.00 kg has three significant figures.

■

2.10 Measurement of Volume

litre

The metric system unit of volume commonly used by chemists is the **litre** (symbol, L). The litre is a little larger than a U.S. quart; 1.000 litre equals 1.057 quarts. The most commonly used fractional unit of a litre is the millilitre: 1 L = 1000 mL, and 946 mL = 1.00 qt. (See Figure 2.3.) This small unit of

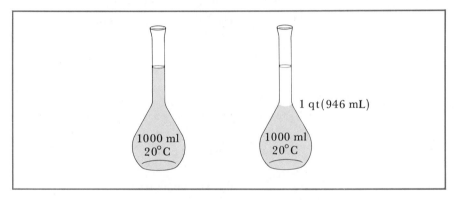

Figure 2.3   Comparison of the volume of a litre and a quart (946 mL = 1 qt).

volume is also commonly referred to as a cubic centimetre, (symbol, cm^3). This is because a litre corresponds to the volume enclosed in a cube measuring exactly 10 cm on an edge. The volume of this cube is determined by multiplying length times width times height—that is, 10 cm $\times$ 10 cm $\times$ 10 cm = 1000 cm^3 (see Figure 2.4). The relationship between a millilitre and a cubic centimetre is that 1 mL equals 1 cm^3 exactly (see Table 2.4).

TABLE 2.4 **Metric units of volume.**

Unit	Symbol	Litre equivalent
Litre	L	1.0 L
Millilitre	mL	0.001 L
Cubic centimetre	cm^3	0.001 L

Examples of volume conversions are given in the problems on the next page.

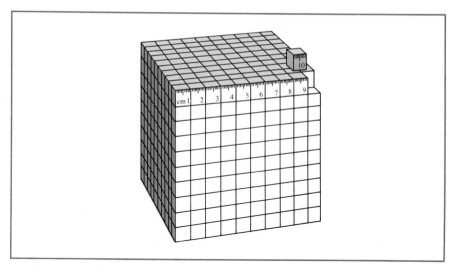

Figure 2.4   The large cube, 10.0 cm on a side, has a volume of 1000 cm³, or 1.0 litre. The small cube on top is 1.0 cm³. The large block contains 1000 of these small cubes.

PROBLEM 2.10   How many millilitres are contained in 3.5 litres?

The conversion factor to change litres to millilitres is 1000 mL/L.

$$3.5 \, \cancel{L} \times \frac{1000 \, \text{mL}}{\cancel{L}} = 3500 \, \text{mL} \quad \text{or} \quad 3.5 \times 10^3 \, \text{mL}$$

Litres may be changed to millilitres by moving the decimal point three places to the right and changing the units to millilitres.

$$1.500 \, \text{L} = 1500 \, \text{mL}$$

∎

PROBLEM 2.11   Cubic centimetres can be calculated by multiplying the length times the width times the height of a solid in centimetres. How many cubic centimetres are in a cube that is 11.1 inches on a side?

First change inches to centimetres. The conversion factor is 2.54 cm/in.

$$11.1 \, \cancel{\text{in.}} \times \frac{2.54 \, \text{cm}}{\cancel{\text{in.}}} = 28.2 \, \text{cm on a side}$$

Then change to cubic volume (length × width × height).

$$28.2 \, \text{cm} \times 28.2 \, \text{cm} \times 28.2 \, \text{cm} = 22{,}426 \, \text{cm}^3 \quad \text{or} \quad 2.24 \times 10^4 \, \text{cm}^3$$

∎

Table 2.5 summarizes the units and conversion factors that are used most often.

TABLE 2.5 Most often used units and their equivalents.

1 km = 1000 m
1 km = 0.62 mile
1 m = 1000 mm
1 m = 39.37 in.
1 cm = 10 mm
2.54 cm = 1 in.
1 g = 1000 mg
1 kg = 1000 g
454 g = 1 lb
1 L = 1000 mL
1 mL = 1 cm^3
946 mL = 1 qt

2.11 Temperature Scales

The *temperature* of a system measures how hot or cold that system is and can be expressed by several different temperature scales. Three commonly used temperature scales are the Celsius (centigrade) scale (pronounced *sell-si-us*), the Kelvin (absolute) scale, and the Fahrenheit scale. A unit of temperature on each of these scales is called a *degree*. The symbol for the degree is ° and it is placed as a superscript after the number and before the temperature scales for Celsius and Fahrenheit. Thus, 100°C means 100 *degrees Celsius*. The degree sign is not used with Kelvin temperatures.

Degrees Celsius (centigrade) = °C

Kelvin (absolute) = K

Degrees Fahrenheit = °F

The Celsius scale is based on dividing the interval between the freezing and boiling temperatures of water into 100 equal parts, or degrees. The freezing point of water is assigned a temperature of 0°C and the boiling point of water a temperature of 100°C. The Kelvin temperature scale is also known as the *absolute temperature scale* because 0 K is the lowest possible temperature theoretically attainable. The Kelvin zero is 273.16 degrees below the Celsius zero. Kelvin degrees are equal in size to Celsius degrees. The freezing point of water on the Kelvin scale is 273.16 K (usually rounded off to 273 K). On the Fahrenheit scale there are 180 degrees between the freezing and boiling temperatures of water. On this scale, the freezing point of water is 32°F and the boiling point is 212°F.

0°C = 273 K = 32°F

The three scales are compared in Figure 2.5. Although absolute zero is the lower limit of temperature on these scales, there is no known upper limit

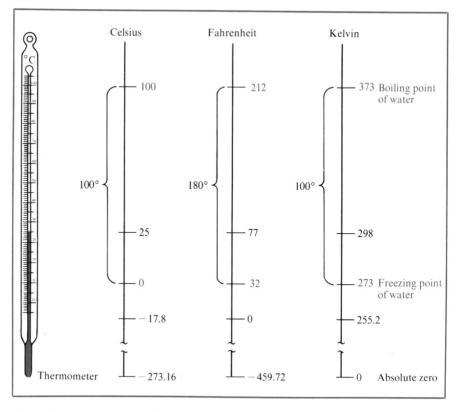

Figure 2.5   Comparison of Celsius, Kelvin, and Fahrenheit temperature scales.

to temperature. (Temperatures of several million degrees are known to exist in the sun and in other stars.)

By examining Figure 2.5 we can see that there are 100 Celsius degrees and 100 Kelvin degrees between the freezing and boiling points of water. But there are 180 Fahrenheit degrees between these two temperatures. Hence, the size of a degree on the Celsius scale is the same as the size of a degree on the Kelvin scale, but the Celsius degree corresponds to 1.8 degrees on the Fahrenheit scale. From these data, mathematical formulas have been derived to convert a temperature on one scale to the corresponding temperature on another scale. These formulas are

$$K = {}^\circ C + 273 \tag{1}$$

$$^\circ F = (1.8 \times {}^\circ C) + 32 \tag{2}$$

$$^\circ C = \frac{({}^\circ F - 32)}{1.8} \tag{3}$$

Interpretation: Formula (1) states that the addition of 273 to the degrees Celsius converts the temperature to Kelvin. Formula (2) states that to obtain the Fahrenheit temperature corresponding to a given Celsius temperature we multiply the degrees Celsius by 1.8 and then add 32. Formula (3) states that to obtain the corresponding Celsius temperature we subtract 32 from the degrees Fahrenheit and then divide this figure by 1.8.

Examples of temperature conversions are given in the following problems.

PROBLEM 2.12 The temperature at which salt (sodium chloride) melts is 800°C. What is this temperature on the Kelvin and Fahrenheit scales?

We need to calculate K from °C, so we use formula (1) above. We also need to calculate °F from °C; for this we use formula (2):

$$K = °C + 273$$

$$K = 800°C + 273 = 1073 \text{ K}$$

$$°F = (1.8 \times °C) + 32$$

$$°F = (1.8 \times 800°C) + 32$$

$$°F = 1440 + 32 = 1472°F \quad \text{or} \quad 1.47 \times 10^3 \text{ °F}$$

$$800°C = 1073 \text{ K} = 1472°F$$

■

PROBLEM 2.13 The temperature for December 1 was 110°F, a new record. Calculate this temperature in degrees Celsius.

Formula (3) applies here:

$$°C = \frac{(°F - 32)}{1.8}$$

$$°C = \frac{(110 - 32)}{1.8} = \frac{78}{1.8} = 43°C$$

■

PROBLEM 2.14 What temperature on the Fahrenheit scale corresponds to −8.0°C? (Be alert to the presence of the minus sign in this problem.)

Use formula (2) for this calculation:

$$°F = (1.8 \times °C) + 32$$

$$°F = [1.8 \times (-8.0)] + 32 = -14.4 + 32$$

$$°F = 18$$

■

Temperatures used in this book are in degrees Celsius (°C) unless specified otherwise.

2.12 Heat and Temperature

heat

temperature

Heat is a form of energy associated with the motion of small particles of matter. Heat is associated with a quantity of energy within a system or a quantity of energy supplied to a system. **Temperature** is a measure of the intensity of heat, or how hot a system is, regardless of its size. Heat always flows from a region of high temperature to one of lower temperature.

calorie
kilocalorie

joule

One unit of heat commonly used in chemical systems is the calorie. A **calorie** (cal) is the quantity of heat required to change the temperature of 1 g of water 1°C, usually measured from 14.5 to 15.5°C. The **kilocalorie** (kcal), also known as the nutritional or large Calorie (spelled with a capital *C* and abbreviated Cal), is equal to 1000 small calories. Another energy unit commonly used in scientific work is the **joule** (pronounced *jool*). One calorie is equal to 4.184 joules (J). The joule is the accepted unit of energy in the SI system. Temperature is measured in degrees and heat quantity is measured in calories or joules.

1 cal = Energy needed to raise the temperature of 1 g of water by 1°C

1 kcal = 1000 cal

1 cal = 4.184 J

The difference in the meanings of the terms *heat* and *temperature* can be seen by this example: Visualize two beakers, A and B. Beaker A contains 100 g of water at 20°C, and beaker B contains 200 g of water also at 20°C. The beakers are now heated until the temperature of the water in each reaches 30°C. The *temperature* of the water in the beakers was raised by exactly the same amount, 10°C. Yet twice as much *heat* (2000 cal) was required by the water in beaker B than was required by the water in beaker A (1000 cal).

heat capacity

The **heat capacity,** also known as *specific heat,* of any substance is the quantity of heat (in calories) required to change the temperature of 1 g of that substance by 1°C. It follows from the definition of the calorie that the heat capacity of water is 1 cal/g°C. The heat capacity of water is high compared to that of most substances. Aluminum and copper, for example, have heat capacities of 0.215 and 0.0921 cal/g°C, respectively (see Table 2.6). The relation of mass, heat capacity, temperature change (Δt), and quantity of heat lost or gained by a system is expressed by this general equation:

(Grams of substance) $\times$ (Heat capacity of substance) $\times \Delta t$ = calories   (4)

Thus, the amount of heat needed to warm 200 g of water from 20°C to 30°C ($\Delta t = 10°C$) can be calculated as follows:

$$200\,g \times \frac{1.00\ cal}{g°C} \times 10°C = 2000\ cal$$

The application of equation (4) to the solution of a more complicated example is illustrated by Problem 2.15.

TABLE 2.6 **Heat capacity of selected substances.**

Substance	Heat capacity (cal/g°C)
Water	1.00
Ethyl alcohol	0.511
Ice	0.492
Aluminum	0.215
Iron	0.113
Copper	0.0921
Gold	0.0312
Lead	0.0305

PROBLEM 2.15   A sample of a metal weighing 200 g is heated to 125°C and then dropped into 500 g of water at 24.0°C. If the final temperature of the water is 29.5°C, what is the heat capacity of the metal? (Assume no heat losses to the surroundings.)

When the metal enters the water it begins to cool, losing heat to the water. At the same time the temperature of the water rises. This process continues until the temperature of the metal and the temperature of the water are equal, at which point (29.5°C) no net flow of heat occurs.

The heat lost or gained by a system is given by equation (4). We use this equation first to calculate the heat gained by the water and then to calculate the heat capacity of the metal.

Temperature rise of the water $(\Delta t) = 5.5°C$ (29.5°C − 24.0°C)

$$\text{Heat gained by the water} = 500 \text{ g} \times \frac{1 \text{ cal}}{\text{g°C}} \times 5.5°C = 2750 \text{ cal}$$

The metal dropped into the water must have a final temperature the same as the water (29.5°C).

Temperature drop by the metal $(\Delta t) = 125°C − 29.5°C = 95.5°C$

Heat lost by the metal = Heat gained by the water (2750 cal)

Rearranging equation (4) we get

$$\text{Heat capacity} = \frac{\text{cal}}{\Delta t \times g}$$

$$\text{Heat capacity of the metal} = \frac{2750 \text{ cal}}{95.5°C \times 200 \text{ g}} = 0.144 \text{ cal/g°C}$$

■

2.13 Tools for Measurement

Common measuring instruments used in chemical laboratories are illustrated in Figures 2.6 and 2.7. A balance is used to measure mass. Balances are obtainable that will weigh objects to the nearest microgram. The choice of the balance

Figure 2.6 (a) A Cent-o-Gram R311 balance has four calibrated horizontal beams, each fitted with a specific movable weight. The weight of the object placed on the pan is determined by moving the weights along the beams until the swinging beam is in balance, as shown by the indicator on the right. *(Courtesy Ohaus Scale Corporation.)* (b) A single-pan, top-loading, rapid-weighing balance with direct readout to the nearest milligram (for example, 125.456 g). *(Courtesy Sartorius Balance Div., Brinkmann Instruments Inc.)* (c) A single-pan analytical balance for high-precision weighing. The precision of this balance is 0.1 mg (0.0001 g). *(Courtesy Mettler Instrument Corporation, Hightstown, N.J.)*

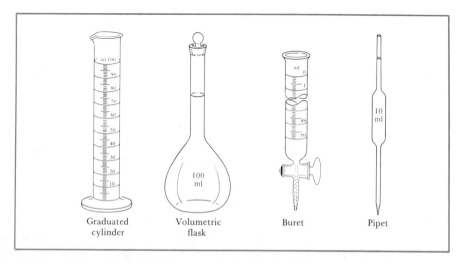

Figure 2.7 Calibrated glassware for measuring the volume of liquids.

depends on the accuracy required and the amount of material being weighed. Three standard balances are shown in Figure 2.6: a four-beam balance with precision up to 0.01 g; a single-pan, top-loading balance with a precision of 0.001 g (1 mg); a single-pan analytical balance with a precision up to 0.0001 g. Automatic-recording balances are also available.

The most common instruments for measuring liquids are the graduated cylinder, volumetric flask, buret, and pipet, which are shown in Figure 2.7. These calibrated pieces are usually made of glass and are available in various sizes.

The common laboratory tool for measuring temperature is a thermometer (see Figure 2.5).

2.14 Density

density

Density (d) is the ratio of the mass of a substance to the volume occupied by that mass; it is the mass per unit of volume and is given by the equation

$$d = \frac{\text{Mass}}{\text{Volume}}$$

Density is a physical characteristic of a substance and may be used as an aid to its identification. When the density of a solid or a liquid is given, the mass is usually expressed in grams and the volume in millilitres or cubic centimetres.

$$d = \frac{\text{Mass}}{\text{Volume}} = \frac{\text{g}}{\text{mL}} \quad \text{or} \quad d = \frac{\text{g}}{\text{cm}^3}$$

Since the volume of a substance (especially liquids and gases) varies with temperature, it is important to state the temperature along with the density. For example, the volume of 1.0000 g of water at 4°C is 1.0000 mL, whereas at 20°C it is 1.0018 mL, and at 80°C it is 1.0290 mL. Density, therefore, also varies with temperature.

The density of water at 4°C is 1.0000 g/mL but at 80°C the density of water is 0.9718 g/mL.

$$d^{4°C} = \frac{1.0000 \text{ g}}{1.0000 \text{ mL}} = 1.0000 \text{ g/mL}$$

$$d^{80°C} = \frac{1.0000 \text{ g}}{1.0290 \text{ mL}} = 0.9718 \text{ g/mL}$$

The density of iron at 20°C is 7.86 g/mL.

$$d^{20°C} = \frac{7.86 \text{ g}}{1.00 \text{ mL}} = 7.86 \text{ g/mL}$$

The density of iron indicates that it is about eight times as heavy as water per unit of volume. The densities of water, sulfur, lead, and gold are compared in Figure 2.8.

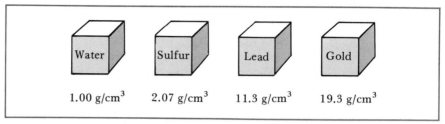

Figure 2.8 Comparison of the masses of 1.00 cm³ (1.00 mL) of water, sulfur, lead, and gold. The mass of each substance per cubic centimetre is its density. (Water is at 0°C; the three solids, at 20°C.)

Densities for liquids and solids are usually represented in terms of grams per millilitre or grams per cubic centimetre. The density of gases, however, is normally expressed in terms of grams per litre (g/L). Unless otherwise stated, gas densities are given for 0°C and 1 atmosphere pressure (discussed further in Chapter 12). Table 2.7 lists the densities of a number of common materials.

TABLE 2.7 Densities of some selected materials.
For comparing densities, the density of water is the reference for solids and liquids; air is the reference for gases.

Liquids and solids		Gases	
Substance	Density (g/mL at 20°C)	Substance	Density (g/L at 0°C)
Wood (Douglas fir)	0.512	Hydrogen	0.090
Ethyl alcohol	0.789	Helium	0.178
Cottonseed oil	0.926	Methane	0.714
Water (4°C)	**1.0000**	Ammonia	0.771
Sugar	1.59	Neon	0.90
Carbon tetrachloride	1.595	Carbon monoxide	1.25
Magnesium	1.74	Nitrogen	1.251
Sulfuric acid	1.84	**Air**	**1.293**
Sulfur	2.07	Oxygen	1.429
Salt	2.16	Hydrogen chloride	1.63
Aluminum	2.70	Argon	1.78
Silver	10.5	Carbon dioxide	1.963
Lead	11.3	Chlorine	3.17
Mercury	13.55		
Gold	19.3		

Suppose that water, carbon tetrachloride, and cottonseed oil are successively poured into a graduated cylinder. The result is a layered three-liquid system (Figure 2.9). Can we predict the order of the liquid layers? Yes, by looking up the liquid densities in Table 2.7. Carbon tetrachloride has the

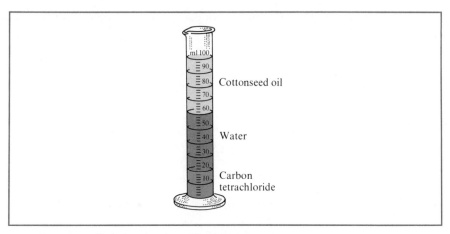

Figure 2.9 Relative density of liquids. When three immiscible (not capable of mixing) liquids are poured together, the liquid with the highest density will be the bottom layer. In the case of cottonseed oil, water, and carbon tetrachloride, cottonseed oil is the top layer.

greatest density (1.595 g/mL) and cottonseed oil has the lowest density (0.926 g/mL). Carbon tetrachloride will, therefore, form the bottom layer and cottonseed oil the top layer. Water, with a density between the other two liquids, will, of course, form the middle layer. This information can also be determined readily by experiment. Add a few millilitres of carbon tetrachloride to a beaker of water. The carbon tetrachloride, being more dense than water, will sink. Cottonseed oil, being less dense than water, will float when added to water. Direct comparisons of density in this manner can be made only with liquids that are *immiscible* (do not dissolve in one another).

The density of air at 0°C is approximately 1.293 g/L. Gases with densities less than this value are said to be "lighter than air." A helium-filled balloon will rise rapidly in air because the density of helium is only 0.178 g/L.

When an insoluble solid object is dropped into water, the object will sink or float—depending on its density. If the object is less dense than water, it will float, displacing a *mass* of water equal to the mass of the object. If the object is more dense than water, it will sink, displacing a *volume* of water equal to the volume of the object. This information can be used to determine the volume (and density) of irregularly shaped objects.

Sample calculations of density problems follow.

PROBLEM 2.16 What is the density of a mineral if 427 g of the mineral occupy a volume of 35.0 mL?

We need to solve for density, so we start by writing the formula for calculating density:

$$d = \frac{\text{Mass}}{\text{Volume}}$$

Then we substitute the data given in the problem into the equation and solve:

$$\text{Mass} = 427\text{ g} \qquad \text{Volume} = 35.0\text{ mL}$$

$$d = \frac{\text{Mass}}{\text{Volume}} = \frac{427\text{ g}}{35.0\text{ mL}} = 12.2\text{ g/mL} \quad \text{(Answer)}$$

■

PROBLEM 2.17 The density of gold is 19.3 g/mL. What is the mass of 25.0 mL of gold?

There are two ways to solve this problem: (1) Solve the density equation for mass, then substitute the density and volume data into the new equation and calculate; (2) use dimensional analysis for the data given.

Method 1. (a) Solve the density equation for mass:

$$d = \frac{\text{Mass}}{\text{Volume}} \qquad \text{Mass} = d \times \text{Volume}$$

(b) Substitute the data and calculate:

$$\text{Mass} = \frac{19.3\text{ g}}{\text{mL}} \times 25.0\text{ mL} = 482\text{ g} \quad \text{(Answer)}$$

Method 2. Dimensional analysis:

$$\textit{Data:} \quad d = \frac{19.3\text{ g}}{\text{mL}} \qquad \text{mL} = 25.0\text{ mL}$$

The conversion of units is

$$\frac{\text{g}}{\text{mL}} \times \text{mL} = \text{g}$$

$$\frac{19.3\text{ g}}{\text{mL}} \times 25.0\text{ mL} = 482\text{ g} \quad \text{(Mass of gold)}$$

■

PROBLEM 2.18 Calculate the volume of 100 g of ethyl alcohol.

From Table 2.7 we see that the density of ethyl alcohol is 0.789 g/mL. This also means that 1 mL of the alcohol weighs 0.789 g (1 mL/0.789 g). The calculation takes this form:

$$\frac{\text{mL}}{\text{g}} \times \text{g} = \text{mL}$$

Substituting the data, we get

$$\frac{1\text{ mL}}{0.789\text{ g}} \times 100\text{ g} = 127\text{ mL of ethyl alcohol}$$

This problem may also be done by solving the density equation for volume and then substituting the data in the new equation.

∎

PROBLEM 2.19   The water level in a graduated cylinder stands at 20.0 mL before and at 26.2 mL after a 16.74 g metal bolt is submerged in the water. (a) What is the volume of the bolt? (b) What is the density of the bolt?

$$\begin{aligned} \text{(a)} \quad 26.2 \text{ mL} &= \text{Volume of water plus bolt} \\ -20.0 \text{ mL} &= \text{Volume of water} \\ \hline 6.2 \text{ mL} &= \text{Volume of bolt} \quad \text{(Answer)} \end{aligned}$$

(b) $d = \dfrac{\text{Mass of bolt}}{\text{Volume of bolt}} = \dfrac{16.74 \text{ g}}{6.2 \text{ mL}} = 2.7 \text{ g/mL}$ (Answer)

∎

2.15 Specific Gravity

specific gravity

The **specific gravity** (sp gr) of a substance is a ratio of the density of that substance to the density of another substance. Water is usually used as the reference standard for solids and liquids:

$$\text{sp gr} = \frac{\text{Density of a liquid or solid}}{\text{Density of water}} \quad \text{or} \quad \frac{\text{Density of a gas}}{\text{Density of air}}$$

Specific gravity has no units because it is calculated by dividing two numbers having the same units. It is a number that compares the density of a liquid or a solid with that of water, or the density of a gas with that of air. Since specific gravity is unitless it tells us how many times as dense or heavy an object is compared to water.

PROBLEM 2.20   What is the specific gravity of mercury with respect to water at 4°C? (Density of water at 4°C is 1.000 g/mL.)

$$\text{sp gr} = \frac{\text{Density of mercury}}{\text{Density of water}} = \frac{13.55 \text{ g/mL}}{1.000 \text{ g/mL}}$$

sp gr of mercury = 13.55

The value for the specific gravity of mercury (13.55) tells us that, per unit volume, mercury is 13.55 times as heavy as water. Do you think that you could readily lift a litre (approximately 1 quart) of mercury?

∎

hydrometer

A **hydrometer** consists of a weighted bulb at the end of a sealed, calibrated tube. This instrument is used to measure the specific gravity of liquids (see Figure 2.10). When a hydrometer is floated in a liquid, the specific gravity is indicated on the scale at the surface of the liquid.

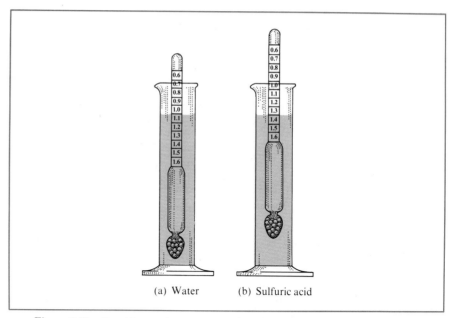

(a) Water   (b) Sulfuric acid

Figure 2.10   Specific gravity determination using hydrometers. The hydrometer in (a) is floating in water, showing a specific gravity of 1.0. The hydrometer in (b) is floating in dilute sulfuric acid (battery acid), showing a specific gravity of 1.3.

QUESTIONS   *An asterisk indicates a more challenging question or problem.*

A. **Review the Meanings of the New Terms Introduced in this Chapter**
   The terms listed in Section A of each set of Questions are new terms defined in the chapter. They appear in boldface type and occur in the chapter in the order listed in Question A.

1. Mass	7. Metre	13. Kilocalorie
2. Weight	8. Kilogram	14. Joule
3. Significant figures	9. Litre	15. Heat capacity
4. Rounding off numbers	10. Heat	16. Density
5. Scientific notation	11. Temperature	17. Specific gravity
6. Metric system, or SI	12. A calorie	18. Hydrometer

B. **Answers to the Following Questions Will Be Found in Tables and Figures**
   1. Use Table 2.2 to determine how many centimetres make up 1 kilometre.
   2. Refer to Figure 2.3. If milk were suddenly to be sold by the litre rather than by the quart, would you expect the price in cents per litre to be higher or lower than the price in cents per quart (assuming the basic value of the milk did not change)?
   3. A water bath is measured as 25°C, which is 75°C below the boiling point of water. If this temperature is measured in degrees Fahrenheit,
      (a) How many Fahrenheit degrees would this be below the boiling point of water?
      (b) What is the ratio of the number of Fahrenheit degrees to the number of Celsius degrees for this temperature range? (See Figure 2.5.)

4. Would more calories be needed to heat 20 g of aluminum or 20 g of gold through a temperature change of 40°C? (Refer to Table 2.6.)
5. Why are the top ends of the pipet and the volumetric flask (see Figure 2.7) narrower than the bulk of these volumetric instruments?
6. Refer to Table 2.7 and describe the arrangement you would see when the following immiscible materials are placed in a 100 mL graduated cylinder: 135.5 g of mercury, 25 g of water, 25 mL of carbon tetrachloride, and a cube of aluminum measuring 1.0 cm on an edge. All are liquids except the aluminum.
7. Arrange the following materials in order of increasing density: aluminum, gold, water, wood, ethyl alcohol, and lead.
8. Would carbon dioxide be a satisfactory lifting gas for a balloon? Explain.
9. Ice floats in water and sinks in ethyl alcohol. What information does this give us about the density of ice? (See Table 2.7.)

C. Review Questions

1. Why is the metric system of weights and measurements more desirable than the American system?
2. What are the abbreviations for the following?

   (a) Gram   (e) Decimetre   (i) Kilolitre
   (b) Milligram   (f) Centimetre   (j) Millilitre
   (c) Kilogram   (g) Cubic centimetre   (k) Angstrom
   (d) Megagram   (h) Nanometre

3. In a number, when is zero significant and when is it not significant?
4. What are the three rules for rounding off a number?
5. Distinguish between heat and temperature.
6. Which of the following statements are correct?

   (a) The prefix *milli* indicates one-millionth of the unit expressed.
   (b) The quantity 10 cm is equal to 100 mm.
   (c) The number 383.263 rounded to four significant figures becomes 383.3.
   (d) The number 8.6453 rounded to three figures becomes 8.64.
   (e) The number of significant figures in the number 29,004 is three.
   (f) The number 0.00652 contains three significant figures.
   (g) The sum of 24.928 g + 2.126 g should contain five significant figures.
   (h) The product of 14.63 cm × 2.50 cm should contain three significant figures.
   (i) One microsecond is 10^{-6} second.
   (j) One thousand metres is a shorter distance than 1000 yards.
   (k) One litre is a larger volume than 1 quart.
   (l) One centimetre is longer than 1 inch.
   (m) One cubic centimetre (cm^3) is equal to 1 millilitre.
   (n) The number 0.002894 expressed in exponential notation is 2.894×10^{-3}.
   (o) $2.0 \times 10^4 \times 6.0 \times 10^6 = 1.2 \times 10^{11}$.
   (p) $5.0 \times 10^5 \times 5.0 \times 10^{-3} = 2.5 \times 10^9$.
   (q) One degree on the Celsius scale is equal to 1 degree on the Kelvin scale and to 1.8 degrees on the Fahreheit scale.
   (r) The direction of heat flow is from cold to hot.
   (s) A calorie is a unit of temperature.
   (t) Temperature is a form of energy.
   (u) The density of water at 4°C is 1.00 g/mL.
   (v) A graduated cylinder would be a more accurate instrument for measuring 10.0 mL of water than would a pipet.
   (w) A hydrometer is an instrument used for measuring the specific gravity of liquids.

D. Review Problems

Significant Figures, Rounding, Exponential Notation, and Mathematical Review

1. How many significant figures are in each of the following numbers?
   - (a) 0.013
   - (b) 20.0
   - (c) 12.3
   - (d) 0.0077
   - (e) 0.0209
   - (f) 517,062
   - (g) 8.40×10^4
   - (h) 5.070×10^9

2. Round off the following numbers to three significant figures:
   - (a) 82.174
   - (b) 3.8826
   - (c) 0.03854
   - (d) 62.15
   - (e) 8.945
   - (f) 216.509
   - (g) 25.453
   - (h) 1.835×10^6

3. Express each of the following numbers in exponential notation:
   - (a) 3,400,000
   - (b) 0.0273
   - (c) 0.88
   - (d) 3066.2
   - (e) 0.00370
   - (f) 20.40
   - (g) 62,000,000
   - (h) 0.000077

4. Solve the following mathematical problems, giving answers to the proper number of significant figures:
   - (a) $18.58 + 2.2 =$
   - (b) $8.44 \times 13.3 =$
   - (c) $8.46 \times 10^2 \times 2.62 \times 10^5 =$
   - (d) $\dfrac{127.6}{5.5} =$
   - (e) $\dfrac{277 \times 5.38}{12.3} =$
   - (f) $347.6 + 1.87 =$
   - (g) $0.0277 \times 15.7 =$
   - (h) $3.07 \times 10^{-2} \times 8.44 \times 10^{-6} =$
   - (i) $\dfrac{0.1372}{81.4} =$
   - (j) $\dfrac{72.4}{466 \times 512} =$

5. Change these fractions to decimals. Express each answer to three significant figures.
   - (a) $\dfrac{5}{8}$
   - (b) $\dfrac{2}{7}$
   - (c) $\dfrac{12}{18}$
   - (d) $\dfrac{11}{22}$

6. Solve each equation for X:
   - (a) $2.43X = 7.5$
   - (b) $\dfrac{X}{17.2} = 8.06$
   - (c) $\dfrac{0.688}{X} = 0.38$
   - (d) $0.388X = 14.2$
   - (e) $\dfrac{X}{0.607} = 12.3$
   - (f) $\dfrac{5.2}{X} = 137$

7. Solve each equation for the unknown:
   - (a) $°C = \dfrac{160 - 32}{1.8}$
   - (b) $°F = 1.8(17) + 32$
   - (c) $K = 25 + 273$
   - (d) $\dfrac{7.8 \text{ g}}{\text{mL}} = \dfrac{26.42 \text{ g}}{\text{Volume}}$

Unit Conversions

8. Make the following conversions, showing mathematical setups:
   - (a) 16.0 cm to m
   - (b) 268 m to km
   - (c) 8.52 cm to mm
   - (d) 125 mm to km
   - (e) 0.525 cm to km
   - (f) 2.5 cm to Å
   - (g) 6.5×10^{-5} m to Å
   - (h) 8.07 m to cm
   - (i) 4.0 km to m
   - (j) 295 mm to cm
   - (k) 42 km to mm
   - (l) 24 nm to cm
   - (m) 0.525 km to cm
   - (n) 4.662 Å to nm
   - (o) 33.6 in. to cm
   - (p) 0.70 mile to in.
   - (q) 405 miles to km
   - (r) 1.00 in.^{2} to cm^2
   - (s) 46.2 m to ft
   - (t) 12.5 km to miles
   - (u) 5.2 in. to mm

9. Make the following conversions, showing mathematical setups:
   - (a) 4.68 g to mg
   - (b) 8.6×10^4 mg to kg
   - (c) 4.54 g to kg
   - (d) 38.4 kg to lb
   - (e) 78 mg to g
   - (f) 0.725 kg to mg
   - (g) 3.2 kg to g
   - (h) 125 lb to g

10. Make the following conversions, showing mathematical setups:
    - (a) 12.3 mL to L
    - (b) 2.25 L to mL
    - (c) 5.4 qt to mL
    - (d) 3.0×10^6 ft^3 to m^3
    - (e) 0.862 L to mL
    - (f) 40.5 L to gal
    - (g) 15 μL to mL
    - (h) 5.60 gal to L

11. An automobile traveling at 55 miles per hour is moving at what speed in (a) kilometres per hour and (b) feet per second?

12. The speed of light in a vacuum is 3.00×10^{10} cm per second. Calculate the speed of light in miles per second.

13. An experiment requires each student to use 7.45 g of sodium chloride. The instructor opens a new 1.00 lb jar of the salt. If 22 students each take exactly the assigned amount of salt, how much should be left in the bottle at the end of the lab period?

14. When the space satellite Voyager I, which gave us new data on the planet Saturn, reaches the planet Neptune in 1989 it will be traveling at an average speed of 13 miles per second. What will be its speed in (a) miles per hour and (b) kilometres per hour?

15. How many kilograms does a 155 lb man weigh?

16. The usual aspirin tablet contains 5.0 grains of aspirin. How many grams of aspirin are in one tablet (1 grain $= \frac{1}{7000}$ lb)?

17. The sun is approximately 93 million miles from the earth. How many seconds will it take light to travel from the sun to the earth if the velocity of light is 3.00×10^{10} cm per second?

18. The average weight of the heart of a human baby is about 1 ounce. What is the weight in milligrams?

*19. More sulfuric acid is manufactured in the United States than any other chemical; the annual production is about 8.1×10^{10} lb. What is the average daily production of sulfuric acid in tons? In kilograms?

20. The price of gold varies greatly and has been as high as $800 per ounce. What is the value of 250 g of gold at $445 per ounce? Gold is priced per troy ounce [1 lb (avoirdupois) $= 14.58$ oz (troy)].

21. An adult ruby-throated hummingbird has an average weight of 3.2 g, whereas an adult California condor may attain a weight of 21 lb. How many times heavier than the hummingbird is the condor?

22. At 41 cents per litre, how much will it cost to fill a 15 gal tank with gasoline?

23. A French automobile manufacturer claims that its sedan uses only 6.0 L of gasoline per 100 km. How many miles per gallon of gasoline could be expected from the car?

24. Calculate the volume, in litres, of a box 80 cm long by 45 cm wide by 60 cm high.

25. At a price of $1.51 per gallon, what will it cost to fill a 50 L tank with gasoline?

26. Calculate the volume, in cubic centimetres, of a box 1.20 m long, 24 cm wide, and 95 mm deep.

27. Wine bottles were recently changed in volume from $\frac{4}{5}$ quart (also called a fifth, from $\frac{1}{5}$ gallon) to 750 mL. Will there be more or less wine in the 750 mL bottle? Show evidence for your answer.

*28. An aquarium has the following dimensions: 30 in. long by 10 in. wide by 20 in. high. How much water will it take to fill the aquarium three-fourths full? Express your answer in both litres and gallons.

*29. Oil spreads in a thin layer on water and is commonly called an "oil slick." How much area in square metres (m^2) will 100 cm^3 of oil cover if it forms a layer 5 Å in thickness?

Temperature Conversions and Heat Capacity

30. Make the following conversions, showing mathematical setups:

    (a) 149°F to °C   (e) 22°C to °F

    (b) 0°F to °C   (f) −16°F to °C

    (c) 0°F to K   (g) 273°C to K

    (d) −12°C to °F   (h) 200 K to °C

31. Normal body temperature for humans is 98.6°F. What is this temperature on the Celsius scale?
32. Which is colder, −100°C or −142°F?
*33. (a) At what temperature are the Fahrenheit and Celsius temperatures exactly equal?
   (b) At what temperature are they numerically equal but opposite in sign?
34. How many calories of heat are required to raise the temperature of 80 g of water from 20°C to 70°C?
35. How many calories are required to raise the temperature of 80 g of iron from 20°C to 70°C?
*36. A 20.0 g piece of a metal at 203°C is dropped into 80.0 g of water at 25.0°C. The water temperature rises to 29.0°C. Calculate the heat capacity (specific heat) of the metal. Assume that all the heat lost by the metal is transferred to the water and no heat is lost to the surroundings.
*37. Assuming no heat losses by the system, what will be the final temperature when 80 g of water at 10°C are mixed with 10 g of water at 80°C?
*38. A 418 g piece of gold at 427°C is dropped into 200 mL of water at 22.0°C. The heat capacity of gold is 0.0312 cal/g°C. Calculate the final temperature of the mixture. (Assume no heat losses to the surroundings.)

Density and Specific Gravity

39. Calculate the density of a liquid if 15.60 mL of the liquid weigh 18.25 g.
40. A 28.0 mL sample of bromine weighs 87.4 g. What is the density of bromine?
41. When a 22.0 g piece of chromium metal was placed into a graduated cylinder containing 25.0 mL of water, the water level rose to 28.1 mL. Calculate the density of the chromium.
42. Concentrated hydrochloric acid has a density of 1.19 g/mL. Calculate the weight, in grams, of 1.00 L of this acid.
43. An empty graduated cylinder weighs 42.817 g. When filled with 50.0 mL of an unknown liquid it weighs 86.773 g. What is the density of the liquid?
44. What weight of mercury (density 13.6 g/mL) will occupy a volume of 50.0 mL?
45. Thirty-five millilitres of ethyl alcohol (density 0.789 g/mL) are added to a graduated cylinder that weighs 44.28 g. What will be the weight of the cylinder plus the alcohol?
46. You are given three cubes, A, B, and C; one is magnesium, one is aluminum, and the third is silver. All three cubes weigh the same, but cube A has a volume of 29.0 mL, cube B has a volume of 18.8 mL, and cube C has a volume of 4.81 mL. Identify cubes A, B, and C.
*47. A cube of aluminum weighs 500 g. What will be the weight of a cube of lead of the same dimensions?
48. Twenty-five millilitres of water at 90°C weigh 24.12 g. Calculate the density of water at this temperature.
49. Calculate (a) the density and (b) the specific gravity of a solid that weighs 160 g and has a volume of 50.0 mL.
50. Five hundred millilitres of gasoline weigh 358 g. Is the specific gravity of gasoline greater or less than that of water? Show evidence for your answer.
51. Which liquid will occupy the greater volume, 100 g of water or 100 g of ethyl alcohol? Explain.
52. A gold bullion dealer advertised a bar of pure gold for sale. The gold bar weighed 3100 g and measured 2.00 cm by 15.0 cm by 6.00 cm. Was the gold bar pure gold? Show evidence for your answer.
53. The largest nugget of gold on record was found in 1872 in New South Wales, Australia, and weighed 93.3 kg. What was the volume of this nugget in cubic centimetres? In litres?

*54. Forgetful Freddie placed 25.0 mL of ethyl alcohol in a graduated cylinder that weighed 89.450 g when empty. When Freddie placed a metal slug weighing 15.434 g into the cylinder, the volume rose to 30.7 mL. Freddie was asked to calculate the density of the alcohol and of the metal slug from his data, but he forgot to obtain the weight of the alcohol. He was told that if he weighed the cylinder containing the alcohol and the slug, he would have enough data for the calculations. He did so, and found it weighed 124.634 g. Calculate the density of the alcohol and of the metal slug.

*55. A solution made by adding 265 g of sulfuric acid to 500 mL of water had a volume of 607 mL.
   (a) What value will a hydrometer read when placed in this solution?
   (b) What volume of concentrated sulfuric acid ($d = 1.84$ g/mL) was added?

3 Properties of Matter

After studying Chapter 3 you should be able to:

1. Understand the terms listed in Question A at the end of the chapter.
2. Identify the three physical states of matter and list the physical properties that characterize each state.
3. Distinguish between the physical and chemical properties of matter.
4. Classify changes undergone by matter as being either physical or chemical changes.
5. Distinguish between substances and mixtures.
6. Distinguish between kinetic and potential energy.
7. State the Law of Conservation of Mass.
8. State the Law of Conservation of Energy.
9. Explain why the laws dealing with the conservation of mass and of energy may be combined into a single more accurate general statement.
10. Calculate the percent composition of compounds from the weights of the elements involved in a chemical reaction or vice versa.

3.1 Matter Defined

matter

The entire universe consists of matter and energy. Every day we come into contact with countless kinds of matter. Air, food, water, rocks, soil, glass, this book—all are different types of matter. Broadly defined, **matter** is *anything* that has mass and occupies space.

Matter may be quite invisible. If an apparently empty test tube is submerged mouth downward in a beaker of water, the water rises only slightly into the tube. The water cannot rise further because the tube is filled with invisible matter—air (see Figure 3.1, page 46).

To the eye matter appears to be continuous and unbroken. However, it is actually discontinuous and is composed of discrete, tiny particles called *atoms*. The particle nature of matter will become evident when we study atomic structure and the properties of gases.

3.2 Physical States of Matter

solid

Matter exists in three physical states: solid, liquid, and gas. A **solid** is characterized by having a definite shape and volume, with particles that cohere rigidly to one another. The shape of a solid may be independent of its container.

Figure 3.1   An apparently empty test tube is submerged, mouth downward, in water. Only a small volume of water rises into the tube, which is actually filled with air. This experiment proves that air, which is matter, occupies space.

For example, a crystal of sulfur has the same shape and volume whether it is placed in a beaker or simply laid on a glass plate.

Most commonly occurring solids, such as salt, sugar, quartz, and metals, are *crystalline.* Crystalline materials exist in regular, recurring internal geometric patterns. Solids such as plastics, glass, and gels, because they do not have any particular regular internal geometric arrangement, are called **amorphous** solids. (*Amorphous* means without shape or form.) Figure 3.2 illustrates three crystalline solids—halite, quartz, and gypsum.

A **liquid** is characterized by having a definite volume, but not a definite shape, with particles that cohere firmly but not rigidly. Although held together by strong attractive forces, the particles are able to move freely but remain in close contact with one another. Particle mobility gives the liquid fluidity and causes it to take the shape of the container in which it is stored. Figure 3.3 shows the same amount of liquid in differently shaped containers.

A **gas** is characterized by having no fixed shape or volume, with particles that are moving independently of one another. The particles in the gaseous state have gained enough energy to overcome the attractive forces holding them together as liquids or solids. A gas presses continuously and in all directions upon the walls of its container. Because of this quality, a gas completely fills a container. The particles of a gas are relatively far apart, compared to those of solids and liquids. The actual volume of the gas particles is usually very small in comparison to the volume of space occupied by the gas. A gas therefore may be compressed into a very small volume or expanded practically indefinitely. Liquids cannot be compressed to any great extent, and solids are even less compressible than liquids.

When a bottle of ammonia solution is opened in one corner of the laboratory, you can soon smell its familiar odor in all parts of the room. The ammonia

amorphous

liquid

gas

Figure 3.2   These three naturally occurring substances are examples of regular geometric formations that are characteristic of crystalline solids: (a) halite (salt); (b) quartz; (c) gypsum.

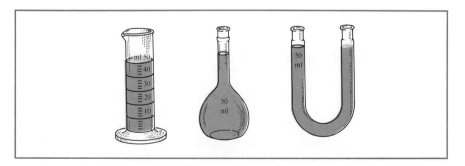

Figure 3.3   Liquids have the property of fluidity and assume the shape of their container, as illustrated in each of the three different calibrated containers.

gas escaping from the solution demonstrates that gaseous particles move freely and rapidly, and tend to permeate the entire area into which they are released.

Although matter is discontinuous, attractive forces exist that hold particles together and give matter its appearance of continuity. These attractive forces are strongest in solids, giving them rigidity; they are weaker in liquids, but still strong enough to hold liquids to definite volumes. In gases the attractive forces are so weak that the particles of a gas are practically independent of one another. Table 3.1 lists a number of common materials that exist as solids, liquids, and gases. Table 3.2 summarizes comparative properties of solids, liquids, and gases.

TABLE 3.1 Common materials in the solid, liquid, and gaseous states of matter at room temperature.

Solids	Liquids	Gases
Aluminum	Alcohol	Acetylene
Copper	Blood	Air
Gold	Gasoline	Butane
Polyethylene	Honey	Carbon dioxide
Salt	Mercury	Chlorine
Sand	Oil	Helium
Steel	Vinegar	Methane
Sulfur	Water	Oxygen

TABLE 3.2 Physical properties of solids, liquids, and gases.

State	Shape	Volume	Particles	Compressibility
Solid	Definite	Definite	Rigidly cohering; tightly packed	Very slight
Liquid	Indefinite	Definite	Mobile; cohering	Slight
Gas	Indefinite	Indefinite	Independent of one another and relatively far apart	High

3.3 Substances and Mixtures

The term *matter* refers to the total concept of material things. There are thousands of distinct and different kinds of matter. Upon closely examining different samples of matter, we can observe them to be either homogeneous or heterogeneous. By homogeneous we mean uniform in appearance when observed by the unaided eye or through a microscope. Matter that has identical properties throughout is **homogeneous.** Matter consisting of two or more physically distinct

homogeneous
heterogeneous
phase

phases is **heterogeneous. A phase** is a homogeneous part of a system separated from other parts by physical boundaries. A system of ice and water is heterogeneous, containing both solid and liquid phases, although each physical state of water is uniform in composition and is homogeneous. Whenever we have a system in which definite boundaries exist between the components, no matter whether they are in the solid, liquid, or gaseous states, the system has more than one phase and is heterogeneous. Thus, when we first put a spoonful of sugar into water, there exist a solid and a liquid phase, and the system is heterogeneous. After the sugar has been stirred and dissolved, the system has only one phase and is homogeneous.

substance

A **substance** is a particular kind of matter that is homogeneous and has a definite, fixed composition. Substances, sometimes known as *pure substances,* occur in two forms: elements and compounds. Several examples of elements and compounds are copper, gold, oxygen, salt, sugar, and water. Elements and compounds are discussed in more detail in Chapter 4.

mixture

Matter that contains two or more substances mixed together is known as a **mixture.** Mixtures are variable in composition and may be either homogeneous or heterogeneous. When sugar is dissolved in water, a sugar solution is formed. All parts of this solution are sweet and contain both substances, sugar and water, uniformly mixed. Solutions are homogeneous mixtures. Air is a homogeneous mixture (solution) of several gases. If we examine ordinary concrete, granite, iron ore, or other naturally occurring mineral deposits, we observe them to be heterogeneous mixtures of several different substances. Of course, it is very easy to prepare a heterogeneous mixture simply by physically mixing two or more substances, such as sugar and salt. We will consider mixtures again in Chapter 4. Figure 3.4 illustrates the relationship between homogeneous and heterogeneous matter.

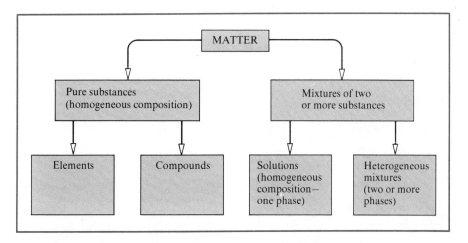

Figure 3.4   Classification of matter: A pure substance is always homogeneous. A mixture always contains two or more substances and may be either homogeneous (a solution) or heterogeneous.

3.4 Properties of Substances

properties

How do we recognize substances? Each substance has a set of **properties** that is characteristic of that substance and gives it a unique identity. Properties are the personality traits of substances and are classified as either physical or chemical.

physical properties

Physical properties are the inherent characteristics of a substance that may be determined without altering the composition of that substance; they are associated with its physical existence. Common physical properties are color, taste, odor, state of matter (solid, liquid, or gas), density, melting point, and boiling point. **Chemical properties** describe the ability of a substance to form new

chemical properties

substances, either by reaction with other substances or by decomposition.

We can select a few of the physical and chemical properties of chlorine as an example. Physically, chlorine is a gas about 2.4 times heavier than air. It is yellowish green in color and has a disagreeable odor. Chemically, chlorine will not burn but will support the combustion of certain other substances. It is used as a bleaching agent, as a disinfectant for water, and in many chlorinated substances such as refrigerants and insecticides. When chlorine combines with the metal sodium, it forms a salt, sodium chloride. These properties, among others, help to characterize and identify chlorine.

Substances are recognized and differentiated by their properties. Table 3.3 lists four substances and tabulates several of their common physical properties. Information about common physical properties, such as given in Table 3.3, is readily available in handbooks of chemistry and physics. Scientists don't pretend to know all the answers or to remember voluminous amounts of data, but it is important for them to know where to look for data in the literature. Handbooks are one of the most widely used resources for scientific data.

No two substances will have identical physical and chemical properties.

TABLE 3.3 **Physical properties of chlorine, water, sugar, and acetic acid.**

Substance	Color	Odor	Taste	Physical state	Boiling point (°C)	Freezing point (°C)
Chlorine	Yellowish green	Sharp, suffocating	Sharp, sour	Gas	−34.6	−101.6
Water	Colorless	Odorless	Tasteless	Liquid	100.0	0.0
Sugar	White	Odorless	Sweet	Solid	Decomposes 170–186	—
Acetic acid	Colorless	Like vinegar	Sour	Liquid	118.0	16.7

3.5 Physical Changes

physical change

Matter can undergo two types of changes, physical and chemical. **Physical changes** are mainly changes in physical properties (such as size, shape, density) or state of matter without an accompanying change in composition. The changing of ice into water and water into steam are physical changes from one state of matter into another. No new substances are formed in these physical changes (see Figure 3.5).

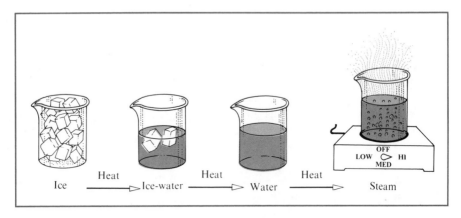

Figure 3.5 Physical changes in the appearance and state of water.

When a platinum wire is heated in a burner flame, the wire will become red-hot. It returns to its original silvery, metallic form after cooling. The platinum undergoes a physical change in appearance while in the flame, but its composition remains the same.

3.6 Chemical Changes

chemical change

In a **chemical change,** new substances are formed that have different properties and composition from the original material. The new substances need not in any way resemble the initial material.

When a clean copper wire is heated in a burner flame, a change in the appearance of the wire is readily noted after it cools. The copper no longer has its characteristic color, but now appears black. The black material is copper(II) oxide, a new substance formed when copper is combined chemically with oxygen in the air during the heating process. The wire before heating was essentially 100% copper, whereas the black copper(II) oxide contains only 79.9% copper, the rest being oxygen (see Figure 3.6). When both platinum and copper are heated under the conditions described, platinum, which does not readily combine with oxygen, changes only physically, but copper changes chemically as well as physically.

Mercury(II) oxide is an orange-red powder that, when subjected to high temperature (500–600°C), decomposes into a colorless gas (oxygen) and a silvery, liquid metal (mercury). The composition of each product, as well as its physical appearance, is noticeably different from that of the starting compound. When mercury(II) oxide is heated in a test tube (see Figure 3.7), small globules of mercury are observed collecting on the cooler part of the tube. Evidence of

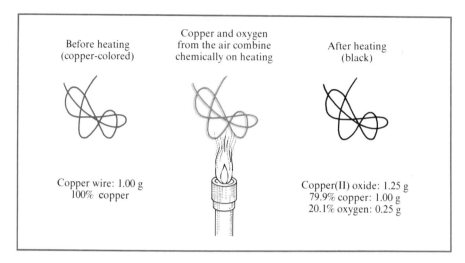

Before heating
(copper-colored)

Copper and oxygen
from the air combine
chemically on heating

After heating
(black)

Copper wire: 1.00 g
100% copper

Copper(II) oxide: 1.25 g
79.9% copper: 1.00 g
20.1% oxygen: 0.25 g

Figure 3.6 Chemical change: formation of copper(II) oxide from copper and oxygen.

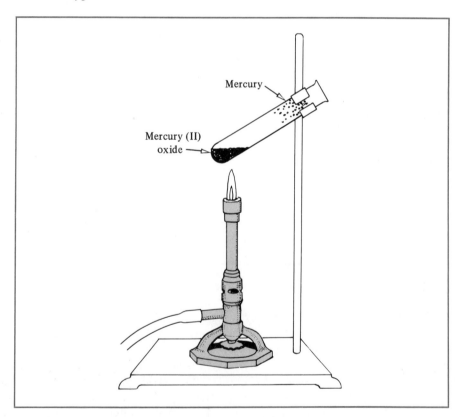

Mercury

Mercury (II)
oxide

Figure 3.7 Heating of mercury(II) oxide causes it to decompose into mercury and oxygen. Observation of the mercury and oxygen with properties different from mercury(II) oxide is evidence that a chemical change has occurred.

the oxygen formed is observed when a glowing wood splint lowered into the tube bursts into flame. Oxygen supports and intensifies the combustion of the wood. From these observations, we can conclude that a chemical change has taken place.

Chemists have devised *chemical equations* as a shorthand method of expressing chemical changes. The two examples of chemical changes just presented may be represented by the following word equations:

$$\text{Copper} + \text{Oxygen} \xrightarrow{\Delta} \text{Copper(II) oxide (Cupric oxide)} \tag{1}$$

$$\text{Mercury(II) oxide} \xrightarrow{\Delta} \text{Mercury} + \text{Oxygen} \tag{2}$$

Equation (1) states that copper plus oxygen when heated produce copper(II) oxide. Equation (2) states that mercury(II) oxide when heated produces mercury plus oxygen. The arrow means "produces"; it points to the products. The delta sign (Δ) represents heat. The starting substances (copper, oxygen, and mercury(II) oxide) are called the *reactants,* and the substances produced (copper(II) oxide, mercury, and oxygen) are called the *products.* In later chapters, equations are presented in a more abbreviated form, with symbols used for each substance.

In most cases a physical change accompanies a chemical change. Table 3.4 lists common physical and chemical changes. In the examples given in the table, you will note that wherever a chemical change occurs, a physical change occurs also. However, wherever a physical change is listed, only a physical change occurs.

TABLE 3.4 Examples of processes involving physical or chemical changes.

Process taking place	Type of change	Accompanying physical changes
Rusting of iron	Chemical	Shiny, bright metal changes to reddish brown rust
Boiling of water	Physical	Liquid changes to vapor
Burning of sulfur in air	Chemical	Yellow solid sulfur changes to gaseous, choking sulfur dioxide
Melting of lead	Physical	Solid changes to liquid
Combustion of gasoline	Chemical	Liquid gasoline burns to gaseous carbon monoxide, carbon dioxide, and water
Cutting of a diamond	Physical	Small diamonds are made from a larger diamond
Sawing of wood	Physical	Smaller pieces of wood plus sawdust are made from a larger piece of wood
Burning of wood	Chemical	Wood burns to ashes and gaseous carbon dioxide and water
Heating of glass	Physical	Solid becomes pliable during heating and the glass may change its shape

3.7 Conservation of Mass

Law of
Conservation
of Mass

The **Law of Conservation of Mass** states that there is no detectable change in the total mass of the substances involved in a chemical change. This law, tested by extensive laboratory experimentation, is the basis for the quantitative weight relationships among reactants and products.

The decomposition of mercury(II) oxide into mercury and oxygen illustrates this law. One hundred grams of mercury(II) oxide decomposes into 92.6 g of mercury and 7.40 g of oxygen.

$$\text{Mercury(II) oxide} \longrightarrow \text{Mercury} + \text{Oxygen}$$
$$\text{100.0 g} \qquad\qquad \text{92.6 g} \qquad \text{7.40 g}$$

| 100 g Reactant | $\longrightarrow$ | 100 g Products |

Sealed within the ordinary photographic flashbulb are fine wires of magnesium (a metal), and oxygen (a gas). When these reactants are energized, they combine chemically, producing magnesium oxide, together with a blinding white light and considerable heat. The chemical change may be represented by this equation:

$$\text{Magnesium} + \text{Oxygen} \longrightarrow \text{Magnesium oxide} + \text{Heat} + \text{Light}$$

When weighed before and after the chemical change, as illustrated in Figure 3.8, the bulb shows no increase or decrease in weight.

Mass of reactants = Mass of products

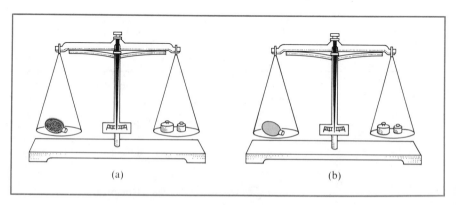

(a)   (b)

Figure 3.8   The flashbulb, containing magnesium and oxygen, weighs the same before (a) and after (b) the bulb is flashed. When the bulb is flashed, a chemical change occurs. The original substances are changed into the white powder, magnesium oxide.

3.8 Energy

energy

From the prehistoric discovery that fire could be used to warm shelters and cook food to the modern-day discovery that nuclear reactors can be used to produce vast amounts of controlled energy, man's progress has been directed by the ability to harness, produce, and use energy. **Energy** is the capacity of matter to do work. Energy exists in several forms; some of the more common forms are mechanical, chemical, electrical, heat, nuclear, and radiant or light energy. Matter can have both potential and kinetic energy.

potential energy

Potential energy is stored energy, or energy an object possesses because of its relative position. For example, a ball located 20 feet above the ground has more potential energy than when located 10 feet above the ground, and will bounce higher when allowed to fall. Water backed up behind a dam represents potential energy that can be converted into useful work in the form of electrical energy. Gasoline represents a source of stored chemical potential energy. When gasoline burns, together with oxygen, heat is released and the substances formed have less potential energy than the gasoline and the oxygen. The heat released is associated with the loss of potential energy of the gasoline.

kinetic energy

Kinetic energy is the energy that matter possesses because of its motion. When the water behind the dam is released and allowed to flow, its potential energy is changed into kinetic energy, which may be used to drive generators and produce electricity. All moving bodies possess kinetic energy. The pressure exerted by a confined gas is due to the kinetic energy of rapidly moving gas particles. We all know the results when two moving vehicles collide—their kinetic energy is expended in the crash that occurs.

Energy may be converted from one form to another form. Some kinds of energy can be converted to other forms easily and efficiently. For example, mechanical energy can be converted to electrical energy with an electric generator at better than 90% efficiency. On the other hand, solar energy has thus far been directly converted to electrical energy at an efficiency of about 15%.

3.9 Energy in Chemical Changes

In all chemical changes, matter either absorbs or releases energy. Chemical changes can be used to produce different forms of energy. Electrical energy to start automobiles is produced by chemical changes in the lead storage battery. Light energy for photographic purposes occurs as a flash during the chemical change in the magnesium flashbulb. Heat and light energies are released from the combustion of fuels. All the energy needed for our life processes—breathing, muscle contraction, blood circulation, and so on—is produced by chemical changes occurring within the cells of the body.

Conversely, energy is used to cause chemical changes. For example, a chemical change occurs in the electroplating of metals when electrical energy is passed through a salt solution in which the metal is submerged. A chemical change also occurs when radiant energy from the sun is used by plants in the process of photosynthesis. And, as we saw, a chemical change occurs when heat causes mercury(II) oxide to decompose into mercury and oxygen. Chemical changes are often used primarily to produce energy rather than to produce new substances. The heat or thrust generated during the combustion of fuels is more important than the new substances formed.

3.10 Conservation of Energy

An energy transformation occurs whenever there is a chemical change. If energy is absorbed during the change, the products will have more chemical or potential energy than the reactants. Conversely, if energy is given off in a chemical change, the products will have less chemical or potential energy than the reactants. Water, for example, can be decomposed in an electrolytic cell. Electrical energy is absorbed in the decomposition, and the products—hydrogen and oxygen— have a greater chemical or potential energy level than that of water. This potential energy is released in the form of heat and light when the hydrogen and oxygen are burned to form water again (see Figure 3.9). Thus, energy can be

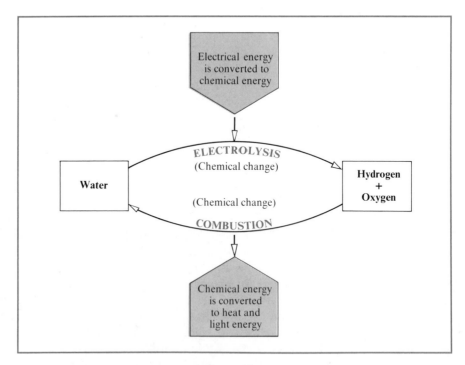

Figure 3.9   Energy transformations during the electrolysis of water and the combustion of hydrogen and oxygen. Electrical energy is converted to chemical energy in the electrolysis, and chemical energy is converted to heat and light energy in the combustion.

changed from one form to another or from one substance to another and therefore is not lost.

The energy changes occurring in many systems have been thoroughly studied by many investigators. No system has been found to acquire energy except at the expense of energy possessed by another system. This is stated in other words as the **Law of Conservation of Energy:** Energy can be neither created nor destroyed, though it may be transformed from one form to another.

Law of Conservation of Energy

3.11 Interchangeability of Matter and Energy

Sections 3.7–3.10 dealt with matter and energy. The two are clearly related; any attempt to deal with one inevitably involves the other. The nature of this relationship eluded the most able scientists until the beginning of the 20th century. Then, in 1905, Albert Einstein (Figure 3.10) presented one of the most original scientific concepts ever devised.

Einstein stated that the quantity of energy (E) equivalent to the mass (m) could be calculated by the equation $E = mc^2$, where m is in grams and c is the velocity of light (3.0×10^{10} cm/s). According to Einstein's equation, whenever energy is absorbed or released by a substance, there must be a loss or gain of mass. Although the energy changes in chemical reactions are measurable and may appear to be large, the amounts are relatively small. The accompanying difference in mass between reactants and products in chemical changes is so

Figure 3.10 Albert Einstein (1879–1955), world-renowned physicist and author of the theory of relativity and the interrelationship between matter and energy: $E = mc^2$. *(Courtesy of The New World Library of World Literature, Inc., The University and Dr. Einstein.)*

small that it cannot be detected by available measuring instruments. According to Einstein's equation, 2.2×10^7 cal (9×10^{14} ergs) of energy are equivalent to 0.000001 g (1 μg) of mass:

$$1 \ \mu\text{g mass} = 2.2 \times 10^7 \text{ cal energy}$$

In a more practical sense, when 2.8×10^3 g of carbon are burned to carbon dioxide, 2.2×10^7 cal of energy are released. Of this very large amount of carbon only about one-millionth of a gram, which is $3.6 \times 10^{-8}\%$ of the starting mass, is converted to energy. Therefore, in practice we may treat the reactants and products of chemical changes as having constant mass. However, because mass and energy are interchangeable, the two laws dealing with the conservation of matter may be combined into a single and generally more accurate statement:

The total amount of mass and energy remains constant during chemical change.

QUESTIONS

A. Review the Meanings of the New Terms Introduced in this Chapter

1. Matter
2. Solid
3. Amorphous
4. Liquid
5. Gas
6. Homogeneous
7. Heterogeneous
8. Phase
9. Substance
10. Mixture
11. Properties
12. Physical properties
13. Chemical properties
14. Physical change
15. Chemical change
16. Law of Conservation of Mass
17. Energy
18. Potential energy
19. Kinetic energy
20. Law of Conservation of Energy

B. Answers to the Following Questions Will Be Found in Tables and Figures

1. Name three liquids listed in Table 3.1 that are mixtures.
2. Which of the gases listed in Table 3.1 is not a pure substance?
3. What physical properties do solids and liquids have in common? (See Table 3.2.)
4. In what physical state will acetic acid exist at 150°C? (See Table 3.3.)
5. In what physical state will water exist at 293 K? (See Table 3.3.)
6. In what physical state will chlorine exist at 100 K? (See Table 3.3.)
7. From Figure 3.1, what evidence can you find that gases occupy space?
8. What effect does the absorption of heat energy have on mercury(II) oxide? (See Figure 3.7.)
9. What physical changes occur to the matter in the flashbulb of Figure 3.8 when the bulb is flashed?

C. Review Questions

1. List three substances in each of the three physical states of matter.
2. Explain why a gas can be compressed and why a liquid cannot be compressed appreciably.
3. In terms of the properties of the ultimate particles, explain why a liquid can be poured but a solid cannot be poured.
4. When the stopper is removed from a partly filled bottle containing solid and liquid acetic acid at 16.7°C, a strong vinegar-like odor is noticeable immediately. How many acetic acid phases must be present in the bottle? Explain.

5. Is the system enclosed in the bottle of Question 4 homogeneous or heterogeneous? Explain.
6. Distinguish between physical and chemical properties of matter.
7. Is a system containing only water necessarily homogeneous? Explain.
8. Is a system containing only one substance necessarily homogeneous? Explain.
9. Distinguish between physical and chemical changes.
10. Classify the following as primarily physical or chemical changes:
    (a) Boiling water   (c) Boiling an egg   (e) Souring milk
    (b) Freezing ice cream   (d) Homogenizing milk
11. Classify the following as primarily physical or chemical changes:
    (a) Formation of a snowflake   (d) Burning of gasoline
    (b) Cake batter being stirred   (e) Cutting a piece of sodium
    (c) Explosion of a firecracker   (f) Decomposition of limestone when heated
12. Reread Section 3.4 and list those properties given for chlorine that are physical and those that are chemical.
13. Cite the evidence demonstrating that the heating of mercury(II) oxide brings about a chemical change.
14. Distinguish between potential and kinetic energy.
15. Is chemical energy potential or kinetic?
16. In an ordinary chemical change, why can we consider that mass is neither lost nor gained (for practical purposes)?
17. When the flashbulb of Figure 3.8 is flashed, energy is given off to the surroundings. Explain why the mass of the bulb appears to be the same after the flashing as it was before, although, according to Einstein, energy is equivalent to mass.
18. Which of the following statements are correct? (Try to answer this question without referring to the text.)
    (a) Liquids are the most compact state of matter.
    (b) Liquids have a definite volume and a definite shape.
    (c) Matter in the solid state is discontinuous — that is, it is made up of discrete particles.
    (d) Wood is homogeneous.
    (e) Wood is a substance.
    (f) Dirt is a substance.
    (g) Seawater, although homogeneous, is considered to be a mixture.
    (h) Any system that consists of two or more phases is heterogeneous.
    (i) A solution, although it contains dissolved material, is considered to be homogeneous.
    (j) Boiling water represents a chemical change because no change of composition occurs.
    (k) All of the following represent chemical change: baking a cake, frying an egg, leaves changing color, iron changing to rust.
    (l) All of the following represent physical changes: breaking a stick, melting wax, folding a napkin, burning hydrogen to form water.
    (m) The normal boiling point of water is 100°C.
    (n) A stretched rubber band possesses kinetic energy.
    (o) An automobile rolling down a hill possesses both kinetic and potential energy.
    (p) When heated in air, metallic copper gains weight.
    (q) When heated in air, mercury(II) oxide loses weight.
    (r) The two types of pure substances are elements and mixtures.

D. Review Problems
1. Calculate the boiling point of chlorine (see Table 3.3) in (a) degrees Fahrenheit, and (b) degrees Kelvin.
2. Bromine boils at 138°F. Convert to degrees Celsius.

3. What weight of mercury can be obtained from 85.0 g of mercury(II) oxide?

4. If a mixture is composed of 20 g of iron filings and 12 g of powdered sulfur, what is the percentage of iron in the mixture?

*5. When 10.5 g of magnesium was heated in air, 17.4 g of magnesium oxide was produced. Given the chemical reaction,

$$\text{Magnesium} + \text{Oxygen} \longrightarrow \text{Magnesium oxide}$$

(a) What weight of oxygen has combined with the magnesium?

(b) What percentage of the magnesium oxide is magnesium?

*6. If a U.S. 25-cent coin weighs about 5.5 g,

(a) How many calories would be released by the complete conversion of a 25-cent coin to energy? (1×10^{-6} g $= 2.2 \times 10^7$ cal)

(b) If 3.03×10^5 cal are needed to heat a gallon of water from room temperature (20°C) to boiling (100°C), how many gallons of water could be heated to the boiling point by the energy from part (a)?

4

Elements
and Compounds

After studying Chapter 4 you should be able to:

1. Understand the terms listed in Question A at the end of the chapter.
2. List in order of abundance the five most abundant elements in the earth's crust, seawater, and atmosphere.
3. List in order of abundance the six most abundant elements in the human body.
4. Write the symbols when given the names or write the names when given the symbols of the common elements listed in Table 4.3.
5. Understand that compounds exist as either molecules or ions.
6. Differentiate among atoms, molecules, and ions.
7. State the Law of Definite Composition.
8. Understand how symbols, including subscripts and parentheses, are used to write chemical formulas.
9. Differentiate between compounds and mixtures.
10. List the characteristics of metals and nonmetals.
11. List the elements that occur as diatomic molecules.
12. Name binary compounds from their formulas.
13. Balance simple chemical equations when the formulas are given.

4.1 Elements

element

All the words in the English dictionary are formed from an alphabet consisting of only 26 letters. All known substances on earth—and most probably in the universe, too—are formed from a sort of "chemical alphabet" consisting of 106 known elements. An **element** is a fundamental or elementary substance that cannot be broken down, by chemical means, to simpler substances. Elements are the basic building blocks of all substances. The elements are numbered in order of increasing complexity beginning with hydrogen, number 1. Of the first 92 elements, 88 are known to occur in nature. The other four—technetium (43), promethium (61), astatine (85), and francium (87)—either do not occur in nature or have only transitory existences resulting from radioactive decay. With the exception of number 94, elements 93–106 are not known to occur naturally, but have been synthesized—usually in very small quantities—in laboratories.

The discovery of trace amounts of element 94 (plutonium) in nature has been reported. Element 106 was reported to have been synthesized in 1974. No elements other than those on the earth have been detected on other bodies in the universe.

Most substances can be decomposed into two or more other simpler substances. We have seen that mercury(II) oxide can be decomposed into mercury and oxygen, and that water can be decomposed into hydrogen and oxygen. Sugar may be decomposed into carbon, hydrogen, and oxygen. Table salt is easily decomposed into sodium and chlorine. An element, however, cannot be decomposed into simpler substances by ordinary chemical changes.

atom

If we could take a small piece of an element, say copper, and divide it and subdivide it into smaller and smaller particles, we finally would come to a single unit of copper that we could no longer divide and still have copper. This ultimate particle, the smallest particle of an element that can exist, is called an **atom.** An atom is also the smallest unit of an element that can enter into a chemical reaction. Chapter 5 describes the smaller subatomic particles that make up atoms, but these particles no longer have the properties of elements.

4.2 Distribution of Elements

Elements are distributed very unequally in nature. Table 4.1 shows that ten of the elements make up about 99% of the weight of the earth's crust, seawater, and the atmosphere (see Figure 4.1). Oxygen, the most abundant of these, constitutes about 50% of this mass. This list does not include the mantle, which

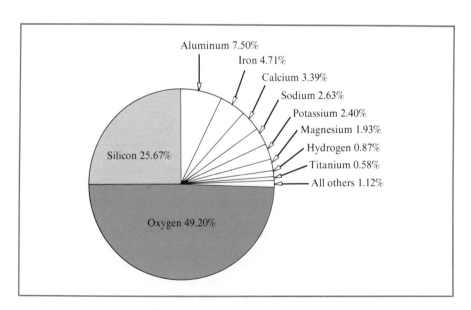

Figure 4.1 Weight percent of the elements in the earth's crust, seawater, and atmosphere.

is rich in iron and magnesium minerals, and the core of the earth, which is believed to be composed of metallic iron and nickel. The average distribution of the elements in the human body is shown in Table 4.2. Note again the high percentage of oxygen.

TABLE 4.1 Distribution of the elements in the earth's crust, seawater, and atmosphere.

Element	Weight percent	Element	Weight percent
Oxygen	49.20	Chlorine	0.19
Silicon	25.67	Phosphorus	0.11
Aluminum	7.50	Manganese	0.09
Iron	4.71	Carbon	0.08
Calcium	3.39	Sulfur	0.06
Sodium	2.63	Barium	0.04
Potassium	2.40	Nitrogen	0.03
Magnesium	1.93	Fluorine	0.03
Hydrogen	0.87		
Titanium	0.58	All others	0.47

TABLE 4.2 Average elemental composition of the human body.

Element	Weight percent
Oxygen	65.0
Carbon	18.0
Hydrogen	10.0
Nitrogen	3.0
Calcium	2.0
Phosphorus	1.0
Traces of several other elements	1.0

4.3 Names of the Elements

The names of the elements came to us from various sources. Many are derived from early Greek, Latin, or German words that generally described some property of the element. For example, iodine is taken from the Greek word *iodes,* meaning violet-like. Iodine, indeed, is violet in the vapor state. The name of the metal bismuth had its origin from the German words *weisse masse,* which means white mass. Miners called it *wismat;* it was later changed to *bismat,* and finally to bismuth. Some elements are named for the location of their discovery —for example, germanium, discovered in 1886 by Clemens Winkler (1838–1904), a German chemist. Others are named in commemoration of famous scientists, such as einsteinium and curium, for Albert Einstein and Marie Curie, respectively.

4.4 Symbols of the Elements

symbol

We all recognize Mr., N.Y., and St. as abbreviations for "mister," "New York," and "street." In like manner, chemists have assigned specific abbreviations to each element; these are called **symbols** of the elements. Fourteen of the elements have a single letter as their symbol, three elements have three-letter symbols, and all the others have two letters. The symbol stands for the element itself, for one atom of the element, and, as we shall see later, for a particular quantity of the element.

Rules governing symbols of elements are as follows:

1. Symbols are composed of one, two, or three letters.
2. If one letter is used, it is capitalized.
3. If two or three letters are used, the first is capitalized and the others are lowercase letters.

Examples: Sulfur S   Barium Ba

The symbols and names of all the elements are given in the table on the inside back cover of this book. Table 4.3 lists the more commonly used symbols. If we examine this table carefully, we note that most of the symbols start with the same letter as the name of the element that is represented. A number of symbols, however, appear to have no connection with the names of the elements they represent (see Table 4.4). These symbols have been carried over from earlier names (usually in Latin) of the elements and are so firmly implanted in the literature that their use is continued today.

Special care must be used in writing symbols. Begin each with a capital letter and use a lowercase second letter if needed. For example, consider Co, the symbol for the element cobalt. If through error CO (capital C and capital O) is written, the two elements carbon and oxygen (the *formula* for carbon mon-

TABLE 4.3 Symbols of the most common elements.

Element	Symbol	Element	Symbol	Element	Symbol
Aluminum	Al	Fluorine	F	Phosphorus	P
Antimony	Sb	Gold	Au	Platinum	Pt
Argon	Ar	Helium	He	Potassium	K
Arsenic	As	Hydrogen	H	Radium	Ra
Barium	Ba	Iodine	I	Silicon	Si
Bismuth	Bi	Iron	Fe	Silver	Ag
Boron	B	Lead	Pb	Sodium	Na
Bromine	Br	Lithium	Li	Strontium	Sr
Cadmium	Cd	Magnesium	Mg	Sulfur	S
Calcium	Ca	Manganese	Mn	Tin	Sn
Carbon	C	Mercury	Hg	Titanium	Ti
Chlorine	Cl	Neon	Ne	Tungsten	W
Chromium	Cr	Nickel	Ni	Uranium	U
Cobalt	Co	Nitrogen	N	Zinc	Zn
Copper	Cu	Oxygen	O		

TABLE 4.4 Symbols of the elements derived from early names. These symbols are in use today, even though they do not appear to correspond to the current name of the element.

Present name	Symbol	Former name
Antimony	Sb	Stibium
Copper	Cu	Cuprum
Gold	Au	Aurum
Iron	Fe	Ferrum
Lead	Pb	Plumbum
Mercury	Hg	Hydrargyrum
Potassium	K	Kalium
Silver	Ag	Argentum
Sodium	Na	Natrium
Tin	Sn	Stannum
Tungsten	W	Wolfram

oxide) are represented instead of the single element cobalt. Another example of the need for care in writing symbols is the symbol Ca for calcium versus Co for cobalt. The letters must be distinct, or else the wrong element may be represented.

A knowledge of symbols is essential for writing chemical formulas and equations. You should begin to learn symbols immediately since they will be used extensively in the remainder of this book and in any future chemistry courses you may take. One way to learn the symbols is to practice a few minutes a day by making side-by-side lists of names and symbols and then covering each list alternately and writing the corresponding name or symbol. Initially, it is a good plan to learn the symbols of the most common elements shown in Table 4.3.

The experiments of alchemists paved the way for the development of chemistry. Alchemists surrounded their work in mysticism, partly by devising a system of symbols known only to practitioners of alchemy (see Figure 4.2). The symbol ℞ (from the Latin *recipe*) is still used in medicine and was established during this time. In the early 1800s the Swedish chemist J. J. Berzelius (1779–1848) made a great contribution to chemistry by devising the present system of symbols using letters of the alphabet.

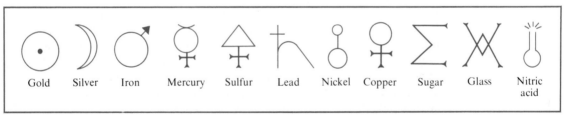

Gold   Silver   Iron   Mercury   Sulfur   Lead   Nickel   Copper   Sugar   Glass   Nitric acid

Figure 4.2   Some typical alchemists' symbols.

4.5 Compounds

compound

A **compound** is a distinct substance that contains two or more elements combined in a definite proportion by weight. Compounds, unlike elements, can be decomposed chemically into simpler substances—that is, into other compounds and/or elements.

The atoms of the elements that form a compound are combined in whole-number ratios, never as fractional parts of an atom. Compounds exist either as molecules or as ions. A **molecule** is the smallest uncharged individual unit of a compound formed by the union of two or more atoms. If we subdivide a drop of water into smaller and smaller particles, we ultimately obtain a single molecule of water consisting of two hydrogen atoms bonded to one oxygen atom. This molecule cannot be further subdivided without destroying the water and forming the elements hydrogen and oxygen. Thus, a molecule of water is the smallest unit of the substance water.

ion

An **ion** is a positive or negative electrically charged atom or group of atoms. The ions in a compound are held together in a crystalline structure by the attractive forces that exist between their positive and negative charges. Compounds consisting of ions do not exist as molecules. Sodium chloride is an example of an ionic compound. Although ionic compounds consist of large aggregates of positive and negative ions, their formulas are normally represented by the simplest ratio of the atoms in the compound. For example, the ratio of ions in sodium chloride is one sodium ion to one chlorine ion and the formula is represented as NaCl. The two types of compound, *molecular* and *ionic,* are illustrated in Figure 4.3.

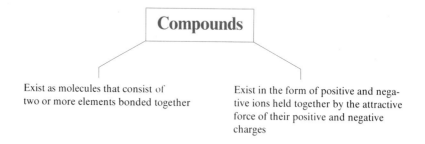

Compounds

Exist as molecules that consist of two or more elements bonded together

Exist in the form of positive and negative ions held together by the attractive force of their positive and negative charges

The compound carbon monoxide (CO) is composed of carbon and oxygen in the ratio of one atom of carbon to one atom of oxygen. Hydrogen chloride (HCl) contains a ratio of one atom of hydrogen to one atom of chlorine. Compounds may contain more than one atom of the same element. Methane (natural gas, CH_4) is composed of a ratio of one atom of carbon to four atoms of hydrogen; ordinary table sugar (sucrose, $C_{12}H_{22}O_{11}$) contains a ratio of 12 atoms of carbon to 22 atoms of hydrogen to 11 atoms of oxygen. These atoms are held together in the compound by *chemical bonds.*

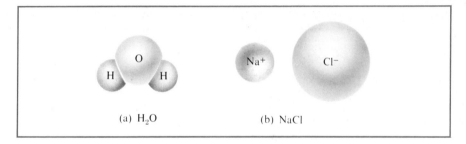

Figure 4.3 Representation of molecular and ionic (nonmolecular) compounds. (a) Two hydrogen atoms combined with an oxygen atom form a molecule of water. (b) A positively charged sodium ion and a negatively charged chloride ion form the compound sodium chloride.

Substance	Each molecule composed of	Formula
Carbon monoxide	1 carbon atom + 1 oxygen atom	CO
Hydrogen chloride	1 hydrogen atom + 1 chlorine atom	HCl
Methane	1 carbon atom + 4 hydrogen atoms	CH_4
Sugar (sucrose)	12 carbon atoms + 22 hydrogen atoms + 11 oxygen atoms	$C_{12}H_{22}O_{11}$
Water	2 hydrogen atoms + 1 oxygen atom	H_2O

There are over 5 million known registered compounds, with no end in sight as to the number that can and will be prepared in the future. Each compound is unique and has characteristic physical and chemical properties. Let us consider in some detail two compounds—water and mercury(II) oxide. Water is a colorless, odorless, tasteless liquid that can be changed to a solid (ice) at 0°C and to a gas (steam) at 100°C. It is composed of two atoms of hydrogen and one atom of oxygen per molecule, which represents 11.2% hydrogen and 88.8% oxygen by weight. Water reacts chemically with sodium to produce hydrogen gas and sodium hydroxide, with lime to produce calcium hydroxide, and with sulfur trioxide to produce sulfuric acid. No other compound has all these exact physical and chemical properties; they are characteristic of water alone.

Mercury(II) oxide is a dense, orange-red powder composed of a ratio of one atom of mercury to one atom of oxygen. Its composition by weight is 92.6% mercury and 7.4% oxygen. When it is heated to temperatures greater than 360°C, a colorless gas, oxygen, and a silvery liquid metal, mercury, are produced. These are specific physical and chemical properties belonging to mercury(II) oxide and to no other substance. Thus, a compound may be identified and distinguished from all other compounds by its characteristic properties.

4.6 Law of Definite Composition of Compounds

Many experiments extending over a long period of time have established that a specific compound always contains the same elements in a fixed proportion by weight. For example, water contains 11.2% hydrogen and 88.8% oxygen by weight. Water will always contain hydrogen and oxygen in this fixed weight ratio. The fact that water contains hydrogen and oxygen in this particular ratio does not mean that hydrogen and oxygen cannot combine in some other ratio. However, the resulting compound will not be water. In fact, hydrogen peroxide is made up of two atoms of hydrogen and two atoms of oxygen per molecule and contains 5.9% hydrogen and 94.1% oxygen by weight; its properties are markedly different from those of water.

	Water	Hydrogen peroxide
Percent H	11.2	5.9
Percent O	88.8	94.1
Atomic composition	2 H + 1 O	2 H + 2 O

Law of Definite Composition

The **Law of Definite Composition** states: A compound always contains two or more elements combined in a definite proportion by weight. The reliability of this law, which in essence states that the composition of a substance will always be the same no matter what its origin or how it is formed, is the cornerstone of the science of chemistry.

4.7 Chemical Formulas

chemical formulas

In a manner similar to the use of symbols for elements, chemists use **chemical formulas** as abbreviations for compounds. The chemical formula represents two or more elements that are in chemical combination. Sodium chloride contains one atom of sodium per atom of chlorine; its formula is NaCl. The formula for water is H_2O; it shows that a molecule of water contains two atoms of hydrogen and one atom of oxygen. Numbers that appear lower and to the right of the symbol of an atom are known as subscripts.

The formula of a compound tells us which elements it is composed of and how many atoms of each element are present in a formula unit. For example, a molecule of sulfuric acid is composed of two atoms of hydrogen, one atom of sulfur, and four atoms of oxygen. We can express this compound as HHSOOOO, but the usual formula for writing sulfuric acid is H_2SO_4. The formula may be expressed verbally as "H-two-S-O-four." Characteristics of chemical formulas are summarized as follows:

1. The formula of a compound contains the symbols of all the elements in the compound.
2. When the formula contains one atom of an element, the symbol of that element represents that one atom. The number one (1) is not used as a subscript to indicate one atom of an element.

3. When the formula contains more than one atom of an element, the number of atoms is indicated by a subscript written to the right of the symbol of that atom. For example, the subscript 2 in H_2O indicates two atoms of H in the formula.

4. When the formula contains more than one of a group of atoms that occurs as a unit, parentheses are placed around the group and the number of units of the group are indicated by a subscript placed to the right of the parentheses. Consider the nitrate group, NO_3^-. In the formula for sodium nitrate, $NaNO_3$, there is only one nitrate group; therefore, no parentheses are needed. In calcium nitrate, $Ca(NO_3)_2$, there are two nitrate groups, and parentheses and the subscript 2 are used to indicate this. There are a total of nine atoms in $Ca(NO_3)_2$—one Ca, two N, and six O atoms. The formula $Ca(NO_3)_2$ is pronounced "C-A-N-O-3-taken-2-times."

5. Formulas written as H_2O, H_2SO_4, $Ca(NO_3)_2$, and $C_{12}H_{22}O_{11}$ show only the number and kind of each atom contained in the compound; they do not show the arrangement of the atoms in the compound or how they are chemically bonded to one another.

Figure 4.4 illustrates how symbols and numbers are used in chemical formulas. There is more extensive use of formulas in later chapters.

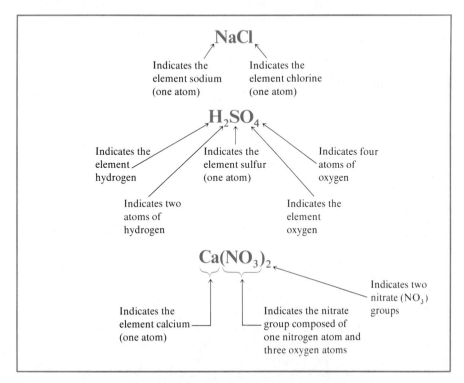

Figure 4.4   Explanation of the formulas NaCl, H_2SO_4, and $Ca(NO_3)_2$.

4.8 Mixtures

Single substances—elements or compounds—seldom occur naturally in the pure state. Air is a mixture of gases; seawater is a mixture containing a variety of dissolved minerals; ordinary soil is a complex mixture of minerals and various organic materials.

How is a mixture distinguished from a pure substance? A mixture (see Section 3.3) always contains two or more substances that can be present in varying concentrations. Let us consider an example of a homogeneous mixture and an example of a heterogeneous mixture. Homogeneous mixtures (solutions) containing either 5% or 10% salt in water can be prepared by simply mixing the correct amounts of salt and water. These mixtures can be separated by boiling away the water, leaving the salt as a residue. The composition of a heterogeneous mixture of sulfur crystals and iron filings can be varied by merely blending in either more sulfur or more iron filings. This mixture can be separated physically, either by using a magnet to attract the iron or by adding carbon disulfide to dissolve the sulfur.

Iron(II) sulfide (FeS) contains 63.5% Fe and 36.5% S by weight. If we mix iron and sulfur in this proportion, do we have iron(II) sulfide? No, it is still a mixture; the iron is still attracted by a magnet. But if this mixture is heated strongly, a chemical change (reaction) occurs, forming iron(II) sulfide. This is a substance whose properties are different from those of iron and sulfur—FeS is neither attracted by a magnet nor dissolved by carbon disulfide. Thus, the properties of the reactants are lost and a compound (a pure substance) is formed.

Mixture of Iron and Sulfur	**Compound of Iron and Sulfur**
Formula: no definite formula; consists of Fe + S	Formula: FeS
Can contain Fe and S in any proportion by weight.	Contains 63.5% Fe and 36.5% S by weight.
Fe and S can be separated by physical means.	Fe and S can be separated only by chemical changes.

Key differences between mixtures and compounds are summarized in Table 4.5.

4.9 Metals, Nonmetals, and Metalloids

metal
nonmetal
metalloid

Three primary classifications of the elements are **metals, nonmetals,** and **metalloids.** Most of the elements are metals. We are familiar with metals because of their widespread use in tools, materials of construction, automobiles, and so on. But nonmetals are equally useful in our everyday life as major components of such items as clothing, food, fuel, glass, plastics, and wood.

TABLE 4.5 **Comparison of mixtures and compounds.**

	Mixture	*Compound*
Composition	May be composed of elements, compounds, or both in variable composition	Composed of two or more elements in a definite, fixed proportion by weight
Separation of components	Separation may be made by simple physical or mechanical means	Elements can be separated by chemical changes only
Identification of components	Components do not lose their identity	A compound does not resemble the elements from which it is formed

The metallic elements are solids at room temperature (mercury is an exception). They have a high luster, are good conductors of heat and electricity, can be rolled or hammered into sheets (they are *malleable*), and can be drawn into wires (they are *ductile*). In addition, most metals have a high melting point and a high density. Metals familiar to most of us are aluminum, chromium, copper, gold, iron, lead, magnesium, mercury, nickel, platinum, silver, tin, and zinc. Other less familiar but still important metals are calcium, cobalt, potassium, sodium, uranium, and titanium.

Metals have little tendency to combine with each other to form compounds. But many metals readily combine with nonmetals such as chlorine, oxygen, and sulfur to form mainly ionic compounds such as metallic chlorides, oxides, and sulfides, respectively. The more active metals are found in nature combined with other elements as minerals. A few of the less active ones—such as copper, gold, and silver—are sometimes found in a native or free state as well.

Nonmetals, unlike metals, are not lustrous, have relatively low melting points and densities, and are generally poor conductors of heat and electricity. Carbon, phosphorus, sulfur, selenium, and iodine are solids; bromine is a liquid; the rest of the nonmetals are gases. Common nonmetals found uncombined in nature are carbon (graphite and diamond), nitrogen, oxygen, sulfur, and the noble gases (helium, neon, argon, krypton, xenon, and radon).

Nonmetals combine with one another to form molecular compounds such as carbon dioxide (CO_2), methane (CH_4), butane (C_4H_{10}), and sulfur dioxide (SO_2). Fluorine, the most reactive nonmetal, combines readily with almost all the other elements.

Several elements (boron, silicon, germanium, arsenic, antimony, tellurium, and polonium) are classified as *metalloids* and have properties that are intermediate between those of metals and those of nonmetals. The intermediate position for these elements is shown in Table 4.6, which classifies all the elements as metals, nonmetals, or metalloids. Certain of the metalloids are the raw materials for the semiconductor devices that make our modern electronics industry possible.

TABLE 4.6 Classification of the elements into metals, metalloids, and nonmetals.

1 H				Key												1 H	2 He
3 Li	4 Be			☐ Metals								5 B	6 C	7 N	8 O	9 F	10 Ne
11 Na	12 Mg			▨ Metalloids								13 Al	14 Si	15 P	16 S	17 Cl	18 Ar
19 K	20 Ca	21 Sc	22 Ti	23 V	24 Cr	25 Mn	26 Fe	27 Co	28 Ni	29 Cu	30 Zn	31 Ga	32 Ge	33 As	34 Se	35 Br	36 Kr
37 Rb	38 Sr	39 Y	40 Zr	41 Nb	42 Mo	43 Tc	44 Ru	45 Rh	46 Pd	47 Ag	48 Cd	49 In	50 Sn	51 Sb	52 Te	53 I	54 Xe
55 Cs	56 Ba	57 La	72 Hf	73 Ta	74 W	75 Re	76 Os	77 Ir	78 Pt	79 Au	80 Hg	81 Tl	82 Pb	83 Bi	84 Po	85 At	86 Rn
87 Fr	88 Ra	89 Ac	104 Unq	105 Unp	106 Unh												

58 Ce	59 Pr	60 Nd	61 Pm	62 Sm	63 Eu	64 Gd	65 Tb	66 Dy	67 Ho	68 Er	69 Tm	70 Yb	71 Lu
90 Th	91 Pa	92 U	93 Np	94 Pu	95 Am	96 Cm	97 Bk	98 Cf	99 Es	100 Fm	101 Md	102 No	103 Lr

Seven of the elements (all nonmetals) occur as *diatomic molecules* (consisting of two atoms). These seven elements, together with their formulas and brief descriptions, are listed in the following table.

Element	Molecular formula	Normal state
Hydrogen	H_2	Colorless gas
Nitrogen	N_2	Colorless gas
Oxygen	O_2	Colorless gas
Fluorine	F_2	Pale yellow gas
Chlorine	Cl_2	Yellow-green gas
Bromine	Br_2	Reddish brown liquid
Iodine	I_2	Grayish black solid

Whether found free in nature or prepared in the laboratory, the molecules of each of these elements contain two atoms. Their formulas, therefore, are always written to show the molecular composition of the element—that is, H_2, N_2, O_2, F_2, Cl_2, Br_2, and I_2.

It is important to understand how symbols are used to designate either an atom or a molecule of an element. Consider the elements hydrogen and oxygen. Hydrogen is found in gases coming from volcanoes and in certain natural gas supplies; it can also be prepared by many different chemical reactions. Regardless of the source, all samples of hydrogen are identical and consist of diatomic molecules. The composition of this molecular hydrogen is expressed by the formula H_2. Free oxygen makes up about 21% (by volume) of the air that we breathe. Whether obtained from the air or prepared by chemical reaction, free oxygen also exists as diatomic molecules; its composition is expressed by the formula O_2. Now consider hydrogen in the compound water, which has the composition expressed by the formula H_2O (or HOH). Water contains neither free hydrogen (H_2) nor free oxygen (O_2); the H_2 part of the formula H_2O simply tells us that each molecule of water contains two hydrogen atoms combined chemically with one oxygen atom. Symbols and subscripts are used in this way to show the molecular composition of elements and to show the composition of compounds.

4.10 Nomenclature and Chemical Equations

NOMENCLATURE

Knowledge of chemical names and of the writing and balancing of chemical equations is vital to the study of chemistry. This section serves only as an introduction to the naming of compounds and writing equations. More complete details of the systematic methods of naming inorganic compounds are given in Chapter 8, and a more detailed explanation of chemical equations is given in Chapter 10. Refer to these two chapters often, as needed. Neither chapter is intended to be studied only in the sequence given in the text; rather, they are common depositories of information on chemical nomenclature and equations.

We have already used such names as hydrogen chloride (HCl), mercury(II) oxide (HgO), magnesium oxide (MgO), and carbon dioxide (CO_2). Note that all four names end in *ide*. This *ide* ending is characteristic of the names of compounds composed of atoms of two different elements. Compounds composed of two elements are called *binary* compounds. Some compounds contain several atoms of the same element (for example, CCl_4, carbon tetrachloride), but as long as there are only two different kinds of atoms, the compound is considered to be binary.

When naming a binary compound consisting of a metal and a nonmetal, the name of the metal is given first, followed by the name of the nonmetal, which is modified to end in *ide* (see Table 4.7).

TABLE 4.7 **Names of selected nonmetals modified to end in *-ide,* for use in naming binary compounds.**

Nonmetal	Modified name used in binary compounds
Oxygen	Oxide
Chlorine	Chloride
Bromine	Bromide
Iodine	Iodide
Sulfur	Sulfide
Nitrogen	Nitride

PROBLEM 4.1 Name the binary compounds having the formula NaCl and MgO.

Metal — Nonmetal
NaCl
Sodium — Name of nonmetal modified to end in *ide* — chloride

The name is sodium chloride.

Metal — Nonmetal
MgO
Magnesium — Name of nonmetal modified to end in *ide* — oxide

The name is magnesium oxide.

There are exceptions to the rule that all names ending in *ide* are binary compounds. The names of some compounds containing more than two elements also end in *ide* (for example, NH_4Cl, ammonium chloride; NaOH, sodium hydroxide). Refer to Section 8.3 for more details on naming binary compounds. Examples of binary compounds with names ending in *ide* follow.

NaCl	Sodium chloride	H_2S	Hydrogen sulfide
CO_2	Carbon dioxide	$AlBr_3$	Aluminum bromide
NaI	Sodium iodide	K_2S	Potassium sulfide
CaF_2	Calcium fluoride	Mg_3N_2	Magnesium nitride

Other names and formulas that you should become familiar with early in your study of chemistry are the commonly used acids and bases shown below. Methods used for naming these compounds are discussed in Chapter 8.

HCl	Hydrochloric acid	NaOH	Sodium hydroxide
H_2SO_4	Sulfuric acid	KOH	Potassium hydroxide
HNO_3	Nitric acid	$Ca(OH)_2$	Calcium hydroxide
$HC_2H_3O_2$	Acetic acid	NH_4OH	Ammonium hydroxide
H_3PO_4	Phosphoric acid		

CHEMICAL EQUATIONS

Chemical changes or reactions result in the formation of substances having compositions different from those of the starting substances. A chemical equation is a shorthand expression for a chemical reaction. Substances in the reaction are represented by their symbols or formulas in the equation. The equation indicates both the reactants (starting substances) and the products, and often shows the conditions necessary to facilitate the chemical change. The reactants are written on the left side and the products on the right side of the equation. An arrow ($\rightarrow$) pointing to the products separates the reactants from the products. A plus sign ($+$) is used to separate one reactant (or product) from another.

Reactants ⟶ Products

We will see in Chapter 5 that every atom has a specific mass. In an equation, the symbols or formulas that represent a substance also represent a specific mass of that substance. Since no detectable change in mass results from a chemical change, the mass of the products must equal the mass of the reactants. In representing a chemical change by an equation, this conservation of mass is attained by balancing the equation. After establishing the correct formulas for the reactants and products, an equation is balanced by placing integral numbers (as needed) in front of the formulas of the substances in the equation. We use these numbers to obtain an equation with the same number of atoms of each kind of element on each side of the equation. When there is the same number of each kind of atom on each side of an equation, the mass of the products is equal to the mass of the reactants.

Consider, again, the reaction of metallic copper heated in air. The chemical change may be represented by the following equations:

$$\text{Copper} + \text{Oxygen} \xrightarrow{\Delta} \text{Copper(II) oxide} \tag{1}$$

$$\text{Cu} + \text{O}_2 \xrightarrow{\Delta} \text{CuO} \quad \text{(Unbalanced)} \tag{2}$$

Copper and oxygen are the reactants, and copper(II) oxide is the product. Equation (2) as written is not balanced because there are two oxygen atoms on the left side and only one on the right side of the equation. We place a 2 in front of Cu and a 2 in front of CuO to obtain the balanced equation (3):

$$2\,\text{Cu} + \text{O}_2 \xrightarrow{\Delta} 2\,\text{CuO} \quad \text{(Balanced)} \tag{3}$$

This balanced equation contains 2 Cu atoms and 2 O atoms on both sides of the equation.

A very important factor to remember when balancing equations is that a correct formula of a substance may not be changed for the convenience of balancing the equation. In the unbalanced equation (2) above, we cannot change the formula of CuO to CuO_2 to balance the equation, even though by so doing we balance the number of atoms of each element on each side of the equation. The formula CuO_2 is not the correct formula for the product. It is also important to be aware that a number in front of a formula multiplies every atom in that formula by that number. Thus,

2 CuO   means 2 Cu atoms and 2 O atoms
3 H_2O means 6 H atoms and 3 O atoms
4 H_2SO_4 means 8 H atoms, 4 S atoms, and 16 O atoms

Once a correct formula is written it must not be changed during the balancing of an equation.

PROBLEM 4.2 How many atoms of each element are in each of these expressions?
(a) CH_4 (b) $2 CH_4$ (c) $3 Ca(OH)_2$

(a) CH_4

1 atom C 4 atoms H

Each molecule of CH_4 contains one C atom and four H atoms.

(b) $2 CH_4$ means two molecules of CH_4. Since one molecule of CH_4 contains one C atom and four H atoms, two CH_4 molecules contain two C atoms and eight H atoms.

(c) $Ca(OH)_2$

1 atom Ca 2 units of OH

Each unit of OH contains one atom of O and one atom of H. Therefore $(OH)_2$ contains two O atoms and two H atoms.

One formula unit of $Ca(OH)_2$ contains one Ca atom, two O atoms, and two H atoms; $3 Ca(OH)_2$ indicates three formula units of $Ca(OH)_2$. Therefore, $3 Ca(OH)_2$ contain three Ca atoms, six O atoms, and six H atoms.

■

PROBLEM 4.3 When methane gas (CH_4) is burned in air, it reacts with oxygen to yield carbon dioxide and water. The equation for the reaction is

$$CH_4 + O_2 \longrightarrow CO_2 + H_2O \quad \text{(Unbalanced)}$$

Balance the equation for this reaction.

Inspecting each side of the equation we observe

Left side: 1 C atom   Right side: 1 C atom
4 H atoms   2 H atoms
2 O atoms   3 O atoms

The numbers of H and O atoms on each side of the equation are not equal so the equation is unbalanced. To balance the equation, first place a 2 in front of the H_2O. This balances the H atoms, giving four H atoms on each side:

$$CH_4 + O_2 \longrightarrow CO_2 + 2 H_2O \quad \text{(Unbalanced)}$$

The oxygen atoms are still unbalanced, with two O atoms on the left and four O atoms on the right. Place a 2 in front of the O_2 and the total equation is balanced:

$$CH_4 + 2 O_2 \longrightarrow CO_2 + 2 H_2O \quad \text{(Balanced)}$$

The balanced equation contains one C, four H, and four O atoms on each side.

■

PROBLEM 4.4 Balance the following equation:

$$Al + Cl_2 \longrightarrow AlCl_3 \quad \text{(Unbalanced)}$$

In this equation, the Cl atoms are not in balance (two Cl atoms on the left and three Cl atoms on the right side of the equation). Placing integral numbers in front of Cl_2 will always result in an even number of Cl atoms on the left side of the equation. Therefore, the number in front of $AlCl_3$ will have to be an even number (2, 4, 6, . . .). The number of Cl atoms needed on each side of the equation is 6, the lowest common multiple between 2 and 3. Placing a 3 in front of Cl_2 and a 2 in front of $AlCl_3$ will give 6 Cl atoms on each side of the equation:

$$Al + 3\,Cl_2 \longrightarrow 2\,AlCl_3 \qquad \text{(Unbalanced)}$$

By balancing the Cl atoms we have unbalanced the Al atoms. But we can correct this by placing a 2 in front of Al to give the total balanced equation:

$$2\,Al + 3\,Cl_2 \longrightarrow 2\,AlCl_3 \qquad \text{(Balanced)}$$

Each side of the balanced equation has two Al and six Cl atoms.

∎

4.11 Laboratory Preparation of Oxygen and Hydrogen

Oxygen and hydrogen can be prepared by many different methods. Two common ways to prepare oxygen in the laboratory are the decomposition of hydrogen peroxide and of potassium chlorate. Hydrogen peroxide decomposes readily at room temperature, but potassium chlorate must be heated for decomposition to occur. For practical purposes a small quantity of manganese dioxide is added to each reaction to increase the rate of decomposition. Since manganese dioxide does not change composition in the reaction but only affects the rate, it is known as a catalyst. Typical laboratory setups are shown in Figure 4.5. Equations for the reactions are

$$\text{Hydrogen peroxide} \xrightarrow[\text{dioxide}]{\text{Manganese}} \text{Water} + \text{Oxygen}$$

$$2\,H_2O_2 \xrightarrow[MnO_2]{} 2\,H_2O + O_2$$

$$\text{Potassium chlorate} \xrightarrow[\text{dioxide}]{\text{Manganese}} \text{Potassium chloride} + \text{Oxygen}$$

$$2\,KClO_3 \xrightarrow[MnO_2]{} 2\,KCl \qquad 3\,O_2$$

Note again that oxygen is a diatomic molecule and its formula is written as O_2. Since oxygen is a gas, special methods must be used to collect and contain it. In the setups shown, it is collected by displacement of water. The fact that oxygen is only slightly soluble in water allows it to be collected in this manner.

Hydrogen is commonly generated in the laboratory by the reaction of a metal with an acid. Many different metals and several different acids can be used. Figure 4.6 shows a typical laboratory setup in which zinc and sulfuric acid react to produce hydrogen and zinc sulfate. Equations for the reaction are

$$\text{Zinc} + \text{Sulfuric acid} \longrightarrow \text{Hydrogen} + \text{Zinc sulfate}$$

$$Zn + H_2SO_4 \longrightarrow H_2 + ZnSO_4$$

The equation is balanced as shown. Hydrogen is a diatomic molecule and its formula is written as H_2. Hydrogen is essentially insoluble in water and is also collected by the displacement of water.

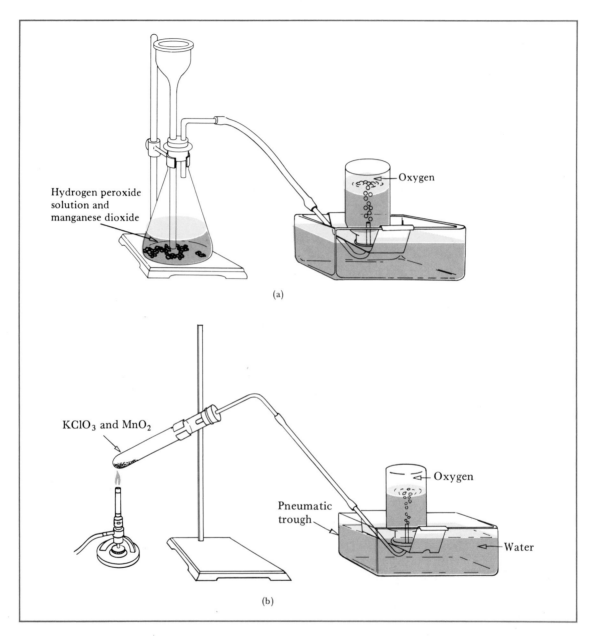

Hydrogen peroxide solution and manganese dioxide

Oxygen

(a)

KClO₃ and MnO₂

Oxygen

Pneumatic trough

Water

(b)

Figure 4.5 Laboratory preparation of oxygen from (a) hydrogen peroxide and (b) potassium chlorate.

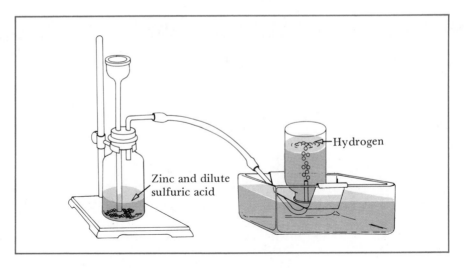

Figure 4.6   Laboratory preparation of hydrogen from zinc and sulfuric acid.

QUESTIONS   **A. Review the Meanings of the New Terms Introduced in this Chapter.**

   1. Element   7. Law of Definite Composition
   2. Atom   8. Chemical formula
   3. Symbol   9. Metal
   4. Compound   10. Nonmetal
   5. Molecule   11. Metalloid
   6. Ion

B. Answers to the Following Questions Will Be Found in Tables and Figures

   1. List, in decreasing order of abundance, the six most abundant elements in the earth's crust, seawater, and atmosphere.
   2. Since there is so much oxygen in the earth's crust, seawater, and atmosphere, is it logical to assume that most of it is present in the seawater as H_2O molecules, considering that in water the mass of oxygen is about eight times that of hydrogen?
   3. Are there more atoms of silicon or hydrogen in the earth's crust, seawater, and atmosphere? Use Table 4.1 and the fact that the mass of a silicon atom is about 28 times that of a hydrogen atom.
   4. List, in decreasing order of abundance, the six most abundant elements in the human body.
   5. Why is the symbol for lead Pb instead of Le?
   6. Make a list of the names of the elements in Table 4.3. Now see how many of the symbols you know by writing the correct symbol after each name.
   7. How many metals are there? Nonmetals? Metalloids? (See Table 4.6.)

C. Review Questions

   1. What does the symbol of an element stand for?
   2. Write down what you believe to be the symbols for the elements phosphorus, aluminum, neon, hydrogen, helium, potassium, magnesium, and manganese. Now look up the correct symbols and rewrite them, comparing the two sets.
   3. Interpret the difference in meanings for each of these pairs:
     (a) Si and SI   (b) Pb and PB

4. Distinguish between an element and a compound.
5. Explain why the Law of Definite Composition does not pertain to mixtures.
6. Does the Law of Definite Composition pertain to an element? Discuss.
7. Distinguish between a chemical formula and a symbol.
8. Given the following list of compounds and their formulas, what elements are present in each compound?
   - (a) Potassium iodide   KI
   - (b) Sodium carbonate Na_2CO_3
   - (c) Aluminum oxide Al_2O_3
   - (d) Calcium bromide $CaBr_2$
   - (e) Carbon tetrachloride CCl_4
   - (f) Magnesium bromide $MgBr_2$
   - (g) Nitric acid HNO_3
   - (h) Barium sulfate $BaSO_4$
   - (i) Aluminum phosphate $AlPO_4$
   - (j) Acetic acid $HC_2H_3O_2$
9. Write the formula for each of the following compounds, the composition of which is given after each name:
   - (a) Zinc oxide   1 atom Zn, 1 atom O
   - (b) Potassium chlorate   1 atom K, 1 atom Cl, 3 atoms O
   - (c) Sodium hydroxide   1 atom Na, 1 atom O, 1 atom H
   - (d) Aluminum bromide   1 atom Al, 3 atoms Br
   - (e) Calcium fluoride   1 atom Ca, 2 atoms F
   - (f) Lead(II) chromate   1 atom Pb, 1 atom Cr, 4 atoms O
   - (g) Ethyl alcohol   2 atoms C, 6 atoms H, 1 atom O
   - (h) Benzene   6 atoms C, 6 atoms H
10. Explain the meaning of each symbol and number in the following formulas:
   - (a) H_2O (c) Na_2SO_4 (e) $C_{12}H_{22}O_{11}$ (sucrose)
   - (b) $AlBr_3$ (d) $Ni(NO_3)_2$
11. How many atoms are represented in each of these formulas?
   - (a) KF   (d) $Ba(ClO_3)_2$ (g) CCl_2F_2 (Freon)
   - (b) $CaCO_3$ (e) $K_2Cr_2O_7$ (h) $Al_2(SO_4)_3$
   - (c) N_2 (f) $NaC_2H_3O_2$ (i) $(NH_4)_2C_2O_4$
12. How many atoms of hydrogen are contained in one molecule of hydrogen?
13. How many atoms of hydrogen and oxygen are contained in one molecule of water? In one molecule of hydrogen peroxide?
14. Write the names and formulas of the elements that exist as diatomic molecules.
15. How many atoms of oxygen are represented in each expression?
   - (a) $4\,H_2O$ (b) $3\,CuSO_4$ (c) H_2O_2 (d) $3\,Fe(OH)_3$ (e) $Al(ClO_3)_3$
16. How many atoms of hydrogen are represented in each expression?
   - (a) $5\,H_2$ (b) $2\,Ba(C_2H_3O_2)_2$ (c) $2\,C_6H_{12}O_6$ (d) $2\,HC_2H_3O_2$
17. Are all mixtures homogeneous? Explain.
18. Classify each of the following materials as an element, compound, or mixture:
   - (a) Air   (c) Sodium chloride   (e) Wine   (g) Sulfuric acid
   - (b) Oxygen   (d) Platinum   (f) Iodine   (h) Crude oil
19. Classify each of the following materials as an element, compound, or mixture:
   - (a) Wood   (d) Fog   (g) Brass
   - (b) Salt   (e) Magnesium iodide   (h) Sodium hydroxide
   - (c) Copper   (f) Silver
20. A white solid, on heating, formed a colorless gas and a yellow solid. Assuming that there was no reaction with the air, is the original solid an element or a compound? Explain.
21. Tabulate the properties that characterize metals and nonmetals.
22. Which of the following are diatomic molecules?
   - (a) H_2 (c) HCl   (e) NO   (g) $MgCl_2$
   - (b) SO_2 (d) H_2O (f) NO_2

23. Name the following binary compounds. Refer to Chapter 8 if necessary.
    (a) AgCl   (c) $MgBr_2$ (e) $SrCl_2$ (g) LiF   (i) H_2S
    (b) HI   (d) CaO   (f) K_2S (h) BN   (j) Al_2O_3
24. Which of the compounds listed in Questions 8 and 9 are binary compounds? What is common to the names of these binary compounds?
25. An atom of silver is represented by the symbol Ag; a hydrogen molecule by the formula H_2; a water molecule by H_2O. Write the expressions to represent:
    (a) Five silver atoms
    (b) Four hydrogen molecules
    (c) Three water molecules
26. Balance these equations (all formulas are correct as written):
    (a) $H_2 + Cl_2 \longrightarrow HCl$
    (b) $Zn + CuSO_4 \longrightarrow Cu + ZnSO_4$
    (c) $HCl + NaOH \longrightarrow NaCl + H_2O$
    (d) $Ca + O_2 \longrightarrow CaO$
    (e) $Fe + HCl \longrightarrow FeCl_2 + H_2$
    (f) $P + I_2 \longrightarrow PI_3$
    (g) $MgO + HCl \longrightarrow MgCl_2 + H_2O$
    (h) $HNO_3 + Ba(OH)_2 \longrightarrow Ba(NO_3)_2 + H_2O$
    (i) $BiCl_3 + H_2S \longrightarrow Bi_2S_3 + HCl$
    (j) $Mg_3N_2 + H_2O \longrightarrow Mg(OH)_2 + NH_3$
27. Balance the following equations, each of which represents a method of preparing oxygen gas:
    (a) $H_2O_2 \longrightarrow H_2O + O_2$
    (b) $KClO_3 \xrightarrow{\Delta} KCl + O_2$
    (c) $KNO_3 \xrightarrow{\Delta} KNO_2 + O_2$
    (d) $Na_2O_2 + H_2O \longrightarrow NaOH + O_2$
    (e) $H_2O \xrightarrow[H_2SO_4]{\text{Electrical energy}} H_2 + O_2$
28. Balance the following equations, each of which represents a method of preparing hydrogen gas:
    (a) $Zn + HCl \longrightarrow ZnCl_2 + H_2$
    (b) $Al + H_2SO_4 \longrightarrow Al_2(SO_4)_3 + H_2$
    (c) $Na + H_2O \longrightarrow NaOH + H_2$
    (d) $C + H_2O \text{ (steam)} \longrightarrow CO + H_2$
    (e) $Fe + H_2O \text{ (steam)} \longrightarrow Fe_3O_4 + H_2$
29. Which of the following statements are correct?
    (a) The smallest unit of an element that can exist and enter into a chemical reaction is called a molecule.
    (b) The basic building blocks of all substances, which cannot be decomposed into simpler substances by ordinary chemical change, are compounds.
    (c) The most abundant element in the earth's crust, seawater, and atmosphere by weight is oxygen.
    (d) The most abundant element in the human body, by weight, is carbon.
    (e) Most of the elements are represented by symbols consisting of one or two letters.
    (f) The symbol for copper is C.
    (g) The symbol for sodium is Na.
    (h) The symbol for potassium is P.
    (i) The symbol for lead is PB.
    (j) Early names for some elements led to unlikely symbols, such as Fe for iron.
    (k) A compound is a distinct substance that contains two or more elements combined in a definite proportion by weight.
    (l) The smallest uncharged individual unit of a compound formed by the union of two or more atoms is called a substance.

(m) An ion is a positive or negative electrically charged atom or group of atoms.

(n) The Law of Definite Composition states that a compound always contains two or more elements combined in a definite proportion by weight.

(o) A chemical formula is a shorthand expression for a chemical reaction.

(p) The formula Na_2CO_3 indicates a total of six atoms, including three oxygen atoms.

(q) A general property of nonmetals is that they are good conductors of heat and electricity.

(r) Metals have the properties of ductility and malleability.

(s) Malleable means that when struck a hard blow the substance will shatter.

(t) Elements that have properties intermediate between metals and nonmetals are called mixtures.

(u) More of the elements are metals than nonmetals.

(v) The nonmetals are located at the upper right of the periodic table of the elements.

(w) Bromine is an element that occurs as diatomic molecules, Br_2.

(x) The binary compound CaS is called calcium sulfate.

(y) The substances on the left side of the arrow in a chemical equation are called reactants and those on the right side are called products.

(z) The equation $HgO \longrightarrow Hg + O_2$ is balanced.

(aa) In preparing oxygen from hydrogen peroxide, the word equation is

$$\text{Hydrogen peroxide} \xrightarrow{\text{Manganese dioxide}} \text{Water} + \text{Oxygen}$$

(bb) The correct formula equation for the preparation of hydrogen from zinc and sulfuric acid is:

$$Zn + H_2SO_4 \longrightarrow H_2$$

D. Review Problems

1. Common table salt, NaCl, contains 39.3% sodium and 60.7% chlorine. What weight of chlorine is present in 35.0 g of salt?

2. Red brass is a homogeneous mixture of 90% copper and 10% zinc. If 30 g of zinc is added to 100 g of red brass, what will be the new composition?

3. Barium oxide, BaO, contains 89.6% barium. What size sample of BaO would contain 12.0 g of barium?

*4. What would be the density of a solution made by mixing 25.0 mL of carbon tetrachloride (CCl_4, $d = 1.595$ g/mL) and 40.0 mL of carbon tetrabromide (CBr_4, $d = 3.420$ g/mL)? Assume that the volume of the mixed liquids is the sum of the two volumes used.

* 5. When 6.00 g of calcium and 6.00 g of sulfur were mixed and reacted to give the compound calcium sulfide (CaS), 1.20 g of sulfur remained unreacted.
    (a) What percentage of the compound is sulfur?
    (b) An atom of which element, Ca or S, has the greater mass? Explain.
    (c) How many grams of sulfur will combine with 30.0 g of calcium?

6. Pure gold is too soft a metal for many uses, so it is alloyed to give it more mechanical strength. One particular alloy is made by mixing 60 g of gold, 8.0 g of silver, and 12 g of copper. What carat gold is this alloy if pure gold is considered to be 24 carat?

*7. Methane, the chief component of natural gas, has the formula CH_4. Each atom of carbon weighs 12 times as much as an atom of hydrogen. Calculate the weight percent of carbon in methane.

8. White gold is a homogeneous solution of 90% gold and 10% palladium. How much gold is present in a bar of white gold weighing 8420 g?

*9. The metal used to make the U.S. nickel coin is an alloy of 75% copper and 25% nickel. What maximum weight of alloy could be produced if only 420 kg of nickel and 1090 kg of copper were on hand?

Review Exercises for Chapters 1-4

CHAPTER 1
INTRODUCTION

True–False: Answer the following as either true or false.
1. Chemistry is the science that deals with the composition of substances and the transformations they undergo.
2. Scientific laws are simple statements of natural phenomena to which no exceptions are known.
3. From 1803 to 1810, John Dalton advanced his atomic theory.
4. A key feature of the scientific method is to plan and do additional experiments to test a hypothesis.

CHAPTER 2
STANDARDS
FOR
MEASUREMENT

True–False: Answer the following as either true or false.
1. A milligram is 0.001 g.
2. A centimetre is longer than a millimetre.
3. If 1 mL = 0.001 litre, then we can use the factor 10^3 mL/litre to convert litres to millilitres.
4. The density of water at 4°C is 1.00 g/mL.
5. As a metric prefix, *kilo* means 1000, or 10^3.
6. One millilitre equals 1 cm^3 exactly.
7. The calorie is a unit of temperature.
8. The measurement 12.200 g contains three significant figures.
9. The answer to 25.2×0.1465 should contain three significant figures.
10. The number 14.0667 rounded off to four digits is 14.07.
11. The answer to $16.215 - 2.32$ should contain three digits.
12. A litre contains 100 mL.
13. 90°C is hotter than 210°F.
14. The units of specific gravity are g/mL.

Multiple Choice: Choose the correct answer to each of the following.
1. One centimetre is equal to how many metres?
   (a) 2.54   (b) 100   (c) 10   (d) 0.01
2. One centimetre is equal to how many inches?
   (a) 0.394   (b) 0.10   (c) 12   (d) 2.54
3. 3.50 ft is how many centimetres?
   (a) 8.90   (b) 16.5   (c) 454   (d) 107
4. The number 0.076 contains how many significant figures?
   (a) 1   (b) 2   (c) 3   (d) 4
5. Express 0.00382 in exponential notation.
   (a) 3.82×10^3 (b) 3.8×10^{-3} (c) 3.82×10^{-2} (d) 3.82×10^{-3}
6. 35°C is equivalent to:
   (a) 273 K   (b) 121°F   (c) 95°F   (d) 1.7°F
7. 257°F is equivalent to:
   (a) 398 K   (b) 111°C   (c) 530 K   (d) 384 K
8. To heat 30 g of water from 20°C to 50°C will require:
   (a) 30 cal   (b) 50 cal   (c) 900 cal   (d) 1500 cal
9. An object has a mass of 50 g and a volume of 4.6 mL. Its density is:
   (a) 0.092 mL/g   (b) 230 g/mL   (c) 9.2 g/mL   (d) 11 g/mL
10. The mass of a block is 8.43 g and its density is 2.35 g/mL. The volume of the block is:
    (a) 3.59 mL   (b) 0.280 mL   (c) 19.8 mL   (d) 2.80 mL
11. The density of copper is 8.92 g/mL. The mass of a piece of copper that has a volume of 7.5 mL is:
    (a) 2.58 g   (b) 67 g   (c) 0.831 g   (d) 1.19 g

12. An empty graduated cylinder weighs 54.772 g. When filled with 50.0 mL of an unknown liquid it weighs 101.074 g. The density of the liquid is:
    (a) 0.926 g/mL   (b) 1.00 g/mL   (c) 2.02 g/mL   (d) 1.845 g/mL
13. The conversion factor to change grams to milligrams is:
    (a) $\dfrac{100 \text{ mg}}{1 \text{ g}}$ (b) $\dfrac{1 \text{ g}}{100 \text{ g}}$ (c) $\dfrac{1 \text{ g}}{1000 \text{ mg}}$ (d) $\dfrac{1000 \text{ mg}}{1 \text{ g}}$
14. $-40°C$ is equivalent to:
    (a) $-40°F$ (b) -40 K (c) $104°F$ (d) 213 K
15. The heat capacity of aluminum is 0.213 cal/g°C. How many calories of energy are required to raise the temperature of 20.0 g Al from 10.0°C to 15.0°C?
    (a) 19 cal   (b) 21 cal   (c) 100 cal   (d) 469 cal

CHAPTER 3
PROPERTIES
OF MATTER

True–False: Answer the following as either true or false.
1. A substance is homogeneous but does not have a fixed composition.
2. A system having more than one phase is heterogeneous.
3. In a chemical change, substances are formed that are entirely different, having different properties and composition from the original material.
4. The Law of Conservation of Energy says that because of the energy shortage, anyone wasting energy can be arrested.
5. The Law of Conservation of Mass states that there is no detectable change in the total mass of the substances involved in a chemical change.
6. The starting substances in a chemical reaction are called the reactants.
7. Plastics, glass, and gels are examples of amorphous solids.
8. Matter that has identical properties throughout is homogeneous.
9. A gas is the least compact of the three states of matter.
10. When a clean copper wire is heated in a burner flame it gains weight.
11. The energy released when hydrogen and oxygen react to form water was stored in the hydrogen and oxygen as chemical or kinetic energy.

Multiple Choice: Choose the correct answer to each of the following.
1. Which of the following is not a physical property?
    (a) Boiling point   (c) Bleaching action
    (b) Physical state   (d) Color
2. Which of the following is a physical change?
    (a) A piece of sulfur is burned.   (c) A rubber band is stretched.
    (b) A firecracker explodes.   (d) A nail rusts.
3. Which of the following is a chemical change?
    (a) Water evaporates.   (c) Rocks are ground to sand.
    (b) Ice melts.   (d) A penny tarnishes.
4. Which of the following is a mixture?
    (a) Water   (c) Sugar solution
    (b) Mercury(II) oxide   (d) Copper(II) oxide
5. When 8.44 g of calcium are heated in air, 11.82 g of calcium oxide are formed. The percentage of oxygen in the compound is:
    (a) 28.6%   (b) 40.0%   (c) 71.4%   (d) 1.40%
6. Mercury(II) sulfide, HgS, contains 86.2% mercury. The weight of HgS that can be made from 40.0 g of mercury is:
    (a) 3448 g   (b) 0.464 g   (c) 34.5 g   (d) 46.4 g

CHAPTER 4
ELEMENTS AND
COMPOUNDS

True–False: Answer the following as either true or false.

1. The basic building blocks of all substances, which cannot be decomposed into simpler substances by ordinary chemical change, are compounds.
2. The smallest particle of an element that can exist and still retain the properties of the element is called an atom.
3. The symbol for silver is Ag.
4. The symbol for nitrogen is Ni.
5. The Law of Definite Composition states that a compound contains two or more elements combined in a definite proportion by weight.
6. The name of $ZnBr_2$ is zinc bromide.
7. Elements that have properties resembling both metals and nonmetals are called mixtures.
8. A molecule is a small, uncharged individual unit of a compound formed by the union of two or more atoms.
9. An ion is a positive or negative electrically charged atom or group of atoms.
10. A chemical formula is a shorthand expression for a chemical reaction.
11. The most abundant element in the earth's crust, seawater, and atmosphere is nitrogen.
12. The symbol for cobalt can be written as Co or CO.
13. Metalloids are elements that can have properties intermediate between those of metals and nonmetals.
14. Compounds exist as either molecules or ions.
15. The main characteristic of a mixture is that it has a definite composition.
16. The characteristic name ending for a binary compound is -*ide*.

Multiple Choice: Choose the correct answer to each of the following.

1. Which of the following is not one of the five most abundant elements by weight in the earth's crust, seawater, and atmosphere?
   (a) Oxygen   (b) Hydrogen   (c) Silicon   (d) Aluminum
2. Which of the following is a compound?
   (a) Lead   (b) Wood   (c) Potassium   (d) Water
3. Which of the following is a mixture?
   (a) Water   (b) Chromium   (c) Wood   (d) Sulfur
4. How many atoms are represented in the formula Na_2CrO_4?
   (a) 3   (b) 5   (c) 7   (d) 8
5. Which of the following is a characteristic of metals?
   (a) Ductile   (b) Easily shattered   (c) Extremely strong   (d) Dull
6. Which of the following is a characteristic of nonmetals?
   (a) Always a gas   (c) Shiny
   (b) Poor conductor of electricity   (d) Combine only with metals
7. When the equation $Al + O_2 \longrightarrow Al_2O_3$ is properly balanced, which of the following terms appears?
   (a) 2 Al   (b) $2 Al_2O_3$ (c) 3 Al   (d) $3 Al_2O_3$
8. Which of the following does not occur as diatomic molecules?
   (a) O_2 (b) I_2 (c) H_2 (d) Na_2
9. Barium iodide, BaI_2, contains 35.1% barium. A 6.50 g sample of barium iodide contains what weight of iodine?
   (a) 4.22 g   (b) 2.28 g   (c) 18.5 g   (d) 64.9 g
10. Which equation is incorrectly balanced?
   (a) $2 KNO_3 \xrightarrow{\Delta} 2 KNO_2 + O_2$
   (b) $H_2O_2 \longrightarrow H_2O + O_2$
   (c) $2 Na_2O_2 + 2 H_2O \longrightarrow 4 NaOH + O_2$
   (d) $2 H_2O \xrightarrow[H_2SO_4]{\text{Electrical energy}} 2 H_2 + O_2$

11. Which of the following is not a binary compound?
    (a) NaI   (b) HClO   (c) K_2S (d) Mg_3N_2
12. Which of the following is the formula for sodium bromide?
    (a) SBr   (b) SoBr   (c) NaBr   (d) NaB
13. When a pure substance was analyzed, it was found to contain carbon and chlorine. This substance must be classified as:
    (a) An element   (c) A compound
    (b) A mixture   (d) Both a mixture and a compound

5

Atomic Theory and Structure

After studying Chapter 5 you should be able to:

1. Understand the terms listed in Question A at the end of the chapter.
2. State the major provisions of Dalton's atomic theory.
3. Give the names, symbols, charges, and relative masses of the three principal subatomic particles.
4. Describe the atom as conceived by Ernest Rutherford after his alpha scattering experiments.
5. Describe the atom as conceived by Niels Bohr.
6. Discuss the contributions to atomic theory made by Dalton, Thomson, Rutherford, Bohr, Chadwick, and Schrödinger.
7. Determine the maximum number of electrons that can exist in a given main energy level.
8. Determine the atomic number, atomic mass, or number of neutrons of any isotope when given any two of these three values.
9. Draw the diagram of any isotope of the first 40 elements, showing the composition of the nucleus and the numbers of electrons in the main energy levels.
10. Give the electron structure ($1s^2 2s^2 2p^6$, etc.) for any of the first 40 elements.
11. Explain what is represented by the electron-dot (Lewis-dot) structure of an element.
12. Draw the electron-dot (Lewis-dot) diagrams for any of the first 20 elements.
13. Name and distinguish among the three isotopes of hydrogen.
14. Convert grams or atoms of an element to moles of that element, and vice versa.
15. Understand the relationships among a mole, Avogadro's number, and a gram-atomic weight of an element.

5.1 Early Thoughts

The structure of matter has long intrigued and engaged the minds of people. The seed of modern atomic theory was sown during the time of the ancient Greek philosophers. About 440 B.C. Empedocles stated that all matter was composed of four "elements"—earth, air, water, and fire. Democritus (about 470–370 B.C.), one of the early atomistic philosophers, thought that all forms of matter were finitely divisible into invisible particles, which he called atoms. He held that atoms were in constant motion and that they combined with one another in various ways. This purely speculative hypothesis was not based on scientific observation. Shortly thereafter, Aristotle (384–322 B.C.) opposed the theory of Democritus and endorsed and advanced the Empedoclean theory. So strong was the influence of Aristotle that his theory dominated the thinking of scientists and philosophers until the beginning of the 17th century. The term *atom* is derived from the Greek word *atomos,* meaning "indivisible."

5.2 Dalton's Atomic Theory

Dalton's atomic theory

More than 2000 years after Democritus, the English schoolmaster John Dalton (1766–1844) revived the concept of atoms and proposed an atomic theory based on facts and experimental evidence. This theory, described in a series of papers published during the period 1803–1810, rested on the idea of a different kind of atom for each element. The essence of **Dalton's atomic theory** may be summed up as follows:

Basicaly to ke Basicaly for Granted, these for Granted, should recognize

1. Elements are composed of minute, indivisible particles called atoms.
2. Atoms of the same element are alike in mass and size.
3. Atoms of different elements have different masses and sizes.
4. Chemical compounds are formed by the union of two or more atoms of different elements.
5. When atoms combine to form compounds, they do so in simple numerical ratios, such as one to one, two to one, two to three, and so on.
6. Atoms of two elements may combine in different ratios to form more than one compound.

Dalton's atomic theory stands as a landmark in the development of chemistry. The major premises of his theory are valid today. However, some of the statements must be modified or qualified because investigations since Dalton's time have shown that (1) atoms are composed of subatomic particles; (2) all the atoms of a specific element do not have the same mass; and (3) atoms, under special circumstances, can be decomposed.

5.3 Subatomic Parts of the Atom

The concept of the atom—a particle so small that it cannot be seen even with the most powerful microscope—and the subsequent determination of its structure stand among the very greatest creative intellectual human achievements.

When we refer to any visible quantity of an element, we are considering a vast number of identical atoms of that element. But when we refer to an atom of an element, we isolate a single atom from the multitude in order to present that element in its simplest form. Figure 5.1 illustrates the hypothetical isolation of a single copper atom from its crystal lattice.

Let us examine this tiny particle we call the atom. The diameter of a single atom ranges from 1 to 5 angstroms (1 Å = 1 × 10⁻⁸ cm). Hydrogen, the smallest atom, has a diameter of about 1 Å. To arrive at some idea of how small an

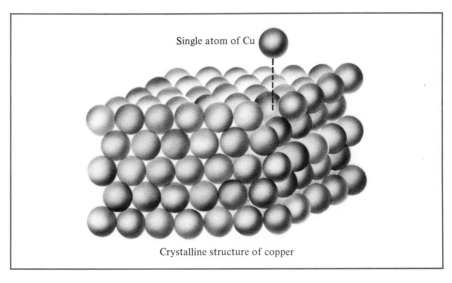

Single atom of Cu

Crystalline structure of copper

Figure 5.1 A single atom of copper compared with copper as it occurs in its regular crystalline lattice structure. Billions of atoms are present in even the smallest strand of copper wire.

atom is, consider this dot (•), which has a diameter of about 1 mm. It would take 10 million hydrogen atoms to form a line stretching across this dot. To carry this size illustration a bit further, 10 million of the 1 millimetre dots laid edge to edge would extend for 10,000 metres, or more than 6 miles! As inconceivably small as atoms are, they contain even more minute particles, the **subatomic particles,** such as electrons, protons, and neutrons.

The experimental discovery of the electron (e^-) was made in 1897 by J. J. Thomson (1856–1940). The **electron** is a particle with a negative electrical charge and a mass of 9.107×10^{-28} g. This mass is 1/1837 the mass of a hydrogen atom and corresponds to 0.0005486 atomic mass unit (amu). One atomic mass unit has a mass of 1.660×10^{-24} g. Although the actual electrical charge of an electron is known, its value is too cumbersome for practical use. The electron, therefore, has been assigned a relative electrical charge of -1. The size of an electron has not been determined exactly, but its diameter is believed to be less than 10^{-12} cm.

Protons were first observed by E. Goldstein (1850–1930) in 1886. However, it was J. J. Thomson who discovered the nature of the proton. He showed that the proton is a particle and he calculated its mass to be about 1837 times that of an electron. The **proton** (p) is a particle with a relative mass of 1 amu and an actual mass of 1.672×10^{-24} g. Its electrical charge ($+1$) is equal in magnitude but of opposite sign to the charge on the electron. The mass of a proton is only very slightly less than that of a hydrogen atom.

The third major subatomic particle was discovered in 1932 by James Chadwick (1891–1974). This particle, the **neutron** (n) bears neither a positive nor a negative charge and has a relative mass of about 1 amu. Its actual mass

subatomic particles

electron

1897
J. J. Thompson

E. Goldstein
1886 *J. J. Thompson*

proton

James Chadwick 1932

neutron

$(1.675 \times 10^{-24}$ g) is only slightly greater than that of a proton. The properties of these three subatomic particles are summarized in Table 5.1.

TABLE 5.1 Properties of electrons, protons, and neutrons.

Particle	Symbol	Electrical charge	Relative mass (amu)	Actual mass (g)
Electron	e^-	-1	1/1837	9.107×10^{-28}
Proton	p	$+1$	1	1.672×10^{-24}
Neutron	n	0	1	1.675×10^{-24}

Nearly all the ordinary chemical properties of matter can be explained in terms of atoms consisting of electrons, protons, and neutrons. The discussion of atomic structure that follows is based on the assumption that atoms contain only these principal subatomic particles. Many other subatomic particles, such as mesons, positrons, neutrinos, pions, muons, and antiprotons, have been discovered. At this time it is not clear whether all these particles are originally present in the atom or whether they are produced by reactions occurring within the nucleus. The field of atomic physics is fascinating and has attracted many young scientists. This interest has resulted in a great deal of research that is producing a long list of additional subatomic particles. Descriptions of the properties of many of these particles are to be found in recent textbooks on atomic physics and in various articles appearing in *Scientific American* over the past several years.

5.4 The Nuclear Atom

The discovery that positively charged particles were present in atoms came soon after the discovery of radioactivity by Henri Becquerel in 1896 (see Chapter 18).

Ernest Rutherford (Figure 5.2) had, by 1907, established that the positively charged alpha particles emitted by certain radioactive elements were ions of the element helium. Rutherford used these alpha particles to establish the nuclear nature of atoms. In some experiments performed in 1911, he directed a stream of positively charged helium ions (alpha particles) at a very thin sheet of gold foil (about 1000 atoms thick). He observed that most of the alpha particles passed through the foil with little or no deflection, but a few of the particles were deflected at large angles, and occasionally one even bounced back from the foil (see Figure 5.3). It was known that like charges repel each other and that an electron with a mass of 1/1837 amu could not have an appreciable effect on the path of a far more massive (4 amu) alpha particle. Rutherford therefore reasoned that each gold atom must contain a positively charged mass occupying a relatively tiny volume and that when an alpha particle approached close enough to this positive mass, it was deflected. Rutherford spoke of this positively charged mass as the *nucleus* of the atom. Since alpha particles are relatively high in mass, the extent of the deflections—remember some actually bounced

Figure 5.2   Ernest Rutherford (1871–1937), British physicist, who identified two of the three principal rays emanating from radioactive substances. His experiments with alpha particles led to the first laboratory transmutation of an element and to his formulation of the nuclear atom. Rutherford was awarded the Nobel Prize in 1908 for his work on transmutation. *(Courtesy Rutherford Museum, McGill University.)*

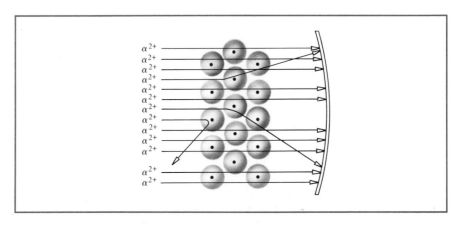

Figure 5.3   Diagram representing Rutherford's experiment on alpha-particle scattering. Positive alpha particles, emanating from a radioactive source, were directed at a thin gold foil. Diagram illustrates the repulsion of the positive alpha particles by the positive nuclei of the metal atoms.

back—indicated to Rutherford that the nucleus is relatively very heavy and dense. (The density of the nucleus of a hydrogen atom, for example, is about 10^{12} g/cm^3, about 1 trillion times the density of water.) Since most of the alpha particles passed through the thousand or so gold atoms without any apparent deflection, Rutherford further concluded that the bulk of an atom consists of empty space.

When we speak of the mass of an atom, we are, for practical purposes, referring primarily to the mass of the nucleus. This is because the nucleus contains all the protons and neutrons, and these represent more than 99.9% of the total mass of any atom (see Table 5.1). By way of illustration, the largest number of electrons known to exist in an atom is 106. The mass of even 106 electrons represents only about 1/17 of the mass of a single proton or neutron. The mass of an atom, therefore, is primarily determined by the mass of its protons and neutrons.

5.5 General Arrangement of Subatomic Particles

nucleus

The alpha particle scattering experiments of Rutherford established that the atom contains a dense, positively charged nucleus. The later work of Chadwick demonstrated that the atom contains neutrons, which are particles with mass but no charge. Light, negatively charged electrons are also present and offset the positive charges in the nucleus. Based on this experimental evidence, a general description of the atom and location of its subatomic particles was devised. Each atom consists of a **nucleus** surrounded by electrons. The nucleus contains protons and neutrons, but electrons are not found in the nucleus. In a neutral atom, the positive charge of the nucleus (due to protons) is exactly offset by the negative electrons. Since the charge on an electron is equal to but of opposite sign to the charge on a proton, a neutral atom must contain exactly the same number of electrons as protons. However, this generalized picture of atomic structure provides no information on the arrangement of electrons within the atom.

5.6 The Bohr Atom

At high temperatures or when subjected to high voltages, elements in the gaseous state give off colored light. Neon signs illustrate this property of matter very well. When passed through the prism or grating of a spectroscope, the light emitted by a gas appears as a set of bright colored lines (band spectra). These colored lines indicate that the light is being emitted only at certain wavelengths or frequencies that correspond to specific colors. Each element possesses a unique set of these spectral lines that is different from the sets of all the other elements.

In 1912–1913, while studying the line spectra of hydrogen (see Figure 5.4), Niels Bohr (1885–1962), a Danish physicist, made a significant contribution to

Figure 5.4   Line spectrum of hydrogen. Each line corresponds to the wavelength of the energy emitted when the electron of a hydrogen atom, which has absorbed energy, falls back to a lower energy level.

the rapidly growing knowledge of atomic structure. His research led him to believe that electrons in an atom exist in specific regions at various distances from the nucleus. He also visualized the electrons as rotating in orbits around the nucleus like planets rotating around the sun.

Bohr's first paper in this field dealt with the hydrogen atom, which he described as a single electron rotating in an orbit about a relatively heavy nucleus (see Figure 5.6). He applied the concept of energy quanta, proposed in 1900 by the German physicist Max Planck (1858–1947), to the observed line spectra of hydrogen. Planck stated that energy is never emitted in a continuous stream, but only in small discrete packets called quanta (from the Latin *quantus,* meaning "how much"). Bohr theorized that there are several possible orbits for electrons at different distances from the nucleus. But an electron had to be in one specific orbit or another; it could not exist between orbits. Bohr also stated that when a hydrogen atom absorbed one or more quanta of energy, its electron would "jump" to another orbit at a greater distance from the nucleus.

Bohr was able to account for spectral lines this way: Each orbit corresponds to a different energy level, the one closest to the nucleus representing the lowest or ground-state energy level. Orbits at increasing distances from the nucleus represent the second, third, fourth energy levels. When an electron "falls" from a high-energy orbit to one of lower energy, a quantum of energy in the form of light is emitted, thus giving rise to a spectral line at a specific frequency. A number of orbits exist, and when electrons "fall" different distances, correspondingly different quanta of energy are emitted, producing the several lines visible in the hydrogen spectra.

Bohr contributed greatly to the advancement of our knowledge of atomic structure by (1) suggesting quantized energy levels for electrons and (2) showing that spectral lines result from the radiation of small increments of energy (Planck's quanta) when electrons shift from one energy level to another.

Much of Bohr's work related to the simplest atom, hydrogen. Difficulties arose when his energy calculations were applied to atoms containing more than one electron. Bohr's concept of the atom has been replaced by the *quantum mechanics* or *wave mechanics* theory, introduced in 1926 by the Austrian physicist Erwin Schrödinger (1887–1961). One of the chief differences between orbitals

the Bohr and quantum mechanics theories is that in quantum mechanics electrons are visualized not as revolving around the nucleus in orbits, but as occupying orbitals. **Orbitals** are cloudlike regions surrounding the nucleus; the electrons are located in these regions. The concept of electrons being in specific energy levels is still retained in the modern theory.

5.7 Energy Levels of Electrons

The discussion in Sections 5.7–5.15 describes the ordered system by which the electrons are distributed within the atoms of all the elements. The classifications given in the periodic table (Chapter 6), as well as the chemical properties of the elements, are dependent on their electronic arrangements.

energy levels
of electrons
electron shells

The electrons in an atom are located at specified distances from the nucleus called **energy levels** or **electron shells.** The main or principal energy levels (n) are numbered, starting with $n = 1$ as the energy level nearest to the nucleus and going to $n = 7$, for the known elements. (Theoretically, there are an infinite number of energy levels.) The energy levels are also identified by the letters K, L, M, N, O, P, Q, with K equivalent to the first energy level, L to the second level, and so on, as follows:

Energy level	n	Letter designation
First	1	K
Second	2	L
Third	3	M
Fourth	4	N
Fifth	5	O
Sixth	6	P
Seventh	7	Q

Each succeeding energy level is located farther away from the nucleus.

There are limits to the number of electrons that can exist in each energy level. The maximum number of electrons that can occupy a specific energy level can be calculated from the formula $2n^2$, where n is the number of the principal energy level. For example, shell K, or energy level 1, can have a maximum of two electrons ($2 \times 1^2 = 2$); shell L, or energy level 2, can have a maximum of eight electrons ($2 \times 2^2 = 8$), and so on. Table 5.2 shows the maximum number of electrons that can exist in each of the first five energy levels.

5.8 Energy Sublevels

The principal energy levels contain sublevels differing slightly in energy from one another. These sublevels are the orbitals in which the electrons are located. There are four such sublevels, which are identified by the letters s, p, d, and f.

TABLE 5.2 Maximum number of electrons that can occupy each principal energy level.

Principal energy level, n	Letter designation	Maximum number of electrons in each energy level, $2n^2$
1	K	$2 \times 1^2 = 2$
2	L	$2 \times 2^2 = 8$
3	M	$2 \times 3^2 = 18$
4	N	$2 \times 4^2 = 32$
5	O	$2 \times 5^2 = 50^a$

^{a}The theoretical value of 50 electrons in energy level 5 has never been attained in any element known to date.

Thus we have s orbitals, p orbitals, d orbitals, and f orbitals. Each type of orbital has a particular spatial arrangement or shape.

In any principal energy level there can be a maximum of one s orbital, three p orbitals, five d orbitals, and seven f orbitals. An electron spins on its own axis in one of only two directions, clockwise or counterclockwise. As a result, only two electrons can occupy the same orbital, one spinning clockwise and the other spinning counterclockwise. When an orbital contains two electrons, the electrons are said to be paired.

Since no more than two electrons can exist in an orbital, the maximum numbers of electrons that can exist in the sublevels are 2 in the s orbital, 6 in the three p orbitals, 10 in the five d orbitals, and 14 in the seven f orbitals.

Type of sublevel	Number of orbitals possible	Number of electrons possible
s	1	2
p	3	6
d	5	10
f	7	14

The order of energy of the sublevels within a specific principal energy level is the following: s electrons are lower in energy than p electrons, which are lower than d electrons, which are lower than f electrons. This may be expressed in the following manner:

Sublevel energy: $s < p < d < f$

The order of energy of the principal energy levels, n, is

Principal energy levels: $1 < 2 < 3 < 4 < 5 < 6 < 7$

Not all principal energy levels contain each and every type of sublevel. To determine what type of sublevels occur in any energy level, we need to know the maximum number of electrons possible in that energy level (see Table 5.2), and we need to use two rules: (1) No more than two electrons can occupy one orbital, and (2) an electron will occupy the lowest possible sublevel. The maximum number of electrons in the first energy level is two; both of these are s electrons. They are designated as $1s^2$, indicating two s electrons in the first energy level. There are no p, d, or f electrons in the first energy level. The s orbital in the second energy level is written as $2s$, in the third energy level as $3s$, and so on. The second energy level, with a maximum of eight electrons, will contain only s and p electrons—namely, a maximum of two s and six p electrons. They are designated as $2s^2 2p^6$. The following diagram illustrates how to read these electron designations:

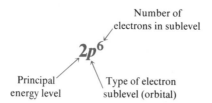

Number of
electrons in sublevel

$$2p^6$$

Principal
energy level

Type of electron
sublevel (orbital)

If each orbital contains two electrons, the second energy level can have four orbitals: one s orbital and three individual p orbitals. These three p orbitals are energetically equivalent to one another and are labeled $2p_x$, $2p_y$, and $2p_z$ to indicate their orientation in space (see Figure 5.5). The symbols $3s^2$, $3p^6$, and $3d^{10}$ illustrate the sublevel breakdown of electrons in the third energy level. From this line of reasoning, we can see that if there are sufficient electrons in an atom, f electrons first appear in the fourth energy level. Table 5.3 shows the type of sublevel electrons and the maximum numbers of orbitals and electrons in each energy level.

Since the $spdf$ atomic orbitals have definite distribution in space, they are represented by particular spatial shapes. At this time we will consider only the s and p orbitals. The s orbitals are spherically symmetrical about the nucleus, as illustrated in Figure 5.5. A $2s$ orbital is a larger sphere than a $1s$ orbital. The

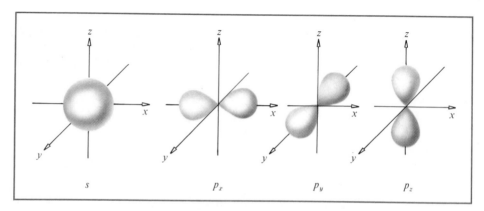

Figure 5.5   Perspective representation of the $s, p_x, p_y,$ and p_z atomic orbitals.

TABLE 5.3 Sublevel electrons in each principal energy level and the maximum number of orbitals and electrons in each energy level.

Energy level	Sublevel electrons	Maximum number of orbitals	Maximum number of electrons
1 (K)	s	1	2
2 (L)	s, p	4	8
3 (M)	s, p, d	9	18
4 (N)	s, p, d, f	16	32
5^a (O)	s, p, d, f	Incomplete	—
6^a (P)	s, p, d	Incomplete	—
7^a (Q)	s	Incomplete	—

[a]Insufficient electrons to complete the shell.

p orbitals (p_x, p_y, p_z) are dumbbell-shaped and are oriented at right angles to one another along the x, y, and z axes in space (see Figure 5.5). An electron has equal probability of being located in either lobe of the p orbital. In illustrations such as Figure 5.5, the boundaries of the orbitals enclose the region of the greatest probability (about 90% chance) of finding an electron. In the ground state of a hydrogen atom, this falls within a sphere having a radius of 0.53 Å.

5.9 Atomic Numbers of the Elements

atomic number

The **atomic number** of an element is the number of protons in the nucleus of an atom of that element. The atomic number is a fundamental characteristic and identifies an atom as being a particular element. The presently known elements are numbered consecutively from 1 to 106 to coincide with the number of protons in their nuclei. Thus, hydrogen, element number 1, has 1 proton; calcium, element number 20, has 20 protons; and uranium, element number 92, has 92 protons in the nucleus. The atomic number tells us not only the amount of positive charge in the nucleus, but also the number of electrons in the neutral atom.

Atomic number = Number of protons in the nucleus

5.10 The Simplest Atom—Hydrogen

The hydrogen atom, consisting of a nucleus containing one proton and an electron orbital containing one electron, is the simplest known atom. (Some hydrogen atoms are known to contain one or two neutrons in their nucleus—see Section 5.16 on isotopes.) The electron in hydrogen occupies an s orbital in the first energy level. This electron does not move in any definite path but rather in a random motion within its orbital, forming an electron cloud about the nucleus.

The diameter of the nucleus is believed to be about 10^{-13} cm, and the diameter of the electron orbital to be about 10^{-8} cm. Hence, the diameter of the electron orbital of a hydrogen atom is about 100,000 times greater than the diameter of the nucleus.

What we have, then, is a positive nucleus surrounded by an electron cloud formed by an electron in an *s* orbital. The net electrical charge on the hydrogen atom is zero; it is called a *neutral atom.* Figure 5.6 shows two methods of representing a hydrogen atom.

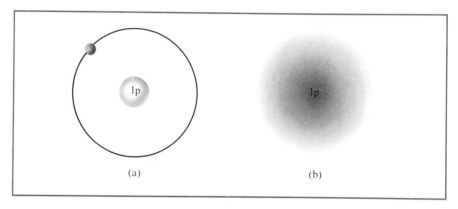

(a)   (b)

Figure 5.6   The hydrogen atom. (a) Illustration of the Bohr description, indicating a discrete electron moving around its nucleus of one proton. (b) Illustration of the modern concept of a hydrogen atom consisting of an electron in an *s* orbital as a cloud of negative charge surrounding the proton in the nucleus.

5.11 Atomic Structure of the First Twenty Elements

Starting with hydrogen, and progressing in order of increasing atomic number to helium, lithium, beryllium, and so on, the atoms of each successive element contain one more proton and one more electron than do the atoms of the preceding element. This sequence continues, without exception, throughout the entire list of known elements and is one of the most impressive examples of order in nature.

The number of neutrons also increases as we progress through the elements. But this number, unlike the number of protons and electrons, does not increase in a perfectly uniform manner as we go from elements of low atomic number to those of higher atomic number. Furthermore, atoms of the same element may contain different numbers of neutrons (see Section 5.16). For example, the predominant isotope of hydrogen contains no neutrons, but two other hydrogen isotopes, containing one and two neutrons, respectively, are known. The predominant isotope of helium (element number 2) has two neutrons, but helium isotopes containing one and four neutrons are also known.

The ground-state (lowest energy state) electronic structures of the first 20 elements fall into a regular pattern. The one hydrogen electron is in the first energy level, and both helium electrons are in the first energy level. The electron structures for hydrogen and helium are written $1s^1$ and $1s^2$, respectively. The maximum number of electrons in the first energy level is two ($2 \times 1^2 = 2$). An atom with three electrons will therefore have an electron in the second energy level, since the first level can contain only two electrons. In lithium (atomic number 3), the third electron is in the $2s$ sublevel of the second energy level. Lithium has the electron structure $1s^2 2s^1$.

In succession, the atoms of beryllium (4), boron (5), carbon (6), nitrogen (7), oxygen (8), fluorine (9), and neon (10) have one more proton and one more electron than the preceding element. Both the first and second energy levels are filled to capacity by the ten electrons of neon, which has two electrons in the first energy level and eight electrons in the second.

H	$1s^1$	B	$1s^2 2s^2 2p^1$	O	$1s^2 2s^2 2p^4$
He	$1s^2$	C	$1s^2 2s^2 2p^2$	F	$1s^2 2s^2 2p^5$
Li	$1s^2 2s^1$	N	$1s^2 2s^2 2p^3$	Ne	$1s^2 2s^2 2p^6$
Be	$1s^2 2s^2$				

Element 11, sodium (Na), has two electrons in the first energy level and eight electrons in the second energy level, with the remaining electron occupying the $3s$ orbital in the third energy level. The electron structure of sodium is $1s^2 2s^2 2p^6 3s^1$. Magnesium (12), aluminum (13), silicon (14), phosphorus (15), sulfur (16), chlorine (17), and argon (18) follow in order, each adding one electron to the third energy level up to argon, which has eight electrons in the M shell.

The placement of the last electron in potassium and calcium, elements number 19 and 20, departs somewhat from the expected order. One might expect that if the third energy level can contain a maximum of 18 electrons (see Table 5.2), electrons would continue to fill this shell until the maximum capacity was reached. However, this is not the case. The $4s$ sublevel is at a lower energy state than the $3d$ sublevel (see Figure 5.7, page 100). Hence, in elements 19 and 20, the last electron is found in the $4s$ level. The electron structure for potassium is $1s^2 2s^2 2p^6 3s^2 3p^6 4s^1$. Calcium has an electron structure similar to that of potassium, except that it has two $4s$ electrons. This break in sequence does not invalidate the formula $2n^2$, which merely prescribes the maximum number of electrons that each shell may contain, but does not state the order in which the shells are filled. Table 5.4 (page 101) shows the electron arrangement of the first 20 elements.

The relative energies of the electron orbitals are shown in Figure 5.7. The order given can be used to determine the electron distribution in the atoms of the elements, although some exceptions to the pattern are known. Suppose we wish to determine the electron structure of a chlorine atom, which has 17 electrons. Following the order in Figure 5.7, we begin by placing two electrons in the $1s$ orbital, then two electrons in the $2s$ orbital, and six electrons in the

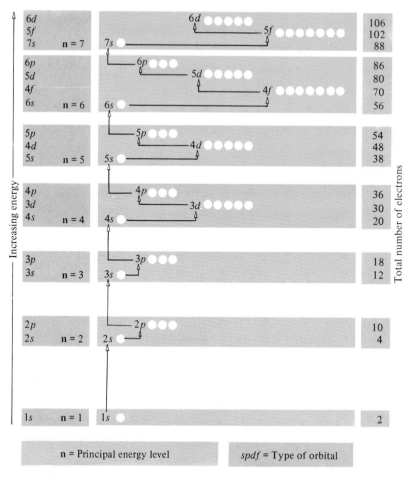

Figure 5.7   Order of filling electron orbitals. Each circle represents an orbital, which can contain two electrons. (Some exceptions to this order are known.)

2p orbitals. We now have used ten electrons. Finally, we place the next two electrons in the 3s orbital and the remaining five electrons in the 3p orbitals, which uses all 17 electrons, giving the electron structure for a chlorine atom as $1s^2 2s^2 2p^6 3s^2 3p^5$. The sum of the superscripts is 17, the number of electrons in the atom. This procedure is summarized as follows:

Order of orbitals to be filled: $1s2s2p3s3p$
Distribution of the 17 electrons in a chlorine atom: $1s^2 2s^2 2p^6 3s^2 3p^5$

PROBLEM 5.1   What is the electron structure of a phosphorus atom?

First determine the number of electrons contained in a phosphorus atom. From the table of atomic weights given on the inside back cover, we see that the atomic number of phosphorus is 15; therefore, each atom contains 15 protons and 15 electrons. Now

TABLE 5.4 Electron arrangement of the first 20 elements.

Element	Symbol	Number of protons (atomic number)	Total number of electrons	Arrangement of electrons, $\mathbf{n} =$ 1 2 3 4
Hydrogen	H	1	1	1
Helium	He	2	2	2
Lithium	Li	3	3	2 1
Beryllium	Be	4	4	2 2
Boron	B	5	5	2 3
Carbon	C	6	6	2 4
Nitrogen	N	7	7	2 5
Oxygen	O	8	8	2 6
Fluorine	F	9	9	2 7
Neon	Ne	10	10	2 8
Sodium	Na	11	11	2 8 1
Magnesium	Mg	12	12	2 8 2
Aluminum	Al	13	13	2 8 3
Silicon	Si	14	14	2 8 4
Phosphorus	P	15	15	2 8 5
Sulfur	S	16	16	2 8 6
Chlorine	Cl	17	17	2 8 7
Argon	Ar	18	18	2 8 8
Potassium	K	19	19	2 8 8 1
Calcium	Ca	20	20	2 8 8 2

tabulate the number of electrons in each principal energy level and sublevel until all 15 electrons are assigned. The order of filling the sublevels is $1s2s2p3s3p$:

Sublevel	Number of e^-	Total e^-
$1s$ orbital	2	2
$2s$ orbital	2	4
$2p$ orbital	6	10
$3s$ orbital	2	12
$3p$ orbital	3	15

Therefore, the electron structure of phosphorus is $1s^2 2s^2 2p^6 3s^2 3p^3$.

■

5.12 Electron Structure of the Elements Beyond Calcium

The elements following calcium have a less regular pattern of adding electrons. The lowest energy level available for the twenty-first electron is the $3d$ level. Thus, scandium (21) has the following electron arrangement: first energy level, two electrons; second energy level, eight electrons; third energy level, nine electrons; fourth energy level, two electrons. The last electron is located in the $3d$ level. The structure for scandium is $1s^2 2s^2 2p^6 3s^2 3p^6 4s^2 3d^1$. The elements

TABLE 5.5   **Electron structure of the elements.**

Element	Atomic number	Electron structure	Element	Atomic number	Electron structure
H	1	$1s^1$	Xe	54	$[Kr]\,5s^24d^{10}5p^6$
He	2	$1s^2$	Cs	55	$[Xe]\,6s^1$
Li	3	$1s^22s^1$	Ba	56	$[Xe]\,6s^2$
Be	4	$1s^22s^2$	La	57	$[Xe]\,6s^25d^1$
B	5	$1s^22s^22p^1$	Ce	58	$[Xe]\,6s^24f^15d^1$
C	6	$1s^22s^22p^2$	Pr	59	$[Xe]\,6s^24f^3$
N	7	$1s^22s^22p^3$	Nd	60	$[Xe]\,6s^24f^4$
O	8	$1s^22s^22p^4$	Pm	61	$[Xe]\,6s^24f^5$
F	9	$1s^22s^22p^5$	Sm	62	$[Xe]\,6s^24f^6$
Ne	10	$1s^22s^22p^6$	Eu	63	$[Xe]\,6s^24f^7$
Na	11	$[Ne]\,3s^1$	Gd	64	$[Xe]\,6s^24f^75d^1$
Mg	12	$[Ne]\,3s^2$	Tb	65	$[Xe]\,6s^24f^9$
Al	13	$[Ne]\,3s^23p^1$	Dy	66	$[Xe]\,6s^24f^{10}$
Si	14	$[Ne]\,3s^23p^2$	Ho	67	$[Xe]\,6s^24f^{11}$
P	15	$[Ne]\,3s^23p^3$	Er	68	$[Xe]\,6s^24f^{12}$
S	16	$[Ne]\,3s^23p^4$	Tm	69	$[Xe]\,6s^24f^{13}$
Cl	17	$[Ne]\,3s^23p^5$	Yb	70	$[Xe]\,6s^24f^{14}$
Ar	18	$[Ne]\,3s^23p^6$	Lu	71	$[Xe]\,6s^24f^{14}5d^1$
K	19	$[Ar]\,4s^1$	Hf	72	$[Xe]\,6s^24f^{14}5d^2$
Ca	20	$[Ar]\,4s^2$	Ta	73	$[Xe]\,6s^24f^{14}5d^3$
Sc	21	$[Ar]\,4s^23d^1$	W	74	$[Xe]\,6s^24f^{14}5d^4$
Ti	22	$[Ar]\,4s^23d^2$	Re	75	$[Xe]\,6s^24f^{14}5d^5$
V	23	$[Ar]\,4s^23d^3$	Os	76	$[Xe]\,6s^24f^{14}5d^6$
Cr	24	$[Ar]\,4s^13d^5$	Ir	77	$[Xe]\,6s^24f^{14}5d^7$
Mn	25	$[Ar]\,4s^23d^5$	Pt	78	$[Xe]\,6s^14f^{14}5d^9$
Fe	26	$[Ar]\,4s^23d^6$	Au	79	$[Xe]\,6s^14f^{14}5d^{10}$
Co	27	$[Ar]\,4s^23d^7$	Hg	80	$[Xe]\,6s^24f^{14}5d^{10}$
Ni	28	$[Ar]\,4s^23d^8$	Tl	81	$[Xe]\,6s^24f^{14}5d^{10}6p^1$
Cu	29	$[Ar]\,4s^13d^{10}$	Pb	82	$[Xe]\,6s^24f^{14}5d^{10}6p^2$
Zn	30	$[Ar]\,4s^23d^{10}$	Bi	83	$[Xe]\,6s^24f^{14}5d^{10}6p^3$
Ga	31	$[Ar]\,4s^23d^{10}4p^1$	Po	84	$[Xe]\,6s^24f^{14}5d^{10}6p^4$
Ge	32	$[Ar]\,4s^23d^{10}4p^2$	At	85	$[Xe]\,6s^24f^{14}5d^{10}6p^5$
As	33	$[Ar]\,4s^23d^{10}4p^3$	Rn	86	$[Xe]\,6s^24f^{14}5d^{10}6p^6$
Se	34	$[Ar]\,4s^23d^{10}4p^4$	Fr	87	$[Rn]\,7s^1$
Br	35	$[Ar]\,4s^23d^{10}4p^5$	Ra	88	$[Rn]\,7s^2$
Kr	36	$[Ar]\,4s^23d^{10}4p^6$	Ac	89	$[Rn]\,7s^26d^1$
Rb	37	$[Kr]\,5s^1$	Th	90	$[Rn]\,7s^26d^2$
Sr	38	$[Kr]\,5s^2$	Pa	91	$[Rn]\,7s^25f^26d^1$
Y	39	$[Kr]\,5s^24d^1$	U	92	$[Rn]\,7s^25f^36d^1$
Zr	40	$[Kr]\,5s^24d^2$	Np	93	$[Rn]\,7s^25f^46d^1$
Nb	41	$[Kr]\,5s^14d^4$	Pu	94	$[Rn]\,7s^25f^6$
Mo	42	$[Kr]\,5s^14d^5$	Am	95	$[Rn]\,7s^25f^7$
Tc	43	$[Kr]\,5s^24d^5$	Cm	96	$[Rn]\,7s^25f^76d^1$
Ru	44	$[Kr]\,5s^14d^7$	Bk	97	$[Rn]\,7s^25f^9$
Rh	45	$[Kr]\,5s^14d^8$	Cf	98	$[Rn]\,7s^25f^{10}$
Pd	46	$[Kr]\,4d^{10}$	Es	99	$[Rn]\,7s^25f^{11}$
Ag	47	$[Kr]\,5s^14d^{10}$	Fm	100	$[Rn]\,7s^25f^{12}$
Cd	48	$[Kr]\,5s^24d^{10}$	Md	101	$[Rn]\,7s^25f^{13}$
In	49	$[Kr]\,5s^24d^{10}5p^1$	No	102	$[Rn]\,7s^25f^{14}$
Sn	50	$[Kr]\,5s^24d^{10}5p^2$	Lr	103	$[Rn]\,7s^25f^{14}6d^1$
Sb	51	$[Kr]\,5s^24d^{10}5p^3$	Unq	104	$[Rn]\,7s^25f^{14}6d^2$
Te	52	$[Kr]\,5s^24d^{10}5p^4$	Unp	105	$[Rn]\,7s^25f^{14}6d^3$
I	53	$[Kr]\,5s^24d^{10}5p^5$	Unh	106	$[Rn]\,7s^25f^{14}6d^4$

Note: For simplicity of expression, symbols of the chemically stable noble gases are used as a portion of the electron structure for the elements beyond neon. For example, the electron structure of a sodium atom, Na, consists of ten electrons, as in neon [Ne], plus a $3s^1$ electron. Detailed electron structures for the noble gases are given in Table 5.6.

following scandium, titanium (22) through zinc (30), continue to add d electrons until the third energy level has its maximum of 18. Two exceptions in the orderly electron addition are chromium (24) and copper (29), the structures of which are given in Table 5.5. The third energy level of the electrons is first completed in the element copper. Electrons continue to add to the fourth energy level, filling the $4p$ orbitals (Ga to Kr). Table 5.5 shows the order of filling of the electron orbitals and the electron configurations of all the known elements.

Problem 5.2 illustrates how to use and interpret Table 5.5

PROBLEM 5.2   What is the electron structure for a sulfur atom (atomic number 16) and for an iron atom (atomic number 26)?

Look in Table 5.5 for the element with atomic number 16 and write down its structure, [Ne] $3s^2 3p^4$. [Ne] is an abbreviated structure for neon, which is $1s^2 2s^2 2p^6$. Therefore, the electron structure for a sulfur atom is $1s^2 2s^2 2p^6 3s^2 3p^4$.

For an iron atom, Table 5.5 shows a structure of [Ar] $4s^2 3d^6$. This means that the electron structure of iron consists of the electron structure for argon plus $4s^2 3d^6$. The table shows that the structure for [Ar] is [Ne] $3s^2 3p^6$, which is equal to $1s^2 2s^2 2p^6 3s^2 3p^6$. Therefore, the electron structure for iron is $1s^2 2s^2 2p^6 3s^2 3p^6 4s^2 3d^6$.

If Table 5.5 is not available, the structure can be determined by tabulating electrons as Problem 5.1. Structures for all the noble gases, He, Ne, Ar, and so on, are also given in Table 5.6.

■

5.13 Diagramming Atomic Structures

Several methods can be used to diagram atomic structures of atoms, depending on what we are trying to illustrate. When we want to show both the nuclear makeup and the total electron structure of each energy level (without orbital detail), we can use a diagram like those shown in Figure 5.8.

A method of diagramming energy sublevels is shown in Figure 5.9 (page 104). Each orbital is represented by a circle ◯. When the orbital contains one electron, an arrow (↾) is placed in the circle. A second arrow, pointing downward (↾), indicates the second electron in that orbital.

cates the second electron in that orbital.

The diagram for hydrogen is ⊙. Helium, with two electrons, is drawn as ⊛; both electrons are $1s$ electrons. The diagram for lithium shows three electrons in two energy levels, $1s^2 2s^1$. All four electrons of beryllium are s electrons,

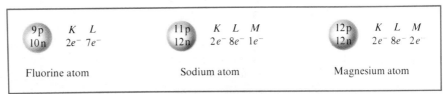

Figure 5.8   Atomic structure diagrams of fluorine, sodium, and magnesium atoms. The number of protons and neutrons in the nucleus are shown; outside the nucleus are shown the number of electrons in each principal energy level.

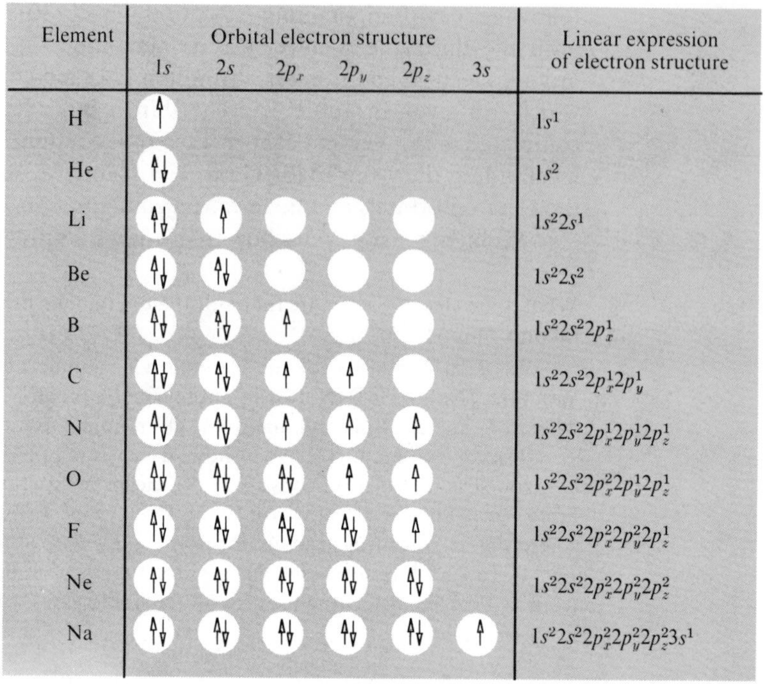

Element	Orbital electron structure						Linear expression of electron structure
	$1s$	$2s$	$2p_x$	$2p_y$	$2p_z$	$3s$	
H	↑						$1s^1$
He	↑↓						$1s^2$
Li	↑↓	↑					$1s^2 2s^1$
Be	↑↓	↑↓					$1s^2 2s^2$
B	↑↓	↑↓	↑				$1s^2 2s^2 2p_x^1$
C	↑↓	↑↓	↑	↑			$1s^2 2s^2 2p_x^1 2p_y^1$
N	↑↓	↑↓	↑	↑	↑		$1s^2 2s^2 2p_x^1 2p_y^1 2p_z^1$
O	↑↓	↑↓	↑↓	↑	↑		$1s^2 2s^2 2p_x^2 2p_y^1 2p_z^1$
F	↑↓	↑↓	↑↓	↑↓	↑		$1s^2 2s^2 2p_x^2 2p_y^2 2p_z^1$
Ne	↑↓	↑↓	↑↓	↑↓	↑↓		$1s^2 2s^2 2p_x^2 2p_y^2 2p_z^2$
Na	↑↓	↑↓	↑↓	↑↓	↑↓	↑	$1s^2 2s^2 2p_x^2 2p_y^2 2p_z^2 3s^1$

Figure 5.9   Energy sublevel electron structures of the elements from hydrogen through sodium. Each electron is indicated by an arrow placed in a circle, which represents an orbital.

$1s^2 2s^2$. Boron has the first p electron, which is located in the $2p_x$ orbital. Since it is energetically more difficult for the next p electron to pair up with the electron in the p_x orbital than to occupy a second p orbital, the second p electron in carbon is located in the $2p_y$ orbital. The third p electron in nitrogen is still unpaired and is found in the $2p_z$ orbital. The next three electrons pair with each of the $2p$ electrons up to the element neon. Also shown in Figure 5.9 are the equivalent linear expressions for these orbital electron structures.

The electrons in successive elements are found in sublevels of increasing energy. The general sequence of increasing energy of subshells is

$$1s\,2s\,2p\,3s\,3p\,4s\,3d\,4p\,5s\,4d\,5p\,6s\,4f\,5d\,6p\,7s\,5f\,6d$$

Figure 5.10 is a useful mnemonic device for writing electron structures. Minor variations from the electron structure predicted by the foregoing general sequence or mnemonic device of Figure 5.10 are found in some atoms. Table 5.5 shows the accepted ground-state electron structures for all the elements.

PROBLEM 5.3   Diagram the electron structure of a zinc atom and of a rubidium atom. Use the $1s^2 2s^2 2p^6$, etc. method.

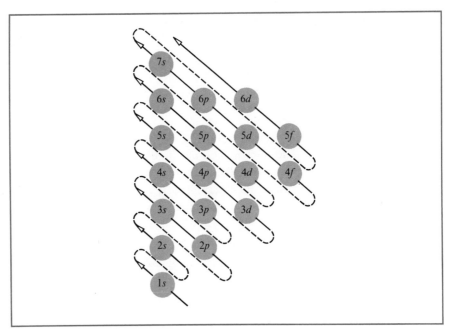

Figure 5.10 Approximate order for placing electrons in energy sublevels. Follow the arrows as indicated. Fill each successive energy sublevel with the proper number of electrons, starting with $1s$ ($s = 2, p = 6, d = 10, f = 14$ electrons), until all the electrons of an atom have been assigned. (There are a few exceptions to this order—for example, chromium and copper among the first 30 elements.)

The atomic number of zinc is 30; therefore, it has 30 protons and 30 electrons in a neutral atom. Using Figure 5.7 or Figure 5.10, tabulate the 30 electrons as follows:

Orbital	Number of e⁻	Total e⁻
$1s$	2	2
$2s$	2	4
$2p$	6	10
$3s$	2	12
$3p$	6	18
$4s$	2	20
$3d$	10	30

The electron distribution in Zn is $1s^2 2s^2 2p^6 3s^2 3p^6 4s^2 3d^{10}$. Check by adding the superscripts, which should equal 30.

The atomic number of rubidium is 37; therefore, it has 37 protons and 37 electrons in a neutral atom. With a little practice, and using either Figure 5.7 or Figure 5.10, the electron structure may be written directly in the linear form. The structure for rubidium is $1s^2 2s^2 2p^6 3s^2 3p^6 4s^2 3d^{10} 4p^6 5s^1$. Check by adding the superscripts, which should equal 37.

5.14 Electron-Dot Representation of Atoms

The electron-dot or Lewis-dot method of diagramming atoms was named after the American chemist G. N. Lewis (1875–1946). It uses the symbol of an element together with dots to represent electrons. The number of dots used represents the number of s and p electrons in the outermost energy level of an atom. The dots are placed around the symbol of the atom. Paired dots represent paired electrons; unpaired dots represent unpaired electrons.

The nucleus and all the electrons other than those in the outermost energy level are called the **kernel** of the atom and are represented by the symbol of the element. For example, $:\dot{B}$ represents a boron atom and tells us that boron has three electrons in its outermost energy level. The kernel of the boron atom includes the nucleus and its $1s^2$ electrons:

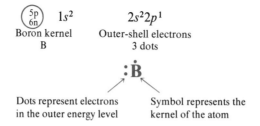

$$1s^2 \qquad 2s^2 2p^1$$

Boron kernel   Outer-shell electrons
B   3 dots

$$:\dot{B}$$

Dots represent electrons   Symbol represents the
in the outer energy level   kernel of the atom

Similarly, $:\ddot{I}\cdot$ indicates an iodine atom, which has seven electrons in its outermost principal energy level. The electron-dot system is used a great deal, not only because of its simplicity, but also because much of the chemistry of the atom is directly associated with the electrons in the outermost energy level. This association is especially true for the first 20 elements and the remaining Group A elements of the periodic table (see Chapter 6). However, electron-dot diagrams are not generally used for the transition elements, since their atoms are filling the d and f orbitals and they all have one or two s electrons in their outer energy levels. Figure 5.11 shows electron-dot diagrams for the elements hydrogen through calcium.

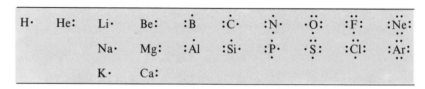

Figure 5.11   Electron-dot diagrams for the first 20 elements. Dots represent electrons in the outermost energy level only. Symbol represents the kernel of the atom.

PROBLEM 5.4   Write the electron-dot structure for a phosphorus atom.

First establish the electron structure for a phosphorus atom. It is $1s^2 2s^2 2p^6 3s^2 3p^3$. Note that there are five electrons in the outermost principal energy level; they are $3s^2 3p^3$. The $1s^2 2s^2 2p^6$ electrons are part of the kernel and are represented by the symbol for phosphorus. Write the symbol for phosphorus, and place the five electrons as dots around it.

$$:\!\overset{\textstyle .}{\text{P}}\!\cdot$$

The $3s^2$ electrons are paired and are represented by the paired dots. The $3p^3$ electrons, which are unpaired, are represented by the single dots.

5.15 The Noble Gases

noble gases

The family of elements consisting of helium, neon, argon, krypton, xenon, and radon is known as the **noble gases.** The electron structure of these gases has particular interest for the chemist. Until 1962, attempts to prepare compounds of these elements met with failure. Because of their supposed inability to enter into chemical combinations, these elements were formerly called *inert* gases. They have also been called the *rare* gases. Because of their chemical inactivity, the electron structure of the noble gases is considered to be extraordinarily stable.

Each of these elements, except helium, has eight electrons in its outer shell (see Table 5.6). This electron structure is such that the *s* and *p* orbitals in the outer shell are filled with paired electrons, an arrangement that is very stable and chemically unreactive.

Recognition of this structure led to the *rule of eight* principle: The elements, through chemical changes, attempt to attain an electron structure of eight electrons in the outermost energy level, a structure identical with that of the chemically stable noble gases. Although the rule of eight principle applies to the chemical behavior of many elements and compounds, it is not universally applicable; some elements do not follow this rule.

All the noble gases are present in the atmosphere. Argon is the most abundant and is found at a concentration of about 1% by volume. The others are present in only trace amounts. Argon was discovered in 1894 by Baron Rayleigh (1842–1919) and Sir William Ramsay (1852–1916). Helium was first observed in the spectrum of the sun during an eclipse in 1868. It was not until 1894 that Ramsay recognized that helium exists in the earth's atmosphere. In 1898 he and his coworker, Morris W. Travers (1872–1961), announced the discovery of neon, krypton, and xenon, having isolated them from liquid air. Friedrich E. Dorn (1848–1916) first identified radon, the heaviest member of the noble gases, as a radioactive gas emanating from the element radium.

Because of its low density and nonflammability, helium has been used for filling balloons and dirigibles. Only hydrogen surpasses helium in lifting power.

Helium mixed with oxygen is used by deep-sea divers for breathing. This mixture reduces the danger of acquiring the "bends" (caisson disease), pains and paralysis suffered by divers on returning from the ocean depths to normal atmospheric pressure. Helium is also used in "heli-arc welding," where it supplies an inert atmosphere for the welding of active metals such as magnesium. Helium is found in some natural gas wells in the southwestern United States. As a liquid, it is used to study the properties of substances at very low temperatures. The boiling point of liquid helium, 4.2 K, is not far above absolute zero.

We are all familiar with the neon sign, in which a characteristic red color is produced when an electric discharge is passed through a tube filled with neon. This color may be modified by mixing the neon with other gases or by changing the color of the glass. Argon is used primarily in gas-filled electric light bulbs and other types of electronic tubes to provide an inert atmosphere for prolonging tube life. Argon is also used in some welding applications where an inert atmosphere is needed. Krypton and xenon have not been used extensively because of their limited availability. Radon is radioactive; it has been used medicinally in the treatment of cancer.

For many years it was believed that the noble gases could not be made to combine chemically with any other element. Then, in 1962, Neil Bartlett at the University of British Columbia, Vancouver, synthesized the first noble gas compound, xenon hexafluoroplatinate, $XePtF_6$. This outstanding discovery opened a new field in the techniques of preparing noble gas compounds and investigating their chemical bonding and properties. Other compounds of xenon as well as compounds of krypton and radon have been prepared. Some of these are XeF_2, XeF_4, XeF_6, $XeOF_4$, $Xe(OH)_6$, and KrF_2.

TABLE 5.6 Arrangement of electrons in the noble gases.
Each gas except helium has eight electrons in the outermost energy level.

Noble gas	Symbol	Electron structure					
		$n = 1$	2	3	4	5	6
Helium	He	$1s^2$					
Neon	Ne	$1s^2$	$2s^22p^6$				
Argon	Ar	$1s^2$	$2s^22p^6$	$3s^23p^6$			
Krypton	Kr	$1s^2$	$2s^22p^6$	$3s^23p^63d^{10}$	$4s^24p^6$		
Xenon	Xe	$1s^2$	$2s^22p^6$	$3s^23p^63d^{10}$	$4s^24p^64d^{10}$	$5s^25p^6$	
Radon	Rn	$1s^2$	$2s^22p^6$	$3s^23p^63d^{10}$	$4s^24p^64d^{10}4f^{14}$	$5s^25p^65d^{10}$	$6s^26p^6$

5.16 Isotopes of the Element

Shortly after Rutherford's conception of the nuclear atom, experiments were performed to determine the masses of individual atoms. These experiments showed that the masses of nearly all atoms were greater than could be accounted for by simply adding up the masses of all the protons and electrons that were known to be present. This fact led to the concept of the neutron, a particle with no charge but with a mass about the same as that of a proton. Since this particle has no charge, it was very difficult to detect, and the existence of the neutron was not proven experimentally until 1932. All atomic nuclei except that of the simplest hydrogen atom are now believed to contain neutrons.

All atoms of a given element have the same number of protons, but experimental evidence has shown that, in most cases, all atoms of a given element do not have identical masses. This is because atoms of the same element may have different numbers of neutrons in their nuclei.

isotopes

Atoms of an element having the same atomic number but different atomic masses are called **isotopes** of that element. Atoms of the isotopes of an element, therefore, have the same number of protons and electrons but different numbers of neutrons.

Three isotopes of hydrogen (atomic number 1) are known. Each has one proton in the nucleus and one electron in the first energy level. The first isotope (protium) does not have a neutron, and its mass is 1; the second isotope (deuterium) has one neutron in the nucleus, and a mass of 2; the third isotope (tritium) has two neutrons, and a mass of 3 (see Figure 5.12).

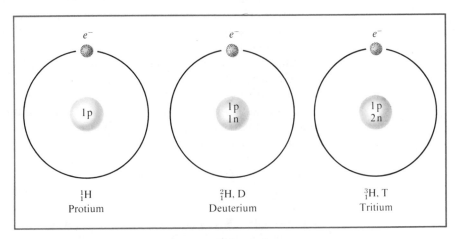

Figure 5.12   The isotopes of hydrogen. The number of protons (p) and neutrons (n) are shown within the nucleus.

The three isotopes of hydrogen may be represented by the symbols $_1^1H$, $_1^2H$, and $_1^3H$, indicating an atomic number of 1 and mass numbers of 1, 2, and 3, respectively. This method of representing atoms is called *isotopic notation.* The subscript number (Z) is the atomic number; the superscript number (A) is **mass number**   the **mass number,** which is the sum of the number of protons and the number of neutrons in the nucleus. The hydrogen isotopes may also be referred to as hydrogen-1, hydrogen-2, and hydrogen-3.

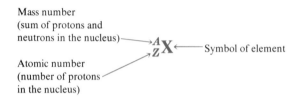

Mass number
(sum of protons and
neutrons in the nucleus) $\longrightarrow {}_Z^A X \longleftarrow$ Symbol of element

Atomic number
(number of protons
in the nucleus)

Two or more isotopes are known for all elements. However, not all isotopes are stable; some are radioactive and are continually decomposing to form other elements (see Chapter 18). For example, of the seven known isotopes of carbon, only two, carbon-12 and carbon-13, are stable. Of the seven known isotopes of oxygen, only three, ${}_8^{16}O$, ${}_8^{17}O$, and ${}_8^{18}O$, are stable. Of the fifteen known isotopes of arsenic, ${}_{33}^{75}As$ is the only one that is stable.

5.17 Atomic Weight

By means of an instrument called a *mass spectrometer,* it is possible to make fairly precise physical determinations of the masses of individual atoms. In the ordinary sense of weighing, single atoms are far too tiny actually to be weighed (the mass of a single hydrogen atom is 1.67×10^{-24} g). If we could magnify the mass of such an atom by a factor of 100 billion, it would still require about 300 million of these magnified hydrogen atoms to equal the weight of a single drop of water (0.05 g).

Because it is not only inconvenient but impractical to express the actual weights of atoms in grams, chemists have devised a useful table of relative atomic weights. The carbon isotope having six protons and six neutrons and designated carbon-12, or ${}_6^{12}C$, was chosen as the standard of reference for atomic weights. The mass of this isotope is assigned a value of exactly 12 atomic mass **atomic mass unit** units (amu). Thus, 1 **atomic mass unit** is defined as $1/12$ the mass of a carbon-12 atom. The atomic weights of all the other atoms are assigned values proportional to the arbitrary mass assigned to the reference isotope carbon-12. A table of atomic weights is given on the inside back cover of this book. Hydrogen atoms, with a mass of about $1/12$ that of a carbon atom, have an average atomic mass of 1.0079 amu on this relative scale. Magnesium atoms, which are about twice as heavy as carbon, have an average mass of 24.305 amu. The average atomic mass of oxygen is 15.9994 amu (usually rounded off to 16.0 for calculations).

Since all elements exist as isotopes having different masses, the atomic weight of an element represents the average relative mass of all the naturally occurring isotopes of that element. The atomic weights of the individual isotopes are approximately whole numbers, because the relative masses of the protons and neutrons are approximately 1.0 amu each. Yet we find that the atomic weights given for many of the elements deviate considerably from whole numbers. For example, the atomic weight of rubidium is 85.4678 amu, that of copper is 63.546 amu, and that of magnesium is 24.305 amu. The deviation of an atomic weight from a whole number is due mainly to the unequal occurrence of the various isotopes of an element. It is also due partly to the difference between the mass of a free proton or neutron and the mass of these same particles in the nucleus. For example, the two principal isotopes of copper are $^{63}_{29}Cu$ and $^{65}_{29}Cu$. It is apparent that copper-63 atoms are the more abundant isotope, since the atomic weight of copper, 63.546 amu, is closer to 63 than to 65 amu. The actual values of the copper isotopes observed by mass spectra determination are shown in the following table:

Isotope	Isotopic mass (amu)	Abundance (%)	Average atomic mass (amu)
$^{63}_{29}Cu$	62.930	69.16	63.546
$^{65}_{29}Cu$	64.928	30.84	

atomic weight

The **atomic weight** of an element is the average relative mass of the isotopes of that element referred to the atomic mass of carbon-12 (exactly 12.0000 amu).

The relationship of mass number and atomic number is such that if we subtract the atomic number from the mass number we obtain the number of neutrons in the nucleus of the atom. Table 5.7 shows the application of this method of determining the number of neutrons. For example, the fluorine atom ($^{19}_{9}F$), atomic number 9, mass number 19, contains 10 neutrons:

Mass number $-$ Atomic number $=$ Number of neutrons

$$19 \quad - \quad 9 \quad = \quad 10$$

The atomic weights given in the table on the inside back cover of this book are values accepted by international agreement. There is no need to memorize atomic weights. In most of the calculations needed in this book, their use to the first decimal place will give results of sufficient accuracy.

TABLE 5.7 Calculation of the number of neutrons in an atom by subtracting the atomic number from the mass number.

	Hydrogen ($^{1}_{1}H$)	Oxygen ($^{16}_{8}O$)	Sulfur ($^{32}_{16}S$)	Fluorine ($^{19}_{9}F$)	Iron ($^{56}_{26}Fe$)
Mass number	1	16	32	19	56
Atomic number	1	8	16	9	26
Number of neutrons	0	8	16	10	30

5.18 The Mole

According to the atomic theory set forth by John Dalton, atoms always combine in whole-number ratios. Since individual atoms certainly cannot be weighed, there was a need to establish some weighable unit for comparing the quantities

**gram-atomic
weight**

of elements involved in chemical reactions. The working unit devised for this purpose is the **gram-atomic weight** (g-at. wt), which is defined as the number of grams of an element *numerically* equal to its atomic weight. Be very careful to note this distinction: The atomic weight of magnesium, for example, as given in the table, is 24.305. This value indicates the average mass of magnesium atoms relative to the carbon-12 isotope. But the gram-atomic weight of magnesium is 24.305 *grams* of magnesium.

It is known that 1 gram-atomic weight of magnesium reacts chemically with 1 gram-atomic weight of sulfur to form magnesium sulfide:

$$Mg + S \longrightarrow MgS$$

Thus, magnesium atoms and sulfur atoms react in a 1:1 atom ratio. It follows, then, that a gram-atomic weight (24.305 g) of magnesium must contain the same number of atoms as a gram-atomic weight (32.06 g) of sulfur.

Since the actual weight of an atom is minute, 1 gram-atomic weight of atoms will contain a large number of individual atoms. The number of atoms in 1 gram-atomic weight of any element has been experimentally determined to be 6.02×10^{23} (that is, 602,000,000,000,000,000,000,000 atoms). This number is known as **Avogadro's number.** Thus, 1 gram-atomic weight (24.305 g) of magnesium contains 6.02×10^{23} magnesium atoms. A gram-atomic weight of any element—for example, 32.06 g of sulfur, 15.9994 g of oxygen, 1.0079 g of hydrogen, and so on—therefore contains Avogadro's number of atoms. Table 5.8 summarizes these relationships.

Avogadro's number

Avogadro's number $= 6.02 \times 10^{23}$

It is difficult to imagine how large Avogadro's number really is, but perhaps the following analogy will help express it: If 10,000 people started to count Avogadro's number and each counted at the rate of 100 numbers per minute each minute of the day, it would take them over 1 trillion (10^{12}) years to count the total number. So, even the minutest amount of matter contains extremely large numbers of atoms. For example, 1 milligram (0.001 g) of sulfur contains 2×10^{19} atoms of sulfur.

TABLE 5.8 Avogadro's number related to the gram-atomic weights of oxygen, hydrogen, sulfur, and magnesium.

Element	Gram-atomic weight	Avogadro's number (Number of atoms/g-at. wt)
Oxygen	15.9994 g	6.02×10^{23}
Hydrogen	1.0079 g	6.02×10^{23}
Sulfur	32.06 g	6.02×10^{23}
Magnesium	24.305 g	6.02×10^{23}

mole

Avogadro's number is the basis for an additional very important quantity used to express a particular number of chemical species—such as atoms, molecules, formula units, ions, or electrons. This quantity is the mole. We define a **mole** (mol) as an amount of a substance containing the same number of formula units as there are atoms in exactly 12 g of carbon-12. Other definitions are used, but they all relate to a mole being Avogadro's number of formula units of a substance. A formula unit is whatever is indicated by the formula of the substance under consideration—for example, Mg, MgS, H_2O, O_2, $^{75}_{33}As$. If 1 mole of atoms contains 6.02×10^{23} atoms and 6.02×10^{23} atoms weigh 1 gram-atomic weight, then 1 gram-atomic weight of atoms is equal to 1 mole of atoms.

1 gram-atomic weight = 1 mole of atoms
= Avogadro's number (6.02×10^{23}) of atoms

Thus, for example, 1 mole of hydrogen atoms represents Avogadro's number of hydrogen atoms and has a mass of 1.0079 g (the gram-atomic weight).

The term *mole* is so commonplace in chemical jargon that chemists use it as freely as the words *atom* and *molecule*. The mole is used in conjunction with many different particles, such as atoms, molecules, ions, and electrons, to represent Avogadro's number of these particles. If we can speak of a mole of atoms, we can also speak of a mole of molecules, a mole of electrons, and a mole of ions, understanding that in each case we mean 6.02×10^{23} formula units of these particles.

1 mole of atoms = 6.02×10^{23} atoms
1 mole of molecules = 6.02×10^{23} molecules
1 mole of ions = 6.02×10^{23} ions

We frequently encounter problems that require interconversions involving the quantities of mass, numbers, and moles of atoms of an element. Conversion factors that may be used for this purpose are

(a) Grams to atoms: $\dfrac{6.02 \times 10^{23} \text{ atoms of the element}}{\text{g-at. wt of the element}}$

(b) Atoms to grams: $\dfrac{\text{g-at. wt of the element}}{6.02 \times 10^{23} \text{ atoms of the element}}$

(c) Grams to moles: (Monatomic elements) $\dfrac{1 \text{ mole of the element}}{1 \text{ g-at. wt of the element}}$

(d) Moles to grams: (Monatomic element) $\dfrac{1 \text{ g-at. wt of the element}}{1 \text{ mole of the element}}$

Sample problems follow.

PROBLEM 5.5   How many moles of iron does 25.0 g of Fe represent?

The problem requires that we change grams of Fe to moles of Fe. We look up the atomic weight of Fe in the atomic weight table and find it to be 55.8. Then we use the proper conversion factor to obtain moles. The conversion factor is (c).

$$\text{Grams Fe} \longrightarrow \text{Moles Fe} \qquad \text{Grams Fe} \times \frac{1 \text{ mole Fe}}{1 \text{ g-at. wt Fe}}$$

$$25.0 \text{ g Fe} \times \frac{1 \text{ mole Fe}}{55.8 \text{ g Fe}} = 0.448 \text{ mole Fe} \quad \text{(Answer)}$$

■

PROBLEM 5.6   How many magnesium atoms does 5.00 g of Mg represent?

The problem requires that we change grams of magnesium to atoms of magnesium.

$$\text{Grams Mg} \longrightarrow \text{Atoms Mg}$$

We find the atomic weight of magnesium to be 24.3 and set up the calculation using conversion factor (a):

$$\text{Grams Mg} \times \frac{6.02 \times 10^{23} \text{ atoms Mg}}{1 \text{ g-at. wt Mg}}$$

$$5.00 \text{ g Mg} \times \frac{6.02 \times 10^{23} \text{ atoms Mg}}{24.3 \text{ g Mg}} = 1.24 \times 10^{23} \text{ atoms Mg} \quad \text{(Answer)}$$

An alternative solution is first to convert grams of magnesium to moles of magnesium, which are then changed to atoms of magnesium (grams Mg $\longrightarrow$ moles Mg $\longrightarrow$ atoms Mg). The calculation setup is

$$5.00 \text{ g Mg} \times \frac{1 \text{ mole Mg}}{24.3 \text{ g Mg}} \times \frac{6.02 \times 10^{23} \text{ atoms Mg}}{1 \text{ mole Mg}} = 1.24 \times 10^{23} \text{ atoms Mg}$$

Thus, there are 1.24×10^{23} atoms of Mg in 5.00 g of Mg.

■

PROBLEM 5.7   What is the mass (in grams) of one atom of carbon?

From the table of atomic weights, we see that the gram-atomic weight of carbon is 12.0 g. The factor needed to convert atoms to grams is conversion factor (b).

$$\text{Atoms C} \longrightarrow \text{Grams C} \qquad \text{Atoms C} \times \frac{1 \text{ g-at. wt C}}{6.02 \times 10^{23} \text{ atoms C}}$$

$$1 \text{ atom C} \times \frac{12.0 \text{ g C}}{6.02 \times 10^{23} \text{ atoms C}} = 1.99 \times 10^{-23} \text{ g C} \quad \text{(Answer)}$$

■

PROBLEM 5.8   How much do 3.01×10^{23} atoms of sodium weigh?

The information needed to solve this problem is the gram-atomic weight of Na (23.0 g) and conversion factor (b).

$$\text{Atoms Na} \longrightarrow \text{Grams Na} \qquad \text{Atoms Na} \times \frac{1 \text{ g-at. wt Na}}{6.02 \times 10^{23} \text{ atoms Na}}$$

$$3.01 \times 10^{23} \text{ atoms Na} \times \frac{23.0 \text{ g Na}}{6.02 \times 10^{23} \text{ atoms Na}} = 11.5 \text{ g Na} \quad \text{(Answer)}$$

■

PROBLEM 5.9   How many grams does 0.252 mole of Cu weigh?

The information needed to solve this problem is the gram-atomic weight of Cu (63.5 g) and conversion factor (d).

$$\text{Moles Cu} \longrightarrow \text{Grams Cu} \qquad \text{Moles Cu} \times \frac{1 \text{ g-at. wt Cu}}{1 \text{ mole Cu}}$$

$$0.252 \text{ mole Cu} \times \frac{63.5 \text{ g Cu}}{1 \text{ mole Cu}} = 16.0 \text{ g Cu} \quad \text{(Answer)}$$

■

PROBLEM 5.10   How many oxygen atoms are present in 1.00 mole of oxygen molecules?

Oxygen is a diatomic molecule with the formula O_2. Therefore, a molecule of oxygen contains two atoms of oxygen.

$$\frac{2 \text{ atoms O}}{1 \text{ molecule O}_2}$$

The sequence of conversions is

$$\text{Moles O}_2 \longrightarrow \text{Molecules O}_2 \longrightarrow \text{Atoms O}$$

Two conversion factors are needed; they are

$$\frac{6.02 \times 10^{23} \text{ molecules O}_2}{1 \text{ mole O}_2} \quad \text{and} \quad \frac{2 \text{ atoms O}}{1 \text{ molecule O}_2}$$

The calculation is

$$1.00 \text{ mole O}_2 \times \frac{6.02 \times 10^{23} \text{ molecules O}_2}{1 \text{ mole O}_2} \times \frac{2 \text{ atoms O}}{1 \text{ molecule O}_2} = 1.20 \times 10^{24} \text{ atoms O}$$

$$\text{(Answer)}$$

■

QUESTIONS   **A. Review the Meanings of the New Terms Introduced in this Chapter**

1. Dalton's atomic theory
2. Subatomic particles
3. Electron
4. Proton
5. Neutron
6. Nucleus
7. Orbital
8. Energy levels of electrons
9. Electron shells
10. Atomic number
11. Kernel of an atom
12. Noble gases
13. Isotopes
14. Mass number
15. Atomic mass unit
16. Atomic weight
17. Gram-atomic weight
18. Avogadro's number
19. Mole

B. **Answers to the Following Questions Will Be Found in Tables and Figures**
   1. What are the atomic numbers of sulfur, iron, gold, and argon?
   2. How many electron orbitals can be present in the third energy level? What are they?
   3. Show the electron structures ($1s^2 2s^2 2p^6$, etc.) for elements of atomic numbers 8, 12, 17, 23, 28, and 33.
   4. Using only Table 5.5, show the electron structures ($1s^2 2s^2 2p^6$, etc.) for elements containing the following number of electrons:
      (a) 9   (b) 26   (c) 31   (d) 39   (e) 52
   5. Diagram the atomic structure of the following atoms: N, Na, Cl, Zn, Zr, I. (See Figure 5.8.)
   6. Show electron-dot structures for C, Mg, Al, Cl, Mn, As, Kr, K.
   7. Explain the meaning of the following symbols: $^{78}_{33}As$, $^{197}_{79}Au$.
   8. List the electrical charge and the relative mass (amu) for the proton, the electron, and the neutron.
   9. In the designation $3d^7$, give the significance of the 3, the d, and the 7.
   10. Sketch the s orbital and the p_x, p_y, and p_z orbitals.
   11. Sketch the hydrogen atom (a) according to the Bohr theory, and (b) according to the modern concept of atomic structure.
   12. In Table 5.5, for elements 71 through 80, which elements do not build upon the previous electron structure in an orderly pattern?
   13. Basing your answer on Table 5.6, tell whether a noble gas structure is achieved each time a principal energy level becomes completely full. Explain.

C. **Review Questions**

Atomic Structure
   1. From the point of view of a chemist, what are the essential differences among a proton, a neutron, and an electron?
   2. Describe the general arrangement of particles in the atom.
   3. What part of the atom contains practically all its mass?
   4. What experimental evidence led Rutherford to conclude each of the following?
      (a) The nucleus of the atom contains most of the atomic mass.
      (b) The nucleus of the atom is positively charged.
      (c) The atom consists of mostly empty space.
   5. What contribution did each of the following scientists make to the atomic theory?
      (a) Dalton   (c) Rutherford   (e) Bohr
      (b) Thomson   (d) Chadwick   (f) Schrödinger
   6. Which of the following statements are correct?
      (a) John Dalton developed an important atomic theory in the early 1800s.
      (b) Dalton said that elements are composed of minute indivisible particles called *atoms.*
      (c) Dalton said that when atoms combine to form compounds, they do so in simple numerical ratios.
      (d) Dalton said that atoms are composed of protons, neutrons, and electrons.
      (e) All of Dalton's theory is still considered valid today.
      (f) Hydrogen is the smallest atom.
      (g) A proton is about 1837 times as heavy as an electron.
      (h) The nucleus of an atom contains protons, neutrons, and electrons.
      (i) The Bohr theory proposed that electrons move around the nucleus in circular orbits.
      (j) Bohr concluded from his experiment that the positive charge and almost all the mass was concentrated in a very small nucleus.

Electron Structure

7. (a) What is an atomic orbital?
   (b) What are the shapes of an *s* orbital and a *p* orbital?
8. Under which conditions can a second electron enter an orbital already containing one electron?
9. What is meant when we say that the electron structure of an atom is in its ground state?
10. List the following electron sublevels in order of increasing energy: 2*s*, 2*p*, 4*s*, 1*s*, 3*d*, 3*p*, 4*p*, 3*s*.
11. How many *s* electrons, *p* electrons, and *d* electrons are possible in any electron shell?
12. How many protons are in the nucleus of an atom of each of these elements: H, B, F, Sc, Ag, U, Br, Sb, and Pb?
13. Give the electron structure ($1s^2 2s^2 2p^6$, etc.) for B, Ti, Zn, Br, and Sr.
14. Why is the eleventh electron of the sodium atom located in the third energy level rather than in the second energy level?
15. Why are the last two electrons in calcium located in the fourth energy level rather than in the third energy level?
16. Which atoms have the following electron structures?
    (a) $1s^2 2s^2 2p^6 3s^2$
    (b) $1s^2 2s^2 2p^5$
    (c) $1s^2 2s^2 2p^6 3s^2 3p^6 4s^2 3d^8$
    (d) $1s^2 2s^2 2p^6 3s^2 3p^6 4s^2 3d^5$
    (e) $1s^2 2s^2 2p^6 3s^2 3p^6 4s^2 3d^{10} 4p^6 5s^1 4d^5$
17. Identify these atoms from their atomic structure diagrams:

    (a) $\left(\begin{smallmatrix} 8p \\ 8n \end{smallmatrix}\right)$ $\begin{smallmatrix} K \\ 2e^- \end{smallmatrix}$ $\begin{smallmatrix} L \\ 8e^- \end{smallmatrix}$ $\begin{smallmatrix} M \\ 6e^- \end{smallmatrix}$

    (b) $\left(\begin{smallmatrix} 27p \\ 32n \end{smallmatrix}\right)$ $\begin{smallmatrix} K \\ 2e^- \end{smallmatrix}$ $\begin{smallmatrix} L \\ 8e^- \end{smallmatrix}$ $\begin{smallmatrix} M \\ 15e^- \end{smallmatrix}$ $\begin{smallmatrix} N \\ 2e^- \end{smallmatrix}$

18. Diagram the atomic structures (as in Question 17) for these atoms:
    (a) $^{27}_{13}Al$ (b) $^{51}_{23}V$ (c) $^{89}_{39}Y$
19. Write electron-dot structures for these atoms: He, B, O, Na, Si, Ar, Ga, Ca, Br, Kr.
20. Which of the following statements are correct?
    (a) In the ground state, electrons tend to occupy orbitals having the lowest possible energy.
    (b) The maximum number of *p* electrons in the first energy level is six.
    (c) A 2*s* electron is in a lower energy state than a 2*p* electron.
    (d) The electron structure for a carbon atom is $1s^2 2s^2 2p^2$.
    (e) The $2p_x$, $2p_y$, and $2p_z$ electron orbitals are all in the same energy state.
    (f) The energy level of a 3*d* electron is higher than that of a 4*s* electron.
    (g) The electron structure for a calcium atom is $1s^2 2s^2 2p^6 3s^2 3p^6 3d^2$.
    (h) There are seven principal energy levels for the known elements.
    (i) The *M* energy level can have a maximum of 18 electrons.
    (j) The number of possible *d* electrons in the third energy level is ten.
    (k) The first *f* electron occurs in the fourth principal energy level.
    (l) The dot structure for an atom of nitrogen is $:\overset{\cdot}{\underset{\cdot}{N}}\cdot$
    (m) The dot structure for an atom of potassium is P·
    (n) Atoms of all the noble gases (except helium) have eight electrons in their outermost energy level.
    (o) A *p* orbital is spherically symmetrical around the nucleus.
    (p) An atom of nitrogen has two electrons in a 1*s* orbital, two electrons in a 2*s* orbital, and one electron in each of three different 2*p* orbitals.
    (q) The maximum number of electrons that can occupy a specific energy level **n** is given by $2n^2$.

Isotopes, Isotopic Notation

21. In what ways are isotopes alike? In what ways are they different?
22. List the similarities and differences in the three isotopes of hydrogen.
23. An atom of an element has an atomic number of 24 and a mass number of 52.
    (a) What is the symbol and name of the element?
    (b) How many protons, neutrons, and electrons are contained in the atom?
24. An atom of an element has a mass number of 201 and has 121 neutrons in its nucleus.
    (a) What is its nuclear charge?
    (b) What is the symbol and name of the element?
25. What is the nuclear composition of the six naturally occurring isotopes of calcium having mass numbers of 40, 42, 43, 44, 46, and 48?
26. Give the isotopic notation ($^{73}_{32}Ge$, etc.) for:
    (a) An atom containing 27 protons, 32 neutrons, and 27 electrons.
    (b) An atom containing 110 neutrons, 74 electrons, and 74 protons.
27. Which of the following statements are correct?
    (a) An element with an atomic number of 29 has 29 protons, 29 neutrons, and 29 electrons.
    (b) An atom of the isotope $^{60}_{26}Fe$ has 34 neutrons in its nucleus.
    (c) $^{2}_{1}H$ is a symbol for the isotope deuterium.
    (d) An atom of $^{31}_{15}P$ contains 15 protons, 16 neutrons, and 31 electrons.
    (e) The proton and the neutron have charges that are approximately equal.
    (f) Isotopes of a given element have the same number of protons but differ in the number of neutrons.
    (g) The three isotopes of hydrogen are called protium, deuterium, and tritium.
    (h) $^{23}_{11}Na$ and $^{24}_{11}Na$ are isotopes.
    (i) $^{24}_{11}Na$ has one more electron than $^{23}_{11}Na$.
    (j) $^{24}_{11}Na$ has one more proton than $^{23}_{11}Na$.
    (k) $^{24}_{11}Na$ has one more neutron than $^{23}_{11}Na$.
    (l) A mole of $^{24}_{11}Na$ contains more atoms than a mole of $^{23}_{11}Na$.
    (m) A mole of $^{24}_{11}Na$ weighs more than a mole of $^{23}_{11}Na$.

Atomic Weights, the Mole

28. Explain why the atomic weights of elements are not whole numbers.
29. Is the isotopic mass of a given isotope ever an exact whole number? Is it always? In answering, consider the masses of $^{12}_{6}C$ and $^{35}_{17}Cl$.
30. What information is needed to calculate the approximate atomic weight of an element?
31. Which of the isotopes of calcium in Question 25 is the most abundant isotope? Can you be sure? Explain your choice.
32. Distinguish between atomic weights expressed in atomic mass units and in grams.
33. What is a mole?
34. Which would weigh more: a mole of potassium atoms, or a mole of gold atoms?
35. Which would contain more atoms: a mole of potassium atoms, or a mole of gold atoms?
36. Which would contain more electrons: a mole of potassium atoms, or a mole of gold atoms?
*37. If the atomic weight scale had been defined differently, with an atom of $^{12}_{6}C$ being defined as weighing 50 amu, would this have any effect on the value of Avogadro's number? Explain.

38. Complete the following statements, supplying the proper quantity:
    (a) A mole of oxygen atoms (O) contains _____ atoms.
    (b) A mole of oxygen molecules (O_2) contains _____ molecules.
    (c) A mole of oxygen molecules (O_2) contains _____ atoms.
    (d) A mole of oxygen atoms (O) weighs _____ grams.
    (e) A mole of oxygen molecules (O_2) weighs _____ grams.
39. Which of the following statements are correct?
    (a) One gram-atomic weight of any element contains 6.02×10^{23} atoms.
    (b) The mass of one atom of chlorine is $\dfrac{35.5 \text{ g}}{6.02 \times 10^{23} \text{ atoms}}$.
    (c) A mole of magnesium atoms (24.3 g) contains the same number of atoms as a mole of sodium atoms (23.0 g).
    (d) A mole of bromine atoms contains 6.02×10^{23} atoms of bromine.
    (e) A mole of chlorine molecules (Cl_2) contains 6.02×10^{23} atoms of chlorine.
    (f) A mole of aluminum atoms weighs the same as a mole of tin atoms.
    (g) A mole of tritium atoms weighs the same as a mole of deuterium atoms.

D. **Review Problems**
    1. Change the following to powers of 10 (scientific notation):
       (a) 51,000   (b) 0.00274   (c) $(0.001)^2$ (d) $(7.0)^3$
    2. Using the formula $2n^2$, calculate the number of electrons that can exist in electron shells K, L, M, N, O, and P. Energy levels: $n = 1, 2, 3, 4, 5,$ and 6.
    3. How many neutrons are in an atom of:
       (a) $^{32}_{16}S$ (b) $^{35}_{17}Cl$ (c) $^{27}_{13}Al$ (d) $^{52}_{24}Cr$ (e) $^{197}_{79}Au$ (f) $^{232}_{90}Th$

    4. Complete the following table with the appropriate data for each isotope given:

Atomic number	Mass number	Symbol of element	Number of protons	Number of neutrons
(a) 5	11			
(b)		Mo		54
(c)	118		50	

    5. How many moles of atoms are contained in the following?
       (a) 15 g Zn   (b) 0.844 g Mg   (c) 492 g Co   (d) 0.077 g Sn
    6. How many atoms are contained in the following?
       (a) 21.6 g Cd   (c) 715 g Fe   (e) 3.8×10^4 g Be
       (b) 0.84 g Sr   (d) 0.067 g Bi   (f) 7.6×10^{-2} g Ti
    7. Calculate the weight, in grams, of one atom of:
       (a) Mercury   (b) Sodium   (c) Bromine   (d) Lead
    8. Complete the following table with the appropriate data for each element given:

Atomic number	Symbol of element	Number of protons	Atomic weight	Gram-atomic weight	Mass (g) of 1 atom
(a)			31.0		
(b) 29					
(c)				137.3 g	
(d)					6.50×10^{-23} g

9. What weight in grams is represented by each of the following?
   (a) 0.00308 mole Ag   (c) 5.00 g-at. wt Se
   (b) 1.8×10^{12} atoms Mn   (d) 28.3 moles W
10. Make the following conversions:
    (a) 9.42 moles of I to grams of I   (d) 9.6×10^{26} atoms of Cd to grams of Cd
    (b) 45 moles of Pb to kilograms   (e) 0.882 mole of V to atoms of V
    of Pb   (f) 35.5 mL Hg $(d = 13.6$ g/mL)
    (c) 48.4 g of S to moles of S   to moles of Hg
11. Make the following conversions:
    (a) 5.8 g of Zr to moles of Zr   (d) 1.72 kg Rb to moles of Rb
    (b) 0.772 mole of C to grams of C   (e) 26 g of Mg to atoms of Mg
    (c) 0.84 mole of Cl_2 to grams of Cl_2 (f) 405 mL of Br_2 $(d = 3.12$ g/mL)
         to moles of Br_2
12. One mole of carbon dioxide (CO_2) contains:
    (a) How many carbon dioxide molecules?
    (b) How many carbon atoms?
    (c) How many oxygen atoms?
13. White phosphorus is one of several forms of phosphorus and exists as a waxy solid consisting of P_4 molecules. How many atoms are present in 0.20 mole of P_4?
14. How many grams of sodium contain the same number of atoms as 8.00 g of potassium?
15. One atom of an unknown element is found to have a mass of 3.15×10^{-22} g. What is the gram-atomic weight of this element?
*16. If a stack of 500 sheets of paper is 4.50 cm high, what will be the height, in metres, of a stack of Avogadro's number of sheets of paper?
17. There are about 4.0 billion (4.0×10^9) people on earth. If 1 mole of dollars were distributed equally among these people, how many dollars would each person receive?
*18. If 20 drops of water equal 1.0 mL (1.0 cm³),
    (a) How many drops of water are there in a cubic mile of water?
    (b) What would be the volume in cubic miles of a mole of drops of water?
*19. A marble is about 1.2 cm in diameter. If a mole of marbles were laid end to end, what distance, in kilometres, would they cover? How many miles is this?
*20. Copper has a density of 8.92 g/cm³. If 1.00 mole of copper were shaped into a cube,
    (a) What would be the volume of the cube?
    (b) What would be the length of one side of the cube?

E. Review Exercises

1. Show the atomic structure of the most abundant isotope of fluorine and of bromine.
2. Which atom would you expect to have the larger volume, chlorine or bromine? Why?
*3. In your own words, describe an atom of phosphorus.
4. Would you expect the nucleus of an atom to have a relatively high or low density? Explain.
5. A scientist investigating the electron structure of an element concluded that the K, L, and M levels were all full, and the N level contained four electrons. What is the element?
*6. Four persons are trying to sell you gold. Person A offers 10.0 g for $305; B offers 1.00 mole for $2533; C offers 1.00 kg for $20,000; and D offers 1.0×10^{21} atoms for $12.63. Assuming they all have truly pure gold, which is the best offer? The offers can be compared on any equivalent basis, but if you wish to relate to commonly quoted prices of gold, 1 troy ounce is equal to 31.1 g.

6

The Periodic Arrangement of the Elements

After studying Chapter 6 you should be able to:

1. Understand the terms listed in Question A at the end of the chapter.
2. Describe briefly the contributions of Döbereiner, Newlands, Mendeleev, Meyer, and Moseley to the development of the periodic law.
3. State the periodic law in its modern form.
4. Explain why there were blank spaces in Mendeleev's periodic table and how he was able to predict the properties of the elements that belonged in those spaces.
5. Indicate the locations of the metals, nonmetals, metalloids, and noble gases in the periodic table.
6. Indicate in the periodic table the areas where the s, p, d, and f orbitals are being filled.
7. Describe how atomic radii vary (a) from left to right in a period and (b) from top to bottom in a family of elements.
8. Describe the changes in outer-level electron structure when (a) moving from left to right in a period and (b) going from top to bottom in a family of elements.
9. List the general characteristics of group or family properties.
10. Predict formulas of simple compounds formed between Group A elements using the periodic table.
11. Point out how the change in electron structure in going from one transition element to the next differs from that in nontransition elements.

6.1 Early Attempts to Classify the Elements

Chemists of the early 19th century had sufficient knowledge of the properties of elements to recognize similarities among groups of elements. J. W. Döbereiner (1780–1849), professor at the University of Jena in Germany, observed the existence of "triads" of similarly behaving elements, in which the middle element had an atomic weight approximating the average of the other two elements. He also noted that many other properties of the central element were approximately the average of the other two elements. Table 6.1 presents comparative data on atomic weight and density for two sets of Döbereiner's triads.

TABLE 6.1 Döbereiner's triads.

Triads	Atomic weight	Density (g/mL at 4°C)
Chlorine	35.5	1.56[a]
Bromine	79.9	3.12
Iodine	126.9	4.95
Average of chlorine and iodine	81.2	3.26
Calcium	40.1	1.55
Strontium	87.6	2.6
Barium	137.3	3.5
Average of calcium and barium	88.8	2.52

[a]Density at −34°C (liquid).

In 1864, J. A. R. Newlands (1837–1898), an English chemist, reported his *Law of Octaves.* In his studies, Newlands observed that when the elements were arranged according to increasing atomic weights, every eighth element had similar properties. (The noble gases had not yet been discovered.) But Newlands' theory was ridiculed by his contemporaries in the Royal Chemical Society, and they refused to publish his work. Many years later, however, Newlands was awarded the highest honor of the society for this important contribution to the development of the periodic law.

In 1869, Dmitri Ivanovitch Mendeleev (1834–1907) of Russia and Lothar Meyer (1830–1895) of Germany independently published period arrangements of the elements that were based on increasing atomic weights. Because his arrangement was published slightly earlier and was in a somewhat more useful form than that of Meyer, Mendeleev's name is usually associated with the modern period table.

6.2 The Periodic Law

Only about 63 elements were known when Mendeleev constructed his periodic table. These elements were arranged so that those with similar chemical properties fitted into columns to form family groups. The arrangement left many gaps between elements. Mendeleev predicted that these spaces would be filled as new elements were discovered. For example, spaces for undiscovered elements were left after calcium, under aluminum, and under silicon. He called these unknown elements eka-boron, eka-aluminum, and eka-silicon. The term *eka* comes from Sanskrit meaning "one" and was used to indicate that the missing element was one place away in the table from the element indicated. Mendeleev even went so far as to predict with high accuracy the physical and chemical properties of these elements yet to be discovered. The three elements, scandium (atomic number 21), gallium (31), and germanium (32) were in fact discovered during Mendeleev's lifetime and were found to have properties agreeing very closely with the predictions that he had made for eka-boron, eka-aluminum, and eka-silicon. The amazing way in which Mendeleev's predictions

TABLE 6.2 Comparison of the properties of eka-silicon predicted by Mendeleev with the properties of germanium.

Property	Mendeleev's predictions in 1871 for eka-silicon (Es)	Observed properties for germanium (Ge)
Atomic weight	72	72.6
Color of metal	Dirty gray	Grayish white
Density	5.5 g/mL	5.47 g/mL
Oxide formula	EsO_2	GeO_2
Oxide density	4.7 g/mL	4.70 g/mL
Chloride formula	$EsCl_4$	$GeCl_4$
Chloride density	1.9 g/mL	1.89 g/mL
Boiling temperature of chloride	Under 100°C	86°C

were fulfilled is illustrated in Table 6.2, which compares the predicted properties of eka-silicon with those of germanium, discovered by the German chemist C. Winkler in 1886.

Two major additions have been made to the periodic table since Mendeleev's time: (1) a new family of elements, the noble gases, was discovered and added; and (2) elements having atomic numbers greater than 92 have been discovered and fitted into the table.

The original table was based on the premise that the properties of the elements are a periodic function of their atomic weights. However, there were some disturbing discrepancies to this basic premise. For example, the atomic weight of argon is greater than that of potassium. Yet potassium had to be placed after argon since argon is certainly one of the noble gases and potassium behaves like the other alkali metals. These discrepancies were resolved by the work of British physicist H. G. J. Moseley (1887–1915) and by the discovery of the existence of isotopes. Moseley noted that the X-ray emission frequencies of the elements increased in a regular, stepwise fashion each time the nuclear charge (atomic number) increased by one unit. This showed that the basis for placing an element in the periodic table should be the atomic number rather than the atomic weight. Atomic weights are the average masses of the naturally occurring mixtures of isotopes of each element. The atomic weight of argon is greater than that of potassium (the next element of higher atomic number) because the average mass of the argon isotopes is greater than the average mass of the potassium isotopes. The current statement of the **periodic law** is:

periodic law

The properties of the elements are a periodic function of their atomic numbers.

As the format of the periodic table is studied, it becomes evident that the periodicity of the properties of the elements is due to the recurring similarities of their electron structures.

TABLE 6.3 Periodic table of the elements.

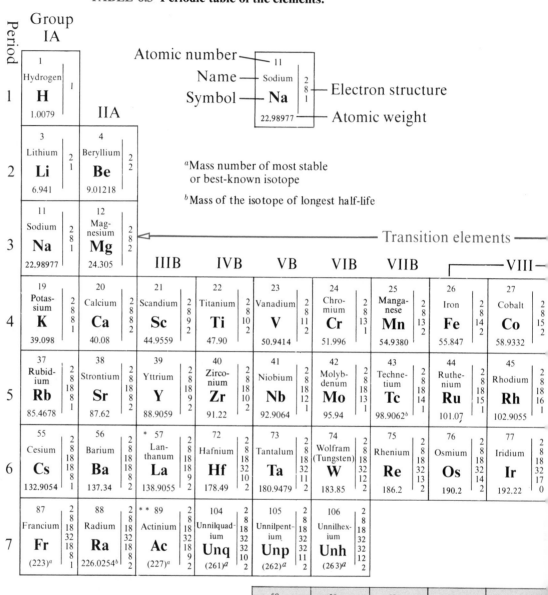

Period	Group IA		IIA		IIIB		IVB		VB		VIB		VIIB		VIII		
1	1 Hydrogen **H** 1.0079	1																
2	3 Lithium **Li** 6.941	2 1	4 Beryllium **Be** 9.01218	2 2														
3	11 Sodium **Na** 22.98977	2 8 1	12 Magnesium **Mg** 24.305	2 8 2														
4	19 Potassium **K** 39.098	2 8 8 1	20 Calcium **Ca** 40.08	2 8 8 2	21 Scandium **Sc** 44.9559	2 8 9 2	22 Titanium **Ti** 47.90	2 8 10 2	23 Vanadium **V** 50.9414	2 8 11 2	24 Chromium **Cr** 51.996	2 8 13 1	25 Manganese **Mn** 54.9380	2 8 13 2	26 Iron **Fe** 55.847	2 8 14 2	27 Cobalt **Co** 58.9332	2 8 15 2
5	37 Rubidium **Rb** 85.4678	2 8 18 8 1	38 Strontium **Sr** 87.62	2 8 18 8 2	39 Yttrium **Y** 88.9059	2 8 18 9 2	40 Zirconium **Zr** 91.22	2 8 18 10 2	41 Niobium **Nb** 92.9064	2 8 18 12 1	42 Molybdenum **Mo** 95.94	2 8 18 13 1	43 Technetium **Tc** 98.9062[b]	2 8 18 14 1	44 Ruthenium **Ru** 101.07	2 8 18 15 1	45 Rhodium **Rh** 102.9055	2 8 18 16 1
6	55 Cesium **Cs** 132.9054	2 8 18 18 8 1	56 Barium **Ba** 137.34	2 8 18 18 8 2	* 57 Lanthanum **La** 138.9055	2 8 18 18 9 2	72 Hafnium **Hf** 178.49	2 8 18 32 10 2	73 Tantalum **Ta** 180.9479	2 8 18 32 11 2	74 Wolfram (Tungsten) **W** 183.85	2 8 18 32 12 2	75 Rhenium **Re** 186.2	2 8 18 32 13 2	76 Osmium **Os** 190.2	2 8 18 32 14 2	77 Iridium **Ir** 192.22	2 8 18 32 17 0
7	87 Francium **Fr** (223)[a]	2 8 18 32 18 8 1	88 Radium **Ra** 226.0254[b]	2 8 18 32 18 8 2	** 89 Actinium **Ac** (227)[a]	2 8 18 32 18 9 2	104 Unnilquadium **Unq** (261)[a]	2 8 18 32 32 10 2	105 Unnilpentium **Unp** (262)[a]	2 8 18 32 32 11 2	106 Unnilhexium **Unh** (263)[a]	2 8 18 32 32 12 2						

Atomic number ——┐
Name ——
Symbol ——
┌— 11
Sodium
Na
22.98977
2
8
1
—— Electron structure
—— Atomic weight

[a]Mass number of most stable or best-known isotope

[b]Mass of the isotope of longest half-life

←———————— Transition elements ————————→

Lanthanide series
6

	58 Cerium **Ce** 140.12	2 8 18 20 8 2	59 Praseodymium **Pr** 140.9077	2 8 18 21 8 2	60 Neodymium **Nd** 144.24	2 8 18 22 8 2	61 Promethium **Pm** (145)[a]	2 8 18 23 8 2	62 Samarium **Sm** 150.4	2 8 18 24 8 2

Actinide series
7

	90 Thorium **Th** 232.0381[b]	2 8 18 32 18 10 2	91 Protactinium **Pa** 231.0359[b]	2 8 18 32 20 9 2	92 Uranium **U** 238.029	2 8 18 32 21 9 2	93 Neptunium **Np** 237.0482	2 8 18 32 22 9 2	94 Plutonium **Pu** (242)[a]	2 8 18 32 23 9 2

Atomic weights are based on carbon-12. Atomic weights in parentheses indicate the most stable or best-known isotope. Slight disagreement exists as to the exact electronic configuration of several of the high-atomic-number elements. Names and symbols for elements 104, 105, and 106 are unofficial.

Noble gases

IIIA	IVA	VA	VIA	VIIA	2 Helium **He** 4.00260 (2)

5 Boron **B** 10.81 (2,3)	6 Carbon **C** 12.011 (2,4)	7 Nitrogen **N** 14.0067 (2,5)	8 Oxygen **O** 15.9994 (2,6)	9 Fluorine **F** 18.99840 (2,7)	10 Neon **Ne** 20.179 (2,8)
13 Aluminum **Al** 26.98154 (2,8,3)	14 Silicon **Si** 28.086 (2,8,4)	15 Phosphorus **P** 30.97376 (2,8,5)	16 Sulfur **S** 32.06 (2,8,6)	17 Chlorine **Cl** 35.453 (2,8,7)	18 Argon **Ar** 39.948 (2,8,8)

IB	IIB

28 Nickel **Ni** 58.71 (2,8,16,2)	29 Copper **Cu** 63.546 (2,8,18,1)	30 Zinc **Zn** 65.38 (2,8,18,2)	31 Gallium **Ga** 69.72 (2,8,18,3)	32 Germanium **Ge** 72.59 (2,8,18,4)	33 Arsenic **As** 74.9216 (2,8,18,5)	34 Selenium **Se** 78.96 (2,8,18,6)	35 Bromine **Br** 79.904 (2,8,18,7)	36 Krypton **Kr** 83.80 (2,8,18,8)
46 Palladium **Pd** 106.4 (2,8,18,18,0)	47 Silver **Ag** 107.868 (2,8,18,18,1)	48 Cadmium **Cd** 112.40 (2,8,18,18,2)	49 Indium **In** 114.82 (2,8,18,18,3)	50 Tin **Sn** 118.69 (2,8,18,18,4)	51 Antimony **Sb** 121.75 (2,8,18,18,5)	52 Tellurium **Te** 127.60 (2,8,18,18,6)	53 Iodine **I** 126.9045 (2,8,18,18,7)	54 Xenon **Xe** 131.30 (2,8,18,18,8)
78 Platinum **Pt** 195.09 (2,8,18,32,17,1)	79 Gold **Au** 196.9665 (2,8,18,32,18,1)	80 Mercury **Hg** 200.59 (2,8,18,32,18,2)	81 Thallium **Tl** 204.37 (2,8,18,32,18,3)	82 Lead **Pb** 207.2 (2,8,18,32,18,4)	83 Bismuth **Bi** 208.9804 (2,8,18,32,18,5)	84 Polonium **Po** (210)a (2,8,18,32,18,6)	85 Astatine **At** (210)a (2,8,18,32,18,7)	86 Radon **Rn** (222)a (2,8,18,32,18,8)

Inner transition elements

63 Europium **Eu** 151.96 (2,8,18,25,8,2)	64 Gadolinium **Gd** 157.25 (2,8,18,25,9,2)	65 Terbium **Tb** 158.9254 (2,8,18,27,8,2)	66 Dysprosium **Dy** 162.50 (2,8,18,28,8,2)	67 Holmium **Ho** 164.9304 (2,8,18,29,8,2)	68 Erbium **Er** 167.26 (2,8,18,30,8,2)	69 Thulium **Tm** 168.9342 (2,8,18,31,8,2)	70 Ytterbium **Yb** 173.04 (2,8,18,32,8,2)	71 Lutetium **Lu** 174.97 (2,8,18,32,9,2)
95 Americium **Am** (243)a (2,8,18,32,25,8,2)	96 Curium **Cm** (247)a (2,8,18,32,25,9,2)	97 Berkelium **Bk** (249)a (2,8,18,32,26,9,2)	98 Californium **Cf** (251)a (2,8,18,32,27,9,2)	99 Einsteinium **Es** (254)a (2,8,18,32,28,9,2)	100 Fermium **Fm** (253)a (2,8,18,32,29,9,2)	101 Mendelevium **Md** (256)a (2,8,18,32,30,9,2)	102 Nobelium **No** (254)a (2,8,18,32,31,9,2)	103 Lawrencium **Lr** (257)a (2,8,18,32,32,9,2)

6.3 Arrangement of the Periodic Table

periodic table

periods of elements

groups or families of elements

The most commonly used **periodic table** is the long form shown in Table 6.3. In this table the elements are arranged horizontally in numerical sequence, according to their atomic numbers; the result is seven horizontal rows called **periods.** Each period, with the exception of the first, starts with an alkali metal and ends with a noble gas. By this arrangement, vertical columns of elements are formed, having identical or similar outer-shell electron structures and thus similar chemical properties. These columns are known as **groups** or **families of elements.**

The heavy zigzag line starting at boron and running diagonally down the table separates the elements into metals and nonmetals. The elements to the right of the line are nonmetallic, and those to the left are metallic. The elements bordering the zigzag line are the metalloids, which show both metallic and nonmetallic properties. With some exceptions, the characteristic electronic arrangement of metals is that their atoms have one, two, or three electrons in the outer energy level, whereas nonmetals have five, six, or seven electrons in the outer energy level.

With this periodic arrangement, the elements fall into blocks according to the sublevel of electrons that is being filled. The grouping of the elements into *spdf* blocks is shown in Table 6.4. The *s* block comprising Groups IA and

TABLE 6.4 Arrangement of the elements into blocks according to the sublevel of electrons being filled in their atomic structure.

IIA has one or two s electrons in its outer energy level. The p block includes Groups IIIA through VIIA and the noble gases (except helium). In these elements, electrons are filling the p sublevel orbitals. The d block includes the transition elements of Groups IB through VIIB and Group VIII. The d sublevels of electrons are being filled in these elements. The f block of elements includes the inner transition series. In the lanthanide series, electrons are filling the $4f$ sublevel. In the actinide series, electrons are filling the $5f$ sublevel.

6.4 Periods of Elements

The number of elements in each period is shown in Table 6.5. The first period contains 2 elements, hydrogen and helium, and coincides with the full K shell of electrons. Period 2 contains 8 elements, starting with lithium and ending with neon. Period 3 also contains 8 elements, sodium to argon. Periods 4 and 5 each contain 18 elements; period 6 has 32 elements; and period 7, which is incomplete, contains the remaining 20 elements.

TABLE 6.5 The number of elements in each period.

Period number	Number of elements	Electron orbitals in each period being filled
1	2	$1s$
2	8	$2s2p$
3	8	$3s3p$
4	18	$4s3d4p$
5	18	$5s4d5p$
6	32	$6s4f5d6p$
7	20	$7s5f6d$

The first three periods are known as *short periods;* the others, as *long periods.* The number of each period corresponds to the outermost energy level in which electrons are located in the neutral atom. For example, the elements in period 1 contain electrons in the first energy level only; period 2 elements contain electrons in the first and second levels; period 3 elements contain electrons in the first, second, and third levels; and so on. Moving horizontally across periods 2 through 6, we find that the properties of the elements vary from strongly metallic at the beginning to nonmetallic at the end of the period. Starting with the third element of the long periods 4, 5, 6, and 7 (scandium, Sc; yttrium, Y; lanthanum, La; actinium, Ac), the inner shells of d and f orbital electrons begin to fill in.

In general, the atomic radii of the elements within a period decrease due to increasing nuclear charge. As the positive charge on the nucleus increases, it exerts a greater attractive force on the electrons, causing the atom to become smaller. Therefore, the size of the atoms becomes progessively smaller from left to right within each period. Because the noble gases do not readily combine with

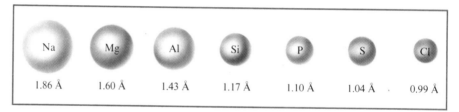

Figure 6.1   Radii of period 3 elements. In general, the size of the atoms in a period decreases with increasing nuclear charge.

other elements to form compounds, the radii of their atoms are not determined in the same comparative manner as for the other elements. However, calculations have shown that the radii of the noble gas atoms are about the same as, or slightly smaller than, the element immediately preceding them. Slight deviations in atomic radii occur in the middle of the long periods of the elements. The elements of period 3 (Figure 6.1) illustrate this principle.

6.5 Groups or Families of Elements

The groups, or families, of elements are designated IA through VIIA, IB through VIIB, VIII, and noble gases. Group A elements are often referred to as the *representative* or *main* groups of elements. They are all filling in electrons in the *s* and *p* orbitals. Group B and VIII elements are known as the *transition elements.* They are filling in electrons in the *d* and *f* orbitals.

The elements comprising each family have similar outer energy-level electron structures. In Group A elements, the number of electrons in the outer energy level is identical to the group number. Group IA is known as the *alkali metal* family. Each atom of this family of elements has one *s* electron in its outer energy level. Group IIA atoms, the *alkaline earth metals,* each have two *s* electrons in their outer energy level. All atoms of the *halogen* family, Group VIIA, have an outer energy-level electron structure of ns^2np^5. The noble gases (except helium) have an outer energy-level structure of ns^2np^6.

Each alkali metal starts a new period of elements in which the *s* electron occupies a principal energy level one greater than in the previous period. As a result, the size of the atoms of this and other families of elements increases from the top to the bottom of the family. The relative size of the alkali metals is illustrated in Figure 6.2.

One major distinction between groups lies in the energy level to which the last electron is added. In the elements of Groups IA through VIIA, IB, IIB, and the noble gases, the last electron is added either to an *s* or to a *p* orbital located in the outermost energy level (copper is an exception; see Table 5.5). In the elements of Groups IIIB through VIIB and VIII, the last electron goes to a *d* or to an *f* orbital located in an inner energy level. For example, the last electron in potassium (a fourth-period, Group IA element) is a 4*s* electron

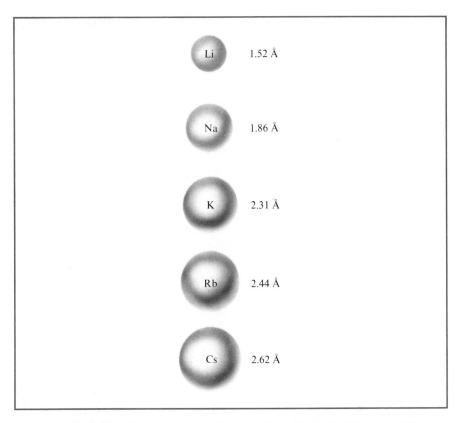

Figure 6.2   The size of the atoms of the alkali metal family. The size of the atoms in a family increases from top to bottom, since each atom progressively contains electrons in a higher energy level.

located in the fourth energy level (an outer shell); in scandium (a fourth-period Group IIIB element) the last electron added is a 3*d* electron located in the third energy level (an inner shell). Figure 6.3 compares the locations of the last electron added in Group A and B types of elements.

Group	Noble gas	IA	IIA	IIIB	IVB
Element	Ar	K	Ca	Sc	Ti
Electron structure	2,8,8	2,8,8,1	2,8,8,2	2,8,9,2	2,8,10,2
Energy level	1,2,3	1,2,3,4	1,2,3,4	1,2,3,4	1,2,3,4
Last electron added	↑	↑	↑	↑	↑

Figure 6.3   Comparison of the placement of the last electron in Group A and Group B elements. The energy level to which the last electron is added is indicated by the arrow—an outer level for Group A elements, an inner level for Group B elements.

The general characteristics of group properties are as follows:

1. The number of electrons in the outer energy level of Groups IA through VIIA, IB, and IIB elements is the same as the group number. The other B groups and Group VIII do not show this characteristic. Each noble gas except helium has eight electrons in its outer energy level.
2. The group number in which an element is located will indicate one of its possible oxidation states, with some exceptions, notably in Group VIII.
3. The groups on the left and in the middle sections of the table tend to be metallic in nature. The groups on the right tend to be nonmetallic.
4. The radii of the elements increase from top to bottom within a particular group (for example, from lithium to francium).
5. Elements at the bottom of a group tend to be more metallic in their properties than those at the top. This tendency is especially noticeable in Groups IVA through VIIA.
6. Elements within an A group have the same number of electrons in their outer shell and show closely related chemical properties.
7. Elements within a B group have some similarity in electron structure and also show some similarities in chemical properties.

6.6 Predicting Formulas by Use of the Periodic Table

The periodic table can be used to predict the formulas of simple compounds. As we shall see in Chapter 7, the chemical properties of the elements are dependent on their electrons. Group A elements ordinarily form compounds using only the electrons in their outer energy level. If we examine Group IA, the alkali metals, we see that all of them have one electron in their outer energy level, all follow a noble gas in the table, and all, except lithium, have eight electrons in their next inner shell. These likenesses suggest that there should be a great deal of similarity in the chemistry of these metals, since their chemical properties are vested primarily in their outer-shell electron. And there is similarity—all readily lose their outer electron and attain a noble gas electron structure. In doing this, they form compounds with similar atomic compositions. For example, the monoxides of Group IA contain two atoms of the alkali metal to one atom of oxygen. Their formulas are Li_2O, Na_2O, K_2O, Rb_2O, Cs_2O, and Fr_2O.

How can we use the periodic table to predict formulas of other compounds? Because of similar electron structures, the elements in a family generally form compounds with the same atomic ratios. This was just shown for the oxides of the Group IA metals. In general, if we know the atomic ratio of a particular compound, say sodium chloride (NaCl), we can predict the atomic ratios and formulas of the other alkali metal chlorides. These formulas are LiCl, KCl, RbCl, CsCl, and FrCl (see Table 6.6).

In a similar way, if we know that the formula of the oxide of hydrogen is H_2O, we predict that the formula of the sulfide will be H_2S because sulfur has

TABLE 6.6 Elements in Group A families have the same outer-shell electron structure and are likely to form compounds with the same atomic ratios. Shown, for example, are the monoxides, chlorides, bromides, and sulfates of the alkali metals.

Electron dot structure	Monoxides	Chlorides	Bromides	Sulfates
Li·	Li_2O	LiCl	LiBr	Li_2SO_4
Na·	Na_2O	NaCl	NaBr	Na_2SO_4
K·	K_2O	KCl	KBr	K_2SO_4
Rb·	Rb_2O	RbCl	RbBr	Rb_2SO_4
Cs·	Cs_2O	CsCl	CsBr	Cs_2SO_4

the same outer-shell electron structure as oxygen. It must be recognized, however, that these are only predictions; it does not necessarily follow that every element in a group will behave like the others, or even that a predicted compound will actually exist. Knowing the formulas for potassium chlorate, bromate, and iodate to be $KClO_3$, $KBrO_3$, and KIO_3, we can correctly predict the corresponding sodium compounds to have the formulas $NaClO_3$, $NaBrO_3$, and $NaIO_3$. Fluorine belongs to the same family of elements (Group VIIA) as chlorine, bromine, and iodine. Therefore, we can predict that the formulas for potassium and sodium fluorates will be KFO_3 and $NaFO_3$. These compounds are not known to exist; however, if they did exist, the formulas could very well be correct, for these predictions are based on comparisons with known formulas and/or similar electron structures.

PROBLEM 6.1   The formula for magnesium sulfate is $MgSO_4$ and that for potassium carbonate is K_2CO_3. Predict the formulas for (a) barium sulfate and (b) lithium carbonate.

(a) Look in the periodic table for the locations of magnesium and barium. They are both in Group IIA. Mg and Ba are in the same family. Since the formula for magnesium sulfate is $MgSO_4$, it is reasonable to predict that the formula for barium sulfate will be $BaSO_4$.

(b) Check the periodic table and locate potassium and lithium in Group IA. Since both elements are in the same family and the formula of potassium carbonate is K_2CO_3, we can predict that the formula for lithium carbonate will be Li_2CO_3.

■

6.7 Transition Elements

transition elements

Elements in Groups IB, IIIB through VIIB, and VIII are known as the **transition elements.** There are four series of transition elements, one in each of the periods 4, 5, 6, and 7. The transition elements are characterized by an increasing number of d or f electrons in an inner shell; they all have either one or two s electrons in their outer shell. In period 4, electrons enter the $3d$ sublevel. In period 5, electrons enter the $4d$ sublevel. The transition elements in period 6 include the lanthanide series (or rare earth elements, La to Lu), in which electrons are entering the 4*f* sublevel. The 5*d* sublevel also fills up in the sixth period. The seventh period of elements is an incomplete period. It includes the actinide series (Ac to Lr) in which electrons are entering the 5*f* and 6*d* sublevels.

All of the transition elements are metals. In the formation of compounds of the transition metals, electrons may come from more than one energy level. For this reason many of these metals form multiple series of compounds. These compounds often occur as some of the most beautifully colored crystals in the entire field of chemistry.

6.8 New Elements

Mendeleev allowed gaps in his orderly periodic table for elements whose discovery he predicted. These elements were actually discovered, as were all the elements up to atomic number 92 that occur naturally on the earth. Fourteen elements beyond uranium (atomic numbers 93–106) have been discovered or synthesized since 1939. All these elements have unstable nuclei and are radioactive. Beyond element 101, the isotopes synthesized thus far have such short lives that chemical identification has not been accomplished.

Intensive research is continuing on the synthesis of still heavier elements. Extending the periodic table beyond the presently known elements, it is predicted that elements 110 to 118 will be very stable, but still radioactive. Element 114 will lie below lead (82) and should be exceptionally stable. Element 118 should be a member of the noble gas family. Elements 119 and 120 should be in Groups IA and IIA, respectively, and have electrons in the 8*s* sublevel.

6.9 Summary

The periodic table has been used for studying the relationships of many properties of the elements. Ionization energies, densities, melting points, atomic radii, atomic volumes, oxidation states, electrical conductance, and electronegativity are just a few of these properties. However, a detailed discussion of all these properties is not practical at this time.

The periodic table is still used as a guide in predicting the synthesis of possible new elements. It presents a very large amount of chemical information in compact form and correlates the properties and relationships of all the elements. The table is so useful that a copy hangs in nearly every chemistry lecture hall and laboratory in the world. Refer to it often.

QUESTIONS   A. **Review the Meanings of the New Terms Introduced in this Chapter**
1. Periodic law
2. Periodic table
3. Periods of elements
4. Groups or families of elements
5. Transition elements

B. Answers to the Following Questions Will Be Found in Tables and Figures
1. How many elements are present in each period?
2. Write the symbols of the alkali metal family in the order of increasing size of their atoms.
3. What similarities do you observe in the elements of Group IIA?
4. Write the symbols for the elements with atomic numbers 18, 36, 54, and 86. What do these elements have in common?
5. Write the names and symbols of the halogens.
6. Write the symbols for the family of elements that have two electrons in their outer energy level.
7. Point out similarities and differences between Group IA and Group IB elements.
8. What similarities do the elements of the lanthanide series possess?
9. Where are the elements with the most metallic characteristics located in the periodic table?

C. Review Questions
1. What were Döbereiner's triads? In what way did they lead to later developments in periodicity?
2. What do you feel is the basis for Newlands' Law of Octaves?
3. Recognizing that the noble gases were not discovered at the time of Newlands' Law of Octaves, could his law be extended as far as the element bromine? Explain.
4. What is meant by the term *periodicity* as applied to the elements?
5. Why are some missing elements in Mendeleev's periodic table considered a victory for his table, rather than a defeat?
6. How does our modern periodic table differ from Mendeleev's?
7. What additional understanding of periodic properties was added by the work of H. G. J. Moseley?
8. State the periodic law in its current form. How is this different from the law as stated by Mendeleev?
9. Find the places in the modern periodic table where elements are not in proper sequence according to atomic weight.
10. How are elements in a period related to one another?
11. How are elements in a group related to one another?
12. What is common about the electron structures of the alkali metals?
13. Why would you expect the elements zinc, cadmium, and mercury to be in the same chemical family?
14. Draw the electron-dot diagrams for Cs, Ba, Tl, Pb, Po, At, and Rn. How do these structures correlate with the group in which each element occurs?
15. Pick the electron structures below that represent elements in the same chemical family:
   (a) $1s^2 2s^1$
   (b) $1s^2 2s^2 2p^4$
   (c) $1s^2 2s^2 2p^2$
   (d) $1s^2 2s^2 2p^6 3s^2 3p^4$
   (e) $1s^2 2s^2 2p^6 3s^2 3p^6$
   (f) $1s^2 2s^2 2p^6 3s^2 3p^6 4s^2$
   (g) $1s^2 2s^2 2p^6 3s^2 3p^6 4s^1$
   (h) $1s^2 2s^2 2p^6 3s^2 3p^6 4s^2 3d^1$
16. Pick the electron structures below that represent elements in the same chemical family:
   (a) $[He] 2s^2 2p^6$
   (b) $[Ne] 3s^1$
   (c) $[Ne] 3s^2$
   (d) $[Ne] 3s^2 3p^3$
   (e) $[Ar] 4s^1 3d^{10}$
   (f) $[Ar] 4s^2 3d^{10} 4p^6$
   (g) $[Ar] 4s^2 3d^5$
   (h) $[Kr] 5s^1 4d^{10}$
17. In the periodic table, calcium, element 20, is surrounded by elements 12, 19, 21, and 38. Which of these have physical and chemical properties most resembling calcium?

18. Oxygen is a gas. Sulfur is a solid. What is it about their electron structures that causes them to be grouped in the same chemical family?

19. Would bromine and iodine be more similar in physical properties or in chemical properties? Give examples of each to illustrate your answer.

20. Classify each of the following elements as metals, nonmetals, or metalloids:
    (a) Potassium   (c) Sulfur   (e) Iodine   (g) Molybdenum
    (b) Plutonium   (d) Antimony   (f) Tungsten   (h) Germanium

21. In which period and group does an electron first appear in a d orbital?

22. How many electrons occur in the outer shell of Group IIIA and IIIB elements? Why are they different?

23. In which groups are transition elements located?

24. How do transition elements differ from other elements?

25. Which element in each of the following pairs has the larger atomic radius?
    (a) Na or K   (b) Na or Mg   (c) O or F   (d) Br or I   (e) Ti or Zr

26. Which element in each of Groups IA–VIIA has the smallest atomic radius?

27. Why does the atomic size increase in going down any family of the periodic table?

28. All the atoms within each Group A family of elements can be represented by the same electron-dot structure. Complete the table, expressing the electron-dot structure for each group. Use E to represent the elements.

Group	IA	IIA	IIIA	IVA	VA	VIA	VIIA
	E·						

29. Letting E be an element in any group, the table represents the possible formulas of such compounds. Following the pattern in the table, write the formulas for the hydrogen and oxygen compounds of:
    (a) Na   (b) Ca   (c) Al   (d) Sn   (e) Sb   (f) Se   (g) Cl

Group	IA	IIA	IIIA	IVA	VA	VIA	VIIA
	EH	EH_2	EH_3	EH_4	EH_3	H_2E	HE
	E_2O	EO	E_2O_3	EO_2	E_2O_5	EO_3	E_2O_7

30. Group IB elements have one electron in their outer shell, as do Group IA elements. Would you expect them to form compounds such as CuCl, AgCl, and AuCl? Explain.

31. The formula for lead(II) bromide is $PbBr_2$; predict formulas for tin(II) and germanium(II) bromides. (See Section 8.3, part 2, for the use of roman numerals in naming compounds.)

32. The formula for sodium sulfate is Na_2SO_4. Write the names and formulas for the other alkali metal sulfates.

33. The formula for calcium bromide is $CaBr_2$. Write formulas for magnesium bromide, strontium bromide, and barium bromide.

34. Why should the discovery of the existence of isotopes have any bearing on the fact that the periodicity of the elements is a function of their atomic numbers and not their atomic weights?

Try to answer Questions 35–37 without referring to the periodic table.

35. The atomic numbers of the noble gases are 2, 10, 18, 36, 54, and 86. What are the atomic numbers for the elements with six electrons in their outer electron shells?

36. Element number 87 is in Group IA, period 7. Describe its outermost energy level. How many energy levels of electrons does it have?

37. If element 36 is a noble gas, in which groups would you expect elements 35 and 37 to occur?

38. Write a paragraph describing the general features of the periodic table.

Periodic Table for Questions 39–42

													A	B	C	
D	E														F	
G				H	I					J					K	
	L	M													N	

Use the skeleton periodic table shown here to answer Questions 39–42. Assume that the fictitious elements A–N have electron configurations consistent with their positions in the table.

39. Give the letters for those elements (A–N) that are (a) alkali metals, (b) alkaline earth metals, (c) halogens, (d) noble gases, (e) transition metals.
40. Rank the following four elements according to the radii of their atoms, from smallest to largest: D, E, F, G.
41. Which metals are most likely to form oxides of the form (a) X_2O and (b) X_2O_3?
42. Write the chemical formulas of the chemical compounds that would be expected to form between elements (a) E and B, and (b) L and C.

43. Which of the following statements are correct?
    (a) Properties of the elements are periodic functions of their atomic numbers.
    (b) There are more nonmetallic elements than metallic elements.
    (c) Metallic properties of the elements increase from left to right across a period.
    (d) Metallic properties of the elements increase from top to bottom in a family of elements.
    (e) Calcium is a member of the alkaline earth family.
    (f) Iron belongs to the alkali metal family.
    (g) Bromine belongs to the halogen family.
    (h) Neon is a noble gas.
    (i) Group A elements do not contain partially filled d or f sublevels.
    (j) An atom of oxygen has a larger volume than an atom of lithium.
    (k) An atom of sulfur is larger than an atom of oxygen.
    (l) An atom of aluminum (Group IIIA) has five electrons in its outer shell.
    (m) If the formula for calcium iodide is CaI_2, then the formula of cesium iodide is CsI_2.
    (n) If the formula of aluminum oxide is Al_2O_3, then the formula of gallium oxide is Ga_2O_3.
    (o) Uranium is a transition element.

(p) The element with the electron configuration $1s^2 2s^2 2p^5$ belongs to Group VA.

(q) The element with the electron configuration $1s^2 2s^2 2p^6 3s^2 3p^1$ belongs to Group IIIA.

(r) A chemical family consists of one of the horizontal rows on the periodic table.

(s) Elements within a family will have similar electron structures and will show some similarities in their chemical properties.

(t) Group A elements are known as the representative elements.

(u) Going from left to right in any period, the representative elements show an increasing number of either s or p electrons.

7

Chemical Bonds—
The Formation of
Compounds from Atoms

After studying Chapter 7 you should be able to:

1. Understand the terms listed in Question A at the end of the chapter.
2. Describe how the ionization energies of the elements vary with respect to (a) position in the periodic table and (b) the removal of successive electrons.
3. Describe (a) the formation of ions by electron transfer and (b) the nature of the chemical bond formed by electron transfer.
4. Show by means of electron-dot structures the formation of an ionic compound from atoms.
5. Describe a crystal of sodium chloride.
6. Predict the formulas of the monatomic ions of Group A metals and nonmetals.
7. Predict the relative sizes of an atom and a monatomic ion for a given element.
8. Describe the covalent bond and predict whether a given covalent bond would be polar or nonpolar.
9. Describe the changes in electronegativity in (a) moving across a period and (b) moving down a group in the periodic table.
10. Draw Lewis electron-dot structures for compounds, monatomic ions, and simple polyatomic ions.
11. Determine from its electron-dot structure whether a molecule is a dipole.
12. Distinguish clearly between ionic and molecular substances.
13. Predict whether the bonding in a compound is primarily ionic or covalent.
14. Distinguish coordinate covalent from covalent bonds in an electron-dot structure.
15. Write the formulas of compounds formed by combining the ions from Tables 7.6 and 7.7 (or from the inside back cover of this book) in the correct ratios.
16. Assign the oxidation number to each element in a compound or ion.
17. Distinguish between oxidation number and ionic charge.
18. Distinguish between oxidation and reduction.
19. Identify in an equation the element that has been oxidized and the element that has been reduced.

7.1 Chemical Bonds

chemical bonds

Except in very rare instances matter does not fly apart spontaneously. It is prevented from doing so by forces acting at the ionic and molecular levels. Through chemical reactions, atoms tend to attain more stable states at lower chemical potential energy levels. Atoms react chemically by losing, gaining, or sharing electrons. Forces arise from electron transferring and electron sharing interactions. Those forces that hold oppositely charged ions together or that bind atoms together in molecules are called **chemical bonds.** There are two principal types of bonds: the ionic or electrovalent bond and the covalent bond. We will study these two bond types and their modifications in this chapter.

7.2 Ionization Energy

Niels Bohr's description of the atom showed that electrons can exist at various energy levels when an atom absorbs energy. If sufficient energy is applied to an atom, it is possible to remove completely (or "knock out") one or more electrons from its structure, thereby forming a positive ion:

$$\text{Atom} + \text{Energy} \longrightarrow \text{Positive ion} + \text{Electron } (e^-)$$

ionization energy

The amount of energy required to remove one electron from an atom and convert it into a positive ion is known as the **ionization energy.** This energy may be expressed in units of kilocalories per mole (kcal/mole), indicating the number of kilocalories required to remove an electron from each atom in a mole of atoms. For example, 314 kcal are required to remove one electron from a mole of hydrogen atoms; 567 kcal are required to remove the first and 1254 kcal to remove the second electron from a mole of helium atoms. Other units, such as joules, are often used to express ionization energy.

Table 7.1 gives ionization energies for the removal of five electrons (where available) from each of several elements. The table shows that it requires in-

TABLE 7.1 Ionization energies for selected elements. Values are expressed in kilocalories per mole, showing energies required to remove up to five electrons per atom. The values shown in color indicate the energy needed to remove an electron from a noble gas electron structure.

Element	Required amounts of energy (kcal/mole)				
	1st e^-	2nd e^-	3rd e^-	4th e^-	5th e^-
H	314				
He	567	1254			
Li	124	1744	2823		
Be	215	420	3548	5020	
B	191	580	874	5980	7843
C	260	562	1104	1487	9034
Ne	497	947	1500	2241	2913
Na	118	1091	1652	2280	3192

creasingly higher amounts of energy to remove the second, third, fourth, and fifth electrons. This is a logical sequence, because as electrons are removed the number of protons and hence the positive charge of the nucleus remain the same and therefore the remaining electrons are held more tightly. The data also show an extra-large increase in the ionization energy when an electron is removed from a noble gas electron structure, indicating the high stability of this structure.

Ionization energies have been experimentally determined for most of the elements. First ionization energies (the energy required to remove the first electron) are given in Table 7.2 (page 140). Figure 7.1 is a graphic plot of the first ionization energies of the first 56 elements, H through Ba.

Certain periodic relationships are noted in Table 7.2 and Figure 7.1. All the alkali metals have relatively low ionization energies, indicating that they each have one electron that is easily removed. Furthermore, the ionization energy decreases from Li to Cs, showing that Cs loses an electron more easily than the other elements. There are two main reasons for this family trend: the electron being removed (1) is farther away from its nucleus and (2) is shielded from its nucleus by more inner shells of electrons as one goes down the family from Li to Cs.

From left to right within a period, the ionization energy, despite some irregularities, gradually increases. The noble gases have relatively high values, confirming the nonreactivity of these elements and the stability of an electron structure containing eight electrons in the outer energy level.

In the discussion on ionization energy, we spoke about the energy needed to remove an electron from an atom or a positive ion. Atoms also have an attraction for electrons, in which case they form negative ions.

$$F + e^- \longrightarrow F^- + Energy$$

The electron affinity of an atom is the amount of energy required to remove an electron from a negative ion. It is a measure of the attraction that an atom has for an electron. Electron affinity values are high for nonmetals and low for metals. The trend is for electron affinities to increase from left to right in any period of the periodic table and increase from bottom to top in a family of elements.

7.3 Electrons in the Outer Shell

Two outstanding properties of the elements are their tendencies (1) to have two electrons in each atomic orbital and (2) to form a stable outer-shell electron structure. For many elements, especially the representative ones, this stable outer shell contains eight electrons and is similar to the outer-shell electron structure of the noble gases. Atoms interact with one another and undergo electron structure changes to attain a state of greater stability. These rearrangements are accomplished by losing, gaining, or sharing electrons with other atoms. For example, a hydrogen atom has a tendency to accept another electron and thus attain an electron structure like that of the stable noble gas helium; a fluorine atom can accommodate one more electron and attain a stable electron structure

TABLE 7.2 First ionization energies of the elements (in kilocalories per mole).

Key

Atomic number → 1
Symbol → H
Ionization energy (kcal/mole) → 314

1 **H** 314																	2 **He** 567
3 **Li** 124	4 **Be** 215											5 **B** 191	6 **C** 260	7 **N** 335	8 **O** 314	9 **F** 402	10 **Ne** 497
11 **Na** 118	12 **Mg** 176											13 **Al** 138	14 **Si** 188	15 **P** 242	16 **S** 239	17 **Cl** 300	18 **Ar** 363
19 **K** 100	20 **Ca** 141	21 **Sc** 151	22 **Ti** 158	23 **V** 155	24 **Cr** 156	25 **Mn** 171	26 **Fe** 182	27 **Co** 181	28 **Ni** 176	29 **Cu** 178	30 **Zn** 217	31 **Ga** 138	32 **Ge** 182	33 **As** 226	34 **Se** 225	35 **Br** 273	36 **Kr** 322
37 **Rb** 96.3	38 **Sr** 131	39 **Y** 147	40 **Zr** 158	41 **Nb** 159	42 **Mo** 164	43 **Tc** 168	44 **Ru** 170	45 **Rh** 172	46 **Pd** 192	47 **Ag** 175	48 **Cd** 207	49 **In** 133	50 **Sn** 169	51 **Sb** 199	52 **Te** 208	53 **I** 241	54 **Xe** 280
55 **Cs** 89.8	56 **Ba** 120	57 **La** 129	72 **Hf** 127	73 **Ta** 182	74 **W** 184	75 **Re** 181	76 **Os** 201	77 **Ir** 207	78 **Pt** 207	79 **Au** 213	80 **Hg** 240	81 **Tl** 141	82 **Pb** 171	83 **Bi** 184	84 **Po** 194	85 **At** 219	86 **Rn** 248
87 **Fr** 88.3	88 **Ra** 122	89 **Ac** 159															

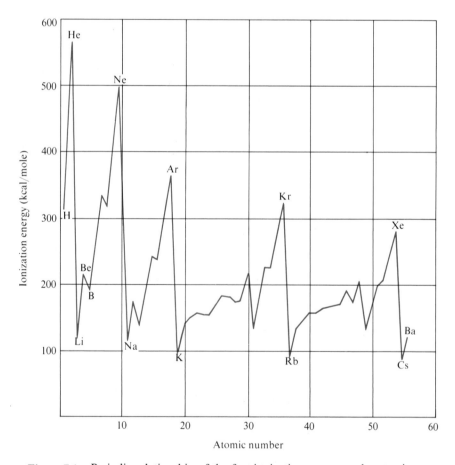

Figure 7.1 Periodic relationship of the first ionization energy to the atomic number of the elements.

like that of neon; a sodium atom tends to lose one electron to attain a stable electron structure like that of neon. This process requires energy.

$$\text{Na}\,(g) + \text{Energy} \longrightarrow \text{Na}^+\,(g) + 1\,e^-$$

valence electrons

The electrons in the outermost shell of an atom are responsible for most of this electron activity and are called the **valence electrons.** In electron-dot formulas of atoms the dots represent the outer-shell electrons and thus also represent the valence electrons. For example, hydrogen has one valence electron; sodium, one; aluminum, three; and oxygen, six. When a rearrangement of these electrons takes place between atoms, a chemical change occurs.

H·	Äl·	·Ö:
One valence electron	Three valence electrons	Six valence electrons

7.4 The Ionic Bond—Transfer of Electrons from One Atom to Another

The chemistry of many of the elements, especially the representative ones, is to attain an outer-shell electron structure like that of the chemically stable noble gases. With the exception of helium, this stable structure is eight electrons in the outer shell (see Table 5.6).

Let us look at the electron structures of sodium and chlorine to see how each element may attain a structure of 8 electrons in its outer shell. A sodium atom has 11 electrons: 2 in the first energy level, 8 in the second energy level, and 1 in the third energy level. Chlorine has 17 electrons: 2 in the first energy level, 8 in the second energy level, and 7 in the third energy level. If a sodium atom transfers or loses its $3s$ electron, its third energy level becomes vacant and it becomes a sodium ion with an electron configuration identical to that of the noble gas neon.

11+ $2\,e^-8\,e^-1\,e^- \longrightarrow$ 11+^{1+} $2\,e^-8\,e^- + 1\,e^-$

Na atom $(1s^22s^22p^63s^1)$ Na$^+$ ion $(1s^22s^22p^6)$

An atom that has lost or gained electrons will have a plus or minus electrical charge, depending on which charged particles, protons or electrons, are in excess. As we have seen, a charged atom or group of atoms is called an *ion*. A positively charged ion is called a *cation* (pronounced cat-i-on); a negative ion is called an *anion* (pronounced an-i-on).

By losing a negatively charged electron, the sodium atom becomes a positively charged particle known as a sodium ion. The charge, $+1$, occurs because the nucleus still contains 11 positively charged protons but the electron orbitals now contain only 10 negatively charged electrons. The charge of the ion is indicated by a plus sign ($+$) and is written as a superscript after the symbol of the element (Na^+).

Determination of the charge of a sodium ion:

Na atom	11 $+$		Na$^+$ ion	11 $+$
	11 e^-			10 e^-
Net charge	0		Net charge	$+1$

A chlorine atom with 7 electrons in the third energy level needs 1 electron to pair up with its 1 unpaired $3p$ electron to attain the stable outer-shell electron structure of argon. By gaining 1 electron, the chlorine atom becomes a chloride ion (Cl^-), a negatively charged particle containing 17 protons and 18 electrons.

17 $+$ $2\,e^-8\,e^-7\,e^-$ $+$ $1\,e^-$ $\longrightarrow$ 17 $+$$^{1-}$ $2\,e^-8\,e^-8\,e^-$

Cl atom ($1s^2 2s^2 2p^6 3s^2 3p^5$)   Cl$^-$ ion ($1s^2 2s^2 2p^6 3s^2 3p^6$)

Determination of the charge of a chloride ion:

Cl atom	17 $+$		Cl$^-$ ion	17 $+$
	17 e^-			18 e^-
Net charge	0		Net charge	-1

Consider the case in which sodium and chlorine atoms react with each other. The $3s$ electron from the sodium atom transfers to the vacant $3p$ orbital in the chlorine atom to form a positive sodium ion and a negative chloride ion. The compound sodium chloride results because the Na^+ and Cl^- ions are strongly attracted to one another by their opposite electrostatic charges. The force holding these two ions together is called an ionic bond.

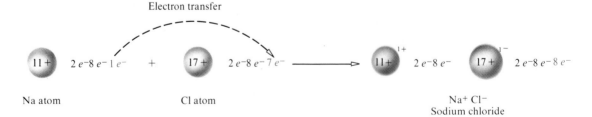

Electron transfer

11 $+$ $2\,e^-8\,e^-1\,e^-$ $+$ 17 $+$ $2\,e^-8\,e^-7\,e^-$ $\longrightarrow$ 11$+$$^{1+}$ $2\,e^-8\,e^-$ 17$+$$^{1-}$ $2\,e^-8\,e^-8\,e^-$

Na atom   Cl atom   Na$^+$ Cl$^-$
                                                                     Sodium chloride

The electron-dot representation of sodium chloride formation is as follows:

$$Na\cdot + \;:\!\ddot{Cl}\!: \longrightarrow Na^+ \;:\!\ddot{Cl}\!:^-$$

The chemical reaction between sodium and chlorine is a very vigorous one. It is highly *exothermic* (evolving heat), liberating 90,200 cal when 1 mole of sodium atoms (23.0 g) combines with 1 mole of chlorine atoms (35.5 g).

Sodium chloride is actually made up of cubic crystals, in which each sodium ion is surrounded by six chloride ions and each chloride ion is surrounded by six sodium ions, except at the crystal surface. A sodium chloride crystal is a regularly arranged aggregate of millions of these ions, but the ratio of sodium to chloride ions is 1:1. The cubic crystalline lattice arrangement of sodium chloride is shown in Figure 7.2.

Figure 7.3 contrasts the relative sizes of sodium and chlorine atoms with those of their ions. The sodium ion is smaller than the atom because (1) the sodium atom has lost its outer shell, consisting of 1 electron, thereby reducing its

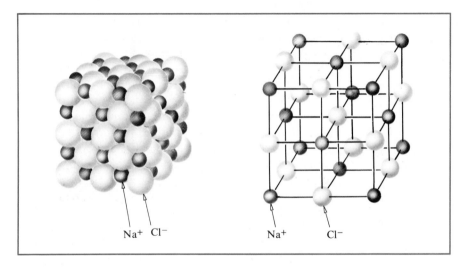

Figure 7.2 Sodium chloride crystal. Diagram represents a small fragment of sodium chloride, which forms cubic crystals. Each sodium ion is surrounded by six chloride ions, and each chloride ion is surrounded by six sodium ions.

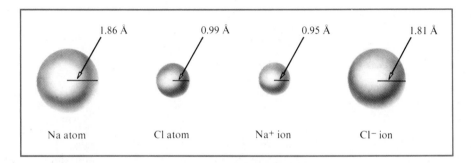

Figure 7.3 Relative sizes (radii) of sodium and chlorine atoms and their ions.

size; and (2) the 10 remaining electrons are now attracted by 11 protons and are thus drawn closer to the nucleus. Conversely, the chloride ion is larger than the atom because it has 18 electrons but only 17 protons. The nuclear attraction on each electron is thereby decreased, allowing the chlorine atom to expand as it forms a chloride ion.

We have seen that when sodium reacts with chlorine, each atom becomes an electrically charged ion. Sodium chloride, like all ionic substances, is held together by the attraction existing between positive and negative charges. An **ionic,** or **electrovalent, bond** is the electrostatic attraction existing between oppositely charged ions.

ionic bond
electrovalent bond

Ionic bonds are formed whenever there is a complete transfer of one or more electrons from one atom to another. The metals, which have relatively little attraction for their valence electrons, tend to form ionic bonds when they combine with nonmetals.

It is important to recognize that electrovalently bonded substances do not exist as molecules. In sodium chloride, for example, the bond does not exist solely between a single sodium ion and a single chloride ion. Each sodium ion in the crystal attracts six near-neighbor negative chloride ions; in turn, each negative chloride ion attracts six near-neighbor positive sodium ions (see Figure 7.2).

A metal will ordinarily have one, two, or three electrons in its outer energy level. In chemical reactions, metal atoms characteristically lose these electrons to become positive ions. A nonmetal, on the other hand, is only a few electrons short of having a complete octet in its outer energy level and thus has a tendency to gain electrons (electron affinity). In reacting with metals, nonmetal atoms characteristically gain one, two, or three electrons, attain the electron structure of a noble gas, and become negative ions. The ions formed by loss of electrons are much smaller than the corresponding metal atoms; the ions formed by gaining electrons are larger than the corresponding nonmetal atoms. The actual dimensions of the atomic and ionic radii of several metals and nonmetals are given in Table 7.3.

TABLE 7.3 Changes in atomic size of selected metals and nonmetals. The metals shown lose electrons to become positive ions. The nonmetals gain electrons to become negative ions.

Atomic radius (Å)		Ionic radius (Å)		Atomic radius (Å)		Ionic radius (Å)	
Li	1.52	Li^+	0.60	F	0.64	F^-	1.36
Na	1.86	Na^+	0.95	Cl	0.99	Cl^-	1.81
K	2.27	K^+	1.33	Br	1.14	Br^-	1.95
Mg	1.60	Mg^{2+}	0.65	O	0.66	O^{2-}	1.40
Al	1.43	Al^{3+}	0.50	S	1.04	S^{2-}	1.84

Study the following examples. Note the loss and gain of electrons between atoms; also note that the ions in each compound have a stable noble gas electron structure.

Example 1:   A magnesium atom of electron structure $1s^2 2s^2 2p^6 3s^2$ must lose two electrons or gain six electrons to reach a stable electron structure. If magnesium reacts with chlorine and each chlorine atom has room for only one electron, two chlorine atoms will be needed to accept the two electrons from one magnesium atom. The compound formed will contain one magnesium ion and two chloride ions. The magnesium ion will have a $+2$ charge, having lost two electrons. Each chloride ion will have a -1 charge. The transfer of electrons from a magnesium atom to two chlorine atoms is shown in the following illustration:

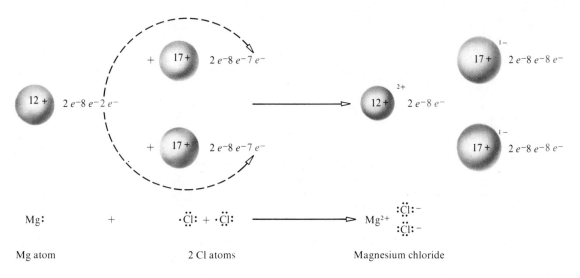

Mg:	+	$\cdot\ddot{C}l:$ + $\cdot\ddot{C}l:$	⟶	Mg²⁺ $\begin{array}{c} :\ddot{C}l:^- \\ :\ddot{C}l:^- \end{array}$
Mg atom		2 Cl atoms		Magnesium chloride

Example 2:   Formation of sodium fluoride, NaF

Na·	+	$\cdot\ddot{F}:$	⟶	Na⁺ $:\ddot{F}:^-$
Sodium atom		Fluorine atom		Sodium fluoride

The fluorine atom, with seven electrons in its outer shell, behaves similarly to a chlorine atom.

Example 3:   Formation of aluminum chloride, $AlCl_3$

$$
\text{Al}\cdot \ + \ \begin{array}{c} \cdot\ddot{C}l: \\[4pt] \cdot\ddot{C}l: \\[4pt] \cdot\ddot{C}l: \end{array} \ \longrightarrow \ \text{Al}^{3+} \begin{array}{c} :\ddot{C}l:^- \\[4pt] :\ddot{C}l:^- \\[4pt] :\ddot{C}l:^- \end{array}
$$

Aluminum atom   Chlorine atoms   Aluminum chloride

Each chlorine atom can accept only one electron. Therefore, three chlorine atoms are needed to combine with the three valence electrons of one aluminum atom. The aluminum atom has lost three electrons to become an aluminum ion, Al^{3+}, with a $+3$ charge.

Example 4:   Formation of magnesium oxide, MgO

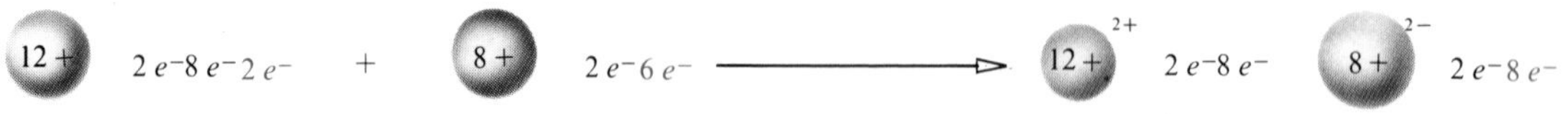

12 + $2e^-8e^-2e^-$	+	8 + $2e^-6e^-$ $\longrightarrow$	12 +$^{2+}$ $2e^-8e^-$	8 +$^{2-}$ $2e^-8e^-$
Mg:	+	$\cdot\ddot{O}$: $\longrightarrow$		Mg^{2+}:$\ddot{O}$:$^{2-}$
Magnesium atom		Oxygen atom		Magnesium oxide

The magnesium atom, with two electrons in the outer energy level, exactly fills the need of two electrons of one oxygen atom. The resulting compound has a ratio of one atom of magnesium to one atom of oxygen. The oxygen (oxide) ion has a -2 charge, having gained two electrons. In combining with oxygen, magnesium behaves the same way as when combining with chlorine—it loses two electrons.

Example 5:   Formation of sodium sulfide, Na_2S

Na· Na^+

    + ·$\ddot{S}$: $\longrightarrow$:$\ddot{S}$:$^{2-}$

Na· Na^+

Sodium   Sulfur   Sodium sulfide
atoms   atom

Two sodium atoms supply the electrons that one sulfur atom needs to make eight electrons in its outer shell.

Example 6:   Formation of aluminum oxide, Al_2O_3

$\ddot{Al}$·   ·$\ddot{O}$:   :$\ddot{O}$:$^{2-}$

                 Al^{3+}

    + ·$\ddot{O}$: $\longrightarrow$:$\ddot{O}$:$^{2-}$

                 Al^{3+}

$\ddot{Al}$·   ·$\ddot{O}$:   :$\ddot{O}$:$^{2-}$

Aluminum   Oxygen   Aluminum oxide
atoms   atoms

One oxygen atom, needing two electrons, cannot accommodate the three electrons from one aluminum atom. One aluminum atom falls one electron short of the four electrons needed by two oxygen atoms. A ratio of two atoms of aluminum to three atoms of oxygen, involving the transfer of six electrons (two to each oxygen atom), gives each atom a stable electron configuration.

Note that, in each of the examples shown, outer shells containing eight electrons were formed in all the negative ions. This formation resulted in the pairing of all the *s* and *p* electrons in these outer shells.

Chemistry would be considerably simpler if all compounds were made by the direct formation of ions as outlined in the examples just given. Unfortunately, this is only one of the two general methods of compound formation. The second general method will be outlined in the sections that follow.

7.5 The Covalent Bond—Sharing of Electrons

covalent bond

The concept of the covalent bond was introduced to the chemical community in 1916 by G. N. Lewis of the University of California at Berkeley. In the millions of compounds known, the covalent bond is the predominant chemical bond. It is the type of bond that is common among the molecules of biological significance. Atoms with similar attraction for electrons do not transfer electrons from one atom to another to form ions. Instead, they form a chemical bond by sharing pairs of electrons between them. A **covalent bond** is formed between two atoms when they share a pair of electrons.

True molecules exist in compounds that are held together by covalent bonds. It is not correct to refer to "a molecule" of sodium chloride or other ionic compounds, since these compounds exist as large aggregates of positive and negative ions. But we can refer to a molecule of hydrogen, chlorine, hydrogen chloride, carbon tetrachloride, sugar, or carbon dioxide, because these compounds contain only covalent bonds and exist in molecular aggregates.

A study of the hydrogen molecule will give a better insight into the nature of the covalent bond and its formation. The formation of a hydrogen molecule, H_2, involves the overlapping and pairing of $1s$ electron orbitals from two hydrogen atoms. This overlapping and pairing is shown in Figure 7.4. Each atom contributes one electron of the pair that is shared jointly by two hydrogen nuclei. The orbital of the electrons now includes both hydrogen nuclei, but probability factors show that the most likely place to find the electrons (the point of highest electron density) is between the two nuclei. The two nuclei are shielded from each other by the pair of electrons, allowing the two nuclei to be drawn very close to each other. (The average bond length between the hydrogen nuclei is 7.4×10^{-9} cm.)

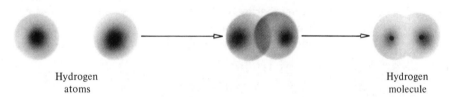

Hydrogen
atoms

Hydrogen
molecule

Figure 7.4 The formation of a hydrogen molecule from two hydrogen atoms. The two $1s$ orbitals overlap, forming the H_2 molecule. In this molecule the two electrons are shared between the atoms, forming a covalent bond.

The tendency for hydrogen atoms to form a molecule is very strong. In the molecule, each electron is attracted by two positive nuclei. This attraction gives the hydrogen molecule a more stable structure than the individual hydrogen atoms had. Experimental evidence of stability is shown by the fact that 104.2 kcal are needed to break the bonds between the hydrogen atoms in 1 mole of hydrogen (2.0 g). The strength of a bond may be determined by the energy required to break it. The energy required to break a covalent bond is known as the *bond dissociation energy*. The following bond dissociation energies illustrate relative bond strengths. (All substances are considered to be in the gaseous state and to form neutral atoms.)

Reaction	Bond dissociation energy (kcal/mole)
$H_2 \longrightarrow 2\,H$	104.2
$N_2 \longrightarrow 2\,N$	226.0
$O_2 \longrightarrow 2\,O$	118.3
$F_2 \longrightarrow 2\,F$	36.6
$Cl_2 \longrightarrow 2\,Cl$	58.0
$Br_2 \longrightarrow 2\,Br$	46.1
$I_2 \longrightarrow 2\,I$	36.1

The formula of chlorine gas is Cl_2. When the two atoms of chlorine combine to form this molecule, the electrons must interact by a method that is different from that shown in the preceding examples. Each chlorine atom would be more stable with eight electrons in its outer shell. But if an electron transfers from one chlorine atom to the other, the first chlorine atom, with only six electrons remaining in its outer shell, would be highly unstable. What actually happens when the two chlorine atoms join together is this: the unpaired $3p$ electron orbital of one chlorine atom overlaps the unpaired $3p$ electron orbital of the other atom, resulting in a pair of electrons that are mutually shared between the two atoms. Each atom furnishes one of the pair of shared electrons. Thus, each atom attains a stable structure of eight electrons by sharing an electron pair with the other atom. The pairing of the p electrons in the formation of a chlorine molecule is illustrated in Figure 7.5. Neither chlorine atom has a positive or negative charge, since both contain the same number of protons and have equal attraction for the pair of electrons being shared. Other examples of molecules in which there is equal sharing of electrons between two atoms are hydrogen, H_2; oxygen, O_2; nitrogen, N_2; fluorine, F_2; bromine, Br_2; and iodine, I_2. Note that two or even three pairs of electrons may be shared between atoms.

H:H	:F̈:F̈:	:B̈r:B̈r:	:Ï:Ï:	:Ö::Ö:	:N:::N:
Hydrogen	Fluorine	Bromine	Iodine	Oxygen	Nitrogen

A practice commonly used in writing molecular structures is to replace by a dash (—) the electron dots used to show a pair of electrons shared between atoms. When two pairs of electrons are shared, two dashes are used; when three

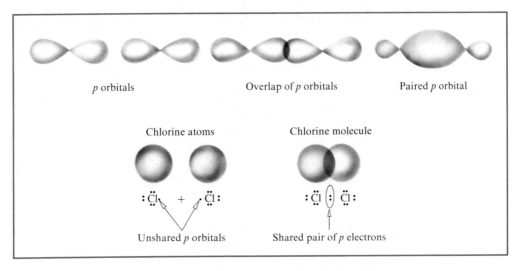

Figure 7.5   Pairing of p electrons in the formation of a chlorine molecule.

pairs of electrons are shared, three dashes are used. One dash (two electrons) is called a single bond; two dashes, a double bond; and three dashes, a triple bond. Thus the six structures just shown can be written this way:

$$ \text{H—H} \qquad :\ddot{\text{F}}—\ddot{\text{F}}: \qquad :\ddot{\text{Br}}—\ddot{\text{Br}}: \qquad :\ddot{\text{I}}—\ddot{\text{I}}: \qquad :\ddot{\text{O}}=\ddot{\text{O}}: \qquad :\text{N}\equiv\text{N}: $$

7.6 Electronegativity

electronegativity

When two different kinds of atoms share a pair of electrons, one atom assumes a partial positive charge and the other assumes a partial negative charge (in respect to each other). This is because the two atoms exert unequal attraction for the pair of shared electrons. The relative attraction that an atom has for the pair of electrons in a covalent bond is called its **electronegativity.** Elements differ in their electronegativities. For example, both hydrogen and chlorine need one electron to form stable electron configurations. They share a pair of electrons in the substance hydrogen chloride, HCl. Chlorine is more electronegative and therefore has a greater attraction for the shared electrons than does hydrogen. As a result, the pair of electrons is displaced toward the chlorine atom, giving it a partial negative charge and leaving the hydrogen atom with a partial positive charge. It should be understood that the electron is not transferred entirely to the chlorine atom, as in the case of sodium chloride, and no ions are formed. The entire molecule, HCl, is electrically neutral.

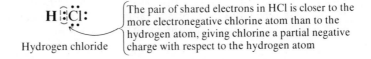

H $\overset{..}{\underset{..}{\text{Cl}}}$:

Hydrogen chloride

The pair of shared electrons in HCl is closer to the more electronegative chlorine atom than to the hydrogen atom, giving chlorine a partial negative charge with respect to the hydrogen atom

The electronegativity, or ability of an atom to attract a pair of shared electrons, depends on several factors: (1) the charge on the nucleus, (2) the distance of the outer electrons from the nucleus, and (3) the amount of shielding of the nucleus by intervening shells of electrons between the outer-shell electrons and the nucleus. A scale of relative electronegativities, in which the most electronegative element, fluorine, is assigned a value of 4.0, was developed by the Nobel laureate (1954 and 1962) Linus Pauling (1901–   ). Table 7.4 shows that the relative electronegativity of the nonmetals is high and that of the metals is low. These electronegativities indicate that atoms of metals have a greater tendency to lose electrons than do atoms of nonmetals, and that nonmetals have a greater tendency to gain electrons. The higher the electronegativity value, the greater the attraction for electrons.

TABLE 7.4 Relative electronegativity of the elements. The electronegativity value is given below the symbol of each element.

Key:
- 9 — Atomic number
- F — Symbol
- 4.0 — Electronegativity

1 **H** 2.1																	2 **He** —
3 **Li** 1.0	4 **Be** 1.5											5 **B** 2.0	6 **C** 2.5	7 **N** 3.0	8 **O** 3.5	9 **F** 4.0	10 **Ne** —
11 **Na** 0.9	12 **Mg** 1.2											13 **Al** 1.5	14 **Si** 1.8	15 **P** 2.1	16 **S** 2.5	17 **Cl** 3.0	18 **Ar** —
19 **K** 0.8	20 **Ca** 1.0	21 **Sc** 1.3	22 **Ti** 1.4	23 **V** 1.6	24 **Cr** 1.6	25 **Mn** 1.5	26 **Fe** 1.8	27 **Co** 1.8	28 **Ni** 1.8	29 **Cu** 1.9	30 **Zn** 1.6	31 **Ga** 1.6	32 **Ge** 1.8	33 **As** 2.0	34 **Se** 2.4	35 **Br** 2.8	36 **Kr** —
37 **Rb** 0.8	38 **Sr** 1.0	39 **Y** 1.2	40 **Zr** 1.4	41 **Nb** 1.6	42 **Mo** 1.8	43 **Tc** 1.9	44 **Ru** 2.2	45 **Rh** 2.2	46 **Pd** 2.2	47 **Ag** 1.9	48 **Cd** 1.7	49 **In** 1.1	50 **Sn** 1.8	51 **Sb** 1.9	52 **Te** 2.1	53 **I** 2.5	54 **Xe** —
55 **Cs** 0.7	56 **Ba** 0.9	57–71 **La-Lu** 1.1–1.2	72 **Hf** 1.3	73 **Ta** 1.5	74 **W** 1.7	75 **Re** 1.9	76 **Os** 2.2	77 **Ir** 2.2	78 **Pt** 2.2	79 **Au** 2.4	80 **Hg** 1.9	81 **Tl** 1.8	82 **Pb** 1.8	83 **Bi** 1.9	84 **Po** 2.0	85 **At** 2.2	86 **Rn** —
87 **Fr** 0.7	88 **Ra** 0.9	89– **Ac-** 1.1–1.7															

7.7 Writing Lewis Electron-Dot Structures

As we have seen, Lewis electron-dot structures are a convenient way of showing the covalent bonds in many molecules or ions of the representative elements. A pair of dots or a dash placed between two atoms is used for each pair of electrons in the covalent bond.

In writing Lewis structures the object is to connect the atoms in a molecule with covalent bonds by rearranging the valence electrons of the atoms so that each atom has eight outer-shell electrons around it. Exceptions to this rule are hydrogen, which requires only two electrons, and several other elements, such as lithium, beryllium, and boron.

Although Lewis structures for many molecules and ions can be written by inspection, the following procedure will be helpful for learning to write these structures:

1. Obtain the total number of electrons to be used in the structure by adding the valence electrons in all of the atoms in the molecule or ion. If you are writing the structure of an ion, add one electron for each negative charge or subtract one electron for each positive charge on the ion. Remember, the number of valence electrons of Group A elements is the same as their group number in the periodic table.

2. Place the atoms in the molecule in a skeletal arrangement and connect adjacent atoms with a single covalent bond (two dots or one dash). Hydrogen, which contains only one bonding electron, can form only one covalent bond. Oxygen atoms are not normally bonded to each other, except in compounds known to be peroxides and in elemental oxygen. Oxygen atoms can have a maximum of two covalent bonds, two single bonds or one double bond.

3. Subtract two electrons for each single bond you used in Step 2 from the total number of electrons calculated in Step 1. Divide this number by 2 to determine how many pairs of electrons are available for use.

4. Place these pairs of electrons (pairs of dots) around each atom (except hydrogen) to give each atom a total of eight electrons around it.

5. If there are too few pairs to give each atom eight electrons, change single bonds between atoms to double or triple bonds by shifting unbonded pairs of electrons as needed.

PROBLEM 7.1   How many valence electrons are in each of these atoms: Cl, H, C, O, N, S, P, I?

You can look in Table 5.5 for the electron structure, or, if the element is in a Group A of the periodic table, the number of valence electrons is equal to the group number.

Atom	Periodic group	Valence electrons
Cl	VIIA	7
H	IA	1
C	IVA	4
O	VIA	6
N	VA	5
S	VIA	6
P	VA	5
I	VIIA	7

PROBLEM 7.2   Write the Lewis structure for water, H_2O.

Step 1. The total number of valence electrons is 8, 2 from the two hydrogen atoms and 6 from the oxygen atom.

Step 2. The two hydrogen atoms are connected to the oxygen atom. Write the skeletal structure:

H   O

H

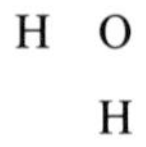

Place two dots between the hydrogen and oxygen atoms to form the covalent bonds:

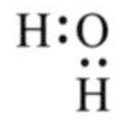

H:O
H

Step 3. Subtract the 4 electrons used in Step 2 from 8 to obtain 4 electrons yet to be used. Divide the 4 electrons by 2 to give two pairs of electrons to be placed.

Step 4. Place the two pairs of electrons around the oxygen atom. Hydrogen atoms cannot accommodate any more electrons.

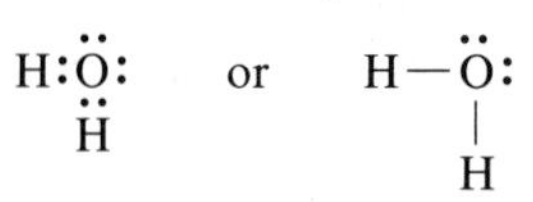

H:Ö: or   H—Ö:
H |
                   H

This is the Lewis structure. Each atom has a noble gas electron structure.

■

PROBLEM 7.3   Write Lewis structures for a molecule of (a) methane, CH_4, and (b) carbon tetrachloride, CCl_4.

(a) **Step 1.** The total number of valence electrons is 8, 1 from each hydrogen atom and 4 from the carbon atom.

Step 2. The skeletal structure contains four H atoms around the C atom. Place 2 electrons between each C and H.

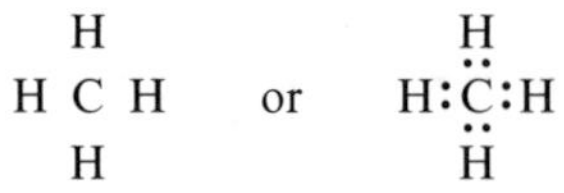

H   H
H C H   or   H:C:H
H   H

Step 3. Subtract the 8 electrons used in Step 2 from 8 to obtain zero electrons yet to be placed. Therefore, the Lewis structure must be as written in Step 2.

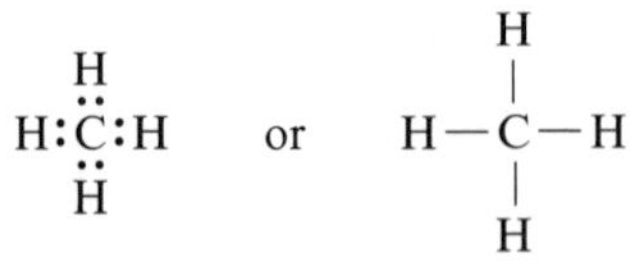

                              H
H |
H:C:H   or   H—C—H
H |
                              H

(b) **Step 1.** The total number of valence electrons to be used is 32, 4 from the carbon atom and 7 from each of the four chlorine atoms.

Step 2. The skeletal structure contains the four Cl atoms around the C atom. Place 2 electrons between the C and each Cl.

$$
\begin{array}{ccc}
& \text{Cl} & \\
\text{Cl} & \text{C} & \text{Cl} \\
& \text{Cl} &
\end{array}
\qquad
\begin{array}{c}
\overset{..}{\text{Cl}} \\
\text{Cl:} \overset{..}{\underset{..}{\text{C}}} \text{:Cl} \\
\text{Cl}
\end{array}
$$

Step 3. Subtract the 8 electrons used in Step 2 from 32, to obtain 24 electrons yet to be placed. Divide 24 by 2 to give 12 pairs of electrons to be placed.

Step 4. Place the 12 pairs of electrons around the Cl atoms so that each Cl atom has 8 electrons around it.

$$
\begin{array}{c}
:\overset{..}{\underset{..}{\text{Cl}}}: \\
:\overset{..}{\underset{..}{\text{Cl}}}: \overset{..}{\underset{..}{\text{C}}} :\overset{..}{\underset{..}{\text{Cl}}}: \\
:\overset{..}{\underset{..}{\text{Cl}}}:
\end{array}
\qquad \text{or} \qquad
\begin{array}{c}
:\overset{..}{\underset{..}{\text{Cl}}}: \\
\mid \\
:\overset{..}{\underset{..}{\text{Cl}}} - \text{C} - \overset{..}{\underset{..}{\text{Cl}}}: \\
\mid \\
:\overset{..}{\underset{..}{\text{Cl}}}:
\end{array}
$$

This is the Lewis structure. CCl_4 contains four covalent bonds.

■

PROBLEM 7.4   Write Lewis structures for (a) carbon dioxide, CO_2, and (b) nitric acid, HNO_3.

(a) **Step 1.** The total number of valence electrons is 16, 4 from the carbon atom and 6 from each oxygen atom.

Step 2. The two oxygen atoms are bonded to the carbon atom. Write the skeletal structure and place 2 electrons between each C and O atom.

$$\text{O:C:O}$$

Step 3. Subtract the 4 electrons used in Step 2 from 16 to obtain 12 electrons yet to be placed. Divide 12 by 2 to give six pairs of electrons to be placed.

Step 4. Place these six pairs of electrons around the C and O atoms. There are several possibilities:

$$
\underset{\text{I}}{:\overset{..}{\underset{..}{\text{O}}}:\text{C}:\overset{..}{\underset{..}{\text{O}}}:}
\qquad
\underset{\text{II}}{:\overset{..}{\underset{..}{\text{O}}}:\text{C}:\overset{..}{\underset{..}{\text{O}}}:}
\qquad
\underset{\text{III}}{:\overset{..}{\underset{..}{\text{O}}}:\overset{..}{\text{C}}:\underset{..}{\text{O}}:}
$$

Step 5. All the atoms do not have 8 electrons around them. Move one pair of unbonded electrons from each O atom in structure I and place one additional pair of electrons between each C and O atom, forming two double bonds:

$$
:\overset{..}{\text{O}}::\text{C}::\overset{..}{\text{O}}:
\qquad \text{or} \qquad
:\overset{..}{\text{O}} = \text{C} = \overset{..}{\text{O}}:
$$

Each atom now has 8 electrons around it. Carbon is sharing four pairs and each oxygen is sharing two pairs of electrons.

(b) **Step 1.** The total number of valence electrons is 24, 1 from the hydrogen atom, 5 from the nitrogen atom, and 6 from each oxygen atom.

Step 2. The three oxygen atoms are bonded to the central nitrogen atom. The hydrogen is bonded to one of the oxygen atoms. Write the skeletal structure and place 2 electrons between each pair of atoms:

$$\begin{array}{c} \text{O} \\ \overset{\cdot\cdot}{} \\ \text{H:O:N:O} \end{array}$$

Step 3. Subtract the 8 electrons used in Step 2 from 24 to obtain 16 electrons yet to be placed. Divide 16 by 2 to give eight pairs of electrons to be placed.

Step 4. Place the eight pairs of electrons around the N and O atoms:

$$\begin{array}{c} \overset{\cdot\cdot}{\text{:O}} \\ \overset{\cdot\cdot}{\text{H:O:N:O:}} \\ \overset{\cdot\cdot}{} \end{array}$$

Step 5. There is still one pair of electrons needed to give all the N and O atoms 8 electrons. Move the unbonded pair of electrons from the N atom and place it between the N and the electron-deficient O atom, making a double bond:

$$\begin{array}{c} \overset{\cdot\cdot}{\text{:O}} \\ \overset{\cdot\cdot}{\text{H:O:N:O:}} \\ \overset{\cdot\cdot}{} \end{array} \quad \text{or} \quad \text{H}-\overset{\cdot\cdot}{\underset{\cdot\cdot}{\text{O}}}-\text{N}=\overset{\cdot\cdot}{\underset{\cdot\cdot}{\text{O}}}$$

This is the Lewis structure.

PROBLEM 7.5 Write the Lewis structure for a sulfate ion, SO_4^{2-}.

Step 1. There are 30 valence electrons in these five atoms (6 in each atom) plus 2 electrons from the −2 charge. This makes 32 electrons to be placed.

Step 2. In the sulfate ion, the sulfur is the central atom, surrounded by the four oxygen atoms. Write the skeletal structure and place two electrons between each pair of atoms:

$$\begin{array}{c} \text{O} \\ \overset{\cdot\cdot}{} \\ \text{O:S:O} \\ \overset{\cdot\cdot}{} \\ \text{O} \end{array}$$

Step 3. Subtract the 8 electrons used in Step 2 from 32 to give 24 electrons yet to be placed. Divide 24 by 2 to give 12 pairs of electrons to be placed.

Step 4. Place the 12 pairs of electrons around the four O atoms and indicate that the sulfate ion has a −2 charge:

$$\begin{array}{c} \overset{\cdot\cdot}{\text{:O:}} \\ \overset{\cdot\cdot}{\text{:O:S:O:}} \\ \overset{\cdot\cdot}{\text{:O:}} \end{array}^{2-} \quad \text{or} \quad \begin{array}{c} \overset{\cdot\cdot}{\text{:O:}} \\ | \\ \text{:O}-\text{S}-\text{O:} \\ | \\ \overset{\cdot\cdot}{\text{:O:}} \end{array}^{2-}$$

This is the Lewis structure. Each atom has 8 electrons around it.

Although many compounds follow the octet rule for covalent bonding, there are numerous exceptions. Sometimes it is impossible to write a structure in which each atom has eight electrons around it. For example, in BF_3 the boron atom has only 6 electrons around it and in SF_6 the sulfur atom has 12 electrons around it.

7.8 Polar Covalent Bonds

We have considered bonds to be either covalent or electrovalent, according to whether electrons are shared between atoms or transferred from one atom to another. In most covalent bonds, the pairs of electrons are not shared equally between the atoms. Such bonds are known as polar covalent bonds.

nonpolar covalent bond

In a **nonpolar covalent bond** the shared pair of electrons is attracted equally by the two atoms. A bond between the same kind of atoms, such as that in hydrogen or chlorine molecules, is nonpolar because the electronegativity difference between identical atoms is zero.

When a covalent bond is formed between two atoms of different electronegativities, the more electronegative atom attracts the shared electron pair toward itself. As a result, the atom with the higher electronegativity acquires a partial negative charge and the other atom acquires a partial positive charge. However, the overall molecule is still neutral. Because of this greater attraction of the electron pair, the bond formed between the two atoms has partial ionic

polar covalent bond

character and is known as a **polar covalent bond.** The resulting molecule is said to be polar. Most bonds are intermediate between the extremes of totally ionic and totally covalent. (See Figure 7.6.)

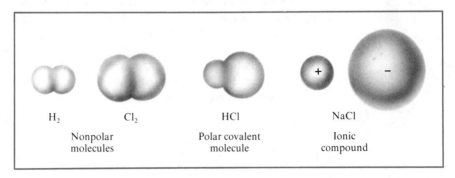

Figure 7.6   Nonpolar, polar covalent, and ionic compounds.

dipole

In two-atom molecules having a polar covalent bond, like HCl, we may say that the molecule is polar and acts as a dipole. In general, a **dipole** is a molecule with a separation of electrostatic charge, causing it to be positive at one end and negative at the other end. A dipole is often written as ⊕ ⊖. A hydrogen chloride molecule is polar and behaves like a small dipole. The HCl dipole may be written as H ⟵⟶ Cl. The arrow points toward the negative end of the dipole. Molecules of H_2O, HBr, and ICl are polar; CH_4, CCl_4, and CO_2 are nonpolar.

$$H \longleftrightarrow Cl \qquad H \longleftrightarrow Br \qquad I \longleftrightarrow Cl \qquad \overset{\displaystyle O}{\underset{H \qquad H}{}}$$

The greater the difference in electronegativity between two atoms, the more polar is the bond between them. As a rule, when the electronegativity difference between the two atoms is greater than 1.7–1.9, the bond between the two atoms will be more ionic than covalent. If the difference in electronegativity is less than 1.7 units, the bond will be essentially covalent. The difference in electronegativity between atoms can also give us a guide to the relative strength of covalent bonds. The greater the difference, the stronger the bond—that is, the more energy required to break the bond. For example, HF has the strongest bond in the series HF, HCl, HBr, and HI, as seen by the bond dissociation data in the following table.

Compound	Electronegativity difference	Bond dissociation energy (kcal/mole)
HF	1.9	135
HCl	0.9	103
HBr	0.7	87.5
HI	0.4	71.4

Care must be taken to distinguish between polar bonds and polar molecules. A covalent bond between different kinds of atoms is always polar. But a molecule containing different kinds of atoms may or may not be polar, depending on its shape or geometry. The HF, HCl, HBr, HI, and ICl molecules just mentioned are all polar because each contains a single polar bond. However, CO_2, CH_4, and CCl_4 are nonpolar molecules despite the fact that all three contain polar bonds. The carbon dioxide molecule, $O{=}C{=}O$, is nonpolar because the three atoms in the molecule are linear and the carbon–oxygen dipoles cancel each other by acting in opposite directions.

$$\overset{\longleftarrow \quad + \quad + \quad \longrightarrow}{O{=}C{=}O}$$

Dipoles in opposite directions

Methane (CH_4) and carbon tetrachloride (CCl_4) are nonpolar because the C—H and C—Cl dipoles form tetrahedral angles and thereby cancel one another (Figure 7.7).

We have said that water is a polar molecule. If the atoms in water were linear, as in carbon dioxide, the two O—H dipoles would cancel each other and the molecule would be nonpolar. However, water is definitely polar and has a nonlinear structure with an angle of 105° between the two O—H bonds (see Figure 7.8).

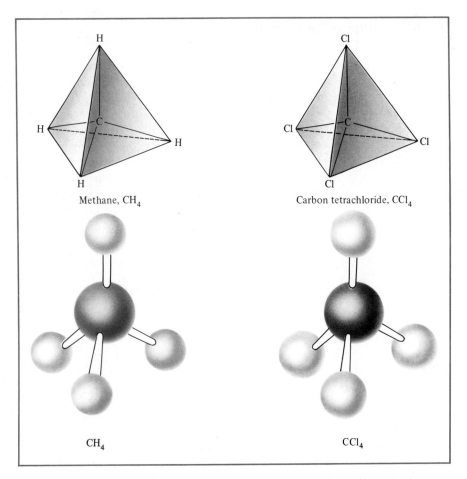

Figure 7.7   Methane and carbon tetrachloride are nonpolar molecules because their polar bonds cancel each other in the tetrahedral arrangement of their atoms. The carbon atom is located in the center of the tetrahedrons.

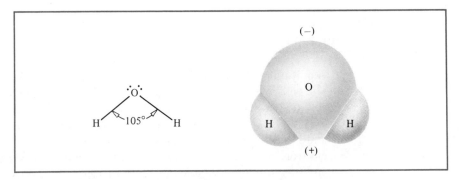

Figure 7.8   The polar water molecule.

7.9 Coordinate Covalent Bonds

coordinate
covalent bond

In Section 7.5, we saw that a covalent bond is formed by the overlapping of electron orbitals between two atoms. The two atoms each furnish an electron to make a pair that is shared between them.

Covalent bonds can also be formed by a single atom furnishing both electrons that are shared between the two atoms. The bond so formed is called a **coordinate covalent,** or **semipolar, bond.** This bond is often designated by an arrow pointing away from the electron donor (for example, A $\longrightarrow$ B). Once formed, a coordinate covalent bond has the same properties as any other covalent bond—it simply is a pair of electrons shared between two atoms.

The electron-dot structures of sulfurous and sulfuric acids show a coordinate covalent bond between the sulfur and the oxygen atoms that are not bonded to hydrogen atoms. The colored dots indicate the electrons of the sulfur atom.

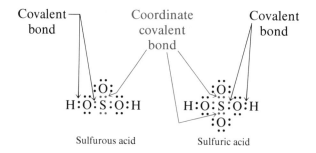

Sulfurous acid   Sulfuric acid

The open (unbonded) pair of electrons on the sulfur atom in sulfurous acid allows room for another oxygen atom with six electrons to fit perfectly into its structure and form sulfuric acid. Other atoms with six electrons in their outer shell, such as sulfur, could also fit into this pattern. The coordinate covalent bond explains the formation of many complex molecules.

7.10 Polyatomic Ions

polyatomic ion

A **polyatomic ion** is a stable group of atoms that has either a positive or a negative charge and behaves as a single unit in many chemical reactions. Sodium sulfate, Na_2SO_4, contains two sodium ions and a sulfate ion. The sulfate ion, SO_4^{2-}, is a polyatomic ion composed of one sulfur atom and four oxygen atoms, and has a charge of -2. One sulfur and four oxygen atoms have a total of 30 electrons in their outer shells. The sulfate ion contains 32 outer-shell electrons and therefore has a charge of -2. In this case, the 2 additional electrons come from the two sodium atoms, which are now sodium ions.

$$Na^+ \left[\begin{array}{c} :\overset{..}{O}: \\ :\overset{..}{O}:\overset{..}{S}:\overset{..}{O}: \\ :\overset{..}{O}: \end{array} \right]^{2-} Na^+ \qquad \left[\begin{array}{c} :\overset{..}{O}: \\ :\overset{..}{O}:\overset{..}{S}:\overset{..}{O}: \\ :\overset{..}{O}: \end{array} \right]^{2-}$$

Sodium sulfate   Sulfate ion

Sodium sulfate has both ionic and covalent bonds. Ionic bonds exist between each of the sodium ions and the sulfate ion. Covalent bonds are present between the sulfur and oxygen atoms within the sulfate ion. One important difference between the ionic and covalent bonds in this compound may be demonstrated by dissolving sodium sulfate in water. It dissolves in water, forming three charged particles—two sodium ions and one sulfate ion per formula unit of sodium sulfate:

$$\underset{\text{Sodium sulfate}}{Na_2SO_4} \xrightarrow{\text{Water}} \underset{\text{Sodium ions}}{2\,Na^+} + \underset{\text{Sulfate ion}}{SO_4^{2-}}$$

The ion SO_4^{2-} remains as a unit, held together by covalent bonds; whereas, where there were ionic bonds, dissociation of the ions took place. Do not think, however, that polyatomic ions are so stable that they cannot be altered. They may indeed be changed into other compounds or ions in certain chemical reactions.

The electron-dot formulas for several common polyatomic ions follow.

$$\left[\begin{array}{c} H \\ H:\overset{..}{N}:H \\ H \end{array} \right]^+ \qquad \left[\begin{array}{c} :\overset{..}{O}: \\ N::O: \\ :\overset{..}{O}: \end{array} \right]^- \qquad \left[\begin{array}{c} :\overset{..}{O}: \\ :\overset{..}{O}:\overset{..}{P}:\overset{..}{O}: \\ :\overset{..}{O}: \end{array} \right]^{3-} \qquad \left[:\overset{..}{O}:H \right]^-$$

Ammonium ion, NH_4^+ Nitrate ion, NO_3^- Phosphate ion, PO_4^{3-} Hydroxide ion, OH^-

7.11 Oxidation Numbers of Atoms

oxidation number

We have seen that atoms can combine to form compounds by losing, gaining, or sharing electrons. The **oxidation number** of an element is a number having a positive, a negative, or a zero value that may be assigned to an atom of that element in a compound. These positive and negative numbers are directly related to the positive and negative charges that result from the transfer of electrons from one atom to another in ionic compounds or from an unequal sharing of electrons between atoms forming covalent bonds. Oxidation numbers are assigned by a somewhat arbitrary system of rules. They are useful for writing formulas, naming compounds, and balancing chemical equations.

In a compound having ionic bonds, the oxidation number of an atom or group of atoms existing as an ion is the same as the *charge of the ion.* Thus, in sodium chloride, NaCl, the oxidation number of sodium is $+1$ and that of chlorine is -1; in magnesium oxide, MgO, the oxidation number of magnesium is $+2$ and that of oxygen is -2; in calcium chloride, $CaCl_2$, the oxidation number of calcium is $+2$ and that of chlorine is -1. The sum of the oxidation numbers of all the atoms in a compound is numerically equal to zero, since a compound is electrically neutral.

For practical purposes it is also convenient to assign oxidation numbers to the individual atoms comprising molecules and polyatomic ions. Here the electrons have not been completely transferred from one atom to another and the assignment cannot be done solely on the basis of ionic charges. However, oxidation numbers can be readily assigned to the atoms in either molecules or polyatomic ions by this general method: For each covalent bond, first assign the shared pair of electrons to the more electronegative atom. Then assign an oxidation number to each atom corresponding to its apparent net charge based (1) on the number of electrons gained or lost and (2) on the fact that the sum of the oxidation numbers must equal zero for a compound or must equal the charge on a polyatomic ion. Consider these substances, H_2, H_2O, CH_4, and CCl_4:

H:H	H:Ö: H	H H:C:H H	:Cl: :Cl: C :Cl: :Cl:
Electrons shared equally	Shared electrons assigned to oxygen	Shared electrons assigned to carbon	Shared electrons assigned to chlorine

In H_2, the pair of electrons is shared equally between the two atoms; therefore, each H is assigned an oxidation number of zero. In H_2O, oxygen is the more electronegative atom and is assigned the two pairs of shared electrons. The oxygen atom now has two additional electrons over the neutral atom, and therefore is assigned an oxidation number of -2. Each hydrogen atom in H_2O has one less electron than the neutral atom and is assigned an oxidation number of $+1$. In CH_4, all four shared pairs of electrons are assigned to the more electronegative carbon atom. The carbon atom then has an additional four electrons and is assigned an oxidation number of -4. Each hydrogen atom has one less electron than the neutral atom and is assigned an oxidation number of $+1$. In CCl_4, one pair of electrons is assigned to each of the four more electronegative chlorine atoms. The carbon atom therefore has four less electrons than the neutral atom and is assigned an oxidation number of $+4$. Each chlorine atom has one additional electron and is assigned an oxidation number of -1.

The following rules govern the assignment of oxidation numbers:

1. The oxidation number of any free element is zero, even when the atoms are combined with themselves (for example, Na, Mg, H_2, O_2).
2. Metals generally have positive oxidation numbers in compounds.
3. The oxidation number of hydrogen in a compound or an ion is $+1$ except in metal hydrides.
4. The oxidation number of oxygen in a compound or an ion is -2 except in peroxides.

5. The oxidation number of a monatomic ion is the same as the charge on the ion.
6. The oxidation number of an atom in a covalent compound is equal to the net apparent charge on the atom after each pair of shared electrons is assigned to the more electronegative element sharing the pair of electrons.
7. The algebraic sum of the oxidation numbers for all the atoms in a compound must equal zero.
8. The algebraic sum of the oxidation numbers for all the atoms in a polyatomic ion must equal the charge on the ion.

The oxidation numbers of many elements are predictable from their position in the periodic table. This is especially true of the Group A elements because the number of electrons in their outer shells corresponds to the group number. Remember that metals lose electrons, becoming positively charged ions. Nonmetals tend to gain electrons, and become negatively charged ions, but they can also share electrons with other atoms to assume a positive or negative oxidation number. Hydrogen can have a $+1$ or -1 oxidation number, depending on the relative electronegativity of the element with which it is combined.

The predictable oxidation numbers of the Group A elements are given in the following table:

	IA	IIA	IIIA	IVA	VA	VIA	VIIA
Oxidation number	$+1$	$+2$	$+3$	$+4$ to -4	-3 to $+5$	-2 to $+6$	-1 to $+7$

Table 7.5 illustrates the use of oxidation numbers to predict formulas of binary compounds from representative members of these groups.

TABLE 7.5 Selected binary hydrogen, oxygen, and chlorine compounds of Group A elements.

	IA	IIA	IIIA	IVA	VA	VIA	VIIA
Hydrogen compound	NaH	CaH_2	AlH_3	CH_4	NH_3	H_2S	HCl
Oxygen compound	Na_2O	CaO	Al_2O_3	CO_2	N_2O_5	SO_3	Cl_2O
Chlorine compound	NaCl	$CaCl_2$	$AlCl_3$	CCl_4	NCl_3	SCl_2	Cl_2

7.12 Oxidation Number Tables

The writing of formulas of compounds and chemical equations is facilitated by a knowledge of oxidation numbers and ionic charges. Table 7.6 lists the names and ionic charges of common monatomic ions. Monatomic ions of Group A

TABLE 7.6 Names, formulas, and charges of selected monatomic ions.

Name	Formula	Oxidation number or charge	Name	Formula	Oxidation number or charge
Aluminum	Al^{3+}	$3+$	Lead(II)	Pb^{2+}	$2+$
Arsenic(III)	As^{3+}	$3+$	Magnesium	Mg^{2+}	$2+$
Barium	Ba^{2+}	$2+$	Manganese(II)	Mn^{2+}	$2+$
Cadmium	Cd^{2+}	$2+$	Mercury(I)	Hg^+	$1+$
Calcium	Ca^{2+}	$2+$	Mercury(II)	Hg^{2+}	$2+$
Chromium(III)	Cr^{3+}	$3+$	Nickel(II)	Ni^{2+}	$2+$
Copper(I)	Cu^+	$1+$	Silver	Ag^+	$1+$
Copper(II)	Cu^{2+}	$2+$	Tin(II)	Sn^{2+}	$2+$
Iron(II)	Fe^{2+}	$2+$	Tin(IV)	Sn^{4+}	$4+$
Iron(III)	Fe^{3+}	$3+$	Zinc	Zn^{2+}	$2+$
Bromide	Br^-	$1-$	Nitride	N^{3-}	$3-$
Chloride	Cl^-	$1-$	Oxide	O^{2-}	$2-$
Fluoride	F^-	$1-$	Sulfide	S^{2-}	$2-$
Iodide	I^-	$1-$			

elements are not given in Table 7.6 because the charges and oxidation numbers of these ions are readily determined from the periodic table. The charges and oxidation numbers of the Groups IA, IIA, and IIIA metal ions are positive and correspond to the group number (for example, Na^+, Ca^{2+}, Al^{3+}). The negative charges and oxidation numbers of Groups VA, VIA, and VIIA monatomic ions can be determined by subtracting 8 from the group number. Sulfur, for example, is in Group VIA, and $6 - 8 = -2$. Therefore, the oxidation number of the sulfide ion (S^{2-}) is -2. All the halogens (F, Cl, Br, I) in binary compounds with metals or hydrogen have an oxidation number of -1.

The names, formulas, and ionic charges of some common polyatomic ions are given in Table 7.7. A more comprehensive list of both monatomic and polyatomic ions is given on the inside back cover of this book. Table 7.8 lists the principal oxidation numbers of common elements that have variable oxidation states.

TABLE 7.7 Names, formulas, and charges of some common polyatomic ions.

Name	Formula	Charge	Name	Formula	Charge
Acetate	$C_2H_3O_2^-$	$1-$	Cyanide	CN^-	$1-$
Ammonium	NH_4^+	$1+$	Dichromate	$Cr_2O_7^{2-}$	$2-$
Arsenate	AsO_4^{3-}	$3-$	Hydroxide	OH^-	$1-$
Bicarbonate	HCO_3^-	$1-$	Nitrate	NO_3^-	$1-$
Bisulfate	HSO_4^-	$1-$	Nitrite	NO_2^-	$1-$
Bromate	BrO_3^-	$1-$	Permanganate	MnO_4^-	$1-$
Carbonate	CO_3^{2-}	$2-$	Phosphate	PO_4^{3-}	$3-$
Chlorate	ClO_3^-	$1-$	Sulfate	SO_4^{2-}	$2-$
Chromate	CrO_4^{2-}	$2-$	Sulfite	SO_3^{2-}	$2-$

TABLE 7.8 Principal oxidation numbers of some common elements that have variable oxidation states.

Element	Oxidation number	Element	Oxidation number
Cu	$+1, +2$	Cl	$-1, +1, +3, +5, +7$
Hg	$+1, +2$	Br	$-1, +1, +3, +5, +7$
Sn	$+2, +4$	I	$-1, +1, +3, +5, +7$
Pb	$+2, +4$	S	$-2, +4, +6$
Fe	$+2, +3$	N	$-3, +1, +2, +3, +4, +5$
Au	$+1, +3$	P	$-3, +3, +5$
Ni	$+2, +3$	C	$-4, +4$
Co	$+2, +3$		
As	$+3, +5$		
Bi	$+3, +5$		
Cr	$+2, +3, +6$		

7.13 Formulas of Electrovalent Compounds

The sum of the oxidation numbers of all the atoms in a compound is zero. This statement applies to all substances, regardless of whether they are electrovalently or covalently bonded. For electrovalently bonded compounds the sum of the charges on all the ions in the compound must also be zero. Hence the formulas of ionic (electrovalently bonded) substances can be determined and written readily. Simply combine the ions in the simplest proportion so that the sum of the ionic charges adds up to zero.

To illustrate: Sodium chloride consists of Na^+ and Cl^- ions. Since $(1+) + (1-) = 0$, these ions combine in a one-to-one ratio, and the formula is written NaCl. Calcium fluoride is made up of Ca^{2+} and F^- ions; one Ca^{2+} and two F^- ions are needed, so the formula is CaF_2. Aluminum oxide is a bit more complicated, because it consists of Al^{3+} and O^{2-} ions. Since 6 is the lowest common multiple of 3 and 2, we have $2(3+) + 3(2-) = 0$; that is, two Al^{3+} ions and three O^{2-} ions are needed. The formula is therefore Al_2O_3.

The foregoing compounds all are made up of monatomic ions. The same procedure is used for polyatomic ions. Consider calcium hydroxide, which is made up of Ca^{2+} and OH^- ions. Since $(2+) + 2(1-) = 0$, one Ca^{2+} and two OH^- ions are needed, and the formula is $Ca(OH)_2$. The parentheses are used to indicate that the formula has two hydroxide ions. It is not correct to write CaO_2H_2 in place of $Ca(OH)_2$ because the identity of the compound would be lost by so doing. Note that the positive ion is written first in formulas. The table at the top of the next page provides examples of formula writing for ionic compounds (see Tables 7.6 and 7.7 or inside the back cover for formulas of common ions).

The sum of the charges on the ions of an electrovalently bonded compound must equal zero.

Name of compound	Ions	Lowest common multiple	Sum of charges on ions	Formula
Sodium bromide	Na^+, Br^-	1	$(1+) + (1-) = 0$	$NaBr$
Potassium sulfide	K^+, S^{2-}	2	$2(1+) + (2-) = 0$	K_2S
Aluminum fluoride	Al^{3+}, F^-	3	$(3+) + 3(1-) = 0$	AlF_3
Zinc sulfate	Zn^{2+}, SO_4^{2-}	2	$(2+) + (2-) = 0$	$ZnSO_4$
Ammonium phosphate	NH_4^+, PO_4^{3-}	3	$3(1+) + (3-) = 0$	$(NH_4)_3PO_4$
Aluminum chromate	Al^{3+}, CrO_4^{2-}	6	$2(3+) + 3(2-) = 0$	$Al_2(CrO_4)_3$

It is not always easy to distinguish between the terms *oxidation number* and *charge.* Oxidation numbers are assigned to atoms according to a set of rules. The charge on an ion is the actual electron excess or deficiency when compared to the neutral atom—or group of atoms, in the case of a polyatomic ion. Oxidation numbers may be assigned to all the atoms, including monatomic ions, in any compound, but only ions have charges. There is no problem with covalently bonded molecules such as methane (CH_4) because they contain no ions. The oxidation number of hydrogen is $+1$, and that of carbon is -4. Electrovalently bonded compounds are apt to be troublesome because the oxidation number and the charge on a monatomic ion have the same numerical value. In sodium chloride (NaCl), the oxidation number of the sodium ion is $+1$ and the charge is $1+$; the oxidation number of the chloride ions is -1 and the charge is $1-$.

In a compound composed of monatomic ions, such as sodium chloride, there is no practical difference in writing the formula regardless of whether oxidation numbers or ionic charges are used. But there is a difference with compounds containing polyatomic ions. Sodium sulfate (Na_2SO_4), for example, consists of two sodium ions (Na^+) and a polyatomic sulfate ion (SO_4^{2-}). The sum of the ionic charges is zero: $2(1+) + (2-) = 0$; it is convenient to make use of this fact in writing the formula. The charge on the polyatomic sulfate ion is not an oxidation number. However, the sum of the oxidation numbers of all the atoms in sodium sulfate is zero: $2(Na^+) + (S^{6+}) + 4(O^{2-}) = 0$; and the sum of the oxidation numbers of all the atoms in the sulfate ion is -2: $(S^{6+}) + 4(O^{2-}) = -2$ (see Section 7.14).

PROBLEM 7.6 Write formulas for (a) calcium chloride, (b) iron(III) sulfide, and (c) aluminum sulfate. Refer to Tables 7.6 and 7.7 as needed.

(a) **Step 1.** From the name we know that calcium chloride is composed of calcium and chloride ions. First write down the formulas of these ions:

Ca^{2+} and Cl^-

Step 2. To write the formula of the compound, combine the smallest numbers of Ca^{2+} and Cl^- ions to give a charge sum equal to zero. In this case, the lowest common multiple of the charges is 2:

$$(Ca^{2+}) + 2(Cl^-) = 0$$

$$(2+) + 2(1-) = 0$$

Therefore, the formula is $CaCl_2$.

(b) Use the same procedure for iron(III) sulfide.

Step 1. Write down the formulas for the iron(III) and sulfide ions:

$$Fe^{3+} \quad \text{and} \quad S^{2-}$$

Step 2. Use the smallest numbers of these ions required to give a charge sum equal to zero. The lowest common multiple of the charges is 6:

$$2(Fe^{3+}) + 3(S^{2-}) = 0$$

$$2(3+) + 3(2-) = 0$$

Therefore, the formula is Fe_2S_3.

(c) Use the same procedure for aluminum sulfate.

Step 1. Write down the formulas for the aluminum and sulfate ions:

$$Al^{3+} \quad \text{and} \quad SO_4^{2-}$$

Step 2. Use the smallest numbers of these ions required to give a charge sum equal to zero. The lowest common multiple of the charges is 6:

$$2(Al^{3+}) + 3(SO_4^{2-}) = 0$$

$$2(3+) + 3(2-) = 0$$

Therefore, the formula is $Al_2(SO_4)_3$. Note the use of parentheses around the SO_4^{2-} ion.

■

7.14 Determining Oxidation Numbers and Ionic Charges from a Formula

If the formula of a compound is known, the oxidation number of an element or the charge on a polyatomic ion in the formula can often be determined by algebraic difference. To begin, you must know the oxidation numbers of a few elements. Excellent ones with which to work are hydrogen, H^+, always $+1$ except in hydrides (a hydride is a compound of hydrogen and a metal); oxygen, O^{2-}, always -2 except in peroxides; and sodium, Na^+, always $+1$. Using the compound sulfuric acid, H_2SO_4, as an example, let us determine the charge of the sulfate ion and the oxidation number of the sulfur atom. The sulfate ion is combined with two hydrogen atoms, each with a $+1$ oxidation number. The sulfate ion must then have a -2 charge for the net charge in the compound to be zero:

$$
\begin{array}{ll}
H^+ & +1 \\
H^+ & +1 \\
SO_4^{2-} & -2 \\
\hline
& 0
\end{array}
$$

To find the oxidation number of sulfur, we proceed as follows:

Step 1. Write the oxidation number of a single atom of hydrogen and a single atom of oxygen below the atoms in the formula.

Step 2. Below this, write the sums of the oxidation numbers of all the H and O atoms: $2(+1) = +2$ and $4(-2) = -8$.

Step 3. Then add the total oxidation numbers of all the atoms, including the sulfur atom, and set them equal to zero: $+2 + S + (-8) = 0$. Solving the equation for S, we determine that the oxidation number of sulfur is $+6$, the value needed to give the sum of zero.

$$
\begin{array}{cccc}
& H_2 & S & O_4 \\
\textbf{Step 1.} & +1 & & -2 \\
\textbf{Step 2.} & 2(+1) = +2 & & 4(-2) = -8 \\
\textbf{Step 3.} & & +2 + S + (-8) = 0 \\
& & S = +6
\end{array}
$$

The oxidation number of sulfur in H_2SO_4 is $+6$.

What is the oxidation number of chromium in sodium dichromate, $Na_2Cr_2O_7$? Using the same method as for H_2SO_4, we have

$$
\begin{array}{cccc}
& Na_2 & Cr_2 & O_7 \\
\textbf{Step 1.} & +1 & & -2 \\
\textbf{Step 2.} & 2(+1) = +2 & & 7(-2) = -14 \\
\textbf{Step 3.} & & +2 + 2\,Cr + (-14) = 0 \\
& & 2\,Cr = +12 \\
& & Cr = +6
\end{array}
$$

The oxidation number of chromium in $Na_2Cr_2O_7$ is $+6$.

The formula of radium chloride is $RaCl_2$. What is the oxidation number of radium? If you remember that the oxidation number of chloride is -1, then the value for radium is $+2$, since one radium ion is combined with two Cl^- ions. If you do not remember the oxidation number of chloride, then you should try to recall the formula of another chloride. One that might come to mind is sodium chloride, NaCl, in which the chloride is -1 because of its combination with one sodium ion of $+1$ This recollection establishes the oxidation number of chloride, which then enables you to calculate the value for radium.

What is the oxidation number of phosphorus in the phosphate ion, PO_4^{3-}? First of all, note that this is a polyatomic ion with a charge of -3. The sum of the oxidation numbers of phosphorus and oxygen must equal -3 and not zero. Four oxygen atoms, each with a -2, give a total of -8. The oxidation number of the phosphorus atom must then be $+5$ ($P - 8 = -3$) for the charge of the ion to be -3.

$$P \quad O_4^{3-}$$
$$-2$$
$$P + 4(-2) = -3$$
$$P = +5$$

The sum of the oxidation numbers of the atoms in a polyatomic ion must equal the charge of the polyatomic ion.

7.15 Oxidation–Reduction: An Introduction

Magnesium burns brilliantly in air, forming magnesium oxide:

$$2\,Mg^0 + O_2^0 \longrightarrow 2\,Mg^{2+}O^{2-}$$

In this reaction, two electrons are transferred from each magnesium atom to the oxygen atoms, resulting in an increase in the oxidation number of magnesium from 0 to $+2$. At the same time, the oxidation number of oxygen has decreased from 0 to -2. Oxidation and reduction have occurred in this reaction. The Mg^0 was oxidized and the O_2^0 was reduced. **Oxidation** is defined as an increase in the oxidation number or oxidation state of an element. **Reduction** is defined as a decrease in the oxidation number or oxidation state of an element. *Oxidation–reduction* occurs as a result of a loss and gain of electrons. Oxidation and reduction occur simultaneously. The element that loses electrons (increases in oxidation number) is oxidized, and the element that gains electrons (decreases in oxidation number) is reduced.

oxidation
reduction

$$2\,Hg^{2+}O^{2-} \xrightarrow{\Delta} 2\,Hg^0 + O_2^0$$

In the decomposition of mercury(II) oxide, the oxidation number of mercury changes from $+2$ to 0; the oxidation number of oxygen changes from -2 to 0. Therefore, oxidation–reduction occurs; oxygen (O^{2-}) is oxidized and mercury (Hg^{2+}) is reduced.

The process of oxidation and reduction involves the transfer of electrons and results in changes in oxidation numbers. The element oxygen is not necessarily involved in this process. For example, in the chemical reaction between sodium and chlorine to form sodium chloride, electrons are transferred from sodium atoms to chlorine atoms:

$$2\,Na^0 + Cl_2^0 \longrightarrow 2\,Na^+Cl^-$$

In this reaction, the oxidation number of sodium increases from 0 to $+1$, and therefore sodium is oxidized. The oxidation number of chlorine decreases from 0 to -1, and, consequently, chlorine is reduced.

A more detailed discussion of oxidation–reduction is given in Chapter 17.

QUESTIONS   **A. Review the Meanings of the New Terms Introduced in this Chapter**

1. Chemical bond
2. Ionization energy
3. Valence electrons
4. Ionic bond
5. Electrovalent bond
6. Covalent bond
7. Electronegativity
8. Nonpolar covalent bond
9. Polar covalent bond
10. Dipole
11. Coordinate covalent bond
12. Polyatomic ion
13. Oxidation number
14. Oxidation
15. Reduction

B. Answers to the Following Questions Will Be Found in Tables and Figures

1. Explain the large increase in ionization energy needed to remove the second electron from a lithium atom compared to that needed to remove the first electron. (See Table 7.1.)
2. Arrange the following elements in order of increasing attraction by which their valence electrons are held in the atom: aluminum, sulfur, silicon, magnesium, chlorine, phosphorus. argon, and sodium.
3. In which general areas of the periodic table are the elements with the lowest and the highest ionization energies and electronegativities located?
4. Using the table of electronegativities (Table 7.4), indicate which element is positive and which is negative in the following compounds:
   (a) NaCl   (e) MgH_2 (i) Cl_2O (m) NO
   (b) SO_2 (f) H_2S (j) PCl_3 (n) CCl_4
   (c) FeS   (g) NH_3 (k) H_2O (o) BF_3
   (d) CaI_2 (h) HCl   (l) OF_2 (p) LiH
5. Classify the bond between the following pairs of atoms as either principally ionic or covalent (use Table 7.4):
   (a) Phosphorus and hydrogen   (d) Hydrogen and chlorine
   (b) Sodium and fluorine   (e) Magnesium and oxygen
   (c) Chlorine and carbon   (f) Hydrogen and sulfur
6. Using the principle employed in Table 7.5, write formulas for:
   (a) The hydrogen compounds of Li, C, I, N, Sr
   (b) The oxygen compounds of Ca, Si, Br, Al, P
   (c) The iodine compounds of Li, Al, C, Sr, S
7. Use the oxidation number tables and determine the formulas for compounds composed of the following ions:
   (a) Sodium and chlorate   (d) Copper(I) and oxide
   (b) Hydrogen and sulfate   (e) Zinc and bicarbonate
   (c) Tin(II) and acetate   (f) Iron(III) and carbonate

C. Review Questions

1. Write an equation representing the change of a fluorine atom to a fluoride ion (F^-).
2. Write an equation representing the change of a calcium atom to a calcium ion (Ca^{2+}).
3. Explain why, in general, the ionization energy decreases from top to bottom in a family of elements.
4. Explain what happens to the electron structures of Mg and Cl atoms when they react to form $MgCl_2$.
5. Why does barium (Ba) have a lower ionization energy than beryllium (Be)?
6. Why is there such a large increase in the ionization energy required to remove the second electron from a sodium atom?
7. How many electrons must be gained or lost for each of the following to achieve a noble gas electron structure?
    (a) A calcium atom   (c) A helium atom   (e) A nitrogen atom
    (b) A sulfur atom   (d) A chloride ion   (f) A potassium atom
8. Explain why potassium forms a K^+ ion but not a K^{2+} ion.
9. What portion of an atom is represented by the kernel?
10. Why does an aluminum ion have a $+3$ charge?
11. Which would be larger, a potassium atom or a potassium ion? Explain.
12. Which would be larger, a bromine atom or a bromide ion? Explain.
13. Which would be larger, a magnesium ion or an aluminum ion? Explain.
14. Which would be larger, Fe^{2+} or Fe^{3+}? Explain.
15. What causes a bond to be polar?
16. How does a coordinate covalent bond differ from an ordinary covalent bond?
17. How does a covalent bond differ from an ionic bond?
18. Classify the bonding in each compound as ionic or covalent:
    (a) H_2O (c) MgO   (e) HCl   (g) NH_3
    (b) NaCl   (d) Br_2 (f) $BaCl_2$ (h) SO_2
19. Draw electron-dot structures for:
    (a) Mg   (d) NH_3 (g) H_2S (j) KI   (m) CS_2
    (b) Br_2 (e) CO_2 (h) BaO   (k) Cl_2O (n) CO
    (c) H_2O (f) HCl   (i) CaF_2 (l) C_3H_8
20. Draw electron-dot structures for:
    (a) Ba^{2+} (c) I^- (e) SO_4^{2-} (g) CN^- (i) ClO_3^-
    (b) Al^{3+} (d) S^{2-} (f) SO_3^{2-} (h) CO_3^{2-} (j) NO_3^-
21. The electron-dot structure for chloric acid is

$$H:\overset{..}{\underset{..}{O}}:\overset{..}{\underset{..}{Cl}}:\overset{..}{\underset{..}{O}}:$$
$$:\overset{}{\underset{..}{O}}:$$

Point out the covalent and coordinate covalent bonds in this structure.
22. Draw the electron-dot structure for ammonia (NH_3). What type of bond is present? Can this molecule form coordinate covalent bonds? If so, how many?
23. Rank these elements from highest electronegativity to lowest: Mg, S, F, H, O, Cs.
24. Classify the following molecules as polar or nonpolar:
    (a) H_2O (b) HBr   (c) CF_4 (d) F_2 (e) CO_2 (f) NH_3
25. Is it possible for a molecule to be nonpolar even though it contains polar covalent bonds? Explain.
26. Why is CO_2 a nonpolar molecule, whereas CO is a polar molecule?
27. Determine the oxidation number of each element in the following:
    (a) HCl   (c) $FeCl_2$ (e) CO_2 (g) NH_3
    (b) MgO   (d) Br_2 (f) Cr_2O_3 (h) SO_3

28. Determine the oxidation number of the element underlined in each formula:
    (a) $\underline{Mn}CO_3$ (c) $K\underline{N}O_3$ (e) $Ba\underline{C}O_3$ (g) $\underline{W}Cl_5$
    (b) $\underline{Sn}F_4$ (d) $K\underline{Mn}O_4$ (f) $\underline{P}Cl_3$ (h) $K_2\underline{Cr}_2O_7$
29. Determine the oxidation number of the element underlined in each formula:
    (a) $\underline{In}I_3$ (c) $Na_2\underline{S}O_4$ (e) $Mg(\underline{N}O_3)_2$ (g) $\underline{Fe}_2(CO_3)_3$
    (b) $K\underline{Cl}O_3$ (d) $\underline{C}_2H_5OH$ (f) $\underline{Sn}O_2$ (h) $Na\underline{Cl}O_4$
30. Write the formula of the compound that would be formed between the given elements:
    (a) Na and I   (c) Al and O   (e) Cs and Cl
    (b) Ba and F   (d) K and S   (f) Sr and Br
31. Write the formula of the compound that would be formed between the given elements:
    (a) Ba and O   (c) Ga and Cl   (e) Li and Si
    (b) H and S   (d) Be and Br   (f) Mg and P
32. Does the fact that two elements combine in a one-to-one atomic ratio mean that their oxidation numbers are both 1? Explain.
33. What might be the formula of a compound formed between elements X and Z where:
    (a) X has 1 electron and Z has 6 electrons in the outer energy level.
    (b) X has 4 electrons and Z has 7 electrons in the outer energy level.
    (c) X has 1 electron and Z has 7 electrons in the outer energy level.
34. Which of these statements are correct? (Try to answer this question without referring to your book.)
    (a)   The amount of energy required to remove one electron from an atom is known as the ionization energy.
    (b)   Metallic elements tend to have relatively low electronegativities.
    (c)   Elements with a high ionization energy tend to have very metallic properties.
    (d)   Sodium and chlorine react to form molecules of NaCl.
    (e)   A chlorine atom has fewer electrons than a chloride ion.
    (f)   The noble gases have a tendency to lose one electron to become positively charged ions.
    (g)   The chemical bonds in a water molecule are ionic.
    (h)   The chemical bonds in a water molecule are polar.
    (i)   Valence electrons are those electrons in the outermost shell of an atom.
    (j)   An atom with eight electrons in its outer shell has all its s and p orbitals filled.
    (k)   Fluorine has the lowest electronegativity of all the elements.
    (l)   Oxygen has a greater electronegativity than carbon.
    (m)   In general, within a group or family, electronegativity increases as atomic size increases.
    (n) Cl_2 is more ionic in character than HCl.
    (o)   A neutral atom with eight electrons in its valence shell must be an atom of a noble gas.
    (p)   A nitrogen atom has four valence electrons.
    (q)   An aluminum atom must lose three electrons to become an aluminum ion, Al^{3+}.
    (r)   A stable group of atoms that has either a positive or a negative charge and behaves as a single unit in many chemical reactions is called a polyatomic ion.
    (s)   Sodium sulfate, Na_2SO_4, has covalent bonds between sulfur and the oxygen atoms, and ionic bonds between sodium ions and the sulfate ion.
    (t)   The water molecule is a dipole.

(u)   In an ethylene molecule, C_2H_4,

$$\begin{matrix} H \\ \\ H \end{matrix} \!\! \diagdown \!\!\!\! \diagup \, C = C \, \diagup \!\!\!\! \diagdown \!\! \begin{matrix} H \\ \\ H \end{matrix}$$

two pairs of electrons are shared between the carbon atoms.

(v)   The octet rule is mainly useful for atoms where only s and p electrons enter into bonding.

(w)   When electrons are transferred from one atom to another, the resulting compound contains ionic bonds.

(x)   A phosphorus atom, $\cdot \ddot{P} \cdot$, needs three additional electrons to attain a stable octet of electrons.

(y)   The simplest compound between oxygen, $\cdot \ddot{O} \cdot$, and fluorine, $: \ddot{F} \cdot$, atoms is FO_2.

(z)   In the molecule $H : \ddot{\underset{..}{Cl}} :$, there are three unshared pairs of electrons.

(aa)   The smaller the difference in electronegativity between two atoms, the more ionic the bond between them will be.

(bb)   In the reaction $2\,H_2O \longrightarrow 2\,H_2 + O_2$, hydrogen is oxidized and oxygen is reduced.

(cc)   Oxidation occurs when an atom loses electrons.

(dd)   In the oxide WO_3, the oxidation number of tungsten is $+6$ and that of oxygen is -2.

(ee)   In the compound $SrCO_3$, the oxidation number of carbon is $+5$.

(ff)   Electron-dot structures are mainly useful for the representative elements.

(gg)   The correct electron-dot structure for NH_3 is $H : \overset{..}{\underset{\underset{\displaystyle H}{\vdots}}{N}} : H$

(hh)   The correct electron-dot structure for CO_2 is $: \overset{..}{\underset{..}{O}} : C : \overset{..}{\underset{..}{O}} :$

(ii)   The correct electron-dot structure for SO_4^{2-} is $: \overset{..}{\underset{..}{O}} : \overset{..}{\underset{..}{O}} : \overset{..}{\underset{..}{S}} : \overset{..}{\underset{..}{O}} : \overset{..}{\underset{..}{O}} :$

(jj)   In period 4 of the periodic table, the element having the lowest ionization energy is Xe.

(kk)   There are four valence electrons in an atom having an electron structure of $1s^2 2s^2 2p^6 3s^2 3p^2$.

(ll)   When an atom of bromine becomes a bromide ion, its size increases.

(mm)   The structures that show that H_2O is a dipole and that CO_2 is not a dipole

are $H - \overset{..}{\underset{..}{O}} - H$ and $: \overset{..}{O} = C = \overset{..}{O} :$

(nn)   The ions Cl^- and S^{2-} have the same electron structure.

D. Review Exercises

1. (a) In terms of electron structure, why is the oxidation number of nitrogen never higher than $+5$ or lower than -3?

   (b) What are the highest and lowest possible oxidation states for bromine?

2. Why do chemical bonds form?

8

Nomenclature of Inorganic Compounds

After studying Chapter 8 you should be able to:

1. Give the name or formula for inorganic binary compounds in which the metal has only one common oxidation state.
2. Give the name or formula for inorganic binary compounds that contain metals of variable oxidation state, using either the Stock System or classical nomenclature.
3. Give the name or formula for inorganic binary compounds that contain two nonmetals.
4. Give the name or formula for binary acids.
5. Give the name or formula for ternary salts.
6. Give the name or formula for ternary inorganic oxy-acids.
7. Given the formula of a salt, write the name and formula of the acid from which the salt may be derived.
8. Give the name or formula for salts that contain more than one positive ion.
9. Give the name or formula for inorganic bases.

Before starting this chapter, it is suggested that you review the interpretation of chemical formulas given in Section 4.7.

8.1 Common, or Trivial, Names

Chemical nomenclature is the system of names that chemists use to identify compounds. When a new substance is formulated, it must be named in order to distinguish it from all other substances. Before chemistry was systematized, a substance was given a name that generally associated it with one of its outstanding physical or chemical properties. For example, *quicksilver* is a common name for mercury; it describes two properties of mercury—a silvery appearance and a quick, liquidlike movement. Nitrous oxide, N_2O, used as an anesthetic in dentistry, has been called *laughing gas* because it induces laughter when inhaled. The name *nitrous oxide* is now giving way to the more systematic name *dinitrogen oxide*. Nonsystematic names are called *common,* or *trivial,* names.

Common names for chemicals are widely used in many industries, since the systematic name frequently is too long or too technical for everyday use. For example, CaO is called *lime,* not *calcium oxide,* by plasterers; photographers refer to $Na_2S_2O_3$ as *hypo,* rather than *sodium thiosulfate;* nutritionists refer to

$C_{27}H_{44}O$ as *vitamin D_3*, not as *9,10-secocholesta-5,7,10(19)-trien-3β-ol*. These common names are chemical nicknames, and, as the vitamin D_3 example shows, there is a practical need for short, common names. Table 8.1 lists the common names, formulas, and chemical names of some familiar substances.

TABLE 8.1 Common names, formulas, and chemical names of some familiar substances.

Common name	Formula	Chemical name
Acetylene	C_2H_2	Ethyne
Lime	CaO	Calcium oxide
Slaked lime	$Ca(OH)_2$	Calcium hydroxide
Water	H_2O	Water
Galena	PbS	Lead(II) sulfide
Alumina	Al_2O_3	Aluminum oxide
Baking soda	$NaHCO_3$	Sodium hydrogen carbonate
Cane or beet sugar	$C_{12}H_{22}O_{11}$	Sucrose
Blue stone, blue vitriol	$CuSO_4 \cdot 5H_2O$	Copper(II) sulfate pentahydrate
Borax	$Na_2B_4O_7 \cdot 10H_2O$	Sodium tetraborate decahydrate
Brimstone	S	Sulfur
Calcite, marble, limestone	$CaCO_3$	Calcium carbonate
Cream of tartar	$KHC_4H_4O_6$	Potassium hydrogen tartrate
Epsom salts	$MgSO_4 \cdot 7H_2O$	Magnesium sulfate heptahydrate
Gypsum	$CaSO_4 \cdot 2H_2O$	Calcium sulfate dihydrate
Grain alcohol	C_2H_5OH	Ethyl alcohol, ethanol
Hypo	$Na_2S_2O_3$	Sodium thiosulfate
Laughing gas	N_2O	Dinitrogen oxide
Litharge	PbO	Lead(II) oxide
Lye, caustic soda	NaOH	Sodium hydroxide
Milk of magnesia	$Mg(OH)_2$	Magnesium hydroxide
Muriatic acid	HCl	Hydrochloric acid
Oil of vitriol	H_2SO_4	Sulfuric acid
Plaster of paris	$CaSO_4 \cdot \frac{1}{2}H_2O$	Calcium sulfate hemihydrate
Potash	K_2CO_3	Potassium carbonate
Pyrites (fool's gold)	FeS_2	Iron disulfide
Quicksilver	Hg	Mercury
Sal ammoniac	NH_4Cl	Ammonium chloride
Saltpeter (chile)	$NaNO_3$	Sodium nitrate
Table salt	NaCl	Sodium chloride
Washing soda	$Na_2CO_3 \cdot 10H_2O$	Sodium carbonate decahydrate
Wood alcohol	CH_3OH	Methyl alcohol, methanol

8.2 Systematic Chemical Nomenclature

The trivial name is not entirely satisfactory to the chemist, who requires a name that will identify the composition of each substance precisely. Therefore, as the number of known compounds increased, it became more and more necessary to develop a scientific, systematic method of identifying compounds by name. The systematic method of naming inorganic compounds considers the compound to be composed of two parts, one positive and one negative. The positive part, which is either a metal, hydrogen, or another positively charged group, is named and written first. The negative part, generally nonmetallic, follows. The names of the elements are modified with suffixes and prefixes to identify the different types or classes of compounds. Thus, the compound composed of sodium ions and chloride ions is named sodium chloride; the compound composed of calcium ions and bromide ions is named calcium bromide; the compound composed of iron(II) ions and chloride ions is named iron(II) chloride (read as "iron-two chloride").

We will consider the naming of acids, bases, salts, and oxides. Refer to Tables 7.6, 7.7, and 7.8 for the names, formulas, and oxidation numbers of ions. For handy, quick reference, the names and formulas of some common ions are given on the inside back cover of this book.

Before you proceed with the naming of different types of compounds, study the following rules:

Rule 1. The usual oxidation number of Group IA elements is $+1$.

Rule 2. The usual oxidation number of Group IIA elements is $+2$.

Rule 3. The usual oxidation number of Group IIIA elements is $+3$.

Rule 4. Elements in all the other periodic groups (except the noble gases) have variable oxidation numbers. Recall, however, that in practically all compounds, combined oxygen has an oxidation number of -2 and the halogens (Group IIIA) have an oxidation number of -1 when combined with metal ions in binary compounds.

8.3 Binary Compounds

Binary compounds contain only two different elements. Their names consist of two parts: the name of the more electropositive element followed by the name of the electronegative element, which is modified to end in *ide*. (The names of nonbinary compounds that use the *ide* ending but are exceptions to the rule are discussed in part 4 of this section.)

1. Binary compounds in which the electropositive element has a fixed oxidation state.   The majority of these compounds contain a metal and a nonmetal. The chemical name is composed of the name of the metal, which is written first, followed by the name of the nonmetal, which has been modified to an identifying stem plus the suffix *ide*. For example, sodium chloride, NaCl, is composed of one atom each of sodium and chlorine. The name of the metal, sodium, is written first and is not modified. The second part of the name is derived from the nonmetal, chlorine, by using the stem *chlor* and adding the ending *ide;* it is named *chloride*. The compound name is sodium chloride.

NaCl

Elements:	Sodium (metal)
	Chlorine (nonmetal)
	name modified to the stem *chlor* + *ide*
Name of compound:	Sodium chloride

Stems of the more common negative-ion-forming elements are shown in the following table.

Symbol	Element	Stem	Binary name ending
B	Boron	Bor	Boride
Br	Bromine	Brom	Bromide
Cl	Chlorine	Chlor	Chloride
F	Fluorine	Fluor	Fluoride
H	Hydrogen	Hydr	Hydride
I	Iodine	Iod	Iodide
N	Nitrogen	Nitr	Nitride
O	Oxygen	Ox	Oxide
P	Phosphorus	Phosph	Phosphide
S	Sulfur	Sulf	Sulfide

Compounds may contain more than one atom of the same element, but as long as they contain only two different elements, and if only one compound of these two elements exists, the name follows the rule for binary compounds:

Examples: $CaBr_2$ Mg_3N_2 KI
                Calcium bromide   Magnesium nitride   Potassium iodide

Table 8.2 shows more examples of compounds with names ending in *ide.*

TABLE 8.2 Examples of compounds with names ending in *ide*.

Formula	Name
$MgBr_2$	Magnesium bromide
Na_2O	Sodium oxide
NaH	Sodium hydride
HCl	Hydrogen chloride
HI	Hydrogen iodide
CaC_2	Calcium carbide
$AlCl_3$	Aluminum chloride
PbS	Lead(II) sulfide
LiI	Lithium iodide
Al_2O_3	Aluminum oxide

2. Binary compounds containing metals of variable oxidation numbers. Two systems are commonly used for compounds in this category. The official system, designated by the International Union of Pure and Applied Chemistry (IUPAC), is known as the *Stock System.* In the Stock System, when a compound contains a metal that can have more than one oxidation number, the oxidation number of the metal in the compound is designated by a roman numeral written in parentheses immediately after the name of the metal. The negative element is treated in the usual manner for binary compounds.

Oxidation number	Roman numeral
+1	(I)
+2	(II)
+3	(III)
+4	(IV)

Examples:

$FeCl_2$	Iron(II) chloride	Fe^{2+}
$FeCl_3$	Iron(III) chloride	Fe^{3+}
CuCl	Copper(I) chloride	Cu^+
$CuCl_2$	Copper(II) chloride	Cu^{2+}

The fact that $FeCl_2$ has two chloride ions, each with -1 charge, establishes that the oxidation number of Fe is $+2$. To distinguish between the two iron chlorides, $FeCl_2$ is named iron(II) chloride and $FeCl_3$ is named iron(III) chloride.

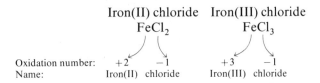

When a metal has only one possible oxidation state, there is no need to distinguish one oxidation state from another, so roman numerals are not needed. Thus, we do not say calcium(II) chloride for $CaCl_2$, but rather calcium chloride, since the oxidation number of calcium is understood to be $+2$.

In classical nomenclature, when the metallic ion has only two oxidation numbers, the name of the metal is modified with the suffixes *ous* and *ic* to distinguish between the two. The lower oxidation state is given the *ous* ending, and the higher one is given the *ic* ending.

Examples:

$FeCl_2$	Ferrous chloride	Fe^{2+}
$FeCl_3$	Ferric chloride	Fe^{3+}
CuCl	Cuprous chloride	Cu^+
$CuCl_2$	Cupric chloride	Cu^{2+}

Table 8.3 (page 178) lists some common metals having more than one oxidation number.

Notice that the *ous–ic* naming system does not give the oxidation state of an element but merely indicates that at least two oxidation states exist. The Stock System avoids any possible uncertainty by clearly stating the oxidation number.

3. Binary compounds containing two nonmetals.   The chemical bond that exists between two nonmetals is predominantly covalent. In a covalent compound, positive and negative oxidation numbers are assigned to the elements according to their electronegativities. The most electropositive element is named

TABLE 8.3 Names and oxidation numbers of some common metal ions that have more than one oxidation number.

Formula	Stock System name	Classical name
Cu^{1+}	Copper(I)	Cuprous
Cu^{2+}	Copper(II)	Cupric
$Hg^{1+}(Hg_2)^{2+}$	Mercury(I)	Mercurous
Hg^{2+}	Mercury(II)	Mercuric
Fe^{2+}	Iron(II)	Ferrous
Fe^{3+}	Iron(III)	Ferric
Sn^{2+}	Tin(II)	Stannous
Sn^{4+}	Tin(IV)	Stannic
As^{3+}	Arsenic(III)	Arsenous
As^{5+}	Arsenic(V)	Arsenic
Sb^{3+}	Antimony(III)	Stibnous
Sb^{5+}	Antimony(V)	Stibnic
Ti^{3+}	Titanium(III)	Titanous
Ti^{4+}	Titanium(IV)	Titanic

first. In a compound between two nonmetals, the element that occurs earlier in the following sequence is written and named first:

B, Si, C, P, N, H, S, I, Br, Cl, O, F

A Latin or Greek prefix is attached to each element to indicate the number of atoms of that element in the molecule. The second element still retains the modified binary ending. The prefix *mono* is generally omitted, except when needed to distinguish between two or more compounds, such as carbon monoxide, CO, and carbon dioxide, CO_2. Some common prefixes and their numerical equivalences follow:

Mono	= 1	Hexa	= 6
Di	= 2	Hepta	= 7
Tri	= 3	Octa	= 8
Tetra	= 4	Nona	= 9
Penta	= 5	Deca	= 10

Here are some examples of compounds that illustrate this system:

CO	Carbon monoxide
CO_2	Carbon dioxide
PCl_3	Phosphorus trichloride
PCl_5	Phosphorus pentachloride
P_2O_5	Diphosphorus pentoxide
CCl_4	Carbon tetrachloride
N_2O	Dinitrogen monoxide
S_2Cl_2	Disulfur dichloride
N_2O_4	Dinitrogen tetroxide
NO	Nitrogen monoxide (nitric oxide)
N_2O_3	Dinitrogen trioxide

$$N_2O_3$$

Dinitrogen   Trioxide

Indicates two   Indicates three
nitrogen atoms   oxygen atoms

4. Exceptions that use *ide* endings.   Three notable exceptions that use the *ide* ending are hydroxides (OH^-), cyanides (CN^-), and ammonium (NH_4^+) compounds. These polyatomic ions, when combined with another element, take the ending *ide* even though more than two elements are present in the compound.

NH_4I Ammonium iodide
$Ca(OH)_2$ Calcium hydroxide
KCN   Potassium cyanide

5. Acids derived from binary compounds.   Certain binary hydrogen compounds, when dissolved in water, form solutions that have *acid* properties. Because of this property, these compounds are given acid names in addition to their regular *ide* names. For example, HCl is a gas and is called *hydrogen chloride,* but its water solution is known as *hydrochloric acid.* Binary acids are composed of hydrogen and one other nonmetallic element. However, not all binary hydrogen compounds are acids. To express the formula of a binary acid, it is customary to write the symbol of hydrogen first, followed by the symbol of the second element (for example, HCl, HBr, H_2S). When we see formulas such as CH_4 or NH_3, we understand that these compounds are not normally considered to be acids.

To name a binary acid, place the prefix *hydro* in front of, and the suffix *ic* after, the stem of the nonmetal. Then add the word *acid.*

	HCl	H_2S
Examples:	*Hydro* chlor/ic acid (hydrochloric acid)	*Hydro* sulfur/ic acid (hydrosulfuric acid)

Acids are hydrogen-containing substances that liberate hydrogen ions when dissolved in water. The same formula is often used to express binary hydrogen compounds such as HCl, regardless of whether they are dissolved in water. Table 8.4 shows examples of binary acids.

TABLE 8.4 Names and formulas of selected binary acids.

Formula	Acid name
HF	Hydrofluoric acid
HCl	Hydrochloric acid
HBr	Hydrobromic acid
HI	Hydriodic acid
H_2S	Hydrosulfuric acid
H_2Se	Hydroselenic acid

8.4 Ternary Compounds

Ternary compounds contain three elements and consist of an electropositive group, which is either a metal or hydrogen, and a polyatomic negative ion. We will consider the naming of compounds in which one of the three elements is oxygen.

In general, in naming ternary compounds the positive group is given first, followed by the name of the negative ion. Rules for naming the positive groups are identical to those used in naming binary compounds. The negative group usually contains two elements: oxygen and a metal or a nonmetal. To name the polyatomic negative ion, add the endings *ite* or *ate* to the stem of the element other than oxygen. Thus, SO_4^{2-} is called *sulfate,* and SO_3^{2-} is called *sulfite.* Note that oxygen is not specifically included in the name, but is understood to be present when the endings *ite* and *ate* are used. The suffixes *ite* and *ate* represent different oxidation states of the element other than oxygen in the polyatomic ion. The *ite* ending represents the lower and the *ate* the higher oxidation state.

ite	*ate*
$\overset{+4}{CaSO_3}$	$\overset{+6}{CaSO_4}$
Calcium sulfite	Calcium sulfate
ite ending indicates lower oxidation state of sulfur in SO_3^{2-}	*ate* ending indicates higher oxidation state of sulfur in SO_4^{2-}

When an element has only one oxidation state, such as C in carbonate, the *ate* ending is used. Examples of ternary compounds and their names are given in Table 8.5.

Ternary oxy-acids.   Inorganic ternary compounds containing hydrogen, oxygen, and one other element are called *oxy-acids.* The element other than hydrogen or oxygen in these acids is usually a nonmetal, but in some cases it can be a metal. The *ous–ic* system is used in naming ternary acids. The suffixes *ous* and *ic* are used to indicate different oxidation states of the element other than hydrogen and oxygen. The *ous* ending again indicates the lower oxidation state and the *ic* ending indicates the higher oxidation state.

To name these acids, we place the ending *ic* or *ous* after the stem of the element other than hydrogen and oxygen, and add the word *acid.* If an element has only one usual oxidation state, the *ic* ending is used. Hydrogen in a ternary oxy-acid is not specifically designated in the acid name but its presence is implied by the use of the word *acid.*

Examples: H_2SO_3 Sulfur/*ous acid*   (S is $+4$)
H_2SO_4 Sulfur/*ic acid*   (S is $+6$)

TABLE 8.5 Names and formulas of selected ternary compounds.

Formula	Name
Na_2SO_3	Sodium sulfite
Na_2SO_4	Sodium sulfate
K_2CO_3	Potassium carbonate
$CaCO_3$	Calcium carbonate
$Al_2(SO_4)_3$	Aluminum sulfate
$KClO_3$	Potassium chlorate
$AlPO_4$	Aluminum phosphate
$FeSO_4$	Iron(II) sulfate or ferrous sulfate
$Fe_2(SO_4)_3$	Iron(III) sulfate or ferric sulfate
$PbCrO_4$	Lead(II) chromate
H_2SO_4	Hydrogen sulfate
HNO_3	Hydrogen nitrate
Cu_2SO_4	Copper(I) sulfate or cuprous sulfate
$CuSO_4$	Copper(II) sulfate or cupric sulfate
Li_3AsO_4	Lithium arsenate
$NaNO_2$	Sodium nitrite
$ZnMoO_4$	Zinc molybdate

Once again, the acid name is associated with the water solution of the pure compound. In the pure state, the usual ternary name may be used. Thus, H_2SO_4 is called both hydrogen sulfate and sulfuric acid.

In cases where there are more than two oxy-acids in a series, the *ous–ic* names are further modified with the prefixes *per* and *hypo*. *Per* is placed before the stem of the element other than hydrogen and oxygen when the element has a higher oxidation number than in the *ic* acid. *Hypo* is used as a prefix before the stem when the element has a lower oxidation number than in the *ous* acid. The use of *per* and *hypo* is illustrated in the oxy-acids of chlorine.

Formula	Name	Oxidation number of chlorine
HClO	Hypochlorous acid	$+1$
$HClO_2$	Chlorous acid	$+3$
$HClO_3$	Chloric acid	$+5$
$HClO_4$	Perchloric acid	$+7$

The electron-dot structures of the oxy-acids of chlorine are

Hypochlorous acid   Chlorous acid   Chloric acid   Perchloric acid

Check the oxidation number of chlorine in each of these oxy-acids using the method for assigning oxidation numbers.

Examples of other ternary oxy-acids and their names are shown in Table 8.6.

TABLE 8.6 **Names and formulas of selected ternary oxy-acids.**

Formula	Acid name	Formula	Acid name
H_2SO_3	Sulfurous acid	HNO_2	Nitrous acid
H_2SO_4	Sulfuric acid	HNO_3	Nitric acid
H_3PO_2	Hypophosphorous acid	$HBrO_3$	Bromic acid
H_3PO_3	Phosphorous acid	HIO_3	Iodic acid
H_3PO_4	Phosphoric acid	H_3BO_3	Boric acid
$HClO$	Hypochlorous acid	$H_2C_2O_4$	Oxalic acid
$HClO_2$	Chlorous acid	$HC_2H_3O_2$	Acetic acid
$HClO_3$	Chloric acid	H_2CO_3	Carbonic acid
$HClO_4$	Perchloric acid		

The endings *ous, ic, ite,* and *ate* are part of classical nomenclature; they are not used in the Stock System to indicate different oxidation states of the elements. These endings are still used, however, in naming many common compounds. The Stock name for H_2SO_4 is tetraoxosulfuric(VI) acid, and that for H_2SO_3 is trioxosulfuric(IV) acid. These Stock names are awkward and are not commonly used.

8.5 Salts

When the hydrogen of an acid is replaced by a metal ion or an ammonium (NH_4^+) ion, the compound formed is classified as a *salt*. Therefore, we can have a series of metal chlorides, bromides, sulfides, sulfates, sulfites, nitrates, phosphates, borates, and so on.

We have already considered the rules for naming some salts, but a comparison of the salt and acid names will reveal definite patterns that are used. The same rules given for naming binary salts are used in naming the positive part of ternary salts. In binary compounds, the usual *ide* ending is given to the negative part of the salt name. In ternary compounds, the *ous* and *ic* endings of the acid names become *ite* and *ate,* respectively, in the salt names, but the names of the stems remain the same.

Ternary oxy-acid		Ternary oxy-salt
ous ending of acid	becomes	*ite* ending in salt
ic ending of acid	becomes	*ate* ending in salt

Acid		Salt	
H_2SO_4	Sulfur/*ic* acid	Na_2SO_4	Sodium sulf/*ate*
H_2SO_3	Sulfur/*ous* acid	$CaSO_3$	Calcium sulf/*ite*
$HClO$	Hypochlor/*ous* acid	$LiClO$	Lithium hypochlor/*ite*
$HClO_4$	Perchlor/*ic* acid	$NaClO_4$	Sodium perchlor/*ate*

Other examples of ternary acids and salts are given in Table 8.7.

TABLE 8.7 Comparison of acid and salt names in ternary oxy-compounds.

Acid	Salt	Name of salt	Acid	Salt	Name of salt
H_2SO_4 Sulfuric acid	Na_2SO_4	Sodium sulfate	H_2SO_3 Sulfurous acid	Na_2SO_3	Sodium sulfite
	$CuSO_4$	Copper(II) sulfate or cupric sulfate		$CaSO_3$	Calcium sulfite
				Ag_2SO_3	Silver sulfite
	$CaSO_4$	Calcium sulfate	H_3PO_3 Phosphorous acid	K_3PO_3	Potassium phosphite
	$Fe_2(SO_4)_3$	Iron(III) sulfate or ferric sulfate		$CrPO_3$	Chromium(III) phosphite
				$Pb_3(PO_3)_2$	Lead(II) phosphite
HNO_3 Nitric acid	KNO_3	Potassium nitrate	H_3PO_4 Phosphoric acid	Li_3PO_4	Lithium phosphate
	$HgNO_3$	Mercury(I) nitrate or mercurous nitrate		$AlPO_4$	Aluminum phosphate
				$Ba_3(PO_4)_2$	Barium phosphate
	$Hg(NO_3)_2$	Mercury(II) nitrate or mercuric nitrate	$HBrO_3$ Bromic acid	$NaBrO_3$	Sodium bromate
	$Fe(NO_3)_2$	Iron(II) nitrate or ferrous nitrate		$Cd(BrO_3)_2$	Cadmium bromate
				$Mg(BrO_3)_2$	Magnesium bromate
	$Al(NO_3)_3$	Aluminum nitrate	HIO_3 Iodic acid	$AgIO_3$	Silver iodate
HNO_2 Nitrous acid	KNO_2	Potassium nitrite		$HgIO_3$	Mercury(I) iodate or mercurous iodate
	$Co(NO_2)_2$	Cobalt(II) nitrite or cobaltous nitrite	$H_2C_2O_4$ Oxalic acid	$Na_2C_2O_4$	Sodium oxalate
	$Mg(NO_2)_2$	Magnesium nitrite		$NaHC_2O_4$	Sodium binoxalate or sodium hydrogen oxalate
$HClO$	$NaClO$	Sodium hypochlorite		ZnC_2O_4	Zinc oxalate
$HClO_2$	$NaClO_2$	Sodium chlorite	$HC_2H_3O_2$ Acetic acid	$NaC_2H_3O_2$	Sodium acetate
$HClO_3$	$NaClO_3$	Sodium chlorate		$NH_4C_2H_3O_2$	Ammonium acetate
$HClO_4$	$NaClO_4$	Sodium perchlorate		$Ca(C_2H_3O_2)_2$	Calcium acetate
H_2CO_3 Carbonic acid	Li_2CO_3	Lithium carbonate		$Fe(C_2H_3O_2)_3$	Iron(III) acetate or ferric acetate
	$CaCO_3$	Calcium carbonate			

8.6 Salts with More Than One Positive Ion

Salts may be formed from acids that contain two or more acid hydrogen atoms by replacing only one of the hydrogen atoms with a metal or by replacing both hydrogen atoms with different metals. Each positive group is named first and then the appropriate salt ending is added.

Acid	Salt	Name of salt
H_2CO_3	$NaHCO_3$	Sodium bicarbonate or sodium hydrogen carbonate
H_2S	$NaHS$	Sodium hydrogen sulfide or sodium bisulfide
H_3PO_4	$MgNH_4PO_4$	Magnesium ammonium phosphate
H_2SO_4	$NaKSO_4$	Sodium potassium sulfate

Note the name *sodium bicarbonate* given in the table. The prefix *bi* is commonly used to indicate a compound in which one of two acid hydrogen atoms has been replaced by a metal. Another example is sodium bisulfate, which has the formula $NaHSO_4$. Table 8.8 shows examples of other salts that contain more than one positive ion.

TABLE 8.8 Names of selected salts that contain more than one positive ion.

Acid	Salt	Name of salt
H_2SO_4	$KHSO_4$	Potassium bisulfate or potassium hydrogen sulfate
H_2SO_3	$Ca(HSO_3)_2$	Calcium bisulfite or calcium hydrogen sulfite
H_2S	NH_4HS	Ammonium hydrogen sulfide or ammonium bisulfide
H_3PO_4	$MgNH_4PO_4$	Magnesium ammonium phosphate
H_3PO_4	NaH_2PO_4	Sodium dihydrogen phosphate
H_3PO_4	Na_2HPO_4	Disodium hydrogen phosphate
$H_2C_2O_4$	KHC_2O_4	Potassium hydrogen oxalate or potassium binoxalate
H_2SO_4	$KAl(SO_4)_2$	Potassium aluminum sulfate
H_2CO_3	$Al(HCO_3)_3$	Aluminum bicarbonate or aluminum hydrogen carbonate

Note that prefixes are also used in chemical nomenclature to give special clarity or emphasis to certain compounds as well as to distinguish between two or more compounds.

Examples: Na_3PO_4 Trisodium phosphate
Na_2HPO_4 Disodium hydrogen phosphate
NaH_2PO_4 Sodium dihydrogen phosphate

8.7 Bases

Inorganic bases contain the hydroxyl group, OH^-, in chemical combination with a metal ion. These compounds are called *hydroxides*. The OH^- group is named as a single ion and is given the ending *ide.* The formulas and names of several common bases follow.

NaOH   Sodium hydroxide
KOH   Potassium hydroxide
NH_4OH Ammonium hydroxide
$Ca(OH)_2$ Calcium hydroxide
$Ba(OH)_2$ Barium hydroxide

We have now looked at methods of naming inorganic acids, bases, salts, and oxides. These four classes are just a handful of the classified chemical compounds. Most other classes fall under the broad field of organic chemistry. A few of these are alcohols, hydrocarbons, ethers, aldehydes, ketones, phenols, and carboxylic acids.

PROBLEM 8.1   Name the compound CaS.

Step 1. From the formula, we know that CaS is a two-element compound and follows the rules for binary compounds.

Step 2. The compound is composed of Ca, a Group IIA metal, and S, a nonmetal. Since Group IIA elements have only one oxidation state, we name the positive part of the compound *calcium*. (You can also look in Table 7.6 for the oxidation number of Ca.)

Step 3. Modify the name of the second element to the identifying stem *sulf* and add the binary ending *ide* to form the name of the negative part, *sulfide*.

Step 4. The name of the compound, therefore, is *calcium sulfide*.

■

PROBLEM 8.2   Name the compound FeS.

Step 1. This compound follows the rules for a binary compound and, like CaS, must be a sulfide.

Step 2. It is a compound of Fe, a metal, and S, a nonmetal. From the oxidation number tables, we see that Fe has variable oxidation numbers. In sulfides, the oxidation number of S is -2. Therefore, the oxidation number of Fe must be $+2$. Thus, the name of the positive part of the compound is *iron(II)* or *ferrous*.

Step 3. We have already determined that the name of the negative part of the compound will be *sulfide*.

Step 4. The name of FeS is *iron(II) sulfide* or *ferrous sulfide*.

■

PROBLEM 8.3   Name the compound PCl_5.

Step 1. Both phosphorus and chlorine are nonmetals, so the rules for naming binary compounds containing two nonmetals apply. Phosphorus is named first since it is less electronegative than chlorine. Therefore, the compound is a chloride.

Step 2. No prefix is needed for phosphorus since there is only one atom of phosphorus per molecule. The prefix *penta* is used with chloride to indicate the five chlorine atoms.

Step 3. The name for PCl_5 is phosphorus pentachloride. The other common chloride is PCl_3, phosphorus trichloride.

■

PROBLEM 8.4   (a) Name the salt KNO_3, and (b) name the acid HNO_3, from which this salt can be derived.

(a) Step 1. From the formula, the compound contains three elements and follows the rules for ternary compounds.

Step 2. The salt is composed of a K^+ ion and a NO_3^- ion. The name of the positive part of the compound is *potassium*.

Step 3. Since it is a ternary salt, the name will end in *ite* or *ate*. From the oxidation number tables, we see that the name of the NO_3^- ion is *nitrate*.

Step 4. The name of the compound is *potassium nitrate*.

(b) The name of the acid follows the rules for ternary oxy-acids. Since the name of the salt KNO_3 ends in *ate*, the name of the corresponding acid will end in *ic acid*. Change the *ate* ending of nitrate to *ic*. Thus, *nitrate* becomes *nitric*, and the name of the acid is *nitric acid*.

■

QUESTIONS   In naming compounds, be careful to use correct spelling. For additional assistance in naming compounds refer to Tables 7.6, 7.7, and 7.8.

1. Write formulas for the following cations (do not forget to include the charges): sodium, magnesium, aluminum, copper(II), iron(II), ferric, lead(II), silver, cobalt(II), barium, hydrogen, mercury(II), tin(II), chromium(III), stannic, manganese(II) bismuth(III).

2. Write formulas for the following anions (do not forget to include the charges): chloride, bromide, fluoride, iodide, cyanide, oxide, hydroxide, sulfide, sulfate, bisulfate, bisulfite, chromate, carbonate, bicarbonate, acetate, chlorate, permanganate, oxalate.

3. Complete the table, filling in each box with the proper formula.

Anions

Cations	Br^-	O^{2-}	NO_3^-	PO_4^{3-}	CO_3^{2-}
K^+	KBr				
Mg^{2+}					
Al^{3+}					
Zn^{2+}				$Zn_3(PO_4)_2$	
H^+					

4. Complete the table, filling in each box with the proper formula.

Anions

Cations	SO_4^{2-}	Cl^-	AsO_4^{3-}	$C_2H_3O_2^-$	CrO_4^{2-}
NH_4^+			$(NH_4)_3AsO_4$		
Ca^{2+}					
Fe^{3+}	$Fe_2(SO_4)_3$				
Ag^+					
Cu^{2+}					

5. State how each of the following is used in naming inorganic compounds: *ide, ous, ic, hypo, per, ite, ate,* roman numerals.

6. Give the name and formula of three salts with *ide* endings that are not binary compounds. Do not use the exact salts used as examples in the chapter.

7. Give the name and formula for an oxy-ion of each: carbon, sulfur, nitrogen, phosphorus, chlorine, and bromine.

8. Give the name and formula for a ternary oxy-acid of each: carbon, sulfur, nitrogen, phosphorus, chlorine, and bromine.

9. Write formulas for the following binary compounds, all of which are composed of nonmetals:
   (a) Carbon monoxide
   (b) Sulfur trioxide
   (c) Carbon tetrabromide
   (d) Phosphorus trichloride
   (e) Nitrogen dioxide
   (f) Dinitrogen pentoxide

10. Name the following binary compounds, all of which are composed of nonmetals:
  (a) CO_2 (c) PCl_5 (e) SO_2 (g) P_2O_5 (i) NF_3
  (b) N_2O (d) CCl_4 (f) N_2O_4 (h) OF_2 (j) CS_2

11. Name the following compounds:
  (a) K_2O (c) CaI_2 (e) Na_3PO_4 (g) $Zn(NO_3)_2$
  (b) NH_4Br (d) $BaCO_3$ (f) Al_2O_3 (h) Ag_2SO_4

12. Write formulas for the following compounds:
  (a) Sodium nitrate (d) Ammonium sulfate
  (b) Magnesium fluoride (e) Silver carbonate
  (c) Barium hydroxide (f) Calcium phosphate

13. Name each of the following compounds by both the Stock (IUPAC) System, and the *ous–ic* system:
  (a) $CuCl_2$ (c) $Fe(NO_3)_2$ (e) SnF_2 (g) $As(C_2H_3O_2)_3$
  (b) $CuBr$ (d) $FeCl_3$ (f) $HgCO_3$ (h) TiI_3

14. Write formulas for the following compounds:
  (a) Tin(IV) bromide (d) Mercuric nitrite
  (b) Copper(I) sulfate (e) Titanic sulfide
  (c) Ferric carbonate (f) Iron(II) acetate

15. Write formulas for the following acids:
  (a) Hydrochloric acid (d) Carbonic acid
  (b) Chloric acid (e) Sulfurous acid
  (c) Nitric acid (f) Phosphoric acid

16. Name the following acids:
  (a) HNO_2 (c) $H_2C_2O_4$ (e) H_3PO_3 (g) HF
  (b) H_2SO_4 (d) HBr (f) $HC_2H_3O_2$ (h) $HBrO_3$

17. Write formulas for the following acids:
  (a) Acetic acid (d) Boric acid
  (b) Hydrofluoric acid (e) Nitrous acid
  (c) Hypochlorous acid (f) Hydrosulfuric acid

18. Name the following acids:
  (a) H_3PO_4 (c) HIO_3 (e) HClO (g) HI
  (b) H_2CO_3 (d) HCl (f) HNO_3 (h) $HClO_4$

19. Name the following compounds:
  (a) $Ba(NO_3)_2$ (d) $MgSO_4$ (g) NiS
  (b) $NaC_2H_3O_2$ (e) $CdCrO_4$ (h) $Sn(NO_3)_4$
  (c) PbI_2 (f) $BiCl_3$ (i) $Ca(OH)_2$

20. Write formulas for the following compounds:
  (a) Silver sulfite (h) Cupric chloride
  (b) Cobalt(II) bromide (g) Potassium permanganate
  (c) Tin(II) hydroxide (j) Barium nitrite
  (d) Aluminum sulfate (k) Sodium peroxide
  (e) Manganese(II) fluoride (l) Ferrous sulfate
  (f) Ammonium carbonate (m) Potassium dichromate
  (g) Chromium(III) oxide (n) Bismuth(III) chromate

21. Write formulas for the following compounds:
  (a) Sodium chromate (h) Cobalt(II) bicarbonate
  (b) Magnesium hydride (i) Sodium hypochlorite
  (c) Nickel(II) acetate (j) Arsenic(V) carbonate
  (d) Calcium chlorate (k) Chromium(III) sulfite
  (e) Lead(II) nitrate (l) Antimony(III) sulfate
  (f) Potassium dihydrogen (m) Sodium oxalate
    phosphate (n) Potassium thiocyanate
  (g) Manganese(II) hydroxide

22. Write the name of each salt and the formula and name of the acid from which the salt may be derived. [*Example:* NiC_2O_4, nickel(II) oxalate, $H_2C_2O_4$, oxalic acid.]

(a) $ZnSO_4$ (f) CoF_2 (k) $Ca(HSO_4)_2$ (p) $BiAsO_4$
(b) $HgCl_2$ (g) $Cr(ClO_3)_3$ (l) $As_2(SO_3)_3$ (q) $Fe(BrO_3)_2$
(c) $CuCO_3$ (h) Ag_3PO_4 (m) $Sn(NO_2)_2$ (r) $(NH_4)_2HPO_4$
(d) $Cd(NO_3)_2$ (i) NiS (n) $FeBr_3$ (s) $NaClO$
(e) $Al(C_2H_3O_2)_3$ (j) $BaCrO_4$ (o) $KHCO_3$ (t) $KMnO_4$

23. Write the chemical formula for each of the following substances:

(a) Baking soda   (f) Potash   (k) Limestone
(b) Lime   (g) Lye   (l) Cane sugar
(c) Epsom salts   (h) Quicksilver   (m) Milk of magnesia
(d) Muriatic acid   (i) Fool's gold   (n) Washing soda
(e) Vinegar   (j) Saltpeter   (o) Grain alcohol

24. Which of these statements are correct?

(a) In FeI_2, the iron is iron(II) since it is combined with two I^- ions.
(b) In Cu_2SO_4, the copper is copper(II) because there are two copper ions.
(c) In Ru_2O_3, we can deduce the oxidation state of Ru as $+3$ because two ions are combined with three oxide ions.
(d) When two nonmetals combine, prefixes of *di-*, *tri-*, *tetra-*, and so on are used to specify how many atoms of each element are in a molecule.
(e) N_2O_4 is called dinitrogen tetroxide.
(f) $Sn(CrO_4)_2$ is called tin dichromate.
(g) In the Stock System of nomenclature, when a compound contains a metal that can have more than one oxidation number, the oxidation number of the metal is designated by a roman numeral written immediately after the name of the metal.
(h) The name of Na_2O is disodium oxide.
(i) The formula for hydrobromic acid is HBr.
(j) HNO_2 is nitric acid.
(k) If the name of an acid ends in *-ous*, the corresponding salt will end in *-ate*.

Review Exercises for Chapters 5–8

CHAPTER 5
ATOMIC
THEORY
AND
STRUCTURE

True–False: Answer the following as either true or false.

1. Dalton's atomic theory states that all atoms are composed of protons, neutrons, and electrons.
2. The electron was discovered by J. J. Thomson.
3. The proton was discovered by James Chadwick in 1932.
4. A p orbital is spherically symmetrical around the nucleus.
5. An atom of nitrogen has two electrons in a $1s$ orbital, two electrons in a $2s$ orbital, and one electron in each of three different $2p$ orbitals.
6. The electron-dot structure for calcium is Ca:
7. An atom of $^{108}_{47}$Ag contains 47 protons, 47 electrons, and 108 neutrons.
8. One atomic mass unit is defined as one-twelfth the mass of a carbon-12 atom.
9. The atomic weight of an element expressed in grams is the gram-atomic weight.
10. A gram-atomic weight of silver is equal to a mole of silver.
11. A mole of silver contains the same number of atoms as a mole of sodium.
12. The proton and the neutron have approximately equal charge.
13. In the ground state, electrons tend to occupy orbitals of the lowest possible energy.
14. The electron-dot structure for potassium is $\cdot \overset{\cdot\cdot}{\underset{\cdot}{\text{P}}} \cdot$
15. The fourth energy level (N) can contain a total of 32 electrons.
16. One gram of tin contains 6.02×10^{23} atoms.
17. The second principal energy level (L) contains only s and p sublevels.
18. The first f sublevel electrons are in the fourth principal energy level.
19. The $4s$ energy sublevel fills before the $3d$ sublevel because the $4s$ is at a lower energy level.
20. A p type energy sublevel can contain six electrons.
21. A d type energy sublevel can contain five orbitals.
22. A million atoms of mercury would be more than a teaspoonful.
23. The listed atomic weight of an element represents the average relative mass of all the naturally occurring isotopes of that element.
24. On a dot structure representation of an element, the symbol of the element is shown surrounded by the number of electrons in the outermost energy level of the atom.
25. One mole of chlorine molecules contains two moles of chlorine atoms.
26. The atoms of two different elements must have different mass numbers.
27. $^{35}_{17}$Cl and $^{37}_{17}$Cl are isotopes of chlorine.
28. The element represented by $1s^2 2s^2 2p^6 3s^2 3p^4$ has an atomic number of 16.
29. The maximum number of electrons in a p orbital is six.
30. An atom with eight electrons in its outer shell is a noble gas.

Multiple Choice: Choose the correct answer to each of the following.

1. The concept of the positive charge and most of the mass concentrated in a small nucleus surrounded by the electrons was the contribution of:
   (a) Dalton   (b) Rutherford   (c) Bohr   (d) Schrödinger.
2. The concept of electrons existing in specific orbits around the nucleus was the contribution of:
   (a) Thomson   (b) Rutherford   (c) Bohr   (d) Schrödinger
3. An equation that leads to a concept of the electron having properties of both a particle and a wave, and forming an electron cloud around the nucleus, was a contribution of:
   (a) Thomson   (b) Rutherford   (c) Bohr   (d) Schrödinger
4. The correct sublevel electron structure for a fluorine atom, F, is:
   (a) $1s^2 2s^2 2p^5$ (b) $1s^2 2s^2 2p^3 3s^2 3p^1$ (c) $1s^2 2s^2 2p^4 3s^1$ (d) $1s^2 2s^2 2p^3$

5. The correct sublevel electron structure for $_{48}$Cd is:

   (a) $1s^2 2s^2 2p^6 3s^2 3p^6 4s^2 3d^{10}$

   (b) $1s^2 2s^2 2p^6 3s^2 3p^6 4s^2 3d^{10} 4p^6 5s^2 4d^{10}$

   (c) $1s^2 2s^2 2p^6 3s^2 3p^6 4s^2 3d^{10} 4p^6 4d^4$

   (d) $1s^2 2s^2 2p^6 3s^2 3p^6 4s^2 4p^6 4d^{10} 5s^2 5d^{10}$

6. The correct sublevel structure of $_{23}$V is:

   (a) [Ar] $4s^2 3d^3$ (b) [Ar] $4s^2 4p^3$ (c) [Ar] $4s^2 4d^3$ (d) [Ar] $4s^5$

7. Which of the following is the correct atomic structure for $^{48}_{22}$Ti?

		K	L	M	N
(a)	22p 26n	2	8	10	2
(c)	26p 22n	2	8	8	4
(b)	22p 48n	2	8	8	4
(d)	22p 26n	2	8	8	4

8. The correct sublevel electron structure for $_{82}$Pb is:

   (a) [Xe] $6s^2 4f^{14} 5d^{10} 6p^2$

   (b) [Xe] $6s^2 5d^{10} 6p^2$

   (c) [Xe] $6s^2 5f^{14} 6p^2$

   (d) [Xe] $6s^2 5f^{14} 6d^{10} 6p^2$

9. An atom of atomic number 53 and mass number 127 contains how many neutrons?

   (a) 53   (b) 74   (c) 127   (d) 180

10. How many electrons are in an atom of $^{40}_{18}$Ar?

    (a) 20   (b) 22   (c) 40   (d) No correct answer given

11. The number of neutrons in an atom of $^{139}_{56}$Ba is:

    (a) 56   (b) 83   (c) 139   (d) No correct answer given

12. The number of electrons in the principal energy level M in an atom having the electron structure $1s^2 2s^2 2p^6 3s^2 3p^2$ is:

    (a) 2   (b) 4   (c) 6   (d) 8

13. The total number of orbitals that contain at least one electron in an atom having the structure $1s^2 2s^2 2p^6 3s^2 3p^2$ is:

    (a) 5   (b) 8   (c) 14   (d) No correct answer given

14. A gram-atomic weight of nickel weighs:

    (a) 6.02×10^{23} g   (b) 28 g   (c) 58.7 g   (d) 14.0 g

15. 4.0 g of oxygen contains:

    (a) 1.5×10^{23} atoms

    (b) 4.0 gram-atomic weights

    (c) 0.50 mole

    (d) 6.02×10^{23} atoms

16. One mole of hydrogen atoms contains:

    (a) 2.0 g   (b) 6.02×10^{23} atoms   (c) 1 atom   (d) 12 g of carbon-12

17. One atom of magnesium weighs:

    (a) 24.3 g   (b) 54.9 g   (c) 12.0 g   (d) 4.04×10^{-23} g

18. Avogadro's number of magnesium atoms:

    (a) Weigh 1.0 g

    (b) Weigh the same as Avogadro's number of sulfur atoms

    (c) Weigh 12.0 g

    (d) Are 1 mole of magnesium atoms

19. Which of the following contains the largest number of moles?

    (a) 1.0 g Li   (b) 1.0 g Na   (c) 1.0 g Al   (d) 1.0 g Ag

20. The number of atoms present in 2.00 g of helium is:

    (a) 3.01×10^{23} (b) 6.02×10^{23} (c) 1.50×10^{23} (d) 2.00×10^{23}

CHAPTER 6
THE PERIODIC
ARRANGEMENT
OF THE
ELEMENTS

True–False: Answer the following as either true or false.

1. The periodic law states that the atomic weights of the elements are a periodic function of their atomic number.

2. The atomic radius of phosphorus, P, will be larger than that of sodium, Na.

3. The atomic radius of barium, Ba, will be larger than that of magnesium, Mg.

4. Elements at the bottom of a group in the periodic table tend to be more metallic in their properties than those at the top.

5. Elements within a group will have similar electron structures and will show some similarities in their chemical properties.
6. The transition elements are characterized by an increasing number of d or f electrons in an inner shell.
7. Based on the periodic table, the formula for aluminum oxide would be Al_2O_3.
8. J. A. R. Newlands observed that when the elements were arranged according to increasing atomic weights, every eighth element had similar properties.
9. The horizontal rows of elements in the periodic table are called periods.
10. The p block of elements contains transition elements.
11. Group IIIA elements all have three electrons in their outer shell.
12. Group B elements are referred to as the representative elements.
13. The element in period 3 having the largest radius is argon.
14. The element in period 3 that is classified as a metalloid is silicon.

Multiple Choice: Choose the correct answer to each of the following.
1. The scientist who noticed the existence of triads of elements with similar properties was:
   (a) Döbereiner   (b) Newlands   (c) Mendeleev   (d) Meyer
2. The chemist who developed a periodic table with the known elements arranged by atomic weight, who left spaces where elements seemed to be missing, and who predicted the properties of the missing elements was:
   (a) Newlands   (b) Mendeleev   (c) Meyer   (d) Moseley
3. The physicist who, through his work in X-ray emission frequencies of elements, established that atomic number was the correct basis for arranging the elements was:
   (a) Döbereiner   (b) Mendeleev   (c) Meyer   (d) Moseley
4. The German scientist who in 1869 independently published a periodic arrangement of elements based on increasing atomic weights similar to Mendeleev's arrangement was:
   (a) Döbereiner   (b) Meyer   (c) Newlands   (d) Moseley
5. Periods IIIA through VIIA plus the noble gases form the area of the periodic table where the electron sublevels being filled are:
   (a) p sublevels   (b) s and p sublevels   (c) d sublevels   (d) f sublevels
6. In moving down an A group on the periodic table, the number of electrons in the outermost energy level:
   (a) Increases regularly   (c) Decreases regularly
   (b) Remains constant   (d) Changes in an unpredictable manner
7. Which of the following would be an incorrect formula?
   (a) NaCl   (b) K_2O (c) AlO   (d) BaO
8. Elements of the noble gas family:
   (a) Form no compounds at all
   (b) Have full outer electron shells
   (c) Have an outer shell electron structure of ns^2np^6 (helium excepted)
   (d) No correct answer given
9. The lanthanide and actinide series of elements are:
   (a) Representative elements   (c) Filling-in d level electrons
   (b) Transition elements   (d) No correct answer given
10. The element having the structure $1s^2 2s^2 2p^6 3s^2 3p^2$ is in Group:
    (a) IIA   (b) IIB   (c) IVA   (d) IVB
11. In Group VA, the element having the smallest atomic radius is:
    (a) Bi   (b) P   (c) As   (d) N
12. In Group IVA, the most metallic element is:
    (a) C   (b) Si   (c) Ge   (d) Sn
13. Which group in the periodic table contains the least reactive elements?
    (a) IA   (b) IIA   (c) IIIA   (d) Noble gases

14. Which group in the periodic table contains the alkali metals?
    (a) IA   (b) IIA   (c) IIIA   (d) IVA
15. An atom of fluorine is smaller than an atom of oxygen. One possible explanation for this is that, compared to oxygen, fluorine has:
    (a) A larger mass number   (c) A greater nuclear charge
    (b) A smaller atomic number   (d) More unpaired electrons

**CHAPTER 7
CHEMICAL
BONDS – THE
FORMATION OF
COMPOUNDS
FROM ATOMS**

True–False: Answer the following as either true or false.
1. The amount of energy required to remove one electron from an atom is known as the ionization energy.
2. It requires less energy to remove a second electron from an atom than to remove the first electron.
3. The first ionization energy of potassium will be greater than that for sodium.
4. The first ionization energy of sulfur will be greater than that for sodium.
5. The electrons in the outermost shell of an element are called the valence electrons.
6. When an atom loses an electron it becomes a negative ion.
7. A bromide ion is smaller than a bromine atom.
8. Crystals of sodium chloride consist of arrays of NaCl molecules.
9. The sharing of a pair of electrons between a positive ion and a negative ion is called an ionic bond.
10. When elements combine by ionic bonding, they normally form molecules.
11. When a single atom furnishes both electrons that are shared between two atoms, the bond is called a coordinate covalent bond.
12. The NH_4^+ ion contains at least one coordinate covalent bond.
13. Oxygen has a greater electronegativity than carbon.
14. Bromine has a greater electronegativity than chlorine.
15. A dipole is a molecule that is electrically unsymmetrical, causing it to be oppositely charged at two points.
16. The water molecule is a dipole.
17. The formula of the aluminum ion is Al^+.
18. A stable group of atoms that has either a positive or a negative charge and behaves as a single unit is called a polyatomic ion.
19. In $KMnO_4$, the oxidation number of Mn is $+7$.
20. A sodium ion is smaller than a sodium atom.
21. In crystals of sodium chloride, each sodium ion is surrounded by six chloride ions, and each chloride ion is surrounded by six sodium ions.
22. Boron has a greater electronegativity than barium.
23. In the reaction $2\ KClO_3 \xrightarrow{\Delta} 2\ KCl + 3\ O_2$, the element reduced is oxygen.
24. Metals tend to lose their valence electrons, forming positively charged ions.
25. When metals with low electronegativity combine with nonmetals of high electronegativity, they tend to form ionic compounds.
26. Covalent bonding is also called ionic bonding.
27. A pair of electrons shared between two atoms constitutes the covalent bond.
28. The ionic bond is the electrostatic attraction existing between oppositely charged ions.

Multiple Choice: Choose the correct answer to each of the following.
1. Which of the following formulas is not correct?
   (a) Na^+ (b) S^- (c) Al^{3+} (d) F^-
2. Which of the following would not include polar covalent bonds?
   (a) CH_4 (b) H_2O (c) CH_3OH (d) Cl_2
3. Which of the following is a dipole?
   (a) HCl   (b) CH_4 (c) H_2 (d) CO_2

4. Which of the following has bonding that is primarily ionic?
   (a) H_2 (b) MgF_2 (c) H_2O (d) CH_4

5. Which of the following is a correct dot structure?

   (a) $:\overset{..}{O}:C:\overset{..}{O}:$ (b) $:\overset{..}{Cl}:\overset{:\overset{..}{Cl}:}{\underset{:\overset{..}{Cl}:}{C}}:\overset{..}{Cl}:$ (c) $\overset{..}{Cl}::\overset{..}{Cl}$ (d) $:N:N:$

6. Which of the following is an incorrect dot structure?

   (a) $H:\overset{\overset{\displaystyle H}{}}{\underset{\displaystyle H}{N}}:H$ (b) $:\overset{..}{\underset{..}{O}}:H$ (c) $H:\overset{\overset{\displaystyle H}{..}}{\underset{\displaystyle H}{C}}:H$ (d) $:N:::N:$

7. Which of the following is an incorrect dot structure for a polyatomic ion?

   (a) $H:\overset{\overset{\displaystyle H}{..}}{\underset{\displaystyle H}{N}}:H$ $^+$ (b) $:\overset{..}{\underset{..}{O}}:\overset{:\overset{..}{O}:}{C}:\overset{..}{\underset{..}{O}}:$ $^{2-}$ (c) $H:\overset{..}{\underset{..}{S}}:$ $^-$ (d) $:\overset{..}{\underset{..}{O}}:\overset{:\overset{..}{O}:}{\underset{:\overset{..}{O}:}{P}}:\overset{..}{\underset{..}{O}}:$ $^{3-}$

8. Which element has seven valence electrons?
   (a) S   (b) Ne   (c) Br   (d) Ag

9. Carbon dioxide, CO_2, is a nonpolar molecule because:
   (a) Oxygen is more electronegative than carbon
   (b) The two oxygen atoms are bonded to the carbon atom
   (c) The molecule has a linear structure with the carbon atom in the middle
   (d) The carbon–oxygen bonds are polar covalent

10. When a magnesium atom participates in a chemical reaction, it is most likely to:
    (a) Lose 1 electron   (c) Lose 2 electrons
    (b) Gain 1 electron   (d) Gain 2 electrons

11. If X represents an element of Group IIIA, what is the general formula for its oxide?
    (a) X_3O_4 (b) X_3O_2 (c) XO   (d) X_2O_3

12. Which particle has the same electron structure as an argon atom?
    (a) Ca^{2+} (b) Cl^0 (c) Na^+ (d) K^0

13. As the difference in electronegativity between two elements decreases, the tendency for the elements to form a covalent bond:
    (a) Increases
    (b) Decreases
    (c) Remains the same
    (d) Sometimes increases and sometimes decreases

14. Which compound forms a tetrahedral molecule?
    (a) NaCl   (b) CO_2 (c) CH_4 (d) H_2O

15. Which compound would show a bent (V-shaped) molecular structure?
    (a) NaCl   (b) CO_2 (c) CH_4 (d) H_2O

16. Which compound has double bonds within its molecular structure?
    (a) NaCl   (b) CO_2 (c) CH_4 (d) H_2O

17. If the chromium(III) ion is Cr^{3+} and the iodide ion is I^-, then the formula for chromium(III) iodide is:
    (a) CrI_3 (b) Cr_3I (c) Cr(III)I   (d) $Cr(III)I_3$

18. If the ammonium ion is NH_4^+, and the carbonate ion is CO_3^{2-}, then the correct formula for ammonium carbonate is:
    (a) NH_4CO_3 (b) $NH_4(CO_3)_2$ (c) $(NH_4)_2CO_3$ (d) $(CO_3)_2NH_4$

19. If the zinc ion is Zn^{2+}, and the acetate ion is $C_2H_3O_2^-$, then the correct formula for zinc acetate is:
    (a) $Zn_2C_2H_3O_2$ (b) $Zn(C_2H_3O_2)_2$ (c) $ZnC_4H_6O_4$ (d) $ZnC_2H_3O_2$

20. If the lead(II) ion is Pb^{2+}, and the phosphate ion is PO_4^{3-}, then the correct formula for lead(II) phosphate is:
(a) Pb_2PO_4 (b) $Pb(II)PO_4$ (c) $Pb_2(PO_4)_3$ (d) $Pb_3(PO_4)_2$

21. In each of the following the oxidation number of the underlined element is given. Which is incorrect?
(a) $Na_2\underline{S}$, -2 (b) $Na_2\underline{S}O_4$, $+6$ (c) $Ba\underline{Cr}O_4$, $+7$ (d) $Ca(\underline{N}O_3)_2$, $+5$

22. In each of the following the oxidation number of the underlined element is given. Which is incorrect?
(a) $Na\underline{N}O_3$, $+6$ (b) $\underline{Na}_2SO_4$, $+1$ (c) $K_2\underline{C}O_3$, $+4$ (d) $H_2\underline{O}_2$, -1

CHAPTER 8
NOMENCLATURE
OF INORGANIC
COMPOUNDS

True–False: Answer the following as either true or false.

1. The name for H_2SO_4 is sulfuric acid.
2. The formula of copper(II) oxide is Cu_2O.
3. The formula for hydrobromic acid is HBr.
4. The name of HNO_2 is nitric acid.
5. The formula of potassium chromate is K_2CrO_4.
6. The formula for barium hydroxide is BaOH.
7. Binary compounds will have names ending in -ide.
8. If the name of an acid ends in -ous, the corresponding salt will end in -ate.
9. Diphosphorus pentoxide is P_2O_5.

In which of the following is the formula correct for the name given?

1. Copper(II) sulfate, $CuSO_4$
2. Ammonium hydroxide, NH_4OH
3. Mercury(I) carbonate, $HgCO_3$
4. Phosphorus triiodide, PI_3
5. Calcium acetate, $Ca(C_2H_3O_2)_2$
6. Hypochlorous acid,   HClO
7. Dichlorine heptoxide, Cl_2O_7
8. Magnesium iodide,   MgI
9. Sulfurous acid, H_2SO_3
10. Potassium manganate, $KMnO_4$
11. Lead(II) chromate, $PbCrO_4$
12. Ammonium bicarbonate, NH_4HCO_3
13. Iron(II) phosphate, $FePO_4$
14. Calcium hydrogen sulfate, $CaHSO_4$
15. Mercury(II) sulfate, $HgSO_4$
16. Dinitrogen pentoxide, N_2O_5
17. Sodium hypochlorite,   NaClO
18. Sodium dichromate, $Na_2Cr_2O_7$
19. Cadmium cyanide, $Ca(CN)_2$
20. Bismuth(III) oxide, Bi_3O_2

21. Carbonic acid, H_2CO_3
22. Silver oxide, Ag_2O
23. Ferric iodide, FeI_2
24. Tin(II) fluoride, TiF_2
25. Carbon monoxide,   CO
26. Phosphoric acid, H_3PO_3
27. Copper(I) sulfide, Cu_2S
28. Hydrosulfuric acid, H_2S
29. Potassium hydroxide,   POH
30. Sodium carbonate, Na_2CO_3
31. Zinc sulfate, $ZnSO_3$
32. Sulfur trioxide, SO_3
33. Tin(IV) nitrate, $Sn(NO_3)_4$
34. Ferrous sulfate, $FeSO_4$
35. Chloric acid,   HCl
36. Aluminum sulfide, Al_2S_3
37. Ammonia, NH_3
38. Acetic acid, $HC_2H_3O_2$
39. Zinc oxide, ZnO_2
40. Calcium hydroxide,   CaOH

9 Quantitative Composition of Compounds

After studying Chapter 9 you should be able to:

1. Understand the terms listed in Question A at the end of the chapter.
2. Determine the formula weight or molecular weight of a compound from its formula.
3. Convert moles (gram-molecular weights or gram-formula weights) to grams, to molecules, or to formula units, and vice versa.
4. Calculate the percentage composition by weight of a compound from its formula.
5. Calculate the percentage composition of a compound from experimental data on combining weights.
6. Explain the relationship between an empirical formula and a molecular formula.
7. Determine the empirical formula of a compound from its percentage composition or combining weights of elements.
8. Calculate the molecular formula of a compound from its empirical formula and molecular weight.

9.1 Formula Weight or Molecular Weight

The quantitative composition of a compound can be determined from its formula, and the formula of a compound can be determined from its quantitative composition. In order to make these determinations, chemists have established a scale of relative masses for atoms known as *atomic weights*. This scale is based on the carbon-12 isotope having a mass of exactly 12 amu (see Section 5.17).

formula weight

molecular weight

Because compounds are composed of atoms, they may be represented by a mass known as the formula weight or the molecular weight. The **formula weight** of a substance is the total mass of all the atoms in the chemical formula of that substance. The **molecular weight** of a substance is the total mass of all the atoms in a molecule of that substance. Formula weight and molecular weight are used interchangeably. However, the term *formula weight* is more inclusive, since it includes both molecular and ionic substances.

9.2 Determination of Molecular Weights from Formulas

If the formula of a substance is known, its formula weight or molecular weight may be determined by adding together the atomic weights of all the atoms in the formula. If more than one atom of any element is present, it must be added as many times as it is used in the formula.

PROBLEM 9.1   The molecular formula for water is H_2O. What is its molecular weight?

Proceed by looking up the atomic weights of H (1.008 amu) and O (15.999 amu) and adding the masses of all the atoms in the formula unit. Water contains two atoms of H and one atom of O. Thus,

$$2 \text{ H atoms} = 2 \times 1.008 = 2.016 \text{ amu}$$
$$1 \text{ O atom} = 1 \times 15.999 = \underline{15.999 \text{ amu}}$$
$$18.015 \text{ amu} = \text{Molecular or formula weight}$$

■

PROBLEM 9.2   Calculate the formula weight of calcium hydroxide, $Ca(OH)_2$.

The formula of this substance contains one atom of Ca and two atoms each of O and H. Thus,

$$1 \text{ Ca atom} = 1 \times 40.08 = 40.08 \text{ amu}$$
$$2 \text{ O atoms} = 2 \times 15.999 = 31.998 \text{ amu}$$
$$2 \text{ H atoms} = 2 \times 1.008 = \underline{2.016 \text{ amu}}$$
$$74.094 \text{ amu} \quad \text{or} \quad 74.09 = \text{Formula weight}$$

The atomic weights of elements are often rounded off to one decimal place to simplify calculations. (However, this simplification cannot be made in the most exacting chemical work.) If we calculate the formula weight of $Ca(OH)_2$ on the basis of one decimal place, we find the value to be 74.1 amu instead of 74.09 amu. The formula weight of $Ca(OH)_2$ would then be calculated as follows:

$$1 \text{ Ca} = 1 \times 40.1 = 40.1 \text{ amu}$$
$$2 \text{ O} = 2 \times 16.0 = 32.0 \text{ amu}$$
$$2 \text{ H} = 2 \times 1.0 = \underline{2.0 \text{ amu}}$$
$$74.1 \text{ amu} = \text{Formula weight}$$

■

PROBLEM 9.3   The formula for barium chloride dihydrate is $BaCl_2 \cdot 2H_2O$. What is its formula weight?

This formula contains one atom of Ba, two atoms of Cl, and two molecules of H_2O. Thus,

$$1 \text{ Ba} = 1 \times 137.3 = 137.3 \text{ amu}$$
$$2 \text{ Cl} = 2 \times 35.5 = 71.0 \text{ amu}$$
$$2 \text{ H}_2\text{O} = 2 \times 18.0 = \underline{36.0 \text{ amu}}$$
$$244.3 \text{ amu} = \text{Formula weight}$$

■

9.3 Gram-Molecular Weight; Gram-Formula Weight; the Mole

**gram-formula
weight**

**gram-molecular
weight**

The quantity of any substance having a mass in grams that is numerically equal to its formula weight is the **gram-formula weight (g-form. wt)** or **gram-molecular weight (g-mol. wt)** of that substance. This quantity represents the weight of 1 mole (6.02×10^{23} formula units or molecules) of the substance. As an illustration, consider the compound hydrogen chloride, HCl. When 1 gram-atomic weight of H (1.00 g representing 6.02×10^{23}, or 1 mole of, H atoms) and 1 gram-atomic weight of Cl (35.5 g representing 6.02×10^{23}, or 1 mole of, Cl atoms) combine, they produce 1 gram-molecular weight of HCl (36.5 g representing 6.02×10^{23}, or 1 mole of, HCl molecules). Since 36.5 g of HCl contain 6.02×10^{23} molecules, we may refer to this quantity as a gram-molecular weight, a gram-formula weight, or simply as a mole of HCl. These relationships are summarized in tabular form for hydrogen chloride.

H	*Cl*	*HCl*
6.02×10^{23} H *atoms*	6.02×10^{23} Cl *atoms*	6.02×10^{23} HCl *molecules*
1 mole H *atoms*	1 mole Cl *atoms*	1 mole HCl *molecules*
1.0 g H	35.5 g Cl	36.5 g HCl
1 g-at. wt H	1 g-at. wt Cl	1 g-mol. wt HCl or
		1 g-form. wt HCl

In dealing with diatomic elements (H_2, O_2, N_2, F_2, Cl_2, Br_2, I_2), special care must be taken to distinguish between a mole of atoms (gram-atomic weight) and a mole of molecules (gram-molecular weight). For example, consider 1 mole of oxygen molecules, which weighs 32.0 g. This quantity is equal to 2 gram-atomic weights of the element oxygen and thus represents 2 moles of oxygen atoms. It is important, therefore, in the case of diatomic elements, to be certain which form we are considering, the atom or the molecule.

Remember that 1 mole of atoms or molecules contains Avogadro's number (6.02×10^{23}) of atoms or molecules, which is also 1 gram-formula weight of atoms or molecules.

For example: 1 mole of Na is 23.0 g Na
1 mole of HCl is 36.5 g HCl
1 mole of Na_2SO_4 is 142.1 g Na_2SO_4

---

1 mole = 6.02×10^{23} molecules or formula units

**1 mole = 1 gram-molecular weight or 1 gram-formula weight
of a compound**

1 mole = 1 gram-atomic weight of a monatomic element

The conversion factors for changing grams of a compound to moles and vice versa are

Grams to moles: $\dfrac{\text{1 mole of a substance}}{\text{1 g-form. wt of the substance}}$

Moles to grams: $\dfrac{\text{1 g-form. wt of a substance}}{\text{1 mole of the substance}}$

PROBLEM 9.4   What is the weight of 1 mole (gram-molecular weight) of sulfuric acid, H_2SO_4?

This problem is solved in a similar manner to Problems 9.1 through 9.3, using atomic weights of the elements in gram units instead of amu. Look up the atomic weights of H, S, and O, and solve.

$$2\,H = 2 \times 1.0 = 2.0\text{ g}$$
$$1\,S = 1 \times 32.1 = 32.1\text{ g}$$
$$4\,O = 4 \times 16.0 = \underline{64.0\text{ g}}$$
$$98.1\text{ g} = \text{Weight of 1 mole (1 g-mol. wt) of } H_2SO_4$$

■

PROBLEM 9.5   How many moles of NaOH are there in 1.00 kg of sodium hydroxide?

First, we know that

$$1\text{ mole} = 1\text{ g-form. wt} = 40.0\text{ g }(23.0 + 16.0 + 1.0\text{ g})\text{ NaOH}$$
$$1\text{ kg} = 1000\text{ g}$$

Now the conversion is

Grams NaOH $\longrightarrow$ Moles NaOH

To convert grams to moles we use the conversion factor

$$\frac{1\text{ mole}}{1\text{ g-form. wt}} = \frac{1\text{ mole NaOH}}{40.0\text{ g NaOH}}$$

The calculation is

$$1000\text{ g NaOH} \times \frac{1\text{ mole NaOH}}{40.0\text{ g NaOH}} = 25.0\text{ moles NaOH}$$

$$1\text{ kg NaOH} = 25.0\text{ moles NaOH}$$

■

PROBLEM 9.6   What is the weight in grams of 5.00 moles of water?

First, we know that

$$1\text{ mole } H_2O = 18.0\text{ g}\quad\text{(Problem 9.1)}$$

The conversion is   Moles $H_2O \longrightarrow$ Grams H_2O

To convert moles to grams, use the conversion factor

$$\frac{1 \text{ g-mol. wt } H_2O}{1 \text{ mole } H_2O} = \frac{18.0 \text{ g } H_2O}{1 \text{ mole } H_2O}$$

The calculation is

$$5.00 \text{ moles } H_2O \times \frac{18.0 \text{ g } H_2O}{1 \text{ mole } H_2O} = 90.0 \text{ g } H_2O \quad \text{(Answer)}$$

■

PROBLEM 9.7 How many molecules of HCl are there in 25.0 g of hydrogen chloride?
From the formula we find that the gram-molecular weight (1 mole) of HCl is 36.5 g. The sequence of conversions is

$$\text{Grams HCl} \longrightarrow \text{Moles HCl} \longrightarrow \text{Molecules HCl}$$

using the conversion factors

$$\frac{1 \text{ mole HCl}}{36.5 \text{ g HCl}} \quad \text{and} \quad \frac{6.02 \times 10^{23} \text{ molecules HCl}}{1 \text{ mole HCl}}$$

$$25.0 \text{ g HCl} \times \frac{1 \text{ mole HCl}}{36.5 \text{ g HCl}} \times \frac{6.02 \times 10^{23} \text{ molecules HCl}}{1 \text{ mole HCl}} = 4.12 \times 10^{23} \text{ molecules HCl}$$

(Answer)

■

9.4 Percentage Composition of Compounds

percentage composition of a compound

Just as each piece of pie represents a percentage of the whole pie, so the mass of each element in a compound represents a percentage of the total mass of that compound. The formula weight may be used to represent the total mass, or 100% of a compound. The **percentage composition of a compound** is the weight percent of each element in the compound. Thus the percentage composition of water, H_2O, is 11.1% H and 88.9% O by weight.

The percentage composition of a compound can be calculated if its formula is known or if the weights of two or more elements that have combined with each other are known or are experimentally determined.

If the formula is known, it is essentially a two-step process to determine the percentage composition:

Step 1. Calculate the gram-formula weight as was done in Sections 9.2 and 9.3.
Step 2. Determine the total weight of each element in the gram-formula weight. Divide the total weight of each element by the gram-formula weight and multiply by 100%. This gives the percentage composition.

$$\frac{\text{Total weight of each element}}{\text{Formula weight}} \times 100\% = \text{Percentage of the element}$$

PROBLEM 9.8   Calculate the percentage composition of sodium chloride, NaCl.

Step 1. First calculate the gram-formula weight of NaCl:

Atomic weights: Na, 23.0; Cl, 35.5

1 Na = 23.0 g
1 Cl = 35.5 g
          58.5 g   (Gram-formula weight)

Step 2. Now calculate the percentage composition. We know there are 23.0 g Na and 35.5 g Cl in 58.5 g NaCl.

Na: $\dfrac{23.0 \text{ g Na}}{58.5 \text{ g NaCl}} \times 100\% =$ 39.3% Na

Cl: $\dfrac{35.5 \text{ g Cl}}{58.5 \text{ g NaCl}} \times 100\% =$ 60.7% Cl
                                                           100.0% total

In any two-component system, if one percentage is known, the other is automatically defined by difference; that is, if Na = 39.3%, then Cl = 100 − 39.3 = 60.7%. However, the calculation of the percentage of each component should be carried out, since this provides a check against possible error. The percentage composition data should add up to 100 ± 0.5%.

■

PROBLEM 9.9   Calculate the percentage composition of potassium chloride, KCl.

Step 1. First calculate the gram-formula weight of KCl:

Atomic weights: K, 39.1; Cl, 35.5

1 K = 39.1 g
1 Cl = 35.5 g
          74.6 g   (Gram-formula weight)

Step 2. Now calculate the percentage composition. We know there are 39.1 g K and 35.5 g Cl in 74.6 g KCl.

K: $\dfrac{39.1 \text{ g K}}{74.6 \text{ g KCl}} \times 100\% =$ 52.4% K

Cl: $\dfrac{35.5 \text{ g Cl}}{74.6 \text{ g KCl}} \times 100\% =$ 47.6% Cl
                                                          100.0% total

KCl contains 52.4% K and 47.6% Cl.

■

Comparing the data calculated for NaCl and for KCl, we see that NaCl contains a higher percentage of Cl by weight, although each compound has a one-to-one atom ratio of Cl to Na and K. The reason for this weight percentage difference is that Na and K do not have the same atomic weights.

It is important to realize that when we compare 1 mole of NaCl with 1 mole of KCl each quantity contains the same number of Cl atoms, namely 1 mole of Cl atoms. However, if we compare equal masses of NaCl and KCl, there will be more Cl atoms in the mass of NaCl since NaCl has a higher weight percent of Cl.

1 mole NaCl contains	*100 g NaCl contains*	*1 mole KCl contains*	*100 g KCl contains*
1 mole Na 1 mole Cl 60.7% Cl	39.3 g Na 60.7 g Cl	1 mole K 1 mole Cl 47.6% Cl	52.4 g K 47.6 g Cl

PROBLEM 9.10   Calculate the percentage composition of potassium sulfate, K_2SO_4.

Step 1. First calculate the gram-formula weight of K_2SO_4:

Atomic weights: K, 39.1; S, 32.1; O, 16.0

$$2 K = 2 \times 39.1 = 78.2 \text{ g}$$
$$1 S = 1 \times 32.1 = 32.1 \text{ g}$$
$$4 O = 4 \times 16.0 = \underline{64.0 \text{ g}}$$
$$174.3 \text{ g}   \text{(Gram-formula weight)}$$

Step 2. Now calculate the percentage composition. We know there are 78.2 g K, 32.1 g S, and 64.0 g O in 174.3 g K_2SO_4.

K: $\dfrac{78.2 \text{ g K}}{174.3 \text{ g K}_2SO_4} \times 100\% = 44.9\% \text{ K}$

S: $\dfrac{32.1 \text{ g S}}{174.3 \text{ g K}_2SO_4} \times 100\% = 18.4\% \text{ S}$

O: $\dfrac{64.0 \text{ g O}}{174.3 \text{ g K}_2SO_4} \times 100\% = \underline{36.7\% \text{ O}}$

$100.0\% \text{ total}$

K_2SO_4 contains 44.9% K, 18.4% S, and 36.7% O.

■

When we have a formula with groups in parentheses, it is helpful to first rewrite the formula, eliminating the parentheses, to indicate the total number of atoms of each element present. For example, $Ca(C_2H_3O_2)_2$ becomes $CaC_4H_6O_4$

and $Fe_2(SO_4)_3$ becomes $Fe_2S_3O_{12}$. Then follow the procedure in Problem 9.10 to calculate the percentage composition.

PROBLEM 9.11   When heated in air, 1.63 g of zinc, Zn, combine with 0.400 g of oxygen, O_2, to form an oxide of zinc. Calculate the percentage composition of the compound formed.

The percentage composition may be calculated on the basis of the individual elements as parts or percentages of the total weight of the compound formed. First, calculate the total weight of the compound formed.

$$\begin{array}{l} 1.63 \ \ \text{g Zn} \\ \underline{0.400 \ \text{g O}} \\ 2.030 \ \text{g} \quad \text{or} \quad 2.03 \ \text{g} = \text{Total weight of product} \end{array}$$

Then divide the weight of each element by the total weight of the compound formed and multiply by 100%.

$$\frac{1.63 \ \cancel{g} \ \text{Zn}}{2.03 \ \cancel{g}} \times 100\% = \ \ 80.3\% \ \text{Zn}$$

$$\frac{0.400 \ \cancel{g} \ \text{O}}{2.03 \ \cancel{g}} \times 100\% = \ \ \underline{19.7\% \ \text{O}}$$
$$100.0\% \ \text{total}$$

The compound formed contains 80.3% Zn and 19.7% O.

■

9.5 Empirical Formula Versus Molecular Formula

empirical formula

The **empirical formula,** or simplest formula, gives the smallest ratio of the atoms that are present in a compound. This formula gives the relative number of atoms of each element in the compound. The empirical formula contains the smallest whole-number ratio that can be derived from the percentages of the different elements in the compound.

molecular formula

The **molecular formula** is the true formula, representing the total number of atoms of each element present in one molecule of a compound. It is entirely possible that two or more substances will have the same percentage composition, yet be distinctly different compounds. For example, acetylene, C_2H_2, is a common gas used in welding; benzene, C_6H_6, is an important solvent obtained from coal tar and is used in the synthesis of styrene and nylon. Both acetylene and benzene contain 92.3% C and 7.7% H. The smallest ratio of C and H corresponding to these percentages is CH (1:1). Therefore, the empirical formula for both acetylene and benzene is CH—even though it is known that the molecular formulas are C_2H_2 and C_6H_6, respectively. It is not uncommon for the molecular formula to be the same as the empirical formula. If the molecular formula is not the same, it will be an integral (whole number) multiple of the empirical formula.

CH   = Empirical formula
$(CH)_2 = C_2H_2 = $ Acetylene (molecular formula)
$(CH)_6 = C_6H_6 = $ Benzene (molecular formula)

Table 9.1 summarizes the data concerning these CH formulas. Table 9.2 shows empirical and molecular formula relationships of other compounds.

TABLE 9.1 Molecular formulas of two compounds having an empirical formula with a 1:1 ratio of carbon and hydrogen atoms.

	Composition		
Formula	%C	%H	Molecular weight
CH (empirical)	92.3	7.7	13.0 (empirical)
C_2H_2 (acetylene)	92.3	7.7	$26.0 \,(2 \times 13.0)$
C_6H_6 (benzene)	92.3	7.7	$78.0 \,(6 \times 13.0)$

TABLE 9.2 Some empirical and molecular formulas.

Compound	Empirical formula	Molecular formula	Compound	Empirical formula	Molecular formula
Acetylene	CH	C_2H_2	Carbon dioxide	CO_2	CO_2
Benzene	CH	C_6H_6	Diborane	BH_3	B_2H_6
Styrene	CH	C_8H_8	Hydrazine	NH_2	N_2H_4
Ethylene	CH_2	C_2H_4	Hydrogen	H	H_2
Formaldehyde	CH_2O	CH_2O	Chlorine	Cl	Cl_2
Acetic acid	CH_2O	$C_2H_4O_2$	Bromine	Br	Br_2
Dextrose	CH_2O	$C_6H_{12}O_6$	Oxygen	O	O_2
Hydrogen chloride	HCl	HCl	Nitrogen	N	N_2

9.6 Calculation of the Empirical Formula

It is possible to establish an empirical formula because (1) the individual atoms in a compound are combined in whole-number ratios and (2) each element has a specific atomic weight.

In order to calculate the empirical formula, we need to know (1) the elements that are combined, (2) their atomic weights, and (3) the ratio by weight or percentage in which they are combined. If elements A and B form a compound, we may represent the empirical formula as A_xB_y, where x and y are small whole numbers that represent the number of atoms of A and B. To write the empirical formula, we must determine x and y.

The solution to this problem requires three or four steps.

Step 1. Assume a definite quantity (usually 100 g) of the compound, if not given, and express the weight of each element in grams.

Step 2. Multiply the grams of each element by the factor 1 mole/1 g-at. wt to convert grams to moles. This conversion gives the relative number of moles of atoms of each element in the quantity assumed. At this point, these numbers will usually not be whole numbers.

Step 3. Divide each of the values obtained in Step 2 by the smallest of these values. If the numbers obtained by this procedure are whole numbers, use them as subscripts in writing the empirical formula. If the numbers obtained are not whole numbers, go on to Step 4.

Step 4. Multiply the values obtained in Step 3 by the smallest number that will convert them to whole numbers. Use these whole numbers as the subscripts in the empirical formula. For example, if the ratio of A to B is 1.0 to 1.5, multiply both numbers by 2 to obtain a ratio of 2 to 3. The empirical formula would then be A_2B_3.

PROBLEM 9.12   Calculate the empirical formula of a compound containing 11.1% hydrogen, H, and 88.9% oxygen, O.

Step 1. Express each element in grams; if we assume that there are 100 g of material, then the percentage of each element is equal to the number of grams of each element in 100 g.

$$H = 11.1\% = \frac{11.1 \text{ g}}{100 \text{ g}}$$

$$O = 88.9\% = \frac{88.9 \text{ g}}{100 \text{ g}}$$

Step 2. Multiply the number of grams of each element by the proper conversion factor to obtain the relative number of moles of atoms:

H: $11.1 \text{ g H} \times \dfrac{1 \text{ mole H atoms}}{1.0 \text{ g H}} = 11.1 \text{ moles H atoms}$

O: $88.9 \text{ g O} \times \dfrac{1 \text{ mole O atoms}}{16.0 \text{ g O}} = 5.56 \text{ moles O atoms}$

The formula could be expressed as $H_{11.1}O_{5.56}$. However, it is customary to use the smallest whole-number ratio of atoms. This ratio is calculated in Step 3.

Step 3. Change these numbers to whole numbers by dividing each by the smallest.

$$H = \frac{11.1 \text{ moles}}{5.56 \text{ moles}} = 2.00 \qquad O = \frac{5.56 \text{ moles}}{5.56 \text{ moles}} = 1.00$$

In this step, the ratio of atoms has not changed, because we divided the number of moles of each element by the same number.

The simplest whole-number ratio of H to O in the compound is 2 to 1.

Empirical formula = H_2O

PROBLEM 9.13   The analysis of a salt showed that it contained 56.58% potassium, K, 8.68% carbon, C, and 34.73% oxygen, O. Calculate the empirical formula for this substance.

Steps 1 and 2. After changing the percentage of each element to grams, calculate the relative number of moles of each element by multiplying by the proper mole/g-at. wt factor.

$$K: \quad 56.58 \text{ g K} \times \frac{1 \text{ mole K atoms}}{39.1 \text{ g K}} = 1.45 \text{ moles K atoms}$$

$$C: \quad 8.68 \text{ g C} \times \frac{1 \text{ mole C atoms}}{12.0 \text{ g C}} = 0.720 \text{ mole C atoms}$$

$$O: \quad 34.73 \text{ g O} \times \frac{1 \text{ mole O atoms}}{16.0 \text{ g O}} = 2.17 \text{ moles O atoms}$$

Step 3. Divide each number of moles by the smallest of these three numbers.

$$K = \frac{1.45 \text{ moles}}{0.720 \text{ mole}} = 2.01$$

$$C = \frac{0.720 \text{ mole}}{0.720 \text{ mole}} = 1.00$$

$$O = \frac{2.17 \text{ moles}}{0.720 \text{ mole}} = 3.01$$

The simplest whole-number ratio of K : C : O is therefore 2 : 1 : 3.

Empirical formula = K_2CO_3

■

PROBLEM 9.14   A sulfide of iron was formed by combining 2.233 g of iron, Fe, with 1.926 g of sulfur, S. What is the empirical formula of the compound?

Steps 1 and 2. Calculate the relative number of moles of each element by multiplying grams of each element by the proper mole/g-at. wt factor.

$$Fe: \quad 2.233 \text{ g Fe} \times \frac{1 \text{ mole Fe atoms}}{55.8 \text{ g Fe}} = 0.0400 \text{ mole Fe atoms}$$

$$S: \quad 1.926 \text{ g S} \times \frac{1 \text{ mole S atoms}}{32.1 \text{ g S}} = 0.0600 \text{ mole S atoms}$$

Step 3. Divide each number of moles by the smaller of the two numbers.

$$Fe = \frac{0.0400 \text{ mole}}{0.0400 \text{ mole}} = 1.00$$

$$S = \frac{0.0600 \text{ mole}}{0.0400 \text{ mole}} = 1.50$$

Step 4. Since we still have not reached a ratio that will give a formula containing whole numbers of atoms, we must double each value to obtain a ratio of 2.00 atoms of Fe to 3.00 atoms of S. Doubling both values does not change the ratio of Fe and S atoms.

Fe: 1.00 × 2 = 2.00
 S: 1.50 × 2 = 3.00

The simplest ratio of Fe to S in the compound is 2 to 3.

Empirical formula = Fe_2S_3

In many of these calculations, results may vary somewhat from an exact whole number. This can be due to experimental errors in obtaining the data from the rounding off of numbers. Calculations that vary by no more than ±0.1 from a whole number can usually be rounded off to the nearest whole number. Deviations greater than about 0.1 unit usually mean that the calculated ratios need to be multiplied by a factor to make them all whole numbers.

■

9.7 Calculation of the Molecular Formula from the Empirical Formula

The molecular formula can be calculated from the empirical formula if the molecular weight is known in addition to data for calculating the empirical formula. The molecular formula, as stated in Section 9.5, will be equal to, or some multiple of, the empirical formula. For example, if the empirical formula of a compound of hydrogen and fluorine is HF, the molecular formula can be expressed as $(HF)_n$, where $n = 1, 2, 3, 4, \ldots$. This means that the molecular formula could be HF, H_2F_2, H_3F_3, H_4F_4, and so on. To determine the molecular formula, we must evaluate n.

$$n = \frac{\text{Molecular weight}}{\text{Empirical formula weight}} = \text{Number of empirical formula units}$$

What we actually calculate is the number of units of the empirical formula that is contained in the molecular formula.

PROBLEM 9.15   A compound of nitrogen and oxygen with a molecular weight of 92.0 was found to have an empirical formula of NO_2. What is its molecular formula?

Step 1. Let n be the number of (NO_2) units in a molecule; then the molecular formula is $(NO_2)_n$.
Step 2. Each (NO_2) unit weighs 46.0 g [14 + (2)(16)]. The gram-molecular weight of $(NO_2)_n$ = 92.0 g and the number of (46.0) units in 92.0 is 2.

$$n = \frac{92.0 \text{ g}}{46.0 \text{ g}} = 2 \quad \text{(Empirical formula units)}$$

Step 3. $2 \times NO_2$ $[(NO_2)_2]$ becomes N_2O_4, which is the molecular formula.
■

PROBLEM 9.16   The hydrocarbon propylene has a gram-molecular weight of 42.0 g/mole and contains 14.3% H and 85.7% C. What is its molecular formula?

Step 1. First calculate the empirical formula:

$$C:\quad 85.7\ g\ C \times \frac{1\ \text{mole C atoms}}{12.0\ g\ C} = 7.14\ \text{moles C atoms}$$

$$H:\quad 14.3\ g\ H \times \frac{1\ \text{mole H atoms}}{1.0\ g\ H} = 14.3\ \text{moles H atoms}$$

Divide each value by the smallest number of moles.

$$C = \frac{7.14\ \text{moles}}{7.14\ \text{moles}} = 1.0$$

$$H = \frac{14.3\ \text{moles}}{7.14\ \text{moles}} = 2.0$$

Empirical formula $= CH_2$

Step 2. Determine the molecular formula from the empirical formula and molecular weight.

Molecular formula $= (CH_2)_n$
Molecular weight $= 42.0$

Each CH_2 unit weighs 14.0 (12 + 2). The number of CH_2 units in 42.0 is 3.

$$n = \frac{42.0}{14.0} = 3 \quad \text{(Empirical formula units)}$$

Step 3. $3 \times CH_2\ [(CH_2)_3]$ becomes C_3H_6, which is the molecular formula.

■

QUESTIONS   A. **Review the Meanings of the New Terms Introduced in this Chapter**
   1. Formula weight
   2. Molecular weight
   3. Gram-formula weight
   4. Gram-molecular weight
   5. Percentage composition of a compound
   6. Empirical formula
   7. Molecular formula

B. **Review Questions**
   1. How are formula weight and molecular weight related to each other? In what respects are they different?
   2. How many molecules are present in 1 g-mol. wt of sulfuric acid, H_2SO_4? How many atoms are present?
   3. What is the relationship between the following?
      (a) Mole and molecular weight   (b) Mole and formula weight
   4. Why is it correct to refer to the weight of 1 mole of sodium chloride, but incorrect to refer to a molecular weight of sodium chloride?
   5. In calculating the empirical formula of a compound from its percentage composition, why do we choose to start with 100 g of the compound?

6. Which of the following statements are correct?
   (a) A mole of sodium and a mole of sodium chloride contain the same number of sodium atoms.
   (b) A compound such as NaCl has a formula weight but no true molecular weight.
   (c) One mole of nitrogen gas weighs 14.0 g.
   (d) The percentage of oxygen is higher in K_2CrO_4 than it is in Na_2CrO_4.
   (e) The number of Cr atoms is the same in a mole of K_2CrO_4 as it is in a mole of Na_2CrO_4.
   (f) Both K_2CrO_4 and Na_2CrO_4 contain the same percentage by weight of Cr.
   (g) A gram-molecular weight of sucrose, $C_{12}H_{22}O_{11}$, contains 1 mole of sucrose molecules.
   (h) Two moles of nitric acid, HNO_3, contain 6 moles of oxygen atoms.
   (i) The empirical formula of sucrose, $C_{12}H_{22}O_{11}$, is CH_2O.
   (j) A hydrocarbon that has a molecular weight of 280 and an empirical formula of CH_2 has a molecular formula of $C_{22}H_{44}$.
   (k) The empirical formula is often called the simplest formula.
   (l) The empirical formula of a compound gives the smallest ratio of the atoms that are present in a compound.
   (m) If the molecular formula and the empirical formula of a compound are not the same, the empirical formula will be an integral multiple of the molecular formula.
   (n) The empirical formula of benzene, C_6H_6, is CH.
   (o) A compound having an empirical formula of CH_2O, and a molecular weight of 60, has a molecular formula of $C_3H_6O_3$.

C. Problems

Formula Weight, Molecular Weight

1. Determine the molecular or formula weight of the following compounds:
   (a) KBr
   (b) Na_2SO_4
   (c) $Pb(NO_3)_2$
   (d) C_2H_5OH
   (e) $HC_2H_3O_2$
   (f) Fe_3O_4
   (g) $C_{12}H_{22}O_{11}$
   (h) $Al_2(SO_4)_3$
   (i) $(NH_4)_2HPO_4$
2. Determine the gram-formula weight of the following compounds:
   (a) NaOH
   (b) Ag_2CO_3
   (c) Cr_2O_3
   (d) $(NH_4)_2CO_3$
   (e) $Mg(HCO_3)_2$
   (f) C_6H_5COOH
   (g) $C_6H_{12}O_6$
   (h) $K_4Fe(CN)_6$
   (i) $BaCl_2 \cdot 2H_2O$

Moles

3. How many moles are contained in each of the following?
   (a) 25.0 g KOH
   (b) 18.0 g Cl_2
   (c) 0.585 g $CaCl_2$
   (d) 12.5 g C_2H_5OH
   (e) 1.77 g $NaNO_3$
   (f) 5.0 lb FeI_3
4. How many moles of atoms are contained in each of the following?
   (a) 8.6 g Al
   (b) 15.3 g I_2
   (c) 2.5×10^{22} atoms Cu
   (d) 3.5×10^{24} molecules N_2
5. Calculate the number of grams contained in each of the following:
   (a) 12.3 moles H_2O
   (b) 0.400 mole $CaCl_2$
   (c) 8.44 moles H_2SO_4
   (d) 6.53×10^{-3} mole $CuSO_4$
   (e) 20.0 moles O_2
   (f) 9.2×10^{22} molecules CH_3OH
6. How many molecules are contained in each of the following?
   (a) 3.0 moles O_2 (b) 0.22 mole CO_2 (c) 5.0 g CH_4 (d) 20.0 g HCl
   How many atoms are present in each of these amounts?

7. What is the weight in grams of each of the following?
   (a) 1 atom of gold
   (b) 1 molecule of water
   (c) 1 atom of neon
   (d) 1 molecule of NH_3
8. How many moles are contained in each of the following?
   (a) 1000 molecules CO_2
   (b) 2.0×10^8 atoms Mn
   (c) 1000 molecules C_6H_6
   (d) 8.5×10^{20} molecules Cl_2
   (e) 4 atoms C
   (f) 1.5×10^{25} molecules H_2O
9. How many atoms of oxygen are contained in each of the following?
   (a) $12 \text{ g } O_2$
   (b) 0.40 mole MgO
   (c) 4.0×10^{22} molecules $C_6H_{12}O_6$
   (d) 5.0 moles SO_2
   (e) $250 \text{ g } CaCO_3$
   (f) 5.0×10^{16} molecules H_2O
10. Calculate the number of:
    (a) Grams of silver in 25.0 g AgCl
    (b) Grams of chlorine in 5.00 g $BaCl_2$
    (c) Grams of nitrogen in 8.52 moles $(NH_4)_3PO_4$
    (d) Grams of oxygen in 9.5×10^{22} molecules SO_3
    (e) Grams of hydrogen in 35.0 g C_3H_8O
11. A solution was made by dissolving 15.0 g of ammonium nitrate, NH_4NO_3, in 250 g of water. How many moles of each compound were used?
*12. A sulfuric acid solution contains 45.0% H_2SO_4 by weight and has a density of 1.35 g/mL. How many moles of the acid are present in 1.00 L of the solution?
*13. A nitric acid solution containing 64.0% HNO_3 by weight has a density of 1.387 g/mL. How many moles of HNO_3 are present in 100 mL of the solution?

Percentage Composition
14. Calculate the percentage composition by weight of the following compounds:
    (a) KBr
    (b) CO_2
    (c) $BaSO_4$
    (d) $NaNO_3$
    (e) $Al(OH)_3$
    (f) Na_2CO_3
15. Calculate the percentage composition by weight of the following compounds:
    (a) ZnI_2
    (b) $NH_4C_2H_3O_2$
    (c) $Cr_2(SO_3)_3$
    (d) $Cr(NO_3)_2$
    (e) $K_2Cr_2O_7$
    (f) $(NH_4)_3PO_4$
16. What is the oxidation number of the first element in each compound in Question 14?
17. Calculate the percentage of iron, Fe, in the following compounds:
    (a) FeO
    (b) Fe_2O_3
    (c) Fe_3O_4
    (d) $K_3Fe(CN)_6$
18. Which of the following chlorides has the highest and which has the lowest percentage of chlorine, Cl, by weight, in its formula?
    (a) KCl
    (b) $BaCl_2$
    (c) $CHCl_3$
    (d) LiCl
19. A 5.88 g sample of manganese, Mn, was converted to an oxide, which was found to weigh 9.30 g. Calculate the percentage composition of the compound formed.
20. A 12.66 g sample of molybdenum, Mo, combined with 16.92 g of sulfur. Calculate the percentage composition of the compound formed.
21. Calculate the percentage of:
    (a) Mercury in Hg_2CO_3
    (b) Oxygen in $Mg(NO_3)_2$
    (c) Nitrogen in NH_4NO_3
    (d) Hydrogen in $HC_2H_3O_2$
22. Answer the following by examination of the formulas. Check your answers by calculations if you wish. Which compound has the:
    (a) Higher percent by weight of hydrogen, H_2O or H_2O_2?
    (b) Lower percent by weight of nitrogen, NO or N_2O_3?
    (c) Higher percent by weight of oxygen, NO_2 or N_2O_4?
    (d) Lower percent by weight of chlorine, NaCl or KCl?
    (e) Higher percent by weight of sulfur, $KHSO_4$ or K_2SO_4?
    (f) Lower percent by weight of chromium, Na_2CrO_4 or $Na_2Cr_2O_7$?

Empirical and Molecular Formulas

23. Calculate the empirical formula of each compound from the percentage compositions given.
    (a) 63.6% N, 36.4% O   (d) 43.4% Na, 11.3% C, 45.3% O
    (b) 46.7% N, 53.3% O   (e) 18.8% Na, 29.0% Cl, 52.3% O
    (c) 30.4% N, 69.6% O   (f) 49.4% K, 20.3% S, 30.3% O

24. Calculate the empirical formula of each compound from the percentage compositions given.
    (a) 44.3% Cu, 55.7% Br   (d) 55.3% K, 14.6% P, 30.1% O
    (b) 28.5% Cu, 71.5% Br   (e) 38.9% Ba, 29.4% Cr, 31.7% O
    (c) 51.9% Cr, 48.1% S   (f) 41.5% Zn, 17.8% N, 40.7% O

25. A sample of tin (Sn) weighing 2.664 g was oxidized and found to have combined with 0.718 g oxygen. Calculate the empirical formula of this oxide of tin.

26. A 7.443 g sample of vanadium (V) combined with oxygen to form 10.949 g of product. Calculate the empirical formula for this compound.

*27. Zinc and sulfur react to form zinc sulfide, ZnS. If we mix 15.3 g of zinc and 9.6 g of sulfur, have we added sufficient sulfur to fully react all the zinc? Show evidence for your answer.

*28. Sodium reacts with oxygen to form sodium oxide, Na_2O. If 70.0 g of sodium and 30.0 g of oxygen are placed in a closed container and reacted, is there sufficient sodium to react with all the oxygen? Show evidence for your answer.

29. Hydroquinone is an organic compound commonly used as a photographic developer. It has a molecular weight of 110 g/mole and a composition of 65.45% C, 5.45% H, and 29.09% O. Calculate the molecular formula of hydroquinone.

30. Fructose is a very sweet natural sugar that is present in honey, fruits, and fruit juices. It has a molecular weight of 180 g/mole and a composition of 40.0% C, 6.7% H, and 53.3% O. Calculate the molecular formula of fructose.

31. Aspirin is well known as a pain reliever (analgesic) and as a fever reducer (antipyretic). It has a molecular weight of 180 and a composition of 60.0% C, 4.48% H, and 35.5% O. Calculate the molecular formula of aspirin.

32. A compound was found to have a composition of 14.43% manganese, 66.65% iodine, and 18.92% water. Calculate the empirical formula of this hydrate. See Table 8.1 for examples of formulas of hydrates.

33. Listed below are the compositions of four different compounds of carbon and hydrogen. Determine both the empirical formula and the molecular formula for each.

	Percentage C	Percentage H	Molecular weight
(a)	74.87	25.13	16.0
(b)	79.89	20.11	30.0
(c)	85.63	14.37	42.0
(d)	93.75	6.25	128

10 Chemical Equations

After studying Chapter 10 you should be able to:

1. Understand the terms listed in Question A at the end of the chapter.
2. Know the format used in setting up chemical equations.
3. Recognize the various symbols commonly used in writing chemical equations.
4. Balance simple chemical equations.
5. Interpret a balanced equation in terms of the relative numbers or amounts of molecules, atoms, grams, or moles of each substance represented.
6. Classify equations as representing combination, decomposition, single-replacement, or double-replacement reactions.
7. Complete and balance equations for simple combination, decomposition, single-replacement, and double-replacement reactions when given the reactants.
8. Distinguish between exothermic and endothermic reactions, and relate the quantity of heat to the amounts of substances involved in the reaction.

10.1 The Chemical Equation

chemical equation

word equation

A **chemical equation** is a shorthand expression for a chemical change or reaction. It shows, among other things, the rearrangement of atoms that are involved in the reaction. A **word equation** states in words, in equation form, the substances involved in a chemical reaction. For example, when mercury(II) oxide is heated, it decomposes to form mercury and oxygen. The word equation for this decomposition is

$$\text{Mercury(II) oxide} + \text{Heat} \longrightarrow \text{Mercury} + \text{Oxygen}$$

From the chemist's point of view, this method of describing a chemical change is inadequate. It is bulky and cumbersome to use and does not give quantitative information. The chemical equation, using symbols and formulas, is a far better way to describe the decomposition of mercury(II) oxide:

$$2\,HgO \xrightarrow{\Delta} 2\,Hg + O_2\uparrow$$

This equation gives all the information from the word equation plus formulas, composition, reactive amounts of all the substances involved in the reaction, and much additional information (see Section 10.4). However, even though a chemical equation provides much quantitative information, it is still not a complete description; it does not tell us how much heat is needed to cause decomposition, what we observe during the reaction, or anything about the rate of reaction. This information must be obtained from other sources or from experimentation.

10.2 Format for Writing Chemical Equations

reactant
product

A chemical equation consists of reactants and products, along with other symbolic terms. The **reactants** are the substances that enter into a chemical change or reaction. The **products** are the substances produced by the reaction. A chemical equation uses the chemical symbols and formulas of the reactants and products and other symbolic terms to represent a chemical reaction. Equations are written according to this general format:

1. The reactants and products are separated by an arrow or other sign indicating equality between reactants and products ($\longrightarrow$, $=$, $\rightleftarrows$).
2. The reactants are placed to the left and the products to the right of the arrow or equality sign. A plus sign ($+$) is placed between reactants and between products when needed.
3. Conditions required to carry out the reaction may, if desired, be placed above or below the arrow or equality sign. For example, a delta sign placed over the arrow ($\xrightarrow{\Delta}$) indicates that heat is supplied to the reaction.
4. Small integral numbers in front of substances (for example, 3 H_2O) are used to balance the equation and to indicate the number of formula units (atoms, molecules, moles, ions) of each substance reacting or being produced. When no number is shown, it is understood that one formula unit of the substance is indicated. A number placed in front of a symbol or formula is called a *coefficient*. Thus, 3 is a coefficient in 3 H_2O.

Symbols commonly used in equations are given in Table 10.1.

TABLE 10.1 Symbols commonly used in chemical equations.

Symbol	Meaning
$\rightarrow$	Yields, produces (points to products)
$=$	Equals; equilibrium between reactants and products
$\rightleftarrows$	Reversible reaction; equilibrium between reactants and products
$\uparrow$	Gas evolved (written after a substance)
$\downarrow$	Solid or precipitate formed (written after a substance)
(s)	Solid (written after a substance)
(l)	Liquid (written after a substance)
(g)	Gas (written after a substance)
Δ	Heat
$+$	Plus or added to
(aq)	Aqueous solution (substance dissolved in water)

10.3 Writing and Balancing Equations

balanced equation

To represent the quantitative relationships of a reaction, the chemical equation must be balanced. A **balanced equation** is one that contains the same number of each kind of atom on each side of the equation. The balanced equation, therefore, obeys the Law of Conservation of Mass.

The ability to balance equations must be acquired by every chemistry student. Simple equations are easy to balance, but some care and attention to detail are required. Clearly, the way to balance an equation is to adjust the number of atoms of each element so that it is the same on each side of the equation. But we must not change a correct formula in order to achieve a balanced equation. Each equation must be treated on its own merits; there is no simple "plug in" formula for balancing equations. The following outline gives a general procedure for balancing equations. Study this outline and refer to it as needed when working examples. There is no substitute for practice in learning to write and balance equations.

1. **Identify the reaction for which the equation is to be written.** Formulate a description or word equation for this reaction (for example, mercury(II) oxide decomposes yielding mercury and oxygen). This, of course, need not be done when the reactants and products are identified and their formulas are given.

2. **Write the unbalanced, or skeleton, equation.** Make sure that the formula for each substance is correct and that the reactants are written to the left and the products to the right of the arrow (for example, $HgO \longrightarrow Hg + O_2$). The correct formulas must be known or ascertained from the periodic table, oxidation numbers, lists of ions, or experimental data.

3. **Balance the equation.** Use the following steps as necessary:
   (a) Count and compare the number of atoms of each element on each side of the equation and determine those that must be balanced.
   (b) Balance each element, one at a time, by placing small whole numbers (coefficients) in front of the formulas containing the unbalanced element. It is usually best to balance metals first, then nonmetals, then hydrogen and oxygen. Select the smallest coefficients that will give the same number of atoms of the element on each side. A coefficient placed before a formula multiplies every atom in the formula by that number (for example, $2\ H_2SO_4$ means two molecules of sulfuric acid and also means four H atoms, two S atoms, and eight O atoms).
   (c) Check all other elements after each individual element is balanced to see if, in balancing one, other elements have become unbalanced. Make adjustments as needed.

(d) Balance polyatomic ions such as SO_4^{2-}, OH^-, and NO_3^-, which remain unchanged from one side of the equation to the other, in the same way as individual atoms.

(e) Do a final check, making sure that each element and/or polyatomic ion is balanced and that the smallest possible set of whole-number coefficients has been used.

$$4\,HgO \longrightarrow 4\,Hg + 2\,O_2 \quad \text{(Incorrect form;}$$
divide each coefficient by 2)

$$2\,HgO \longrightarrow 2\,Hg + O_2 \quad \text{(Correct form)}$$

The following problems show stepwise sequences leading to balanced equations. Study each one carefully.

PROBLEM 10.1 Write the balanced equation for the reaction that takes place when magnesium metal is burned in air to produce magnesium oxide. Magnesium combines with oxygen in the air.

1. Word equation:

Magnesium + Oxygen $\longrightarrow$ Magnesium oxide

2. Skeleton equation:

$Mg + O_2 \longrightarrow MgO$ (Unbalanced)

3. Balance:
(a) Magnesium is balanced; oxygen is not balanced. There are two O atoms on the left side and one on the right side.
(b) Place the coefficient 2 before MgO to balance the O atoms:

$Mg + O_2 \longrightarrow 2\,MgO$ (Unbalanced)

(c) Now Mg is not balanced. There is one Mg atom on the left side and two on the right side. Place a 2 before Mg:

$2\,Mg + O_2 \longrightarrow 2\,MgO$ (Balanced)

(d) *Check:* Each side has two Mg and two O atoms.

PROBLEM 10.2 Methane, CH_4, undergoes complete combustion to produce carbon dioxide and water. Write the balanced equation for this reaction.

1. Word equation:

Methane + Oxygen $\longrightarrow$ Carbon dioxide + Water

2. Skeleton equation:

$CH_4 + O_2 \longrightarrow CO_2 + H_2O$ (Unbalanced)

3. **Balance:**
   (a) Carbon is balanced; hydrogen and oxygen are not balanced.
   (b) Balance H atoms by placing a 2 before H_2O:

   $$CH_4 + O_2 \longrightarrow CO_2 + 2\,H_2O \quad \text{(Unbalanced)}$$

   Each side of the equation has four H atoms; oxygen is still not balanced. Place a 2 before O_2 to balance the oxygen atoms:

   $$CH_4 + 2\,O_2 \longrightarrow CO_2 + 2\,H_2O \quad \text{(Balanced)}$$

   (c) *Check:* The equation is correctly balanced; it has one C atom, four O atoms, and four H atoms on each side.

■

PROBLEM 10.3   Oxygen and potassium chloride are formed by heating potassium chlorate. Write a balanced equation for this reaction.

1. **Word equation:**

   Potassium chlorate $\xrightarrow{\Delta}$ Potassium chloride + Oxygen

2. **Skeleton equation:**

   $$KClO_3 \xrightarrow{\Delta} KCl + O_2 \quad \text{(Unbalanced)}$$

3. **Balance:**
   (a) Potassium and chlorine are balanced. Oxygen is unbalanced (three O atoms on the left side and two on the right side).
   (b) How many oxygen atoms are needed? The subscripts of oxygen (3 and 2) in $KClO_3$ and O_2 have a common denominator of 6. Therefore, coefficients of $KClO_3$ and O_2 are needed to give six oxygen atoms on each side. Place a 2 before $KClO_3$ and a 3 before O_2 to give six O atoms on each side:

   $$2\,KClO_3 \xrightarrow{\Delta} KCl + 3\,O_2 \quad \text{(Unbalanced)}$$

   Now K and Cl are not balanced. Place a 2 before KCl, which balances both K and Cl at the same time:

   $$2\,KClO_3 \xrightarrow{\Delta} 2\,KCl + 3\,O_2 \quad \text{(Balanced)}$$

   (c) *Check:* Each side contains two K, two Cl, and six O atoms.

■

PROBLEM 10.4   Balance by starting with the word equation given.

1. **Word equation:**

   Silver nitrate + Hydrogen sulfide $\longrightarrow$ Silver sulfide + Nitric acid

2. **Skeleton equation:**

   $$AgNO_3 + H_2S \longrightarrow Ag_2S + HNO_3 \quad \text{(Unbalanced)}$$

3. **Balance:**
   (a) Ag and H are unbalanced.
   (b) Place a 2 in front of $AgNO_3$ to balance Ag:

   $$2\,AgNO_3 + H_2S \longrightarrow Ag_2S + HNO_3 \quad \text{(Unbalanced)}$$

   (c) This leaves H and NO_3^- unbalanced. Balance by placing a 2 in front of HNO_3:

   $$2\,AgNO_3 + H_2S \longrightarrow Ag_2S + 2\,HNO_3 \quad \text{(Balanced)}$$

   (d) In this example, N and O atoms are balanced by balancing the NO_3^- ion as a unit.
   (e) *Check:* Each side has two Ag, two H, and one S atom. Also, each side has two NO_3^- ions.

   ■

PROBLEM 10.5   Balance by starting with the word equation given.

1. **Word equation:**

   Aluminum hydroxide + Sulfuric acid $\longrightarrow$ Aluminum sulfate + Water

2. **Skeleton equation:**

   $$Al(OH)_3 + H_2SO_4 \longrightarrow Al_2(SO_4)_3 + H_2O \quad \text{(Unbalanced)}$$

3. **Balance:**
   (a) All elements are unbalanced.
   (b) Balance Al by placing a 2 in front of $Al(OH)_3$. Treat the unbalanced SO_4^{2-} ion as a unit and balance by placing a 3 before H_2SO_4. Note that Step 3 (d) may sometimes be combined with Step 3 (b) (see pages 213–214).

   $$2\,Al(OH)_3 + 3\,H_2SO_4 \longrightarrow Al_2(SO_4)_3 + H_2O \quad \text{(Unbalanced)}$$

   Balance the unbalanced H and O by placing a 6 in front of H_2O:

   $$2\,Al(OH)_3 + 3\,H_2SO_4 \longrightarrow Al_2(SO_4)_3 + 6\,H_2O \quad \text{(Balanced)}$$

   (c) *Check:* Each side has 2 Al, 12 H, 3 S, and 18 O atoms.

   ■

PROBLEM 10.6   When butane undergoes complete combustion (burns), it reacts with oxygen to form carbon dioxide and water. Write the balanced equation for this reaction.

1. **Word equation:**

   Butane + Oxygen $\longrightarrow$ Carbon dioxide + Water

2. **Skeleton equation:**

   $$C_4H_{10} + O_2 \longrightarrow CO_2 + H_2O \quad \text{(Unbalanced)}$$

3. **Balance:**

  (a) All elements are unbalanced.

  (b) In this equation it is best to balance oxygen last, since it occurs in more than one product. First, balance C by placing a 4 in front of CO_2:

  $$C_4H_{10} + O_2 \longrightarrow 4\,CO_2 + H_2O \quad \text{(Unbalanced)}$$

  Balance H by placing a 5 in front of H_2O:

  $$C_4H_{10} + O_2 \longrightarrow 4\,CO_2 + 5\,H_2O \quad \text{(Unbalanced)}$$

  Oxygen remains unbalanced. The oxygen atoms on the right side are fixed, since $4\,CO_2$ and $5\,H_2O$ are balanced. When we try to balance the O atoms, we find that there is no integer (whole number) that can be placed in front of O_2 to bring about a balance. The equation can be balanced if we use $6\frac{1}{2}\,O_2$ and then double the coefficients of each substance, including the $6\frac{1}{2}\,O_2$, to obtain the balanced equation:

  $$2\,C_4H_{10} + 13\,O_2 \longrightarrow 8\,CO_2 + 10\,H_2O \quad \text{(Balanced)}$$

  An alternative procedure is to rewrite the last unbalanced equation, doubling all the coefficients except that of O_2:

  $$2\,C_4H_{10} + O_2 \longrightarrow 8\,CO_2 + 10\,H_2O$$

  Now balance the O_2—the result is the same balanced equation as above.

  (c) *Check:* Each side now has 8 C, 20 H, and 26 O atoms.

■

10.4 What Information Does an Equation Give Us?

The meaning of a formula has been considerably expanded since formulas were introduced in Chapter 4. Depending on the context in which it is used, a formula can have several meanings. For example, the formula H_2O can be used to indicate the following:

$$H_2O = \begin{cases} \text{1 molecule of water} \\ \text{2 hydrogen atoms plus 1 oxygen atom} \\ \text{1 molecular weight of water} \\ \text{1 mole of water} \\ 6.02 \times 10^{23} \text{ molecules of water} \\ \text{18.0 grams of water} \end{cases}$$

Formulas used in equations can be expressed in any of these units, moles being the most commonly used unit. For example, in the reaction of hydrogen and oxygen to form water,

$$2\,H_2 + O_2 \longrightarrow 2\,H_2O$$

the $2\,H_2$ can represent 2 moles or 2 molecules of hydrogen; the O_2 can represent 1 mole or 1 molecule of oxygen; the $2\,H_2O$ can represent 2 moles or 2 molecules of water. One way this equation is commonly read is as follows: 2 moles of H_2 react with 1 mole of O_2 to give 2 moles of H_2O.

Interpreting the information given in an equation is important if we are to gain the full benefit of its use in evaluating a chemical reaction. The balanced equation tells us the following:

1. What the reactants are and what the products are
2. The formulas of the reactants and products
3. The number of molecules or formula units of reactants and products in the reaction
4. The number of atoms of each element involved in the reaction
5. The number of molecular weights or formula weights of each substance used or produced
6. The number of moles of each substance
7. The number of gram-molecular weights or gram-formula weights of each substance used or produced
8. The number of grams of each substance used or produced

Consider the following equation:

$$H_2(g) + Cl_2(g) \longrightarrow 2\,HCl(g)$$

This equation states that hydrogen gas reacts with chlorine gas to produce hydrogen chloride, also a gas. Let us summarize all the information relating to the equation. The information that can be stated about the relative amount of each substance, with respect to all other substances in the balanced equation, is written below its formula in the following equation:

$H_2(g)$ +	$Cl_2(g)$ $\longrightarrow$	$2\,HCl(g)$
Hydrogen	*Chlorine*	*Hydrogen chloride*
1 molecule	1 molecule	2 molecules
2 atoms	2 atoms	2 atoms H + 2 atoms Cl
1 mol. wt	1 mol. wt	2 mol. wt
1 mole	1 mole	2 moles
2.0 g	71.0 g	2 × 36.5 g (73.0 g)

Let us try another equation. When propane gas is burned in air, the products are carbon dioxide, CO_2, and water, H_2O. The equation and its interpretation are as follows:

$C_3H_8(g)$ +	$5\,O_2(g)$ $\longrightarrow$	$3\,CO_2(g)$ +	$4\,H_2O(g)$
Propane	*Oxygen*	*Carbon dioxide*	*Water*
1 molecule	5 molecules	3 molecules	4 molecules
3 atoms C	10 atoms O	3 atoms C	8 atoms H
8 atoms H		6 atoms O	4 atoms O
1 mol. wt	5 mol. wt	3 mol. wt	4 mol. wt
1 mole	5 moles	3 moles	4 moles
44.0 g	5 × 32.0 (160.0 g)	3 × 44.0 (132.0 g)	4 × 18.0 (72.0 g)

These data are very useful in calculating quantitative relationships that exist among substances in a chemical reaction. For example, if we react 2 moles of hydrogen (twice as much as is indicated by the equation) with 2 moles of chlorine, we can expect to obtain 4 moles, or 146 g, of hydrogen chloride, as a product. We will study this phase of using equations in more detail in the next chapter.

10.5 Types of Chemical Equations

Chemical equations represent chemical changes or reactions. To be of any significance, an equation must represent an actual or possible reaction. Part of the problem in writing equations is determining the products formed. There is no sure method of predicting products, nor do we have time to carry out experimentally all the reactions we may wish to consider. Therefore, we must use data reported in the writings of other workers, certain rules to aid in our predictions, and the atomic structure and combining capacities of the elements to help us predict the formulas of the products of a chemical reaction. The final proof of the existence of any reaction, of course, is in the actual observation of the reaction in the laboratory (or elsewhere).

Reactions are classified into types to assist in writing equations and to aid in predicting other reactions. Many chemical reactions fit one or another of the four principal reaction types that are discussed in the following paragraphs. Reactions are also classified as oxidation–reduction. Special methods are used to balance complex oxidation–reduction equations (see Chapter 17).

combination or synthesis reaction

1. Combination or synthesis reaction. In this type of reaction, direct union or combination of two substances produces one new substance. Oxidation–reduction is involved in some, but not all, combination reactions. The general form of the equation is

$$A + B \longrightarrow AB$$

where A and B are either elements or compounds and AB is a compound. The formula of the compound in many cases can be determined from a knowledge of the oxidation numbers of the reactants in their combined states. Some reactions that fall into this category are the following:

(a) Metal + Oxygen $\longrightarrow$ Metal oxide
$$2\,Mg + O_2 \xrightarrow{\Delta} 2\,MgO$$
$$4\,Al + 3\,O_2 \xrightarrow{\Delta} 2\,Al_2O_3$$

(b) Nonmetal + Oxygen $\longrightarrow$ Nonmetal oxide
$$S + O_2 \xrightarrow{\Delta} SO_2$$
$$N_2 + O_2 \xrightarrow{\Delta} 2\,NO$$

(c) Metal + Nonmetal $\longrightarrow$ Salt
$$2\,Na(s) + Cl_2(g) \longrightarrow 2\,NaCl(s)$$
$$2\,Al(s) + 3\,Br_2(l) \longrightarrow 2\,AlBr_3(s)$$

(d) Metal oxide + Water $\longrightarrow$ Base (metal hydroxide)

$Na_2O(s) + H_2O(l) \longrightarrow 2\ NaOH(aq)$

$CaO + H_2O \longrightarrow Ca(OH)_2$

(e) Nonmetal oxide + Water $\longrightarrow$ Oxyacid

$SO_3(g) + H_2O(l) \longrightarrow H_2SO_4(aq)$

$N_2O_5(s) + H_2O(l) \longrightarrow 2\ HNO_3(aq)$

decomposition reaction

2. Decomposition reaction.   In this type of reaction a single substance is decomposed or broken down into two or more different substances. The reaction may be considered the reverse of combination. The starting material must be a compound, and the products may be elements or compounds. Oxidation–reduction is involved in some, but not all, decomposition reactions. The general form of the equation is

$$AB \longrightarrow A + B$$

Predicting the products of a decomposition reaction can be difficult and requires an understanding of each individual reaction. Heating oxygen-containing compounds often results in decomposition. Some reactions that fall into this category are the following:

(a) Metal oxides—some metal oxides decompose to yield the free metal plus oxygen. Others give a lower oxide, and some are very stable, resisting decomposition by heating.

$2\ HgO(s) \xrightarrow{\Delta} 2\ Hg(l) + O_2(g)$

$2\ PbO_2(s) \xrightarrow{\Delta} 2\ PbO(s) + O_2(g)$

(b) Carbonates and bicarbonates decompose to yield CO_2 when heated.

$CaCO_3(s) \xrightarrow{\Delta} CaO(s) + CO_2(g)$

$2\ NaHCO_3(s) \xrightarrow{\Delta} Na_2CO_3(s) + H_2O(g) + CO_2(g)$

(c) Miscellaneous.

$2\ KClO_3(s) \xrightarrow{\Delta} 2\ KCl(s) + 3\ O_2(g)$

$2\ NaNO_3(s) \longrightarrow 2\ NaNO_2(s) + O_2(g)$

$2\ H_2O(l) \xrightarrow{\text{Electrical energy}} 2\ H_2(g) + O_2(g)$

single-replacement or substitution reaction

3. Single-replacement or substitution reaction.   In this type of reaction one simple substance (element) reacts with a compound substance to form a new simple substance and a new compound, one element displacing another in a compound. Oxidation–reduction is always present in single-replacement reactions. The general form of the equation is

$$A + BC \longrightarrow B + AC$$

If A is a metal, A will replace B to form AC, providing A is a more reactive metal than B. If A is a halogen, it will replace C to form BA, providing A is a more reactive halogen than C.

A brief list, in descending order of reactivity, of selected metals (and hydrogen) and the halogens follows. Any metal on the list will react with and replace the ions of those metals that appear anywhere to the right of it on the list. For example, zinc metal will replace hydrogen from a hydrochloric acid solution. But copper metal, which is to the right of hydrogen on the list and thus less reactive than hydrogen, will not replace hydrogen from a hydrochloric acid solution.

Metals: K, Ca, Na, Mg, Al, Zn, Fe, Sn, Pb, H, Cu, Ag, Hg, Au
Halogens: F_2, Cl_2, Br_2, I_2

Some reactions that fall into this category follow:

(a) Metal + Acid $\longrightarrow$ Hydrogen + Salt
$$Zn(s) + 2\,HCl(aq) \longrightarrow H_2(g) + ZnCl_2(aq)$$
$$2\,Al(s) + 3\,H_2SO_4(aq) \longrightarrow 3\,H_2(g) + Al_2(SO_4)_3(aq)$$

(b) Metal + Water $\longrightarrow$ Hydrogen + Metal hydroxide or metal oxide
$$2\,Na(s) + 2\,H_2O \longrightarrow H_2(g) + 2\,NaOH(aq)$$
$$Ca(s) + 2\,H_2O \longrightarrow H_2(g) + Ca(OH)_2(aq)$$
$$3\,Fe(s) + 4\,H_2O(g) \longrightarrow 4\,H_2(g) + Fe_3O_4(s)$$
$$\text{Steam}$$

(c) Metal + Salt $\longrightarrow$ Metal + Salt
$$Fe(s) + CuSO_4(aq) \longrightarrow Cu(s) + FeSO_4(aq)$$
$$Cu(s) + 2\,AgNO_3(aq) \longrightarrow 2\,Ag(s) + Cu(NO_3)_2(aq)$$

(d) Halogen + Halogen salt $\longrightarrow$ Halogen + Halogen salt
$$Cl_2(g) + 2\,NaBr(aq) \longrightarrow Br_2(l) + 2\,NaCl(aq)$$
$$Cl_2(g) + 2\,KI(aq) \longrightarrow I_2(s) + 2\,KCl(aq)$$

double-replacement or metathesis reaction

4. **Double-replacement or metathesis reaction.** In this type of reaction, two compounds react with each other to produce two different compounds. Oxidation–reduction does not occur in double-replacement reactions. The general form of the equation is

$$AB + CD \longrightarrow AD + CB$$

This reaction may be thought of as an exchange of positive and negative groups, where the positive group (A) of the first reactant combines with the negative group (D) of the second reactant, and the positive group (C) of the second reactant combines with the negative group (B) of the first reactant. In writing the formulas of the products, we must take into account the oxidation numbers of the combining groups. Some reactions that fall into this category are:

(a) Neutralization of an acid and a base
Acid + Base $\longrightarrow$ Salt + Water
$$HCl(aq) + NaOH(aq) \longrightarrow NaCl(aq) + H_2O$$
$$H_2SO_4(aq) + Ba(OH)_2(aq) \longrightarrow BaSO_4\!\downarrow + 2\,H_2O$$

(b) Formation of an insoluble precipitate
$$BaCl_2(aq) + 2\,AgNO_3(aq) \longrightarrow 2\,AgCl\downarrow + Ba(NO_3)_2(aq)$$
$$FeCl_3(aq) + 3\,NH_4OH(aq) \longrightarrow Fe(OH)_3\downarrow + 3\,NH_4Cl(aq)$$

(c) Metal oxide + Acid $\longrightarrow$ Salt + Water
$$CuO(s) + 2\,HNO_3(aq) \longrightarrow Cu(NO_3)_2(aq) + H_2O$$
$$CaO(s) + 2\,HCl(aq) \longrightarrow CaCl_2(aq) + H_2O$$

(d) Formation of a gas
$$H_2SO_4(l) + NaCl(s) \longrightarrow NaHSO_4(s) + HCl\uparrow$$
$$2\,HCl(aq) + ZnS(s) \longrightarrow ZnCl_2(aq) + H_2S\uparrow$$
$$2\,HCl(aq) + Na_2CO_3(s) \longrightarrow 2\,NaCl(aq) + H_2O(l) + CO_2\uparrow$$

All substances that we attempt to react may not react, or the conditions under which they react may not be present. For example, mercury(II) oxide does not decompose until it is heated; magnesium does not burn in air or oxygen until the temperature is raised to the point at which it begins to react. When silver is placed in a solution of copper(II) sulfate, no reaction takes place; however, when a strip of copper is placed in a solution of silver nitrate, the single-replacement reaction as given in 3(c) above takes place because copper is a more reactive metal than silver. The successful prediction of the products of a reaction is not always easy. The ability to predict products correctly comes with knowledge and experience. Although you may not be able to predict many reactions at this point, as you continue you will find that reactions can be categorized, and that prediction of the products thereby becomes easier—if not always certain.

PROBLEM 10.7   Write the equation for the reaction between aqueous solutions of hydrobromic acid and potassium hydroxide.

First write the formulas for the reactants. They are HBr and KOH. Then classify the type of reaction that would occur between them. Since both reactants are compounds, and since one is an acid and one is a base, the reaction will be of the neutralization type:

Acid + Base $\longrightarrow$ Salt + Water

Now rewrite the equation by putting down the formulas for the known substances:

$$HBr(aq) + KOH(aq) \longrightarrow Salt + H_2O$$

In this reaction, which is a double-replacement type, the H^+ from the acid combines with the OH^- from the base to form water. The salt must be composed of the other two ions, K^+ and Br^-. We determine the formula of the salt to be KBr from the fact that K is a $+1$ ion and Br is a -1 ion. The final balanced equation is

$$HBr(aq) + KOH(aq) \longrightarrow KBr(aq) + H_2O$$

■

PROBLEM 10.8   Complete and balance the equation for the reaction between aqueous solutions of barium chloride and sodium sulfate.

First determine the formulas for the reactants. They are $BaCl_2$ and Na_2SO_4. Then classify these substances as acids, bases, or salts. Both substances are salts. Since both substances are compounds, the reaction looks as if it will be of the double-replacement type. Start to write the equation with the reactants:

$$BaCl_2(aq) + Na_2SO_4(aq) \longrightarrow$$

If the reaction is double replacement, Ba^{2+} will be written combined with SO_4^{2-} and Na^+ with Cl^- as the products. The balanced equation is

$$BaCl_2(aq) + Na_2SO_4(aq) \longrightarrow BaSO_4 + 2\,NaCl$$

The final step is to determine the nature of the products, which controls whether or not the reaction will take place. If both products are soluble, we may merely have a mixture of all the possible products in solution. But if an insoluble precipitate is formed, the reaction will definitely occur. We know from experience that NaCl is fairly soluble in water, but what about $BaSO_4$? The Solubility Table in Appendix IV can give us this information. From this table we see that $BaSO_4$ is insoluble in water, so it will be a precipitate in the reaction. Thus the reaction will occur, a white precipitate will form, and the equation is

$$BaCl_2(aq) + Na_2SO_4(aq) \longrightarrow BaSO_4\downarrow + 2\,NaCl(aq)$$

■

There is a great deal yet to learn about which substances react with each other, how they react, and what conditions are necessary to bring about their reaction. It is possible to make accurate predictions concerning the occurrence of proposed reactions. Such predictions require, in addition to appropriate data, a good knowledge of thermodynamics. But the study of this subject is usually reserved for advanced courses in chemistry and physics. Even without the formal use of thermodynamics, your knowledge of such generalities as the four reaction types just cited, the periodic table, atomic structure, oxidation numbers, and so on, can be put to good use in predicting reactions and in writing equations. Indeed, such applications serve to make chemistry an interesting and fascinating study.

10.6 Heat in Chemical Reactions

Energy changes alway accompany chemical reactions. One reason why reactions may occur is that the products attain a lower, more stable energy state than the reactants. For the products to attain this more stable state, energy must be liberated and given off to the surroundings as heat (or as heat and work). When a solution of a base is neutralized by the addition of an acid, the liberation of heat energy is signaled by an immediate rise in the temperature of the solution. When an automobile engine burns gasoline, heat is certainly liberated, and, at the same time, part of the liberated energy does the work of moving the automobile.

exothermic reaction

endothermic reaction

Reactions are either exothermic or endothermic. **Exothermic reactions** liberate heat; **endothermic reactions** absorb heat. In an exothermic reaction, heat is a product and may be written on the right side of the equation for the reaction. In an endothermic reaction, heat can be regarded as a reactant and is written on the left side of the equation. Examples indicating heat in an exo-thermic and an endothermic reaction follow.

$$H_2(g) + Cl_2(g) \longrightarrow 2\,HCl(g) + 44.2\,kcal \quad \text{(Exothermic)}$$

$$N_2(g) + O_2(g) + 43.2\,kcal \longrightarrow 2\,NO(g) \quad \text{(Endothermic)}$$

heat of reaction

The quantity of heat produced by a reaction is known as the **heat of reaction.** The units used can be calories or kilocalories. Consider the reaction represented by this equation:

$$C(s) + O_2(g) \longrightarrow CO_2(g) + 94.0\,kcal$$

When the heat liberated is expressed as part of the equation, the substances are expressed in units of moles. Thus, when 1 mole (12.0 g) of C combines with 1 mole (32.0 g) of O_2, 1 mole (44.0 g) of CO_2 is formed and 94.0 kcal of heat are liberated. Assuming that coal is 90% C and the combustion product is only CO_2, 6.4×10^9 calories would be released when 1 ton of coal is burned. In this reac-tion, as in many others, the heat or energy is more useful than the chemical products. At 1 calorie per gram per degree, 6.4×10^9 calories is sufficient energy to heat about 21,000 gallons of water from room temperature (20°C) to 100°C.

As another example, aluminum metal is produced by electrolyzing alu-minum oxide:

$$2\,Al_2O_3 + 779\,kcal \longrightarrow 4\,Al + 3\,O_2$$

For each mole of Al_2O_3 the equivalent of 389.5 kcal must be supplied as elec-trical energy. Since only 54 g of aluminum are obtained from 1 mole of Al_2O_3, this means that each ton of aluminum produced requires more than 6.56×10^6 kcal of energy.

Be careful not to confuse an exothermic reaction that merely requires heat (activation energy) to get it started with a truly endothermic process. The com-bustion of magnesium is highly exothermic, yet magnesium must be heated to a fairly high temperature in air before combustion begins. Once started, however, the combustion reaction goes very vigorously until either the magne-sium or the available supply of oxygen is exhausted. The electrolytic decom-position of water to hydrogen and oxygen is highly endothermic. If the electric current is shut off when this process is going on, the reaction stops instantly. The relative energy levels of reactants and products in exothermic and in endo-thermic processes are presented graphically in Figure 10.1.

In reaction (a) of Figure 10.1, the products are at a lower energy level than the reactants. Energy (heat) was given off to the surroundings and the reaction is exothermic. In reaction (b), the products are at a higher energy level than the reactants. Energy has therefore been absorbed and the reaction is endothermic.

Examples of endothermic and exothermic processes that can be demon-strated easily in the laboratory are shown in Figure 10.2. When dissolving am-monium chloride, NH_4Cl, in water, we observe an endothermic process. The temperature changes from 24.5°C to 18.1°C when 10 g of NH_4Cl are added

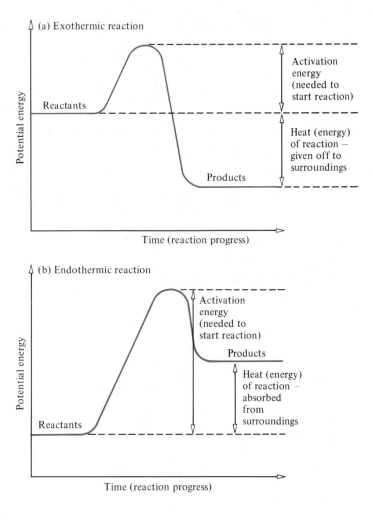

Figure 10.1 Energy states in exothermic and endothermic reactions.

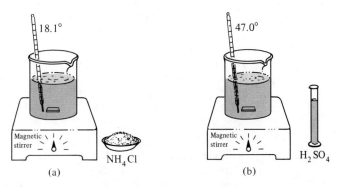

Figure 10.2 Endothermic and exothermic processes.

to 100 mL of water. This energy, in the form of heat, is taken from the immediate surroundings, the water, causing the salt solution to become colder. In the second example, we observe a temperature change of from 24.8°C to 47.0°C when 10 mL of concentrated sulfuric acid, H_2SO_4, are dissolved in 100 mL of water. This is an exothermic process. In both examples, the temperature changes are large enough to be detected by touching the containers.

QUESTIONS

A. **Review the Meanings of the New Terms Introduced in this Chapter**
   1. Chemical equation
   2. Word equation
   3. Reactant
   4. Product
   5. Balanced equation
   6. Combination reaction
   7. Decomposition reaction
   8. Single-replacement reaction
   9. Double-replacement or metathesis reaction
   10. Exothermic reaction
   11. Endothermic reaction
   12. Heat of reaction

B. Study Table 10.1 so that you will be familiar with the most common symbols used in equations.

C. **Review Questions**
   1. Balance the following equations:
      (a) $H_2 + O_2 \longrightarrow H_2O$
      (b) $H_2 + Br_2 \longrightarrow HBr$
      (c) $Ca + O_2 \longrightarrow CaO$
      (d) $H_2O_2 \longrightarrow H_2O + O_2$
      (e) $Ba(ClO_3)_2 \xrightarrow{\Delta} BaCl_2 + O_2$
      (f) $H_2SO_4 + NaOH \longrightarrow H_2O + Na_2SO_4$
      (g) $NH_4I + Cl_2 \longrightarrow NH_4Cl + I_2$
      (h) $CrCl_3 + AgNO_3 \longrightarrow Cr(NO_3)_3 + AgCl$
      (i) $Al_2(CO_3)_3 \xrightarrow{\Delta} Al_2O_3 + CO_2$
      (j) $Al + C \xrightarrow{\Delta} Al_4C_3$
   2. Balance the following equations:
      (a) $SO_2 + O_2 \longrightarrow SO_3$
      (b) $Al + MnO_2 \xrightarrow{\Delta} Mn + Al_2O_3$
      (c) $Na + H_2O \longrightarrow NaOH + H_2$
      (d) $AgNO_3 + Ni \longrightarrow Ni(NO_3)_2 + Ag$
      (e) $Bi_2S_3 + HCl \longrightarrow BiCl_3 + H_2S$
      (f) $PbO_2 \xrightarrow{\Delta} PbO + O_2$
      (g) $LiAlH_4 \xrightarrow{\Delta} LiH + Al + H_2$
      (h) $KI + Br_2 \longrightarrow KBr + I_2$
      (i) $K_3PO_4 + BaCl_2 \longrightarrow KCl + Ba_3(PO_4)_2$
      (j) $C_4H_{10} + O_2 \longrightarrow CO_2 + H_2O$
   3. Balance the following equations:
      (a) $Na_2CO_3 + HCl \longrightarrow NaCl + H_2O + CO_2$
      (b) $KClO_3 \xrightarrow{\Delta} KCl + O_2$
      (c) $P + Br_2 \longrightarrow PBr_3$
      (d) $CuSO_4 \cdot 5H_2O \xrightarrow{\Delta} CuSO_4 + H_2O$
      (e) $HC_2H_3O_2 + O_2 \longrightarrow CO_2 + H_2O$
      (f) $Li + O_2 \longrightarrow Li_2O$
      (g) $(NH_4)_2Cr_2O_7 \xrightarrow{\Delta} N_2 + Cr_2O_3 + H_2O$

(h) $C_2H_2 + O_2 \longrightarrow CO_2 + H_2O$

(i) $Fe(OH)_3 + H_2SO_4 \longrightarrow Fe_2(SO_4)_3 + H_2O$

(j) $C_3H_5(NO_3)_3 \xrightarrow{\Delta} N_2 + O_2 + CO_2 + H_2O$
Nitroglycerine

4. Change the following word equations into formula equations and balance them:

    (a) Copper + Sulfur $\xrightarrow{\Delta}$ Copper(I) sulfide

    (b) Phosphoric acid + Calcium hydroxide $\longrightarrow$ Calcium phosphate + Water

    (c) Silver oxide $\xrightarrow{\Delta}$ Silver + Oxygen

    (d) Iron(III) chloride + Sodium hydroxide $\longrightarrow$
Iron(III) hydroxide + Sodium chloride

    (e) Nickel(II) phosphate + Sulfuric acid $\longrightarrow$
Nickel(II) sulfate + Phosphoric acid

    (f) Zinc carbonate + Hydrochloric acid $\longrightarrow$
Zinc chloride + Water + Carbon dioxide

    (g) Silver nitrate + Aluminum chloride $\longrightarrow$
Silver chloride + Aluminum nitrate

5. Change the following word equations into formula equations and balance them:

    (a) Water $\longrightarrow$ Hydrogen + Oxygen

    (b) Acetic acid + Potassium hydroxide $\longrightarrow$ Potassium acetate + Water

    (c) Phosphorus + Iodine $\longrightarrow$ Phosphorus triiodide

    (d) Aluminum + Copper(II) sulfate $\longrightarrow$ Copper + Aluminum sulfate

    (e) Ammonium sulfate + Barium chloride $\longrightarrow$
Ammonium chloride + Barium sulfate

    (f) Sulfur tetrafluoride + Water $\longrightarrow$ Sulfur dioxide + Hydrogen fluoride

    (g) Chromium(III) carbonate $\longrightarrow$ Chromium(III) oxide + Carbon dioxide

6. Complete and balance the equations for these combination reactions:

    (a) $K + O_2 \longrightarrow$ (c) $CO_2 + H_2O \longrightarrow$

    (b) $Al + Cl_2 \longrightarrow$ (d) $CaO + H_2O \longrightarrow$

7. Complete and balance the equations for these decomposition reactions:

    (a) $HgO \xrightarrow{\Delta}$ (c) $MgCO_3 \xrightarrow{\Delta}$

    (b) $NaClO_3 \xrightarrow{\Delta}$ (d) $PbO_2 \xrightarrow{\Delta} PbO +$

8. Complete and balance the equations for these single-replacement reactions:

    (a) $Zn + H_2SO_4 \longrightarrow$ (c) $Mg + AgNO_3 \longrightarrow$

    (b) $AlI_3 + Cl_2 \longrightarrow$ (d) $Al + CoSO_4 \longrightarrow$

9. Complete and balance the equations for these double-replacement reactions:

    (a) $ZnCl_2 + KOH \longrightarrow$ (d) $(NH_4)_3PO_4 + Ni(NO_3)_2 \longrightarrow$

    (b) $CuSO_4 + H_2S \longrightarrow$ (e) $Ba(OH)_2 + HNO_3 \longrightarrow$

    (c) $Zn(OH)_2 + HCl \longrightarrow$ (f) $(NH_4)_2S + HCl \longrightarrow$

10. Complete and balance the equations for these reactions:

    (a) $H_2 + I_2 \longrightarrow$ (e) $SO_2 + H_2O \longrightarrow$

    (b) $CaCO_3 \xrightarrow{\Delta}$ (f) $BaCl_2 \cdot 2H_2O \xrightarrow{\Delta}$

    (c) $Mg + H_2SO_4 \longrightarrow$ (g) $Ca + H_2O \longrightarrow$

    (d) $FeCl_2 + NaOH \longrightarrow$ (h) $Bi(NO_3)_3 + H_2S \longrightarrow$

11. Complete and balance the equations for the following reactions:

    (a) $Ba + O_2 \longrightarrow$ (f) $C + O_2 \longrightarrow$

    (b) $NaHCO_3 \xrightarrow{\Delta} Na_2CO_3 +$ (g) $Al(ClO_3)_3 \xrightarrow{\Delta} O_2 +$

    (c) $Ni + CuSO_4 \longrightarrow$ (h) $CuBr_2 + Cl_2 \longrightarrow$

    (d) $MgO + HCl \longrightarrow$ (i) $SbCl_3 + (NH_4)_2S \longrightarrow$

    (e) $H_3PO_4 + KOH \longrightarrow$ (j) $NaNO_3 \xrightarrow{\Delta} NaNO_2 +$

12. What is the purpose of balancing equations?

13. What is represented by the numbers (coefficients) that are placed in front of the formulas in a balanced equation?

14. Interpret the following chemical reactions in terms of the number of moles of each reactant and product:

    (a) $MgBr_2 + 2\,AgNO_3 \longrightarrow Mg(NO_3)_2 + 2\,AgBr$

    (b) $N_2 + 3\,H_2 \longrightarrow 2\,NH_3$

    (c) $2\,CH_3CH(OH)CH_3 + 9\,O_2 \longrightarrow 6\,CO_2 + 8\,H_2O$
        Isopropyl alcohol

15. Interpret each of the following equations in terms of the relative number of moles of each substance involved and indicate whether the reaction is exothermic or endothermic:

    (a) $2\,Na + Cl_2 \longrightarrow 2\,NaCl + 196.4\,kcal$

    (b) $PCl_5 + 22.2\,kcal \longrightarrow PCl_3 + Cl_2$

16. Write balanced equations for each of these reactions, including the heat term:

    (a) Lime, CaO, is converted to slaked lime, $Ca(OH)_2$, by reaction with water. The reaction liberates 15.6 kcal of heat for each mole of lime reacted.

    (b) Photosynthesis in plants produces glucose, $C_6H_{12}O_6$, and oxygen from carbon dioxide and water. The energy, which is absorbed from sunlight, amounts to 673 kcal per mole of glucose formed.

17. In the reaction $H_2(g) + Br_2(g) \longrightarrow 2\,HBr(g) + 24.7\,kcal$, the net heat liberated is a result of breaking the bonds of H_2 and Br_2 molecules and forming the bonds of 2 HBr molecules. Energy is absorbed in breaking bonds and energy is released in forming bonds. Use the bond dissociation energy data given in Sections 7.5 and 7.8 to verify that this reaction is exothermic by 24.7 kcal.

18. Which of the following statements are correct?

    (a) The coefficients in front of the formulas in a balanced chemical equation give the relative number of moles of the reactants and products in the reaction.

    (b) A balanced chemical equation is one that has the same number of moles on each side of the equation.

    (c) In a chemical equation, the symbol $\xrightarrow{\Delta}$ indicates that the reaction is exothermic.

    (d) A chemical change that absorbs heat energy is said to be endothermic.

    (e) In the reaction $H_2 + Cl_2 \longrightarrow 2\,HCl$, 100 molecules of HCl are produced for every 50 molecules of H_2 reacted.

    (f) The symbol (aq) after a substance in an equation means that the substance is in a water solution.

    (g) The equation $H_2O \longrightarrow H_2 + O_2$ can be balanced by placing a 2 in front of H_2O.

    (h) In the equation $3\,H_2 + N_2 \longrightarrow 2\,NH_3$ there are fewer moles of product than there are moles of reactants.

    (i) The total number of moles of reactants and products represented by this equation is 5 moles:
        $Mg + 2\,HCl \longrightarrow MgCl_2 + H_2$

    (j) One mole of glucose, $C_6H_{12}O_6$, contains 6 moles of carbon atoms.

    (k) The reactants are the substances produced by the chemical reaction.

    (l) In a balanced equation each side of the equation contains the same number of atoms of each element.

    (m) When a precipitate is formed in a chemical reaction, it can be indicated in the equation with the symbol $\downarrow$ immediately before the formula of the substance precipitated.

    (n) When a gas is evolved in a chemical reaction, it can be indicated in the equation with the symbol $\uparrow$ immediately following the formula of the gas.

    (o) According to the equation $3\,H_2 + N_2 \longrightarrow 2\,NH_3$, 4 moles of NH_3 will be formed when 6 moles of H_2 and 2 moles of N_2 react.

    (p) In Figure 10.1 the products of an exothermic reaction are at a lower energy state than the reactants.

11

Calculations from Chemical Equations

After studying Chapter 11 you should be able to:

1. Understand the terms listed in Question A at the end of the chapter.
2. Give mole ratios for any two substances involved in a chemical reaction.
3. Outline the mole or mole-ratio method for making stoichiometric calculations.
4. Calculate the number of moles of a desired substance obtainable from a given number of moles of a starting substance in a chemical reaction (mole to mole calculations).
5. Calculate the weight of a desired substance obtainable from a given number of moles of a starting substance in a chemical reaction and vice versa (mole to weight and weight to mole calculations).
6. Calculate the weight of a desired substance involved in a chemical reaction from a given weight of a starting substance (weight to weight calculations).
7. Deduce the limiting reagent or reactant when given the amounts of starting substances and then calculate the moles or weight of desired substance obtainable from a given chemical reaction (limiting reagent calculations).
8. Apply theoretical yield, actual yield, and percentage yield to any of the foregoing types of problems, or calculate theoretical and percentage yields of a chemical reaction.

This chapter shows the quantitative relationship between reactants and products in chemical reactions and also reviews and correlates such concepts as molecular weight, the molecule, the mole, and balancing equations.

11.1 A Short Review

1. Molecular weight or formula weight. The molecular weight is the sum of the atomic weights of all the atoms in a molecule. The formula weight is the sum of the atomic weights of all the atoms in a given formula of a compound or an ion. The terms *molecular weight* and *formula weight* are commonly used interchangeably.

2. Relationship between molecule and mole. A molecule is the smallest unit of a molecular substance (e.g., Cl_2) and a mole is Avogadro's number, 6.02×10^{23}, of molecules of that substance. A mole of chlorine (Cl_2) has the same number of molecules as a mole of carbon dioxide, a mole of water, or a mole of any other molecular substance. When we relate molecules to gram-molecular weight, 1 g-mol. wt = 1 mole, or 6.02×10^{23} molecules.

In addition to molecular substances, the term *mole* may refer to any chemical species. It represents a quantity in grams equal to the formula weight (1 g-form. wt) and may be applied to atoms, ions, electrons, and formula units of nonmolecular substances. For example, a mole of water consists of 18.0 g of water, or 6.02×10^{23} molecules; a mole of sodium is 23.0 g of sodium, or 6.02×10^{23} atoms; a mole of chloride ions is 35.5 g of chloride ions, or 6.02×10^{23} of these ions; and a mole of sodium chloride (a nonmolecular substance) is 58.5 g or 6.02×10^{23} formula units of sodium chloride.

$$1 \text{ mole} = \begin{cases} 1 \text{ g-mol. wt} = 6.02 \times 10^{23} \text{ molecules} \\ 1 \text{ g-form. wt} = 6.02 \times 10^{23} \text{ formula units} \\ 1 \text{ g-at. wt} = 6.02 \times 10^{23} \text{ atoms} \\ 1 \text{ g-ionic wt} = 6.02 \times 10^{23} \text{ ions} \end{cases}$$

Other useful mole relationships are

$$\text{Number of moles} = \frac{\text{Grams of a substance}}{\text{Gram-formula weight of the substance}}$$

$$\text{Number of moles} = \frac{\text{Grams of a monatomic element}}{\text{Gram-atomic weight of the element}}$$

$$\text{Number of moles} = \frac{\text{Number of molecules}}{6.02 \times 10^{23} \text{ molecules/mole}}$$

Two other useful equalities may be derived algebraically from each of these mole relationships. What are they?

3. Balanced equations.   In using equations for calculations of mole–weight–volume relationships between reactants and products, the equations must be balanced. Remember that the number in front of a formula in a balanced equation represents the number of moles of that substance reacting in the chemical change.

11.2 Calculations from Chemical Equations: The Mole Method

stoichiometry

In chemical work it is often necessary to calculate the amount of a substance that is produced from, or needed to react with, a given quantity of another substance. The area of chemistry that deals with the quantitative relationships among reactants and products in a chemical reaction is known as **stoichiometry.** There are several methods for solving stoichiometric problems. However, it is my firm belief that the *mole* or *mole-ratio method* is best and gives the student a better understanding of the relationships of the reacting species in a chemical reaction.

mole ratio

A **mole ratio** is a ratio between the number of moles of any two species involved in a chemical reaction. For example, in the reaction

$$2\,H_2 \;+\; O_2 \;\longrightarrow\; 2\,H_2O$$

2 moles   1 mole   2 moles

there are six mole ratios that apply only to this reaction:

$$\frac{2\text{ moles }H_2}{1\text{ mole }O_2} \qquad \frac{2\text{ moles }H_2}{2\text{ moles }H_2O} \qquad \frac{1\text{ mole }O_2}{2\text{ moles }H_2}$$

$$\frac{1\text{ mole }O_2}{2\text{ moles }H_2O} \qquad \frac{2\text{ moles }H_2O}{2\text{ moles }H_2} \qquad \frac{2\text{ moles }H_2O}{1\text{ mole }O_2}$$

The mole ratio is used as a conversion factor, to convert the moles of one substance to moles of another substance in a reaction.

Since stoichiometric problems are encountered throughout the entire field of chemistry, it is profitable to master this general method for their solution. The mole method makes use of three simple basic operations:

A. Conversion of the starting substance to moles (if it is not given in moles).
B. Calculation of the moles of the desired substance obtainable from the available moles of starting substance. This is done by multiplying the moles of starting substance (from A) by the mole ratio. This mole ratio is taken from the balanced equation and is the number of moles of the desired substance over the number of moles of the starting substance.
C. Conversion of the moles of desired substance (from B) to whatever units are required.

Like learning to balance chemical equations, learning to make stoichiometric calculations requires some practice. A detailed step-by-step description of the general method, together with a variety of worked examples, is given in the following paragraphs. Study this material and apply the method to the problems at the end of this chapter.

Step 1. Use a balanced equation. Write a balanced equation for the chemical reaction in question or check to see that the equation given is balanced.

Step 2. Determine the number of moles of starting substance. Identify the starting substance from the data given in the statement of the problem. When the starting substance is given in moles, use it in that form; if it is not in moles, convert the quantity of the starting substance to moles.

Step 3. Determine the mole ratio of the desired substance to the starting substance. The number of moles of each substance in the balanced equation is indicated by the coefficient in front of each substance. Use these coefficients to set up the mole ratio:

$$\text{Mole ratio} = \frac{\text{Moles of desired substance in the equation}}{\text{Moles of starting substance in the equation}}$$

Step 4. Calculate the number of moles of the desired substance. Multiply the number of moles of starting substance (from Step 2) by the mole ratio (from Step 3) to obtain the number of moles of desired substance:

$$\begin{pmatrix} \text{Moles of desired} \\ \text{substance} \end{pmatrix} = \underbrace{\begin{pmatrix} \text{Moles of starting} \\ \text{substance} \end{pmatrix}}_{\text{From Step 2}} \times \underbrace{\frac{\begin{pmatrix} \text{Moles of desired} \\ \text{substance in} \\ \text{the equation} \end{pmatrix}}{\begin{pmatrix} \text{Moles of starting} \\ \text{substance in} \\ \text{the equation} \end{pmatrix}}}_{\text{Mole ratio from Step 3}}$$

Note that the units of moles of starting substance cancel out in the numerator and the denominator.

Step 5. Calculate the desired substance in the units asked for in the problem. If the answer is requested in moles, the problem is finished in Step 4. If units other than moles are wanted, multiply the moles of the desired substance from Step 4 by the appropriate factor to convert moles to the units required. For example, if grams of the desired substance are wanted,

$$\begin{pmatrix} \text{Grams of desired} \\ \text{substance} \end{pmatrix} = \underbrace{\begin{pmatrix} \text{Moles of desired} \\ \text{substance} \end{pmatrix}}_{\text{From Step 4}} \times \begin{pmatrix} \text{Gram-formula weight} \\ \dfrac{\text{of desired substance}}{\text{1 mole of}} \\ \text{desired substance} \end{pmatrix}$$

These steps are summarized in Figure 11.1.

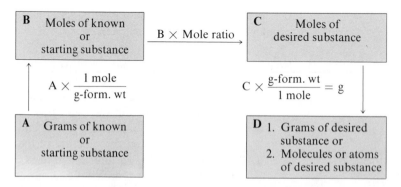

Figure 11.1 Basic steps in using the mole ratio to convert moles of one substance to moles of another substance in a chemical reaction. Problems can start at either Block A or Block B. There is no direct path between Blocks A and D.

11.3 Mole–Mole Calculations

The object of mole–mole calculations is to calculate the number of moles of one substance that reacts with, or is produced from, a given number of moles of another substance. Illustrative problems follow.

PROBLEM 11.1 How many moles of water will be produced from 3.5 moles of magnesium hydroxide, $Mg(OH)_2$, according to the following reaction?

$$Mg(OH)_2 + H_2CO_3 \longrightarrow MgCO_3 + 2\,H_2O$$

1 mole   1 mole   1 mole   2 moles

When the equation is balanced, it states that 2 moles of water will be produced from 1 mole of $Mg(OH)_2$. Even though we can quickly ascertain that 7.0 moles of water will be formed from 3.5 moles of $Mg(OH)_2$, the mole method for solving the problem is shown here. This method of working with mole ratios will be very helpful in solving later problems.

Step 1. The equation given is balanced.

Step 2. The moles of starting substance are 3.5 moles of $Mg(OH)_2$. The conversion needed is

$$\text{Moles } Mg(OH)_2 \longrightarrow \text{Moles } H_2O$$

Step 3. From the balanced equation, set up the mole ratio between the two substances in question, placing the number of moles of the substance being sought in the numerator and the number of moles of the starting substance in the denominator. The number of moles, in each case, is the same as the coefficient in front of the substance in the balanced equation.

$$\text{Mole ratio} = \frac{2 \text{ moles } H_2O}{1 \text{ mole } Mg(OH)_2} \quad \text{(From equation)}$$

Step 4. Multiply the 3.5 moles of $Mg(OH)_2$ given in the problem by this mole ratio:

$$3.5 \text{ moles } Mg(OH)_2 \times \frac{2 \text{ moles } H_2O}{1 \text{ mole } Mg(OH)_2} = 7.0 \text{ moles } H_2O \quad \text{(Answer)}$$

Again note the use of units. The moles of $Mg(OH)_2$ cancel, leaving the answer in units of moles of H_2O.
∎

PROBLEM 11.2 How many moles of ammonia can be produced from 8.00 moles of hydrogen reacting with nitrogen?

Step 1. First, we need the balanced equation:

$$3\,H_2 + N_2 \longrightarrow 2\,NH_3$$

Step 2. The moles of starting substance are 8.00 moles of hydrogen. The conversion needed is

$$Moles\ H_2 \longrightarrow Moles\ NH_3$$

Step 3. The balanced equation states that we get 2 moles of NH_3 for every 3 moles of H_2 that react. Set up the mole ratio of desired substance (NH_3) to starting substance (H_2):

$$Mole\ ratio = \frac{2\ moles\ NH_3}{3\ moles\ H_2}$$

Step 4. Multiplying the 8.00 moles of starting H_2 by this mole ratio, we get

$$8.00\ \cancel{moles\ H_2} \times \frac{2\ moles\ NH_3}{3\ \cancel{moles\ H_2}} = 5.33\ moles\ NH_3 \quad \text{(Answer)}$$

■

PROBLEM 11.3   Given the balanced equation

$$\underset{\text{1 mole}}{K_2Cr_2O_7} + \underset{\text{6 moles}}{6\ KI} + 7\ H_2SO_4 \longrightarrow Cr_2(SO_4)_3 + 4\ K_2SO_4 + \underset{\text{3 moles}}{3\ I_2} + 7\ H_2O$$

calculate (a) the number of moles of potassium dichromate ($K_2Cr_2O_7$) that will react with 2.0 moles of potassium iodide (KI); (b) the number of moles of iodine (I_2) that will be produced from 2.0 moles of potassium iodide.

After the equation is balanced, we are concerned only with $K_2Cr_2O_7$, KI, and I_2, and we can ignore all the other substances. The equation states that 1 mole of $K_2Cr_2O_7$ will react with 6 moles of KI to produce 3 moles of I_2.

(a) Calculate the number of moles of $K_2Cr_2O_7$:

Step 1. The equation given is balanced.
Step 2. The moles of starting substance are 2.0 moles of KI. The conversion needed is

$$Moles\ KI \longrightarrow Moles\ K_2Cr_2O_7$$

Step 3. Set up the mole ratio of desired substance to starting substance:

$$Mole\ ratio = \frac{1\ mole\ K_2Cr_2O_7}{6\ moles\ KI} \quad \text{(From equation)}$$

Step 4. Multiply the moles of starting material by this ratio to obtain the answer:

$$2.0\ \cancel{moles\ KI} \times \frac{1\ mole\ K_2Cr_2O_7}{6\ \cancel{moles\ KI}} = 0.33\ mole\ K_2Cr_2O_7 \quad \text{(Answer)}$$

(b) Calculate the number of moles of I_2.

Steps 1 and 2. The equation given is balanced and the moles of starting substance are 2.0 moles KI as in part (a). The conversion needed is

$$Moles\ KI \longrightarrow Moles\ I_2$$

Step 3. Set up the mole ratio of desired substance to starting substance:

$$\text{Mole ratio} = \frac{3 \text{ moles } I_2}{6 \text{ moles } KI} \quad \text{(From equation)}$$

Step 4. Multiply the moles of starting material by this ratio to obtain the answer.

$$2.0 \text{ moles } KI \times \frac{3 \text{ moles } I_2}{6 \text{ moles } KI} = 1.0 \text{ mole } I_2 \quad \text{(Answer)}$$

Thus, 2.0 moles KI will react with 0.33 mole $K_2Cr_2O_7$ to produce 1.0 mole I_2.

■

PROBLEM 11.4 How many molecules of water can be produced by reacting 0.010 mole of oxygen with hydrogen?

The sequence of conversions needed in the calculation is

$$\text{Moles } O_2 \longrightarrow \text{Moles } H_2O \longrightarrow \text{Molecules } H_2O$$

Step 1. First, we write the balanced equation:

$$2 H_2 + O_2 \longrightarrow 2 H_2O$$
$$\quad\quad\quad\; 1 \text{ mole} \quad\; 2 \text{ moles}$$

Step 2. The number of moles of starting substance is 0.010 mole O_2.

Step 3. Set up the mole ratio of desired substance to starting substance:

$$\text{Mole ratio} = \frac{2 \text{ moles } H_2O}{1 \text{ mole } O_2} \quad \text{(From equation)}$$

Step 4. Multiplying 0.010 mole of oxygen by this ratio, we obtain

$$0.010 \text{ mole } O_2 \times \frac{2 \text{ moles } H_2O}{1 \text{ mole } O_2} = 0.020 \text{ mole } H_2O$$

Step 5. Since the problem asks for molecules instead of moles of H_2O, we must convert moles to molecules. Use the conversion factor (6.02×10^{23} molecules)/mole.

$$0.020 \text{ mole } H_2O \times \frac{6.02 \times 10^{23} \text{ molecules}}{\text{mole}} = 1.2 \times 10^{22} \text{ molecules } H_2O$$

We should note that 0.020 mole is still quite a large number of water molecules.

■

11.4 Mole–Weight and Weight–Weight Calculations

The object of these types of problems is to calculate the weight of one substance that reacts with, or is produced from, a given number of moles or a given weight of another substance in a chemical reaction. The mole ratio is used to convert from moles of starting substance to moles of desired substance.

PROBLEM 11.5   What weight of hydrogen can be produced by reacting 6.0 moles of aluminum with hydrochloric acid?

First calculate the number of moles of hydrogen produced, using the mole-ratio method, and then calculate the weight of hydrogen by multiplying the moles of hydrogen by its weight per mole. The sequence of conversions in the calculation is

$$\text{Moles Al} \longrightarrow \text{Moles H}_2 \longrightarrow \text{Grams H}_2$$

Step 1. The balanced equation is

$$2\,\text{Al}(s) + 6\,\text{HCl}(aq) \longrightarrow 2\,\text{AlCl}_3(aq) + 3\,\text{H}_2(g)$$
$$\text{2 moles} \qquad\qquad\qquad\qquad\qquad\qquad \text{3 moles}$$

Step 2. The moles of starting substance are 6.0 moles of aluminum.
Steps 3 and 4. Calculate moles of H_2.

$$\text{Mole ratio} = \frac{3 \text{ moles H}_2}{2 \text{ moles Al}}$$

$$6.0 \text{ moles Al} \times \frac{3 \text{ moles H}_2}{2 \text{ moles Al}} = 9.0 \text{ moles H}_2$$

Step 5. Convert moles of H_2 to grams [g = moles × (g/mole)]:

$$9.0 \text{ moles H}_2 \times \frac{2.0 \text{ g H}_2}{1 \text{ mole H}_2} = 18 \text{ g H}_2$$

We see that 18 g of H_2 can be produced by reacting 6.0 moles of Al with HCl. The following setup combines all the steps into one continuous calculation:

$$6.0 \text{ moles Al} \times \frac{3 \text{ moles H}_2}{2 \text{ moles Al}} \times \frac{2.0 \text{ g H}_2}{1 \text{ mole H}_2} = 18 \text{ g H}_2$$

∎

PROBLEM 11.6   What weight of carbon dioxide is produced by the complete combustion of 100 g of the hydrocarbon pentane, C_5H_{12}?

The sequence of conversions in the calculation is

$$\text{Grams C}_5\text{H}_{12} \longrightarrow \text{Moles C}_5\text{H}_{12} \longrightarrow \text{Moles CO}_2 \longrightarrow \text{Grams CO}_2$$

Formula weights: C_5H_{12}, 72.0; CO_2, 44.0

Step 1. The balanced equation is

$$\text{C}_5\text{H}_{12} + 8\,\text{O}_2 \longrightarrow 5\,\text{CO}_2 + 6\,\text{H}_2\text{O}$$
$$\text{1 mole} \qquad\qquad\quad \text{5 moles}$$

Step 2. The starting substance is 100 g of C_5H_{12}. Convert 100 g of C_5H_{12} to moles:

$$100 \text{ g C}_5\text{H}_{12} \times \frac{1 \text{ mole C}_5\text{H}_{12}}{72.0 \text{ g C}_5\text{H}_{12}} = 1.39 \text{ moles C}_5\text{H}_{12}$$

Steps 3 and 4. Calculate the moles of CO_2 by the mole-ratio method:

$$1.39 \; \text{moles } C_5H_{12} \times \frac{5 \text{ moles } CO_2}{1 \text{ mole } C_5H_{12}} = 6.95 \text{ moles } CO_2$$

Step 5. Convert moles of CO_2 to grams:

$$\text{Moles } CO_2 \times \frac{\text{g-form. wt } CO_2}{1 \text{ mole } CO_2} = \text{Grams } CO_2$$

$$6.95 \; \text{moles } CO_2 \times \frac{44.0 \text{ g } CO_2}{1 \text{ mole } CO_2} = 306 \text{ g } CO_2 \quad \text{(Answer)}$$

We see that 306 g of CO_2 are produced from the complete combustion of 100 g of C_5H_{12}. The calculation in a continuous setup is

$$\boxed{100 \text{ g } C_5H_{12}} \times \boxed{\frac{1 \text{ mole } C_5H_{12}}{72.0 \text{ g } C_5H_{12}}} \times \boxed{\frac{5 \text{ moles } CO_2}{1 \text{ mole } C_5H_{12}}} \times \boxed{\frac{44.0 \text{ g } CO_2}{1 \text{ mole } CO_2}} = 306 \text{ g } CO_2$$

$$\text{Grams } C_5H_{12} \longrightarrow \text{Moles } C_5H_{12} \longrightarrow \text{Moles } CO_2 \longrightarrow \text{Grams } CO_2$$

PROBLEM 11.7 How many grams of nitric acid, HNO_3, are required to produce 8.75 g of dinitrogen monoxide, N_2O, according to the following equation?

$$4 \, Zn(s) + \underset{\text{10 moles}}{10 \, HNO_3(aq)} \longrightarrow 4 \, Zn(NO_3)_2(aq) + \underset{\text{1 mole}}{N_2O(g)} + 5 \, H_2O(l)$$

In this problem, the amount of product is given and we are asked to calculate the amount of reactant required to produce that product. The calculation is no different than in Problem 11.6. The starting substance for the calculation is 8.75 g N_2O. The grams of HNO_3 need to be calculated. We shall do this problem in a continuous calculation setup.

Step 1. The equation for the reaction is balanced.
Step 2. The starting substance is 8.75 g N_2O. The sequence of conversions needed is

$$\text{Grams } N_2O \longrightarrow \text{Moles } N_2O \longrightarrow \text{Moles } HNO_3 \longrightarrow \text{Grams } HNO_3$$

Formula weights: N_2O, 44.0; HNO_3, 63.0

Steps 3–5. The calculation is

$$8.75 \text{ g } N_2O \times \frac{1 \text{ mole } N_2O}{44.0 \text{ g } N_2O} \times \frac{10 \text{ moles } HNO_3}{1 \text{ mole } N_2O} \times \frac{63.0 \text{ g } HNO_3}{1 \text{ mole } HNO_3} = 125 \text{ g } HNO_3$$

$$\text{Grams } N_2O \longrightarrow \text{Moles } N_2O \longrightarrow \text{Moles } HNO_3 \longrightarrow \text{Grams } HNO_3$$

Thus, 125 g HNO_3 are required to produce 8.75 g N_2O in this reaction.

11.5 Limiting Reagent and Yield Calculations

limiting reagent

In many chemical processes the quantities of the reactants used are such that the moles of one reactant are in excess of the moles of a second reactant in the reaction. The amount of the product(s) formed in such a case will be dependent on the reactant that is not in excess. Thus, the reactant that is not in excess is known as the **limiting reagent,** since it limits the amount of product that can be formed.

As an example, consider the case where solutions containing 1.0 mole of sodium hydroxide and 1.5 moles of hydrochloric acid are mixed:

$$NaOH + HCl \longrightarrow NaCl + H_2O$$
$$\text{1 mole} \quad \text{1 mole} \quad \text{1 mole} \quad \text{1 mole}$$

According to the equation it is possible to obtain 1.0 mole of NaCl from 1.0 mole of NaOH, and 1.5 moles of NaCl from 1.5 moles of HCl. However, we cannot have two different yields of NaCl from the reaction. When 1.0 mole of NaOH and 1.5 moles of HCl are mixed, there is insufficient NaOH to react with all of the HCl. Therefore, HCl is the reagent in excess and NaOH is the limiting reagent. Since the NaCl formed is dependent on the limiting reagent, the amount of NaCl formed will be 1.0 mole. Since 1.0 mole of NaOH reacts with 1.0 mole of HCl, 0.5 mole of HCl remains unreacted.

$$\left.\begin{array}{l}\text{1.0 mole NaOH} \\ \text{1.5 moles HCl}\end{array}\right\} \longrightarrow \begin{array}{l}\text{1.0 mole NaCl} \\ \text{1.0 mole H}_2\text{O}\end{array} + 0.5 \text{ mole HCl unreacted}$$

Problems giving the amounts of two reactants are generally of the limiting reagent type. Several methods can be used to identify the limiting reagent in a chemical reaction. The most direct method is to calculate the amount of product that can be formed from each starting substance. The substance that gives the smaller amount of product is the limiting reagent. The other substance will be in excess and some of it will remain unreacted at the end of the reaction. Another method is to calculate the moles of each starting substance first. Compare the ratio of the calculated moles to the mole ratio of these substances in the equation to see which substance is limiting and which is in excess.

Using the first method, two steps are needed to solve problems in which a limiting reagent controls the amount of product formed:

1. Calculate the number of moles of each starting substance.
2. Calculate the number of moles or grams of product that can be formed from the number of moles of each starting substance. The substance that gives the least amount of product is the limiting reagent.

PROBLEM 11.8   How many grams of silver bromide, AgBr, can be formed when solutions containing 50.0 g of $MgBr_2$ and 100 g of $AgNO_3$ are mixed together?

First identify the limiting reagent and the reagent in excess.

$$MgBr_2(aq) + 2\ AgNO_3(aq) \longrightarrow 2\ AgBr\downarrow + Mg(NO_3)_2(aq)$$

Formula Weights: $MgBr_2$, 184.1; $AgNO_3$, 169.9; AgBr, 187.8

Step 1. Moles of $MgBr_2$ and $AgNO_3$:

$$50.0\ g\ MgBr_2 \times \frac{1\ mole\ MgBr_2}{184.1\ g\ MgBr_2} = 0.272\ mole\ MgBr_2$$

$$100\ g\ AgNO_3 \times \frac{1\ mole\ AgNO_3}{169.9\ g\ AgNO_3} = 0.589\ mole\ AgNO_3$$

Step 2. Calculate the grams of AgBr from the moles of each reagent.

Moles reagent $\longrightarrow$ Moles AgBr $\longrightarrow$ Grams AgBr

(a) Grams AgBr from $MgBr_2$:

$$0.272\ mole\ MgBr_2 \times \frac{2\ moles\ AgBr}{1\ mole\ MgBr_2} \times \frac{187.8\ g\ AgBr}{1\ mole\ AgBr} = 102\ g\ AgBr$$

(b) Grams AgBr from $AgNO_3$:

$$0.589\ mole\ AgNO_3 \times \frac{2\ moles\ AgBr}{2\ moles\ AgNO_3} \times \frac{187.8\ g\ AgBr}{1\ mole\ AgBr} = 111\ g\ AgBr$$

The limiting reagent is $MgBr_2$, since it gives less AgBr. $AgNO_3$ is in excess. The yield is 102 g AgBr. The final mixture will contain 102 g AgBr, $Mg(NO_3)_2$, and some unreacted $AgNO_3$.

■

PROBLEM 11.9 How many moles of Fe_3O_4 can be obtained by reacting 16.8 g Fe with 10.0 g H_2O?

$$3\ Fe(s) + 4\ H_2O(g) \xrightarrow{\Delta} Fe_3O_4(s) + 4\ H_2(g)$$

This is a typical problem in which one of the starting substances will control or limit the yield of product.

Step 1. Calculate the moles of each starting substance.

Grams of reagent $\longrightarrow$ Moles of reagent

Formula weights: Fe, 55.8; H_2O, 18.0

$$16.8\ g\ Fe \times \frac{1\ mole\ Fe}{55.8\ g\ Fe} = 0.301\ mole\ Fe$$

$$10.0\ g\ H_2O \times \frac{1\ mole\ H_2O}{18.0\ g\ H_2O} = 0.556\ mole\ H_2O$$

Step 2. Calculate the moles of Fe_3O_4 from the moles of each reactant.

$$\text{Moles reagent} \longrightarrow \text{Moles } Fe_3O_4$$

This calculation involves the mole ratios from the equation:

$$0.301 \text{ mole Fe} \times \frac{1 \text{ mole } Fe_3O_4}{3 \text{ moles Fe}} = 0.100 \text{ mole } Fe_3O_4$$

$$0.556 \text{ mole } H_2O \times \frac{1 \text{ mole } Fe_3O_4}{4 \text{ moles } H_2O} = 0.139 \text{ mole } Fe_3O_4$$

Clearly, the limiting reagent is Fe and the yield of product is 0.100 mole Fe_3O_4.

∎

The quantities of the products that we have been calculating from equations represent the maximum yield (100%) of product according to the reaction represented by the equation. Many reactions fail to give a 100% yield of product. The main reason for this failure are the side reactions that give products other than the main product and the fact that many reactions are reversible. In addition, some product may be lost in handling and transferring from one vessel to another. The yield calculated from the chemical equation is commonly known as the **theoretical yield,** which is the maximum amount of product that can be produced according to the equation. The **actual yield** is the amount of product that we finally obtain.

theoretical yield
actual yield

percentage yield

The **percentage yield** is the ratio of the actual yield to the theoretical yield multiplied by 100%. Both yields must have the same units.

$$\frac{\text{Actual yield}}{\text{Theoretical yield}} \times 100\% = \text{Percentage yield}$$

For example, if the yield (theoretical) calculated for a reaction is 14.8 g and the amount of product obtained is 9.25 g, the percentage yield is

$$\frac{9.25 \text{ g}}{14.8 \text{ g}} \times 100\% = 62.5\%$$

PROBLEM 11.10 Carbon tetrachloride, CCl_4, was prepared by reacting 100 g of chlorine with excess carbon disulfide. Calculate the percentage yield if 65.0 g of CCl_4 were obtained from the reaction.

$$CS_2 + 3 Cl_2 \longrightarrow CCl_4 + S_2Cl_2$$

In this problem we first need to calculate the quantity of CCl_4 (theoretical yield) that can be formed according to the equation. Then we can compare this amount with the 65.0 g CCl_4 obtained (actual yield) to calculate the percentage yield.

Step 1. Calculate the grams of CCl_4 from the equation using the mole-ratio method. The sequence of conversions is

Grams $Cl_2 \longrightarrow$ Moles $Cl_2 \longrightarrow$ Moles $CCl_4 \longrightarrow$ Grams CCl_4

Formula weights: Cl_2, 71.0; CCl_4, 154.0

$$100 \text{ g } \cancel{Cl_2} \times \frac{1 \text{ mole } \cancel{Cl_2}}{71.0 \text{ g } \cancel{Cl_2}} \times \frac{1 \text{ mole } \cancel{CCl_4}}{3 \text{ moles } \cancel{Cl_2}} \times \frac{154 \text{ g } CCl_4}{1 \text{ mole } \cancel{CCl_4}} = 72.3 \text{ g } CCl_4$$

Step 2. According to the equation, 72.3 g CCl_4 is the maximum amount or theoretical yield of CCl_4 possible from 100 g Cl_2. The actual yield is 65.0 g CCl_4.

$$\text{Percentage yield} = \frac{65.0 \text{ g}}{72.3 \text{ g}} \times 100\% = 89.9\%$$

■

When solving problems, you will achieve better results if at first you do not take shortcuts. Solve each problem by writing down the data in a logical, orderly manner. Make certain that the equations are balanced and that the computations are accurate and expressed to the correct number of significant figures. Remember that units are very important; a number without units has little meaning. Finally, an electronic calculator will save many hours of tedious computations.

QUESTIONS

A. Review the Meanings of the New Terms Introduced in this Chapter
  1. Stoichiometry
  2. Mole ratio
  3. Limiting reagent
  4. Theoretical yield
  5. Actual yield
  6. Percentage yield

B. Review Problems

In some of the following problems, equations have not been balanced.

  1. Calculate the number of moles in each of the following quantities:
     (a) 15.0 g KNO_3
     (b) 4.82 g $Ca(NO_3)_2$
     (c) 5.4×10^3 g $(NH_4)_2C_2O_4$
     (d) 1.20 kg $NaHCO_3$
     (e) 385 mg $ZnCl_2$
     (f) 35 millimoles $NaOH$
     (g) 8.4×10^{24} molecules CO_2
     (h) 400 mL ethyl alcohol, C_2H_5OH ($d = 0.789$ g/mL)
     (i) 35.0 mL H_2SO_4 solution ($d = 1.727$ g/mL, 80.0% H_2SO_4 by weight)
  2. Calculate the number of grams in each of the following quantities:
     (a) 4.88 moles $Fe(OH)_3$
     (b) 0.0844 mole $NiSO_4$
     (c) 388 kg $CaCO_3$
     (d) 0.0400 mole $HC_2H_3O_2$
     (e) 25.3 moles NH_3
     (f) 0.913 mole Bi_2S_3
     (g) 28 millimoles HCl
     (h) 2.7×10^{21} molecules glucose, $C_6H_{12}O_6$
     (i) 600 mL of liquid Br_2 ($d = 3.119$ g/mL)
     (j) 80 mL K_2CrO_4 solution ($d = 1.175$ g/mL, 20.0% K_2CrO_4 by weight)

3. Which contains the larger number of molecules?
   (a) 5.0 g H_2O or 5.0 g H_2O_2 (b) 20.0 g HCl or 60.0 g $C_6H_{12}O_6$
4. Given the equation

$$2\,C_3H_7OH + 9\,O_2 \longrightarrow 6\,CO_2 + 8\,H_2O$$

set up the correct mole ratio to calculate:
   (a) Moles O_2 from moles C_3H_7OH
   (b) Moles C_3H_7OH from moles O_2
   (c) Moles O_2 from moles CO_2
   (d) Moles H_2O from moles C_3H_7OH
   (e) Moles CO_2 from moles H_2O
5. Given the equation

$$4\,NH_3 + 3\,O_2 \longrightarrow 2\,N_2 + 6\,H_2O$$

   (a) How many moles of O_2 are required for the reaction of 5.0 moles of NH_3?
   (b) How many moles of water will be produced from the reaction of 0.442 mole NH_3?
   (c) How many grams of N_2 can be formed from the reaction of 2.00 moles NH_3?
   (d) How many grams of NH_3 are needed to produce 400 g N_2?
6. Given the equation

$$4\,FeS_2 + 11\,O_2 \longrightarrow 2\,Fe_2O_3 + 8\,SO_2$$

   (a) How many moles of Fe_2O_3 can be made from 1.00 mole FeS_2?
   (b) How many moles of O_2 are required to react with 3.3 moles FeS_2?
   (c) If the reaction produces 2.11 moles Fe_2O_3, how many moles of SO_2 are produced?
   (d) How many grams of SO_2 can be formed from 0.622 mole FeS_2?
   (e) If the reaction produces 25.3 g SO_2, how many moles of O_2 were reacted?
   (f) How many grams of FeS_2 are needed to produce 121 g Fe_2O_3?
7. An early method of producing chlorine was by the reaction of pyrolusite, MnO_2, and hydrochloric acid. How many moles of HCl will react with 1.52 moles of MnO_2? (Balance the equation first.)

$$MnO_2(s) + HCl(aq) \longrightarrow Cl_2(g) + MnCl_2(aq) + H_2O$$

8. Given the reaction

$$Zn + HCl \longrightarrow ZnCl_2 + H_2$$

200 g of zinc was dropped into a beaker of hydrochloric acid. After the reaction ceased, 35 g of unreacted zinc remained in the beaker.
   (a) How many moles of hydrogen gas were produced?
   (b) What weight of HCl was reacted?
9. Given the balanced equation

$$PCl_5 + 4\,H_2O \longrightarrow H_3PO_4 + 5\,HCl$$

   (a) What is the theoretical yield (in grams) of HCl that can be produced from 750 g of PCl_5?
   *(b) If the yield of H_3PO_4 is only 85% of theoretical, what weight of phosphoric acid would be obtained from 95 g of phosphorus pentachloride?

10. In a blast furnace, iron(III) oxide reacts with coke (carbon) to produce molten iron and carbon monoxide:

$$Fe_2O_3 + C \longrightarrow 2\,Fe + 3\,CO$$

What weight of iron would be formed from 250 kg of Fe_2O_3?

11. What weight of steam and iron must react to produce 350 g of magnetic iron oxide, Fe_3O_4?

$$3\,Fe(s) + 4\,H_2O(g) \longrightarrow Fe_3O_4(s) + 4\,H_2(g)$$

*12. Ethyl alcohol, C_2H_5OH, also called grain alcohol, can be made by the fermentation of sugar, which often comes from starch in grain:

$$\underset{\text{Glucose}}{C_6H_{12}O_6} \longrightarrow \underset{\text{Ethyl alcohol}}{2\,C_2H_5OH} + 2\,CO_2$$

If an 82% yield of ethyl alcohol is obtained,

(a) What weight of ethyl alcohol will be produced from 850 g of glucose?
(b) What weight of glucose should be used to produce 415 g C_2H_5OH?

13. Ethane gas, C_2H_6, burns in air (i.e., reacts with the oxygen in air) to form carbon dioxide and water:

$$2\,C_2H_6 + 7\,O_2 \longrightarrow 4\,CO_2 + 6\,H_2O$$

(a) How many moles of O_2 are needed for the complete combustion of 10.0 moles of ethane?
(b) How many grams of CO_2 are produced for each 6.00 g of H_2O produced?
(c) What weight of CO_2 will be produced by the combustion of 85.0 g of C_2H_6?

*14. Both $CaCl_2$ and $MgCl_2$ react with $AgNO_3$ to precipitate $AgCl$. When solutions containing equal weights of $CaCl_2$ and $MgCl_2$ are reacted, which salt will produce the larger mass of $AgCl$? Show proof.

*15. Carbon disulfide, CS_2, can be made from coke, C, and sulfur dioxide, SO_2:

$$5\,C + 2\,SO_2 \longrightarrow CS_2 + 4\,CO_2$$

If the yield of CS_2 is 82% of theoretical, what weight of coke is needed to produce 950 g of CS_2?

*16. Acetylene, C_2H_2, can be manufactured by the reaction of water and calcium carbide, CaC_2:

$$CaC_2 + 2\,H_2O \longrightarrow C_2H_2\uparrow + Ca(OH)_2$$

When 38.5 g of commercial grade (impure) calcium carbide were reacted, 0.440 mole of C_2H_2 was produced. Assuming that all of the CaC_2 was reacted to C_2H_2, what is the percent of CaC_2 in the commercial grade material?

17. In the following equations, determine which reactant is the limiting reagent and which reactant is in excess. The amounts mixed together are shown below each reactant. Show evidence for your answers.

(a) $\underset{12.0\,g}{KOH} + \underset{16.0\,g}{HNO_3} \longrightarrow KNO_3 + H_2O$

(b) $\underset{6.0\,g}{2\,NaOH} + \underset{10.0\,g}{H_2SO_4} \longrightarrow Na_2SO_4 + 2\,H_2O$

(c) $\underset{25.0\,g}{2\,Bi(NO_3)_3} + \underset{8.0\,g}{3\,H_2S} \longrightarrow Bi_2S_3 + 6\,HNO_3$

(d) $\underset{15.0\,g}{3\,Fe} + \underset{8.0\,g}{4\,H_2O} \longrightarrow Fe_3O_4 + 4\,H_2$

18. The reaction for the combustion of propane, C_3H_8, is:

$$C_3H_8 + 5\,O_2 \longrightarrow 3\,CO_2 + 4\,H_2O$$

(a) If 5.0 moles C_3H_8 and 5.0 moles O_2 are reacted, how many moles of CO_2 can be produced?

(b) If 3.0 moles C_3H_8 and 20.0 moles O_2 are reacted, how many moles of CO_2 can be produced?

(c) If 20.0 moles C_3H_8 and 3.0 moles O_2 are reacted, how many moles of CO_2 can be produced?

*(d) If 2.0 moles C_3H_8 and 14.0 moles O_2 are placed in a closed container, and they react to completion (until one reactant is completely used up), what compounds are present in the container after the reaction, and how many moles of each compound are present?

(e) If 20.0 g C_3H_8 and 20.0 g O_2 are reacted, how many grams of CO_2 can be produced?

(f) If 20.0 g C_3H_8 and 80.0 g O_2 are reacted, how many grams of CO_2 can be produced?

(g) If 20.0 g C_8H_8 and 200 g O_2 are reacted, how many grams of CO_2 can be produced?

19. Aluminum reacts with bromine to form aluminum bromide:

$$2\,Al + 3\,Br_2 \longrightarrow 2\,AlBr_3$$

(a) If 5.0 moles Al and 5.0 moles Br_2 are reacted, what weight of $AlBr_3$ can be produced?

(b) If 30 g Al and 100 g Br_2 are reacted, what weight of $AlBr_3$ can be produced?

(c) If 30 g Al and 100 g Br_2 are reacted, and 44.2 g $AlBr_3$ product are recovered, what is the percentage yield for the reaction?

20. Methyl alcohol, CH_3OH, is made by reacting carbon monoxide and hydrogen in the presence of certain metal oxide catalysts. How much alcohol can be obtained by reacting 40.0 g CO and 10.0 g H_2?

$$CO(g) + 2\,H_2(g) \longrightarrow CH_3OH(l)$$

21. Iron was reacted with a solution containing 400 g of copper(II) sulfate. The reaction was stopped after 1 hour, and 144 g of copper were obtained. Calculate the percentage yield of copper obtained.

$$Fe(s) + CuSO_4(aq) \longrightarrow Cu(s) + FeSO_4(aq)$$

*22. An astronaut excretes about 2500 g of water a day. If lithium oxide, Li_2O, is used in the spaceship to absorb this water, how many kilograms of Li_2O must be carried for a 30-day space trip for three astronauts?

$$Li_2O + H_2O \longrightarrow 2\,LiOH$$

*23. Most commercial hydrochloric acid is prepared by the reaction of concentrated sulfuric acid with sodium chloride:

$$H_2SO_4 + 2\,NaCl \longrightarrow Na_2SO_4 + 2\,HCl$$

How many kilograms of concentrated H_2SO_4, 96% H_2SO_4 by weight, are required to produce 20.0 L of concentrated hydrochloric acid ($d = 1.20$ g/mL, 42.0% HCl by weight)?

*24. What weight, in grams, of air is required to complete the combustion of 250 g of phosphorus to diphosphorus pentoxide, assuming the air to be 23% oxygen by weight?

*25. 12.82 g of a mixture of $KClO_3$ and NaCl are heated strongly. The $KClO_3$ reacts according to the following equation:

$$2 \, KClO_3(s) \longrightarrow 2 \, KCl(s) + 3 \, O_2(g)$$

The NaCl does not undergo any reaction. After the heating, the residue (KCl and NaCl) weighs 10.08 g. Assuming that all the loss of weight represents loss of oxygen gas, calculate the percentage of $KClO_3$ in the original mixture.

26. Phosphine, PH_3, can be prepared by the hydrolysis of calcium phosphide, Ca_3P_2:

$$Ca_3P_2 + 6 \, H_2O \longrightarrow 3 \, Ca(OH)_2 + 2 \, PH_3$$

Based on this equation, which of the following statements are correct?
(a) One mole of Ca_3P_2 produces 2 moles of PH_3.
(b) One gram of Ca_3P_2 produces 2 g of PH_3.
(c) Three moles of $Ca(OH)_2$ are produced for each 2 moles of PH_3 produced.
(d) The mole ratio between phosphine and calcium phosphide is

$$\frac{2 \text{ moles } PH_3}{1 \text{ mole } Ca_3P_2}$$

(e) When 2.0 moles Ca_3P_2 and 3.0 moles H_2O react, 4.0 moles PH_3 can be formed.
(f) When 2.0 moles Ca_3P_2 and 15.0 moles H_2O react, 6.0 moles $Ca(OH)_2$ can be formed.
(g) When 200 g Ca_3P_2 and 100 g H_2O react, Ca_3P_2 is the limiting reagent.
(h) When 200 g Ca_3P_2 and 100 g H_2O react, the theoretical yield of PH_3 is 57.4 g.

27. The equation representing the reaction used for the commercial preparation of hydrogen cyanide is

$$2 \, CH_4 + 3 \, O_2 + 2 \, NH_3 \longrightarrow 2 \, HCN + 6 \, H_2O$$

Based on this equation, which of the statements below are correct?
(a) Three moles of O_2 are required for 2 moles of NH_3.
(b) Twelve moles of HCN are produced for every 16 moles of O_2 that react.
(c) The mole ratio between H_2O and CH_4 is

$$\frac{6 \text{ moles } H_2O}{2 \text{ moles } CH_4}$$

(d) When 12 moles of HCN are produced, 4 moles of H_2O will also be formed.
(e) When 10 moles CH_4, 10 moles O_2, and 10 moles NH_3 are mixed and reacted, O_2 is the limiting reagent.
(f) When 3 moles each of CH_4, O_2, and NH_3 are mixed and reacted, 3 moles of HCN will be produced.

12

The Gaseous State of Matter

After studying Chapter 12 you should be able to:

1. Understand the terms listed in Question A at the end of the chapter.
2. State the principal assumptions of the Kinetic-Molecular Theory (KMT).
3. Estimate the relative rates of effusion of two gases of known molecular weight.
4. Sketch and explain the operation of a mercury barometer.
5. Tell what two factors determine gas pressure in a vessel of fixed volume.
6. Work problems involving (a) Boyle's and (b) Charles' gas laws.
7. State what is meant by standard temperature and pressure (STP).
8. Give the equation for the combined gas law that deals with the pressure, volume, and temperature relationships expressed in Boyle's, Charles', and Gay-Lussac's gas laws.
9. Use Dalton's Law of Partial Pressures and the combined gas laws to calculate the dry STP volume of a gas collected over water.
10. State Avogadro's hypothesis.
11. Understand the mole–weight–volume relationship of gases.
12. Determine the density of any gas at STP.
13. Determine the molecular weight of a gas from its density at a known temperature and pressure.
14. Make mole-to-volume, weight-to-volume, and volume-to-volume stoichiometric calculations from a balanced chemical equation.
15. Define an ideal gas.
16. State two valid reasons why real gases may deviate from the behavior predicted for an ideal gas.
17. Solve problems involving the ideal gas equation.

12.1 General Properties of Gases

In Chapter 3, solids, liquids, and gases are described in a brief outline. In this chapter we will consider the behavior of gases in greater detail.

Of the three states of matter, gases are the least compact and most mobile. A solid has a rigid structure and its particles remain in essentially fixed positions. When a solid absorbs sufficient heat, it melts and changes into a liquid. Melting occurs because the molecules (or ions) have absorbed enough energy to break out of the rigid crystal lattice structure of the solid. The molecules or ions in the liquid are more energetic than they were in the solid, as shown by their increased mobility. Molecules in the liquid state are *coherent*—that is, they cling to one another. When the liquid absorbs additional heat, the more energetic molecules break away from the liquid surface and go into the gaseous state. Gases represent the most mobile state of matter. Gas molecules move with very high velocities and have high kinetic energy (KE). The average velocity of hydrogen molecules at 0°C is over 1600 metres (1 mile) per second. Because of the high velocities of their molecules, mixtures of gases are uniformly distributed within the container in which they are confined.

A quantity of a substance occupies a much greater volume as a gas than does a like quantity of the substance as a liquid or a solid. For example, 1 mole of water (18 g) has a volume of 18 mL at 4°C. This same amount of water would occupy about 22,400 mL in the gaseous state—more than a 1200-fold increase in volume. We may assume from this difference in volume that (1) gas molecules are relatively far apart, (2) gases are capable of being greatly compressed, and (3) the volume occupied by a gas is mostly empty space.

12.2 The Kinetic-Molecular Theory

Kinetic-Molecular Theory (KMT)

Careful scientific studies of the behavior and properties of gases were begun in the 17th century by Robert Boyle (1627–1691). This work was carried forward by many investigators after Boyle. The accumulated data were used in the second half of the 19th century to formulate a general theory to explain the behavior and properties of gases, called the **Kinetic-Molecular Theory (KMT).** The KMT has since been extended to cover, in part, the behavior of liquids and solids. It ranks today with the atomic theory as one of the greatest generalizations of modern science.

ideal, or perfect, gas

The KMT is based on the motion of particles, particularly gas molecules. A gas that behaves exactly as outlined by the theory is known as an **ideal,** or **perfect, gas.** Actually, there are no ideal gases, but under certain conditions of temperature and pressure, gases approach ideal behavior, or at least show only small deviations from it. Under extreme conditions, such as very high pressure and low temperature, real gases may deviate greatly from ideal behavior. For example, at low temperature and high pressure many gases become liquids.

The principal assumptions of the Kinetic-Molecular Theory are

1. Gases consist of tiny (submicroscopic) molecules.
2. The distance between molecules is large compared to the size of the molecules themselves. The volume occupied by a gas consists mostly of empty space.
3. Gas molecules have no attraction for one another.
4. Gas molecules move in straight lines in all directions, colliding frequently with one another and with the walls of the container.
5. No energy is lost by the collision of a gas molecule with another gas molecule or with the walls of the container. All collisions are perfectly elastic.

6. The average kinetic energy for molecules is the same for all gases at the same temperature, and its value is directly proportional to the Kelvin temperature.

Let us consider the facts supporting the theory. Assumption 1 is based on the size of atoms and molecules already established in previous chapters. Assumptions 2 and 3 are based on the comparison of volumes occupied by equal masses of the solid, liquid, and gaseous states of a substance and the fact that gases continue to expand and completely fill any size container. Assumption 4, that gases are in constant motion, is shown by the fact that gases exert pressure, expand into larger containers, and diffuse.

diffusion   The property of **diffusion,** the ability of two or more gases to mix spontaneously, also supports the assumption that gas molecules have very little attraction for one another. The diffusion of gases may be illustrated by use of the apparatus shown in Figure 12.1. Two large flasks, one containing reddish brown bromine vapors and the other dry air, are connected by a side tube. When the stopcock between the flasks is opened, the bromine and air will diffuse into each other. After standing awhile, both flasks will contain bromine and air.

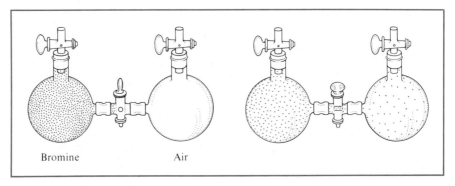

Bromine   Air

Figure 12.1   Diffusion of gases. When the stopcock between the two flasks is opened, colored bromine molecules can be seen diffusing into the flask containing air.

Figure 12.2 (page 250) shows that a gas exerts the same pressure at all parts of a container. The three gauges, located at different parts of the cylinder, show the same pressure.

With billions of molecules present in even a very small mass of gas, it is safe to assume that there will be collisions between these molecules as well as collisions with the walls of the container. Assumption 5 is borne out by the fact that gases do not change temperature upon standing (external causes excepted). This shows that the molecules do not suffer loss of energy by collisions. Although one molecule may transfer energy to another molecule in a collision, the average or total energy of the system remains the same. The kinetic energy (KE)

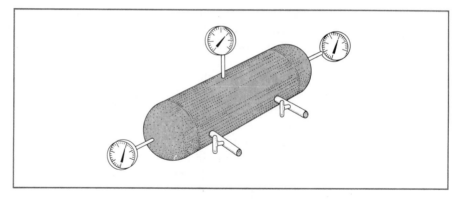

Figure 12.2 A gas moves in all directions and exerts the same pressure in all directions.

of a molecule is one-half of its mass times its velocity squared. It is expressed by the equation

$$KE = \frac{1}{2}mv^2$$

where m is the mass and v is the velocity of the molecule.

Experimental evidence shows that 2.0 g H_2 (1 mole) and 32.0 g O_2 (1 mole) in containers of equal volume at the same temperature exert the same pressure. This evidence supports assumption 6—that the kinetic energy for all gases is the same at the same temperature—and leads us to reason that the molecules of different gases, because of differing masses, will have different average velocities. The *relative* molecular velocities of different gases can be calculated from their kinetic energies. For example, the mass of any oxygen molecule is 32 amu and that of hydrogen is 2 amu. From the Kinetic-Molecular Theory, we have

$$KE \text{ of } H_2 = KE \text{ of } O_2$$

$$\frac{1}{2}m_{H_2}v^2_{H_2} = \frac{1}{2}m_{O_2}v^2_{O_2}$$

$$\frac{1}{2} \times 2 \times v^2_{H_2} = \frac{1}{2} \times 32 \times v^2_{O_2}$$

$$\frac{v^2_{H_2}}{v^2_{O_2}} = \frac{16}{1}$$

Taking the square root of both sides of the equation, we have

$$\frac{v_{H_2}}{v_{O_2}} = \frac{4}{1}$$

These calculations show that the average velocity of a hydrogen molecule is four times greater than that of an oxygen molecule.

effusion

Effusion is a process by which gas molecules pass through a very small orifice (opening) from a container at higher pressure to one at lower pressure. The rates of effusion of different gases are directly proportional to their molecular velocities. Inspection of the foregoing equations shows that molecular velocities —and therefore the rates of effusion—of different gases are inversely proportional to the square roots of their molecular weights. This principle was first introduced by the Scottish chemist Thomas Graham (1805–1869) and is known

Graham's law of effusion

as **Graham's law of effusion:** The rates of effusion of different gases are inversely proportional to the square roots of their molecular weights (or densities).

$$\frac{\text{Rate of effusion of gas A}}{\text{Rate of effusion of gas B}} = \sqrt{\frac{\text{mol. wt B}}{\text{mol. wt A}}}$$

A major application of Graham's law occurred during World War II with the separation of the isotopes of uranium-235 (U-235) and uranium-238 (U-238). Naturally occurring uranium consists of 0.7% U-235, 97.3% U-238, and a small amount of U-234. However, only U-235 is useful as fuel for nuclear reactors and atomic bombs. So the isotopes had to be separated in order to obtain a sufficient quantity of U-235 for use.

Uranium was first changed to uranium hexafluoride, UF_6, a white solid that readily goes into the gaseous state. The gaseous mixture of $^{235}UF_6$ and $^{238}UF_6$ was then allowed to effuse through porous walls. Although the effusion rate of the lighter gas is only slightly faster than that of the heavier one,

$$\frac{\text{Effusion rate } ^{235}UF_6}{\text{Effusion rate } ^{238}UF_6} = \sqrt{\frac{352}{349}} = 1.0043$$

the separation and enrichment of U-235 was accomplished by subjecting the gaseous mixture to several thousand stages of effusion.

The properties of an ideal gas are independent of the molecular constitution of the gas. Mixtures of gases also obey the Kinetic-Molecular Theory if the gases in the mixture do not enter into a chemical reaction with one another.

12.3 Measurement of Pressure of Gases

pressure

Pressure is defined as force per unit area. Do gases exert pressure? Yes. When a rubber balloon is inflated with air, it stretches and maintains an abnormally large size because the pressure on the inside is greater than that on the outside. Pressure results from the collisions of gas molecules with the walls of the balloon (see Figure 12.3). When the gas is released, the force or pressure of the air escaping from the small neck propels the balloon in a rapid, irregular path. If the balloon is inflated until it bursts, the gas escaping all at once causes a small explosive noise. This pressure that gases display can be measured; it can also be transformed into useful work. Steam under pressure, as used in the steam locomotive, played an important role in the early development of the United States. Compressed steam is used today to generate at least part of the electricity for

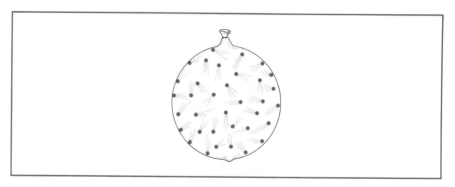

Figure 12.3   Pressure resulting from the collisions of gas molecules with the walls of the balloon keep the balloon inflated.

many cities. Compressed air is used to operate many different kinds of mechanical equipment.

The mass of air surrounding the earth is called the *atmosphere.* It is composed of about 78% nitrogen, 21% oxygen, and 1% argon and other minor constituents by volume (see Table 12.1). The outer boundary of the atmosphere is not known precisely, but more than 99% of the atmosphere is below an altitude of 20 miles (32 km). Thus, the concentration of gas molecules in the atmosphere decreases with altitude, and at about 4 miles there is insufficient oxygen to sustain human life. The gases in the atmosphere exert a pressure known as **atmospheric pressure.** The pressure exerted by a gas depends on the number of molecules of gas present, the temperature, and the volume in which the gas is confined. Gravitational forces confine the atmosphere relatively close to the earth and act to prevent air molecules from flying off into outer space. Thus, the atmospheric pressure at any point is due to the weight of the atmosphere pressing downward at that point.

atmospheric pressure

TABLE 12.1   Average composition of normal dry air.

Gas	Percent by volume	Gas	Percent by volume
N_2	78.08	He	0.0005
O_2	20.95	CH_4	0.0002
Ar	0.93	Kr	0.0001
CO_2	0.033	Xe, H_2, and N_2O	Trace
Ne	0.0018		

The pressure of a gas can be measured with a pressure gauge, a manometer, or a **barometer.** A mercury barometer is commonly used in the laboratory to measure atmospheric pressure. A simple barometer of this type may be prepared by completely filling a long tube with pure, dry mercury and inverting the open end into an open dish of mercury. If the tube is longer than 760 mm, the mercury level will drop to a point at which the column of mercury in the tube

barometer

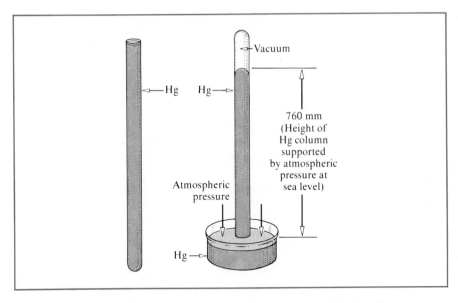

Figure 12.4 Preparation of a mercury barometer. The full tube of mercury at the left is inverted and placed in a dish of mercury.

is just supported by the pressure of the atmosphere. If the tube is properly prepared, a vacuum will exist above the mercury column. The weight of mercury, per unit area, is equal to the pressure of the atmosphere. The column of mercury is supported by the pressure of the atmosphere, and the height of the column is a measure of this pressure (see Figure 12.4). The mercury barometer was invented in 1643 by the Italian physicist E. Torricelli (1608–1647), for whom the unit of pressure *torr* was named.

1 atmosphere

Air pressure is measured and expressed in many units. The standard atmospheric pressure or simply **1 atmosphere** is the pressure exerted by a column of mercury 760 mm high at a temperature of 0°C. The abbreviation for atmosphere is atm. The normal pressure of the atmosphere at sea level is 1 atm or 760 mm Hg. Other units for expressing pressure are inches of mercury, centimetres of mercury, torr, the millibar (mbar), and pounds per square inch (lb/in.²). The meteorologist uses inches of mercury in reporting atmospheric pressure. The values of these units equivalent to 1 atm are summarized in Table 12.2 (1 atm ≡ 760 mm Hg ≡ 76 cm Hg ≡ 29.9 in. Hg ≡ 760 torr ≡ 1013 mbar ≡ 14.7 lb/in.²).

Atmospheric pressure varies with altitude. The average pressure at Denver, Colorado, 1.61 km (1 mile) above sea level, is 63 cm Hg (0.83 atm). Atmospheric pressure is 0.5 atm at about 5.5 km (3.4 miles) altitude. Other liquids besides mercury may be employed for barometers, but they are not as useful as mercury because of the difficulty of maintaining a vacuum above the liquid and because of impractical heights of the liquid column. For example, a pressure of 1 atm will support a column of water about 10,336 mm (33.9 ft) high.

**TABLE 12.2 Pressure units
equivalent to 1 atmosphere.**

1 atm
760 mm Hg
76 cm Hg
760 torr
1013 mbar
29.9 in. Hg
14.7 lb/in.^{2}

PROBLEM 12.1   The average atmospheric pressure at Walnut, California, is 740 mm Hg. Calculate this pressure in (a) torr and (b) atmospheres.

This problem can be solved using conversion factors relating one unit of pressure to another. From Table 12.2 we have

$$1 \text{ atm} = 760 \text{ mm Hg} = 760 \text{ torr}$$

(a) To convert mm Hg to torr, use the conversion factor 760 torr/760 mm Hg:

$$740 \text{ mm Hg} \times \frac{760 \text{ torr}}{760 \text{ mm Hg}} = 740 \text{ torr}$$

(b) To convert mm Hg to atm, use the conversion factor 1 atm/760 mm Hg:

$$740 \text{ mm Hg} \times \frac{1 \text{ atm}}{760 \text{ mm Hg}} = 0.934 \text{ atm}$$

12.4 Dependence of Pressure on Number of Molecules and Temperature

Pressure is produced by gas molecules colliding with the walls of a container. At a specific temperature and volume, the number of collisions depends on the number of gas molecules present. The number of collisions may be increased by increasing the number of gas molecules present. If we double the number of molecules, the frequency of collisions and the pressure should double. We find, for an ideal gas, that this doubling is actually what happens. The pressure, therefore, when the temperature and volume are kept constant, is directly proportional to the number of moles or molecules of gas present. Figure 12.5 illustrates this.

A good example of this molecule–pressure relationship may be observed on an ordinary cylinder of compressed gas that is equipped with a pressure gauge. When the valve is opened, gas escapes from the cylinder. The volume of the cylinder is constant and the decrease in quantity of gas is registered by a drop in pressure indicated on the gauge.

The pressure of a gas in a fixed volume also varies with temperature. When the temperature is increased, the kinetic energy of the molecules increases, causing more frequent collisions of the molecules with the walls of the container. This results in a pressure increase (see Figure 12.6).

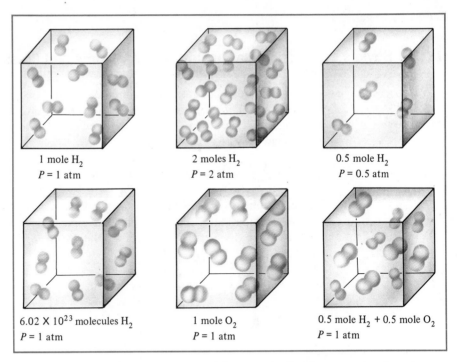

Figure 12.5 The pressure exerted by a gas is directly proportional to the number of molecules present. In each case shown, the volume is 22.4 L and the temperature is 0°C.

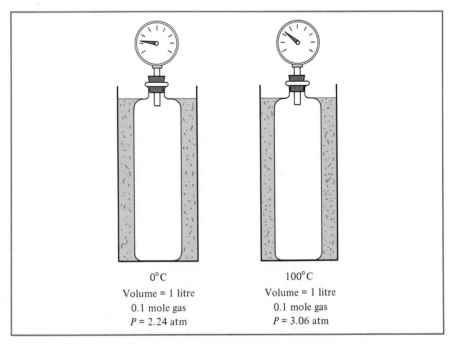

Figure 12.6 The pressure of a gas in a fixed volume increases with increasing temperature. The increased pressure is due to more frequent collisions of the gas molecules with the walls of the container at the higher temperature.

12.5 Boyle's Law—The Relationship of the Volume and Pressure of a Gas

Boyle's law

Robert Boyle demonstrated experimentally that, at constant temperature (T), the volume (V) of a fixed mass of a gas is inversely proportional to the pressure (P). This relationship of P and V is known as **Boyle's law.** Mathematically, Boyle's law may be expressed as

$$V \propto \frac{1}{P} \quad \text{(Mass and temperature are constant)}$$

This equation says that the volume varies inversely with the pressure, at constant mass and temperature. When the pressure on a gas is increased, its volume will decrease, and vice versa. The inverse relationship of pressure and volume is shown graphically in Figure 12.7.

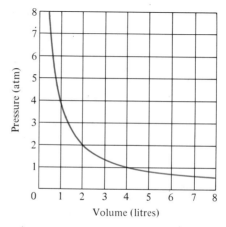

Figure 12.7 Graph of pressure versus volume showing the inverse PV relationship of an ideal gas.

Boyle demonstrated that when he doubled the pressure on a specific quantity of a gas, keeping the temperature constant, the volume was reduced to one-half the original volume; when he tripled the pressure on the system, the new volume was one-third the original volume; and so on. His demonstration shows that the product of volume and pressure is constant if the temperature is not changed:

$$PV = \text{Constant} \quad \text{or} \quad PV = k \quad \text{(Mass and } T \text{ are constant)}$$

Let us demonstrate this law by taking a cylinder of gas with a movable piston, so that the volume may be varied by changing the external pressure (see Figure 12.8). We assume that there is no change in temperature or in the number of molecules. Let us start with a volume of 1000 mL and a pressure of 1 atm. When we change the pressure to 2 atm, the gas molecules are crowded closer together and the volume is reduced to 500 mL. When we increase the pressure to 4 atm, the volume becomes 250 mL.

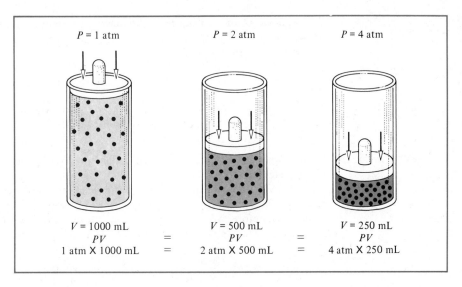

Figure 12.8 The effect of pressure on the volume of a gas.

Note that the product of the pressure times the volume in each case is the same number, substantiating Boyle's law. We may then say that

$$P_1 V_1 = P_2 V_2$$

where $P_1 V_1$ is the pressure–volume product at one set of conditions and $P_2 V_2$, at another set of conditions. In each case, the new volume may be calculated by multiplying the starting volume by a ratio of the two pressures involved. Of course, the ratio of pressures used must reflect the direction in which the volume should change. When the pressure is changed from 1 atm to 2 atm, the ratio to be used is 1 atm/2 atm. Now we can verify the results given in Figure 12.8:

(a) Starting volume: 1000 mL; pressure change (1 atm → 2 atm)

$$1000 \text{ mL} \times \frac{1 \text{ atm}}{2 \text{ atm}} = 500 \text{ mL}$$

(b) Starting volume: 1000 mL; pressure change (1 atm → 4 atm)

$$1000 \text{ mL} \times \frac{1 \text{ atm}}{4 \text{ atm}} = 250 \text{ mL}$$

(c) Starting volume: 500 mL; pressure change (2 atm → 4 atm)

$$500 \text{ mL} \times \frac{2 \text{ atm}}{4 \text{ atm}} = 250 \text{ mL}$$

In summary, a change in the volume of a gas due to a change in pressure may be calculated by multiplying the original volume by a ratio of the two pressures. If the pressure is increased, the ratio should have the smaller pressure in the numerator and the larger pressure in the denominator. If the pressure is decreased, the larger pressure should be in the numerator and the smaller pressure should be in the denominator.

$$\text{New volume} = \text{Original volume} \times \text{Ratio of pressures}$$

Examples of problems based on Boyle's law follow. If no mention is made of temperature, assume that it remains constant.

PROBLEM 12.2   What volume will 2.50 L of a gas occupy if the pressure is changed from 760 mm Hg to 630 mm Hg?

First we must determine whether the pressure is being increased or decreased. In this case it is being decreased. This decrease in pressure should result in an increase in volume. Therefore, we need to multiply 2.50 L by a ratio of the pressures, which will give us an increase in volume. This ratio is 760 mm Hg/630 mm Hg. The calculation is

$$V = 2.50 \text{ L} \times \frac{760 \text{ mm Hg}}{630 \text{ mm Hg}} = 3.02 \text{ L} \quad \text{(New volume)}$$

Alternatively, an algebraic approach may be used, solving $P_1 V_1 = P_2 V_2$ for V_2:

$$V_2 = V_1 \times \frac{P_1}{P_2} = 2.50 \text{ L} \times \frac{760 \text{ mm Hg}}{630 \text{ mm Hg}} = 3.02 \text{ L}$$

where $V_1 = 2.50 \text{ L}$, $P_1 = 760 \text{ mm Hg}$, and $P_2 = 630 \text{ mm Hg}$.

PROBLEM 12.3   A given mass of hydrogen occupies 40.0 L at 760 mm Hg pressure. What volume will it occupy at 5 atm pressure?

Since the units of the two pressures are not the same they must be made the same; otherwise, the units will not cancel in the final calculation. Since the pressure is increased, the volume should decrease. Therefore, we need to multiply 40.0 L by a ratio of the pressures that will give us a decrease in volume.

Step 1. Convert 760 mm Hg to atmospheres by multiplying by the conversion factor 1 atm/760 mm Hg:

$$760 \text{ mm Hg} \times \frac{1 \text{ atm}}{760 \text{ mm Hg}} = 1 \text{ atm}$$

Step 2. Set up a ratio of the pressures that will give a volume decrease:

$$\frac{1 \text{ atm}}{5 \text{ atm}}$$

Step 3. Multiply the volume (40.0 L) by this pressure ratio:

$$V = 40.0 \text{ L} \times \frac{1 \text{ atm}}{5 \text{ atm}} = 8.00 \text{ L} \quad \text{(Answer)}$$

PROBLEM 12.4   A gas occupies a volume of 200 mL at 400 mm Hg pressure. To what pressure must the gas be subjected in order to change the volume to 75.0 mL?

In order to reduce the volume from 200 mL to 75.0 mL, it will be necessary to increase the pressure. In the same way we calculated volume change affected by a change in pressure, we must multiply the original pressure by a ratio of the two volumes. The volume ratio in this case should be 200 mL/75.0 mL. The calculation is

$$\text{New pressure} = \text{Original pressure} \times \text{Ratio of volumes}$$

$$P = 400 \text{ mm Hg} \times \frac{200 \text{ mL}}{75.0 \text{ mL}} = 1067 \text{ mm Hg } (1.07 \times 10^3 \text{ mm Hg}) \quad \text{(New pressure)}$$

Algebraically, $P_1 V_1 = P_2 V_2$ may be solved for P_2:

$$P_2 = P_1 \times \frac{V_1}{V_2} = 400 \text{ mm Hg} \times \frac{200 \text{ mL}}{75.0 \text{ mL}} = 1.07 \times 10^3 \text{ mm Hg}$$

where $P_1 = 400$ mm Hg, $V_1 = 200$ mL, and $V_2 = 75.0$ mL.

In problems of this type, it is good practice to check the answers to see if they are consistent with the given facts. For example, if the data indicate that the pressure is increased, the final volume should be smaller than the initial volume.

■

12.6 Charles' Law—The Effect of Temperature on the Volume of a Gas

The effect of temperature on the volume of a gas was observed in about 1787 by the French physicist J. A. C. Charles (1746–1823). Charles found that various gases expanded by the same fractional amount when heated through the same temperature interval. Later it was found that if a given volume of any gas initially at 0°C was cooled by 1°C, the volume decreased by $\frac{1}{273}$; if cooled by 2°C, by $\frac{2}{273}$; if cooled by 20°C, by $\frac{20}{273}$; and so on. Since each degree of cooling reduced the volume by $\frac{1}{273}$, it was apparent that any quantity of any gas would have zero volume, if it could only be cooled to −273°C. Of course, no real gas can be cooled to −273°C for the simple reason that it liquefies before that temperature is reached. However, −273°C (more precisely −273.16°C) is referred to as *absolute zero;* this temperature is the zero point on the Kelvin (absolute) temperature scale. It is the temperature at which the volume of an ideal, or perfect, gas would become zero.

The volume–temperature relationship for gases is shown graphically in Figure 12.9. Experimental data show the graph to be a straight line that, when extrapolated, crosses the temperature axis at −273.16°C, or absolute zero.

Charles' law

In modern form, **Charles' law** states that at constant pressure the volume of a fixed weight of any gas is directly proportional to the absolute temperature. Mathematically, Charles' law may be expressed as

$$V \propto T \quad (P \text{ is constant})$$

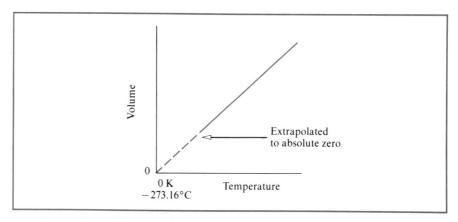

Figure 12.9 Volume–temperature relationship of gases. Extrapolated portion of the graph is shown by the broken line.

which means that the volume of a gas varies directly with the absolute temperature when the pressure remains constant. In equation form Charles' law may also be written as

$$V = kT \qquad \text{or} \qquad \frac{V}{T} = k \quad \text{(At constant pressure)}$$

where k is a constant for a fixed weight of the gas. If the absolute temperature of a gas is doubled, the volume will double. (A capital T is usually used for absolute temperature, K, and a small t for °C.)

To illustrate, let us return to the gas cylinder with the movable or free-floating piston (see Figure 12.10). Assume that the cylinder labeled (a) contains a quantity of gas and the pressure on it is 1 atm. When the gas is heated, the molecules move faster and their kinetic energy increases. This action should increase the number of collisions per unit of time and thereby the pressure. However, the increased internal pressure will cause the piston to rise to a level at which the internal and external pressures again equal 1 atm, as we see in cylinder (b). The net result is an increase in volume due to an increase in temperature.

Another equation relating the volume of a gas at two different temperatures is

$$\frac{V_1}{T_1} = \frac{V_2}{T_2} \quad \text{(Constant } P\text{)}$$

where V_1 and T_1 are one set of conditions and V_2 and T_2 are another set of conditions.

A simple experiment showing the variation of the volume of a gas with temperature is illustrated in Figure 12.11. A flask to which a balloon is attached is immersed in either ice water or hot water. In ice water the volume is reduced, as shown by the collapse of the balloon; in hot water the gas expands and the balloon increases in size.

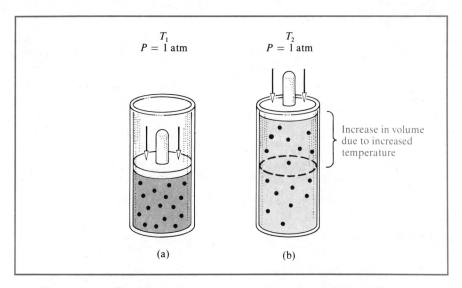

Figure 12.10 The effect of temperature on the volume of a gas. The gas in cylinder (a) is heated from T_1 to T_2. With the external pressure constant at 1 atm, the free-floating piston rises, resulting in an increased volume, as shown in cylinder (b).

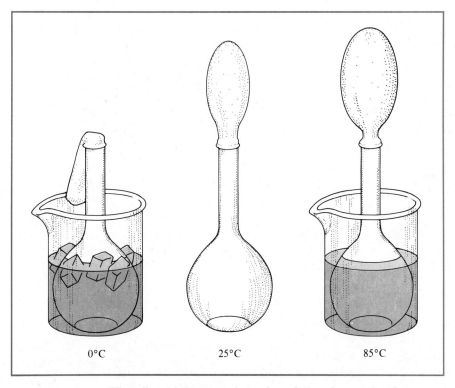

Figure 12.11 The effect of temperature on the volume of a gas. A volume decrease occurs when a flask to which a balloon is attached is immersed in ice water; the volume increases when the flask is immersed in hot water.

The calculation of changes in volume due to changes in temperature involves two basic steps: (1) changing the temperatures to K and (2) multiplying the original volume by a ratio of the initial and final temperatures. If the temperature is increased, the higher temperature is placed in the numerator of the ratio and the lower temperature is placed in the denominator. If the temperature is decreased, the lower temperature is placed in the numerator of the ratio and the higher temperature is placed in the denominator.

New volume = Original volume × Ratio of temperatures (K)

Problems based on Charles' law follow.

PROBLEM 12.5 Three litres of hydrogen at −20°C are allowed to warm to a room temperature of 27°C. What is the volume at room temperature if the pressure remains constant?

Step 1. Change °C to K.

$$°C + 273 = K$$
$$-20°C + 273 = 253 \text{ K}$$
$$27°C + 273 = 300 \text{ K}$$

Step 2. Since the temperature is increased, the volume should increase. The original volume should be multiplied by the temperature ratio of 300 K/253 K. The calculation is

$$V = 3.00 \text{ L} \times \frac{300\,K}{253\,K} = 3.56 \text{ L} \quad \text{(New volume)}$$

To obtain the answer by algebra, solve $V_1/T_1 = V_2/T_2$ for V_2:

$$V_2 = V_1 \times \frac{T_2}{T_1} = 3.00 \text{ L} \times \frac{300\,K}{253\,K} = 3.56 \text{ L}$$

where $V_1 = 3.00$ L, $T_1 = 253$ K, and $T_2 = 300$ K.

PROBLEM 12.6 If 20.0 L of nitrogen are cooled from 100°C to 0°C, what is the new volume?
Since no mention is made of pressure, assume that there is no pressure change.

Step 1. Change °C to K.

$$100°C + 273 = 373 \text{ K}$$
$$0°C + 273 = 273 \text{ K}$$

Step 2. The ratio of temperature to be used is 273 K/373 K, since the final volume should be smaller than the original volume. The calculation is

$$V = 20.0 \text{ L} \times \frac{273\,K}{373\,K} = 14.6 \text{ L} \quad \text{(New volume)}$$

Three variables—pressure, P; volume, V; and temperature, T—are needed to describe a fixed amount of a gas. Boyle's law, $PV = k$, relates pressure and volume at constant temperature; Charles' law, $V = kT$, relates volume and temperature at constant pressure. A third relationship involving pressure and temperature at constant volume is also known and is stated: The pressure of a fixed weight of a gas, at constant volume, is directly proportional to the Kelvin temperature. In equation form, the relationship is

$$P = kT \quad \text{(At constant volume)}$$

This relationship is a modification of Charles' law and is sometimes called Gay-Lussac's law.

We may summarize the effect of changes in pressure, temperature, and quantity of a gas as follows:

1. In the case of a fixed or constant volume,
   (a) when the temperature is increased, the pressure increases.
   (b) when the quantity of a gas is increased, the pressure increases
      (T remaining constant).
2. In the case of a variable volume,
   (a) when the external pressure is increased, the volume decreases
      (T remaining constant).
   (b) when the temperature of a gas is increased, the volume increases
      (P remaining constant).
   (c) when the quantity of a gas is increased, the volume increases
      (P and T remaining constant).

12.7 Standard Temperature and Pressure

standard conditions
standard
temperature and
pressure (STP)

In order to compare volumes of gases, common reference points of temperature and pressure were selected and called **standard conditions** or **standard temperature and pressure** (abbreviated **STP**). Standard temperature is 273 K (0°C) and standard pressure is 1 atm, or 760 mm Hg. For purposes of comparison, volumes of gases are usually changed to STP conditions.

Standard temperature = 273 K or 0°C
Standard pressure = 1 atm or 760 mm Hg

12.8 Combined Gas Laws (Simultaneous Changes in Pressure, Volume, and Temperature)

When both temperature and pressure change at the same time, the new volume may be calculated by multiplying the initial volume by the correct ratios of both pressure and temperature, as follows:

$$\text{Final volume} = \text{Initial volume} \times \left(\begin{array}{c}\text{Ratio of}\\ \text{pressures}\end{array}\right) \times \left(\begin{array}{c}\text{Ratio of}\\ \text{temperatures}\end{array}\right)$$

This equation combines both Boyle's and Charles' laws, and the same considerations for the pressure and the temperature ratios should be used in the calculation. There are four possible variations:

1. Both T and P cause an increase in volume.
2. Both T and P cause a decrease in volume.
3. T causes an increase and P causes a decrease in volume.
4. T causes a decrease and P causes an increase in volume.

The P, V, and T relationships for a given weight of any gas, in fact, may be expressed as a single equation, $PV/T = k$. For problem solving, this equation is usually written

$$\frac{P_1V_1}{T_1} = \frac{P_2V_2}{T_2}$$

where P_1V_1/T_1 are the initial conditions and P_2V_2/T_2 are the final conditions.

This equation may be solved for any one of the six variables represented and is very generally useful in dealing with the pressure–volume–temperature relationships of gases. Note that when T is constant ($T_1 = T_2$), Boyle's law is represented; when P is constant ($P_1 = P_2$), Charles' law is represented; and when V is constant ($V_1 = V_2$), the modified Charles' or Gay-Lussac's law is represented.

PROBLEM 12.7   Given 20.0 L of ammonia gas at 5°C and 730 mm Hg pressure, calculate the volume at 50°C and 800 mm Hg.

In order to get a better look at the data, tabulate the initial and final conditions:

	Initial	Final
V	20.0 L	V_2
T	5°C	50°C
P	730 mm Hg	800 mm Hg

Step 1. Change °C to K:

$$5°C + 273 = 278 \text{ K}$$
$$50°C + 273 = 323 \text{ K}$$

Step 2. Set up ratios of T and P:

$$T \text{ ratio} = \frac{323 \text{ K}}{278 \text{ K}} \qquad \text{(Increase in } T \text{ should increase } V)$$

$$P \text{ ratio} = \frac{730 \text{ mm Hg}}{800 \text{ mm Hg}} \qquad \text{(Increase in } P \text{ should decrease } V)$$

Step 3. The calculation is

$$V_2 = 20.0 \text{ L} \times \frac{730 \text{ mm Hg}}{800 \text{ mm Hg}} \times \frac{323 \text{ K}}{278 \text{ K}} = 21.2 \text{ L}$$

The algebraic solution is:

Solve $\quad \dfrac{P_1 V_1}{T_1} = \dfrac{P_2 V_2}{T_2} \quad$ for V_2 by multiplying both sides of the equation

by T_2/P_2 and rearranging to obtain

$$V_2 = \frac{V_1 \times P_1 \times T_2}{P_2 \times T_1}$$

Tabulate the known values:

$V_1 = 20.0 \text{ L}$ $\qquad\qquad$ $V_2 = ?$
$T_1 = 5°\text{C} + 273 = 278 \text{ K}$ $\qquad$ $T_2 = 50°\text{C} + 273 = 323 \text{ K}$
$P_1 = 730 \text{ mm Hg}$ $\qquad\qquad$ $P_2 = 800 \text{ mm Hg}$

Substitute these values in the equation and calculate the value of V_2:

$$V_2 = \frac{20.0 \text{ L} \times 730 \text{ mm Hg} \times 323 \text{ K}}{800 \text{ mm Hg} \times 278 \text{ K}} = 21.2 \text{ L}$$

■

PROBLEM 12.8 To what temperature (°C) must 10.0 L of nitrogen at 25°C and 700 mm Hg be heated in order to have a volume of 15.0 L and a pressure of 760 mm Hg?
This problem is conveniently handled by an algebraic solution.

Solve $\quad \dfrac{P_1 V_1}{T_1} = \dfrac{P_2 V_2}{T_2} \quad$ for T_2 to obtain $\quad T_2 = \dfrac{T_1 \times P_2 \times V_2}{P_1 \times V_1}$

Tabulate the known values:

$P_1 = 700 \text{ mm Hg}$ $\qquad\qquad$ $P_2 = 760 \text{ mm Hg}$
$V_1 = 10.0 \text{ L}$ $\qquad\qquad$ $V_2 = 15.0 \text{ L}$
$T_1 = 25°\text{C} + 273 = 298 \text{ K}$ $\qquad$ $T_2 = ?$

Substitute these known values in the equation and evaluate T_2:

$$T_2 = \frac{298 \text{ K} \times 760 \text{ mm Hg} \times 15.0 \text{ L}}{700 \text{ mm Hg} \times 10.0 \text{ L}} = 485 \text{ K}$$

Since the problem asks for °C, we must subtract 273 from the K answer:

$485 \text{ K} - 273 = 212°\text{C}$ (Answer)

■

PROBLEM 12.9   The volume of a gas-filled balloon is 50.0 L at 20°C and 742 mm Hg. What volume will it occupy at standard temperature and pressure (STP)? Tabulate the data.

	Initial	Final
V	50.0 L	V_2
T	20°C	273 K
P	742 mm Hg	760 mm Hg

Step 1. STP conditions are 273 K and 760 mm Hg. First change °C to K:

$$20°C + 273 = 293 \text{ K}$$

Step 2. Set up ratios of T and P:

$$T \text{ ratio} = \frac{273 \text{ K}}{293 \text{ K}} \qquad [\text{Decrease in } T \text{ (293 K to 273 K) should decrease } V]$$

$$P \text{ ratio} = \frac{742 \text{ mm Hg}}{760 \text{ mm Hg}} \quad [\text{Increase in } P \text{ (742 mm to 760 mm) should decrease } V]$$

Step 3. The calculation is

$$V_2 = 50.0 \text{ L} \times \frac{273 \text{ K}}{293 \text{ K}} \times \frac{742 \text{ mm Hg}}{760 \text{ mm Hg}}$$

$$V_2 = 45.5 \text{ L}$$

The problem may also be done by solving $\dfrac{P_1 V_1}{T_1} = \dfrac{P_2 V_2}{T_2}$ for V_2 and substituting PVT values in the new equation.

■

12.9 Dalton's Law of Partial Pressures

Dalton's Law of Partial Pressures

If gases behave according to the Kinetic–Molecular Theory, there should be no difference in their pressure–volume–temperature relationships, whether the gas molecules are all the same or different. This similarity in behavior is the basis for an understanding of **Dalton's Law of Partial Pressures,** which states that the total pressure of a mixture of gases is the sum of the partial pressures exerted by each gas in the mixture. Each gas in the mixture exerts a pressure independent of the other gases present. These pressures are called *partial pressures.* Thus, if we have a mixture of three gases, A, B, and C, exerting 50 mm, 150 mm, and 400 mm Hg pressure, respectively, the total pressure will be 600 mm Hg.

$$P_{\text{Total}} = P_A + P_B + P_C$$
$$P_{\text{Total}} = 50 \text{ mm Hg} + 150 \text{ mm Hg} + 400 \text{ mm Hg} = 600 \text{ mm Hg}$$

We can see an application of Dalton's law in the collection of gases over water. When oxygen is prepared in the laboratory, it is commonly collected

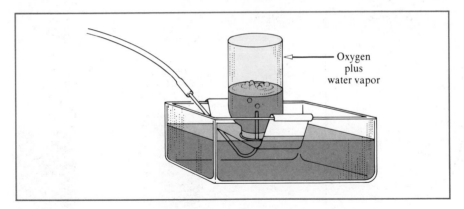

Figure 12.12 Oxygen collected over water.

over water (see Figure 12.12). The O_2, collected by the downward displacement of water, is not pure but contains water vapor mixed with it. When the water level is adjusted to be the same inside and outside the bottle, the pressure of the oxygen plus water vapor inside the bottle is equal to the atmospheric pressure:

$$P_{atm} = P_{O_2} + P_{H_2O}$$
$$P_{O_2} = P_{atm} - P_{H_2O}$$

To determine the amount of O_2 or any other gas collected over water, we must subtract the pressure of the water vapor from the total pressure of the gas. The vapor pressure of water at various temperatures is tabulated in Appendix II.

An illustrative problem follows.

PROBLEM 12.10 A 500 mL sample of oxygen, O_2, was collected over water at 23°C and 760 mm Hg pressure. What volume will the dry O_2 occupy at 23°C and 760 mm Hg? The vapor pressure of water at 23°C is 21.0 mm Hg.

To solve this problem, we must first find the pressure of the O_2 alone by subtracting the pressure of the water vapor present.

$$P_{Total} = 760 \text{ mm Hg} = P_{O_2} + P_{H_2O}$$
$$P_{O_2} = 760 \text{ mm Hg} - 21.0 \text{ mm Hg} = 739 \text{ mm Hg}$$

Thus, the pressure of dry O_2 is 739 mm Hg.

The problem is now of the Boyle's law type. It is treated as if we had 500 mL of dry O_2 at 739 mm Hg pressure, which is then changed to 760 mm Hg pressure, with the temperature remaining constant. The calculation is

$$V = 500 \text{ mL} \times \frac{739 \cancel{\text{ mm Hg}}}{760 \cancel{\text{ mm Hg}}} = 486 \text{ mL dry } O_2$$

This means that 486 mL of the 500 mL mixture of O_2 and water vapor is pure O_2. Figure 12.13 (page 268) depicts the pressure and volume changes involved in this problem.

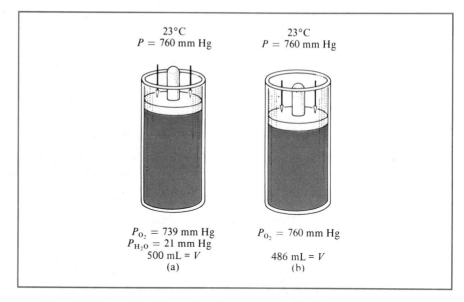

Figure 12.13   A 500 mL sample of oxygen was collected over water at 23°C and 760 mm Hg pressure. The original gas collected is shown in cylinder (a). When the water vapor is removed (b), the volume is reduced. The external pressure, being greater than the pressure of the oxygen, forces the cylinder lid downward until the pressure of the oxygen is 760 mm Hg. The volume the dry oxygen would occupy is 486 mL.

12.10 Avogadro's Hypothesis

Gay-Lussac's Law of Combining Volumes of Gases

Early in the 19th century, J. L. Gay-Lussac (1778–1850) of France studied the volume relationships of reacting gases. His results, published in 1809, were summarized in a statement known as **Gay-Lussac's Law of Combining Volumes of Gases:** *When measured at the same temperature and pressure, the ratios of the volumes of reacting gases are small whole numbers.* Thus, H_2 and O_2 combine to form water vapor in a volume ratio of 2 to 1 (Figure 12.14); H_2 and Cl_2 react to form HCl in a volume ratio of 1 to 1; H_2 and N_2 react to form NH_3 in a volume ratio of 3 to 1.

Avogadro's hypothesis

Two years later, in 1811, Amedeo Avogadro of Italy (1776–1856) used the Law of Combining Volumes of Gases to make a simple but very significant and far-reaching generalization concerning gases. **Avogadro's hypothesis** states:

Equal volumes of different gases at the same temperature and pressure contain the same number of molecules.

This hypothesis was a real breakthrough in understanding the nature of gases. (1) It offered a rational explanation of Gay-Lussac's Law of Combining

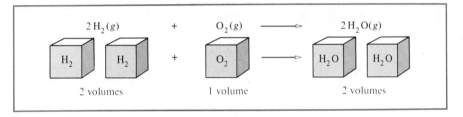

Figure 12.14   Gay-Lussac's Law of Combining Volumes of Gases applied to the reaction of hydrogen and oxygen. When measured at the same temperature and pressure, hydrogen and oxygen react in a volume ratio of 2 to 1.

Volumes of Gases and indicated the diatomic nature of such elemental gases as hydrogen, chlorine, and oxygen; (2) it provided a method for determining the molecular weights of gases and for comparing the densities of gases of known molecular weight (see Section 12.11); and (3) it afforded a firm foundation for the development of the Kinetic-Molecular Theory.

On a volume basis, hydrogen and chlorine react thus:

$$\text{Hydrogen} + \text{Chlorine} \longrightarrow \text{Hydrogen chloride}$$
$$\text{1 volume} \qquad \text{1 volume} \qquad\qquad \text{2 volumes}$$

By Avogadro's hypothesis, equal volumes of hydrogen and chlorine must contain the same number of molecules. Therefore, hydrogen molecules react with chlorine molecules in a 1:1 ratio. Since two volumes of hydrogen chloride are produced, one molecule of hydrogen and one of chlorine must produce two molecules of hydrogen chloride. Therefore, each hydrogen molecule and each chlorine molecule is made up of two atoms. The coefficients of the balanced equation for the reaction give the correct ratios for volumes, molecules, and moles of reactants and products:

$$H_2 \quad + \quad Cl_2 \quad \longrightarrow \quad 2\,HCl$$

H_2	Cl_2	$2\,HCl$
1 volume	1 volume	2 volumes
1 molecule	1 molecule	2 molecules
1 mole	1 mole	2 moles

By like reasoning, oxygen molecules must contain at least two atoms because one volume of oxygen reacts with two volumes of hydrogen to produce two volumes of water vapor.

The volume of a gas depends on the temperature, the pressure, and the number of gas molecules. Two or more gases at the same temperature have the same average kinetic energy. If these gases occupy the same volume, they will exhibit the same pressure. Such a system of identical PVT properties can only be produced by the same number of molecules having the same average kinetic energy.

12.11   Mole–Weight–Volume Relationship of Gases

molar volume of a gas

A mole of any gas contains 6.02×10^{23} molecules (Avogadro's number). However, the volume occupied by a mole of a gas varies with temperature and pressure. It has been experimentally determined that 1 mole of any gas occupies a volume of 22.4 litres at standard temperature and pressure. This volume, 22.4 litres, is known as the **molar volume of a gas.** The molar volume is a cube about 28.2 cm (11.1 in.) on a side (see Figure 12.15). The gram-molecular weights of several gases, each occupying 22.4 L at STP, are also shown in Figure 12.15.

One mole of a gas occupies 22.4 litres at STP.

The molar volume is useful for determining the molecular weight of a gas or of substances that can be easily vaporized into gases. If the weight and the volume of a gas at STP are known, we can calculate its molecular weight. For example, 1 litre of pure oxygen at STP weighs 1.429 g. The molecular weight of oxygen may be calculated by multiplying the weight of 1 litre by 22.4 litres/mole.

$$\frac{1.429 \text{ g}}{\cancel{L}} \times \frac{22.4 \cancel{L}}{\text{mole}} = 32.0 \text{ g/mole} \quad \text{(g-mol. wt)}$$

If the weight and volume are at other than standard conditions, we first change the volume to STP and then calculate the molecular weight. Note that we do not correct the weight to standard conditions—only the volume.

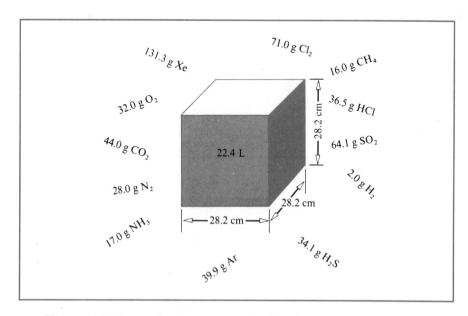

Figure 12.15   One mole of a gas occupies 22.4 litres at STP. The weight given for each gas is its gram-molecular weight (1 mole).

The molar volume, 22.4 L/mole, is used as a conversion factor to convert grams per litre to grams per mole and also to convert litres to moles. The two conversion factors are

$$\frac{22.4 \text{ L}}{1 \text{ mole}} \quad \text{and} \quad \frac{1 \text{ mole}}{22.4 \text{ L}}$$

These conversions must be done at STP except under certain special circumstances. Examples of problems follow.

PROBLEM 12.11 If 2.00 L of a gas measured at STP weigh 3.23 g, what is the molecular weight of the gas? The unit of molecular weight is g/mole; the conversion is from

$$\frac{g}{L} \longrightarrow \frac{g}{\text{mole}}$$

The starting amount is $\dfrac{3.23 \text{ g}}{2.00 \text{ L}}$

The conversion factor is $\dfrac{22.4 \text{ L}}{\text{mole}}$

The calculation is

$$\frac{3.23 \text{ g}}{2.00 \cancel{L}} \times \frac{22.4 \cancel{L}}{\text{mole}} = 36.2 \text{ g/mole} \quad \text{(g-mol. wt)}$$

∎

PROBLEM 12.12 Measured at 40°C and 630 mm Hg, 691 mL of ethyl ether weigh 1.65 g. Calculate the gram-molecular weight of ethyl ether.

In order to use 22.4 L/mole, we must first correct the volume to standard conditions:

740 mm Hg $\longrightarrow$ 760 mm Hg (Standard pressure)
40°C (313 K) $\longrightarrow$ 273 K   (Standard temperature)

$$V = 691 \text{ mL} \times \frac{273 \cancel{K}}{313 \cancel{K}} \times \frac{630 \cancel{\text{ mm Hg}}}{760 \cancel{\text{ mm Hg}}}$$

V at (STP) = 500 mL = 0.500 L

The weight of the gas has not been altered by correcting the volume to STP, so that 500 mL at STP weigh 1.65 g. The conversion is from g/L to g/mole.

$$\frac{1.65 \text{ g}}{0.500 \cancel{L}} \times \frac{22.4 \cancel{L}}{\text{mole}} = 73.9 \text{ g/mole} \quad \text{(g-mol. wt)}$$

∎

12.12 Density and Specific Gravity of Gases

The density, d, of a gas is its mass per unit volume, which is generally expressed in grams per litre (g/L) as follows:

$$d = \frac{\text{Mass}}{\text{Volume}} = \frac{\text{g}}{\text{L}}$$

Because the volume of a gas depends on temperature and pressure, both of these should be given when stating the density of a gas. The volume of solids and liquids is hardly affected by changes in pressure and is changed only by a small degree when the temperature is varied. Increasing the temperature from 0°C to 50°C will reduce the density of a gas by about 18% if the gas is allowed to expand. In comparison, a 50°C rise in the temperature of water (0°C → 50°C) will change its density by less than 0.2%.

The density of a gas at any temperature and pressure may be determined by calculating the weight of gas present in 1 litre. At STP, in particular, the density may be calculated by multiplying the gram-molecular weight of the gas by 1 mole/22.4 L.

$$d \text{ (at STP)} = \text{g-mol. wt} \times \frac{1 \text{ mole}}{22.4 \text{ L}}$$

$$\text{g-mol. wt} = d \text{ (at STP)} \times \frac{22.4 \text{ L}}{\text{mole}}$$

PROBLEM 12.13   Calculate the density of Cl_2 at STP.

$$\text{g-mol. wt of } Cl_2 = 71.0 \text{ g/mole}$$

$$d = \frac{71.0 \text{ g}}{\text{mole}} \times \frac{1 \text{ mole}}{22.4 \text{ L}} = 3.17 \text{ g/L}$$

∎

The specific gravity (sp gr) of a gas is the ratio of the mass of any volume of the gas to the mass of an equal volume of some reference gas. Specific gravities of gases are commonly quoted in reference to air = 1.00. The actual mass of air at STP is 1.29 g/L, which is the density of air.

The specific gravity can be calculated by dividing the density of a gas by the density of air. Both gases must be at the same temperature and pressure.

$$\text{sp gr} = \frac{\text{Density of a gas}}{\text{Density of air}}$$

The specific gravity of Cl_2, for example, is

$$\text{sp gr of } Cl_2 = \frac{\text{Density of } Cl_2}{\text{Density of air}} = \frac{3.17 \text{ g/L}}{1.29 \text{ g/L}} = 2.46$$

This indicates that Cl_2 is 2.46 times as heavy as air. Table 12.3 lists the density and specific gravity of some common gases.

TABLE 12.3 Density and specific gravity of common gases at STP.

Gas	Molecular weight	Density (g/L at STP)	Specific gravity (air = 1.00)
H_2	2.0	0.090	0.070
CH_4	16.0	0.714	0.553
NH_3	17.0	0.760	0.589
C_2H_2	26.0	1.16	0.899
HCN	27.0	1.21	0.938
CO	28.0	1.25	0.969
N_2	28.0	1.25	0.969
O_2	32.0	1.43	1.11
H_2S	34.1	1.52	1.18
HCl	36.5	1.63	1.26
F_2	38.0	1.70	1.32
CO_2	44.0	1.96	1.52
C_3H_8	44.0	1.96	1.52
O_3	48.0	2.14	1.66
SO_2	64.1	2.86	2.22
Cl_2	71.0	3.17	2.46

12.13 Calculations from Chemical Equations Involving Gases

1. Mole–volume (gas) and weight–volume (gas) calculations. Stoichiometric problems involving gas volumes can be solved by the general mole-ratio method outlined in Chapter 11. The factors 1 mole/22.4 L and 22.4 L/1 mole are used as needed for converting volume to moles and moles to volume, respectively. These conversion factors are used under the assumptions that the gases are at STP and behave as ideal gases. In practice, gases are measured at other than STP conditions, and the volumes are converted to STP for stoichiometric calculations.

In a balanced equation, the number in front of the formula of a gaseous substance represents the number of moles or molar volumes (22.4 L at STP) of that substance.

The following are examples of typical problems involving gases and chemical equations.

PROBLEM 12.14   What volume of oxygen (at STP) can be formed from 0.500 mole of potassium chlorate?

Step 1. Write the balanced equation:

$$2 \, KClO_3 \longrightarrow 2 \, KCl + 3 \, O_2 \uparrow$$

2 moles   3 moles

Step 2. The starting amount in moles is 0.500 mole $KClO_3$. The conversion is from

Moles $KClO_3 \longrightarrow$ Moles $O_2 \longrightarrow$ Litres O_2

Step 3. Calculate the moles of O_2, using the mole-ratio method:

$$0.500 \text{ mole KClO}_3 \times \frac{3 \text{ moles O}_2}{2 \text{ moles KClO}_3} = 0.750 \text{ mole O}_2$$

Step 4. Convert moles of O_2 to litres of O_2. The moles of a gas at STP are converted to litres by multiplying by the molar volume, 22.4 L per mole:

$$0.750 \text{ mole O}_2 \times \frac{22.4 \text{ L}}{\text{mole}} = 16.8 \text{ L O}_2 \quad \text{(Answer)}$$

Setting this up in a continuous calculation, we obtain

$$0.500 \text{ mole KClO}_3 \times \frac{3 \text{ moles O}_2}{2 \text{ moles KClO}_3} \times \frac{22.4 \text{ L O}_2}{\text{mole O}_2} = 16.8 \text{ L O}_2$$

■

PROBLEM 12.15 How many grams of zinc must react with sulfuric acid to produce 1000 mL of hydrogen gas at STP?

Step 1. The balanced equation is

$$\underset{\text{1 mole}}{\text{Zn}} + \text{H}_2\text{SO}_4 \longrightarrow \text{ZnSO}_4 + \underset{\text{1 mole}}{\text{H}_2\uparrow}$$

Step 2. Moles of H_2: The equation states that 1 mole of H_2 is produced from 1 mole of Zn; 1000 mL of H_2 equals 1 litre of H_2 and represents a fraction of a mole.

$$1000 \text{ mL H}_2 \times \frac{1 \text{ L H}_2}{1000 \text{ mL H}_2} \times \frac{1 \text{ mole}}{22.4 \text{ L}} = 0.0446 \text{ mole H}_2$$

Step 3. Convert to moles of Zn:

$$0.0446 \text{ mole H}_2 \times \frac{1 \text{ mole Zn}}{1 \text{ mole H}_2} = 0.0446 \text{ mole Zn}$$

Step 4. Convert to grams of Zn:

$$0.0446 \text{ mole Zn} \times \frac{65.4 \text{ g Zn}}{\text{mole Zn}} = 2.92 \text{ g Zn} \quad \text{(Answer)}$$

The continuous calculation setup is

$$1000 \text{ mL H}_2 \times \frac{1 \text{ litre}}{1000 \text{ mL}} \times \frac{1 \text{ mole}}{22.4 \text{ L}} \times \frac{1 \text{ mole Zn}}{1 \text{ mole H}_2} \times \frac{65.4 \text{ g Zn}}{\text{mole Zn}} = 2.92 \text{ g Zn}$$

Millilitres $H_2 \longrightarrow$ Litres $H_2 \longrightarrow$ Moles $H_2 \longrightarrow$ Moles $Zn \longrightarrow$ Grams Zn

■

PROBLEM 12.16 What volume of hydrogen, collected at 30°C and 700 mm Hg pressure, will be formed by reacting 50.0 g of aluminum with hydrochloric acid?

$$2 \text{ Al} + 6 \text{ HCl} \longrightarrow 2 \text{ AlCl}_3 + 3 \text{ H}_2$$
2 moles   3 moles

In this problem, the volume of H_2 is first calculated from the equation, as we have done before. But, because the volume calculated by use of the equation is at STP, it must be changed to the conditions at which the gas is collected.

Step 1. Calculate litres of H_2 at STP using the mole-ratio method:

$$\text{Grams Al} \longrightarrow \text{Moles Al} \longrightarrow \text{Moles H}_2 \longrightarrow \text{Litres H}_2$$

$$50.0 \text{ g Al} \times \frac{1 \text{ mole Al}}{27.0 \text{ g Al}} \times \frac{3 \text{ moles H}_2}{2 \text{ moles Al}} \times \frac{22.4 \text{ L H}_2}{1 \text{ mole H}_2} = 62.2 \text{ L H}_2 \quad \text{(at STP)}$$

Step 2. Calculate the volume of H_2 at 30°C and 700 mm Hg pressure:

6.22 L $\longrightarrow$ New volume
273 K $\longrightarrow$ 303 K
760 mm Hg $\longrightarrow$ 700 mm Hg

$$\text{Volume} = 62.2 \text{ L} \times \frac{303 \text{ K}}{273 \text{ K}} \times \frac{760 \text{ mm Hg}}{700 \text{ mm Hg}} = 75.0 \text{ L H}_2 \quad \text{(Answer)}$$

■

2. Volume–volume calculations. When all substances in a reaction are in the gaseous state, simplifications in the calculation can be made based on Avogadro's hypothesis that gases under identical conditions of temperature and pressure contain the same number of molecules and occupy the same volume. Using this same hypothesis, we can state that, under the same conditions of temperature and pressure, the volumes of gases reacting are proportional to the number of moles of the gases in the balanced equation. Consider the reaction

$$H_2(g) + Cl_2(g) \longrightarrow 2 \text{ HCl}(g)$$

1 mole	1 mole	2 moles
22.4 L	22.4 L	2 × 22.4 L
1 volume	1 volume	2 volumes
Y volume	Y volume	2 Y volumes

In this reaction, 22.4 L of hydrogen will react with 22.4 L of chlorine to give $2 \times 22.4 = 44.8$ L of hydrogen chloride gas. This is true because these volumes are equivalent to the number of reacting moles in the equation. Therefore, Y volume of H_2 will combine with Y volume of Cl_2 to give $2\,Y$ volume of HCl. For example, 100 L of H_2 react with 100 L of Cl_2 to give 200 L of HCl; if the 100 L of H_2 and Cl_2 are at 50°C, they will give 200 L of HCl at 50°C. When the temperature and pressure before and after a reaction are the same, calculation of volumes can be done without correcting the volumes to STP.

For reacting gases: volume–volume relationships are the same as mole–mole relationships.

PROBLEM 12.17   What volume of oxygen will react with 150 L of hydrogen to form water vapor? What volume of water vapor will be formed?

Assume that both reactants and products are measured at the same conditions. Let us compare the two methods for solving this problem, using the mole method first and then the principle of reacting volumes.

$$2\,H_2(g) + O_2(g) \longrightarrow 2\,H_2O(g)$$
$$\text{2 moles} \qquad \text{1 mole} \qquad\qquad \text{2 moles}$$

Volume of O_2 and H_2O by the mole method:

$$\text{Litres } H_2 \longrightarrow \text{Moles } H_2 \longrightarrow \text{Moles } O_2 \text{ or } H_2O \longrightarrow \text{Litres } O_2 \text{ or } H_2O$$

$$150\,\cancel{L}\,\cancel{H_2} \times \frac{1\,\cancel{\text{mole}}}{22.4\,\cancel{L}} \times \frac{1\,\cancel{\text{mole}}\,O_2}{2\,\cancel{\text{moles}}\,\cancel{H_2}} \times \frac{22.4\,L}{\cancel{\text{mole}}} = 75.0\,L\,O_2$$

$$150\,\cancel{L}\,\cancel{H_2} \times \frac{1\,\cancel{\text{mole}}}{22.4\,\cancel{L}} \times \frac{2\,\cancel{\text{moles}}\,H_2O}{2\,\cancel{\text{moles}}\,\cancel{H_2}} \times \frac{22.4\,L}{\cancel{\text{mole}}} = 150\,L\,H_2O(g)$$

Calculation by reacting volumes:

$$\begin{array}{ccccc}
2\,H_2(g) & + & O_2(g) & \longrightarrow & 2\,H_2O(g) \\
\text{2 moles} & & \text{1 mole} & & \text{2 moles} \\
2 \times 22.4\,L & & 22.4\,L & & 2 \times 22.4\,L \\
\text{2 volumes} & & \text{1 volume} & & \text{2 volumes} \\
150\,L & & \boxed{1/2 \times 150\,L = 75\,L} & & \boxed{2/2 \times 150\,L = 150\,L}
\end{array}$$

Thus, 150 L of H_2 will react with 75 L of O_2 to produce 150 L of $H_2O(g)$. The calculation by reacting volumes, which may be done by inspection, is certainly simpler and more direct.

■

PROBLEM 12.18   The equation for the preparation of ammonia is

$$3\,H_2 + N_2 \xrightarrow{\;400°C\;} 2\,NH_3$$

Assuming that the reaction goes to completion,

    (a) What volume of H_2 will react with 50 L of N_2?
    (b) What volume of NH_3 will be formed from 50 L of N_2?
    (c) What volume of N_2 will react with 100 mL of H_2?
    (d) What volume of NH_3 will be produced from 100 mL of H_2?
    (e) If 600 mL of H_2 and 400 mL of N_2 are sealed in a flask and allowed to react, what amounts of H_2, N_2, and NH_3 are in the flask at the end of the reaction?

The answers to parts (a)–(d) are shown in the boxes and can be determined from the equation by inspection, using the principle of reacting volumes.

$$3\,H_2 \;+\; N_2 \;\longrightarrow\; 2\,NH_3$$

3 volumes   1 volume   2 volumes

(a) $\boxed{150\,L}$ 50 L

(b)   50 L $\boxed{100\,L}$

(c) 100 mL $\boxed{33.3\,mL}$

(d) 100 mL $\boxed{66.7\,mL}$

(e) Volume ratio from the equation $= \dfrac{3 \text{ volumes } H_2}{1 \text{ volume } N_2}$

Volume ratio used $= \dfrac{600 \text{ mL } H_2}{400 \text{ mL } N_2} = \dfrac{3 \text{ volumes } H_2}{2 \text{ volumes } N_2}$

Comparing these two ratios, we see that an excess of N_2 is present in the gas mixture. Therefore, the reagent limiting the amount of NH_3 that can be formed is H_2:

$$3\,H_2 \;+\; N_2 \;\longrightarrow\; 2\,NH_3$$

600 mL   200 mL   400 mL

In order to have a $3:1$ ratio of volumes reacting, 600 mL of H_2 will react with 200 mL of N_2 to produce 400 mL of NH_3, leaving 200 mL of N_2 unreacted. At the end of the reaction, the flask will contain 400 mL of NH_3 and 200 mL of N_2.

■

12.14 Ideal Gas Equation

We have used four variables in calculations involving gases: the volume, V; the pressure, P; the absolute temperature, T; and the number of molecules or moles, n. Combining these variables into a single expression, we obtain

$$V \propto \frac{nT}{P} \quad \text{or} \quad V = \frac{nRT}{P}$$

where R is a proportionality constant known as the *ideal gas constant*. The equation is commonly written as

$$PV = nRT$$

ideal gas equation

and is known as the **ideal gas equation.** This equation states in a single expression what we have considered earlier in our discussions—that the volume of a gas varies directly with the number of gas molecules and the absolute temperature and varies inversely with the pressure. The value and units of R depend on the units of P, V, and T. We can calculate one value of R by taking 1 mole of a gas at STP conditions. Solve the equation for R:

$$R = \frac{PV}{nT} = \frac{1 \text{ atm} \times 22.4 \text{ L}}{1 \text{ mole} \times 273 \text{ K}} = 0.0821 \frac{\text{L-atm}}{\text{mole-K}}$$

The units of R in this case are litre-atmospheres (L-atm) per mole-K.

The ideal gas equation can be used to calculate any one of the four variables if the other three are known.

When the value of $R = 0.0821$ L-atm/mole-K,

P is in atmospheres;　n is in moles;
V is in litres;　T is in Kelvin.

PROBLEM 12.19　What pressure will be exerted by 0.400 mole of a gas in a 5.00 L container at 17°C?

First, solve the ideal gas equation for P:

$$PV = nRT \quad \text{or} \quad P = \frac{nRT}{V}$$

Then substitute the data in the problem into the equation and solve:

$$P = \frac{0.400 \; \text{mole} \times 0.0821 \; \text{L-atm} \times 290 \; \text{K}}{5.00 \; \text{L} \times \text{mole-K}} = 1.90 \text{ atm} \quad \text{(Answer)}$$

■

PROBLEM 12.20　How many moles of oxygen gas are in a 50.0 L tank at 22°C if the pressure gauge reads 2000 lb/in.²?

First change to pressure in atmospheres. Then solve the ideal gas equation for n (moles), and substitute the data in the equation to complete the calculation.

Step 1. Pressure in atmospheres:

$$\frac{2000 \; \text{lb}}{\text{in.}^2} \times \frac{1 \text{ atm}}{14.7 \; \text{lb/in.}^2} = 136 \text{ atm}$$

Step 2. Solve for moles using the ideal gas equation:

$$PV = nRT \quad \text{or} \quad n = \frac{PV}{RT}$$

$$n = \frac{136 \; \text{atm} \times 50.0 \; \text{L}}{(0.0821 \; \text{L-atm/mole-K}) \times 295 \; \text{K}} = 281 \text{ moles O}_2 \quad \text{(Answer)}$$

The ideal gas equation can also be used for problems involving a specific mass of gas by substituting the mass–mole relationship, $n = $ g/g-mol. wt, into the equation.

■

All the gas laws are based on the behavior of an ideal gas—that is, a gas with a behavior that is described exactly by the gas laws for all possible values of P, V, and T. Most real gases actually do behave as predicted by the gas laws over a fairly wide range of temperatures and pressures. However, when conditions are such that the gas molecules are crowded closely together (high pressure and/or low temperature), they show marked deviations from ideal behavior. Deviations occur because molecules have finite volumes and also exhibit intermolecular attractions. This results in less compressibility at high pressures and greater compressibility at low temperatures than predicted by the gas laws. Many gases become liquids at high pressure and low temperature.

QUESTIONS **A. Review the Meanings of the New Terms Introduced in this Chapter**
1. Kinetic-Molecular Theory (KMT)
2. Ideal, or perfect, gas
3. Diffusion
4. Effusion
5. Graham's law of effusion
6. Pressure
7. Atmospheric pressure
8. Barometer
9. 1 atmosphere
10. Boyle's law
11. Charles' law
12. Standard conditions
13. Standard temperature and pressure (STP)
14. Dalton's Law of Partial Pressures
15. Gay-Lussac's Law of Combining Volumes of Gases
16. Avogadro's hypothesis
17. Molar volume of a gas
18. Ideal gas equation

B. Answers to the Following Questions Will Be Found in Tables and Figures
1. What evidence is used to show diffusion in Figure 12.1? If hydrogen, H_2, and oxygen, O_2, are in the two bulbs, how could we prove that diffusion had taken place?
2. Why do all the pressure gauges in Figure 12.2 indicate the same pressure?
3. How does the air pressure inside the balloon shown in Figure 12.3 compare to the air pressure outside the balloon? Explain.
4. According to Table 12.1, what two gases are the major constituents of dry air?
5. How does the pressure represented by 1 torr compare in magnitude to the pressure represented by 1 mm Hg? See Table 12.2.
6. In which container illustrated by Figure 12.6 are the molecules of gas moving faster? Assume both gases to be hydrogen.
7. In Figure 12.7, what gas pressure corresponds to a volume of 4 L?
8. How do the data illustrated in Figure 12.7 substantiate Boyle's law?
9. What effect would you observe in Figure 12.10 if T_2 were lower than T_1?
10. In the diagram shown in Figure 12.12, is the pressure of the oxygen plus water vapor inside the bottle equal to, greater than, or less than the atmospheric pressure outside the bottle? Explain.
11. List five gases in Table 12.3 that are denser than air. Explain the basis for your selections.
12. Explain why the pressure of oxygen, P_{O_2}, has changed from 739 mm Hg to 760 mm Hg in Figure 12.13.

C. Review Questions
1. What are the basic assumptions of the Kinetic-Molecular Theory?
2. Arrange the following gases, all at standard temperature, in order of increasing relative molecular velocities: H_2, CH_4, Rn, N_2, F_2, He. What is your basis for determining the order?
3. How do the average kinetic energies of the molecules in Question 2 compare?
4. What are the four parameters used to describe the behavior of a gas?
5. What are the characteristics of an ideal gas?

6. Under what condition of temperature, high or low, is a gas least likely to exhibit ideal behavior? Explain.

7. Under what conditions of pressure, high or low, is a gas least likely to exhibit ideal behavior? Explain.

8. Compare, at the same temperature and pressure, equal volumes of H_2 and O_2 as to:
   (a) Number of molecules
   (b) Mass
   (c) Number of moles
   (d) Average kinetic energy of the molecules
   (e) Rate of effusion
   (f) Density

9. How does the Kinetic-Molecular Theory account for the behavior of gases as described by:
   (a) Boyle's law?
   (b) Charles' law?
   (c) Dalton's Law of Partial Pressures?

10. In which physical state—solid, liquid, or gas—do water molecules at 0°C have the highest average kinetic energy?

11. What is the reason for referring gases to STP?

12. Maintaining constant pressure, what effect does heating a mole of N_2 gas have on the following?
   (a) Its density
   (b) Its specific gravity
   (c) Its mass
   (d) The average kinetic energy of its molecules
   (e) The average velocity of its molecules
   (f) The number of N_2 molecules in the sample

13. Assuming ideal gas behavior, which of the following statements are correct? (Try to answer without referring to your text.)
   (a) The pressure exerted by a gas at constant volume is independent of the temperature of the gas.
   (b) At constant temperature, increasing the pressure exerted on a gas sample will cause a decrease in the volume of the gas sample.
   (c) At constant pressure, the volume of a gas is inversely proportional to the absolute temperature.
   (d) At constant temperature, doubling the pressure on a gas sample will cause the volume of the gas sample to decrease to one-half its original volume.
   (e) Compressing a gas at constant temperature will cause its density and mass to increase.
   (f) Equal volumes of CO_2 and CH_4 gases at the same temperature and pressure contain:
      (1) The same number of molecules
      (2) The same mass
      (3) The same densities
      (4) The same number of moles
      (5) The same number of atoms
   (g) At constant temperature, the average kinetic energy of O_2 molecules at 200 atm pressure is greater than the average kinetic energy of O_2 molecules at 100 atm pressure.
   (h) According to Charles' law, the volume of a gas becomes zero at −273°C.
   (i) One litre of O_2 gas at STP has the same mass as 1 L of O_2 gas at 273°C and 2 atm pressure.
   (j) The volume occupied by a gas depends only on its temperature and pressure.
   (k) In a mixture containing O_2 molecules and N_2 molecules, the O_2 molecules, on the average, are moving faster than the N_2 molecules.
   (l) $PV = k$ is a statement of Charles' law.
   (m) If the temperature of a sample of gas is increased from 25°C to 50°C, the volume of the gas will increase by 100%.
   (n) One mole of chlorine, Cl_2, at 20°C and 600 mm Hg pressure contains 6.02×10^{23} molecules.

(o) Although a nitrogen molecule is 14 times as heavy as a hydrogen molecule, they both have the same kinetic energy at the same temperature.

(p) When the pressure on a sample of gas is halved, with the temperature remaining constant, the density of the gas is also halved.

(q) When the temperature of a sample of gas is increased at constant pressure, the density of the gas will decrease.

D. Review Problems

1. The barometer reads 725 mm Hg. Calculate the corresponding pressure in
   (a) atmospheres   (d) torrs
   (b) inches of Hg   (e) millibars
   (c) lb/in.^{2}

2. Express the following pressures in atmospheres:
   (a) 18 mm Hg   (b) 7000 cm Hg   (c) 745 torr   (d) 0.760 m Hg

3. A gas occupies a volume of 300 mL at 500 mm Hg pressure. What will be its volume if the pressure is changed to (a) 700 mm Hg; (b) 300 mm Hg; (c) 1.00 atm, with the temperature remaining constant?

4. A 500 mL sample of a gas is at a pressure of 640 mm Hg. What must be the pressure, at constant temperature, if the volume is changed to (a) 800 mL; (b) 350 mL?

5. At constant temperature, what pressure would be required to compress 2500 L of hydrogen gas at 1.0 atm pressure into a 30 L tank?

6. Given 5.00 L of N_2 gas at $-25°C$, what volume will the nitrogen occupy at (a) 0.0°C; (b) 0.0°F; (c) 200 K; (d) 325 K? (Assume constant pressure.)

7. Given a sample of a gas at 22°C, at what temperature would the volume of the gas sample be doubled, the pressure remaining constant?

*8. A gas sample at 22°C and 740 mm Hg pressure is heated until its volume is doubled. What pressure would restore the sample to its original volume?

9. A gas occupies a volume of 210 mL at 27°C and 740 mm Hg pressure. Calculate the volume the gas would occupy at (a) STP; (b) 150°C and 650 mm Hg pressure.

10. What volume would 2.30 L of H_2 gas at STP occupy at 70°C and 830 mm Hg pressure?

11. What pressure will 700 mL of a gas at STP exert when its volume is 200 mL at 30°C?

12. What volume will 1.5 moles of a gas occupy at STP?

13. An expandable balloon contains 1200 L of He at 0.950 atm pressure and 18°C. At an altitude of 22 miles (temperature 2°C and pressure 4 mm Hg), what will be the volume of the balloon?

14. At 27°C and 750 mm Hg pressure, what will be the volume of 1.3 moles of Ne?

15. What volume will a mixture of 4.00 moles of H_2 and 0.500 mole of CO_2 occupy at STP?

16. Eighty moles of H_2 are in a steel cylinder at a pressure of 1500 lb/in.^{2}. How many moles of H_2 are in the cylinder when the pressure reads 850 lb/in.^{2}? How many grams of H_2 were initially in the cylinder?

17. How many grams of H_2 are present in 2500 mL of H_2 at STP?

18. A gas has a density at STP of 1.30 g/L. What is its molecular weight?

19. Calculate the density of the following gases at STP:
    (a) Rn   (b) He   (c) SO_2 (d) CH_4

20. Calculate the density of:
    (a) Cl_2 gas at STP
    (b) Cl_2 gas at 27°C and 1 atm pressure

*21. At what temperature (°C) will the density of methane, CH_4, be 1.0 g/L at 1.0 atm pressure?

22. At STP 340 mL of a gas has a mass of 0.740 g. What is the gram-molecular weight of the gas?

23. What volume will each of the following occupy at STP?
    (a) 1.0 mole of H_2S
    (b) 17.05 g of H_2S
    (c) 1.20×10^{24} molecules of H_2S

24. An equilibrium mixture contains H_2 at 600 mm Hg pressure, N_2 at 200 mm Hg pressure, and O_2 at 300 mm Hg pressure.
    (a) What is the total pressure of the gases in the system?
    *(b) If the volume of the container is 1.0 L and the temperature is 30°C, what is the total number of moles of gases in the system?

*25. How many gas molecules are present in 400 mL of N_2O at 40°C and 400 mm Hg pressure? How many atoms are present? What would be the volume of the sample at STP? Is the density of the gas sample at STP greater than the density at the initial temperature and pressure?

26. What would be the partial pressure of O_2 gas collected over water at 20°C and 740 mm Hg pressure? (Check Appendix II for the vapor pressure of water.)

27. How many grams of methane, CH_4, would be present in 1000 mL of methane collected over water at 22°C and 745 mm Hg pressure?

28. (a) What volume of hydrogen at STP can be produced by reacting 6.30 moles of Al with sulfuric acid? The equation is

$$2\,Al(s) + 3\,H_2SO_4(aq) \longrightarrow Al_2(SO_4)_3(aq) + 3\,H_2(g)$$

(b) How many moles of H_2SO_4 react with the aluminum?

29. Given the equation,

$$4\,NH_3(g) + 5\,O_2(g) \longrightarrow 4\,NO(g) + 6\,H_2O(g)$$

(a) How many moles of NH_3 are required to produce 7.0 moles of NO?
(b) How many moles of NH_3 will react with 7.0 moles of O_2?
(c) How many grams of NO can be made from 12 moles of O_2 and 10 moles of NH_3?
(d) At constant temperature and pressure how many litres of NO can be made by the reaction of 500 mL of O_2?
(e) At constant temperature and pressure, what is the maximum volume, in litres, of NO that can be made from 2.0 L of NH_3 and 2.0 L of O_2?
(f) How many grams of O_2 must react to produce 50 L of NO measured at STP?
*(g) How many grams of NH_3 must react to produce a total of 20 L of products, NO plus H_2O, measured at STP?

30. Given the equation,

$$4\,FeS_2(s) + 11\,O_2(g) \xrightarrow{\Delta} 2\,Fe_2O_3(s) + 8\,SO_2(g)$$

(a) How many litres of O_2, measured at STP, will react with 0.500 kg of FeS_2?
(b) How many litres of SO_2, measured at STP, will be produced from 0.500 kg of FeS_2?

*31. Assume that the reaction $2\,CO(g) + O_2(g) \longrightarrow 2\,CO_2(g)$ goes to completion. When 12 moles of CO and 8.0 moles of O_2 react in a closed 10 L vessel,
    (a) How many moles of CO, O_2, and CO_2 are present at the end of the reaction?
    (b) What will be the total pressure in the flask at 0°C?

32. In the preparation of hydrogen by the following reaction, at constant tempera-ture and pressure, what volume of H_2 is produced per cubic foot of CH_4 reacted? What volume of carbon monoxide is also produced as a by-product?

$$2 CH_4(g) + 2 H_2O(g) \longrightarrow 2 CO(g) + 6 H_2(g)$$

33. Given the reaction,

$$CH_4(g) + 2 O_2(g) \longrightarrow CO_2(g) + 2 H_2O(g)$$

   (a) How many litres of CH_4 will react with 30.0 L of air (21% O_2), all measure-ments made at STP?
   (b) How does the volume of the system, before and after reaction, compare?

*34. Six (6.00) litres of O_2, measured at STP, were obtained by the decomposition of the $KClO_3$ in a 30.0 g mixture of KCl and $KClO_3$:

$$2 KClO_3(s) \overset{\Delta}{\longrightarrow} 2 KCl(s) + 3 O_2(g)$$

   What is the percentage by weight of $KClO_3$ in the mixture?

35. Using the ideal gas equation, $PV = nRT$, calculate:
   (a) The volume of 0.410 mole of H_2 at 47°C and 1.2 atm pressure
   (b) The number of grams in 16.0 L of CH_4 at 27°C and 500 mm Hg pressure
   *(c) The density of CO_2 at 2.00 atm pressure and -10°C
   *(d) The molecular weight of a gas having a density of 2.00 g/L at 27°C and 740 mm Hg pressure.
   Hints for (c) and (d): n = moles = grams/molecular weight, and density = grams/volume.

E. Review Exercises

1. Suggest reasons why gases deviate from ideal behavior.
2. Under what conditions of temperature and pressure would deviations from ideal gas behavior be most pronounced?
3. What major exception can you visualize where the pressures exerted by mixtures of gases are not in accord with Dalton's Law of Partial Pressures?
*4. (a) What are the relative rates of effusion of methane, CH_4, and helium, He?
   (b) If these two gases are simultaneously introduced into opposite ends of a 100 cm tube and allowed to diffuse toward each other, at what distance from the helium end will molecules of the two gases meet?
*5. A gas has a percent composition by weight of 79.89% carbon and 20.11% hydro-gen. At STP the density of the gas is 1.34 g/L. What is the molecular formula of the gas?
*6. A single smelter or coal-burning power plant can release several hundred tons of the pollutant sulfur dioxide, SO_2, into the atmosphere daily. Air oxidation fol-lowed by hydrolysis converts the SO_2 to sulfuric acid, H_2SO_4. What volume of SO_2 measured at STP would be needed to produce 100 metric tons of H_2SO_4? (1 metric ton = 1000 kg) The pertinent equations are:

$$2 SO_2(g) + O_2(g) \longrightarrow 2 SO_3(g) \quad \text{and} \quad SO_3(g) + H_2O(g) \longrightarrow H_2SO_4(g)$$

Review Exercises for Chapters 9–12

CHAPTER 9
QUANTITATIVE
COMPOSITION
OF
COMPOUNDS

True–False: Answer the following as either true or false.

1. A compound such as NaCl has a formula weight but no true molecular weight.
2. One mole of $HC_2H_3O_2$ weighs 60.0 g.
3. If the molecular formula and empirical formula of a compound are not the same, the empirical formula will be an integral multiple of the molecular formula.
4. The empirical formula of a compound gives the smallest ratio of the atoms that are present in a compound.
5. A mole of magnesium and a mole of magnesium oxide, MgO, contain the same number of magnesium atoms.
6. The number of sulfur atoms is the same in 1 mole of Na_2SO_4 as in 1 mole of K_2SO_4.
7. The number of sulfur atoms is the same in 1 g of Na_2SO_4 as in 1 g of K_2SO_4.
8. There are 14 moles of chlorine atoms in 3.5 moles of CCl_4.
9. A mole of CO_2 contains more molecules than a mole of CO.
10. A compound that has a carbon to hydrogen ratio of 1:2 can have a molecular weight of 48.0.

Multiple Choice: Choose the correct answer to each of the following.

1. The formula weight of $Ba(NO_3)_2$ is:
   (a) 199.3   (b) 261.3   (c) 247.3   (d) 167.3
2. A sample of 16 g of O_2:
   (a) Is 1 mole of O_2 (c) Is 0.50 molecule of O_2
   (b) Contains 6.02×10^{23} molecules   (d) Is 0.50 g-mol. wt of O_2
3. 2.00 moles of CO_2:
   (a) Weigh 2.00 g
   (b) Contain 1.20×10^{24} molecules
   (c) Weigh 56.0 g
   (d) Contain 6.00 molecular weights of CO_2
4. In Ag_2CO_3, the percentage by weight of:
   (a) Carbon is 43.5%   (c) Oxygen is 17.4%
   (b) Silver is 64.2%   (d) Oxygen is 21.9%
5. The empirical formula of the compound containing 31.0% Ti and 69.0% Cl is:
   (a) TiCl   (b) $TiCl_2$ (c) $TiCl_3$ (d) $TiCl_4$
6. A compound contains 54.3% C, 5.6% H, and 40.1% Cl. The empirical formula is:
   (a) CH_3Cl (b) C_2H_5Cl (c) $C_2H_4Cl_2$ (d) C_4H_5Cl
7. A compound contains 40.0% C, 6.7% H, and 53.3% O. The molecular weight is 60.0. The molecular formula is:
   (a) $C_2H_3O_2$ (b) C_3H_8O (c) C_2HCl (d) $C_2H_4O_2$
8. How many chlorine atoms are in 4.0 moles of PCl_3?
   (a) 3   (b) 7.2×10^{24} (c) 12.0   (d) 2.4×10^{24}
9. What is the weight of 4.53 moles of Na_2SO_4?
   (a) 142.1 g   (b) 644 g   (c) 31.4 g   (d) 3.19×10^{-2} g
10. The percentage composition of Mg_3N_2 is:
    (a) 72.2% Mg, 27.8% N   (c) 83.9% Mg, 16.1% N
    (b) 63.4% Mg, 36.6% N   (d) No correct answer given
11. How many grams of oxygen are contained in 0.500 mole of Na_2SO_4?
    (a) 16.0 g   (b) 32.0 g   (c) 64.0 g   (d) No correct answer given
12. The empirical formula of a compound is CH. If the molecular weight of this compound is 78.0, then the molecular formula is:
    (a) C_2H_2 (b) C_5H_{18} (c) C_6H_6 (d) No correct answer given

CHAPTER 10
CHEMICAL
EQUATIONS

True–False: Answer the following as either true or false.

1. In balancing an equation, we change the formulas of compounds to make the number of atoms on each side of the equation balance.
2. The equation $N_2 + 3 H_2 \rightleftharpoons 2 NH_3$ can be interpreted as saying that 1 mole of N_2 reacts with 3 moles of H_2 to form 2 moles of NH_3.
3. The equation $N_2 + 3 H_2 \rightleftharpoons 2 NH_3$ can be interpreted as saying that 1 g of N_2 reacts with 3 g of H_2 to form 2 g of NH_3.
4. The substances on the right side of a chemical equation are called the products.
5. When a gas is evolved in a chemical reaction, it can be indicated in the equation with the symbol ↑ or (g) immediately following the formula of the gas.
6. A balanced chemical equation is one that has the same number of moles on each side of the equation.
7. The coefficients in front of the formulas in a balanced chemical equation give the relative number of moles of the reactants and products in the reaction.
8. Water is formed in a neutralization reaction.
9. When carbonates or bicarbonates react with acids, carbon monoxide is formed.
10. In the reaction

$$Cu(s) + 2 AgNO_3(aq) \longrightarrow Cu(NO_3)_2(aq) + 2 Ag(s)$$

Cu replaces Ag^+ because copper is a more reactive element than silver.

Multiple Choice: Choose the correct answer to each of the following.

1. The reaction $BaCl_2 + (NH_4)_2CO_3 \longrightarrow BaCO_3 + 2 NH_4Cl$ is an example of:
   (a) Combination
   (b) Decomposition
   (c) Single replacement
   (d) Double replacement

2. The reaction $2 Al + 3 Br_2 \longrightarrow 2 AlBr_3$ is an example of:
   (a) Combination
   (b) Single replacement
   (c) Decomposition
   (d) Double replacement

3. When the equation $PbO_2 \xrightarrow{\Delta} PbO + O_2$ is balanced, one of the terms in the balanced equation is:
   (a) PbO_2 (b) $3 O_2$ (c) $3 PbO$ (d) O_2

4. When the equation $Cr_2S_3 + HCl \longrightarrow CrCl_3 + H_2S$ is balanced, one of the terms in the balanced equation is:
   (a) $3 HCl$ (b) $CrCl_3$ (c) $3 H_2S$ (d) $2 Cr_2S_3$

5. When the equation $F_2 + H_2O \longrightarrow HF + O_2$ is balanced, a term in the balanced equation is:
   (a) $2 HF$ (b) $3 O_2$ (c) $4 HF$ (d) $4 H_2O$

6. When the equation $NH_4OH + H_2SO_4 \longrightarrow$
   is completed and balanced, one of the terms in the balanced equation is:
   (a) NH_4SO_4 (b) $2 H_2O$ (c) H_2OH (d) $2 (NH_4)_2SO_4$

7. When the equation $H_2 + V_2O_5 \longrightarrow V +$
   is completed and balanced, a term in the balanced equation is:
   (a) $2 V_2O_5$.   (b) $3 H_2O$ (c) $2 V$ (d) $8 H_2$

8. When the equation $Fe_2(SO_4)_3 + Ba(OH)_2 \longrightarrow$ +
   is completed and balanced, a term in the balanced equation is:
   (a) $Ba_2(SO_4)_3$ (b) $2 Fe(OH)$ (c) $2 Fe_2(SO_4)_3$ (d) $2 Fe(OH)_3$

9. For the reaction $2 H_2 + O_2 \longrightarrow 2 H_2O + 136.8$ kcal, which of the following is not true?
   (a) The reaction is exothermic.
   (b) 136.8 kcal of heat are liberated for each mole of water formed.
   (c) Two moles of hydrogen react with 1 mole of oxygen.
   (d) 136.8 kcal of heat are liberated for each 2 moles of hydrogen reacted.
10. When a nonmetal oxide reacts with water,
   (a) A base is formed.   (c) A salt is formed.
   (b) An acid is formed.   (d) A nonmetal oxide is formed.

CHAPTER 11
CALCULATIONS
FROM
CHEMICAL
EQUATIONS

True–False: Answer the following as either true or false.
1. Stoichiometry is the section of chemistry involving calculations based on mass and mole relationships of substances in chemical reactions.
2. In a limiting-reagent problem, you determine which reactant has the fewest moles available.
3. The maximum amount of product that can be produced according to the equation, from the amounts of reactants supplied, is the theoretical yield.

Multiple Choice: Choose the correct answer to each of the following.
1. 20.0 g of Na_2CO_3 is how many moles?
   (a) 1.89   (b) 2.12×10^3 (c) 212   (d) 0.189
2. What is the weight in grams of 0.30 mole of $BaSO_4$?
   (a) 7.0×10^3 (b) 0.13   (c) 70   (d) 700.20
3. How many molecules are in 5.8 g of acetone, C_3H_6O?
   (a) 0.10   (b) 6.0×10^{22} (c) 3.5×10^{24} (d) 6.02×10^{23}

Problems 4 through 10 refer to the reaction,

$$2 C_2H_4 + 6 O_2 \longrightarrow 4 CO_2 + 4 H_2O$$

The following molecular weights may be needed for these problems: $C_2H_4 = 28.0$, $O_2 = 32.0$, $CO_2 = 44.0$, $H_2O = 18.0$

4. If 6.0 moles of CO_2 are produced, how many moles of O_2 were reacted?
   (a) 4.0   (b) 7.5   (c) 9.0   (d) 15.0
5. How many moles of O_2 are required for the complete reaction of 45 g of C_2H_4?
   (a) 1.3×10^2 (b) 0.64   (c) 112.5   (d) 4.8
6. If 18.0 g of CO_2 are produced, what weight of H_2O is produced?
   (a) 7.36 g   (b) 3.68 g   (c) 9.0 g   (d) 14.7 g
7. How many moles of CO_2 can be produced by the reaction of 5.0 moles of C_2H_4 and 12.0 moles O_2?
   (a) 4.0 moles   (b) 5.0 moles   (c) 8.0 moles   (d) 10.0 moles
8. How many moles of CO_2 can be produced by the reaction of 0.480 mole of C_2H_4 and 1.08 moles O_2?
   (a) 0.240 mole   (b) 0.960 mole   (c) 0.720 mole   (d) 0.864 mole
9. How many grams of CO_2 could be produced from 2.0 g of C_2H_4 and 5.0 g O_2?
   (a) 5.5 g   (b) 4.6 g   (c) 7.6 g   (d) 6.3 g
10. If 14.0 g of C_2H_4 is reacted, and the actual yield of H_2O is 7.84 g, the percentage yield in the reaction is:
   (a) 0.56%   (b) 43.6%   (c) 87.1%   (d) 56.0%

CHAPTER 12
THE GASEOUS
STATE OF
MATTER

True–False: Answer the following as either true or false.

1. The average kinetic energy of molecules of a gas increases with increased temperature.
2. Pressure is defined as force per unit area.
3. One torr is equal to 760 mm Hg.
4. One mole of hydrogen gas and 1 mole of oxygen gas, each in a box of equal volume, and each at the same temperature, will exert the same pressure.
5. Boyle's law states that the volume of a gas is inversely proportional to the pressure.
6. Charles' law states that the pressure of a gas is directly proportional to the temperature.
7. One mole of any gas always occupies 22.4 litres.
8. The specific gravity of a gas is the ratio of the mass of any volume of the gas to the mass of an equal volume of some reference gas at the same temperature and pressure.
9. Avogadro's hypothesis states that equal volumes of different gases at the same temperature and pressure contain the same weights of gas.
10. An ideal gas is one whose behavior is described exactly by the gas laws for all possible values of P, V, and T.
11. According to Avogadro, there are 6.02×10^{23} molecules of gas in a litre box at STP.
12. When the temperature of a gas is decreased at fixed volume, the pressure decreases.
13. At constant temperature and pressure, N_2 will diffuse more rapidly than O_2.
14. A barometer is an instrument used to measure the weight of a certain quantity of mercury.
15. The volume of a gas is dependent on the number of gas molecules, the pressure, the absolute temperature, and the gas constant, R.

Multiple Choice: Choose the correct answer to each of the following.

1. Which of the following is not one of the principal assumptions of the Kinetic-Molecular Theory?
   (a) All collisions of gaseous molecules are perfectly elastic.
   (b) A mole of any gas occupies 22.4 L at STP.
   (c) Gas molecules have no attraction for each other.
   (d) The average kinetic energy for molecules is the same for all gases at the same temperature.
2. Which of the following is not equal to 1.00 atm pressure?
   (a) 760 cm Hg   (b) 29.9 in. Hg   (c) 760 mm Hg   (d) 760 torr
3. If the pressure on 45 mL of gas is changed from 600 torr to 800 torr, the new volume will be:
   (a) 60 mL   (b) 34 mL   (c) 0.060 L   (d) 22.4 L
4. The volume of a gas is 300 mL at 740 torr, 25°C. If the pressure remains constant and the temperature is raised to 100°C, the new volume will be:
   (a) 240 mL   (b) 1.20 L   (c) 376 mL   (d) 410 mL
5. The volume of a dry gas is 4.00 L at 15°C and 745 torr. What volume will it occupy at 40°C and 700 torr?
   (a) 4.63 L   (b) 3.46 L   (c) 3.92 L   (d) 4.08 L
6. A sample of Cl_2 occupies 8.50 L at 80°C and 740 mm Hg. What volume will it occupy at STP?
   (a) 10.7 L   (b) 6.75 L   (c) 11.3 L   (d) 6.40 L
7. What volume will 8.00 g of O_2 gas occupy at 45°C and 2.00 atm?
   (a) 0.462 L   (b) 104 L   (c) 9.62 L   (d) 3.26 L
8. The density of NH_3 gas at STP is:
   (a) 0.759 g/mL   (b) 0.759 g/L   (c) 1.32 g/mL   (d) 1.32 g/L

9. The ratio of the relative rate of diffusion of methane, CH_4, to sulfur dioxide, SO_2, is:

   (a) $\dfrac{64}{16}$ (b) $\dfrac{16}{64}$ (c) $\dfrac{1}{4}$ (d) $\dfrac{2}{1}$

10. Measured at 65°C and 500 mm Hg, 3.21 L of a gas weighed 3.5 g. The molecular weight of this gas is:

    (a) 21 g/mole   (b) 46 g/mole   (c) 24 g/mole   (d) 130 g/mole

11. Box A contains O_2 (mol. wt = 32.0) at a pressure of 200 mm Hg. Box B, which is identical to Box A in volume, contains twice as many molecules of CH_4 (mol. wt = 16.0) as the molecules of O_2 in Box A. The temperatures of the gases are identical. The pressure in Box B is:

    (a) 100 mm Hg   (b) 200 mm Hg   (c) 400 mm Hg   (d) 800 mm Hg

12. A 300 mL sample of oxygen, O_2, was collected over water at 23°C and 725 mm Hg pressure. If the vapor pressure of water at 23°C is 21.0 mm Hg, the volume of dry O_2 at STP would be:

    (a) 256 mL   (b) 351 mL   (c) 341 mL   (d) 264 mL

13. A tank containing 0.01 mole of neon and 0.04 mole of helium shows a pressure of 1 atm. What is the partial pressure of neon in the tank?

    (a) 1 atm   (b) 0.01 atm   (c) 0.2 atm   (d) 0.5 atm

14. How many litres of NO_2 (at STP) can be produced from 25.0 g of Cu reacting with concentrated nitric acid?

$$Cu(s) + 4\,HNO_3(aq) \longrightarrow Cu(NO_3)_2(aq) + 2\,H_2O(l) + 2\,NO_2\uparrow$$

    (a) 4.4 L   (b) 8.8 L   (c) 17.6 L   (d) 22.4 L

15. How many litres of butane vapor are required to produce 2.0 L of CO_2 at STP?

$$2\,C_4H_{10}(g) + 13\,O_2(g) \longrightarrow 8\,CO_2(g) + 10\,H_2O(g)$$
Butane

    (a) 2.0 L   (b) 4.0 L   (c) 8.0 L   (d) 0.50 L

16. What volume of CO_2 (at STP) can be produced when 15.0 g of C_2H_6 and 50.0 g of O_2 are reacted?

$$2\,C_2H_6(g) + 7\,O_2(g) \longrightarrow 4\,CO_2(g) + 6\,H_2O(g)$$

    (a) 20.0 L   (b) 22.4 L   (c) 35.0 L   (d) 61.2 L

13

Water and the Properties of Liquids

After studying Chapter 13 you should be able to:

1. Understand the terms listed in Question A at the end of the chapter.
2. Describe a water molecule with respect to electron-dot structure, bond angle, and polarity.
3. Make sketches showing hydrogen bonding (a) between water molecules and (b) between hydrogen fluoride molecules.
4. Explain the effect of hydrogen bonding on the physical properties of water.
5. Determine whether a compound will or will not form hydrogen bonds.
6. Complete and balance equations showing the formation of water (a) from hydrogen and oxygen, (b) by neutralization, and (c) by combustion of hydrogen-containing compounds.
7. Complete and balance equations for (a) the reaction of water with Na, K, and Ca; (b) the reaction of steam with Zn, Al, Fe, and C; and (c) the reaction of water with halogens.
8. Identify metal oxides as basic anhydrides and write balanced equations for their reactions with water.
9. Identify nonmetal oxides as acid anhydrides and write balanced equations for their reactions with water.
10. Deduce the formula of the acid anhydride or the basic anhydride when given the formula of the corresponding acid or base.
11. Identify, name, and write equations for the complete dehydration of hydrates.
12. Distinguish clearly between peroxides and ordinary oxides.
13. Discuss the occurrence of ozone and its effects on humans.
14. Outline the processes needed to prepare a potable water supply from a contaminated river source.
15. Describe how water may be softened by distillation, chemical precipitation, ion exchange, and demineralization—including chemical equations where appropriate.
16. Explain the process of evaporation from the standpoint of kinetic energy.
17. Relate vapor pressure data or vapor pressure curves of different substances to their relative rates of evaporation and to their relative boiling points.
18. Explain what is happening in the different segments of the time–temperature phase diagram of water.

13.1 Occurrence of Water

Water is our most common natural resource; it covers about 70% of the earth's surface. Not only is it found in the oceans and seas, in lakes, rivers, streams, and in glacial ice deposits; it is also always present in the atmosphere and in cloud formations.

About 97% of the earth's water is in the oceans. This is saline water that contains vast amounts of dissolved minerals. The world's fresh water comprises the other 3%, of which about two-thirds is locked up in polar ice caps and glaciers. The remaining fresh water is found in ground water, lakes, and the atmosphere. More than 70 elements have been detected in the mineral content of seawater. Only four of these—chlorine, sodium, magnesium, and bromine—are now commercially obtained from the sea.

Water is an essential constituent of all living matter. It is the most abundant compound in the human body, making up about 70% of total body weight. About 92% of blood plasma is water; about 80% of muscle tissue is water; and about 60% of the red blood cells are water. Water is more important than food in the sense that a person can survive much longer without food than without water.

13.2 Physical Properties of Water

Water is a colorless, odorless, tasteless liquid with a melting point of 0°C and a boiling point of 100°C at 1 atm pressure. Two additional physical properties of matter are introduced with the study of water: heat of fusion and heat of vaporization. **Heat of fusion** is the amount of heat required to change 1 g of a solid into a liquid at its melting point. The heat of fusion of water is 80 cal/g. The temperature of the solid–liquid system does not change during the absorption of this heat. The heat of fusion is the energy used in breaking down the crystalline lattice of ice from a solid to a liquid. **Heat of vaporization** is the amount of heat required to change 1 g of liquid to a vapor at its normal boiling point. The value for water is 540 cal/g. Once again, there is no change in temperature during the absorption of this heat. The heat of vaporization is the energy needed to overcome the attractive forces between molecules in changing them from the liquid to the gaseous state. The values for water for both the heat of fusion and the heat of vaporization are relatively high compared to those for other substances; these high values indicate that strong attractive forces are acting between the molecules.

heat of fusion

heat of vaporization

Ice and water exist together in equilibrium at 0°C, as shown in Figure 13.1. When ice at 0°C melts, it absorbs 80 cal/g in changing into a liquid; the temperature remains at 0°C. In order to refreeze the water, we have to remove 80 cal/g from the liquid at 0°C.

Both boiling water and steam are shown in Figure 13.2 to have a temperature of 100°C. It takes 100 cal to heat 1 g of water from 0°C to 100°C, but water at its boiling point absorbs 540 cal/g in changing to steam. Although boiling

0°C

Figure 13.1   Ice and water in equilibrium at 0°C.

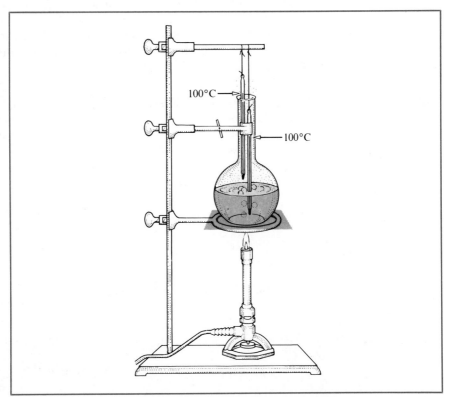

100°C

100°C

Figure 13.2   Boiling water and steam in equilibrium at 100°C.

water and steam are both at the same temperature, steam contains considerably more heat per gram and can cause more severe burns than hot water. The physical properties of water are tabulated and compared with those of other hydrogen compounds of Group VIA elements in Table 13.1.

TABLE 13.1 Physical properties of water and other hydrogen compounds of Group VIA elements.

Formula	Color	Molecular weight	Melting point (°C)	Boiling point, 760 mm Hg (°C)	Heat of fusion (cal/g)	Heat of vaporization (cal/g)
H_2O	Colorless	18.0	0.00	100.0	80.0	540
H_2S	Colorless	34.1	−85.5	−60.3	16.7	131
H_2Se	Colorless	81.0	−65.7	−41.3	7.4	57.0
H_2Te	Colorless	129.6	−51	−2.3	−	42.8

The maximum density of water is 1.000 g/mL at 4°C. Water has the unusual property of contracting in volume as it is cooled to 4°C and then expanding when cooled from 4°C to 0°C. (Most liquids contract in volume all the way down to the point at which they solidify.) Therefore, 1 g of water occupies a volume greater than 1 mL at all temperatures above and below 4°C. There is, on the other hand, a large increase (about 9%) in volume when water is changed from a liquid at 0°C to a solid (ice) at 0°C. The density of ice at 0°C is 0.915 g/mL, which means that ice, being less dense than water, will float in water.

PROBLEM 13.1 How many calories of energy are needed to change 10.0 g of ice at 0°C to 20°C?
Ice will absorb 80 cal/g (heat of fusion) in going from a solid at 0°C to a liquid at 0°C. Then an additional 1 cal/g is needed to raise the temperature for each 1°C.
Calories needed to melt the ice:

$$10.0 \, \cancel{g} \times \frac{80 \text{ cal}}{\cancel{g}} = 800 \text{ cal}$$

Calories needed to heat the water from 0°C to 20°C:

$$10.0 \, \cancel{g} \times \frac{1 \text{ cal}}{\cancel{g} \, °\cancel{C}} \times 20°\cancel{C} = 200 \text{ cal}$$

Thus, 800 cal + 200 cal = 1000 cal are needed.

∎

PROBLEM 13.2 How many calories of energy are needed to change 20.0 g of water at 20°C to steam at 100°C?
Calories needed to heat the water from 20°C to 100°C:

$$20.0 \, \cancel{g} \times \frac{1 \text{ cal}}{\cancel{g} \, °\cancel{C}} \times 80°\cancel{C} = 1600 \text{ cal}$$

Calories needed to change water at 100°C to steam at 100°C:

$$20.0 \not{g} \times \frac{540 \text{ cal}}{\not{g}} = 10,800 \text{ cal}$$

Thus, 1600 cal + 10,800 cal = 12,400 cal are needed.

■

13.3 Structure of the Water Molecule

A single water molecule consists of two hydrogen atoms and one oxygen atom. Each H atom is attached to the O atom by a single covalent bond. This bond is formed by the overlap of the 1s orbital of hydrogen with an unpaired 2p electron orbital of oxygen. The average distance between the two nuclei is known as the *bond length*. The O—H bond length in water is 0.96 Å. The water molecule is nonlinear and has a V-shaped structure with an angle of about 105° between the two bonds (see Figure 13.3).

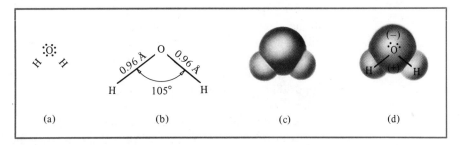

(a)   (b)   (c)   (d)

Figure 13.3   Diagrams of a water molecule: (a) electron distribution, (b) bond angle and O—H bond length, (c) molecular orbital structure, and (d) dipole representation.

Oxygen is the second most electronegative element. As a result, the two covalent OH bonds in water are polar. If the three atoms in a water molecule were aligned in a linear structure such as H+→O←+H, the two polar bonds would be acting in equal and opposite directions and the molecule would be nonpolar. However, water is a highly polar molecule. Therefore, it does not have a linear structure. When atoms are bonded together in a nonlinear fashion, the angle formed by the bonds is called the *bond angle*. In water the HOH bond angle is 105°. The two polar covalent bonds and the bent structure result in the oxygen atom having a partial negative charge and each hydrogen atom having a partial positive charge. The polar nature of water is responsible for many of its properties, including its behavior as a solvent.

13.4 The Hydrogen Bond

Table 13.1 compares the physical properties of H_2O, H_2S, H_2Se, and H_2Te. From this comparison it is apparent that four physical properties of water—

melting point, boiling point, heat of fusion, and heat of vaporization—are extremely high and do not fit the trend relative to the molecular weights of the four compounds. If the properties of water followed the normal progression shown by the other three compounds, we would expect the melting point of water to be below $-85°C$ and the boiling point to be below $-60°C$.

Why does water have these anomalous physical properties? The answer is that liquid water molecules are linked together by hydrogen bonds. A **hydrogen bond** is a chemical bond that is formed between polar molecules that contain hydrogen covalently bonded to a small, highly electronegative atom such as fluorine, oxygen, or nitrogen (F—H, O—H, N—H). The bond is actually the dipole–dipole attraction of polar molecules.

hydrogen bond

Elements that have significant hydrogen bonding ability are F, O, and N.

What is a hydrogen bond, or H-bond? Because a hydrogen atom has only one electron, it can form only one covalent bond. When it is attached to a strong electronegative atom such as oxygen, a hydrogen atom will also be attracted to an oxygen atom of another molecule, forming a bond (or bridge) between the two molecules. Water has two types of bonds: covalent bonds that exist between hydrogen and oxygen atoms within a molecule, and hydrogen bonds that exist between hydrogen and oxygen atoms in different water molecules.

Hydrogen bonds are *intermolecular* bonds; that is, they are formed between atoms in different molecules. They are somewhat ionic in character because they are formed by electrostatic attraction. Hydrogen bonds are much weaker than the ionic or covalent bonds that unite atoms to atoms to form compounds. Despite their weakness, hydrogen bonds are of great chemical importance.

Figure 13.4, part (a), shows two water molecules linked by a hydrogen bond, and part (b) shows six water molecules linked together by hydrogen bonds. A dash (—) is used for the covalent bond and a dotted line (----) for the hydrogen bond. In water, one molecule is linked to another through hydrogen bonds, forming a three-dimensional aggregate of water molecules. This molecular bonding effectively gives water the properties of a much larger, heavier molecule, explaining in part its relatively high melting point, boiling point, heat of fusion, and heat of vaporization. As water is heated and energy is absorbed, hydrogen bonds are continually being broken until at 100°C, with the absorption of an additional 540 cal/g, water separates into individual molecules, going into the gaseous state. Sulfur, selenium, and tellurium are not sufficiently electronegative for their hydrogen compounds to behave like water. As a result, H-bonding in H_2S, H_2Se, and H_2Te is only of small consequence (if any) to their physical properties.

Fluorine, the most electronegative element, forms the strongest hydrogen bonds. This bonding is strong enough to link hydrogen fluoride molecules together as *dimers*, H_2F_2, or as higher, $(HF)_n$, molecular units. The dimer

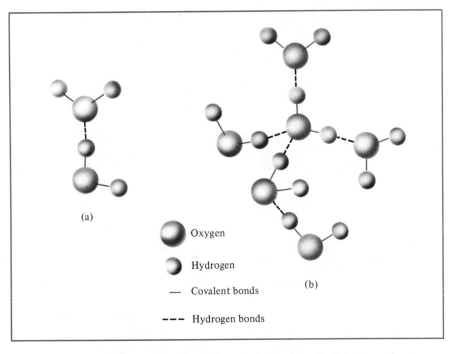

Figure 13.4 Hydrogen bonding: Water in the liquid and solid states exists as aggregates in which the water molecules are linked together by hydrogen bonds.

structure may be represented in this way:

$$H\diagdown_{F}\text{-...}H\diagup^{F}$$
$$\diagdown\text{H-bond}$$

The existence of salts, such as KHF_2 and NH_4HF_2, verifies the hydrogen fluoride (bifluoride) structure, HF_2^- $(F-H\text{---}F)^-$, where one H atom is bonded to two F atoms through one covalent bond and one hydrogen bond.

PROBLEM 13.3   Would you expect hydrogen bonding to occur between molecules in the following substances?

(a)
$$\begin{array}{cc} H & H \\ | & | \\ H-C-C-O-H \\ | & | \\ H & H \end{array}$$
Ethyl alcohol

(b)
$$\begin{array}{cc} H & H \\ | & | \\ H-C-O-C-H \\ | & | \\ H & H \end{array}$$
Dimethyl ether

(a) There should be hydrogen bonding in ethyl alcohol because one hydrogen atom is bonded to an oxygen atom.

$$\begin{array}{cccc} H & H & & H & H & H \\ | & | & & | & | & | \\ H-C-C-O-H\text{---}O-C-C-H \\ | & | & \uparrow & | & | \\ H & H & \text{H-bond} & H & H \end{array}$$

(b) There is no hydrogen bonding in dimethyl ether because all the hydrogen atoms are bonded only to carbon atoms.

Both ethyl alcohol and dimethyl ether have the same molecular weight (46.0). Although both compounds have the same molecular formula, C_2H_6O, ethyl alcohol has a much higher boiling point (78.4°C) than dimethyl ether (-23.7°C), as a result of the hydrogen bonding between molecules.

■

Hydrogen bonding can occur between two different atoms that are capable of forming H-bonds. Thus, we may have an $O\cdots H-N$ or $O-H\cdots N$ linkage in which the hydrogen atom forming the H-bond is between an oxygen and a nitrogen atom. This form of the H-bond exists in certain types of protein molecules and many biologically active substances.

13.5 Formation of Water and Chemical Properties of Water

Water is very stable to heat; it decomposes to the extent of only about 1% at temperatures up to 2000°C. Pure water is a nonconductor of electricity. But when a small amount of sulfuric acid or sodium hydroxide is added, the solution is readily decomposed into hydrogen and oxygen by an electric current. Two volumes of hydrogen are produced for each volume of oxygen:

$$2\,H_2O(l) \xrightarrow[\text{H}_2\text{SO}_4\text{ or NaOH}]{\text{Electrical energy}} 2\,H_2(g) + O_2(g)$$

Formation. Water is formed when hydrogen burns in air. Pure hydrogen burns very smoothly in air, but mixtures of hydrogen and air (or oxygen) are dangerous and explode when ignited. The reaction is strongly exothermic:

$$2\,H_2(g) + O_2(g) \longrightarrow 2\,H_2O(g) + 115.6\,\text{kcal}$$

Water is produced by a variety of other reactions, especially by (1) acid–base neutralizations, (2) combustion of hydrogen-containing materials, and (3) metabolic oxidation in living cells.

1. $HCl(aq) + NaOH(aq) \longrightarrow NaCl(aq) + H_2O$

2. $2\,C_2H_2(g) + 5\,O_2(g) \longrightarrow 4\,CO_2(g) + 2\,H_2O(g) + 289.6\,\text{kcal}$
   Acetylene

   $CH_4 + 2\,O_2 \longrightarrow CO_2 + 2\,H_2O + 192\,\text{kcal}$
   Methane

3. $C_6H_{12}O_6 + 6\,O_2 \xrightarrow{\text{Enzymes}} 6\,CO_2 + 6\,H_2O + 673\,\text{kcal}$
   Glucose

The combustion of acetylene shown in (2) is strongly exothermic and is capable of producing very high temperatures. It is used in oxygen–acetylene torches to cut and weld steel and other metals. Methane is known as natural gas and is commonly used as fuel for heating and cooking. The reaction of glucose with oxygen shown in (3) is the reverse of photosynthesis. It is the overall reaction by which living cells obtain needed energy by metabolizing glucose to carbon dioxide and water.

Reactions of water with metals and nonmetals.   The reactions of metals with water at different temperatures show that these elements vary greatly in their reactivity. Metals such as sodium, potassium, and calcium react with cold water to produce hydrogen and a metal hydroxide. A small piece of sodium added to water melts from the heat produced by the reaction, forming a silvery metal ball, which rapidly flits back and forth on the surface of the water. One must use caution when experimenting with this reaction, since the hydrogen produced is frequently ignited by the sparking of the sodium, and it will explode, spattering sodium. Potassium reacts even more vigorously than sodium. Calcium sinks in water and liberates a gentle stream of hydrogen. The equations for these reactions are

$$2\,Na(s) + 2\,H_2O(l) \longrightarrow H_2\uparrow + 2\,NaOH(aq)$$

$$2\,K(s) + 2\,H_2O(l) \longrightarrow H_2\uparrow + 2\,KOH(aq)$$

$$Ca(s) + 2\,H_2O(l) \longrightarrow H_2\uparrow + Ca(OH)_2(aq)$$

Zinc, aluminum, and iron do not react with cold water but will react with steam at high temperatures, forming hydrogen and a metallic oxide. The equations are

$$Zn(s) + H_2O(steam) \longrightarrow H_2\uparrow + ZnO(s)$$

$$2\,Al(s) + 3\,H_2O(steam) \longrightarrow 3\,H_2\uparrow + Al_2O_3(s)$$

$$3\,Fe(s) + 4\,H_2O(steam) \longrightarrow 4\,H_2\uparrow + Fe_3O_4(s)$$

Copper, silver, and mercury are examples of metals that do not react with cold water or steam to produce hydrogen. We conclude from these reactions that sodium, potassium, and calcium are chemically more reactive than zinc, aluminum, and iron, which are more reactive than copper, silver, and mercury.

Certain nonmetals react with water under various conditions. For example, fluorine reacts violently with cold water, producing hydrogen fluoride and free oxygen. The reactions of chlorine and bromine are much milder, producing what is commonly known as "chlorine water" and "bromine water," respectively. Chlorine water contains HCl, HOCl, and dissolved Cl_2; the free chlorine gives it a yellow-green color. Bromine water contains HBr, HOBr, and dissolved Br_2; the free bromine gives it a red-brown color. Steam passed over hot coke (carbon) produces a mixture of carbon monoxide and hydrogen that is known as "water gas." Since water gas is combustible, it is useful as a fuel. It is also the starting material for the commercial production of several alcohols. The equations for these reactions are

$$2 F_2(g) + 2 H_2O(l) \longrightarrow 4 HF(aq) + O_2(g)$$

$$Cl_2(g) + H_2O(l) \longrightarrow HCl(aq) + HOCl(aq)$$

$$Br_2(l) + H_2O(l) \longrightarrow HBr(aq) + HOBr(aq)$$

$$C(s) + H_2O(g) \xrightarrow{1000°C} CO(g) + H_2(g)$$

basic anhydride

Reactions of water with metal and nonmetal oxides. Metal oxides that react with water to form bases are known as **basic anhydrides.** Examples are

$$CaO(s) + H_2O \longrightarrow \underset{\text{Calcium hydroxide}}{Ca(OH)_2(aq)}$$

$$Na_2O(s) + H_2O \longrightarrow \underset{\text{Sodium hydroxide}}{2 NaOH(aq)}$$

Certain metal oxides, such as CuO and Al_2O_3, do not form basic solutions because the oxides are insoluble in water.

acid anhydride

Nonmetal oxides that react with water to form acids are known as **acid anhydrides.** Examples are

$$CO_2 + H_2O \rightleftharpoons \underset{\text{Carbonic acid}}{H_2CO_3}$$

$$SO_2(g) + H_2O \rightleftharpoons \underset{\text{Sulfurous acid}}{H_2SO_3(aq)}$$

$$N_2O_5(s) + H_2O \longrightarrow \underset{\text{Nitric acid}}{2 HNO_3(aq)}$$

The word *anhydrous* means "without water." An anhydride is a metal oxide or a nonmetal oxide derived from a base or an oxy-acid by the removal of water. To determine the formula of an anhydride, the elements of water, H_2O, are removed from an acid or base formula until all the hydrogen is removed. The formula of the anhydride then consists of the remaining metal or nonmetal and oxygen atoms. In calcium hydroxide, removal of water as indicated leaves CaO as the anhydride:

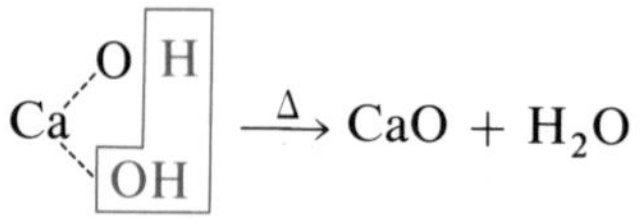

$$Ca \overset{\text{O}|\text{H}}{\underset{|\text{OH}}{}} \xrightarrow{\Delta} CaO + H_2O$$

In sodium hydroxide, H_2O cannot be removed from one formula unit, so two formula units of NaOH must be used, leaving Na_2O as the formula of the anhydride:

$$\begin{matrix} NaO & \boxed{H} \\ Na & \boxed{OH} \end{matrix} \xrightarrow{\Delta} Na_2O + H_2O$$

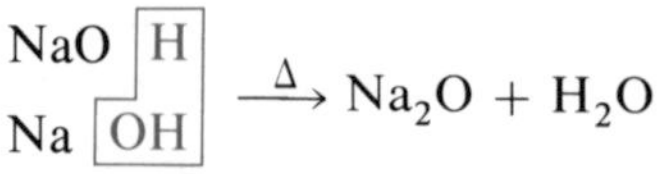

The removal of water from sulfuric acid, H_2SO_4, gives the acid anhydride SO_3:

$$H_2SO_4 \xrightarrow{\Delta} SO_3 + H_2O$$

The foregoing are examples of typical reactions of water but are by no means a complete list of the known reactions of water.

13.6 Hydrates

hydrate
water of hydration
water of
crystallization

When certain salt solutions are allowed to evaporate to dryness, some water molecules remain as part of the crystalline salt that is left after evaporation is complete. Solids that contain water molecules as part of their crystalline structure are known as **hydrates.** Water in a hydrate is known as **water of hydration** or **water of crystallization.**

Formulas for hydrates are expressed by first writing the usual anhydrous (without water) formula for the compound, then adding a dot, followed by the number of water molecules present. An example is $BaCl_2 \cdot 2H_2O$. This formula tells us that each formula unit of this salt contains one barium ion, two chloride ions, and two water molecules. A crystal of the salt contains many of these units in its crystalline lattice.

In naming hydrates, we first name the compound exclusive of the water and then add the term *hydrate,* with the proper prefix representing the number of water molecules in the formula. For example, $BaCl_2 \cdot 2H_2O$ is called *barium chloride dihydrate.* Hydrates are true compounds and follow the Law of Definite Composition. The formula weight of $BaCl_2 \cdot 2H_2O$ is 244.3; it contains 56.20% barium, 29.06% chlorine, and 14.74% water.

Water molecules in hydrates are bonded by electrostatic forces between polar water molecules and the positive or negative ions of the compound. These forces are not as strong as covalent or ionic chemical bonds. As a result, water of crystallization may be removed by moderate heating of the compound. A partially dehydrated or completely anhydrous compound may result. When $BaCl_2 \cdot 2H_2O$ is heated, it loses its water at about 100°C:

$$BaCl_2 \cdot 2H_2O \xrightarrow{100°C} BaCl_2 + 2\,H_2O\uparrow$$

When a solution of copper(II) sulfate $(CuSO_4)$ is allowed to evaporate, beautiful blue crystals containing 5 moles of water per mole of $CuSO_4$ are formed. The formula for this hydrate is $CuSO_4 \cdot 5H_2O$; it is called copper(II) sulfate pentahydrate or cupric sulfate pentahydrate. When $CuSO_4 \cdot 5H_2O$ is heated, water is lost and a pale green-white powder, anhydrous $CuSO_4$, is formed:

$$CuSO_4 \cdot 5H_2O \xrightarrow{250°C} CuSO_4 + 5\,H_2O\uparrow$$

When water is added to anhydrous copper(II) sulfate, the foregoing reaction is reversed and the salt turns blue again. Because of this outstanding color change, anhydrous copper(II) sulfate has been used as an indicator to detect small amounts of water. The formation of the hydrate is noticeably exothermic.

The formula for plaster of paris is $(CaSO_4)_2 \cdot H_2O$. When mixed with the proper quantity of water, plaster of paris forms a dihydrate and sets to a hard mass. It is, therefore, useful for making patterns for the reproduction of art objects, molds, and surgical casts. The chemical reaction is

$$(CaSO_4)_2 \cdot H_2O(s) + 3\ H_2O(l) \longrightarrow 2\ CaSO_4 \cdot 2H_2O(s)$$

The occurrence of hydrates is commonplace in salts. Table 13.2 lists a number of common hydrates.

TABLE 13.2 Selected hydrates.

Hydrate	Name
$CaCl_2 \cdot 2H_2O$	Calcium chloride dihydrate
$Ba(OH)_2 \cdot 8H_2O$	Barium hydroxide octahydrate
$MgSO_4 \cdot 7H_2O$	Magnesium sulfate heptahydrate
$SnCl_2 \cdot 2H_2O$	Tin(II) chloride dihydrate
$CoCl_2 \cdot 6H_2O$	Cobalt(II) chloride hexahydrate
$Na_2CO_3 \cdot 10H_2O$	Sodium carbonate decahydrate
$(NH_4)_2C_2O_4 \cdot H_2O$	Ammonium oxalate monohydrate
$NaC_2H_3O_2 \cdot 3H_2O$	Sodium acetate trihydrate
$Na_2B_4O_7 \cdot 10H_2O$	Sodium tetraborate decahydrate
$Na_2S_2O_3 \cdot 5H_2O$	Sodium thiosulfate pentahydrate

13.7 Hygroscopic Substances; Deliquescence; Efflorescence

hygroscopic substances

Many anhydrous salts and other substances readily absorb water from the atmosphere. Such substances are said to be **hygroscopic.** This property can be observed in the following simple experiment: Spread a weighed 10–20 g sample of anhydrous copper(II) sulfate on a watch glass and set it aside so that the salt is exposed to the air. Then weigh the sample periodically for 24 hours, noting the increase in weight and the change in color. Water is absorbed from the atmosphere, forming the blue pentahydrate $CuSO_4 \cdot 5H_2O$.

deliquescence

Some compounds continue to absorb water beyond the hydrate stage to form solutions. A substance that absorbs water from the air until it forms a solution is said to be **deliquescent.** A few granules of anhydrous calcium chloride or pellets of sodium hydroxide exposed to the air will appear moist in a few minutes, and within an hour will absorb enough water to form a puddle of solution. Phosphorus pentoxide (P_2O_5) picks up water so rapidly that it cannot be weighed accurately unless it is weighed in an anhydrous atmosphere.

Compounds that absorb water are very useful as drying agents (desiccants). Refrigeration systems must be kept dry with such agents or the moisture will freeze and clog the tiny orifices in the mechanism. Bags of drying agents are often enclosed in packages containing iron or steel parts to absorb moisture and prevent rusting. Anhydrous calcium chloride, magnesium sulfate, sodium sulfate, calcium sulfate, silica gel, and phosphorus pentoxide are some of the compounds commonly used for drying liquids and gases containing small amounts of moisture.

efflorescence

The process by which crystalline materials spontaneously lose water when exposed to the air is known as **efflorescence.** Glauber's salt ($Na_2SO_4 \cdot 10H_2O$), a transparent crystalline salt, loses water when exposed to the air. One can actually observe these well-defined, large crystals crumbling away as they lose water, forming a white, noncrystalline-appearing powder. From our discussion of the decomposition of hydrates, we can predict that heat will increase the rate of efflorescence. The rate also depends on the concentration of moisture in the air. A dry atmosphere will allow the process to take place more rapidly.

13.8 Hydrogen Peroxide

Although both water and hydrogen peroxide are compounds of hydrogen and oxygen, their properties are very different. A hydrogen peroxide molecule (H_2O_2) is composed of two H atoms and two O atoms. Its composition by weight is 94.1% oxygen and 5.9% hydrogen. Pure hydrogen peroxide is a pale blue liquid that has a melting point of $-0.9°C$, boils at $151°C$, and has a density of 1.44 g/mL at 20°C. It is miscible with water in all proportions. Water solutions of hydrogen peroxide are slightly acid. The structure of hydrogen peroxide may be represented as

$$
\overset{\text{H}}{\underset{\text{H}}{:\!\ddot{O}\!:\!\ddot{O}\!:}} \qquad \text{or} \qquad \overset{\text{H}}{\underset{\text{H}}{:\ddot{O}-.\ddot{O}.}}
$$

Hydrogen peroxide is a common, useful source of oxygen, since it decomposes easily to give oxygen and water:

$$2\,H_2O_2(l) \longrightarrow 2\,H_2O(l) + O_2\!\uparrow + 46.0\text{ kcal}$$

This decomposition is accelerated by heat and light, but may be minimized by storing peroxide solutions in brown bottles, keeping them cold, and adding stabilizers. The decomposition may also be accelerated by catalysts such as manganese dioxide.

The peroxide group, like an oxygen molecule (O_2), contains two oxygen atoms linked by a covalent bond. It has a -2 oxidation number and is written O_2^{2-} or $:\!\ddot{O}\!:\!\ddot{O}\!:^{2-}$; each O atom is considered to have an oxidation number of -1. Metal dioxides also contain two O atoms, but each is bonded individually to the metal ion. Thus, a metal dioxide contains two oxide ($:\!\ddot{O}\!:^{2-}$) ions.

Some discretion must be used when working with peroxide formulas. From their formulas, BaO_2 and TiO_2 appear to be similar compounds, but BaO_2 is a peroxide consisting of a $+2$ barium ion and a -2 peroxide ion whereas titanium dioxide is an oxide consisting of a $+4$ titanium ion and two -2 oxide ions.

Peroxides may be distinguished from dioxides chemically because they generally yield H_2O_2 or O_2 when treated with acids or water. Two examples are the reactions of barium peroxide and sodium peroxide:

$$BaO_2(s) + H_2SO_4(aq) \longrightarrow H_2O_2(aq) + BaSO_4(s)$$

$$Na_2O_2(s) + 2 H_2O(l) \longrightarrow H_2O_2(aq) + 2 NaOH(aq)$$

A 3% solution of H_2O_2, which is commonly available at drugstores, is used as an antiseptic to cleanse open wounds. Somewhat stronger H_2O_2 solutions are widely used as bleaching agents for cotton, wood, and hair. For certain oxidation processes, the chemical industry uses a 30% solution. Concentrations of 85% and higher are used for oxidizing fuels in rocket propulsion. These highly concentrated solutions are extremely sensitive to decomposition and represent a fire hazard if allowed to come into contact with organic material.

Ozone is another compound containing multiple oxygen linkages. One molecule, O_3, contains three atoms of oxygen:

:Ö::Ö: ·Ö·:Ö:
            Ö

Oxygen   Ozone

Ozone can be prepared by passing air or oxygen through an electrical discharge:

$$3 O_2 \xrightarrow[\text{discharge}]{\text{Electrical}} 2 O_3 - 68.4 \text{ kcal}$$

The characteristic pungent odor of ozone is noticeable in the vicinity of electrical machines and power transmission lines. Ozone is formed in the atmosphere during electrical storms and by the photochemical action of ultraviolet radiation on a mixture of nitrogen dioxide and oxygen. Areas with high air pollution are subject to high atmospheric ozone concentrations.

Ozone is not a desirable low-altitude constituent of the atmosphere, since it is known to cause extensive plant damage, cracking of rubber, and the formation of eye-irritating substances. Concentrations of ozone greater than 0.1 part per million (ppm) of air cause coughing, choking, headache, fatigue, and reduced resistance to respiratory infection. Concentrations between 10 and 20 ppm are fatal to humans.

High-energy radiation from the sun also converts oxygen in the upper atmosphere (stratosphere) to ozone, forming a protective ozone layer in the stratosphere:

$$O_2 \xrightarrow{\text{Sunlight}} \underset{\text{Oxygen atoms}}{O + O}$$

$$O_2 + O \longrightarrow O_3$$

Ultraviolet radiation from the sun is highly damaging to living tissue of plants and animals. The ozone layer, however, shields the earth by absorbing ultraviolet radiation and thus prevents most of this lethal radiation from reaching the earth's surface. The reaction that occurs is the reverse of the preceding one:

$$O_3 \xrightarrow[\text{radiation}]{\text{Ultraviolet}} O_2 + O + \text{Heat}$$

Scientists have become concerned about a growing hazard to the ozone layer. Fluorocarbon propellants, such as the Freons CCl_3F and CCl_2F_2, which were used in aerosol spray cans and are used in refrigeration and air conditioning units, are stable compounds and remain unchanged in the lower atmosphere. When these fluorocarbons are carried by convection currents to the stratosphere, they absorb ultraviolet radiation and produce chlorine atoms, which in turn react to destroy part of the ozone layer. The following sequence of steps has been proposed for this reaction:

$$CCl_3F \xrightarrow[\text{radiation}]{\text{Ultraviolet}} \cdot CCl_2F + \underset{\substack{\text{Chlorine} \\ \text{atom}}}{Cl}$$

$$Cl + O_3 \longrightarrow ClO + O_2 + \text{Heat}$$

$$ClO + O \longrightarrow Cl + O_2$$

With the amount of fluorocarbons in the atmosphere already, scientists predict that up to 15% of the ozone layer could be destroyed by the year 2000. This could result in damaging effects on life and major climate changes due to more heat passing into the lower atmosphere.

allotropy

Many elements exist in two or more molecular or crystalline forms. This phenomenon is known as **allotropy** (from the Greek *allotropia*, meaning "variety"). The individual forms of an element are known as allotropic forms or allotropes. Oxygen (O_2) and ozone (O_3) are allotropic forms of the element oxygen. Two other common elements that exhibit allotropy are sulfur and carbon. Allotropic forms of sulfur are rhombic and monoclinic sulfur. Diamond and graphite are allotropic forms of carbon.

13.9 Natural Waters

Natural fresh waters are not pure, but contain dissolved minerals, suspended matter, and sometimes harmful bacteria. The water supplies of large cities are usually drawn from rivers or lakes. Such water is generally unsafe to drink without treatment. To make such water potable (that is, safe to drink), it is treated by some or all of the following processes:

1. **Screening.**   Removal of relatively large objects, such as trash, fish, and so on.
2. **Flocculation and sedimentation.**   Chemicals, usually lime and alum (aluminum sulfate), are added to form a flocculent jellylike precipitate of aluminum hydroxide. This precipitate traps most of the suspended

fine matter in the water and carries it to the bottom of the sedimentation basin.

3. **Sand filtration.**   Water is drawn from the top of the sedimentation basin and passed downward through fine sand filters. Nearly all the remaining suspended matter and bacteria are removed by the sand filters.

4. **Aeration.**   Water is drawn from the bottom of the sand filters and is aerated by spraying. The purpose of this process is to remove objectionable odors and tastes.

5. **Disinfection.**   In the final stage, chlorine gas is injected into the water to kill harmful bacteria before it is distributed to the public. Ozone is also used in some countries to disinfect water. In emergencies, water may be disinfected by simply boiling for a few minutes.

If the drinking water of children contains an optimum amount of fluoride ion, their teeth will be more resistant to decay. Therefore, in many communities NaF or Na_2SiF_6 is added to the water supply to bring the fluoride ion concentration up to the optimum level of about 1.0 ppm. Excessively high concentrations of fluoride ion can cause mottling of the teeth.

Water that contains dissolved calcium and magnesium salts is called *hard water.* One drawback of hard water is that ordinary soap does not lather well in it; the soap reacts with the calcium and magnesium ions to form an insoluble greasy scum. However, synthetic soaps, known as detergents or syndets, are available; they have excellent cleaning qualities and do not form precipitates with hard water. Hard water is also undesirable because it causes "boiler scale" to form on the walls of water heaters and steam boilers, which greatly reduces their efficiency.

Four techniques used to "soften" hard water are distillation, chemical precipitation, ion exchange, and demineralization. In distillation, the water is boiled and the steam thus formed is then condensed to a liquid again, leaving the minerals behind in the distilling vessel. Figure 13.5 illustrates a simple laboratory distillation apparatus. Commercial stills are available that are capable of producing hundreds of litres of distilled water per hour.

Calcium and magnesium ions are precipitated from hard water by adding sodium carbonate and lime. Insoluble calcium carbonate and magnesium hydroxide are precipitated and are removed by filtration or sedimentation.

In the ion-exchange method, used in many households, hard water is effectively softened as it is passed through a bed or tank of zeolite. Zeolite is a complex sodium aluminum silicate. In this process, sodium ions replace objectionable calcium and magnesium ions, and the water is thereby softened:

$$Na_2Zeolite(s) + Ca^{2+}(aq) \longrightarrow CaZeolite(s) + 2\,Na^+(aq)$$

The zeolite is regenerated by back-flushing with concentrated sodium chloride solution, reversing the foregoing reaction.

The sodium ions that are present in water softened either by chemical precipitation or by the zeolite process are not objectionable to most users of soft water.

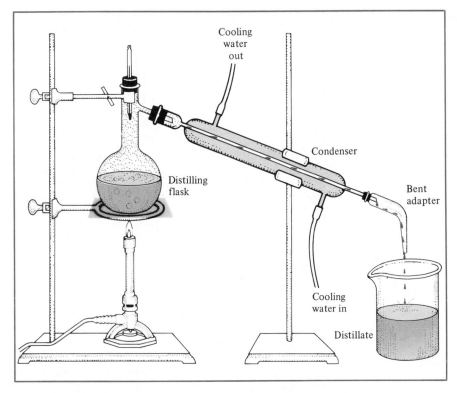

Figure 13.5 Simple laboratory setup for distillation of liquids.

In demineralization, both cations and anions are removed by a two-stage ion-exchange system. Special synthetic organic resins are used in the ion-exchange beds. In the first stage, metal cations are replaced by hydrogen ions. In the second stage, anions are replaced by hydroxide ions. The hydrogen and hydroxide ions react, and essentially pure, mineral-free water leaves the second stage (see Figure 13.6, page 306).

The oceans represent an inexhaustible source of water; however, seawater contains about 3.5 lb of salts per 100 lb of water. This 35,000 ppm of dissolved salts makes seawater unfit for agricultural and domestic uses. Water that contains less than 1000 ppm of salts is considered reasonably good for drinking, and potable (fresh) water is already being obtained from the sea in many parts of the world. Continuous research is being done in an effort to make usable water from the oceans more abundant and economical.

13.10 Water Pollution

Polluted water was formerly thought of as water that was unclear, had a bad odor or taste, and contained disease-causing bacteria. However, such factors as increased population, industrial requirements for water, atmospheric pollution, and use of pesticides have greatly modified the problem of water pollution.

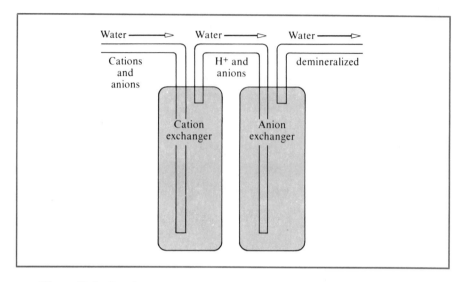

Figure 13.6 Demineralization of water: Water is passed through two beds of synthetic resin. In the cation exchanger, metal ions are exchanged for hydrogen ions. In the anion exchanger, anions are exchanged for hydroxide ions. The H^+ and OH^- ions react to form water, giving essentially pure, demineralized water.

Many of the "newer" pollutants are not removed or destroyed by the usual water-treatment processes. For example, among the 66 organic compounds found in the drinking water of a major city on the Mississippi River, 3 are labeled slightly toxic, 17 moderately toxic, 15 very toxic, 1 extremely toxic, and 1 supertoxic. Two are known carcinogens (cancer-producing agents), 11 are suspect, and 3 are metabolized to carcinogens. The United States Public Health Service classifies water pollutants under eight broad categories. These are shown in Table 13.3.

Many outbreaks of disease or poisoning such as typhoid, dysentery, and cholera have been attributed directly to drinking water. Rivers and streams are a natural means for municipalities to dispose of their domestic and industrial waste products. Much of this water is used again by people downstream, and then discharged back into the water source. Then another community still farther downstream draws the same water and discharges its own wastes. Thus, along waterways such as the Mississippi and Delaware rivers, water is withdrawn and discharged many times. If this water is not properly treated, harmful pollutants will build up, causing epidemics of various diseases.

Hazardous waste products are unavoidable in the manufacture of many products that we use in our everyday life. One common way to dispose of these wastes is to place them in toxic waste dumps. What has been found after many years of disposing of wastes in this manner is that toxic substances have seeped into the ground-water deposits. As a result many people have become ill and water wells have been closed until satisfactory methods of detoxifying this water are found. This is a serious problem, since one-half the United States population

TABLE 13.3 Classification of water pollutants.

Type of pollutant	Examples
Oxygen-demanding wastes	Decomposable organic wastes from domestic sewage and industrial wastes of plant and animal origin
Infectious agents	Bacteria, viruses, and other organisms from domestic sewage, animal wastes, and animal process wastes
Plant nutrients	Principally compounds of nitrogen and phosphorus
Organic chemicals	Large numbers of chemicals synthesized by industry; pesticides; chlorinated organic compounds
Other minerals and chemicals	Inorganic chemicals from industrial operations, mining, oil field operations, and agriculture
Radioactive substances	Waste products from mining and processing of radioactive materials, airborne radioactive fallout, increased use of radioactive materials in hospitals and research
Heat from industry	Large quantities of heated water returned to water bodies from power plants and manufacturing facilities after use for cooling purposes
Sediment from land erosion	Solid matter washed into streams and oceans by erosion, rain, and water runoff

gets its drinking water from ground water. To clean up the thousands of industrial dumps, find new methods of disposing of wastes, and develop safe dumping grounds will be very costly.

Mercury and its compounds have long been known to be highly toxic. Mercury gets into the body primarily through the foods we eat. Although it is not an essential mineral for the body, mercury accumulates in the blood, kidneys, liver, and brain tissues. Mercury in the brain causes serious damage to the central nervous system. The sequence of events that have led to incidents of mercury poisoning is as follows: Mercury and its compounds are used in many industries and in agriculture, primarily as a fungicide in the treatment of seeds. One of the largest uses is in the electrochemical conversion of sodium chloride brines to chlorine and sodium hydroxide, as represented by this equation:

$$2\,NaCl + 2\,H_2O \xrightarrow{\text{Electrolysis}} Cl_2 + 2\,NaOH + H_2$$

Although no mercury is shown in the chemical equation, it is used in the process for electrical contact, and small amounts are discharged along with spent brine solutions. Thus, considerable quantities of mercury, in low concentrations, have been discharged into lakes and other surface waters from the effluents of these manufacturing plants. The mercury compounds discharged into the water are converted by bacterial action and other organic compounds to methyl mercury, $(CH_3)_2Hg$, which then accumulates in the bodies of fish. Several major episodes of mercury poisoning that have occurred in the past years were the result of eating mercury-contaminated fish. The best way to control this contaminant is at the source, and much has been done since 1970 to eliminate the discharge of mercury in industrial wastes. In 1976, the Environmental Protection Agency banned the use of all mercury-containing insecticides and fungicides.

Many other major water pollutants have been recognized and steps have been taken to eliminate them. Three that pose serious problems are lead, detergents, and chlorine-containing organic compounds. Lead poisoning, for example, has been responsible for many deaths in past years. The toxic action of lead in the body is the inhibition of the enzyme necessary for the production of hemoglobin in the blood. The normal intake of lead into the body is through food. However, extraordinary amounts of lead can be ingested from water running through lead pipes and by using lead-containing ceramic containers for storage of food and beverages.

Keeping our lakes and rivers free from pollution is a very costly and complicated process. However, it has been clearly demonstrated that waterways rendered so polluted that the water is neither fit for human use nor able to sustain marine life can be successfully restored.

13.11 Evaporation

When beakers of water, ethyl ether, and ethyl alcohol are allowed to stand uncovered in an open room, the volumes of these liquids gradually decrease. The process by which this takes place is called *evaporation.*

Attractive forces exist between molecules in the liquid state. All these molecules, however, do not have the same kinetic energy. Molecules that have greater than average kinetic energy may overcome the attractive forces and break away from the surface of the liquid, flying off and becoming a gas. **Evaporation** is the escape of molecules from the liquid state to the gas or vapor state.

evaporation

In evaporation, molecules of higher than average kinetic energy escape from a liquid, leaving it cooler than it was before they escaped. For this reason, evaporation of perspiration is one way the human body cools itself and keeps its temperature constant. When volatile liquids such as ethyl chloride (C_2H_5Cl) are sprayed on the skin, they evaporate rapidly, cooling the area by removing heat. The numbing effect of the low temperature produced by evaporation of ethyl chloride allows it to be used as a local anesthetic for minor surgery.

Solids such as iodine, camphor, naphthalene (moth balls), and, to a small extent, even ice, will go directly from the solid to the gaseous state, bypassing the liquid state. This change is a form of evaporation and is called **sublimation.**

sublimation

$$\text{Liquid} \xrightarrow{\text{Evaporation}} \text{Vapor}$$

$$\text{Solid} \xrightarrow{\text{Sublimation}} \text{Vapor}$$

13.12 Vapor Pressure

When a liquid evaporates in a closed system as shown in Figure 13.7, part (b), some of the molecules in the vapor or gaseous state strike the surface and return to the liquid state by the process of *condensation.* The rate of condensation increases until it is equal to the rate of evaporation. At this point, the space above the liquid is said to be saturated with vapor, and an equilibrium, or steady state, exists between the liquid and the vapor. The equilibrium equation is

$$\text{Liquid} \xrightleftharpoons[\text{Condensation}]{\text{Evaporation}} \text{Vapor}$$

This equilibrium is dynamic; both processes—evaporation and condensation—are taking place, even though one cannot visually observe or measure a change. The number of molecules leaving the liquid in a given time interval is equal to the number of molecules returning to the liquid.

At the point of equilibrium, the molecules in the vapor exert a pressure like any other gas. The pressure exerted by a vapor in equilibrium with its liquid is known as the **vapor pressure** of the liquid. The vapor pressure may be thought of as an internal pressure, a measure of the escaping tendency that molecules have to go from the liquid to the vapor state. The vapor pressure of a liquid is independent of the amount of liquid and vapor present, but it increases as

vapor pressure

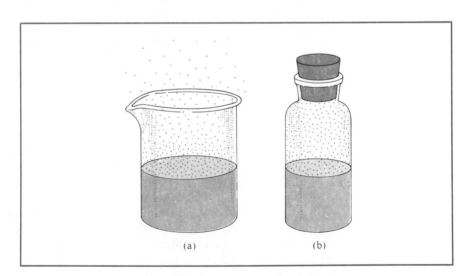

Figure 13.7   (a) Molecules in an open beaker may evaporate from the liquid and be dispersed into the atmosphere. Under this condition, evaporation will continue until all the liquid is gone. (b) Molecules leaving the liquid are confined to a limited space. With time, the concentration in the vapor phase will increase to a point at which an equilibrium between liquid and vapor is established.

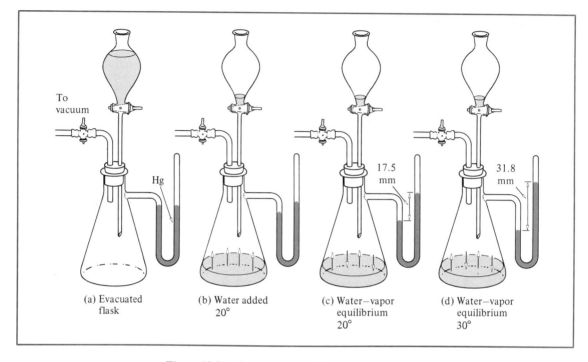

To
vacuum

Hg

17.5
mm

31.8
mm

(a) Evacuated
flask

(b) Water added
20°

(c) Water–vapor
equilibrium
20°

(d) Water–vapor
equilibrium
30°

Figure 13.8 Measurement of the vapor pressure of water at 20°C and 30°C. In flask (a) the system is evacuated. The mercury manometer attached to the flask shows equal pressure in both legs. In (b), water has been added to the flask and begins to evaporate, exerting pressure as indicated by the manometer. In (c), when equilibrium is established, the pressure inside the flask remains constant at 17.5 mm Hg. In (d) the temperature is changed to 30°C and equilibrium is reestablished with the vapor pressure at 31.8 mm Hg.

the temperature rises (see Table 13.4). Figure 13.8 illustrates a liquid–vapor equilibrium and the measurement of its vapor pressure.

When equal volumes of water, ethyl ether, and ethyl alcohol are placed in separate beakers and allowed to evaporate at the same temperature, we observe that the ether evaporates faster than the alcohol, which in turn evaporates faster than the water. This order of evaporation is consistent with the fact that ether has a higher vapor pressure at any particular temperature than ethyl alcohol or water. One reason for this higher vapor pressure is that there is less attraction between ether molecules than there is between alcohol molecules or between water molecules. The vapor pressures of these three compounds at various temperatures are compared in Table 13.4.

volatile

Substances that evaporate readily are said to be **volatile.** A volatile liquid has a relatively high vapor pressure at room temperature. Ethyl ether is a very volatile liquid; water is not too volatile; mercury, which has a vapor pressure of 0.0012 mm Hg at 20°C, is essentially a nonvolatile liquid. Most substances that are normally solids are nonvolatile (solids that sublime are exceptions).

TABLE 13.4 **The vapor pressure of water, ethyl alcohol, and ethyl ether at various temperatures.**

Temperature (°C)	Vapor pressure (mm Hg)		
	Water	Ethyl alcohol	Ethyl ether[a]
0	4.6	12.2	185.3
10	9.2	23.6	291.7
20	17.5	43.9	442.2
30	31.8	78.8	647.3
40	55.3	135.3	921.3
50	92.5	222.2	1276.8
60	152.9	352.7	1729.0
70	233.7	542.5	2296.0
80	355.1	812.6	2993.6
90	525.8	1187.1	3841.0
100	760.0	1693.3	4859.4
110	1074.6	2361.3	6070.1

[a]Note that the vapor pressure of ethyl ether at temperatures of 40°C and higher exceeds standard pressure, 760 mm Hg. This indicates that the substance has a low boiling point and therefore should be stored in a cool place in a tightly sealed container.

13.13 Boiling Point

The boiling temperature of a liquid is associated with its vapor pressure. We have seen that the vapor pressure increases as the temperature increases. When the internal or vapor pressure of a liquid becomes equal to the external pressure, the liquid boils. (By external pressure we mean the pressure of the atmosphere above the liquid.) The boiling temperature of a pure liquid remains constant as long as the external pressure does not vary.

The boiling point of water is 100°C. Table 13.4 shows that the vapor pressure of water at 100°C is 760 mm Hg, a figure we have seen many times before. The significant fact here is that the boiling point is the temperature at which the vapor pressure of the water or other liquid is equal to standard, or atmospheric, pressure at sea level. These relationships lead to the following definition: The **boiling point** is the temperature at which the vapor pressure of a liquid is equal to the external pressure above the liquid.

boiling point

We can readily see that a liquid has an infinite number of boiling points. When we give the boiling point of a liquid, we should also state the pressure. When we express the boiling point without stating the pressure, we mean it to be the **standard** or **normal boiling point** at standard pressure (760 mm Hg). Using Table 13.4 again, we see that the normal boiling point of ethyl ether is between 30°C and 40°C, and for ethyl alcohol it is between 70°C and 80°C, because, for each compound, 760 mm Hg pressure lies within these stated temperature ranges. At the normal boiling point, 1 g of a liquid changing to a vapor (gas) absorbs an amount of energy equal to its heat of vaporization (see Table 13.5).

standard or normal boiling point

TABLE 13.5 Physical properties of ethyl chloride, ethyl ether, ethyl alcohol, and water.

	Boiling point (°C)	Melting point (°C)	Heat of vaporization (cal/g)	Heat of fusion (cal/g)
Ethyl chloride	13	−139	92.5	—
Ethyl ether	34.6	−116	83.9	—
Ethyl alcohol	78.4	−112	204.3	24.9
Water	100.0	0	540	80

The boiling point at various pressures may be evaluated by plotting the data of Table 13.4 on the graph in Figure 13.9, where temperature is plotted horizontally along the x axis and vapor pressure is plotted vertically along the y axis. The resulting curves are known as **vapor pressure curves.** Any point on these curves represents a vapor–liquid equilibrium at a particular temperature and pressure. We may find the boiling point at any pressure by tracing a hori-

vapor pressure curve

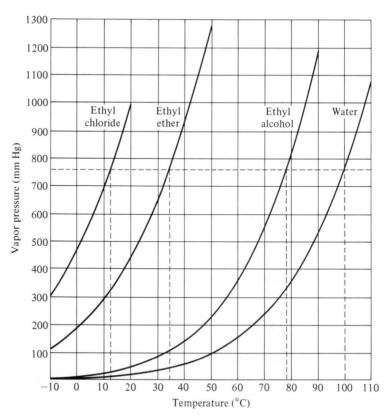

Figure 13.9 Vapor pressure–temperature curves for ethyl chloride, ethyl ether, ethyl alcohol, and water.

zontal line from the designated pressure to a point on the vapor pressure curve. From this point we draw a vertical line to obtain the boiling point on the temperature axis. Four such points are shown in Figure 13.9; they represent the normal boiling points of the four compounds at 760 mm Hg pressure.

See if you can verify from the graph that the boiling points of ethyl chloride, ethyl ether, ethyl alcohol, and water at 600 mm Hg pressure are 8.5°C, 28°C, 73°C, and 93°C, respectively. By reversing this process, you can ascertain at what pressure a substance will boil at a specific temperature. The boiling point is one of the most commonly used physical properties for characterizing and identifying substances.

13.14 Freezing Point or Melting Point

freezing or melting point

As heat is removed from a liquid, the liquid becomes colder and colder, until a temperature is reached at which it begins to solidify. A liquid changing into a solid is said to be *freezing*, or *solidifying*. When a solid is heated continually, a temperature is reached at which the solid begins to liquefy. A solid changing into a liquid is said to be *melting*. The temperature at which the solid phase of a substance is in equilibrium with its liquid phase is known as the **freezing point** or **melting point** of that substance. The equilibrium equation is

$$\text{Solid} \underset{\text{Freezing}}{\overset{\text{Melting}}{\rightleftharpoons}} \text{Liquid}$$

When a solid is slowly and carefully heated so that a solid–liquid equilibrium is maintained, the temperature will remain constant as long as both phases are present. One gram of a solid in changing into a liquid absorbs an amount of energy equal to its *heat of fusion* (see Table 13.5). The melting point is another physical property that is commonly used for characterizing substances.

The most common example of a solid–liquid equilibrium is ice and water (see Figure 13.1). In a well-stirred system of ice and water, the temperature remains at 0°C as long as both phases are present. The melting point is subject to changes in pressure, but is hardly affected unless the pressure change is very large.

It has been known for a long time that dissolved substances markedly decrease the freezing point of a liquid. For example, salt–water–ice equilibrium mixtures may be obtained at temperatures as low as −20°C, 20 degrees below the usual freezing point of water.

If, after all the solid has been melted, the liquid is heated continually, the temperature will rise until the liquid boils. The temperature will remain constant at the boiling point until all the liquid has boiled away. One gram of a liquid in changing into a gas at the normal boiling point absorbs an amount of energy equal to its *heat of vaporization* (see Table 13.5). The whole process of heating a substance (water) is illustrated graphically in Figure 13.10, which is called a *phase diagram*. In the diagram line *AB* represents the solid being heated and line *BC* represents the time during which the solid is melting and is in equilibrium

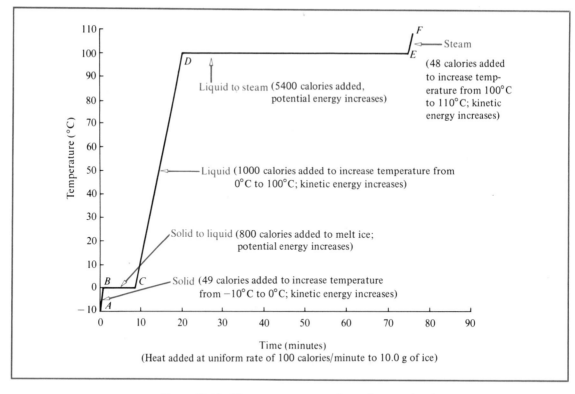

Figure 13.10 Time–temperature phase diagram for the absorption of heat by a substance from the solid state to the gaseous state. Using water as an example, the interval *AB* represents the ice phase; *BC* interval, the melting of ice to water; *CD* interval, the elevation of the temperature of water from 0°C to 100°C; *DE* interval, the boiling of water to steam; and *EF* interval, the heating of steam.

with the liquid. Along line *CD* the liquid absorbs heat and finally, at point *D*, boils and continues to boil at a constant temperature (line *DE*). In the interval *EF*, all the water exists as steam and is being further heated or superheated.

QUESTIONS   A. **Review the Meanings of the New Terms Introduced in this Chapter**

1. Heat of fusion	11. Efflorescence
2. Heat of vaporization	12. Allotropy
3. Hydrogen bond	13. Evaporation
4. Basic anhydride	14. Sublimation
5. Acid anhydride	15. Vapor pressure
6. Hydrate	16. Volatile
7. Water of hydration	17. Boiling point
8. Water of crystallization	18. Standard or normal boiling point
9. Hygroscopic substances	19. Vapor pressure curve
10. Deliquescence	20. Freezing or melting point

B. **Answers to the Following Questions Will Be Found in Tables and Figures**
   1. Compare the potential energy of the two states of water shown in Figure 13.1.
   2. In what state (solid, liquid, or gas) would H_2S, H_2Se, and H_2Te be at 0°C? (See Table 13.1.)
   3. The two thermometers in Figure 13.2 read 100°C. What is the pressure of the atmosphere?
   4. Draw a diagram of a water molecule and point out the areas that are the negative and positive ends of the dipole.
   5. If the water molecule were linear, with all three atoms in a straight line rather than in the shape of a V, as shown in Figure 13.3, what effect would this have on the physical properties of water?
   6. Based on Table 13.2, how do we specify 1, 2, 3, 4, 5, 6, 7, and 8 molecules of water in the formulas of hydrates?
   7. Would the distillation setup in Figure 13.5 be satisfactory for separating salt and water? Ethyl alcohol and water? Explain.
   8. If the liquid in the flask in Figure 13.5 is ethyl alcohol and the atmospheric pressure is 543 mm Hg, what temperature would show on the thermometer? (Use Figure 13.9.)
   9. If water were placed in both containers in Figure 13.7, would both have the same vapor pressure at the same temperature? Explain.
   10. In Figure 13.7, in which case, (a) or (b), will the atmosphere above the liquid reach a point of saturation?
   11. Suppose that a solution of ethyl ether and ethyl alcohol were placed in the closed bottle in Figure 13.7. Use Figure 13.9 for information on the substances.
       (a) Would both substances be present in the vapor?
       (b) If the answer to part (a) is yes, which would have more molecules in the vapor?
   12. In Figure 13.8, if 50% more water had been added in part (b), what equilibrium vapor pressure would have been observed in (c)?
   13. At approximately what temperature would each of the substances listed in Table 13.5 boil when the pressure is 30 mm Hg?
   14. Use the graph in Figure 13.9 to find the following:
       (a) The boiling point of water at 500 mm Hg pressure
       (b) The normal boiling point of ethyl alcohol
       (c) The boiling point of ethyl ether at 0.50 atm
   15. Consider Figure 13.10.
       (a) Why is line BC horizontal? What is happening in this interval?
       (b) What phases are present in the interval BC?
       (c) When heating is continued after point C, another horizontal line, DE, is reached at a higher temperature. What does this line represent?

C. **Review Questions**
   1. List six physical properties of water.
   2. What condition is necessary for water to have its maximum density? What is its maximum density?
   3. Account for the fact that an ice–water mixture remains at 0°C until all the ice is melted, even though heat is applied to it.
   4. Which contains less heat, ice at 0°C or water at 0°C? Explain.
   5. Why does ice float in water? Would ice float in ethyl alcohol ($d = 0.789$ g/mL)? Explain.
   6. If water molecules were linear instead of bent, would the heat of vaporization be higher or lower? Explain.
   7. The heat of vaporization for ethyl ether is 83.9 cal/g and that for ethyl alcohol is 204.3 cal/g. Which of these compounds has hydrogen bonding? Explain.

8. Would there be more or less H-bonding if water molecules were linear instead of bent? Explain.

9. Which would show hydrogen bonding, ammonia (NH_3) or methane (CH_4)? Explain.

10. In which condition are there fewer hydrogen bonds between molecules: water at 40°C or water at 80°C?

11. Write equations to show how the following metals react with water: aluminum, calcium, iron, sodium, zinc. State the conditions for each reaction.

12. Is the formation of hydrogen and oxygen from water an exothermic or endothermic reaction? How do you know?

13. (a) What is an anhydride?
    (b) What type of compound will be an acid anhydride?
    (c) What type of compound will be a basic anhydride?

14. (a) Write the formulas for the anhydrides of the following acids:
       H_2SO_3, H_2SO_4, HNO_3, $HClO_4$, H_2CO_3
    (b) Write the formulas for the anhydrides of the following bases:
       $NaOH$, KOH, $Ba(OH)_2$, $Ca(OH)_2$, $Mg(OH)_2$

15. Complete and balance the following equations:
    (a) $Ba(OH)_2 \xrightarrow{\Delta}$
    (b) $CH_3OH + O_2 \longrightarrow$
        Methyl alcohol
    (c) $Rb + H_2O \longrightarrow$
    (d) $SnCl_2 \cdot 2H_2O \xrightarrow{\Delta}$
    (e) $HNO_3 + NaOH \longrightarrow$
    (f) $Li_2O + H_2O \longrightarrow$
    (g) $KOH \xrightarrow{\Delta}$
    (h) $Ba + H_2O \longrightarrow$
    (i) $Cl_2 + H_2O \longrightarrow$
    (j) $SO_3 + H_2O \longrightarrow$
    (k) $H_2SO_3 + KOH \longrightarrow$
    (l) $CO_2 + H_2O \longrightarrow$

16. Is the conversion of oxygen to ozone an exothermic or an endothermic reaction? How do you know?

17. Write formulas for an oxygen atom, an oxygen molecule, an ozone molecule, and a peroxide ion. How many electrons are in a peroxide ion? How many in an oxygen molecule?

18. How does ozone in the stratosphere protect the earth from excessive ultraviolet radiation?

19. Name each of the following hydrates:
    (a) $BaBr_2 \cdot 2H_2O$
    (b) $AlCl_3 \cdot 6H_2O$
    (c) $FePO_4 \cdot 4H_2O$
    (d) $MgNH_4PO_4 \cdot 6H_2O$
    (e) $FeSO_4 \cdot 7H_2O$
    (f) $SnCl_4 \cdot 5H_2O$

20. Explain how anhydrous copper(II) sulfate ($CuSO_4$) can act as an indicator for moisture.

21. Compare the types of bonds in metal dioxides and metal peroxides.

22. Distinguish between deionized water and:
    (a) Hard water   (b) Soft water   (c) Distilled water

23. How can soap function to make soft water from hard water? What objections are there to using soap for this purpose?

24. What substance is commonly used to destroy bacteria in water?

25. What chemical, other than chlorine or chlorine compounds, can be used to disinfect water for domestic use?

26. Some organic pollutants in water can be oxidized by dissolved molecular oxygen. What harmful effect can result from this depletion of oxygen in the water?

27. Why should you not drink liquids that are stored in ceramic containers, especially unglazed ones?

28. Write the chemical equation showing how magnesium ions are removed by a zeolite water softener.

29. Write an equation to show how hard water containing calcium chloride ($CaCl_2$) is softened by using sodium carbonate (Na_2CO_3).

30. The vapor pressure at 20°C is given for the following compounds:

Methyl alcohol	96 mm Hg	Water	17.5	mm Hg
Acetic acid	11.7 mm Hg	Carbon tetrachloride	91	mm Hg
Benzene	74.7 mm Hg	Mercury	0.0012	mm Hg
Bromine	173 mm Hg	Toluene	23	mm Hg

    (a) Arrange these compounds in their order of increasing rate of evaporation.
    (b) Which substance listed would have the highest boiling point, and which would have the lowest?

31. Explain why rubbing alcohol, warmed to body temperature, still feels cold when applied to your skin.
32. Suggest a method whereby water could be made to boil at 50°C.
33. If a dish of water initially at 20°C is placed in a living room maintained at 20°C, the water temperature will fall below 20°C. Explain.
34. Explain why a higher temperature is obtained in a pressure cooker than in an ordinary cooking pot.
35. What is the relationship between vapor pressure and boiling point?
36. On the basis of the Kinetic-Molecular Theory, explain why vapor pressure increases with temperature.
37. Why does water have such a relatively high boiling point?
38. The boiling point of ammonia (NH_3) is -33.4°C and that of sulfur dioxide (SO_2) is -10.0°C. Which will have the higher vapor pressure at -40°C?
39. Explain what is occurring physically when a substance is boiling.
40. Explain why HF (bp 19.4°C) has a higher boiling point than HCl (bp -85°C), whereas F_2 (bp -188°C) has a lower boiling point than Cl_2 (bp -34°C).
41. Which of the following statements are correct?
    (a) The process of a substance changing directly from a solid to a gas is called sublimation.
    (b) When water is decomposed, the volume ratio of H_2 to O_2 is $2:1$, but the mass ratio of H_2 to O_2 is $1:8$.
    (c) Hydrogen sulfide is a larger molecule than water.
    (d) The changing of ice into water is an exothermic process.
    (e) Water and hydrogen fluoride are both nonpolar molecules.
    (f) The main use of hydrogen peroxide is as an oxidizing agent.
    (g) $H_2O_2 \longrightarrow 2\,H_2O + O_2$ represents a balanced equation for the decomposition of hydrogen peroxide.
    (h) Steam at 100°C can cause more severe burns than liquid water at 100°C.
    (i) The density of water is independent of temperature.
    (j) Liquid A boils at a lower temperature than liquid B. This indicates that liquid A has a lower vapor pressure than liquid B at any particular temperature.
    (k) Water boils at a higher temperature in the mountains than at sea level.
    (l) No matter how much heat you put under an open pot of pure water on a stove, you cannot heat the water above its boiling point.
    (m) The vapor pressure of a liquid at its boiling point is equal to the prevailing atmospheric pressure.
    (n) The normal boiling temperature of water is 273°C.
    (o) The pressure exerted by a vapor in equilibrium with its liquid is known as the vapor pressure of the liquid.
    (p) Sodium, potassium, and calcium each react with water to form hydrogen gas and a metal hydroxide.
    (q) Calcium oxide reacts with water to form calcium hydroxide and hydrogen gas.
    (r) Carbon dioxide is the hydride of carbonic acid.

(s) Water in a hydrate is known as water of hydration or water of crystalliza-
    tion.
(t) A substance that spontaneously loses its water of hydration when exposed
    to the air is said to be efflorescent.
(u) A substance that absorbs water from the air until it forms a solution is
    deliquescent.
(v) Distillation is effective for softening water because the minerals boil away,
    leaving soft water behind.
(w) The original source of mercury-contaminated fish is industrial pollution.
(x) Disposal of toxic industrial wastes in toxic waste dumps has been found
    to be a very satisfactory long-term solution to the problem of what to do
    with these wastes.
(y) The amount of heat needed to change 1 mole of ice at $0°C$ to a liquid at $0°C$
    is 1.44 kcal.

D. Review Problems

1. How many moles of compound are in 100 g of each of these hydrates?
    (a) $CuSO_4 \cdot 5H_2O$ (b) $FeI_2 \cdot 4H_2O$
2. How many moles of water can be obtained from 100 g of each of these hydrates?
    (a) $CuCl_2 \cdot 2H_2O$ (b) $CoBr_2 \cdot 6H_2O$
3. When a person purchases epsom salts, $MgSO_4 \cdot 7H_2O$, what percentage of the
    compound is water?
4. How many calories are needed to change 220 g of water at $20°C$ to steam at
    $100°C$?
5. How many calories of energy must be removed in the freezer to change 85 g
    of water at $24°C$ to ice at $0°C$?
6. How many calories are required to change 125 g ice at $0°C$ to steam at $100°C$?
7. The molar heat of vaporization is the number of calories required to change
    1 mole of a substance from liquid to vapor at its boiling point.
    (a) What is the molar heat of vaporization of water?
    (b) How many calories would be required to change 5.00 moles of water at $30°C$
        to steam at $100°C$?
*8. Suppose 80 g of ice at $0°C$ are added to 300 g of water at $25°C$. Is this sufficient
    ice to lower the temperature of the system to $0°C$ and still have ice remaining?
    Show evidence for your answer.
*9. Suppose 50 g of steam at $100°C$ are added to 300 g of water at $25°C$. Is there
    sufficient steam to heat all the water to $100°C$ and still have steam remaining?
    Show evidence for your answer.
10. What weight of water must be decomposed to produce 30.0 L of oxygen at STP?
11. Compare the volume occupied by 1.00 mole of liquid water at $0°C$ and 1.00 mole
    of water vapor at STP.
12. (a) How many moles of oxygen can be obtained by the decomposition of 8.0
        moles of hydrogen peroxide?
    (b) What volume will this much oxygen occupy at $35°C$ and 710 mm Hg
        pressure?
13. What volume of oxygen at STP can be obtained by decomposing 800 g of 3.0%
    hydrogen peroxide?
*14. Magnesium carbonate, $MgCO_3$, forms a hydrate containing 39.1% water of
    hydration. Calculate the formula of this hydrate.
15. How many grams of $NaC_2H_3O_2 \cdot 3H_2O$ need to be decomposed to obtain
    35.0 mL of water ($d = 1.00$ g/mL)?
16. How many grams of water will react with each of the following?
    (a) 1.00 mole Na   (d) 1.00 mole SO_2
    (b) 1.00 mole Ca   (e) 1.00 g CaO
    (c) 1.00 g K   (f) 1.00 g N_2O_5

17. Suppose 1.00 mole of water evaporates in 1 day. How many water molecules, on the average, leave the liquid each second?

*18. A quantity of sulfuric acid is added to 100 mL of water. The final volume of the solution is 122 mL and it has a density of 1.26 g/mL. What weight of acid was added? Assume the density of the water is 1.00 g/mL.

19. A mixture of 70.0 mL of hydrogen and 50.0 mL of oxygen is ignited by a spark to form water.
    (a) Does any gas remain unreacted? Which one, H_2 or O_2?
    (b) What volume of which gas (if any) remains unreacted? (Assume the same conditions before and after the reaction.)

20. What is the empirical formula of a hydrate found to contain 56.1% $ZnSO_4$ and 43.9% water?

E. **Review Exercises**

1. Can ice be colder than 0°C? Explain.
2. Why does a boiling liquid maintain a constant temperature when heat is continually being added?
3. At what temperature will copper have a vapor pressure of 760 mm Hg?

*4. Why does a lake freeze from the top down?

*5. What water temperature would you theoretically expect to find at the bottom of a very deep lake? Explain.

6. What reasons can you give for two compounds having approximately the same molecular weight but very different boiling points?

14

Solutions

After studying Chapter 14 you should be able to:

1. Understand the terms listed in Question A at the end of the chapter.
2. Describe the different types of solutions that are possible based on the three states of matter.
3. List the general properties of solutions.
4. Outline the solubility rules for common mineral substances.
5. Describe and illustrate the process by which an ionic substance like sodium chloride dissolves in water.
6. Tell how temperature changes affect the solubilities of solids and gases in liquids.
7. Tell how changes in pressure affect the solubility of a gas in a liquid.
8. Identify and discuss the variables that affect the rate at which a solid dissolves in a liquid.
9. Determine by using a solubility graph or table whether a given solution is unsaturated, saturated, or supersaturated at a given temperature.
10. Calculate the weight percent or volume percent composition of a solution from appropriate data.
11. Calculate the amount of solute in a given quantity of a solution when given the weight percent or volume percent composition.
12. Calculate the molarity of a solution when given the volume of solution and moles or grams of solute.
13. Calculate the weight of a substance needed to prepare a solution of specified volume and molarity.
14. Determine the resulting molarity when a given volume of a solution of known molarity is diluted with a specified volume of water or is mixed with a solution of different molarity.
15. Apply stoichiometric principles to chemical reactions in which the amounts of reactants used are given in units of volume and molarity.
16. Understand the concepts of equivalent weight and normality and do calculations involving these concepts.
17. Relate the effect of a solute on the vapor pressure of a solvent to the freezing point and boiling point of a solution.
18. Calculate the freezing point or boiling point of a solution from the appropriate concentration data.
19. Calculate molality and molecular weight of a solute from freezing point or boiling point and weight concentration data.

14.1 Components of a Solution

solution

solute
solvent

The term **solution** is used in chemistry to describe a system in which one or more substances are homogeneously mixed or dissolved in another substance. A simple solution has two components, a solute and a solvent. The **solute** is the substance that is dissolved. The **solvent** is the dissolving agent and usually makes up the greater proportion of the solution. For example, when salt is dissolved in water to form a solution, salt is the solute and water is the solvent. Complex solutions containing more than one solute and/or more than one solvent are common.

14.2 Types of Solutions

From the three states of matter—solid, liquid, and gas—it is possible to have nine different types of solutions: solid dissolved in solid, solid dissolved in liquid, solid dissolved in gas, liquid dissolved in liquid, and so on. Of these, the most common solutions are solid dissolved in a liquid, liquid dissolved in liquid, gas dissolved in a liquid, and gas dissolved in a gas.

14.3 General Properties of Solutions

A true solution is one in which the dissolved solute is molecular or ionic in size, generally in the range of 1 to 10 angstrom units (10^{-8} to 10^{-7} cm). The properties of a true solution are as follows:

1. It is a homogeneous mixture of two or more substances, solute and solvent.
2. It has a variable composition.
3. The dissolved solute is either molecular or ionic in size.
4. It may be either colored or colorless but is usually transparent.
5. The solute remains uniformly distributed throughout the solution and will not settle out with time.
6. A solution has the same chemical composition, the same chemical properties, and the same physical properties in every part.
7. The solute generally may be separated from the solvent by purely physical means (for example, by evaporation).

These properties are illustrated by water solutions of sugar and of potassium permanganate. Suppose that we prepare two sugar solutions, the first containing 10 g of sugar added to 100 mL of water and the second containing 20 g of sugar added to 100 mL of water. Each solution is stirred until all the solute dissolves, demonstrating that we may vary the composition of a solution. Every portion of the solution has the same sweet taste because the sugar molecules are uniformly distributed throughout. If confined so that no solvent is lost, the solution will taste and appear the same a week or a month later. The properties of the solution are unaltered after the solution is passed through filter paper. But by carefully evaporating the water, we may recover the sugar from the solution.

To observe the dissolving of potassium permanganate ($KMnO_4$) we affix a few crystals of $KMnO_4$ to paraffin wax or rubber cement at the end of a glass rod and submerge the entire rod, with the wax–permanganate end up, in a cylinder of water. Almost at once the beautiful purple color of dissolved permanganate ions (MnO_4^-) appears at the top of the rod and streams to the bottom of the cylinder as the crystals dissolve. The purple color at first is mostly at the bottom of the cylinder because potassium permanganate is denser than water. But after a while the purple color disperses until it is evenly distributed throughout the solution. This demonstrates that molecules and ions move about freely and spontaneously (diffuse) in a liquid or solution. Once the solution is formed, it is permanent; the solute does not settle out.

The permanence of a solution is explained by the Kinetic-Molecular Theory (KMT). All matter, according to the KMT, is in some kind of motion at all temperatures above absolute zero, and the intensity of motion increases with increasing temperature. Once prepared, a solution remains homogeneous because of this constant random or thermal motion of the solute and solvent particles. This constant random motion is responsible for diffusion in liquids and gases.

14.4 Solubility

solubility

We use the term **solubility** to describe the amount of solute that will dissolve in a specific amount of solvent. For example, 36.0 g of sodium chloride (NaCl) will dissolve in 100 g of water at 20°C. We say, then, that the solubility of NaCl in water is 36.0 g per 100 g of water at 20°C.

Solubility is often used in a relative way. We say that a substance is very soluble, moderately soluble, slightly soluble, or insoluble. Although these terms do not accurately indicate how much solute will dissolve, they are frequently used to describe the solubility of a substance qualitatively.

Two other terms often used to describe solubility are *miscible* and *im-*

miscible

miscible. Liquids that are capable of mixing and forming a solution are **miscible;** those that do not form solutions or are generally insoluble in each other are

immiscible

immiscible. Methyl alcohol and water are miscible in each other in all proportions. Carbon tetrachloride and water are immiscible, forming two separate layers when they are mixed. Miscible and immiscible systems are illustrated in Figure 14.1 (page 324).

The general rules for the solubility of common mineral substances are given in Table 14.1. The solubilities of over 200 compounds are given in the Solubility Table in Appendix IV. Solubility data for thousands of compounds may be found by consulting standard reference sources.*

*Two commonly used handbooks are *Lange's Handbook of Chemistry,* 12th ed. (New York: McGraw-Hill, 1979), and *Handbook of Chemistry and Physics,* 62nd ed. (Cleveland: Chemical Rubber Co., 1981).

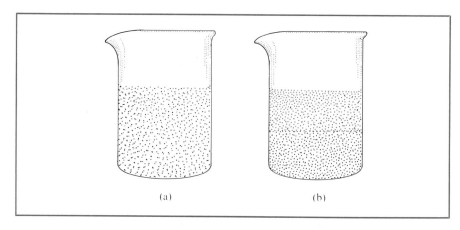

Figure 14.1 Miscible and immiscible systems: (a) miscible: H_2O and CH_3OH; (b) immiscible: H_2O and CCl_4. In a miscible system, a solution is formed, consisting of a single phase with the solute and solvent uniformly dispersed. An immiscible system is heterogeneous, and, in the case of two liquids, forms two liquid layers.

TABLE 14.1 General solubility rules for common mineral substances.[a]

Class	Solubility in cold water[b]
Nitrates	All nitrates are soluble.
Acetates	All acetates are soluble.
Chlorides ⎫ Bromides ⎬ Iodides ⎭	All chlorides, bromides, and iodides are soluble except those of Ag, Hg(I), and Pb(II); $PbCl_2$ and $PbBr_2$ are slightly soluble in hot water.
Sulfates	All sulfates are soluble except those of Ba, Sr, and Pb; Ca and Ag sulfates are slightly soluble.
Carbonates ⎫ Phosphates ⎬	All carbonates and phosphates are insoluble except those of Group IA metals and ammonium compounds. Many bicarbonates and acid phosphates are soluble.
Hydroxides	All hydroxides are insoluble except those of the Group IA metals and NH_4OH; $Ba(OH)_2$ and $Ca(OH)_2$ are slightly soluble.
Sodium salts ⎫ Potassium salts ⎬ Ammonium salts ⎭	All common salts of these ions are soluble.
Sulfides	All sulfides are insoluble except those of the Group IA metals and ammonium compounds.

[a]When we say a substance is soluble, we mean that the substance is reasonably soluble. All substances have some solubility in water, although the amount of solubility may be very small; the solubility of silver iodide, for example, is about 1×10^{-8} mole AgI per litre H_2O.
[b]There are, of course, exceptions to these rules.

The quantitative expression of the amount of dissolved solute in a particular quantity of solvent is known as the **concentration of a solution.** Several methods of expressing concentration will be described in Section 14.7.

14.5 Factors Related to Solubility

The entire concept of predicting solubilities is, at best, very complex and difficult. There are many variables, such as size of ions, charge on ions, interaction between ions, interaction between solute and solvent, and temperature, all of which bear upon the problem. Because of the factors involved, there are many exceptions to the general rules of solubility given in Table 14.1. However, the rules are very useful, because they do apply to a good many of the more common compounds that we encounter in the study of chemistry. Keep in mind that these are rules, not laws, and are therefore subject to exceptions. Fortunately, the solubility of a solute is relatively easy to determine experimentally. Four factors related to solubility are discussed in the following paragraphs.

1. The nature of the solute and solvent. The old adage that "like dissolves like" has merit, in a general way. Polar substances tend to be more miscible, or soluble, with other polar substances. Nonpolar substances tend to be miscible with other nonpolar substances and less miscible with polar substances. Thus, mineral acids, bases, and salts, which are polar, tend to be much more soluble in water, which is polar, than in solvents such as ether, carbon tetrachloride, or benzene, which are essentially nonpolar. Sodium chloride, a very polar substance, is soluble in water, slightly soluble in ethyl alcohol (less polar than water), and insoluble in ether and benzene. Pentane (C_5H_{12}), a nonpolar substance, is only slightly soluble in water but is very soluble in benzene and ether.

At the molecular level, the formation of a solution from two nonpolar substances, such as carbon tetrachloride and benzene, can be visualized as a process of simple mixing. The nonpolar molecules, having little tendency to either attract or repel one another, easily intermingle to form a homogeneous mixture.

Solution formation between ionic or polar substances is much more complex. For example, the process by which sodium chloride dissolves in water is illustrated in Figure 14.2. Water molecules are very polar and are attracted to other polar molecules or ions. When salt crystals (NaCl) are put into water, polar water molecules become attracted to the sodium and chloride ions on the crystal surfaces and weaken the attraction between Na^+ and Cl^- ions. The positive end of the water dipole is attracted to the Cl^- ions, and the negative end of the water dipole is attracted to the Na^+ ions. The weaker attraction permits the ions to move apart, making room for more water dipoles. Thus, the surface ions are surrounded by water molecules, becoming hydrated ions, $Na^+(aq)$ and $Cl^-(aq)$, and slowly diffuse away from the crystals as dissolved ions in solution.

$$NaCl(s) \xrightarrow{H_2O} Na^+(aq) + Cl^-(aq)$$

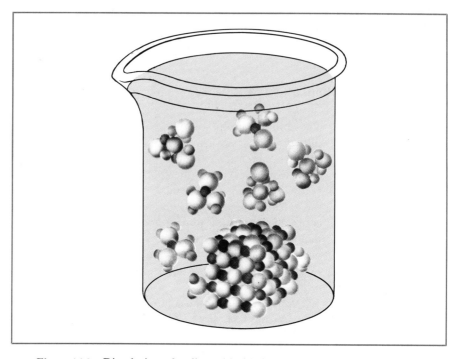

Figure 14.2   Dissolution of sodium chloride in water. Polar water molecules are attracted to Na^+ and Cl^- ions in the salt crystal, weakening the attraction between the ions. As the attraction between the ions weakens, the ions move apart and become surrounded by water dipoles. The hydrated ions slowly diffuse away from the crystal to become dissolved in solution.

Examination of the data in Table 14.2 reveals some of the complex questions relating to solubility. For example, why are lithium halides, except for lithium fluoride (LiF), more soluble than sodium and potassium halides? Why, indeed, are the solubilities of LiF and sodium fluoride (NaF) so low in comparison to those of the other salts? Why does not the solubility of LiF, NaF, and NaCl increase proportionately with temperature, as the solubilities of the other salts do? Sodium chloride is appreciably soluble in water, but is insoluble in concentrated hydrochloric acid (HCl) solution. On the other hand, LiF and NaF are not very soluble in water but are quite soluble in hydrofluoric acid (HF) solution —why? These questions will not be answered directly here, but it is hoped that your curiosity will be aroused to the point that you will do some reading and research on the properties of solutions.

2. The effect of temperature on solubility.   There is no general rule for changes in solubility of solids in liquids with a change in temperature. Most solutes have a limited solubility in a specific amount of solvent at a fixed temperature. Most solids are more soluble in water at higher temperatures (see Figure 14.3). Some solids show very little increase in solubility with increasing tem-

TABLE 14.2 Solubility of alkali metal halides in water.

Salt	Solubility (g salt/100 g H_2O)	
	0°C	*100°C*
LiF	0.12	0.14 (at 35°C)
LiCl	67	127.5
LiBr	143	266
LiI	151	481
NaF	4	5
NaCl	35.7	39.8
NaBr	79.5	121
NaI	158.7	302
KF	92.3 (at 18°C)	Very soluble
KCl	27.6	57.6
KBr	53.5	104
KI	127.5	208

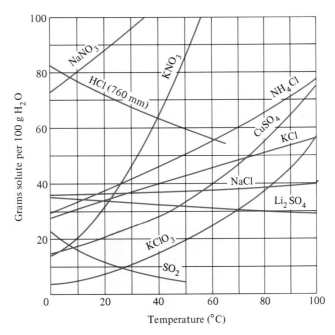

Figure 14.3 Solubility of various compounds in water.

perature (see NaCl in Figure 14.3), and others show a decrease in solubility with increasing temperature (see Li_2SO_4 in Figure 14.3).

The solubility of a gas in a liquid always decreases with increasing temperature (see HCl and SO_2 in Figure 14.3). The tiny bubbles formed when water is heated are due to the decreased solubility of air at higher temperatures.

3. The effect of pressure on solubility.   Small changes in pressure have little effect on the solubility of solids in liquids but have a marked effect on the solubility of gases in liquids. The solubility of a gas in a liquid is directly proportional to the pressure of that gas above the solution. Thus, the amount of a gas that is dissolved in solution will double if the pressure of that gas over the solution is doubled. For example, carbonated beverages contain dissolved carbon dioxide at pressures greater than atmospheric pressure. When a bottle of carbonated soda is opened, the pressure is immediately reduced to the atmospheric pressure, and the excess dissolved carbon dioxide bubbles out of the solution.

4. Rate of dissolving.   The rate at which a solid solute dissolves is affected by (a) particle size of the solute, (b) temperature, (c) agitation or stirring, and (d) concentration of the solution.

    (a) *Particle size.*   A solid can dissolve only at the surface that is in contact with the solvent. Because the surface-to-volume ratio increases as size decreases, smaller crystals dissolve faster than larger ones. For example, if a salt crystal 1 cm on a side (6 cm² surface area) is divided into 1000 cubes, each 0.1 cm on a side, the total surface of the smaller cubes is 60 cm² — a tenfold increase in surface area (see Figure 14.4).

    (b) *Temperature.*   In most cases, the rate of dissolving of a solid increases with temperature. This is because of kinetic effects. The solvent molecules, moving more rapidly at higher temperatures, strike the solid surfaces more often and harder, causing the rate of dissolving to increase.

    (c) *Agitation or stirring.*   The effect of agitation or stirring is kinetic. When a solid is first put into water, the only solvent with which it comes

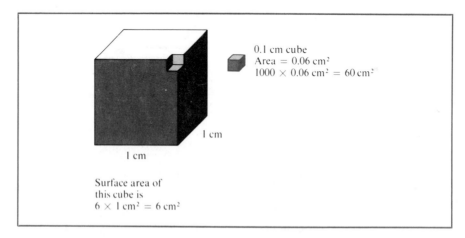

Figure 14.4   Surface area of crystals: A crystal 1 cm on a side has a surface area of 6 cm². Subdivided into 1000 smaller crystals, each 0.1 cm on a side, the total surface area is increased to 60 cm².

in contact is in the immediate vicinity. As the solid dissolves, the amount of dissolved solute around the solid becomes more and more concentrated and the rate of dissolving slows down. If the mixture is not stirred, the dissolved solute diffuses very slowly throughout the entire solution; weeks may pass before the solid is entirely dissolved. Through stirring, the dissolved solute is distributed rapidly throughout the solution and more solvent is brought into contact with the solid, causing it to dissolve more rapidly.

(d) *Concentration of the solution.* When the solute and solvent are first mixed, the rate of dissolving is at its maximum. As the concentration of the solution increases and the solvent becomes more nearly saturated with the solute, the rate of dissolving decreases greatly. The rate of dissolving is pictured graphically in Figure 14.5. Note that about 17 g dissolve in the first 5 minute interval but only about 1 g dissolves in the fourth 5 minute interval. Although different solutes show different rates, the rate of dissolving always becomes very slow as the concentration approaches the saturation point.

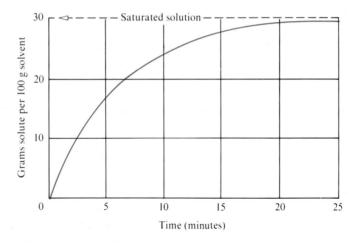

Figure 14.5 Rate of dissolution of a solid solute in a solvent. The rate is maximum at the beginning and decreases as the concentration approaches the saturation point.

14.6 Solutions: A Reaction Zone

Many solids must be put in solution in order to undergo appreciable chemical reactions. We can easily write the equation for the double-replacement reaction between sodium chloride and silver nitrate:

$$NaCl + AgNO_3 \longrightarrow AgCl + NaNO_3$$

But suppose we mix solid NaCl and solid AgNO$_3$ and look for a chemical change. Some reaction may occur, but if it does, it is quite slow and essentially undetectable. In fact, the crystalline structures of NaCl and AgNO$_3$ are so different that we can separate them by tediously picking out the two different kinds of crystals from the mixture. But if we dissolve the sodium chloride and silver nitrate separately in water and mix the two solutions, we observe the immediate formation of a white, curdy precipitate of silver chloride.

Molecules or ions must come into intimate contact or collide with one another in order to react. In the foregoing example, the two solids did not react because the ions were securely locked within the crystal structures. But when the sodium chloride and silver nitrate are dissolved, their crystal lattices are broken down and the ions become mobile. When the two solutions are mixed, the mobile Ag$^+$ and Cl$^-$ ions come into contact and react to form insoluble AgCl, which precipitates out of solution. The soluble Na$^+$ and NO$_3^-$ ions remain in solution but form the crystalline salt NaNO$_3$ when the water is evaporated:

$$NaCl(aq) + AgNO_3(aq) \longrightarrow AgCl\downarrow + NaNO_3(aq)$$

$$(Na^+ + Cl^-) + (Ag^+ + NO_3^-) \xrightarrow{H_2O} AgCl\downarrow + (Na^+ + NO_3^-)$$

Sodium chloride solution   Silver nitrate solution   Silver chloride   Sodium nitrate in solution

The mixture of the two solutions provided a zone or space in which the Ag$^+$ and Cl$^-$ ions could react. (See Chapter 15 for discussion of ionic reactions.)

Solutions also function as diluents in reactions in which the undiluted reactants would combine with each other too violently. Moreover, a solution of known concentration provides a convenient method for delivering specific amounts of reagents.

14.7 Concentration of Solutions

The concentration of a solution gives us information concerning the amount of solute dissolved in a unit volume of solution. Because reactions are often conducted in solution, it is important to understand the methods of expressing concentration and know how to prepare solutions of particular concentrations.

DILUTE AND CONCENTRATED SOLUTIONS

When we say that a solution is *dilute* or *concentrated,* we are expressing, in a relative way, the amount of solute present. One gram of salt and 2 g of salt in solution are both dilute solutions when compared to the same volume of a solution containing 20 g of salt. Ordinary concentrated hydrochloric acid (HCl) contains 12 moles of HCl per litre of solution. In some laboratories, the dilute acid is made by mixing equal volumes of water and the concentrated acid. In other laboratories, the concentrated acid is diluted with two or three volumes of water, depending on its use. The term **dilute solution,** then, describes a solution that contains a relatively small amount of dissolved solute. Conversely, a **concentrated solution** contains a relatively large amount of dissolved solute.

dilute solution
concentrated solution

SATURATED, UNSATURATED, AND SUPERSATURATED SOLUTIONS

At a specific temperature there is a limit to the amount of solute that will dissolve in a given amount of solvent. When this limit is reached, the resulting solution is said to be *saturated*. For example, when we put 40.0 g KCl into 100 g H₂O at 20°C, we find that 34.0 g KCl dissolve and 6.0 g KCl remain undissolved. The solution formed is a saturated solution of KCl.

Two processes are occurring simultaneously in a saturated solution. The solid is dissolving into solution and, at the same time, the dissolved solute is crystallizing out of solution. This may be expressed as

Solute (undissolved) $\rightleftharpoons$ Solute (dissolved)

When these two opposing processes are occurring at the same rate, the amount of solute in solution is constant and a condition of equilibrium is established between dissolved and undissolved solute. A **saturated solution** contains dissolved solute in equilibrium with undissolved solute.

saturated solution

It is especially important to state the temperature of a saturated solution. A solution that is saturated at one temperature may not be saturated at another. If the temperature of a saturated solution is changed, the equilibrium is disturbed, and the amount of dissolved solute will change to reestablish the equilibrium.

A saturated solution may be either dilute or concentrated, depending on the solubility of the solute. A saturated solution can be conveniently prepared by dissolving the solute at a temperature somewhat higher than room temperature. The amount of solute in solution should be in excess of its solubility at room temperature. When the solution cools, the excess solute crystallizes, leaving the solution saturated. In this case, the solute must be more soluble at higher temperatures and must not form a supersaturated solution. Examples expressing the solubility of saturated solutions at two different temperatures are given in Table 14.3.

unsaturated
solution

A solution containing less solute per unit of volume than does its corresponding saturated solution is said to be **unsaturated.** In other words, more solute can be dissolved into an unsaturated solution without altering other conditions. Consider again a solution made from 40 g of KCl and 100 g of water at

TABLE 14.3 Saturated solutions at 20°C and 50°C.

Solute	Solubility (g solute/100 g H₂O)	
	20°C	*50°C*
NaCl	36.0	37.0
KCl	34.0	42.6
NaNO₃	88.0	114.0
KClO₃	7.4	19.3
AgNO₃	222.0	455.0
C₁₂H₂₂O₁₁	203.9	260.4

20°C (see Table 14.3). The solution formed will certainly be saturated and there will be about 6 g of undissolved salt, because the maximum amount of KCl that can dissolve in 100 g of water at 20°C is 34 g. If the solution is now heated and maintained at 50°C, all the salt will dissolve and, in fact, more can be dissolved. Thus, the solution at 50°C is unsaturated.

supersaturated solution

In some instances, solutions can be prepared that contain more solute than that of the saturated solution at a particular temperature. Such solutions are said to be **supersaturated.** However, we must qualify this definition by noting that a supersaturated solution is unstable. Disturbances such as jarring, stirring, scratching the walls of the container, or dropping in a "seed" crystal cause the supersaturation to break. When a supersaturated solution is disturbed, the excess solute crystallizes out rapidly, returning the solution to a saturated state.

Supersaturated solutions, while not easy to prepare, may be formed from selected substances by dissolving, in warm solvent, an amount of solute greater than that needed for a saturated solution at room temperature. The warm solution is then allowed to cool very slowly. With the proper solute and careful work, a supersaturated solution will result. Two substances commonly used to demonstrate this property are sodium thiosulfate pentahydrate, $Na_2S_2O_3 \cdot 5H_2O$, and sodium sulfate, Na_2SO_4 (from a saturated solution at 30°C).

PROBLEM 14.1   Will a solution made by adding 2.5 g $CuSO_4$ to 10 g H_2O be saturated or unsaturated at 20°C?

To answer this question we first need to know the solubility of $CuSO_4$ at 20°C. From Figure 14.3 we see that the solubility of $CuSO_4$ at 20°C is about 21 g per 100 g H_2O. This is equivalent to 2.1 g $CuSO_4$ per 10 g H_2O.

Since 2.5 g per 10 g H_2O is greater than 2.1 g per 10 g H_2O the solution will be saturated. There will be 0.4 g $CuSO_4$ undissolved.

■

WEIGHT PERCENT SOLUTION

The weight percent of a solute in a solution can have two meanings: (a) percentage of the solute by weight of the solution, or (b) percentage of the solute by volume of the solution. The percentage by weight is the one most commonly used and the one we will use in this book. *Percentage by weight* means that for a given weight of solution, a certain percentage of that weight is solute. Suppose that we take a bottle from the reagent shelf that reads "Sodium hydroxide, NaOH, 10%" (see Figure 14.6). This means that for every 100 g of this solution we use, 10 g will be NaOH and 90 g will be water. (Note that this is 100 g and not 100 mL of solution.) We could also make this same concentration of solution by dissolving 2 g of NaOH in 18 g of water. The mathematical equation for calculating weight percent is

$$\text{Weight percent} = \frac{\text{g solute}}{\text{g solute} + \text{g solvent}} \times 100\% = \frac{\text{g solute}}{\text{g solution}} \times 100\%$$

Illustrative problems follow.

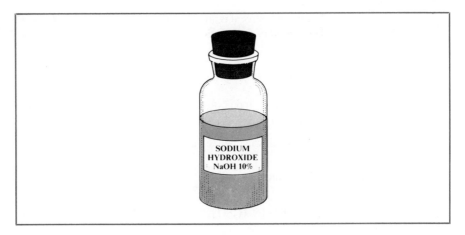

Figure 14.6 Weight percent concentration of solutions. The bottle contains 10 g of NaOH per 90 g of H_2O.

PROBLEM 14.2 What is the weight percent of sodium hydroxide in a solution that is made by dissolving 8.00 g of NaOH in 50.0 g of H_2O?

Grams of solute (NaOH) = 8.00 g

Grams of solvent (H_2O) = 50.0 g

$$\frac{8.00 \text{ g NaOH}}{8.00 \text{ g NaOH} + 50.0 \text{ g H}_2\text{O}} \times 100\% = 13.8\% \text{ NaOH solution}$$

■

PROBLEM 14.3 What weights of potassium chloride (KCl) and water are needed to make 250 g of 5.00% solution?

The percentage expresses the weight of KCl in the solution: 5.00% of the solution is KCl, 95.0% is H_2O.

250 g = Total weight of solution

5.00% of 250 g = 0.0500 × 250 g = 12.5 g KCl   (Solute)

250 g − 12.5 g = 237.5 g H_2O

Dissolving 12.5 g KCl in 237.5 g H_2O gives a 5.00% KCl solution.

■

PROBLEM 14.4 A 34.0% sulfuric acid solution has a density of 1.25 g/mL. How many grams of H_2SO_4 are contained in 1.00 L of this solution?

Since H_2SO_4 is the solute, we first solve the weight percent equation for grams of solute:

$$\text{Weight percent} = \frac{\text{g solute}}{\text{g solution}} \times 100\%$$

$$\text{g solute} = \frac{\text{Weight percent} \times \text{g solution}}{100\%}$$

The weight percent is given in the problem. We need to determine the grams of solution. The weight of the solution can be calculated from the density data.

Convert density (g/mL) to grams:

$$1.00 \text{ L} = 1000 \text{ mL}$$

$$\frac{1.25 \text{ g}}{\text{mL}} \times 1000 \text{ mL} = 1250 \text{ g} \quad \text{(g solution)}$$

Now we can calculate the grams of solute.

$$\text{g solute} = \frac{34.0\% \times 1250 \text{ g}}{100\%} = 425 \text{ g H}_2\text{SO}_4$$

1.00 L of 34.0% H_2SO_4 contains 425 g H_2SO_4.

∎

The student should note that the concentration expressed as weight percent is independent of the formula of the solute.

VOLUME PERCENT

Solutions that are formulated from two liquids are often expressed as *volume percent* with respect to the solute. The label on a bottle of ordinary rubbing alcohol reads "Isopropyl alcohol, 70% by volume." Such a solution could be made by mixing 70 mL of alcohol and 30 mL of water. If we assume that these volumes are additive (which they are not, exactly), 1 litre of 70% isopropyl alcohol by volume will contain 700 mL of the alcohol.

$$\text{Volume percent} = \frac{\text{Volume of liquid in question}}{\text{Total volume of solution}} \times 100\%$$

MOLARITY

Weight percent solutions do not equate or express the number of formula or molecular weights of the solute in solution. For example, 1000 g of 10% NaOH solution contain 100 g of NaOH; 1000 g of 10% KOH solution contain 100 g of KOH. In terms of moles of NaOH and KOH, these solutions contain

$$\text{Moles NaOH} = 100 \text{ g NaOH} \times \frac{1 \text{ mole NaOH}}{40.0 \text{ g NaOH}} = 2.50 \text{ moles NaOH}$$

$$\text{Moles KOH} = 100 \text{ g KOH} \times \frac{1 \text{ mole KOH}}{56.1 \text{ g KOH}} = 1.78 \text{ moles KOH}$$

From these figures, we see that the two 10% solutions do not contain the same number of moles of NaOH and KOH. Yet 1 mole of each of these two bases will neutralize the same amount of acid. As a result, we find that a 10% NaOH solution has more reactive alkali than a 10% KOH solution.

We need a method expressing concentration that will easily identify how many moles or formula weights of solute are present per unit of volume of solution. For this purpose, the molar method of expressing concentration is used.

1 molar solution

A **1 molar solution** contains 1 mole of solute per litre of solution. For example, to make a 1 molar solution of sodium hydroxide (NaOH), we dissolve 40.0 g of NaOH (1 mole) in water and dilute the solution with more water to a volume of 1 litre. The solution contains 1 mole of the solute in 1 litre of solution and is said to be 1 molar (1 M) in concentration. Figure 14.7 illustrates the preparation of a 1 molar solution. Note that the volume of the solute and the solvent together is 1 litre.

molarity

The concentration of a solution may, of course, be varied by using more or less solute or solvent; but in any case, the **molarity** of a solution is the number of moles of solute per litre of solution. A capital M is the abbreviation for

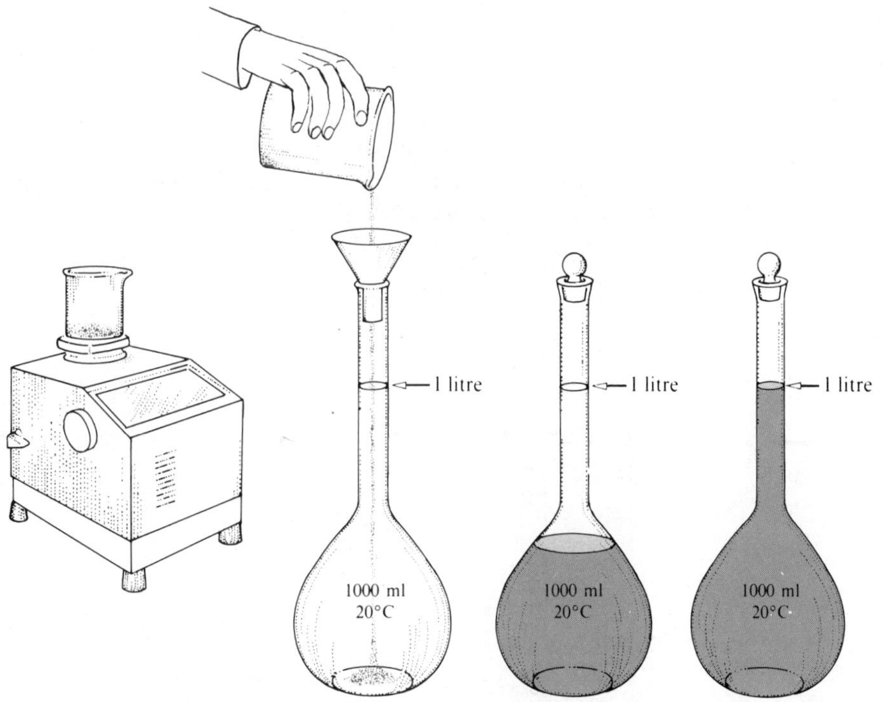

1 litre	1 litre	1 litre	
1000 ml	1000 ml	1000 ml	
20°C	20°C	20°C	
(a) Weigh 1 mole of solute	(b) Transfer weighed solute to a 1 litre volumetric flask	(c) Dissolve in solvent	(d) Add solvent to the 1 litre mark and mix thoroughly

Figure 14.7 Preparation of a 1 molar solution.

molarity. The units of molarity are moles per litre. The expression "2.0 M NaOH" means a 2.0 molar solution of NaOH (2.0 moles, or 80.0 g, of NaOH dissolved in 1 litre of solution).

$$\text{Molarity} = M = \frac{\text{Number of moles of solute}}{\text{Litres of solution}} = \frac{\text{Moles}}{\text{Litre}}$$

Two other useful equations can be derived from this equation; they are

$$\text{Moles} = M \times \text{Litres} \qquad \text{Litres} = \frac{\text{Moles}}{M}$$

Flasks that are calibrated to contain specific volumes at a particular temperature are used to prepare solutions of a desired concentration. These *volumetric flasks* have a calibration mark on the neck to indicate accurately the measured volume.

Suppose we want to make 500 mL of a 1 M solution. This solution can be prepared by weighing 0.5 mole of the solute and diluting with water in a 500 mL volumetric flask. The molarity will be

$$M = \frac{0.5 \text{ mole solute}}{0.5 \text{ L solution}} = 1 \text{ molar}$$

Thus, you can see that it is not necessary to have a litre of solution to express molarity. All we need to know is the number of moles of dissolved solute and the volume of solution (see Figure 14.8). Thus, 0.001 mole NaOH in 10 mL of solution is 0.1 M:

$$\frac{0.001 \text{ mole}}{10 \text{ mL}} \times \frac{1000 \text{ mL}}{\text{L}} = 0.1 \ M$$

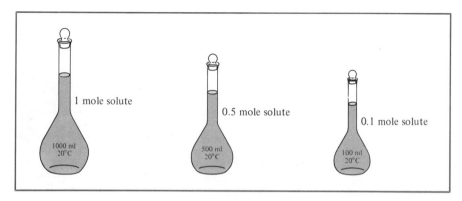

Figure 14.8 Molar solutions. The molarity of the solutions in each of these volumetric flasks is 1 M.

When we stop to think that a balance is not calibrated in moles but in grams, we see that we really need to incorporate grams into the molarity formula. This is done by using the relationship

$$\text{Number of moles} = \frac{\text{Grams}}{\text{Formula wt}}$$

Substituting this relationship into our expression for molarity, we get

$$M = \frac{\text{Moles}}{\text{Litre}} = \frac{\text{Grams of solute}}{\text{Formula wt solute} \times \text{Litres of solution}} = \frac{\text{Grams}}{\text{Formula wt} \times \text{Litres}}$$

Using this formula, we can weigh any amount of a solute that has a known formula, dilute it to any volume, and calculate the molarity of the solution.

The molarities of the concentrated acids commonly used in the laboratory are

HCl	12 M	$HC_2H_3O_2$	17 M
HNO_3	16 M	H_2SO_4	18 M

Illustrative problems follow.

PROBLEM 14.5 What is the molarity of a solution containing 1.4 moles of acetic acid ($HC_2H_3O_2$) in 250 mL of solution?

Use the equation $\quad M = \dfrac{\text{Moles}}{\text{Litre}}$

We know the number of moles of $HC_2H_3O_2$ (1.4 moles), but we need litres of solution. Convert 250 mL to litres:

$$250 \text{ mL} \times \frac{1 \text{ litre}}{1000 \text{ mL}} = 0.250 \text{ L}$$

Then,

$$M = \frac{1.4 \text{ moles}}{0.250 \text{ L}} = \frac{5.6 \text{ moles}}{\text{L}} = 5.6 \ M \quad \text{(Answer)}$$

■

PROBLEM 14.6 What is the molarity of a solution made by dissolving 2.00 g of potassium chlorate ($KClO_3$) in enough water to make 150 mL of solution?

Use the equation $\quad M = \dfrac{\text{Moles}}{\text{Litre}}$

We know the volume is 150 mL (0.150 L). We first need to convert 2.00 g $KClO_3$ to moles $KClO_3$:

1 mole $KClO_3$ weighs 122.6 g (39.1 + 35.5 + 48.0).

$$2.00 \text{ g } \cancel{KClO_3} \times \frac{1 \text{ mole } KClO_3}{122.6 \text{ g } \cancel{KClO_3}} = 1.63 \times 10^{-2} \text{ mole } KClO_3$$

Then,

$$M = \frac{1.63 \times 10^{-2} \text{ mole } KClO_3}{0.150 \text{ L}} = \frac{0.109 \text{ mole}}{\text{L}} = 0.109 \text{ } M \quad \text{(Answer)}$$

This problem may also be solved using the unit conversion method. The steps in the conversions must lead to units of moles/litre.

$$\frac{\text{g } KClO_3}{\text{mL}} \longrightarrow \frac{\text{g } KClO_3}{\text{L}} \longrightarrow \frac{\text{moles } KClO_3}{\text{L}} = M$$

$$\frac{2.00 \text{ g } \cancel{KClO_3}}{150 \text{ } \cancel{mL}} \times \frac{1000 \text{ } \cancel{mL}}{\text{L}} \times \frac{1 \text{ mole}}{122.6 \text{ g } \cancel{KClO_3}} = \frac{0.109 \text{ mole}}{\text{L}} = 0.109 \text{ } M$$

■

PROBLEM 14.7   How many grams of potassium hydroxide are required to prepare 600 mL of 0.450 M KOH solution?

The conversion is   Litres $\longrightarrow$ Moles $\longrightarrow$ Grams

Data: Volume = 0.600 L $M = \dfrac{0.450 \text{ mole}}{\text{L}}$

1 mole of KOH weighs 56.1 g (39.1 + 16.0 + 1.0).

The calculation is

$$0.600 \text{ } \cancel{L} \times \frac{0.450 \text{ mole}}{\cancel{L}} \times \frac{56.1 \text{ g KOH}}{\cancel{\text{mole}}} = 15.1 \text{ g KOH} \quad \text{(Answer)}$$

■

PROBLEM 14.8   What volume of 0.250 M solution can be prepared from 16.0 g of potassium carbonate (K_2CO_3)?

Use the equation $\text{Litres} = \dfrac{\text{Moles}}{M}$

Step 1. Convert 16.0 g K_2CO_3 to moles K_2CO_3.

1 mole K_2CO_3 weighs 138.2 g (78.2 + 12.0 + 48.0).

$$16.0 \text{ g } \cancel{K_2CO_3} \times \frac{1 \text{ mole } K_2CO_3}{138.2 \text{ g } \cancel{K_2CO_3}} = 0.116 \text{ mole } K_2CO_3$$

Step 2. Substitute into the equation and solve:

$$\text{Litres} = \frac{0.116 \text{ } \cancel{\text{mole}} \text{ } K_2CO_3}{0.250 \dfrac{\cancel{\text{mole}}}{\text{L}}} = 0.464 \text{ L}$$

By the unit conversion method the calculation is

$$16.0 \text{ g } K_2CO_3 \times \frac{1 \text{ mole } K_2CO_3}{138.1 \text{ g } K_2CO_3} \times \frac{1 \text{ L}}{0.250 \text{ mole } K_2CO_3} = 0.463 \text{ L}$$

Thus, a 0.250 M solution can be made by dissolving 16.0 g of K_2CO_3 in water and diluting to 463 mL.

■

PROBLEM 14.9 How many millilitres of 2.0 M HCl will react with 28.0 g of NaOH?

Step 1. Write and balance the equation for the reaction:

$$HCl(aq) + NaOH(aq) \longrightarrow NaCl(aq) + H_2O(aq)$$

The equation states that 1 mole of HCl reacts with 1 mole of NaOH.

Step 2. Calculate the number of moles of NaOH in 28.0 g of NaOH:

1 mole NaOH weighs 40.0 g (23.0 + 16.0 + 1.0).

$$28.0 \text{ g } NaOH \times \frac{1 \text{ mole } NaOH}{40.0 \text{ g } NaOH} = 0.700 \text{ mole NaOH}$$

28.0 g NaOH = 0.700 mole NaOH

Step 3. Solve for moles and volume of HCl needed. From Steps 1 and 2 we see that 0.700 mole of NaOH will react with 0.700 mole of HCl, because the ratio of moles reacting is 1:1. We also know that 2.0 M HCl contains 2.0 moles of HCl per litre; therefore, the volume that contains 0.700 mole of HCl will be less than 1 L.

Moles NaOH $\longrightarrow$ Moles HCl $\longrightarrow$ Litres HCl $\longrightarrow$ Millilitres HCl

$$0.700 \text{ mole } NaOH \times \frac{1 \text{ mole } HCl}{1 \text{ mole } NaOH} \times \frac{1 \text{ L HCl}}{2 \text{ moles } HCl} = 0.350 \text{ L HCl}$$

0.350 L HCl $\times$ 1000 mL/L = 350 mL HCl

Therefore, 350 mL of 2.0 M HCl contain 0.700 mole of HCl and will react with 0.700 mole, or 28.0 g, of NaOH.

■

PROBLEM 14.10 Calculate the number of moles of nitric acid in 325 mL of 16 M HNO_3 solution.

Use the equation   Moles = Litres $\times$ M

Substitute the data given in the problem and solve:

$$\text{Moles} = 0.325 \text{ L} \times \frac{16 \text{ moles } HNO_3}{L} = 5.2 \text{ moles } HNO_3 \quad \text{(Answer)}$$

■

Dilution problems. Chemists often find it necessary to dilute solutions from one concentration to another by adding more solvent to the solution. If a solution is diluted by adding pure solvent, the volume of the solution increases but the number of moles of solute in the solution remains the same. Thus, the moles/litre (molarity) of the solution decreases. It is important to read a problem carefully to distinguish between two types of problems: (1) how much solvent must be added to dilute a solution to a particular concentration and (2) to what volume a solution must be diluted to prepare a solution of a particular concentration.

PROBLEM 14.11   Calculate the molarity of a sodium hydroxide solution that is prepared by mixing 100 mL of 0.20 M NaOH with 150 mL of water.

This is a dilution problem. If we double the volume of a solution by adding water, we cut the concentration in half. Since the addition of 150 mL of water will more than double the volume, the concentration of the diluted solution should be less than 0.10 M. In the dilution, the moles of NaOH remain constant; the molarity and volume change.

To solve this problem, (1) calculate the moles of NaOH in the original solution, and (2) divide the moles of NaOH by the final volume of the solution to obtain the new molarity.

Step 1. Calculate the moles of NaOH in the original solution.

$$\text{Moles} = \text{Litres} \times M = 0.100\ \cancel{L} \times \frac{0.20\ \text{mole NaOH}}{1\ \cancel{L}} = 0.020\ \text{mole NaOH}$$

Step 2. Calculate the new molarity. The final volume of the solution is 250 mL (0.250 L).

100 mL (original) + 150 mL H_2O added

Calculate the new molarity using $M = \dfrac{\text{Moles}}{\text{Litre}}$

$$M = \frac{0.020\ \text{mole NaOH}}{0.250\ \text{L}} = 0.080\ M\ \text{NaOH}\quad \text{(Answer)}$$

Alternate Solution: When the moles of solute in a solution before and after dilution are the same, then the moles before and after dilution may be set equal to each other:

$$\text{Moles}_1 = \text{Moles}_2$$

where moles$_1$ = moles before dilution and moles$_2$ = moles after dilution. Then

$$\text{Moles}_1 = \text{Litres}_1 \times M_1 \qquad \text{Moles}_2 = \text{Litres}_2 \times M_2$$

$$\text{Litres}_1 \times M_1 = \text{Litres}_2 \times M_2$$

When both volumes are in the same units, a more general statement can be made:

$$V_1 \times M_1 = V_2 \times M_2$$

For this problem: $V_1 = 100\ \text{mL}$ $V_2 = 250\ \text{mL}$

$M_1 = 0.020\ M$ $M_2 = M_2$ (Unknown)

Then,

$$100\ \text{mL} \times 0.20\ M = 250\ \text{mL} \times M_2$$

Solving for M_2, we get

$$M_2 = \frac{100 \text{ mL} \times 0.20 \, M}{250 \text{ mL}} = 0.080 \, M \text{ NaOH}$$

PROBLEM 14.12 How many millilitres of water should be added to 150 mL of 0.420 M NaCl solution to make the solution 0.200 M?

This is a dilution problem and we can use the formula $V_1 M_1 = V_2 M_2$ to calculate the final volume of the solution. Subtracting the original volume from the final volume gives us the amount of water to be added.

$$V_1 M_1 = V_2 M_2 \quad \text{or} \quad V_2 = \frac{V_1 M_1}{M_2}$$

$$V_1 = 150 \text{ mL} \qquad M_1 = 0.420 \, M \qquad M_2 = 0.200 \, M \qquad V_2 = \text{Final volume}$$

$$V_2 = \frac{150 \text{ mL} \times 0.420 \, M}{0.200 \, M} = 315 \text{ mL}$$

The amount of water to be added is 165 mL (315 mL − 150 mL).

PROBLEM 14.13 What weight of silver chloride (AgCl) will be precipitated by adding sufficient silver nitrate ($AgNO_3$) to react with 1500 mL of 0.400 M $BaCl_2$ (barium chloride) solution?

$$2 \text{ AgNO}_3(aq) + \underset{\text{1 mole}}{\text{BaCl}_2(aq)} \longrightarrow \underset{\text{2 moles}}{2 \text{ AgCl}\downarrow} + \text{Ba(NO}_3)_2(aq)$$

This is a stoichiometry problem. The fact that $BaCl_2$ is in solution means that we first need to convert the volume and concentration of the $BaCl_2$ solution to moles of $BaCl_2$.

Step 1. Determine the number of moles of $BaCl_2$ in 1500 mL of 0.400 M solution:

Use the equation $\qquad$ Moles = Litres $\times M$

1500 mL = 1.500 L

$$1.500 \, \cancel{L} \times \frac{0.400 \text{ mole BaCl}_2}{\cancel{L}} = 0.600 \text{ mole BaCl}_2$$

Step 2. Calculate the moles of AgCl formed by using the mole-ratio method:

$$0.600 \, \cancel{\text{mole BaCl}_2} \times \frac{2 \text{ moles AgCl}}{1 \, \cancel{\text{mole BaCl}_2}} = 1.20 \text{ moles AgCl}$$

Step 3. Convert the moles of AgCl to grams:

1 mole of AgCl weighs 143.4 g (107.9 + 35.5).

$$1.20 \, \cancel{\text{moles AgCl}} \times \frac{143.4 \text{ g AgCl}}{\cancel{\text{mole AgCl}}} = 172 \text{ g AgCl} \quad \text{(Answer)}$$

normality
1 normal solution

Normality is another way of expressing the concentration of a solution. It is based on an alternative chemical unit of mass called the *equivalent weight*. The **normality** of a solution is the concentration expressed as the number of equivalent weights (equivalents) of solute per litre of solution. A **1 normal** (1 *N*) **solution** contains 1 equivalent weight of solute per litre of solution. Normality is widely used in analytical chemistry because it simplifies many of the calculations involving solution concentration.

Every substance may be assigned an equivalent weight equal either to its formula weight or to some small integral fractional part of its formula weight (that is, the formula weight divided by 2, 3, 4, and so on). To gain an understanding of the meaning of equivalent weight, let us start by considering these two reactions:

$$HCl(aq) + NaOH(aq) \longrightarrow NaCl(aq) + H_2O$$
$$\underset{(36.5\,g)}{\overset{1\,mole}{}} \qquad \underset{(40.0\,g)}{\overset{1\,mole}{}}$$

$$H_2SO_4(aq) + 2\,NaOH(aq) \longrightarrow Na_2SO_4(aq) + 2\,H_2O$$
$$\underset{(98.1\,g)}{\overset{1\,mole}{}} \qquad \underset{(80.0\,g)}{\overset{2\,moles}{}}$$

We note first that 1 mole of hydrochloric acid (HCl) reacts with 1 mole of sodium hydroxide (NaOH) and 1 mole of sulfuric acid (H_2SO_4) reacts with 2 moles of NaOH. If we make 1 molar solutions of these acids, 1 litre of 1 *M* HCl will react with 1 litre of 1 *M* NaOH, and 1 litre of 1 *M* H_2SO_4 will react with 2 litres of 1 *M* NaOH. From this, we can see that H_2SO_4 has twice the chemical capacity of HCl when reacting with NaOH. We can, however, adjust these acid solutions to be equal in reactivity by dissolving only 0.5 mole of H_2SO_4 per litre of solution. By doing this, we find that we are required to use 49.0 g of H_2SO_4 per litre (instead of 98.1 g of H_2SO_4 per litre) to make a solution that is equivalent to one made from 36.5 g of HCl per litre. These weights, 49.0 g of H_2SO_4 and 36.5 g of HCl, are chemically equivalent and are known as the equivalent weights of these substances, because they react with the same amount of NaOH (40.0 g). The equivalent weight of HCl is equal to its formula weight, but that of H_2SO_4 is one-half its formula weight. Table 14.4 summarizes these relationships.

TABLE 14.4 Comparison of molar and normal solutions of HCl and H_2SO_4 reacting with NaOH.

	Formula weight	Concentration	Volumes that react	Equivalent weight	Concentration	Volumes that react
HCl	36.5	1 *M*	1 litre	36.5	1 *N*	1 litre
NaOH	40.0	1 *M*	1 litre	40.0	1 *N*	1 litre
H_2SO_4	98.1	1 *M*	1 litre	49.0	1 *N*	1 litre
NaOH	40.0	1 *M*	2 litres	40.0	1 *N*	1 litre

Expressions for normality follow. Notice the similarity to the molar solution definition. N is the abbreviation for normality.

$$\text{Normality} = N = \frac{\text{Number of equivalents of solute}}{1 \text{ litre of solution}} = \frac{\text{Equivalents}}{\text{Litre}}$$

where

$$\text{Number of equivalents of solute} = \frac{\text{Grams of solute}}{\text{Equivalent weight (eq wt) of solute}}$$

Then

$$N = \frac{\text{g solute}}{\text{eq wt solute} \times \text{L solution}} = \frac{\text{g}}{\text{eq wt} \times \text{L}}$$

Thus, 1 litre of solution containing 36.5 g of HCl would be 1 N, and 1 litre of solution containing 49.0 g of H_2SO_4 would also be 1 N. A solution containing 98.1 g of H_2SO_4 (1 mole) per litre would be 2 N when reacting with NaOH in the foregoing equation.

Consider the following reactions, in which an excess of HCl is present. Hydrogen actually exists as H_2 molecules, but for convenience in considering the data, the hydrogen produced is shown as the number of moles (atomic weights) of hydrogen atoms (H^0) released per mole of metal reacting. Table 14.5 summarizes the pertinent data for these reactions.

$$Na(s) + HCl(aq) \longrightarrow NaCl(aq) + H^0(g)$$

$$Ca(s) + 2\,HCl(aq) \longrightarrow CaCl_2(aq) + 2\,H^0(g)$$

$$Al(s) + 3\,HCl(aq) \longrightarrow AlCl_3(aq) + 3\,H^0(g)$$

In each reaction, the equivalent weight of the reacting metal is the weight that reacts with 1 equivalent weight of the acid, liberates 1 mole of H atoms (1.0 g H),

TABLE 14.5 Equivalent weights of sodium, calcium, and aluminum in reaction with hydrochloric acid.

Metal	Atomic weight (1 mole)	Number of moles of hydrogen atoms liberated per mole of metal	Equivalent weight of metal
Na	23.0	1	$\frac{23.0}{1} = 23.0$
Ca	40.0	2	$\frac{40.1}{2} = 20.0$
Al	27.0	3	$\frac{27.0}{3} = 9.0$

23.0 g Na, 20.0 g Ca, and 9.0 g Al each reacting with hydrochloric acid produce 1.0 g H and are, therefore, equivalent to one another.

or involves the transfer of 1 mole of electrons in the reaction. One mole of Na metal lost 1 electron per atom in going to NaCl; 1 mole of Ca metal lost 2 electrons per atom in going to $CaCl_2$; 1 mole of Al metal lost 3 electrons per atom in going to $AlCl_3$. In each reaction, 1 mole of H^+ gained 1 electron per atom in going to free hydrogen.

$$\text{eq wt} = \frac{\text{at. wt Na}}{1} = \frac{\text{at. wt Ca}}{2} = \frac{\text{at. wt Al}}{3} = \frac{\text{at. wt H}}{1}$$

equivalent weight   Two definitions of **equivalent weight** can now be stated:

1. The equivalent weight is that weight of a substance that will react with, combine with, contain, replace, or in any other way be equivalent to 1 mole of hydrogen atoms (1.0 g H).
2. In oxidation–reduction reactions, the equivalent weight is that weight of a substance that loses or gains Avogadro's number of electrons.

The equivalent weight of a substance may be variable; its value is dependent on the reaction that the substance is undergoing. Consider the reactions represented by these equations:

$$NaOH(aq) + H_2SO_4(aq) \longrightarrow NaHSO_4(aq) + H_2O \tag{1}$$

$$2\,NaOH(aq) + H_2SO_4(aq) \longrightarrow Na_2SO_4(aq) + 2\,H_2O \tag{2}$$

In equation (1), 1 mole of sulfuric acid furnishes 1 mole of hydrogen ions. Therefore, the equivalent weight of sulfuric acid is its formula weight, namely 98.1 g. But in equation (2), 1 mole of H_2SO_4 furnishes 2 moles of hydrogen ions. Therefore, the equivalent weight of the sulfuric acid is one-half of its formula weight, or 49.0 g.

14.8 Colligative Properties of Solutions

When two solutions are prepared, one containing 1 mole (60.0 g) of urea (NH_2CONH_2) and the other containing 1 mole (342 g) of sucrose ($C_{12}H_{22}O_{11}$) in 1 kg of water, the freezing point of each solution is $-1.86°C$, not $0°C$ as in pure water. Urea and sucrose are very different substances, yet each lowers the freezing point of the water by the same amount. The only thing apparently common to these two solutions is that each contains 1 mole (6.02×10^{23} molecules) of solute and 1 kg of solvent. In fact, if we dissolved 1 mole of any other solute (provided that it is one that does not produce ions in solution) in 1 kg of water, the freezing point of the resulting solution would be $-1.86°C$.

This leads us to conclude that the freezing point depression for a solution containing 6.02×10^{23} solute molecules (particles) and 1 kg of water is a constant, namely $1.86°C$. Freezing point depression is a general property of solutions: When solutions are prepared from 1 kg of any solvent and 1 mole of any (non-ionized) solute, the freezing point will be lower than that of the pure solvent.

TABLE 14.6 **Freezing point depression and boiling point elevation constants of selected solvents.**

Solvent	Freezing point of pure solvent (°C)	Freezing point depression constant, K_f $\left(\dfrac{°C, kg\ solvent}{mole\ solute}\right)$	Boiling point of pure solvent (°C)	Boiling point elevation constant, K_b $\left(\dfrac{°C, kg\ solvent}{mole\ solute}\right)$
Water	0.00	1.86	100.0	0.52
Acetic acid	16.6	3.90	118.5	3.07
Benzene	5.5	5.1	80.1	2.53
Camphor	178	40	208.2	5.95

colligative properties

Furthermore, the amount by which the freezing point is depressed is the same for all solutions made with a given solvent; that is, each solvent shows a characteristic *freezing point depression constant*. Freezing point depression constants for several solvents are given in Table 14.6.

The solution formed by the addition of a nonvolatile solute to a solvent has a lower freezing point, a higher boiling point, and a lower vapor pressure than that of the pure solvent. All these effects are related and are known as colligative properties. The **colligative properties** are properties that depend only upon the number of solute atoms or molecules in a solution and not on the nature of those atoms or molecules. Freezing point depression, boiling point elevation, and vapor pressure lowering are colligative properties of solutions.

The colligative properties of a solution can be considered in terms of vapor pressure. The vapor pressure of a pure liquid depends on the tendency of molecules to escape from its surface. Thus, if 10% of the molecules in a solution are nonvolatile solute molecules, the vapor pressure of the solution is 10% lower than that of the pure solvent. The vapor pressure is lower because the surface of the solution contains 10% nonvolatile molecules and 90% volatile solvent molecules. A liquid boils when its vapor pressure equals the pressure of the atmosphere. Thus, we can see that the solution just described as having a lower vapor pressure will have a higher boiling point than the pure solvent. The solution with a lowered vapor pressure does not boil until it has been heated above the boiling point of the solvent (see Figure 14.9, page 346). Each solvent has its own characteristic *boiling point elevation constant* (see Table 14.6). The boiling point elevation constant is based on a solution that contains 1 mole of solute particles per kilogram of solvent. For example, the boiling point elevation for a solution containing 1 mole of solute particles per kilogram of water is 0.52°C. This means that this water solution will boil at 100.52°C.

The freezing behavior of a solution can also be considered in terms of lowered vapor pressure. Figure 14.9 shows the vapor pressure relationships of ice, water, and a solution containing 1 mole of solute per kilogram of water. The freezing point of water is at the intersection of the water and ice vapor pressure curves; that is, at the point where water and ice have the same vapor pressure. Because the vapor pressure of water is lowered by the solute, the vapor pressure curve of the solution does not intersect the vapor pressure curve of ice until the

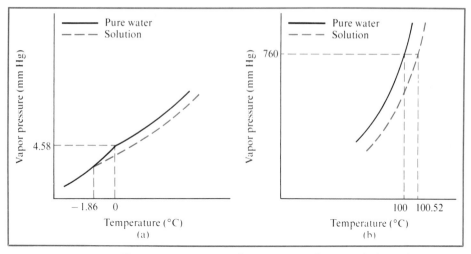

Figure 14.9   Vapor pressure curves of pure water and water solutions, showing (a) freezing point depression and (b) boiling point elevation effects. (Concentration: 1 mole of solute per kilogram of water.)

solution has been cooled below the freezing point of pure water. Thus, it is necessary to cool the solution below 0°C in order to form ice.

The foregoing discussion dealing with freezing point depressions is restricted to *un-ionized* substances. The discussion of boiling point elevations is restricted to *nonvolatile* and un-ionized substances. The colligative properties of ionized substances (electrolytes; see Chapter 15) are not explored at this point.

Some practical applications involving colligative properties are (1) use of salt–ice mixtures to provide low freezing temperatures for homemade ice cream, (2) use of salt or calcium chloride to melt ice from streets, (3) use of ethylene glycol and water mixtures as antifreeze in automobile radiators (ethylene glycol also raises the boiling point of radiator fluid and thus allows the engine to operate at a higher temperature).

Both the freezing point depression and the boiling point elevation are directly proportional to the number of moles of solute per kilogram of solvent. When we deal with the colligative properties of solutions, another concentration expression, *molality,* is used. The **molality** (*m*) of a solute is the number of moles of solute per kilogram of solvent:

molality

$$m = \frac{\text{moles solute}}{\text{kg solvent}}$$

Note that a lowercase *m* is used for molality concentrations, whereas a capital *M* is used for molarity. The difference between molality and molarity is that molality refers to moles of solute *per kilogram of solvent,* whereas molarity refers to moles of solute *per litre of solution.* For un-ionized substances, the colligative properties of a solution are directly proportional to its molality.

The following equations show the relationship of freezing point depression, molecular weight, and solution concentration. The symbol Δt_f indicates the change in the freezing point of the solution with respect to the freezing point of the pure solvent; K_f is the freezing point depression constant of the solvent.

$$\Delta t_f = K_f m \qquad \text{or} \qquad m = \frac{\Delta t_f}{K_f}$$

$$\Delta t_f = K_f \times \frac{\text{moles solute}}{\text{kg solvent}} = K_f \times \frac{\text{g solute}}{\text{mol. wt solute}} \times \frac{1}{\text{kg solvent}}$$

These equations are commonly used to calculate the freezing points of solutions and the molecular weights of compounds. For boiling point elevation calculations, substitute Δt_b for Δt_f and K_b for K_f, where Δt_b is the observed boiling point elevation and K_b is the boiling point elevation constant of the solvent.

PROBLEM 14.14 What is the molality (m) of a solution prepared by dissolving 2.70 g CH_3OH in 25.0 g H_2O?

Since   Molality $= \dfrac{\text{moles solute}}{\text{kg solvent}}$,   the conversion is

$$\frac{2.70 \text{ g } CH_3OH}{25.0 \text{ g } H_2O} \longrightarrow \frac{\text{Moles } CH_3OH}{25.0 \text{ g } H_2O} \longrightarrow \frac{\text{Moles } CH_3OH}{1 \text{ kg } H_2O}$$

The formula weight for CH_3OH is 32.0 (12.0 + 4.0 + 16.0).

$$\frac{2.70 \text{ g } CH_3OH}{25.0 \text{ g } H_2O} \times \frac{1 \text{ mole } CH_3OH}{32.0 \text{ g } CH_3OH} \times \frac{1000 \text{ g } H_2O}{1 \text{ kg } H_2O} = \frac{3.38 \text{ moles } CH_3OH}{1 \text{ kg } H_2O}$$

The molality is 3.38 m.

■

PROBLEM 14.15 A solution is made by dissolving 100 g of ethylene glycol ($C_2H_6O_2$) in 200 g of water. What is the freezing point of this solution?

To calculate the freezing point of the solution, we first need to calculate Δt_f, the change in freezing point.

Use the equation $\Delta t_f = K_f \times \dfrac{\text{moles solute}}{\text{kg solvent}}$

K_f (for water): $\dfrac{1.86°C \text{ kg solvent}}{\text{mole solute}}$ (from Table 14.6)

Moles solute: $100 \text{ g } C_2H_6O_2 \times \dfrac{1 \text{ mole } C_2H_6O_2}{62.0 \text{ g } C_2H_6O_2} = 1.61 \text{ moles } C_2H_6O_2$

kg solvent: $200 \text{ g } H_2O \times \dfrac{1 \text{ kg}}{1000 \text{ g}} = 0.200 \text{ kg } H_2O$

$$\Delta t_f = \frac{1.86°C \text{ kg } H_2O}{\text{mole } C_2H_6O_2} \times \frac{1.61 \text{ mole } C_2H_6O_2}{0.200 \text{ kg } H_2O} = 15.0°C$$

Since 15.0°C is the freezing point depression, it must be subtracted from 0°C, the freezing point of the pure solvent. Therefore, the freezing point of the solution is −15°C. The calculation can also be done using the equation

$$\Delta t_f = K_f \times \frac{\text{g solute}}{\text{mol. wt solute}} \times \frac{1}{\text{kg solvent}}$$

■

PROBLEM 14.16   A solution made by dissolving 3.25 g of a compound of unknown molecular weight in 100.0 g of water has a freezing point of −1.46°C. What is the molecular weight of the compound?

Solve the equation

$$\Delta t_f = K_f \times \frac{\text{g solute}}{\text{mol. wt solute}} \times \frac{1}{\text{kg solvent}}$$

for molecular weight:

$$\text{mol. wt solute} = K_f \times \frac{\text{g solute}}{\Delta t_f} \times \frac{1}{\text{kg solvent}}$$

A freezing point of −1.46°C in an aqueous solution means that $\Delta t_f = 1.46$°C, since pure water freezes at 0°C.

$$K_f = \frac{1.86°\text{C kg H}_2\text{O}}{\text{mole solute}} \qquad \text{kg H}_2\text{O} = 0.100 \text{ kg H}_2\text{O}$$

$$\text{mol. wt} = \frac{1.86°\text{C kg H}_2\text{O}}{\text{mole solute}} \times \frac{3.25 \text{ g}}{1.46°\text{C}} \times \frac{1}{0.100 \text{ kg H}_2\text{O}} = 41.4 \text{ g/mole} \quad \text{(Answer)}$$

■

QUESTIONS   A. **Review the Meanings of the New Terms Introduced in this Chapter**

1. Solution	11. Unsaturated solution
2. Solute	12. Supersaturated solution
3. Solvent	13. 1 molar solution
4. Solubility	14. Molarity
5. Miscible	15. Normality
6. Immiscible	16. 1 normal solution
7. Concentration of a solution	17. Equivalent weight
8. Dilute solution	18. Colligative properties
9. Concentrated solution	19. Molality
10. Saturated solution	

B. **Answers to the Following Questions Will Be Found in Tables and Figures**

1. Make a sketch indicating the orientation of water molecules (a) about a single sodium ion and (b) about a single chloride ion in solution. (See Figure 14.2.)
2. Which of the following salts are insoluble: $CaCO_3$, $Pb(OH)_2$, $ZnCl_2$, Hg_2Cl_2, $(NH_4)_3AsO_4$, $Bi(NO_3)_3$, AgI, K_3PO_4, $NaBr$, $PbSO_4$? (See Table 14.1.)

3. What is different in the solubility trend of the potassium halides compared to that of the lithium halides and the sodium halides? (See Table 14.2.)
4. What is the solubility, in grams of solute per 100 g H_2O, of (a) $KClO_3$ at 60°C; (b) HCl at 20°C; (c) Li_2SO_4 at 80°C; and (d) KNO_3 at 0°C? (See Figure 14.3.)
5. List one substance shown in Figure 14.3 whose solubility decreases with increased temperature.
6. Which substance, KNO_3 or NH_4Cl, shows the greater increase in solubility with increased temperature? (See Figure 14.3.)
7. What would be the total surface area if the 1 cm cube in Figure 14.4 were cut into cubes 0.01 cm on a side?
8. Would the volumetric flask shown in Figure 14.7 be satisfactory for preparing normal solutions? Explain.
9. How are the equivalent weights of the metals in Table 14.5 related to their electron structure?
10. How many gram-equivalent weights are there in 1 mole of H_2SO_4? (See Table 14.4.)
11. Which solution, a 2 molal solution in benzene or a 1 molal solution in camphor, will show the greater freezing point depression? (See Table 14.6.)

C. Review Questions

1. Name and distinguish between the two components of a solution.
2. Is it always apparent in a solution which component is the solute, for example, in a solution of a liquid in a liquid?
3. Explain why the solute does not settle out of a solution.
4. Is it possible to have one solid dissolved in another? Explain.
5. An aqueous solution of KCl is colorless, $KMnO_4$ is purple, and $K_2Cr_2O_7$ is orange. What color would you expect of an aqueous solution of $Na_2Cr_2O_7$? Explain.
6. Explain why carbon tetrachloride will dissolve benzene but will not dissolve sodium chloride.
7. Some drinks like tea are consumed either hot or cold, whereas others like Coca-Cola are drunk only cold. Why?
8. Why is air considered to be a solution?
9. In which will a teaspoonful of sugar dissolve more rapidly, 200 mL of iced tea or 200 mL of hot coffee? Explain in terms of the KMT.
10. What is the effect of pressure on the solubility of gases in liquids? Solids in liquids?
11. Why do smaller particles dissolve faster than large ones?
12. In a saturated solution containing undissolved solute, solute is continually dissolving but the concentration of the solution remains unchanged. Explain.
13. Explain why there is no apparent reaction when crystals of $AgNO_3$ and NaCl are mixed, but a reaction is apparent immediately when solutions of $AgNO_3$ and NaCl are mixed.
14. What do we mean when we say that concentrated nitric acid (HNO_3) is 16 molar?
15. Will 1 litre of 1 molar NaCl contain more chloride ions than 0.5 litre of 1 molar $MgCl_2$? Explain.
16. What disadvantages are there in expressing the concentration of solutions as dilute or concentrated?
17. Explain how concentrated H_2SO_4 can be both 18 molar and 36 normal in concentration.
18. Describe how you would prepare 750 mL of 5 molar NaCl solution.
19. Arrange the following bases (in descending order), according to the volume of each that will react with 1 litre of 1 M HCl: (a) 1 M NaOH; (b) 1.5 M Ca(OH)$_2$; (c) 2 M KOH; (d) 0.6 M Ba(OH)$_2$.

20. Explain in terms of vapor pressure why the boiling point of a solution containing a nonvolatile solute is higher than that of the pure solvent.

21. Explain why the freezing point of a solution is lower than the freezing point of the pure solvent.

22. Which would be colder, a glass of water and crushed ice or a glass of Seven-Up and crushed ice? Explain.

23. When water and ice are mixed, the temperature of the mixture is 0°C. But if methyl alcohol and ice are mixed, a temperature of −10°C is readily attained. Explain why the two mixtures show such different temperature behavior.

24. Which would be more effective in lowering the freezing point of 500 g of water?

    (a) 100 g of sucrose ($C_{12}H_{22}O_{11}$) or 100 g of ethyl alcohol (C_2H_5OH)

    (b) 100 g of sucrose or 20.0 g of ethyl alcohol

    (c) 20.0 g of ethyl alcohol or 20.0 g of methyl alcohol (CH_3OH)

25. Is the molarity of a 5 molal aqueous solution of NaCl greater or less than 5 molar? Explain.

26. Explain the fact that when hydrogen chloride gas, HCl, is bubbled into a saturated sodium chloride solution, NaCl crystallizes out of solution.

27. Express, in terms of its molarity, the normality of an H_2SO_4 solution used to titrate NaOH solutions. (Assume both hydrogens react.)

28. Which of the following statements are correct?

    (a) A solution is a homogeneous mixture.

    (b) It is possible for the same substance, in two different solutions, to be the solvent in one solution and the solute in the other.

    (c) A solute can be removed from a solution by filtration.

    (d) Saturated solutions are always concentrated solutions.

    (e) If a solution of sugar in water is allowed to stand undisturbed for a long period of time, the sugar will gradually settle to the bottom of the container.

    (f) The proportions in which two liquids are mixed sometimes determine whether the resulting mixture is miscible or immiscible.

    (g) Gases are generally more soluble in hot water than in cold water.

    (h) It is impossible to prepare a two-phase liquid mixture from two liquids that are miscible with each other in all proportions.

    (i) A solution that is 10% NaCl by weight always contains 10 g of NaCl.

    (j) Small changes in pressure have little effect on the solubility of solids in liquids but a marked effect on the solubility of gases in liquids.

    (k) How fast a solute dissolves depends mainly on the size of the solute particles, the temperature of the solvent, and the degree of agitation or stirring taking place.

    (l) In order to have a 1 molar solution you must have 1 mole of solute dissolved in sufficient solvent to give 1 litre of solution.

    (m) Dissolving 1 mole of NaCl in 1 litre of water will give a 1 molar solution.

    (n) One mole of solute in 1 litre of solution has the same concentration as 0.1 mole of solute in 100 mL of solution.

    (o) When 100 mL of 0.200 M HCl is diluted to 200 mL volume by the addition of water, the resulting solution is 0.100 M and contains one-half the number of moles of HCl as were in the original solution.

    (p) Fifty millilitres of 0.1 M H_2SO_4 will neutralize the same volume of 0.1 M NaOH as 100 mL of 0.1 M HCl.

    (q) Fifty millilitres of 0.1 N H_2SO_4 will neutralize the same volume of 0.1 M NaOH as 100 mL of 0.1 M HCl.

    (r) An aqueous solution that freezes below 0°C will have a normal boiling point below 100°C.

    (s) The colligative properties of a solution depend on the number of solute particles dissolved in solution.

D. Review Problems

1. How many more grams of NaBr will dissolve in 100 g of water at 100°C than at 0°C? (See Table 14.2)

2. How many grams of $KClO_3$ would be required to saturate 150 g of water at 50°C? (See Table 14.3.)

3. How many grams of a solution that is 15.0% by weight $AgNO_3$ would contain the following?
   (a) 30.0 g of $AgNO_3$ (b) 0.400 mole of $AgNO_3$

4. Calculate the weight percent of the following solutions:
   (a) 50.0 g NaCl + 200.0 g H_2O
   (b) 0.50 mole $HC_2H_3O_2$ + 3.0 moles H_2O
   (c) 1.0 molal solution of $C_{12}H_{22}O_{11}$ in water

5. How much solute is present in each of the following?
   (a) 75 g of 5.0% KCl solution
   (b) 150 g of 15.0% NaCl solution
   *(c) A solution that contains 100.0 g of water, and is 7.0% by weight sodium bicarbonate, $NaHCO_3$

6. Physiological saline solutions used in intravenous injections have a concentration of 0.90% NaCl by weight.
   (a) How many grams of NaCl are needed to prepare 300 g of this solution?
   *(b) How much water must evaporate from this solution to give a solution that is 9.0% NaCl by weight?

7. A solution is made from 45.0 g of KNO_3 and 150 g of H_2O. How many grams of water must evaporate to give a saturated solution of KNO_3 in water at 20°C? (See Figure 14.3.)

8. How many grams of solution, 10% NaOH by weight, are required to neutralize 100 mL of a 1.0 M HCl solution?

*9. How many grams of solution, 10% NaOH by weight, are required to neutralize 500 g of a 1.0 molal solution of HCl?

*10. A sugar syrup solution contains 22.0% sugar, $C_{12}H_{22}O_{11}$, by weight and has a density of 1.09 g/mL.
   (a) How many grams of sugar are in 1.0 L of this syrup?
   (b) What is the molarity of this solution?
   (c) What is the molality of this solution?

11. At 20°C an aqueous solution of H_2SO_4 that is 35% H_2SO_4 by weight has a density of 1.26 g/mL.
   (a) How many grams of H_2SO_4 are present in 1.00 L of this solution?
   (b) What volume of this solution will contain 500 g of H_2SO_4?

12. Calculate the molarity of the following solutions:
   (a) 0.10 mole solute in 200 mL of solution
   (b) 3.0 moles of NaCl in 0.650 L of solution
   (c) 0.015 mole of HCl in 10 mL of solution
   (d) 0.25 mole $BaCl_2 \cdot 2H_2O$ in 593 mL of solution

13. Calculate the molarity of the following solutions:
   (a) 53.0 g Na_2CO_3 in 1.00 L of solution
   (b) 400 g of $C_6H_{12}O_6$ in 800 mL of solution
   (c) 1.00 g $Al_2(SO_4)_3$ in 2.00 L of solution
   (d) 0.0170 g $Ca(NO_3)_2$ in 1.00 mL of solution

14. Calculate the number of moles of solute in each of the following solutions:
   (a) 25.0 L of 1.0 M LiCl solution   (c) 249 mL of 0.0010 M NaOH
   (b) 25.0 mL of 6.0 M H_2SO_4 (d) 5000 mL of 3.1 M $AgNO_3$

15. Calculate the grams of solute in each of the following solutions:
   (a) 100 L of 1.0 M NaCl   (c) 160 mL of 18 M H_2SO_4
   (b) 0.025 L of 10.0 M HCl   (d) 7.00 mL of 7.00 M $Na_2C_2O_4$

16. How many millilitres of 0.371 M KCl solution will contain the following?
    (a) 0.430 mole of KCl   (c) 20.0 g of KCl
    (b) 10.0 moles of KCl   *(d) 71.0 g of chloride ion, Cl$^-$

*17. What is the molarity of a sulfuric acid solution if the solution is 35.0% H_2SO_4 by weight and has a density of 1.26 g/mL?

18. What will be the molarity of the resulting solutions made by mixing the following? (Assume volumes are additive.)
    (a) 100 mL 12 M HCl + 100 mL H_2O
    (b) 70.0 mL 0.60 M $ZnSO_4$ + 500 mL H_2O
    (c) 100 mL 1.0 M HCl + 100 mL 2.0 M HCl

19. Calculate the volume of concentrated reagent required to prepare the diluted solutions indicated:
    (a) 12 M HCl to prepare 500 mL of 6.0 M HCl
    (b) 15 M NH_3 to prepare 40 mL of 6.0 M NH_3
    (c) 16 M HNO_3 to prepare 100 mL of 1.0 M HNO_3
    (d) 18 M H_2SO_4 to prepare 350 mL of 10.0 N H_2SO_4

20. To what volume must a solution of 50.0 g of H_2SO_4 in 600 mL of water be diluted to give a 0.10 M solution?

21. What will be the molarity of each of the solutions made by mixing 250 mL of 0.75 M H_2SO_4 with (a) 200 mL H_2O, (b) 200 mL of 0.70 M H_2SO_4, and (c) 400 mL of 1.50 M H_2SO_4?

22. How many millilitres of water must be added to 200 mL of 1.40 M HCl to make a solution that is 0.500 M HCl?

23. Given the balanced equation,

$$6\,FeCl_2 + K_2Cr_2O_7 + 14\,HCl \longrightarrow 6\,FeCl_3 + 2\,CrCl_3 + 2\,KCl + 7\,H_2O$$

    (a) How many moles of KCl will be produced from 1.0 mole of $FeCl_2$?
    (b) How many moles of $CrCl_3$ will be produced from 2.0 moles of $FeCl_2$?
    (c) How many moles of $FeCl_2$ will react with 0.040 mole of $K_2Cr_2O_7$?
    (d) How many millilitres of 0.060 M $K_2Cr_2O_7$ will react with 0.025 mole of $FeCl_2$?
    (e) How many millilitres of 6.0 M HCl will react with 10.0 mL of 8.0 M $FeCl_2$?

24. $2\,KMnO_4 + 16\,HCl \longrightarrow 2\,MnCl_2 + 5\,Cl_2 + 8\,H_2O + 2\,KCl$

    Calculate the following using the above equation:
    (a) The moles of Cl_2 produced from 0.50 mole of $KMnO_4$
    (b) The moles of HCl required to react with 1.0 L of 2.0 M $KMnO_4$
    (c) The millilitres of 6.0 M HCl required to react with 500 mL of 0.40 M $KMnO_4$
    (d) The litres of Cl_2 gas at STP produced by the reaction of 50.0 mL of 5.0 M HCl

25. $BaCl_2(aq) + K_2CrO_4(aq) \longrightarrow BaCrO_4(s) + 2\,KCl(aq)$

    Using the above equation, calculate the following:
    (a) The grams of $BaCrO_4$ that can be obtained from 100 mL of 0.300 M $BaCl_2$
    (b) The volume of 1.0 M $BaCl_2$ solution needed to react with 50.0 mL of 0.300 M K_2CrO_4 solution

26. (a) How many moles of hydrogen will be liberated from 200 mL of 2.00 M HCl reacting with an excess of magnesium? The equation is

$$Mg(s) + 2\,HCl(aq) \longrightarrow MgCl_2(aq) + H_2(g)$$

(b) How many litres of hydrogen gas, H_2, measured at 27°C and 745 mm Hg, will be obtained? [*Hint:* Use the ideal gas equation.]

*27. What is the molarity of an HCl solution, 130 mL of which, when treated with excess magnesium, liberates 2.50 L of H_2 gas measured at STP?

28. On an equal mass basis, which base, $Mg(OH)_2$ or $Al(OH)_3$, is more effective in neutralizing excess stomach acid, HCl?

29. Calculate the equivalent weight of the acid and base in each of the following reactions:
(a) $HCl + NaOH \longrightarrow NaCl + H_2O$
(b) $2\,HCl + Ba(OH)_2 \longrightarrow BaCl_2 + 2\,H_2O$
(c) $H_2SO_4 + Ca(OH)_2 \longrightarrow CaSO_4 + 2\,H_2O$
(d) $H_2SO_4 + KOH \longrightarrow KHSO_4 + H_2O$
(e) $H_3PO_4 + 2\,LiOH \longrightarrow Li_2HPO_4 + 2\,H_2O$

30. Calculate the equivalent weight of the metal in each of the following reactions:
(a) $2\,K + 2\,H_2O \longrightarrow 2\,KOH + H_2\uparrow$
(b) $Mg + H_2O \overset{\Delta}{\longrightarrow} MgO + H_2\uparrow$
(c) $2\,Fe + 6\,HBr \longrightarrow 2\,FeBr_3 + 3\,H_2\uparrow$
(d) $Zn + CuSO_4 \longrightarrow ZnSO_4 + Cu$
(e) $Cu_2O + H_2 \longrightarrow 2\,Cu + H_2O$
(f) $2\,Ga + 3\,Cl_2 \longrightarrow 2\,GaCl_3$

31. (a) What is the molality of a solution of 1.20 g of urea, CH_4ON_2, in 26.0 g of water?
(b) At what temperature would the solution freeze?

*32. What would be (a) the boiling point and (b) the molality of an aqueous sugar ($C_{12}H_{22}O_{11}$) solution that freezes at $-4.4°C$?

33. (a) What is the freezing point of a solution containing 50.0 g of ethylene glycol, $C_2H_6O_2$, in 150.0 g of water?
(b) What is the boiling point of this solution?
(c) What is the molality of this solution?

34. What is (a) the molality, (b) the freezing point, and (c) the boiling point of a solution containing 1.68 g of naphthalene, $C_{10}H_8$, in 18.4 g of benzene?

*35. The freezing point of a solution of 9.00 g of an unknown compound dissolved in 40.0 g of acetic acid is 13.2°C. Calculate the molecular weight of the compound.

*36. A solution of 3.84 g of C_4H_2N (empirical formula) in 250 g of benzene depresses the freezing point of benzene 0.614°C. What is the molecular formula of the compound?

37. A solution of 6.20 g of $C_2H_6O_2$ in water has a freezing point of $-0.186°C$. How many grams of H_2O are in the solution?

38. What (a) weight and (b) volume of ethylene glycol ($C_2H_6O_2$, density = 1.11 g/mL) should be added to 12.0 L of water in an automobile radiator to protect it from freezing at $-15.0°C$? (c) To what temperature Fahrenheit will the radiator be protected?

15

Ionization.
Acids, Bases, Salts

After studying Chapter 15 you should be able to:

1. Understand the terms listed in Question A at the end of the chapter.
2. Define an acid and a base in terms of Arrhenius, Brønsted–Lowry, and Lewis theories.
3. Identify conjugate acid–base pairs in an equation.
4. When given the reactants, complete and balance equations for the reactions of acids with metals, bases, metal oxides, and carbonates.
5. When given the reactants, complete and balance equations for the reaction of an amphoteric hydroxide with either a strong acid or a strong base.
6. Write balanced equations for the reaction of sodium hydroxide or potassium hydroxide with zinc and with aluminum.
7. Classify common compounds as electrolytes or nonelectrolytes.
8. Write equations for the dissociation or ionization of acids, bases, and salts in water.
9. Given the molarity of a salt solution, determine the molarity of the ions in solution.
10. Understand the distinction between strong and weak electrolytes.
11. Describe and write equations for the ionization of water.
12. Calculate the pH of a solution when given the hydrogen ion concentration.
13. Determine the hydrogen ion concentration from integral values of pH.
14. Calculate the molarity or volume of an acid or base solution from appropriate titration data.
15. Write balanced un-ionized, total ionic, and net ionic equations for neutralization and other reactions.

15.1 Acids and Bases

The word *acid* is derived from the Latin *acidus,* meaning "sour" or "tart," and is also related to the Latin word *acetum,* meaning "vinegar." Vinegar has been known since antiquity as the product of the fermentation of wine and of apple cider. The sour constituent of vinegar is acetic acid ($HC_2H_3O_2$).

Some of the characteristic properties commonly associated with acids are the following: Water solutions of acids are sour to the taste and are capable of changing the color of litmus, a vegetable dye, from blue to red. Water solutions of nearly all acids are able to react with: (1) metals such as zinc and magnesium to produce hydrogen gas; (2) bases to produce water and a salt; and

(3) carbonates to produce carbon dioxide. These properties are due to hydrogen ions, H^+, released by the acid in a water solution.

Classically, a *base* is a substance capable of liberating hydroxide ions, OH^-, in water solution. Hydroxides of the alkali metals (Group IA) and alkaline earth metals (Group IIA), such as $LiOH$, $NaOH$, KOH, $Ca(OH)_2$, and $Ba(OH)_2$, are the most common inorganic bases. Water solutions of bases are called *alkaline solutions* or *basic solutions.* They have the following properties: a bitter or caustic taste; a slippery, soapy feeling; the ability to change litmus from red to blue; and the ability to interact with acids to form a salt and water.

Several theories have been proposed to answer the question "What are acids and bases?" One of the earliest, most significant of these theories was advanced in a doctoral thesis in 1884 by Svante Arrhenius (1859–1927), a Swedish scientist, who stated that an acid is a hydrogen-containing substance that dissociates to produce hydrogen ions, and that a base is a hydroxide-containing substance that dissociates to produce hydroxide ions in aqueous solutions. Arrhenius postulated that the hydrogen ions were produced by the dissociation of acids in water, and that hydroxide ions were produced by the dissociation of bases in water.

$$HA \longrightarrow H^+ + A^-$$
Acid

$$MOH \longrightarrow M^+ + OH^-$$
Base

Thus, an acid solution contains an excess of hydrogen ions and a base solution contains an excess of hydroxide ions.

In 1923, the Brønsted–Lowry proton transfer theory was introduced by J. N. Brønsted, a Danish chemist (1879–1947), and T. M. Lowry, an English chemist (1874–1936). This theory states that an acid is a proton donor and a base is a proton acceptor. A proton is a hydrogen ion, H^+ (a hydrogen atom minus its electron).

Consider the reaction of hydrogen chloride gas with water to form hydrochloric acid:

$$HCl(g) + H_2O(l) \longrightarrow H_3O^+(aq) + Cl^-(aq) \tag{1}$$

In the course of the reaction, HCl donates, or gives up, a proton to form a Cl^- ion and H_2O accepts a proton to form the H_3O^+ ion. Thus, HCl is an acid and H_2O is a base, according to the Brønsted–Lowry theory.

A hydrogen ion, H^+, is nothing more than a bare proton and does not exist by itself in an aqueous solution. In water, a proton combines with a polar water molecule to form a hydrated hydrogen ion, H_3O^+ [$H(H_2O)^+$], commonly called a **hydronium ion.** The proton is attracted to a polar water molecule, forming a bond with one of the two pairs of unshared electrons:

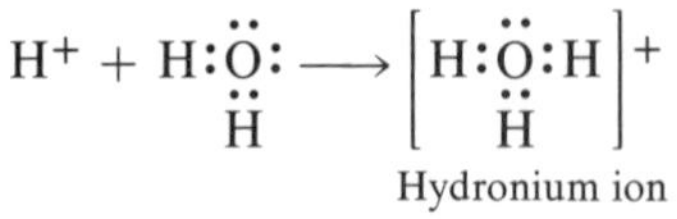

hydronium ion

$$H^+ + H\!:\!\overset{\cdot\cdot}{\underset{\cdot\cdot}{O}}\!:\ \longrightarrow \left[H\!:\!\overset{\cdot\cdot}{\underset{\cdot\cdot}{O}}\!:\!H\right]^+$$
$$\ H\phantom{:\overset{\cdot\cdot}{\underset{\cdot\cdot}{O}}:\ \longrightarrow\ \ }H$$

Hydronium ion

Note the electron structure of the hydronium ion. For simplicity of expression in equations, we often use H^+ instead of H_3O^+, with the explicit understanding that H^+ is always hydrated in solution.

Whereas the Arrhenius theory is restricted to aqueous solutions, the Brønsted–Lowry approach has application in all media and has become the more important theory when the chemistry of substances in solutions other than water is studied. Ammonium chloride (NH_4Cl) is a salt, yet its water solution has an acidic reaction. From this test we must conclude that NH_4Cl has acidic properties. The Brønsted–Lowry explanation shows that the ammonium ion, NH_4^+, is a proton donor, and water is the proton acceptor:

$$\underset{\text{Acid}}{NH_4^+} \rightleftharpoons \underset{\text{Base}}{NH_3} + \underset{\text{Acid}}{H^+} \tag{2}$$

$$\underset{\text{Acid}}{NH_4^+} + \underset{\text{Base}}{H_2O} \longrightarrow \underset{\text{Acid}}{H_3O^+} + \underset{\text{Base}}{NH_3} \tag{3}$$

The Brønsted–Lowry theory also applies to certain cases where no solution is involved. For example, in the reaction of hydrogen chloride and ammonia gases, HCl is the proton donor and NH_3 is the base. [Remember that "(g)" after a formula in equations stands for a gas.]

$$\underset{\text{Acid}}{HCl(g)} + \underset{\text{Base}}{NH_3(g)} \longrightarrow \underset{\text{Acid}}{NH_4^+} + \underset{\text{Base}}{Cl^-} \tag{4}$$

In equations (1), (3), and (4), a conjugate acid and base are produced as products. The formulas of a conjugate acid–base pair differ by one proton (H^+). In equation (1) the conjugate acid–base pairs are HCl–Cl^- and H_3O^+–H_2O. Cl^- is the conjugate base of HCl, and HCl is the conjugate acid of Cl^-. H_2O is the conjugate base of H_3O^+, and H_3O^+ is the conjugate acid of H_2O.

$$\overset{\text{Conjugate acid–base pair}}{\overbrace{HCl(g) + H_2O(l)}} \longrightarrow Cl^-(aq) + H_3O^+(aq)$$

Conjugate acid–base pair

In equation (4) the conjugate acid–base pairs are HCl–Cl^- and NH_4^+–NH_3.

PROBLEM 15.1 Write the formula for (a) the conjugate base of H_2O and HNO_3, and (b) the conjugate acid of SO_4^{2-} and $C_2H_3O_2^-$. The difference between an acid or a base and its conjugate is one proton, H^+.

(a) To write the conjugate base of an acid, remove one proton from the acid formula. Thus,

$$H_2O \xrightarrow{-H^+} OH^- \quad \text{(Conjugate base)}$$

$$HNO_3 \xrightarrow{-H^+} NO_3^- \quad \text{(Conjugate base)}$$

Note that by removing a H^+, the conjugate base becomes more negative than the acid by one minus charge.

(b) To write the conjugate acid of a base, add one proton to the formula of the base. Thus,

$$SO_4^{2-} \xrightarrow{+H^+} HSO_4^- \quad \text{(Conjugate acid)}$$

$$C_2H_3O_2^- \xrightarrow{+H^+} HC_2H_3O_2 \quad \text{(Conjugate acid)}$$

In each case the conjugate acid becomes more positive than the base by one positive charge due to the addition of H^+.

■

A more general concept of acids and bases was introduced by Gilbert N. Lewis (1875–1946). The Lewis theory deals with the way in which a substance with an unshared pair of electrons reacts in an acid–base type of reaction. According to this theory a base is any substance that has an unshared pair of electrons (electron-pair donor) and an acid is any substance that will attach itself to or accept a pair of electrons. In the reaction

$$H^+ + :\overset{\displaystyle H}{\underset{\displaystyle H}{N}}:H \longrightarrow H:\overset{\displaystyle H}{\underset{\displaystyle H}{N}}:H^+$$

Acid   Base

H^+ is a Lewis acid and $:NH_3$ is a Lewis base. According to the Lewis theory, substances other than proton donors (for example, BF_3) behave as acids:

$$F:\overset{\displaystyle F}{\underset{\displaystyle F}{B}} + :\overset{\displaystyle H}{\underset{\displaystyle H}{N}}:H \longrightarrow F:\overset{\displaystyle F}{\underset{\displaystyle F}{B}}:\overset{\displaystyle H}{\underset{\displaystyle H}{N}}:H$$

Acid   Base

The Lewis and Brønsted–Lowry bases are identical, because to accept a proton a base must have an unshared pair of electrons.

The three theories are summarized in Table 15.1. These theories explain how acid–base reactions occur. We will generally use the theory that best

TABLE 15.1 Summary of acid–base definitions according to Arrhenius, Brønsted–Lowry, and G. N. Lewis theories.

Theory	Acid	Base
Arrhenius	A hydrogen-containing substance that produces hydrogen ions in aqueous solution	A hydroxide-containing substance that produces hydroxide ions in aqueous solution
Brønsted–Lowry	A proton (H^+) donor	A proton (H^+) acceptor
Lewis	Any species that will bond to an unshared pair of electrons (electron-pair acceptor)	Any species that has an unshared pair of electrons (electron-pair donor)

explains the reaction that is under consideration. Most of our examples will refer to aqueous solutions. It is important to realize that in an aqueous acidic solution, the H^+ ion concentration is always greater than the OH^- ion concentration. And, vice versa, in an aqueous basic solution, the OH^- ion concentration is always greater than the H^+ ion concentration. When the H^+ and OH^- ion concentrations in a solution are equal, the solution is neutral; that is, it is neither acidic nor basic.

15.2 Reactions of Acids

In aqueous solutions it is the H^+ or H_3O^+ ions that are responsible for the characteristic reactions of acids. All the following reactions are in an aqueous medium.

Reaction with metals.   Acids react with metals that lie above hydrogen in the activity series of elements to produce hydrogen and a salt (see Section 17.5).

Acid + Metal $\longrightarrow$ Hydrogen + Salt

$2\ HCl(aq) + Ca(s) \longrightarrow H_2\uparrow + CaCl_2(aq)$

$H_2SO_4(aq) + Mg(s) \longrightarrow H_2\uparrow + MgSO_4(aq)$

$6\ HC_2H_3O_2(aq) + 2\ Al(s) \longrightarrow 3\ H_2\uparrow + 2\ Al(C_2H_3O_2)_3(aq)$

Acids such as nitric acid (HNO_3) are oxidizing substances (see Chapter 17) and react with metals to produce water instead of hydrogen. For example,

$3\ Zn(s) + 8\ HNO_3(\text{dilute}) \longrightarrow 3\ Zn(NO_3)_2(aq) + 2\ NO(g) + 4\ H_2O$

Reaction with bases.   The interaction of an acid and a base is called a *neutralization reaction.* In aqueous solutions, the products of this reaction are water and a salt.

Acid + Base $\longrightarrow$ Salt + Water

$HBr(aq) + KOH(aq) \longrightarrow KBr(aq) + H_2O$

$2\ HNO_3(aq) + Ca(OH)_2(aq) \longrightarrow Ca(NO_3)_2(aq) + 2\ H_2O$

$2\ H_3PO_4(aq) + 3\ Ba(OH)_2(aq) \longrightarrow Ba_3(PO_4)_2\downarrow + 6\ H_2O$

Reaction with metal oxides.   This reaction is closely related to that of an acid with a base. With an aqueous acid, the products are water and a salt.

Acid + Metal oxide $\longrightarrow$ Salt + Water

$2\ HCl(aq) + Na_2O(s) \longrightarrow 2\ NaCl(aq) + H_2O$

$H_2SO_4(aq) + MgO(s) \longrightarrow MgSO_4(aq) + H_2O$

$6\ HCl(aq) + Fe_2O_3(s) \longrightarrow 2\ FeCl_3(aq) + 3\ H_2O$

Reaction with carbonates.   Many acids react with carbonates to produce carbon dioxide, water, and a salt. Carbonic acid (H_2CO_3) is not the product because it is unstable and decomposes into water and carbon dioxide.

Acid + Carbonate $\longrightarrow$ Salt + Water + Carbon dioxide

$2\,HCl(aq) + Na_2CO_3(aq) \longrightarrow 2\,NaCl(aq) + H_2O + CO_2\uparrow$

$H_2SO_4(aq) + MgCO_3(s) \longrightarrow MgSO_4(aq) + H_2O + CO_2\uparrow$

15.3 Reactions of Bases

The OH^- ions are responsible for the characteristic reactions of bases. All the following reactions are in an aqueous medium.

Reaction with acids.   Bases react with acids to produce a salt and water. See the discussion of the reaction of acids with bases in Section 15.2.

amphoteric

Amphoteric hydroxides.   Hydroxides of certain metals, such as zinc, aluminum, and chromium, are **amphoteric;** that is, they are capable of reacting as either an acid or a base. When treated with a strong acid, they behave like bases; when reacted with a strong base, they behave like acids.

$Zn(OH)_2(s) + 2\,HCl(aq) \longrightarrow ZnCl_2(aq) + 2\,H_2O$

$Zn(OH)_2(s) + 2\,NaOH(aq) \longrightarrow Na_2ZnO_2(aq) + 2\,H_2O$

Reaction of NaOH and KOH with certain metals.   Some amphoteric metals react directly with the strong bases sodium hydroxide and potassium hydroxide to produce hydrogen and a salt.

Base + Metal $\longrightarrow$ Salt + Hydrogen

$2\,NaOH(aq) + Zn(s) \longrightarrow Na_2ZnO_2(aq) + H_2\uparrow$

$6\,KOH(aq) + 2\,Al(s) \longrightarrow 2\,K_3AlO_3(aq) + 3\,H_2\uparrow$

Reaction with salts.   Bases will react with many salts in solution due to the formation of insoluble metal hydroxides.

Base + Salt $\longrightarrow$ Metal hydroxide$\downarrow$ + Salt

$2\,NaOH(aq) + MnCl_2(aq) \longrightarrow Mn(OH)_2\downarrow + 2\,NaCl(aq)$

$3\,NH_4OH(aq) + FeCl_3(aq) \longrightarrow Fe(OH)_3\downarrow + 3\,NH_4Cl(aq)$

$2\,KOH(aq) + CuSO_4(aq) \longrightarrow Cu(OH)_2\downarrow + K_2SO_4(aq)$

15.4 Salts

Salts are very abundant in nature. Most of the rocks and minerals of the earth's mantle are salts of one kind or another. Huge quantities of dissolved salts also exist in the oceans. Salts may be considered to be compounds that have been derived from acids and bases. They consist of positive metal or ammonium ions (H^+ excluded) combined with negative nonmetal ions (OH^- and O^{2-} excluded). The positive ion is the base counterpart and the nonmetal ion is the acid counterpart:

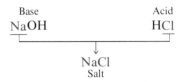

Salts are generally crystalline and have high melting and boiling points.

From a single acid such as hydrochloric acid (HCl), we may produce many chloride salts by replacing the hydrogen with a metal ion (for example, NaCl, KCl, RbCl, $CaCl_2$, $NiCl_2$). The number of known salts greatly exceeds the number of known acids and bases. Salts are ionic compounds. If the hydrogen atoms of a binary acid are replaced by a nonmetal, the resulting compound has covalent bonding and is therefore not considered to be a salt (for example, PCl_3, S_2Cl_2, Cl_2O, NCl_3, ICl).

A review of Chapter 8 on the nomenclature of acids, bases, and salts may be beneficial at this point.

15.5 Electrolytes and Nonelectrolytes

Some of the most convincing evidence as to the nature of chemical bonding within a substance is the ability (or lack of ability) of a water solution of the substance to conduct electricity.

It can be readily shown that solutions of certain substances are conductors of electricity. A simple apparatus to demonstrate conductivity consists of a pair of electrodes connected to a voltage source through a light bulb and a switch (Figure 15.1, page 362). When the switch is closed and the medium between the electrodes is a conductor of electricity, the light bulb glows. When chemically pure water is placed in the beaker and the switch is closed, the light does not glow, indicating that water is a nonconductor. When we dissolve a small amount of sugar in water and test the resulting solution, the light does not glow, showing that a sugar solution is also a nonconductor. But when a small amount of salt, NaCl, is dissolved in water and this solution is tested, the light glows brilliantly. Thus, the salt solution conducts electricity. A fundamental difference exists between the chemical bonding of sugar and that of salt. Sugar is a covalently bonded (molecular) substance; salt is an electrovalently bonded (ionic) substance.

electrolyte
nonelectrolyte

Substances whose aqueous solutions are conductors of electricity are called **electrolytes.** Substances whose solutions are nonconductors are known as **nonelectrolytes.** The classes of compounds that are electrolytes are acids, bases, and salts. Solutions of certain oxides are also conductors because they form an acid or a base when dissolved in water. One major difference between electrolytes

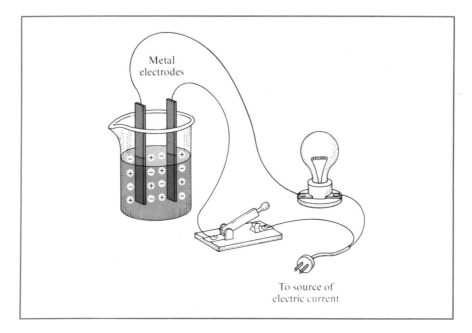

Figure 15.1   A simple conductivity apparatus for testing electrolytes and non-electrolytes in solution. If the solution contains an electrolyte, the light will glow when the switch is closed.

and nonelectrolytes is that electrolytes exist as ions or are capable of producing ions in solution, whereas nonelectrolytes do not have this property. Solutions that contain a sufficient number of ions will conduct an electric current. Although pure water is a nonconductor, many city water supplies contain enough dissolved ionic matter to cause the light to glow dimly when tested in a conductivity apparatus. Table 15.2 lists some common electrolytes and nonelectrolytes.

Acids, bases, and salts are electrolytes.

TABLE 15.2 Representative electrolytes and nonelectrolytes.

Electrolytes	Nonelectrolytes
H_2SO_4	$C_{12}H_{22}O_{11}$ (sugar)
HCl	C_2H_5OH (ethyl alcohol)
HNO_3	$C_2H_4(OH)_2$ (ethylene glycol)
NaOH	$C_3H_5(OH)_3$ (glycerol)
$HC_2H_3O_2$	CH_3OH (methyl alcohol)
NH_4OH	$CO(NH_2)_2$ (urea)
K_2SO_4	O_2
$NaNO_3$	

15.6 Dissociation and Ionization of Electrolytes

Arrhenius received the 1903 Nobel Prize in chemistry for his work on electrolytes. He stated that a solution conducts electricity because the solute dissociates immediately upon dissolving into electrically charged particles called *ions*. The movement of these ions toward oppositely charged electrodes causes the solution to be a conductor. According to his theory, solutions that are relatively poor conductors contain electrolytes that are partly dissociated. Arrhenius also believed that ions exist in solution whether or not there is an electric current. In other words, the electric current does not cause the formation of ions. Positive ions, attracted to the cathode, are cations; and negative ions, attracted to the anode, are anions.

Positive ions are called cations. Negative ions are called anions.

We have seen that sodium chloride crystals consist of sodium and chloride ions held together by ionic bonds. When placed in water, these ions are attracted by polar water molecules, which surround each ion as it dissolves. In water, the salt dissociates, forming hydrated sodium and chloride ions (see Figure 15.2). The sodium and chloride ions in solution are bonded to a specific number of water dipoles and have less attraction for each other than they had in the crystalline state. The equation representing this dissociation is

$$NaCl(s) + (x + y)H_2O \longrightarrow Na^+(H_2O)_x + Cl^-(H_2O)_y$$

A simplified dissociation equation in which the water is omitted but understood to be present is

$$NaCl \longrightarrow Na^+ + Cl^-$$

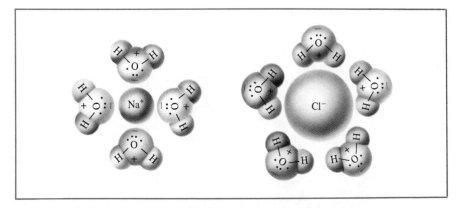

Figure 15.2 Hydrated sodium and chloride ions. When sodium chloride dissolves in water, each Na^+ and Cl^- ion becomes surrounded by water molecules. The negative end of the water dipole is attracted to the Na^+ ion, and the positive end is attracted to the Cl^- ion.

It is important to remember that sodium chloride exists in an aqueous solution as hydrated ions and not as NaCl units, although the formula as such is very often used in equations.

The chemical reactions of salts in solution are the reactions of their ions. For example, when sodium chloride and silver nitrate react and form a precipitate of silver chloride, only the Ag^+ and Cl^- ions participate in the reaction. The Na^+ and NO_3^- remain as ions in solution.

$$Ag^+ + Cl^- \longrightarrow AgCl\downarrow$$

In many cases, the number of molecules of water associated with a particular ion is known. For example, the blue color of the copper(II) ion is due to the hydrated ion $Cu(H_2O)_4^{2+}$. The hydration of ions can be demonstrated in a striking way with cobalt(II) chloride. When cobalt(II) chloride hexahydrate ($CoCl_2 \cdot 6H_2O$) is dissolved in water, a pink solution forms due to the $Co(H_2O)_6^{2+}$ ions. If concentrated hydrochloric acid is added to this pink solution, the color gradually changes to blue. If water is now added to the blue solution, the color gradually changes to pink again. These color changes are due to the exchange of water molecules and chloride ions on the cobalt ion. The complex ion $CoCl_4^{2-}$ is blue. Thus, the hydration of the cobalt ion is a reversible reaction (see Chapter 16). The equilibrium equation representing these changes is

$$\underset{\text{Pink}}{Co(H_2O)_6^{2+}} + 4\,Cl^- \rightleftharpoons \underset{\text{Blue}}{CoCl_4^{2-}} + 6\,H_2O$$

Ionization is the formation of ions; it may occur as a result of a chemical reaction of certain substances with water. Glacial acetic acid (100% $HC_2H_3O_2$) is a liquid that behaves as a nonelectrolyte when tested by the method described in Section 15.5. But a water solution of acetic acid conducts an electric current, as indicated by the dull-glowing light of the conductivity apparatus. The equation for the reaction with water, forming hydronium and acetate ions, is

$$\underset{\text{Acid}}{HC_2H_3O_2} + \underset{\text{Base}}{H_2O} \rightleftharpoons \underset{\text{Acid}}{H_3O^+} + \underset{\text{Base}}{C_2H_3O_2^-}$$

or, in the simplified equation,

$$HC_2H_3O_2 \rightleftharpoons H^+ + C_2H_3O_2^-$$

In this ionization reaction, water serves not only as a solvent but also as a base according to the Brønsted–Lowry theory.

The bond in hydrogen chloride is predominantly covalent, but when HCl is dissolved in water, it reacts, forming hydronium and chloride ions:

$$HCl(g) + H_2O(l) \longrightarrow H_3O^+(aq) + Cl^-(aq)$$

When a hydrogen chloride solution is tested for conductivity, the light glows brilliantly, indicating many ions in the solution.

In each of the foregoing two reactions with water, ionization occurs, producing ions in solution. The necessity for water in the ionization process may be demonstrated by dissolving hydrogen chloride in a nonpolar solvent such as benzene, and testing the solution for conductivity. The solution fails to conduct electricity, indicating that no ions are produced.

The terms *dissociation* and *ionization* are often used interchangeably to describe processes taking place in water. But, strictly speaking, the two are different. In the **dissociation** of a salt, the salt already exists as ions. When dissolved in water, the ions separate or dissociate and increase in mobility. In the **ionization** process, ions are actually produced by the reaction of a compound with water.

dissociation
ionization

Electrolytes composed of two ions per formula unit dissociate to give two ions in solution; electrolytes composed of three ions per formula unit dissociate to give three ions in solution; and so on. The dissociation equations for several electrolytes follow. In all cases, the ions are actually hydrated.

$NaOH \xrightarrow{H_2O} Na^+ + OH^-$ 2 ions in solution per formula unit

$Ca(OH)_2 \xrightarrow{H_2O} Ca^{2+} + 2\,OH^-$ 3 ions in solution per formula unit

$Na_2SO_4 \xrightarrow{H_2O} 2\,Na^+ + SO_4^{2-}$ 3 ions in solution per formula unit

$AlCl_3 \xrightarrow{H_2O} Al^{3+} + 3\,Cl^-$ 4 ions in solution per formula unit

$Fe_2(SO_4)_3 \xrightarrow{H_2O} 2\,Fe^{3+} + 3\,SO_4^{2-}$ 5 ions in solution per formula unit

One mole of NaCl will give 1 mole of Na^+ ions and 1 mole of Cl^- ions in solution, assuming complete dissociation of the salt. One mole of $CaCl_2$ will give 1 mole of Ca^{2+} ions and 2 moles of Cl^- ions in solution. With complete dissociation, only the ions are in solution; there is no NaCl or $CaCl_2$ in solution.

$NaCl \xrightarrow{H_2O} Na^+ + Cl^-$
1 mole   1 mole   1 mole

$CaCl_2 \xrightarrow{H_2O} Ca^{2+} + 2\,Cl^-$
1 mole   1 mole   2 moles

PROBLEM 15.2   What is the molarity of each ion in a solution of (a) 2.0 M NaCl, and (b) 0.75 M K_2SO_4? (Assume complete dissociation.)

(a) According to the dissociation equation,

$NaCl \xrightarrow{H_2O} Na^+ + Cl^-$
1 mole   1 mole   1 mole

The concentration of Na^+ is equal to that of NaCl (1 mole NaCl $\longrightarrow$ 1 mole Na^+) and the concentration of Cl^- is also equal to that of NaCl. Therefore, the concentrations of the ions in 2.0 M NaCl are 2.0 M Na^+ and 2.0 M Cl^-.

(b) According to the dissociation equation,

$K_2SO_4 \xrightarrow{H_2O} 2\,K^+ + SO_4^{2-}$
1 mole   2 moles   1 mole

The concentration of K^+ is twice that of K_2SO_4 and the concentration of SO_4^{2-} is equal to that of K_2SO_4. Therefore, the concentrations of the ions in 0.75 M K_2SO_4 are 1.5 M K^+ and 0.75 M SO_4^{2-}.

■

We have learned that when 1 mole of sucrose, a nonelectrolyte, is dissolved in 1000 g of water, the solution freezes at $-1.86°C$. When 1 mole of NaCl is dissolved in 1000 g of water, the freezing point of the solution is not $-1.86°C$, as might be expected, but is closer to $-3.72°C$ (-1.86×2). The reason for the lower freezing point is that 1 mole of NaCl in solution produces 2 moles of particles ($2 \times 6.02 \times 10^{23}$ ions) in solution. Thus, the freezing point lowering by 1 mole of NaCl is essentially equivalent to that produced by 2 moles of a non-electrolyte. An electrolyte such as $CaCl_2$, which yields three ions in water, gives a freezing point depression about three times that of a nonelectrolyte. These freezing point data provide additional evidence that electrolytes dissociate when dissolved in water.

15.7 Strong and Weak Electrolytes

strong electrolyte
weak electrolyte

Electrolytes are classified as either strong or weak, depending on the degree or extent of dissociation or ionization. **Strong electrolytes** are essentially 100% ionized in solution; **weak electrolytes** are considerably less ionized (assuming 0.1 M solutions). Most electrolytes are either strong or weak, with a small number being classified as moderately strong or weak. Most salts are strong electrolytes. Acids and bases that are strong electrolytes (highly ionized) are called *strong acids* and *strong bases*. Acids and bases that are weak electrolytes (slightly ionized) are called *weak acids* and *weak bases*.

For equivalent concentrations, solutions of strong electrolytes contain many more ions than solutions of weak electrolytes. As a result, solutions of strong electrolytes are better conductors of electricity. Consider the two solutions, 1 M HCl and 1 M $HC_2H_3O_2$. Hydrochloric acid is almost 100% ionized; acetic acid is about 1% ionized. Thus, HCl is a strong acid and $HC_2H_3O_2$ is a weak acid. Hydrochloric acid has about 100 times as many hydronium ions in solution as acetic acid, making the HCl solution much more acidic.

One can distinguish between strong and weak electrolytes experimentally by using the apparatus described in Section 15.5 A 1 M HCl solution causes the light to glow brilliantly, but a 1 M $HC_2H_3O_2$ solution causes only a dull glow. In a similar fashion, the strong base sodium hydroxide (NaOH) may be distinguished from the weak base ammonium hydroxide (NH_4OH). The ionization of a weak electrolyte in water is represented by an equilibrium equation showing that both the un-ionized and ionized forms are present in solution. In the equilibrium equation of $HC_2H_3O_2$ and its ions, the equilibrium is far to the left, since relatively few hydrogen and acetate ions are present in solution:

$$HC_2H_3O_2(aq) \rightleftharpoons H^+ + C_2H_3O_2^-$$

We have previously used a double arrow in an equation to represent reversible processes in the equilibrium between dissolved and undissolved solute in a saturated solution. A double arrow ($\rightleftharpoons$) is also used in the ionization equation of soluble weak electrolytes to indicate that the solution contains a considerable amount of the un-ionized compound in equilibrium with its ions in solution.

(See Section 16.1 for a discussion of reversible reactions.) A single arrow is used to indicate that the electrolyte is essentially all in the ionic form in the solution. For example, nitric acid is a strong acid; nitrous acid is a weak acid. Their ionization equations in water may be indicated as

$$HNO_3(aq) \xrightarrow{H_2O} H^+ + NO_3^- \quad \text{(Strong acid)}$$

$$HNO_2(aq) \xrightleftharpoons{H_2O} H^+ + NO_2^- \quad \text{(Weak acid)}$$

Practically all soluble salts, acids such as sulfuric, nitric, and hydrochloric acids, and bases such as sodium, potassium, calcium, and barium hydroxides are strong electrolytes. Weak electrolytes include numerous other acids and bases such as acetic acid, nitrous acid, carbonic acid, and ammonium hydroxide. The terms *strong acid, strong base, weak acid,* and *weak base* refer to whether an acid or base is a strong or weak electrolyte. A list of strong and weak electrolytes is given in Table 15.3.

TABLE 15.3 Strong and weak electrolytes.

Strong electrolytes	Weak electrolytes
Almost all soluble salts	$HC_2H_3O_2$
	H_2CO_3
H_2SO_4	HNO_2
HNO_3	H_2SO_3
HCl	H_2S
HBr	$H_2C_2O_4$
$HClO_4$	H_3BO_3
NaOH	HClO
KOH	NH_4OH
$Ca(OH)_2$	HF
$Ba(OH)_2$	

15.8 Ionization of Water

The more we study chemistry, the more intriguing the little molecule of water appears. Water ionizes to a small degree. Two equations commonly used to show how water ionizes are

$$\underset{\text{Acid}}{H_2O} + \underset{\text{Base}}{H_2O} \rightleftharpoons \underset{\text{Acid}}{H_3O^+} + \underset{\text{Base}}{OH^-}$$

and

$$H_2O \rightleftharpoons H^+ + OH^-$$

The first equation represents the Brønsted–Lowry concept, with water reacting as both an acid and a base, forming a hydronium ion and a hydroxide ion. The second equation is a simplified version, indicating that water ionizes to give a hydrogen and a hydroxide ion. Actually, the proton, H^+, is hydrated and exists as a hydronium ion. In either case, equal molar amounts of acid and base are produced so that water is neutral, having neither H^+ nor OH^- ions in excess. The ionization of water at 25°C produces an H^+ ion concentration of 1.0×10^{-7} mole per litre and an OH^- ion concentration of 1.0×10^{-7} mole per litre. These concentrations are usually expressed as

$$[H^+] \text{ or } [H_3O^+] = 1.0 \times 10^{-7} \text{ mole/L}$$
$$[OH^-] = 1.0 \times 10^{-7} \text{ mole/L}$$

These figures mean that about two out of every billion water molecules are ionized. This amount of ionization, small as it is, is a significant factor in the behavior of water in many chemical reactions.

The square brackets, [ ], indicate that the concentration is in moles per litre. Thus, $[H^+]$ means the concentration of H^+ in moles per litre.

15.9 Introduction to pH

The acidity of an aqueous solution depends on the concentration of hydrogen or hydronium ions. The acidity of solutions involved in a chemical reaction is often critically important, especially for biochemical reactions. The pH scale of acidity was devised to fill the need for a simple, convenient numerical way to state the acidity of a solution. Values on the pH scale are obtained by mathematical conversion of H^+ ion concentrations to pH by these expressions:

$$pH = \log\frac{1}{[H^+]} \quad \text{or} \quad -\log[H^+]$$

pH

where $[H^+] = H^+$ or H_3O^+ ion concentration in moles per litre. The **pH** is defined as the logarithm (log) of the reciprocal of the H^+ or H_3O^+ ion concentration in moles per litre. The scale itself is based on the H^+ concentration in water at 25°C. At this temperature, water has an H^+ concentration of 1×10^{-7} mole/L and is calculated to have a pH of 7.

$$pH = \log\frac{1}{[H^+]} = \log\frac{1}{[1 \times 10^{-7}]} = \log 1 \times 10^7 = 7$$

The pH of pure water at 25°C is 7 and is said to be neutral; that is, it is neither acidic nor basic, because the concentrations of H^+ and OH^- are equal. Solutions that contain more H^+ ions than OH^- ions have pH values less than 7, and solutions that contain fewer H^+ ions than OH^- ions have pH values greater than 7.

pH < 7.00 Acidic solution

pH = 7.00 Neutral solution

pH > 7.00 Basic solution

When $[H^+] = 1 \times 10^{-5}$ mole/L, pH = 5 (Acidic)
When $[H^+] = 1 \times 10^{-9}$ mole/L, pH = 9 (Basic)

Instead of saying that the hydrogen ion concentration in a solution is 1×10^{-5} mole/L, it is customary to say that the pH of the solution is 5. The smaller the pH value, the more acidic the solution (see Figure 15.3).

pH

Acidic   Basic

0 ▬▬▬▬▬▬ 7 ▬▬▬▬▬▬ 14

▲

Neutral

Figure 15.3   The pH scale of acidity and basicity.

A solution of a strong acid is more acidic (has more H^+) than a weak acid at the same molarity. The pH of 0.1 M HCl is 1.00 and that of 0.1 M $HC_2H_3O_2$ is 2.87, indicating that hydrochloric acid is a stronger acid than acetic acid. The $[H^+]$ and thus the pH varies with the degree of dilution of a solution. The following comparative data show that although acetic acid is a weak acid, its pH approaches that of hydrochloric acid (100% ionized) as the solution becomes more dilute; this indicates that a higher percentage of acetic acid molecules ionize as the solution becomes more dilute:

HCl solution	pH	HC₂H₃O₂ solution	pH
0.100 M	1.00	0.100 M	2.87
0.0100 M	2.00	0.0100 M	3.37
0.00100 M	3.00	0.00100 M	3.90

The pH scale, along with its interpretation, is given in Table 15.4 on page 370. Note that a change of only 1 pH unit means a tenfold increase or decrease in H^+ ion concentration. A simplified method of determining pH from $[H^+]$ is as follows:

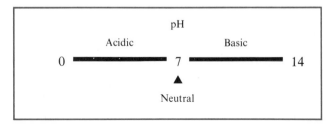

$$[H^+] = 1 \times 10^{-5}$$

When this number   pH is this number (5)
is exactly 1   pH = 5

$$[H^+] = 2 \times 10^{-5}$$

When this number   pH is between this number and
is between 1 and 10   next lower number (4 and 5)
                       pH = 4.7

TABLE 15.4 The pH scale for expressing acidity.

$[H^+]$ (mole/litre)	pH	
$1.0 = 1 \times 10^0$	0	↑
$0.1 = 1 \times 10^{-1}$	1	
1×10^{-2}	2	
$0.001 = 1 \times 10^{-3}$	3	Increasing acidity
1×10^{-4}	4	
$0.00001 = 1 \times 10^{-5}$	5	
1×10^{-6}	6	
$0.0000001 = 1 \times 10^{-7}$	7	Neutral
1×10^{-8}	8	
$0.000000001 = 1 \times 10^{-9}$	9	
1×10^{-10}	10	Increasing basicity
$0.00000000001 = 1 \times 10^{-11}$	11	
1×10^{-12}	12	
$0.0000000000001 = 1 \times 10^{-13}$	13	
1×10^{-14}	14	↓

TABLE 15.5 The pH of some common solutions.

Solution	pH
Gastric juice	0.9
0.1 M HCl	1.0
Lemon juice	2.3
Vinegar	2.8
Orange juice	2.8
0.1 M $HC_2H_3O_2$	2.9
Tomato juice	4.1
Urine	6.0
Milk	6.6
Pure water (25°C)	7.0
Blood	7.4
Household ammonia	10.5
1 M NaOH	14.0

Table 15.5 lists the pH of some common solutions.

Calculation of the pH value corresponding to any H^+ ion concentration requires the use of logarithms. However, if you are not familiar with the use of logarithms (abbreviated logs), the simplified log scale in Figure 15.4 can be used to estimate the logarithms of various numbers.

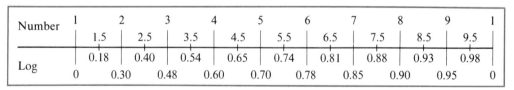

Figure 15.4   Simplified logarithm scale. For example, the logarithm (log) of 5 is 0.70; the logarithm of 7.5 is 0.88.

Let us see how to use this log scale in calculating the pH of a solution with $[H^+] = 2 \times 10^{-5}$:

$$[H^+] = \textcircled{2} \times 10^{-\textcircled{5}}$$

pH = This number minus the log of this number
(which must be between 1 and 10)

pH = $5 - \log 2$

From the log scale, log 2 = 0.30. Thus, pH = $5 - 0.30 = 4.70$ (Answer)

PROBLEM 15.3   What is the pH of a solution having a $[H^+]$ of (a) 1×10^{-11}, and (b) 6×10^{-4}?

(a) $[H^+] = 1 \times 10^{-11}$
   $pH = 11$

(b) $[H^+] = 6 \times 10^{-4}$
   $pH = 4 - \log 6$

From Figure 15.4, the log of 6 is 0.78; therefore,

$pH = 4 - 0.78 = 3.22$

■

The measurement and control of pH is extremely important in many fields of science and technology. The proper soil pH is necessary to grow certain types of plants successfully. The pH of certain foods is too acid for some diets. Many biological processes are delicately controlled pH systems. The pH of human blood is regulated to very close tolerances by the uptake or release of H^+ by mineral ions such as HCO_3^- and CO_3^{2-}. Changes in the pH of the blood by as little as 0.4 pH unit result in death.

Compounds with colors that change at particular pH values are used as indicators in acid–base reactions. For example, phenolphthalein, an organic compound, is colorless in acid solution and changes to pink at a pH of 8.3. When a solution of sodium hydroxide is added to a hydrochloric acid solution containing phenolphthalein, the change in color (from colorless to pink) indicates that all the acid is neutralized. Commercially available pH test paper, such as shown in Figure 15.5, contains chemical indicators. The indicator in the paper takes on different colors when wetted with solutions of different pH. Thus, the pH of a solution can be estimated by placing a drop on the test paper and comparing the color of the test paper with a color chart calibrated at different pH values. Electronic pH meters of the type shown in Figure 15.6 are used for making rapid and highly precise pH determinations.

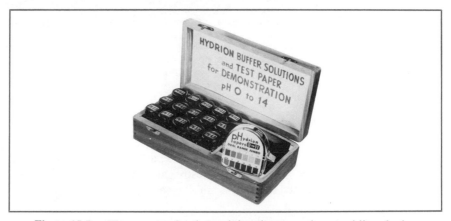

Figure 15.5   pH test paper for determining the approximate acidity of solutions. *(Courtesy Micro Essential Laboratory, Inc.)*

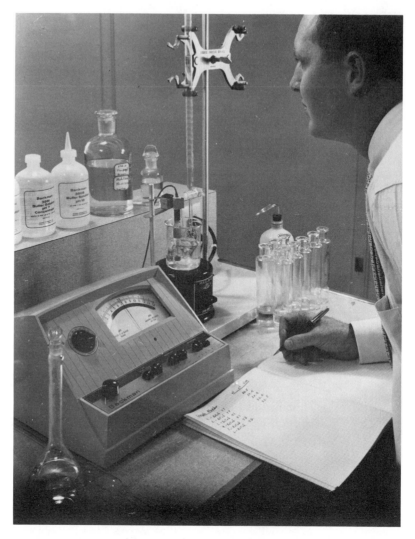

Figure 15.6   An electronic pH meter: Accurate measurements may be made by meters of this type. The scale is calibrated to read in both pH units and millivolts. *(Courtesy Beckman Instruments, Inc. Zeromatic is a registered trademark.)*

15.10 Neutralization

neutralization

The reaction of an acid and a base to form a salt and water is known as **neutralization.** We have seen this reaction before, but now, in the light of what we have learned about ions and ionization, let us reexamine what occurs during neutralization.

Consider the reaction that occurs when solutions of sodium hydroxide and hydrochloric acid are mixed. The ions present initially are Na^+ and OH^- from the base and H^+ and Cl^- from the acid. The products, sodium chloride and water, exist as Na^+ and Cl^- ions and H_2O molecules. A chemical equation representing this reaction is:

$$HCl(aq) + NaOH(aq) \longrightarrow NaCl(aq) + H_2O \tag{5}$$

This equation, however, does not show that HCl, NaOH, and NaCl exist as ions in solution. The following total ionic equation gives a much better representation of the reaction:

$$(H^+ + Cl^-) + (Na^+ + OH^-) \longrightarrow Na^+ + Cl^- + H_2O \tag{6}$$

spectator ions

Equation (6) shows that the Na^+ and Cl^- ions did not react; they are still ions in solution. These ions are **spectator ions** because they were present but did not take part in the reaction. The only reaction that occurred was that between the H^+ and OH^- ions. Therefore, the equation for the neutralization can be written as this net ionic equation:

$$\underset{\text{Acid}}{H^+} + \underset{\text{Base}}{OH^-} \longrightarrow \underset{\text{Water}}{H_2O} \tag{7}$$

This simple net ionic equation (7) represents not only the reaction of sodium hyroxide and hydrochloric acid but also the reaction of any acid with any base in an aqueous solution. The driving force of a neutralization reaction is the ability of an H^+ ion and an OH^- ion to react and form a molecule of un-ionized water.

titration

The amount of acid, base, or other species in a sample may be determined by titration. **Titration** is the process of measuring the volume of one reagent required to react with a measured weight or volume of another reagent. Let us consider the titration of an acid with a base. A measured volume of acid of unknown concentration is placed into a flask and a few drops of an indicator solution are added. Base solution of known concentration is slowly added from a buret to the acid until the indicator changes color (Figure 15.7, page 374). The indicator selected is one that changes color when the stoichiometric quantity (according to the equation) of base has been added to the acid. At this point, known as the *end point of the titration,* the titration is complete and the volume of base used to neutralize the acid is read from the buret. The concentration or amount of acid in solution can be calculated from the titration data and the chemical equation for the reaction. Illustrative problems follow.

PROBLEM 15.4   Suppose that 42.00 mL of 0.15 M NaOH solution are required to titrate 100 mL of hydrochloric acid solution. What is the molarity of the acid solution?

The equation for the reaction is

$$NaOH(aq) + HCl(aq) \longrightarrow NaCl(aq) + H_2O$$

In this neutralization, NaOH and HCl react in a 1:1 mole ratio. Therefore, the moles of HCl in solution are equal to the moles of NaOH required to react with it. First we calculate the moles of NaOH used, and from this value, the moles of HCl.

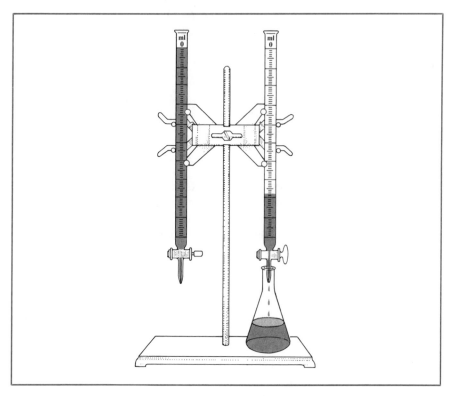

Figure 15.7 Graduated burets are used in titrations for neutralization of acids and bases as well as for many other volumetric determinations. Figure 15.6 illustrates a titration using a pH meter as indicator.

Data: 42.00 mL of 0.15 *M* NaOH   100 mL HCl
         Molarity of acid = *M*   (Unknown)

Moles of NaOH:

   Moles = Litres × *M*   42.00 mL = 0.04200 L

   $0.04200 \cancel{L} \times \dfrac{0.15 \text{ mole NaOH}}{\cancel{L}} = 0.0063 \text{ mole NaOH}$

Since NaOH and HCl react in a 1:1 mole ratio, 0.0063 mole of HCl was present in the 100 mL of HCl solution. Therefore, the molarity of the HCl is

   $M = \dfrac{\text{Moles}}{\text{Litre}} = \dfrac{0.0063 \text{ mole HCl}}{0.100 \text{ L}} = 0.063 \ M \text{ HCl}$ (Answer)

■

PROBLEM 15.5   Suppose that 42.00 mL of 0.15 *M* NaOH solution are required to titrate 100 mL of sulfuric acid (H_2SO_4) solution. What is the molarity of the acid solution?

The equation for the reaction is

$$2\,NaOH(aq) + H_2SO_4(aq) \longrightarrow Na_2SO_4(aq) + 2\,H_2O$$

The same amount of base (0.0063 mole of NaOH) is used in this titration as in Problem 15.4. However, the mole ratio of acid to base in the H_2SO_4 reaction is 1:2. The moles of H_2SO_4 reacted can be calculated by using the mole-ratio method.

Data: 42.00 mL of 0.15 M NaOH = 0.0063 mole NaOH

$$0.0063 \text{ mole NaOH} \times \frac{1 \text{ mole } H_2SO_4}{2 \text{ moles NaOH}} = 0.00315 \text{ mole } H_2SO_4$$

Therefore, 0.00315 mole of H_2SO_4 was present in 100 mL of H_2SO_4 solution. The molarity of the H_2SO_4 is

$$M = \frac{\text{Moles}}{\text{Litre}} = \frac{0.00315 \text{ mole } H_2SO_4}{0.100 \text{ L}} = 0.0315\ M\ H_2SO_4 \quad \text{(Answer)}$$

15.11 Writing Ionic Equations

In Section 15.10, we wrote the reaction of hydrochloric acid and sodium hydroxide in three different equations. Equation (5) was the un-ionized equation; equation (6), the total ionic equation; and equation (7), the net ionic equation.

un-ionized equation
total ionic equation

In the **un-ionized equation,** compounds are written in their molecular or normal formula expressions. In the **total ionic equation,** compounds are written in the form in which they are predominantly present: strong electrolytes as ions in solution; and nonelectrolytes, weak electrolytes, precipitates, and gases in their molecular or un-ionized forms. In the **net ionic equation,** only those molecules or ions that have changed are included in the equation; ions or molecules that do not change are omitted. Up to now, we have been concerned only with balancing the individual elements when we balanced equations. Because we are using ions, which are electrically charged, net ionic equations are often not neutral in charge and end up with a net electrical charge. The net electrical charge of an ionic equation, as well as its atoms, should be in balance. Therefore, a balanced ionic equation will have the same net electrical charge on both sides of the equation, whether it is zero, positive, or negative.

net ionic equation

Study the following examples. All reactions are in aqueous solution.

Example 1

$$HNO_3(aq) + KOH(aq) \longrightarrow KNO_3(aq) + H_2O \qquad \text{Un-ionized equation}$$

$$(H^+ + NO_3^-) + (K^+ + OH^-) \longrightarrow (K^+ + NO_3^-) + H_2O$$
$$\text{Total ionic equation}$$

$$H^+ + OH^- \longrightarrow H_2O \qquad\qquad\qquad \text{Net ionic equation}$$

HNO_3, KOH, and KNO_3 are soluble, strong electrolytes. K^+ and NO_3^- are spectator ions, have not changed, and are not included in the net ionic equation. Water is a nonelectrolyte and is written in the molecular form.

Example 2

$$2\,AgNO_3(aq) + BaCl_2(aq) \longrightarrow 2\,AgCl{\downarrow} + Ba(NO_3)_2(aq)$$

$$(2\,Ag^+ + 2\,NO_3^-) + (Ba^{2+} + 2\,Cl^-) \longrightarrow 2\,AgCl{\downarrow} + (Ba^{2+} + 2\,NO_3^-)$$

$$Ag^+ + Cl^- \longrightarrow AgCl{\downarrow} \qquad \text{Net ionic equation}$$

Although silver chloride (AgCl) is an ionic salt, it is written in the un-ionized form here because most of the Ag^+ and Cl^- ions are no longer in solution but have formed a precipitate of AgCl. Ba^{2+} and NO_3^- are spectator ions.

Example 3

$$Na_2CO_3(aq) + H_2SO_4(aq) \longrightarrow Na_2SO_4(aq) + H_2O + CO_2{\uparrow}$$

$$(2\,Na^+ + CO_3^{2-}) + (2\,H^+ + SO_4^{2-}) \longrightarrow (2\,Na^+ + SO_4^{2-}) + H_2O + CO_2{\uparrow}$$

$$CO_3^{2-} + 2\,H^+ \longrightarrow H_2O + CO_2{\uparrow} \qquad \text{Net ionic equation}$$

Carbon dioxide (CO_2) is a gas and evolves from solution, and Na^+ and SO_4^{2-} are spectator ions.

Example 4

$$HC_2H_3O_2(aq) + NaOH(aq) \longrightarrow NaC_2H_3O_2(aq) + H_2O$$

$$HC_2H_3O_2 + (Na^+ + OH^-) \longrightarrow (Na^+ + C_2H_3O_2^-) + H_2O$$

$$HC_2H_3O_2 + OH^- \longrightarrow C_2H_3O_2^- + H_2O \qquad \text{Net ionic equation}$$

Acetic acid ($HC_2H_3O_2$), a weak acid, is written in the molecular form, but sodium acetate ($NaC_2H_3O_2$), a soluble salt, is written in the ionic form. The Na^+ ion is the only spectator ion in this reaction. Both sides of the net ionic equation have a -1 electrical charge.

Example 5

$$Mg(s) + 2\,HCl(aq) \longrightarrow MgCl_2(aq) + H_2{\uparrow}$$

$$Mg + (2\,H^+ + 2\,Cl^-) \longrightarrow (Mg^{2+} + 2\,Cl^-) + H_2{\uparrow}$$

$$Mg + 2\,H^+ \longrightarrow Mg^{2+} + H_2{\uparrow} \qquad \text{Net ionic equation}$$

The net electrical charge on both sides of the equation is $+2$.

Example 6

$$H_2SO_4(aq) + Ba(OH)_2(aq) \longrightarrow BaSO_4\downarrow + 2\,H_2O$$

$$(2\,H^+ + SO_4^{2-}) + (Ba^{2+} + 2\,OH^-) \longrightarrow BaSO_4\downarrow + 2\,H_2O$$

$$2\,H^+ + SO_4^{2-} + Ba^{2+} + 2\,OH^- \longrightarrow BaSO_4\downarrow + 2\,H_2O$$

Net ionic equation

Barium sulfate ($BaSO_4$) is a highly insoluble salt. If we conduct this reaction using the apparatus described in Section 15.5, the light, which glows brightly at first, will be extinguished when the reaction is complete, because essentially no ions are left in solution. The $BaSO_4$ precipitates out of solution, and water is a nonconductor of electricity.

Here is a list of rules to observe when writing ionic equations:

1. Strong electrolytes are written in their ionic form.
2. Weak electrolytes are written in their molecular or un-ionized form.
3. Nonelectrolytes are written in their molecular form.
4. Insoluble substances, precipitates, and gases are written in their molecular or un-ionized forms.
5. The net ionic equation should include only those substances that have undergone a chemical change.
6. Equations must be balanced, both in atoms and in electrical charge.

QUESTIONS

A. Review the Meanings of the New Terms Introduced in this Chapter

1. Hydronium ion
2. Amphoteric
3. Electrolyte
4. Nonelectrolyte
5. Dissociation
6. Ionization
7. Strong electrolyte
8. Weak electrolyte
9. pH
10. Neutralization
11. Spectator ions
12. Titration
13. Un-ionized equation
14. Total ionic equation
15. Net ionic equation

B. Answer to the Following Questions Will Be Found in Tables and Figures

1. Since a hydrogen ion and a proton are identical, what differences exist between the Arrhenius and Brønsted–Lowry definitions of an acid? (See Table 15.1.)
2. According to Figure 15.1, what type of substance must be in solution in order for the bulb to light?
3. Which of the following classes of compounds are electrolytes: acids, alcohols, bases, salts? (See Table 15.2.)
4. What two differences are apparent in the arrangement of water molecules about the hydrated ions as depicted in Figure 15.2?
5. Between what whole number pH range would a hydrogen ion concentration of 0.003 M fall? (See Table 15.4.)
6. Which is the more acidic, tomato juice or blood? (See Table 15.5.)

C. Review Questions

1. According to each of the three acid–base theories (Arrhenius, Brønsted–Lowry, and Lewis), define (a) an acid, (b) a base, and (c) neutralization.

2. For each of the acid–base theories referred to in Question 1, write an equation illustrating the neutralization of an acid with a base.
3. Identify the conjugate acid–base pairs in the following equations:
   (a) $HCl + NH_3 \longrightarrow NH_4^+ + Cl^-$
   (b) $HCO_3^- + OH^- \rightleftharpoons CO_3^{2-} + H_2O$
   (c) $HCO_3^- + H_3O^+ \rightleftharpoons H_2CO_3 + H_2O$
   (d) $HC_2H_3O_2 + H_2O \rightleftharpoons H_3O^+ + C_2H_3O_2^-$
   (e) $HC_2H_3O_2 + H_2SO_4 \rightleftharpoons H_2C_2H_3O_2^+ + HSO_4^-$
   (f) The two-step ionization of sulfuric acid,
   $$H_2SO_4 + H_2O \longrightarrow H_3O^+ + HSO_4^-$$
   $$HSO_4^- + H_2O \rightleftharpoons H_3O^+ + SO_4^{2-}$$
4. Write the electron-dot structure for (a) bromide ion, (b) hydroxide ion, and (c) cyanide ion. Why are these ions considered to be bases according to the Brønsted–Lowry and Lewis acid–base theories?
5. Complete and balance the following equations:
   (a) $Mg(s) + HCl(aq) \longrightarrow$
   (b) $BaO(s) + HBr(aq) \longrightarrow$
   (c) $Al(s) + H_2SO_4(aq) \longrightarrow$
   (d) $Na_2CO_3(aq) + HCl(aq) \longrightarrow$
   (e) $Fe_2O_3(s) + HBr(aq) \longrightarrow$
   (f) $Ca(OH)_2(aq) + H_2CO_3(aq) \longrightarrow$
   (g) $NaOH(aq) + HBr(aq) \longrightarrow$
   (h) $KOH(aq) + HCl(aq) \longrightarrow$
   (i) $Ca(OH)_2(aq) + HI(aq) \longrightarrow$
   (j) $Al(OH)_3(s) + HBr(aq) \longrightarrow$
6. Into what three classes of compounds do electrolytes generally fall?
7. Which of the following compounds are electrolytes? Consider each substance to be mixed with water.
   (a) HCl
   (b) CO_2
   (c) $CaCl_2$
   (d) $C_{12}H_{22}O_{11}$ (sugar)
   (e) C_3H_7OH (rubbing alcohol)
   (f) CCl_4 (insoluble)
   (g) $NaHCO_3$ (baking soda)
   (h) N_2 (insoluble gas)
   (i) $AgNO_3$
   (j) $HCOOH$ (formic acid)
   (k) $RbOH$
   (l) K_2CrO_4
8. Name each compound listed in Table 15.3.
9. A solution of HCl in water conducts an electric current while a solution of HCl in benzene does not. Explain this behavior in terms of ionization and chemical bonding.
10. How do salts exist in their crystalline structure? What occurs when they are dissolved in water?
11. An aqueous methyl alcohol, CH_3OH, solution does not conduct an electric current, but a solution of sodium hydroxide, $NaOH$, does. What does this information tell us about the OH group in the alcohol?
12. Why does molten sodium chloride conduct electricity?
13. Explain the difference between dissociation of ionic compounds and ionization of molecular compounds.
14. Distinguish between strong and weak electrolytes.
15. Explain why ions are hydrated in aqueous solutions.
16. Indicate, by simple equations, how the following substances dissociate or ionize in water:
   (a) $Cu(NO_3)_2$ (c) HNO_2 (e) NH_4Br (g) $NaClO_3$
   (b) $HC_2H_3O_2$ (d) $LiOH$ (f) K_2SO_4 (h) K_3PO_4
17. What is the main distinction between water solutions of strong and weak electrolytes?

18. What are the relative concentrations of $H^+(aq)$ and $OH^-(aq)$ in (a) a neutral solution, (b) an acid solution, and (c) a basic solution?

19. Write the net ionic equation for the reaction of an acid with a base in an aqueous solution.

20. The solubility of hydrogen chloride gas in water, a polar solvent, is much greater than its solubility in benzene, a nonpolar solvent. How can you account for this?

21. Pure water, containing both acid and base ions, is neutral. Why?

22. Rewrite the following unbalanced equations, changing them into balanced net ionic equations. All reactions are in water solution.

(a) $K_2SO_4(aq) + Ba(NO_3)_2(aq) \longrightarrow KNO_3(aq) + BaSO_4(s)$

(b) $CaCO_3(s) + HCl(aq) \longrightarrow CaCl_2(aq) + CO_2(aq) + H_2O$

(c) $Mg(s) + HC_2H_3O_2(aq) \longrightarrow Mg(C_2H_3O_2)_2(aq) + H_2(g)$

(d) $H_2S(g) + CdCl_2(aq) \longrightarrow CdS(s) + HCl(aq)$

(e) $CaCl_2(aq) + Na_2C_2O_4(aq) \longrightarrow CaC_2O_4(s) + NaCl(aq)$

23. In each of the following pairs which solution is more acidic? (All are water solutions.)

(a) 1 molar HCl or 1 molar H_2SO_4?

(b) 1 molar HCl or 1 molar $HC_2H_3O_2$?

(c) 1 molar HCl or 2 molar HCl?

(d) 1 normal H_2SO_4 or 1 molar H_2SO_4?

24. How does a hydronium ion differ from a hydrogen ion?

25. Arrange, in decreasing order of freezing points, 1 molal aqueous solutions of HCl, $HC_2H_3O_2$, $C_{12}H_{22}O_{11}$ (sucrose), and $CaCl_2$. (List the one with the highest freezing point first.)

26. At $100°C$ the H^+ concentration in water is about 1×10^{-6} mole /L, about 10 times that of water at $25°C$. At which temperature is (a) the pH of water the greatest, (b) the hydrogen ion (hydronium ion) concentration the highest, and (c) the water neutral?

27. A 1 molal solution of acetic acid in water freezes at a lower temperature than a 1 molal solution of ethyl alcohol, C_2H_5OH, in water. Explain.

28. At the same cost per pound, which alcohol, CH_3OH or C_2H_5OH, would be more economical to purchase as an antifreeze for your car?

29. Which of the following statements are correct?

(a) The Arrhenius theory of acids and bases is restricted to aqueous solutions.

(b) The Brønsted–Lowry theory of acids and bases is restricted to solutions other than aqueous solutions.

(c) All substances that are acids according to the Brønsted–Lowry theory will also be acids by the Lewis theory.

(d) All substances that are acids according to the Lewis theory will also be acids by the Brønsted–Lowry theory.

(e) An electron-pair donor is a Lewis acid.

(f) All Arrhenius acid–base neutralization reactions can be represented by a single net ionic equation.

(g) The formation of a coordinate covalent bond represents neutralization according to the Lewis theory of acids and bases.

(h) In the auto-ionization of water, $2 H_2O \rightleftharpoons H_3O^+ + OH^-$, the H_3O^+ and the OH^- constitute a conjugate acid–base pair.

(i) In the reaction in part (h), H_2O is both the acid and the base.

(j) An amphoteric substance may undergo neutralization by reaction with either an acid or a base.

(k) A solution of pH 3 is 100 times more acidic than a solution of pH 5.

(l)   In general, ionic substances when placed in water will give a solution capable of conducting an electric current.
(m) The terms *dissociation* and *ionization* are synonymous.
(n)   Crystalline NaCl and gaseous HCl both ionize and dissociate when dissolved in water.
(o)   The terms *strong acid, strong base, weak acid,* and *weak base* refer to whether an acid or base solution is concentrated or dilute.
(p)   pH is defined as the negative logarithm of the molar concentration of H^+ ions (or H_3O^+ ions).
(q)   All reactions may be represented by net ionic equations.
(r)   One mole of $CaCl_2$ contains more anions than cations.
(s)   It is possible to boil seawater at a lower temperature than that required to boil pure water (both at the same pressure).
(t)   It is possible to have a neutral aqueous solution whose pH is not 7.

D. Review Problems

1. Calculate the molarity of the ions present in each of the following salt solutions. Assume each salt to be 100% dissociated.
   (a) 0.010 M NaCl
   (b) 10.0 M $NaKSO_4$
   (c) 0.55 M $ZnBr_2$
   (d) 1.25 M $Al_2(SO_4)_3$
   (e) 0.10 M $CaCl_2$
   (f) 18.0 g KI in 500 mL of solution
   (g) 1000 g $(NH_4)_2SO_4$ in 20.0 L of solution
   (h) 0.0100 g $Mg(ClO_3)_2$ in 1.00 mL of solution

2. In Problem 1, how many grams of each ion would be present in 100 mL of each solution?

3. What is the concentration of Ca^{2+} ions in a solution of CaI_2 having an I^- ion concentration of 0.720 M?

4. What is the molar concentration of all ions present in a solution prepared by mixing the following?
   (a) 20.0 mL of 1.0 M NaCl and 30.0 mL of 1.0 M NaCl
   (b) 20.0 mL of 1.0 M HCl and 20.0 mL of 1.0 M NaOH
   (c) 50.0 mL of 2.0 M KCl and 50.0 mL of 1.0 M $CaCl_2$
   *(d) 100.0 mL of 0.30 M KOH and 100.0 mL of 0.60 M HCl
   (e) 35.0 mL of 0.20 M $Ba(OH)_2$ and 35.0 mL of 0.20 M H_2SO_4
   (f) 1.00 L of 1.0 M NaCl and 500 mL of 2.0 M $AgNO_3$

   (Neglect the concentration of H^+ and OH^- from water. Also, assume volumes of solutions are additive.)

5. How many millilitres of 0.60 M HCl can be made by diluting 500 mL of 12 M HCl with water?

6. Given the data for the following six titrations, calculate the molarity of the HCl in titrations (a), (b), and (c), and the molarity of he NaOH in titrations (d), (e), and (f).

	mL HCl	Molarity HCl	mL NaOH	Molarity NaOH
(a)	40.13	M HCl	37.70	0.874
(b)	19.00	M HCl	33.66	0.404
(c)	27.25	M HCl	18.00	0.777
(d)	37.19	0.145	31.91	M NaOH
(e)	48.04	0.516	24.02	M NaOH
(f)	13.13	1.243	39.39	M NaOH

7. If 29.26 mL of 0.430 M HCl neutralizes 17.40 mL of Ba(OH)$_2$ solution, what is the molarity of the Ba(OH)$_2$ solution? The reaction is

$$Ba(OH)_2(aq) + 2\,HCl(aq) \longrightarrow BaCl_2(aq) + 2\,H_2O$$

8. What volume (in millilitres) of 0.300 M HCl will neutralize (a) 50.0 mL of 0.100 M Ca(OH)$_2$, and (b) 10.0 g of Al(OH)$_3$? The equations are:
   (a) $2\,HCl(aq) + Ca(OH)_2(aq) \longrightarrow CaCl_2(aq) + 2\,H_2O$
   (b) $3\,HCl(aq) + Al(OH)_3(s) \longrightarrow AlCl_3(aq) + 3\,H_2O$

9. A sample of pure sodium carbonate weighing 0.420 g was dissolved in water and neutralized with 42.4 mL of hydrochloric acid. Calculate the molarity of the acid:

$$Na_2CO_3(aq) + 2\,HCl(aq) \longrightarrow 2\,NaCl(aq) + CO_2(g) + H_2O$$

10. How many moles of solute and solvent are used in preparing a solution containing 140 g of benzene, C_6H_6, and 667 g of carbon tetrachloride, CCl_4?

*11. (a) What volume of each component should be used to prepare the solution in Problem 10?
   (b) If this solution is 3.11 molar in benzene, what is the density of the solution? (Densities: benzene = 0.879 g/mL; carbon tetrachloride = 1.595 g/mL.)

*12. What volume of H$_2$ gas, measured at 27°C and 700 mm Hg pressure, can be obtained by reacting 4.00 g of zinc metal with (a) 100 mL of 0.450 M HCl, and (b) 200 mL of 0.450 M HCl? The equation is:

$$Zn(s) + 2\,HCl(aq) \longrightarrow ZnCl_2(aq) + H_2(g)$$

13. Calculate the pH of solutions having the following H$^+$ ion concentrations:
   (a) 0.01 M (d) $1 \times 10^{-7}\,M$
   (b) 1.0 M (e) 0.50 M
   (c) $6.5 \times 10^{-9}\,M$ (f) 0.00010 M

14. Calculate the pH of the following:
   (a) Orange juice, $2.1 \times 10^{-4}\,M$ H$^+$ (c) Black coffee, $6.2 \times 10^{-5}\,M$ H$^+$
   (b) Vinegar, $1.5 \times 10^{-3}\,M$ H$^+$ (d) Limewater, $3.4 \times 10^{-11}\,M$ H$^+$

*15. A solution of 75 g of acetic acid, $HC_2H_3O_2$, in 250 g of water has a density of 1.03 g/mL. What is the (a) volume of the solution, (b) molarity of the solution, and (c) the molality of the solution?

*16. What is (a) the weight percent, (b) the molarity, and (c) the molality of 125 mL of an aqueous solution having a density of 1.203 g/mL and containing 30.0 g of BaCl$_2$?

17. What volume of concentrated (18.0 M) sulfuric acid must be used to prepare 50.0 L of 4.0 M solution?

18. How many grams of silver iodide, AgI, will be precipitated when 10.0 mL of 1.00 M KI are mixed with (a) 19.0 mL of 0.500 M AgNO$_3$, and (b) 21.0 mL of 0.500 M AgNO$_3$?

19. Two (2.0) grams of NaOH are added to 250 mL of 0.10 M HCl. Will the resulting solution be acidic or basic? Show evidence for your answer.

16

Chemical Equilibrium

After studying Chapter 16 you should be able to:

1. Understand the terms listed in Question A at the end of the chapter.
2. Describe a reversible reaction.
3. Explain why the rate of the forward reaction decreases and the rate of the reverse reaction increases as a chemical reaction approaches equilibrium.
4. State and understand the qualitative effect of Le Chatelier's principle.
5. Tell how the rate of a chemical reaction is affected by the following: (a) changes in concentration of reactants, (b) changes in pressure on gaseous reactants, (c) changes in temperature, and (d) the presence of a catalyst.
6. Write the equilibrium constant expression for a chemical reaction from a balanced equation.
7. Calculate the K_{eq} when given the concentration of reactants and products in equilibrium.
8. Calculate the concentration of one substance in an equilibrium when given the equilibrium constant and the concentrations of all the other substances.
9. Calculate the ionization constant for a weak acid from appropriate data.
10. Calculate the concentrations of all the chemical species in a solution of a weak acid when given the percent ionization or the ionization constant.
11. Compare the relative strengths of acids by using their ionization constants.
12. Using the ion product constant for water, K_w, calculate $[H^+]$, $[OH^-]$, pH, or pOH when given any one of these quantities.
13. Calculate the solubility product constant, K_{sp}, of a slightly soluble salt when given its solubility, or vice versa.
14. Calculate the solubility of a salt when given the K_{sp} value.
15. Compare the relative solubilities of salts when given their solubility products.
16. Explain how a buffer solution is able to counteract the addition of small amounts of either H^+ or OH^- ions.
17. Explain the relative energy diagram of a reaction in terms of activation energy, exothermic or endothermic reaction, and the effect of a catalyst.

16.1 Reversible Reactions

In the preceding chapters we have treated chemical reactions mainly as reactants going to products. However, many reactions do not go to completion. One reason why reactions do not go to completion is that many of them are reversible; that is, when the products are formed, they react to produce the starting reactants.

We have encountered several reversible systems. One is the vaporization of a liquid by heating and subsequent condensation by cooling:

$$\text{Liquid} + \text{Heat} \longrightarrow \text{Vapor}$$

$$\text{Vapor} + \text{Cooling} \longrightarrow \text{Liquid}$$

Another is the crystallization of an aqueous salt solution, which may be considered the reverse of the dissolving and dissociation of the salt:

$$NaCl(s) \xrightarrow{\text{H}_2\text{O}} Na^+(aq) + Cl^-(aq) \quad \text{(Dissociation)}$$

$$Na^+(aq) + Cl^-(aq) \longrightarrow NaCl(s) \quad \text{(Crystallization)}$$

Weak electrolytes are ionized to a small degree because of the reversible reaction of their ions to form the un-ionized compound. A 1 M solution of acetic acid illustrates this behavior:

$$HC_2H_3O_2(aq) + H_2O(l) \longrightarrow H_3O^+(aq) + C_2H_3O_2^-(aq)$$
$$\text{(Forward reaction, 1\%)}$$

$$H_3O^+(aq) + C_2H_3O_2^-(aq) \longrightarrow HC_2H_3O_2(aq) + H_2O(l)$$
$$\text{(Reverse reaction, 99\%)}$$

These two reactions may be represented by a single equation with a double arrow, $\rightleftarrows$, to indicate that they are taking place at the same time:

$$HC_2H_3O_2(aq) + H_2O(l) \rightleftarrows H_3O^+(aq) + C_2H_3O_2^-(aq)$$
(This single equation represents both the forward and reverse reactions.)

The interconversion of nitrogen dioxide (NO_2) and dinitrogen tetroxide (N_2O_4) offers visible evidence of the reversibility of a reaction. NO_2 is a reddish brown gas that changes, with cooling, to N_2O_4, a yellow liquid boiling at 21.2°C and to a colorless solid (N_2O_4) melting at -11.2°C. The reaction is reversible by heating N_2O_4.

$$2\ NO_2(g) \xrightarrow{\text{Cooling}} N_2O_4(l)$$

$$N_2O_4(l) \xrightarrow{\text{Heating}} 2\ NO_2(g)$$

This reversible reaction can readily be demonstrated by sealing samples of NO_2 in two tubes and placing one tube in warm water and the other in ice water (see Figure 16.1). Heating promotes disorder or randomness in a system, so we would expect more NO_2, a gas, to be present at higher temperatures.

reversible reaction   A **reversible reaction** is one in which the products formed in a chemical reaction also react to produce the original reactants. Both the forward reaction and the reverse reaction occur simultaneously. The forward reaction is called *the reaction to the right,* and the reverse reaction is called *the reaction to the left.* A double arrow is used in the equation to indicate that the reaction is reversible.

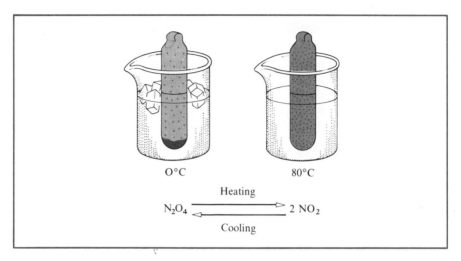

Figure 16.1.   Reversible reaction of nitrogen dioxide (NO_2) and dinitrogen tetroxide (N_2O_4). More of the reddish brown NO_2 molecules are visible in the tube that is heated than in the tube that is cooled.

16.2 Rates of Reaction

chemical kinetics

Every reaction has a rate, or speed, at which it proceeds. Some are fast and some are extremely slow. The study of reaction rates and reaction mechanisms is known as **chemical kinetics.**

The speed of a reaction is variable and depends on the concentration of the reacting species, the temperature, the presence of catalytic agents, and the nature of the reactants. Consider the hypothetical reaction

$$A + B \longrightarrow C + D \quad \text{(Forward reaction)}$$

$$C + D \longrightarrow A + B \quad \text{(Reverse reaction)}$$

where a collision between A and B is necessary for a reaction to occur. The rate at which A and B react depends on the concentration or the number of A and B molecules present; it will be fastest, for a fixed set of conditions, when they are first mixed. As the reaction proceeds, the number of A and B molecules available for reaction decreases, and the rate of reaction slows down. If the reaction is reversible, the speed of the reverse reaction is zero at first, and gradually increases as the concentrations of C and D increase. As the number of A and B molecules decreases, the forward rate slows down, because A and B cannot find one another as often in order to accomplish a reaction. To counteract this diminishing rate of reaction, an excess of one reagent is often used to keep the reaction from becoming unreasonably slow. Collisions between molecules may be likened to the scooters or "dodge'ems" found at amusement parks. When many cars are on the floor, collisions occur frequently, but if only a few cars are present, collisions can usually be avoided.

16.3 Chemical Equilibrium

equilibrium

chemical equilibrium

Any system at **equilibrium** represents a dynamic state in which two or more opposing processes are taking place at the same time and at the same rate. A chemical equilibrium is a dynamic system in which two or more chemical reactions are going on at the same time and at the same rate. When the rate of the forward reaction is exactly equal to the rate of the reverse reaction, a condition of **chemical equilibrium** exists (see Figure 16.2). The concentrations of the products and the reactants are not changing and the system appears to be at a standstill because the products are reacting at the same rate at which they are being formed.

Chemical equilibrium:

Rate of forward reaction = Rate of reverse reaction

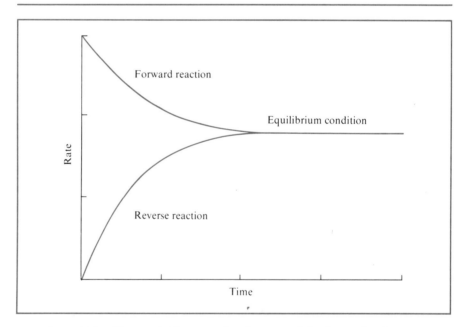

Figure 16.2. The graph illustrates that the rates of the forward and reverse reactions become equal at some point in time. The forward reaction rate decreases as a result of decreasing amounts of reactants. The reverse reaction rate starts at zero and increases as the amount of product increases. When the two rates become equal, a state of chemical equilibrium has been reached.

A saturated salt solution is in a condition of equilibrium:

$$NaCl(s) \rightleftharpoons Na^+(aq) + Cl^-(aq)$$

At equilibrium, salt crystals are continuously dissolving, and Na^+ and Cl^- ions are continuously crystallizing. Both processes are occurring at the same rate.

The ionization of weak electrolytes represents another common chemical equilibrium system:

$$HC_2H_3O_2(aq) + H_2O(l) \rightleftharpoons H_3O^+(aq) + C_2H_3O_2^-(aq)$$

In this reaction, the equilibrium is established in a 1 M solution when the forward reaction has gone about 1%—that is, when only 1% of the acetic acid molecules in solution have ionized. Therefore, only a relatively few ions are present, and the acid behaves as a weak electrolyte. In any acid–base equilibrium system, the position of equilibrium favored is toward the weaker conjugate acid and base. In the ionization of acetic acid, $HC_2H_3O_2$ is a weaker acid than H_3O^+, and H_2O is a weaker base than $C_2H_3O_2^-$.

The reactions represented by

$$H_2(g) + I_2(g) \rightleftharpoons 2\ HI(g)$$

provide another good example of a chemical equilibrium. Theoretically, 1.00 mole of hydrogen should react with 1.00 mole of iodine to yield 2.00 moles of hydrogen iodide. Actually, when 1.00 mole of H_2 and 1.00 mole of I_2 are reacted at 700 K, only 1.58 moles of HI are present when equilibrium is attained. Since 1.58 is 79% of the theoretical yield of 2.00 moles of HI, the forward reaction is only 79% complete at equilibrium. The equilibrium mixture will also contain 0.21 mole each of unreacted H_2 and I_2 because only 79% has reacted (1.00 mole − 0.79 mole = 0.21 mole).

$$\underset{\substack{1.00 \\ \text{mole}}}{H_2} + \underset{\substack{1.00 \\ \text{mole}}}{I_2} \xrightarrow{700\ K} \underset{\substack{2.00 \\ \text{moles}}}{2\ HI}$$ (This would represent the condition if the reaction were 100% complete; 2.00 moles of HI would be formed and no H_2 and I_2 would be left unreacted.)

$$\underset{\substack{0.21 \\ \text{mole}}}{H_2} + \underset{\substack{0.21 \\ \text{mole}}}{I_2} \underset{}{\overset{700\ K}{\rightleftharpoons}} \underset{\substack{1.58 \\ \text{moles}}}{2\ HI}$$ (This represents the actual equilibrium attained starting with 1.00 mole each of H_2 and I_2. It shows that the forward reaction is only 79% complete.)

16.4 Principle of Le Chatelier

principle of
Le Chatelier

In 1888, the French chemist Henri Le Chatelier (1850–1936) set forth a simple, far-reaching generalization on the behavior of equilibrium systems. This generalization is known as the **principle of Le Chatelier** and states: If the conditions of a system in equilibrium are altered, then processes occur in the system that tend to counteract the change. In other words, if a stress is applied to a system in equilibrium, the system will behave in such a way as to relieve that stress and restore equilibrium, but under a new set of conditions.

The application of Le Chatelier's principle helps us to predict the effect of changing conditions in chemical reactions. We will examine the effect of changes in concentration, temperature, and pressure.

16.5 Effect of Concentration on Reaction Rate and Equilibrium

The way in which the rate of a chemical reaction depends on the concentration of the reactants must be determined experimentally. Many simple, one-step reactions occur as the result of a collision between two molecules or ions. The rate of such one-step reactions can be altered by changing the concentration of the reactants or products. An increase in concentration of the reactants provides more individual reacting species for collisions and results in an increase in the rate of reaction.

An equilibrium is disturbed when the concentration of one or more of its components is changed. As a result, the concentration of all the species will change and a new equilibrium mixture will be established. Consider the hypothetical equilibrium represented by the equation

$$A + B \rightleftharpoons C + D$$

where A and B react in one step to form C and D. When the concentration of B is increased, the following occurs:

1. The rate of the reaction to the right (forward) increases. This rate is proportional to the concentration of A times the concentration of B.
2. Therefore, the rate to the right is greater than the rate to the left.
3. Reactants A and B are used faster than they are produced; C and D are produced faster than they are used.
4. After a period of time, rates to the right and left become equal, and the system is again in equilibrium.
5. In the new equilibrium, the concentration of A is less and the concentrations of C and D are greater than in the original equilibrium.

Conclusion: The equilibrium has shifted to the right.

Applying this change in concentration to the equilibrium mixture of 1.00 mole of hydrogen and 1.00 mole of iodine from Section 16.3, we find that when an additional 0.20 mole of I_2 is added, the yield of HI is 85% (1.70 moles) instead of 79%. A comparison of the two systems after the new equilibrium mixture is reached follows.

Original equilibrium	*New equilibrium*
1.00 mole H_2 + 1.00 mole I_2	1.00 mole H_2 + 1.20 moles I_2
Yield: 79% HI	Yield: 85% HI (based on H_2)
Equilibrium mixture contains:	Equilibrium mixture contains:
1.58 moles HI	1.70 moles HI
0.21 mole H_2	0.15 mole H_2
0.21 mole I_2	0.35 mole I_2

Analyzing this new system, we see that when the 0.20 mole I_2 was added, the equilibrium shifted to the right in order to counteract the change in I_2 concentration. Some of the H_2 reacted with added I_2 and produced more HI, until an equilibrium mixture was established again. When I_2 was added, the concentration of I_2 increased, the concentration of H_2 decreased, and the concentration of HI increased. What do you think would be the effects of adding (a) more H_2 or (b) more HI?

The equation

$$Fe^{3+}(aq) + SCN^-(aq) \rightleftharpoons Fe(SCN)^{2+}(aq)$$

Pale yellow   Colorless   Red

represents an equilibrium that is used in certain analytical procedures as an indicator because of the readily visible, intense red color of the complex $Fe(SCN)^{2+}$ ion. A very dilute solution of iron(III), Fe^{3+}, and thiocyanate, SCN^-, is light red. When the concentration of Fe^{3+} or SCN^- is increased, the equilibrium shift to the right is observed by an increase in the intensity of the color, resulting from the formation of additional $Fe(SCN)^{2+}$.

If either Fe^{3+} or SCN^- is removed from solution, the equilibrium will shift to the left, and the solution will become lighter in color. When Ag^+ is added to the solution, a white precipitate of silver thiocyanate (AgSCN) is formed, thus removing SCN^- ion from the equilibrium:

$$Ag^+(aq) + SCN^-(aq) \longrightarrow AgSCN\downarrow$$

The system accordingly responds to counteract the change in SCN^- concentration by shifting the equilibrium to the left. This shift is evident by a decrease in the intensity of the red color due to a decreased concentration of $Fe(SCN)^{2+}$.

Let us now consider the effect of changing the concentrations in the equilibrium mixture of chlorine water. The equilibrium equation is

$$Cl_2(aq) + 2 H_2O \rightleftharpoons HOCl(aq) + H_3O^+ + Cl^-(aq)$$

The variation in concentrations and the equilibrium shifts are tabulated in the following table. An X in the second or third column indicates the reagent that is increased or decreased. The fourth column indicates the direction of the equilibrium shift.

| Reagent | Concentration | | Equilibrium shift |
	Increase	Decrease	
Cl_2	—	X	Left
H_2O	X	—	Right
HOCl	X	—	Left
H_3O^+	—	X	Right
Cl^-	X	—	Left

Consider the equilibrium in a 0.1 M acetic acid solution:

$$HC_2H_3O_2 + H_2O \rightleftharpoons H_3O^+ + C_2H_3O_2^-$$

In this solution, the concentration of the hydronium ion (H_3O^+), which is a measure of the acidity, is 1.34×10^{-3} mole/litre, corresponding to a pH of 2.87. What will happen to the acidity when 0.1 mole of sodium acetate ($NaC_2H_3O_2$) is added to 1 litre of 0.1 M $HC_2H_3O_2$? When $NaC_2H_3O_2$ dissolves, it dissociates into sodium ions (Na^+) and acetate ions ($C_2H_3O_2^-$). The acetate ion from the salt is a common ion to the acetic acid equilibrium system and increases the total acetate ion concentration in the solution. As a result, the equilibrium shifts to the left, decreasing the hydronium ion concentration and lowering the acidity of the solution. Evidence of this decrease in acidity is shown by the fact that the pH of a solution that is 0.1 M in $HC_2H_3O_2$ and 0.1 M in $NaC_2H_3O_2$ is 4.74. The pH of several different solutions of $HC_2H_3O_2$ and $NaC_2H_3O_2$ is shown in the table that follows. Each time the acetate ion is increased, the pH increases, showing a further shift in the equilibrium toward un-ionized acetic acid.

Solution	pH
1 litre 0.1 M $HC_2H_3O_2$	2.87
1 litre 0.1 M $HC_2H_3O_2$ + 0.1 mole $NaC_2H_3O_2$	4.74
1 litre 0.1 M $HC_2H_3O_2$ + 0.2 mole $NaC_2H_3O_2$	5.05
1 litre 0.1 M $HC_2H_3O_2$ + 0.3 mole $NaC_2H_3O_2$	5.23

A secondary reaction of the acetate ion, which also aids in reducing the acidity, is its reaction with water, forming un-ionized $HC_2H_3O_2$ and a hydroxide (OH^-) ion:

$$C_2H_3O_2^- + H_2O \rightleftharpoons HC_2H_3O_2 + OH^-$$

The OH^- ion produced reacts with H_3O^+ to decrease the acidity. Proof that sodium acetate produces OH^- ions in solution is shown by the fact that a 0.1 M $NaC_2H_3O_2$ solution is alkaline, having a pH of 8.87.

In summary, we can say that when the concentration of a reagent on the left side of an equation is increased, the equilibrium shifts to the right. When the concentration of a reagent on the right side of an equation is increased, the equilibrium shifts to the left. In accordance with Le Chatelier's principle, the equilibrium always shifts in the direction that tends to reduce the concentration of the added reactant.

16.6 Effect of Pressure on Reaction Rate and Equilibrium

Changes in pressure significantly affect the reaction rate only when one or more of the reactants or products is a gas. In these cases the effect of increasing the pressure of the reacting gases is equivalent to increasing their concentrations. In the reaction

$$CaCO_3(s) \overset{\Delta}{\rightleftharpoons} CaO(s) + CO_2(g)$$

calcium carbonate decomposes into calcium oxide and carbon dioxide when heated above 825°C. Increasing the pressure of the equilibrium system by adding CO_2 or by decreasing the volume speeds up the reverse reaction and causes the equilibrium to shift to the left. The increased pressure gives the same effect as that caused by increasing the concentration of CO_2, the only gaseous substance in the reaction.

We have seen that when the pressure on a gas is increased, its volume is decreased. In a system composed entirely of gases, an increase in pressure will cause the reaction and the equilibrium to shift to the side that contains the smaller volume or smaller number of moles. This is because the increase in pressure is partially relieved by the system's shifting its equilibrium toward the side in which the substances occupy the smaller volume.

Prior to World War I, Fritz Haber in Germany invented the first major process for the fixation of nitrogen. In the Haber process nitrogen and hydrogen are reacted together in the presence of a catalyst at moderately high temperature and pressure to produce ammonia. The catalyst consists of iron and iron oxide with small amounts of potassium and aluminum oxides. For this process, Haber received the Nobel Prize in chemistry in 1918.

$$N_2(g) + 3 H_2(g) \rightleftharpoons 2 NH_3(g) + 22.1 \text{ kcal}$$

1 mole	3 moles	2 moles
1 volume	3 volumes	2 volumes

The left side of the equation in the Haber process represents four volumes of gas combining to give two volumes of gas on the right side of the equation. An increase in the total pressure on the system shifts the equilibrium to the right. This increase in pressure results in a higher concentration of both reactants and products. Since there are fewer moles of NH_3 than moles of N_2 and H_2, the equilibrium shifts to the right when the pressure is increased.

Ideal conditions for the Haber process are 200°C and 1000 atm pressure. However, at 200°C the rate of reaction is very slow, and at 1000 atm extraordinarily heavy equipment is required. As a compromise the reaction is run at 400–600°C and 200–350 atm pressure, which gives a reasonable yield at a reasonable rate. The effect of pressure on the yield of ammonia at one particular temperature is shown in Table 16.1.

TABLE 16.1 The effect of pressure in the conversion of H_2 and N_2 to NH_3 at 450°C. The starting ratio of H_2 to N_2 is 3 moles to 1 mole.

Pressure (atm)	Yield of NH_3 (%)	Pressure (atm)	Yield of NH_3 (%)
10	2.04	300	35.5
30	5.80	600	53.4
50	9.17	1000	69.4
100	16.4		

When the total number of gaseous molecules on both sides of an equation is the same, a change in pressure does not cause an equilibrium shift. The following reaction is an example:

$$N_2(g) \quad + \quad O_2(g) \; \rightleftharpoons \; 2\,NO(g)$$

1 mole	1 mole	2 moles
1 volume	1 volume	2 volumes
6.02×10^{23}	6.02×10^{23}	$2 \times 6.02 \times 10^{23}$
molecules	molecules	molecules

When the pressure on this system is increased, the rate of both the forward and the reverse reactions will increase because of the higher concentrations of N_2, O_2, and NO. But the equilibrium will not shift, because the increase in concentration of molecules is the same on both sides of the equation and the decrease in volume is the same on both sides of the equation.

PROBLEM 16.1   What effect would increasing the pressure have on the position of equilibrium in the following reactions?

(a) $2\,SO_2(g) + O_2(g) \rightleftharpoons 2\,SO_3(g)$

(b) $H_2(g) + Cl_2(g) \rightleftharpoons 2\,HCl(g)$

(c) $N_2O_4(l) \rightleftharpoons 2\,NO_2(g)$

(a) The equilibrium will shift to the right because the substances on the right have a smaller volume than those on the left.

(b) There will be no effect on the equilibrium position because the volumes (or moles) of gases on both sides of the equation are the same.

(c) The equilibrium will shift to the left because $N_2O_4(l)$ occupies a much smaller volume than does $2\,NO_2(g)$.

■

16.7 Effect of Temperature on Reaction Rate and Equilibrium

An increase in temperature is generally accompanied by an increased rate of reaction. Molecules at elevated temperatures are more energetic and have more kinetic energy; thus, their collisions are more likely to result in a reaction. However, we cannot assume that the rate of a desired reaction will keep increasing indefinitely as the temperature is raised. High temperatures may cause the destruction or decomposition of the reactants and products or may initiate reactions other than the one desired. For example, when calcium oxalate (CaC_2O_4) is heated to 500°C, it decomposes into calcium carbonate and carbon monoxide:

$$CaC_2O_4(s) \xrightarrow{\;500°C\;} CaCO_3(s) + CO\uparrow$$

If calcium oxalate is heated to 850°C, the products are calcium oxide, carbon monoxide, and carbon dioxide:

$$CaC_2O_4(s) \xrightarrow{850°C} CaO(s) + CO\uparrow + CO_2\uparrow$$

When heat is applied to a system in equilibrium, the reaction that absorbs heat is favored. When the process, as written, is endothermic, the forward reaction is increased. When the reaction is exothermic, the reverse reaction is favored. In this sense heat may be treated as a reactant in endothermic reactions or as a product in exothermic reactions. Therefore, temperature is analogous to concentration when applying Le Chatelier's principle to heat effects on a chemical reaction.

Hot coke (C) is a very good reducing agent. In the reaction

$$C(s) + CO_2(g) + Heat \rightleftharpoons 2\, CO\uparrow$$

very little, if any, CO is formed at room temperature. At 1000°C, the equilibrium mixture contains about an equal number of moles of CO and CO_2. At higher temperatures, the equilibrium shifts to the right, increasing the yield of CO. The reaction is endothermic and, as can be seen, the equilibrium is shifted to the right at higher temperatures.

Phosphorus trichloride reacts with dry chlorine gas to form phosphorus pentachloride. The reaction is exothermic:

$$PCl_3(l) + Cl_2(g) \rightleftharpoons PCl_5(s) + 21\ kcal$$

Heat must continually be removed during the reaction to obtain a good yield of the product. According to the principle of Le Chatelier, heat will cause the product, PCl_5, to decompose, re-forming PCl_3 and Cl_2. The equilibrium mixture at 200°C contains 52% PCl_5, and at 300°C it contains 3% PCl_5, verifying that heat causes the equilibrium mixture to shift to the left.

When the temperature of a system is raised, the rate of reaction increases because of increased kinetic energy and more frequent collisions of the reacting species. In a reversible reaction, the rate of both the forward and the reverse reactions is increased by an increase in temperature; however, the reaction that absorbs heat increases to a greater extent, and the equilibrium shifts to favor that reaction. The following examples illustrate these effects:

$$4\, HCl(g) + O_2(g) \rightleftharpoons 2\, H_2O(g) + 2\, Cl_2(g) + 28.4\ kcal \qquad (1)$$

$$H_2(g) + Cl_2(g) \rightleftharpoons 2\, HCl(g) + 44.2\ kcal \qquad (2)$$

$$CH_4(g) + 2\, O_2(g) \rightleftharpoons CO_2(g) + 2\, H_2O(l) + 212.8\ kcal \qquad (3)$$

$$N_2O_4(l) + 14\ kcal \rightleftharpoons 2\, NO_2(g) \qquad (4)$$

$$2\, CO_2(g) + 135.2\ kcal \rightleftharpoons 2\, CO(g) + O_2(g) \qquad (5)$$

$$H_2(g) + I_2(g) + 12.4\ kcal \rightleftharpoons 2\, HI(g) \qquad (6)$$

Reactions (1), (2), and (3) are exothermic; an increase in temperature will cause the equilibrium to shift to the left. Reactions (4), (5), and (6) are endothermic; an increase in temperature will cause the equilibrium to shift to the right.

16.8 Effect of Catalysts on Reaction Rate and Equilibrium

catalyst

A **catalyst** is a substance that influences the speed of a reaction and may be recovered essentially unchanged at the end of the reaction. A catalyst does not shift the equilibrium of a reaction; it affects only the speed at which the equilibrium is reached. If a catalyst does not affect the equilibrium, then it follows that it must affect the speed of both the forward and the reverse reactions equally.

The reaction between phosphorus trichloride (PCl_3) and sulfur is highly exothermic, but it is so slow that very little product, thiophosphoryl chloride ($PSCl_3$), is obtained, even after prolonged heating. When a catalyst, such as aluminum chloride ($AlCl_3$), is added, the reaction is complete in a few seconds:

$$PCl_3(l) + S(s) \xrightarrow{\text{AlCl}_3} PSCl_3(l)$$

We have already demonstrated that manganese dioxide used as a catalyst increases the rates of decomposition of both potassium chlorate and hydrogen peroxide.

Catalysts are extremely important to industrial chemistry. Hundreds of chemical reactions that are otherwise too slow to be of practical value have been put to commercial use once a suitable catalyst was found. But it is in the area of biochemistry that catalysts are of supreme importance. Nearly all the chemical reactions associated with all forms of life are completely dependent on biochemical catalysts known as *enzymes*.

16.9 Equilibrium Constants

Law of
Mass Action

The **Law of Mass Action** states that the rate of a chemical reaction is proportional to the concentration of the reacting species. This simply means that the higher the concentration of reactants, the more frequently they collide and form products. For the equilibrium system in which A and B react in one step to give C and D,

$$A + B \rightleftharpoons C + D$$

the rates of the forward and reverse reactions can be expressed as

$$\text{Rate}_f = k_f \times \text{Concentration of A} \times \text{Concentration of B}$$

$$\text{Rate}_r = k_r \times \text{Concentration of C} \times \text{Concentration of D}$$

where Rate_f is the rate of the forward reaction, Rate_r is the rate of the reverse reaction, and k_f and k_r are the proportionality rate constants. For dilute solutions, the unit of concentration is moles per litre; for gases, it is either moles per litre or pressure. To simplify the rate expressions, we place the formula of each substance in brackets to indicate the concentration of each substance. The concentrations of A, B, C, and D are then expressed as [A], [B], [C], and [D], respectively.

At equilibrium,

$$\text{Rate}_f = \text{Rate}_r$$

Then

$$\text{Rate}_f = k_f[A][B]$$
$$\text{Rate}_r = k_r[C][D]$$

$$\frac{k_f}{k_r} = \frac{[C][D]}{[A][B]}$$

equilibrium constant, K_{eq}

Since both k_f and k_r are constants, k_f/k_r is also a constant and is known as the **equilibrium constant,** abbreviated K_{eq}:

$$K_{eq} = \frac{[C][D]}{[A][B]}$$

This expression reads as follows: The equilibrium constant, K_{eq}, is equal to the product of the concentration of C and the concentration of D divided by the product of the concentration of A and the concentration of B. Consider the equilibrium

$$2\,A \rightleftharpoons C + D$$

The forward reaction is dependent on the collision of two A molecules. Therefore,

$$\text{Rate}_f = k_f[A][A] = k_f[A]^2$$

$$\text{Rate}_r = k_r[C][D]$$

$$K_{eq} = \frac{k_f}{k_r} = \frac{[C][D]}{[A]^2}$$

In this equilibrium, the equilibrium constant is equal to the product of the concentrations of C and D divided by the concentration of A squared.

For the general reaction

$$n\,A + m\,B \rightleftharpoons p\,C + q\,D$$

where n, m, p, and q are the small whole numbers in the balanced equation, the equilibrium constant expression is

$$K_{eq} = \frac{[C]^p[D]^q}{[A]^n[B]^m}$$

Observe that the concentration of each substance is raised to a power that is the same as its numerical coefficient in the balanced equation. It is conventional to place the concentrations of the products (the substances on the right side of the equation as written) in the numerator and the concentrations of the reactants in the denominator.

If the molar concentrations of all the species in an equilibrium are known, the K_{eq} can be calculated by substituting the concentrations into the equilibrium constant expression.

PROBLEM 16.2   Calculate the K_{eq} for the following reaction:

$$PCl_5(g) \rightleftharpoons PCl_3(g) + Cl_2(g)$$

At 300°C the concentrations are: $PCl_5 = 0.030$ mole/L; $PCl_3 = 0.97$ mole/L; $Cl_2 = 0.97$ mole/L. First write the K_{eq} expression:

$$K_{eq} = \frac{[PCl_3][Cl_2]}{[PCl_5]}$$

Substitute the respective concentrations into this equation and solve:

$$K_{eq} = \frac{[0.97][0.97]}{[0.030]} = 31$$

This K_{eq} is considered to be a fairly large value, indicating that at 300°C the decomposition of PCl_5 proceeds far to the right.

■

16.10 Ionization Constants

ionization
constant, K_i

As a first application of an equilibrium constant, let us consider the constant for acetic acid in solution. Because it is a weak acid, an equilibrium is established between molecular $HC_2H_3O_2$ and its ions in solution. The constant is called the **ionization constant,** K_i, a special type of equilibrium constant. The concentration of water in the solution does not change appreciably, so we may use the following simplified equation to set up the constant:

$$HC_2H_3O_2 \rightleftharpoons H^+ + C_2H_3O_2^-$$

The ionization constant expression is

$$K_i = \frac{[H^+][C_2H_3O_2^-]}{[HC_2H_3O_2]}$$

It states that the ionization constant, K_i, is equal to the product of the hydrogen ion $[H^+]$ concentration and the acetate ion $[C_2H_3O_2^-]$ concentration divided by the concentration of the un-ionized acetic acid $[HC_2H_3O_2]$.

At 25°C, a 0.1 M $HC_2H_3O_2$ solution is 1.34% ionized and has a hydrogen ion concentration of 1.34×10^{-3} mole/L. From this information, we can calculate the ionization constant for acetic acid.

A 0.10 M solution contains 0.10 mole of acetic acid per litre. Of this 0.10 mole, only 1.34%, or 1.34×10^{-3} mole, is ionized. This gives a H^+ ion concentration of 1.34×10^{-3} mole/L. Since each molecule of acid that ionizes yields one H^+ and one $C_2H_3O_2^-$, the concentration of $C_2H_3O_2^-$ ions is also 1.34×10^{-3} mole/L. This ionization leaves $0.10 - 0.00134 = 0.09866$ mole/L of un-ionized acetic acid.

	Initial concentration (moles/L)	Equilibrium concentration (moles/L)
$[HC_2H_3O_2]$	0.10	0.09866
$[H^+]$	0	0.00134
$[C_2H_3O_2^-]$	0	0.00134

Substituting these concentrations in the equilibrium expression, we obtain the value for K_i:

$$K_i = \frac{[H^+][C_2H_3O_2^-]}{[HC_2H_3O_2]}$$

$$= \frac{[1.34 \times 10^{-3}][1.34 \times 10^{-3}]}{[0.09866]} = 1.8 \times 10^{-5}$$

The magnitude of an equilibrium constant indicates the extent to which the forward and reverse reactions take place. If the equilibrium constant is very large, the forward reaction goes nearly to completion and the reverse reaction does not occur appreciably. If the equilibrium constant is very small, the reverse reaction goes nearly to completion and the forward reaction goes only to a small extent. The K_i for acetic acid, 1.8×10^{-5}, is small and indicates that the position of the equilibrium is far toward the un-ionized acetic acid. In fact, a 0.1 M acetic acid solution is 98.66% un-ionized.

Once the K_i for acetic acid is established, it can be used to describe any system containing H^+, $C_2H_3O_2^-$, and $HC_2H_3O_2$ in equilibrium at 25°C. The ionization constants for several other weak acids are listed in Table 16.2.

TABLE 16.2 Ionization constants (K_i) of weak acids at 25°C.

Acid	Formula	K_i
Acetic	$HC_2H_3O_2$	1.8×10^{-5}
Benzoic	$HC_7H_5O_2$	6.3×10^{-5}
Carbolic (phenol)	HC_6H_5O	1.3×10^{-10}
Cyanic	$HCNO$	2.0×10^{-4}
Formic	$HCHO_2$	1.8×10^{-4}
Hydrocyanic	HCN	4.0×10^{-10}
Hypochlorous	$HClO$	3.5×10^{-8}
Nitrous	HNO_2	4.5×10^{-4}

PROBLEM 16.3   What is the ionization constant expression for nitrous acid?
First, write the simplified ionization equation:

$$HNO_2(aq) \rightleftarrows H^+ + NO_2^-   \text{(Simplified)}$$

The format of the ionization constant expression, K_i, is the concentrations of the products divided by the concentration of the reactant. Thus,

$$K_i = \frac{[H^+][NO_2^-]}{[HNO_2]}   \text{(Answer)}$$

∎

PROBLEM 16.4   What is the H^+ ion concentration in a 0.50 M $HC_2H_3O_2$ solution? The ionization constant, K_i, for $HC_2H_3O_2$ is 1.8×10^{-5}.
To solve this problem, first write the equilibrium equation and the K_i expression:

$$HC_2H_3O_2 \rightleftarrows H^+ + C_2H_3O_2^-$$

$$K_i = \frac{[H^+][C_2H_3O_2^-]}{[HC_2H_3O_2]} = 1.8 \times 10^{-5}$$

We know that the initial concentration of $HC_2H_3O_2$ is 0.50 M. We also know from the ionization equation that one $C_2H_3O_2^-$ is produced for every H^+ produced. That is, the $[H^+]$ and the $[C_2H_3O_2^-]$ are equal. To solve, let $Y = [H^+]$, which also equals the $[C_2H_3O_2^-]$. The un-ionized $[HC_2H_3O_2]$ remaining will then be $0.50 - Y$, the starting concentration minus the amount that ionized.

$$[H^+] = [C_2H_3O_2^-] = Y   \quad [HC_2H_3O_2] = 0.50 - Y$$

Substituting these values into the K_i expression, we obtain

$$K_i = \frac{(Y)(Y)}{0.50 - Y} = \frac{Y^2}{0.50 - Y} = 1.8 \times 10^{-5}$$

An exact solution of this equation for Y requires the use of a mathematical equation known as the quadratic equation. However, an approximate solution is readily obtained if we assume that Y is small and can be neglected compared to 0.50. Then $0.50 - Y$ will be equal to approximately 0.50. The equation now becomes

$$\frac{Y^2}{0.50} = 1.8 \times 10^{-5}$$

$$Y^2 = 1.8 \times 10^{-5} \times 0.50 = 0.90 \times 10^{-5} = 9.0 \times 10^{-6}$$

Taking the square root of both sides of the equation, we get

$$Y = \sqrt{9.0 \times 10^{-6}} = 3.0 \times 10^{-3} \text{ mole/L}$$

Thus, $[H^+]$ is approximately 3.0×10^{-3} mole/L in a 0.50 M $HC_2H_3O_2$ solution. The exact solution to this problem, using the quadratic equation, gives a value of 2.99×10^{-3} mole/L for $[H^+]$, showing that we were justified in neglecting Y compared to 0.50.

∎

PROBLEM 16.5 Calculate the percent ionization in a 0.50 M $HC_2H_3O_2$ solution.

The percent ionized may be calculated from the following equation:

$$\frac{\text{Concentration of } [H^+] \text{ or } [C_2H_3O_2^-]}{\text{Initial concentration of } [HC_2H_3O_2]} \times 100\% = \text{Percent ionized}$$

To solve this problem we first need to calculate $[H^+]$. This has already been done in Problem 16.4 for a 0.50 M solution.

$$[H^+] = 3.0 \times 10^{-3} \text{ mole/L in a } 0.50 \; M \text{ solution}$$

This $[H^+]$ represents a fractional amount of the initial 0.50 M $HC_2H_3O_2$. Therefore,

$$\frac{3.0 \times 10^{-3} \text{ mole/L}}{0.50 \text{ mole/L}} \times 100\% = 0.60\% \text{ ionized}$$

A 0.50 M $HC_2H_3O_2$ solution is 0.60% ionized.

■

16.11 Ion Product Constant for Water

We have seen that water ionizes to a slight degree. This ionization is represented by these equilibrium equations:

$$H_2O + H_2O \rightleftharpoons H_3O^+ + OH^- \tag{7}$$

$$H_2O \rightleftharpoons H^+ + OH^- \tag{8}$$

Equation (7) is the more accurate representation of the equilibrium since free protons (H^+) do not exist in water. Equation (8) is a simplified and widely used representation of the water equilibrium. The actual concentration of H^+ produced in pure water is very minute and amounts to only 1×10^{-7} mole/L at 25°C. In pure water,

$$[H^+] = [OH^-] = 1 \times 10^{-7} \text{ mole/L}$$

since both ions are produced in equal molar amounts, as shown in equation (8).

The $H_2O \rightleftharpoons H^+ + OH^-$ equilibrium exists in water and in all water solutions. A special equilibrium constant called the **ion product constant for water, K_w**, applies to this equilibrium. The constant K_w is defined as the product of the H^+ ion concentration and the OH^- ion concentration, each in moles per litre:

ion product constant for water
K_w

$$K_w = [H^+][OH^-]$$

The numerical value of K_w is 1×10^{-14}, since for pure water at 25°C

$$K_w = [H^+][OH^-] = [1 \times 10^{-7}][1 \times 10^{-7}] = 1 \times 10^{-14}$$

The value of K_w for all water solutions at 25°C is the constant 1×10^{-14}. It is important to realize that as the concentration of one of these ions, H^+ or OH^-,

TABLE 16.3   Relationship of H⁺ and OH⁻ concentrations in water solutions.

$[H^+]$	$[OH^-]$	K_w	pH	pOH
1×10^{-2}	1×10^{-12}	1×10^{-14}	2	12
1×10^{-4}	1×10^{-10}	1×10^{-14}	4	10
2×10^{-6}	5×10^{-9}	1×10^{-14}	5.7	8.3
1×10^{-7}	1×10^{-7}	1×10^{-14}	7	7
1×10^{-9}	1×10^{-5}	1×10^{-14}	9	5

increases, the other decreases. However, the product of $[H^+]$ and $[OH^-]$ always equals the constant 1×10^{-14}. This relationship can be seen in the examples shown in Table 16.3. If the concentration of one ion is known, the concentration of the other can be calculated from the K_w expression.

$$K_w = [H^+][OH^-] \qquad [H^+] = \frac{K_w}{[OH^-]} \qquad [OH^-] = \frac{K_w}{[H^+]}$$

PROBLEM 16.6   What is the concentration of (a) H^+ and (b) OH^- in a 0.001 M HCl solution? Assume that HCl is 100% ionized.

(a) Since all the HCl is ionized, $H^+ = 0.001$ mole/L.

$$\text{HCl} \longrightarrow \text{H}^+ \; + \; \text{Cl}^-$$
0.001 M 0.001 M 0.001 M

$[H^+] = 1 \times 10^{-3}$ (Answer)

(b) To calculate the $[OH^-]$ in this solution, use the following equation and substitute the values for K_w and $[H^+]$:

$$[OH^-] = \frac{K_w}{[H^+]}$$

$$[OH^-] = \frac{1 \times 10^{-14}}{1 \times 10^{-3}} = 1 \times 10^{-11} \text{ mole/L}   \text{(Answer)}$$

PROBLEM 16.7   What is the pH of a 0.01 M NaOH solution? Assume that NaOH is 100% ionized. Since all the NaOH is ionized, $OH^- = 0.01$ mole/L or 1×10^{-2} mole/L.

$$\text{NaOH} \longrightarrow \text{Na}^+ + \text{OH}^-$$
0.01 M 0.01 M 0.01 M

In order to find the pH of the solution, we must first calculate the H^+ concentration. Use the following equation and substitute the values for K_w and $[OH^-]$.

$$[H^+] = \frac{K_w}{[OH^-]}$$

$$[H^+] = \frac{1 \times 10^{-14}}{1 \times 10^{-2}} = 1 \times 10^{-12} \text{ mole/L}$$

$$pH = \log\frac{1}{[H^+]} = \log\frac{1}{10^{-12}} = \log 10^{12} = 12 \quad \text{(Answer)}$$

The pH can also be calculated by the method shown in Section 15.9:

$$[H^+] = 1 \times 10^{-12}$$

$$pH = 12 - \log 1 = 12 - 0 = 12 \quad \text{(Answer)}$$

■

Just as pH is used to express the acidity of a solution, pOH is used to express the basicity of an aqueous solution. The pOH is related to the OH^- ion concentration in the same way that the pH is related to the H^+ ion concentration:

$$pOH = \log\frac{1}{[OH^-]} \quad \text{or} \quad -\log[OH^-]$$

Thus, a solution in which $[OH^-] = 1 \times 10^{-2}$, as in Problem 16.7, will have pOH = 2.

In pure water, where $[H^+] = 1 \times 10^{-7}$ and $[OH^-] = 1 \times 10^{-7}$, the pH is 7 and the pOH is 7. The sum of the pH and pOH is always 14.

$$pH + pOH = 14$$

This relationship holds in all aqueous solutions and is illustrated in the examples in Table 16.3.

16.12 Solubility Product Constant

solubility product constant, K_{sp}

The **solubility product constant,** abbreviated K_{sp}, is another application of the equilibrium constant. It is the equilibrium constant of a slightly soluble salt. The following example illustrates how K_{sp} is evaluated.

The solubility of silver chloride (AgCl) in water is 1.3×10^{-5} mole/L at 25°C. The equation for the equilibrium between AgCl and its ions in solution is

$$AgCl(s) \rightleftharpoons Ag^+ + Cl^-$$

The equilibrium constant expression is

$$K_{eq} = \frac{[Ag^+][Cl^-]}{[AgCl(s)]}$$

The amount of solid AgCl does not affect the equilibrium system provided that some is present. In other words, the concentration of solid silver chloride is constant, whether 1 mg or 10 g of the salt are present. Therefore, the product obtained by multiplying the two constants K_{eq} and [AgCl(s)] is also a constant. This constant is the solubility product constant, K_{sp}.

$$K_{eq} \times [AgCl(s)] = [Ag^+][Cl^-] = K_{sp}$$

$$K_{sp} = [Ag^+][Cl^-]$$

The K_{sp} is equal to the product of the Ag^+ ion and the Cl^- ion concentrations, each in moles per litre. When 1.3×10^{-5} mole/L of AgCl dissolves, it produces 1.3×10^{-5} mole/L each of Ag^+ and Cl^-. From these concentrations, the K_{sp} can be evaluated.

$$[Ag^+] = 1.3 \times 10^{-5} \text{ mole/L} \qquad [Cl^-] = 1.3 \times 10^{-5} \text{ mole/L}$$

$$K_{sp} = [Ag^+][Cl^-] = [1.3 \times 10^{-5}][1.3 \times 10^{-5}] = 1.7 \times 10^{-10}$$

Once the K_{sp} value for AgCl is established, it can be used to describe other systems containing Ag^+ and Cl^-. For example, if silver nitrate, $AgNO_3$, is added to a saturated AgCl solution until the Ag^+ concentration is 0.10 M, what will be the Cl^- ion concentration remaining in solution? The addition of $AgNO_3$ puts Ag^+ ions into solution and causes the AgCl equilibrium to shift to the left, reducing the Cl^- ion concentration. This process of increasing the concentration of one of the ions in an equilibrium, thereby causing the other ion to decrease in concentration, is known as the **common ion effect.**

common ion effect

We use the K_{sp} to calculate the Cl^- ion concentration remaining in solution. The K_{sp} is constant at a particular temperature and remains the same no matter how we change the concentration of the species involved.

$$K_{sp} = [Ag^+][Cl^-] = 1.7 \times 10^{-10} \qquad [Ag^+] = 0.10 \text{ mole/L}$$

We then substitute the concentration of Ag^+ ion into the K_{sp} expression and calculate:

$$[0.10][Cl^-] = 1.7 \times 10^{-10}$$

$$[Cl^-] = \frac{1.7 \times 10^{-10}}{0.10}$$

$$[Cl^-] = 1.7 \times 10^{-9} \text{ mole/L}$$

This calculation shows a 10,000-fold reduction of Cl^- ions in solution. It illustrates that Cl^- ions may be quantitatively removed from solution with an excess of Ag^+ ions. The equilibrium equations and the K_{sp} expressions for several other substances follow.

$$AgBr(s) \rightleftharpoons Ag^+ + Br^- \qquad\qquad K_{sp} = [Ag^+][Br^-]$$

$$BaSO_4(s) \rightleftharpoons Ba^{2+} + SO_4^{2-} \qquad\qquad K_{sp} = [Ba^{2+}][SO_4^{2-}]$$

$$Ag_2CrO_4(s) \rightleftharpoons 2\,Ag^+ + CrO_4^{2-} \qquad\qquad K_{sp} = [Ag^+]^2[CrO_4^{2-}]$$

$$CuS(s) \rightleftharpoons Cu^{2+} + S^{2-} \qquad\qquad K_{sp} = [Cu^{2+}][S^{2-}]$$

$$Mn(OH)_2(s) \rightleftharpoons Mn^{2+} + 2\,OH^- \qquad\qquad K_{sp} = [Mn^{2+}][OH^-]^2$$

$$Fe(OH)_3(s) \rightleftharpoons Fe^{3+} + 3\,OH^- \qquad\qquad K_{sp} = [Fe^{3+}][OH^-]^3$$

TABLE 16.4 Solubility product constants (K_{sp}) at 25°C.

Compound	K_{sp}
AgCl	1.7×10^{-10}
AgBr	5×10^{-13}
AgI	8.5×10^{-17}
$AgC_2H_3O_2$	2×10^{-3}
Ag_2CrO_4	1.9×10^{-12}
$BaCrO_4$	8.5×10^{-11}
$BaSO_4$	1.5×10^{-9}
CaF_2	3.9×10^{-11}
CuS	9×10^{-45}
$Fe(OH)_3$	6×10^{-38}
PbS	7×10^{-29}
$PbSO_4$	1.3×10^{-8}
$Mn(OH)_2$	2.0×10^{-13}

Table 16.4 lists K_{sp} values for these and several other substances. As in any equilibrium constant, the concentration of each substance in the K_{sp} expressions is raised to a power that is the same number as its numerical coefficient in the balanced equilibrium equation, for example, $[Ag^+]^2$ in Ag_2CrO_4.

When the product of the molar concentration of the ions in solution, each raised to its proper power, is greater than the K_{sp} for that substance, precipitation should occur. If the ion product is less than the K_{sp} value, no precipitation will occur.

PROBLEM 16.8   Write K_{sp} expressions for AgI and PbI_2, both of which are slightly soluble salts.

First write the equilibrium equations:

$$AgI(s) \rightleftharpoons Ag^+ + I^-$$

$$PbI_2(s) \rightleftharpoons Pb^{2+} + 2\,I^-$$

Since the concentration of the solid crystals is constant, the K_{sp} equals the product of the molar concentrations of the ions in solution. In the case of PbI_2, the $[I^-]$ must be squared.

$$K_{sp} = [Ag^+][I^-]$$

$$K_{sp} = [Pb^{2+}][I^-]^2$$

■

PROBLEM 16.9   The K_{sp} value for lead sulfate is 1.3×10^{-8}. Calculate the solubility of $PbSO_4$ in grams per litre.

First write the equilibrium equation and the K_{sp} expression:

$$PbSO_4 \rightleftharpoons Pb^{2+} + SO_4^{2-}$$

$$K_{sp} = [Pb^{2+}][SO_4^{2-}] = 1.3 \times 10^{-8}$$

Since the lead sulfate that is in solution is completely dissociated, the concentration of $[Pb^{2+}]$ or $[SO_4^{2-}]$ is equal to the solubility of $PbSO_4$ in moles per litre.

Let $Y = [Pb^{2+}] = [SO_4^{2-}]$

Substitute Y into the K_{sp} equation and solve.

$$[Pb^{2+}][SO_4^{2-}] = [Y][Y] = 1.3 \times 10^{-8}$$

$$Y^2 = 1.3 \times 10^{-8}$$

$$Y = 1.14 \times 10^{-4} \, mole/L$$

The solubility of $PbSO_4$, therefore, is 1.14×10^{-4} mole/L. Now convert moles/litre to grams/litre:

1 mole of $PbSO_4$ weighs 303.3 g (207.2 + 32.1 + 64.0)

$$\frac{1.14 \times 10^{-4} \, \text{mole}}{L} \times \frac{303.3 \, g}{\text{mole}} = 3.5 \times 10^{-2} \, g/L$$

The solubility of $PbSO_4$ is 3.5×10^{-2} g/L.

■

16.13 Buffer Solutions—The Control of pH

The control of pH within narrow limits is critically important in many chemical applications and vitally important in many biological systems. For example, human blood must be maintained between a pH of 7.35 and 7.45 for the efficient transport of oxygen from the lungs to the cells. This narrow pH range is maintained by buffer systems within the blood.

buffer solution

A solution that resists changes in pH when diluted or when small amounts of acid or base are added is called a **buffer solution.** Two common types of buffer solutions are (1) a weak acid together with a salt of that weak acid and (2) a weak base together with a salt of that weak base.

The action of a buffer system can be understood by considering a solution of acetic acid and sodium acetate. The weak acid, $HC_2H_3O_2$, is mostly un-ionized and is in equilibrium with its ions in solution. The salt, $NaC_2H_3O_2$, is completely ionized.

$$HC_2H_3O_2(aq) \rightleftharpoons H^+(aq) + C_2H_3O_2^-(aq)$$

$$NaC_2H_3O_2(aq) \longrightarrow Na^+(aq) + C_2H_3O_2^-(aq)$$

Since the salt is completely ionized, the solution contains a much higher concentration of acetate ions than would be present if only acetic acid were in solution. The acetate ion represses the ionization of acetic acid and also reacts with water, causing the solution to have a higher pH (be more basic) than an acetic acid solution (see Section 16.5). Thus, a 0.1 M acetic acid solution has a pH of 2.87, but a solution that is 0.1 M in acetic acid and 0.1 M in sodium acetate has a pH of 4.74.

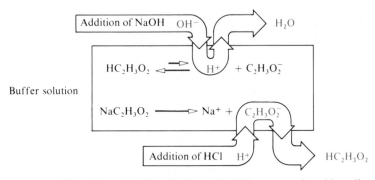

Figure 16.3   The effect of adding HCl and NaOH to an acetic acid–sodium acetate buffer solution. The added H$^+$ from HCl is removed from solution by forming un-ionized acetic acid. The added OH$^-$ from NaOH is removed by reacting with H$^+$ to form water.

A buffer solution has a built-in mechanism that counteracts the effect of adding acid or base. Consider the effect of adding HCl or NaOH to an acetic acid–sodium acetate buffer. When HCl is added, the acetate ions of the buffer combine with the H$^+$ ions of HCl to form un-ionized acetic acid, thus neutralizing the added acid and maintaining the approximate pH of the solution. When NaOH is added, the H$^+$ ions in the buffer combine with the OH$^-$ ions to form water. Additional acetic acid then ionizes (Le Chatelier's principle) to restore the H$^+$ ions and maintain the approximate pH of the solution. The action of this buffer system in counteracting added acid or added base is illustrated in Figure 16.3.

Data comparing the changes in pH caused by adding HCl and NaOH to pure water and to an acetic acid–sodium acetate buffer solution are shown in Table 16.5.

There are a number of buffer systems in the human body. One of these, the bicarbonate–carbonic acid buffer, $HCO_3^- – H_2CO_3$, maintains the blood plasma at a pH of 7.4. The phosphate system, $HPO_4^{2-} – H_2PO_4^-$, is an important buffer in the red blood cells as well as in other places in the body.

TABLE 16.5 Changes in pH caused by the addition of HCl and NaOH to pure water and to an acetic acid–sodium acetate buffer solution.

Solution	pH	*Change in pH*
H$_2$O (1000 mL)	7	—
H$_2$O + 0.01 mole HCl	2	5
H$_2$O + 0.01 mole NaOH	12	5
Buffer solution (1000 mL),		
0.10 M HC$_2$H$_3$O$_2$ + 0.10 M NaC$_2$H$_3$O$_2$	4.74	—
Buffer + 0.01 mole HCl	4.66	0.08
Buffer + 0.01 mole NaOH	4.83	0.09

16.14 Mechanism of Reactions

mechanism
of a reaction

How a reaction occurs—that is, the manner or route by which it proceeds—is known as the **mechanism of the reaction.** The mechanism shows us the path, or course, the atoms and molecules take to arrive at the products. Our aim here is not to study the mechanisms themselves but to show that chemical reactions occur by specific routes.

When hydrogen and iodine are mixed at room temperature, we observe no appreciable reaction. In this case, the reaction takes place as a result of a collision between an H_2 and an I_2 molecule, but at room temperature the collisions do not result in reaction because the molecules do not have sufficient energy to react with each other. We might say that an energy barrier to reaction exists. As heat is added, the kinetic energy of the molecules increases. When molecules of H_2 and I_2 having sufficient energy collide, an intermediate, known as the **activated complex,** is formed. The amount of energy needed to form the activated complex is known as the **activation energy.** The complex, H_2I_2, is in a metastable form and has an energy level higher than that of the reactants or that of the product. It can decompose to form either the reactants or the product. Three steps constitute the mechanism of the reaction: (1) collision of an H_2 and an I_2 molecule; (2) formation of the activated complex, H_2I_2; and (3) decomposition to the product, HI. The various steps in the formation of HI are shown in Figure 16.4. Figure 16.5 illustrates the energy relationships in this reaction.

activated complex
activation energy

The reaction of hydrogen and chlorine proceeds by a different mechanism. When H_2 and Cl_2 are mixed and kept in the dark, essentially no product is formed. But if the mixture is exposed to sunlight or ultraviolet radiation, it reacts very rapidly. The overall reaction is

$$H_2(g) + Cl_2(g) \longrightarrow 2\,HCl(g)$$

free radical

This reaction proceeds by what is known as a *free radical mechanism.* A **free radical** is a neutral atom or group containing one or more unpaired electrons.

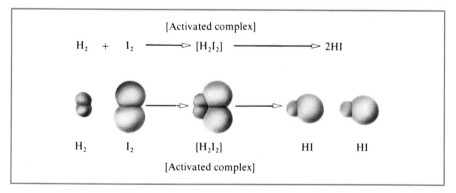

Figure 16.4 Mechanism of the reaction between hydrogen and iodine: H_2 and I_2 molecules of sufficient energy unite, forming the intermediate activated complex, which decomposes to the product, hydrogen iodide.

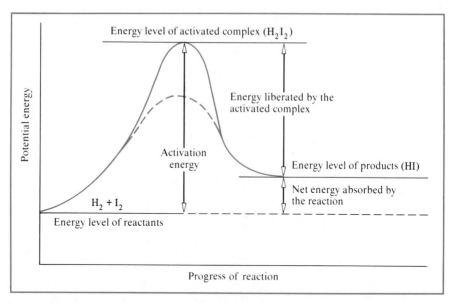

Figure 16.5   Relative-energy diagram for the reaction between hydrogen and iodine.

$$H_2 + I_2 \longrightarrow [H_2I_2] \longrightarrow 2\,HI$$

Energy equal to the activation energy is put into the system in forming the activated complex, H_2I_2. When this complex decomposes, it liberates energy, forming the product. In this case, the product is at a higher energy level than the reactants, indicating that the reaction is endothermic and that energy is absorbed during the reaction. The dotted line represents the effect that a catalyst would have on the reaction. The catalyst lowers the activation energy, thereby increasing the rate of the reaction.

Both atomic chlorine ($:\!\overset{..}{\underset{..}{Cl}}\!\cdot$) and atomic hydrogen ($H\cdot$) have an unpaired electron and are free radicals. The reaction occurs in three steps:

Step 1. *Initiation:*

$$:\!\overset{..}{\underset{..}{Cl}}\!:\!\overset{..}{\underset{..}{Cl}}\!: \; + \; h\nu \longrightarrow \quad :\!\overset{..}{\underset{..}{Cl}}\!\cdot \; + \; :\!\overset{..}{\underset{..}{Cl}}\!\cdot$$
Chlorine free radicals

In this step a chlorine molecule absorbs energy in the form of a photon, $h\nu$, of light or ultraviolet radiation. The energized chlorine molecule then splits into two chlorine free radicals.

Step 2. *Propagation:*

$$:\!\overset{..}{\underset{..}{Cl}}\!\cdot \; + \; H\!:\!H \longrightarrow HCl \quad + \quad H\cdot$$
Hydrogen
free radical

$$H\cdot \; + \; :\!\overset{..}{\underset{..}{Cl}}\!:\!\overset{..}{\underset{..}{Cl}}\!: \longrightarrow HCl + :\!\overset{..}{\underset{..}{Cl}}\!\cdot$$

This step begins when a chlorine free radical reacts with a hydrogen molecule to produce a molecule of hydrogen chloride and a hydrogen free radical. The hydrogen radical next reacts with another chlorine molecule to form hydrogen chloride and another chlorine free radical. This chlorine radical can repeat the process by reacting with another hydrogen molecule, and the reaction continues to propagate itself in this manner until one or both of the reactants are used up. Almost all of the product is formed in this step.

Step 3. *Termination:*

$$:\ddot{C}l\cdot + :\ddot{C}l\cdot \longrightarrow Cl_2$$

$$H\cdot + H\cdot \longrightarrow H_2$$

$$H\cdot + :\ddot{C}l\cdot \longrightarrow HCl$$

Hydrogen and chlorine free radicals can react in any of the three ways shown. Unless further activation occurs, the formation of hydrogen chloride will terminate when the radicals form molecules. In an exothermic reaction such as that between hydrogen and chlorine, there is usually enough heat and light energy available to maintain the supply of free radicals, and the reaction will continue until at least one reactant is exhausted.

QUESTIONS

A. Review the Meanings of the New Terms Introduced in this Chapter

1. Reversible reaction
2. Chemical kinetics
3. Equilibrium
4. Chemical equilibrium
5. Principle of Le Chatelier
6. Catalyst
7. Law of Mass Action
8. Equilibrium constant, K_{eq}
9. Ionization constant, K_i
10. Ion product constant for water, K_w
11. Solubility product constant, K_{sp}
12. Common ion effect
13. Buffer solution
14. Mechanism of a reaction
15. Activated complex
16. Activation energy
17. Free radical

B. Answers to the Following Questions Will Be Found in Tables and Figures

1. How would you expect the two tubes in Figure 16.1 to appear if both are at 25°C?
2. Is the reaction $N_2O_4 \rightleftharpoons 2\,NO_2$ (Figure 16.1) exothermic or endothermic?
3. At equilibrium how do the forward and reverse reaction rates compare? (See Figure 16.2.)
4. Would the reaction of 30 moles of H_2 and 10 moles of N_2 produce a greater yield of NH_3 carried out in a 1 L or a 2 L vessel? (See Table 16.1.)
5. Of the acids listed in Table 16.2, which ones are stronger than acetic acid, and which are weaker?
6. For each of the solutions in Table 16.3, what is the sum of the pH plus the pOH? What would be the pOH of a solution whose pH was -1?
7. Using Table 16.4, tabulate the relative order of molar solubilities of AgCl, AgBr, AgI, $AgC_2H_3O_2$, $PbSO_4$, $BaSO_4$, $BaCrO_4$, and PbS. List the most soluble first.
8. Which compound in each of the following pairs has the greater molar solubility? (See Table 16.4.)
   (a) $Mn(OH)_2$ or Ag_2CrO_4 (b) $BaCrO_4$ or Ag_2CrO_4

9. Using Figure 16.3 and Table 16.5, explain how the acetic acid–sodium acetate buffer system maintains its pH when 0.01 mole of HCl is added to 1 L of the buffer solution.
10. How would Figure 16.5 be altered if the reaction were exothermic?

C. Review Questions

1. Express the following reversible systems in equation form:
   (a) A mixture of ice and liquid water at $0°C$
   (b) Liquid water and vapor at $100°C$ in a pressure cooker
   (c) Crystals of Na_2SO_4 in a saturated aqueous solution of Na_2SO_4
   (d) A closed system containing boiling sulfur dioxide, SO_2
2. Explain why a precipitate of NaCl forms when hydrogen chloride gas is passed into a saturated aqueous solution of NaCl.
3. Why does the rate of a reaction usually increase when the concentration of one of the reactants is increased?
4. Consider the following system at equilibrium:

$$4 NH_3(g) + 3 O_2(g) \rightleftharpoons 2 N_2(g) + 6 H_2O(g) + 366 \text{ kcal}$$

   (a) Is the reaction exothermic or endothermic?
   (b) If the system's state of equilibrium is disturbed by the addition of O_2, in which direction, left or right, must reaction occur to reestablish equilibrium? After the new equilibrium has been established, how will the final molar concentrations of NH_3, O_2, N_2, and H_2O compare (increase or decrease) to their concentrations before the addition of the O_2?
   (c) If the system's state of equilibrium is disturbed by the addition of heat, in which direction will reaction occur, left or right, to reestablish equilibrium?
5. Consider the following system at equilibrium:

$$N_2(g) + 3 H_2(g) \rightleftharpoons 2 NH_3(g) + 22.1 \text{ kcal}$$

Complete the following table. Indicate changes in moles and concentrations by entering I, D, N, or ? in the table. (I = increase, D = decrease, N = no change, ? = insufficient information to determine.)

Change or stress imposed on the system at equilibrium	Direction of reaction, left or right, to reestablish equilibrium	Change in number of moles			Change in molar concentrations		
		N_2	H_2	NH_3	N_2	H_2	NH_3
(a) Add N_2							
(b) Remove H_2							
(c) Decrease volume of reaction vessel							
(d) Increase volume of reaction vessel							
(e) Increase temperature							
(f) Add catalyst							
(g) Add both H_2 and NH_3							

6. If pure hydrogen iodide, HI, is placed in a vessel at 700 K, will it decompose? Explain.

7. For the equations that follow, tell in which direction, left or right, the equilibrium will shift to when the following changes are made: the temperature is increased; the pressure is increased by decreasing the volume of the reaction vessel; a catalyst is added.

(a) $3\,O_2(g) + 64.8\,kcal \rightleftharpoons 2\,O_3(g)$

(b) $CH_4(g) + Cl_2(g) \rightleftharpoons CH_3Cl(g) + HCl(g) + 26.4\,kcal$

(c) $2\,NO(g) + 2\,H_2(g) \rightleftharpoons N_2(g) + 2\,H_2O(g) + 159\,kcal$

(d) $2\,SO_3(g) + 47\,kcal \rightleftharpoons 2\,SO_2(g) + O_2(g)$

(e) $4\,NH_3(g) + 3\,O_2(g) \rightleftharpoons 2\,N_2(g) + 6\,H_2O(g) + 366\,kcal$

8. Explain why an increase in temperature causes the rate of reaction to increase.

9. Give a word description of how equilibrium is reached when the substances A and B are first mixed and react as

$$A + B \rightleftharpoons C + D$$

10. With dilution, aqueous solutions of acetic acid, $HC_2H_3O_2$, show increased ionization. For example, a 1.0 M solution of acetic acid is 0.42% ionized, whereas a 0.10 M solution is 1.34% ionized. Explain this behavior using the ionization equation and equilibrium principles.

11. A 1.0 M solution of acetic acid ionizes less and has a higher concentration of H^+ ions than a 0.10 M acetic acid solution. Explain this behavior. (See Question 10 for data.)

12. Write the equilibrium constant expression for each of the following reactions:

(a) $H_2(g) + I_2(g) \rightleftharpoons 2\,HI(g)$

(b) $N_2(g) + 3\,H_2(g) \rightleftharpoons 2\,NH_3(g)$

(c) $PCl_5(g) \rightleftharpoons PCl_3(g) + Cl_2(g)$

(d) $HClO_2(aq) \rightleftharpoons H^+(aq) + ClO_2^-(aq)$

(e) $NH_4OH(aq) \rightleftharpoons NH_4^+(aq) + OH^-(aq)$

(f) $4\,NH_3(g) + 5\,O_2(g) \rightleftharpoons 4\,NO(g) + 6\,H_2O(g)$

13. What would cause two separate samples of pure water to have slightly different pH values?

14. Why are the pH and pOH equal in pure water?

15. What effect will increasing the H^+ ion concentration of a solution have upon (a) pH, (b) pOH, (c) $[OH^-]$, and (d) K_w?

16. Write the solubility product expression (K_{sp}) for each of the following substances:

(a) CuS   (c) $PbBr_2$ (e) $Fe(OH)_3$ (g) CaF_2

(b) $BaSO_4$ (d) Ag_3AsO_4 (f) Sb_2S_5 (h) $Ba_3(PO_4)_2$

17. Explain why silver acetate, $AgC_2H_3O_2$, is more soluble in nitric acid than in water. [*Hint:* Write the equilibrium equation first and then consider the effect of the acid on the acetate ion.] What would happen if hydrochloric acid, HCl, were used in place of nitric acid?

18. For the reaction $2\,SO_3(g) + 47\,kcal = 2\,SO_2(g) + O_2(g)$, what effect does increasing pressure have on (a) the reaction rate and (b) the equilibrium concentration of SO_3? Answer this question by considering the following ways by which the pressure of the system might be increased:

(a) Decrease the volume of the reaction vessel.

(b) Put some additional SO_3 in the reaction vessel.

(c) Put some He, an inert gas, in the reaction vessel.

(d) Increase the temperature of the system.

19. Dissolution of sodium acetate, $NaC_2H_3O_2$, in pure water gives a basic solution. Why? [*Hint:* A small amount of $HC_2H_3O_2$ is formed.]

20. One of the important pH-regulating systems in the blood consists of a carbonic acid–sodium bicarbonate buffer.

$$H_2CO_3(aq) \rightleftharpoons H^+(aq) + HCO_3^-(aq)$$

$$NaHCO_3(aq) \longrightarrow Na^+(aq) + HCO_3^-(aq)$$

Explain how this buffer resists changes in pH when (a) excess acid and (b) excess base get into the bloodstream.

21. Which of the following statements are correct?
    (a) In a reaction at equilibrium the concentrations of reactants and products are equal.
    (b) A catalyst increases the concentrations of products present at equilibrium.
    (c) A catalyst increases reaction rates by changing reaction mechanisms.
    (d) A catalyst lowers the activation energy of a reaction by equal amounts for both the forward and the reverse reactions.
    (e) If an increase in temperature causes an increase in the concentration of products present at equilibrium, then the reaction is exothermic.
    (f) The magnitude of an equilibrium constant is independent of the reaction temperature.
    (g) The magnitude of an equilibrium constant depends on the nature of the reactants and products and on the temperature at which the reaction occurs.
    (h) A large equilibrium constant for a reaction indicates that the reaction, at equilibrium, favors products over reactants.
    (i) The amount of product obtained at equilibrium is proportional to how fast equilibrium is attained.
    (j) The study of reaction rates and reaction mechanisms is known as chemical kinetics.
    (k) For the reaction $CaCO_3(s) \overset{\Delta}{\rightleftharpoons} CaO(s) + CO_2(g)$, increasing the pressure of CO_2 present at equilibrium will cause the reaction to shift left.
    (l) The greater the value of the equilibrium constant, the greater the proportion of products present at equilibrium.

Statements m–s pertain to the equilibrium system,

$$2 NO(g) + O_2(g) \rightleftharpoons 2 NO_2(g) + 27 \text{ kcal}$$

    (m) The reaction as shown is endothermic.
    (n) Increasing the temperature will cause the equilibrium to shift left.
    (o) Increasing the temperature will increase the magnitude of the equilibrium constant, K_{eq}.
    (p) Decreasing the volume of the reaction vessel will shift the equilibrium to the right and decrease the concentrations of the reactants.
    (q) Removal of some of the O_2 will cause, as the system readjusts, an increase in the concentrations of the reactants.
    (r) High temperatures and pressures favor increased yields of NO_2.
    (s) The equilibrium constant expression for the reaction is

$$K_{eq} = \frac{[NO_2]^2}{[NO]^2[O_2]}$$

    (t) A solution with an H^+ ion concentration of 1×10^{-5} mole/litre has a pOH of 9.
    (u) An aqueous solution that has an OH^- ion concentration of 1×10^{-4} mole/litre has an H^+ ion concentration of 1×10^{-10} mole/litre.
    (v) $K_w = [H^+][OH^-] = 1 \times 10^{-14}$ and $pK_w = pH + pOH = 14$.
    (w) As solid $BaSO_4$ is added to a saturated solution of $BaSO_4$, the magnitude of its K_{sp} increases.

D. Review Problems

1. What is the maximum amount (moles) of HI that can be obtained from a reaction mixture containing 2.30 moles of I_2 and 2.10 moles of H_2?

2. (a) How many moles of hydrogen iodide, HI, will be produced when 2.00 moles of H_2 and 2.00 moles of I_2 are reacted at 700 K? (Reaction is 79% complete.)
   (b) Addition of 0.27 mole of I_2 to the system increases the yield of HI to 85%. How many moles of H_2, I_2, and HI are now present?
   (c) From the data in part (a), calculate K_{eq} for the reaction at 700 K.

*3. Six (6.00) grams of hydrogen, H_2, and 200 g of iodine, I_2, are reacted at 500 K. After equilibrium is reached, analysis shows that there are 64.0 g of HI in the flask. How many moles of H_2, I_2, and HI are present in this equilibrium mixture?

*4. How many moles of I_2 are in equilibrium with 0.10 mole of H_2 and 0.050 mole of HI at 500 K in a 1 litre flask? ($H_2 + I_2 \rightleftharpoons 2$ HI; $K_{eq} = 0.17$.)

5. If the velocity of a reaction doubles for every 10° rise in temperature, how much faster will the reaction proceed at 100°C than at 30°C?

6. Calculate the ionization constant for each of the following monoprotic acids. Each acid ionizes as follows: $HA \rightleftharpoons H^+ + A^-$.

Acid	Acid concentration	$[H^+]$
Hypochlorous, HOCl	0.10 M	5.9×10^{-5} mole/L
Propanoic, $HC_3H_5O_2$	0.20 M	1.6×10^{-3} mole/L
Hydrocyanic, HCN	0.50 M	1.4×10^{-5} mole/L

7. Calculate (a) the H^+ ion concentration, (b) the pH, and (c) the percent ionization of a 0.20 M solution of $HC_2H_3O_2$. ($K_i = 1.8 \times 10^{-5}$.)

8. A 1.0 M solution of a weak acid, HA, is 0.61% ionized. Calculate the ionization constant, K_i, for the acid.

9. A 0.25 M solution of a weak acid, HA, has a pH of 5. Calculate the ionization constant, K_i, for the acid.

10. Calculate the percent ionization and pH of solutions of $HC_2H_3O_2$ having the following molarities: (a) 1.0 M, (b) 0.10 M, and (c) 0.010 M. ($K_i = 1.8 \times 10^{-5}$.)

11. Given the following solubility data, calculate the solubility product constant for each substance.
    (a) $BaSO_4$, 3.9×10^{-5} mole/L
    (b) Ag_2CrO_4, 7.8×10^{-5} mole/L
    (c) ZnS, 3.5×10^{-12} mole/L
    (d) $Pb(IO_3)_2$, 4.0×10^{-5} mole/L
    (e) AgCl, 0.0019 g/L
    (f) $CaSO_4$, 0.67 g/L
    (g) $Zn(OH)_2$, 2.33×10^{-4} g/L
    (h) Ag_3PO_4, 6.73×10^{-3} g/L

*12. Calculate the molar solubility for each of the following substances:
    (a) FeS, $K_{sp} = 4 \times 10^{-19}$
    (b) $NiCO_3$, $K_{sp} = 1.4 \times 10^{-7}$
    (c) Ag_2S, $K_{sp} = 5.5 \times 10^{-51}$
    (d) $Fe(OH)_2$, $K_{sp} = 8.0 \times 10^{-16}$

13. Calculate, for each of the substances in Question 12, the solubility in grams per 100 mL of water.

14. The K_{sp} of CaF_2 is 3.9×10^{-11}. Calculate (a) the molar concentrations of Ca^{2+} and F^- in a saturated solution, and (b) the grams of CaF_2 that will dissolve in 100 mL of water.

*15. The following pairs of solutions are mixed. Show by calculation whether or not a precipitate will form.
    (a) 100 mL of 0.010 M Na_2SO_4 and 100 mL of 0.001 M $Pb(NO_3)_2$
    (b) 50.0 mL of $1.0 \times 10^{-4} M$ $AgNO_3$ and 100 mL of $1.0 \times 10^{-4} M$ NaCl
    (c) 1.0 g $Ca(NO_3)_2$ in 150 mL H_2O and 250 mL of 0.01 M NaOH
    (K_{sp} $PbSO_4 = 1.3 \times 10^{-8}$; K_{sp} AgCl $= 1.7 \times 10^{-10}$;
    K_{sp} $Ca(OH)_2 = 1.3 \times 10^{-6}$.)

16. Calculate the H^+ ion concentration and the pH of buffer solutions that are 0.25 M in $HC_2H_3O_2$ and contain sufficient sodium acetate to make the $C_2H_3O_2^-$ ion concentration equal to (a) 0.10 M; (b) 0.25 M. ($K_i\, HC_2H_3O_2 = 1.8 \times 10^{-5}$.)

17. (a) When 1.0 mL of 1.0 M HCl is added to 50 mL of 1.0 M NaCl, the H^+ ion concentration changes from $1 \times 10^{-7}\, M$ to $2.0 \times 10^{-2}\, M$.

(b) When 1.0 mL of 1.0 M HCl is added to 50 mL of a buffer solution that is 1.0 M in $HC_2H_3O_2$ and 1.0 M in $NaC_2H_3O_2$, the H^+ ion concentration changes from $1.8 \times 10^{-5}\, M$ to $1.9 \times 10^{-5}\, M$.

Calculate the initial pH and the pH change in each solution (log 1.8 = 0.26; log 1.9 = 0.78; log 2.0 = 0.30).

18. Calculate the pH and the pOH of the following solutions:

(a) 0.00010 M HCl

(b) 0.010 M NaOH

(c) 0.0025 M NaOH

(d) 0.10 M HClO ($K_i = 3.5 \times 10^{-8}$)

(e) Saturated $Fe(OH)_2$ solution ($K_{sp} = 8.0 \times 10^{-16}$)

Review Exercises for Chapters 13–16

CHAPTER 13
WATER AND THE
PROPERTIES OF
LIQUIDS

True–False: Answer the following as either true or false.

1. Sodium, potassium, and calcium each react with water to form hydrogen gas and a metal hydroxide.
2. Calcium oxide reacts with water to form calcium hydroxide and hydrogen gas.
3. Water in a hydrate is known as water of hydration or water of crystallization.
4. Substances that can spontaneously lose their water of hydration when exposed to air are said to be hygroscopic.
5. A substance that absorbs water from the air until it forms a solution is said to be deliquescent.
6. A chemical method of softening water is to add lime (CaO) and sodium carbonate, which will precipitate calcium carbonate and magnesium hydroxide.
7. Evaporation is the escape of molecules from the liquid state to the vapor state.
8. Sublimation is the change from the vapor to the liquid state.
9. $CuSO_4 \cdot 5H_2O$ is named copper(II) sulfate pentahydrate.
10. The pressure exerted by a vapor in equilibrium with its liquid is known as the vapor pressure of the liquid.
11. The boiling point of a liquid is the temperature at which the vapor pressure of the liquid is equal to the external pressure above the liquid.
12. Of two liquids, the one having the higher vapor pressure at a given temperature will have the higher normal boiling point.
13. In treating city water supplies, fluorine gas is often added to prevent tooth decay.
14. In treating city water supplies, objectionable odors and tastes are best removed by spraying the water into the air.
15. In treating city water supplies, chlorine is injected into the water to kill harmful bacteria before it is distributed to the public.
16. Fluorocarbons, which were used in aerosol spray cans, and are used in refrigeration and air conditioners, escape to the stratosphere, and react to destroy part of the ozone layer.
17. The original source of mercury-contaminated fish is industrial pollution.
18. Substances that evaporate readily are said to be volatile.
19. The phenomenon in which an element can exist in two or more molecules or crystalline forms is known as allotropy.
20. Metal oxides that react with water to form bases are known as basic anhydrides.
21. Nonmetal oxides that react with water to form acids are known as acid anhydrides.
22. SO_2 is the anhydride of H_2SO_4.
23. When P_2O_5 reacts with water, it would be expected to make the solution basic.
24. When hard water is softened in a zeolite water softener, the dissolved minerals are replaced in the water by zeolite molecules.
25. Substances that readily absorb water from the atmosphere are said to be hygroscopic.

Multiple Choice: Choose the correct answer to each of the following.

1. The heat of fusion of water is:
   (a) 1.0 cal/g   (b) 80 cal/g   (c) 540 cal/g   (d) 540 cal/mole
2. The heat of varporization of water is:
   (a) 1.0 cal/g   (b) 80 cal/g   (c) 540 cal/g   (d) 540 cal/mole
3. The heat capacity of water is:
   (a) 1.0 cal/g °C   (b) 80 cal/g °C   (c) 540 cal/g °C   (d) 18 cal/g °C
4. The density of water at 4°C is:
   (a) 1.0 g/mL   (b) 80 g/mL   (c) 18.0 g/mL   (d) 14.7 lb/in.³
5. SO_2 can be properly classified as a(n)
   (a) basic anhydride   (b) hydrate   (c) anhydrous salt   (d) acid anhydride

6. When compared to H_2S, H_2Se, and H_2Te, water is found to have the highest boiling point. This is because it:
   (a) has the lowest molecular weight
   (b) is the smallest molecule
   (c) has the highest bonding
   (d) will form hydrogen bonds better than the others

7. In which of the following molecules will hydrogen bonding be important?

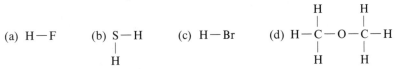

   (a) H—F   (b) S—H   (c) H—Br   (d) H—C—O—C—H

8. Which of the following is an incorrect equation?
   (a) $H_2SO_4 + 2\,NaOH \longrightarrow Na_2SO_4 + 2\,H_2O$
   (b) $C_2H_6 + O_2 \longrightarrow 2\,CO_2 + 3\,H_2$
   (c) $2\,H_2O \xrightarrow[H_2SO_4]{Electrolysis} 2\,H_2 + O_2$
   (d) $Ca + 2\,H_2O \longrightarrow H_2 + Ca(OH)_2$

9. Which of the following is an incorrect equation?
   (a) $C + H_2O(g) \xrightarrow{1000°C} CO(g) + H_2(g)$
   (b) $CaO + H_2O \longrightarrow Ca(OH)_2$
   (c) $NO_2 + H_2O \longrightarrow HNO_3$
   (d) $Cl_2 + H_2O \longrightarrow HCl + HOCl$

10. A correct Lewis dot structure of hydrogen peroxide is:
   (a) $:\!\overset{..}{\underset{..}{O}}\!:\!\overset{..}{\underset{..}{O}}\!:H$ (b) $:\!\overset{..}{O}\!:\!\overset{..}{\underset{..}{O}}\!:H$ H   (c) $:\!\overset{..}{\underset{..}{O}}\!:\!\overset{..}{O}\!:H$ H   (d) $:\!\overset{..}{\underset{..}{O}}\!:H^-$

11. How many calories are required to change 85 g of water at 25°C to steam at 100°C?
   (a) 5.2×10^4 cal   (b) 4.8×10^4 cal   (c) 1.4×10^6 cal   (d) 3.4×10^2 cal

12. A chunk of 0°C ice weighing 145 g is dropped into 75 g of water at 62°C. The heat of fusion of water is 80 cal/g. The result, after thermal equilibrium is attained, will be:
   (a) 87 g of ice and 133 g of liquid water, all at 0°C
   (b) 58 g of ice and 162 g of liquid water, all at 0°C
   (c) 220 g of water at 7°C
   (d) 220 g of water at 17°C

**CHAPTER 14
SOLUTIONS**

True–False: Answer the following as either true or false.

1. Solubility describes the amount of one substance that will dissolve into another substance.
2. When solid NaCl dissolves in water, the sodium ions attract the positive end of the water dipole.
3. Bromine is more soluble in polar water than in nonpolar carbon tetrachloride because the polar water molecules help it form ions.
4. For most solids or gases dissolved in a liquid, an increase in temperature results in an increase in solubility.
5. An increase in temperature almost always increases the rate at which a solid will dissolve in a liquid.
6. A supersaturated solution contains dissolved solute in equilibrium with undissolved solute.
7. A 1 molar solution contains 1 mole of solute per litre of solution.
8. When a solute is dissolved in a solvent, the freezing point of the solution will be higher than that of the pure solvent.

9. Freezing point depression, boiling point elevation, and vapor pressure lowering are colligative properties of solutions.
10. Liquids that mix with water in all proportions are usually ionic in solution or are polar substances.
11. Liquids that do not mix are said to be miscible.
12. As a general rule, sodium and potassium salts of the common ions are soluble in water.
13. The vapor pressure of the solvent in a solution is lower than the vapor pressure of the pure solvent.
14. Salt water has a higher normal boiling point than distilled water.
15. One definition of equivalent weight is the weight of a substance that will combine with, contain, replace, or in any other way be equivalent to 1 g-at. wt of hydrogen.
16. One mole of H_2SO_4 will react with twice as much NaOH as 1 mole of HCl.
17. A solution containing a nonvolatile solute has a lower boiling point than the pure solvent as a result of having a lower vapor pressure than the pure solvent.
18. A sulfuric acid solution with a molarity of 4.0 M would have a normality of 2.0 N.

Multiple Choice: Choose the correct answer to each of the following.
1. Which of the following is not a general property of solutions?
   (a) It is a homogeneous mixture of two or more substances.
   (b) It has a variable composition.
   (c) The dissolved solute breaks down to individual molecules.
   (d) The solution has the same chemical composition, the same chemical properties, and the same physical properties in every part.
2. If NaCl is soluble in water to the extent of 36.0 g NaCl/100 g H_2O at 20°C, then a solution at 20°C containing 45 g NaCl/150 g H_2O would be:
   (a) Dilute   (b) Saturated   (c) Supersaturated   (d) Unsaturated
3. If 5.00 g NaCl are dissolved in 25.0 g of water, the percentage of NaCl by weight is:
   (a) 16.7   (b) 20.0   (c) 0.20   (d) No correct answer given
4. How many grams of 9.0% $AgNO_3$ solution will contain 5.3 g $AgNO_3$?
   (a) 47.7   (b) 0.58   (c) 59   (d) No correct answer given
5. The molarity of a solution containing 2.5 moles of acetic acid, $HC_2H_3O_2$, in 400 mL of solution is:
   (a) 0.062 M (b) 1.0 M (c) 0.103 M (d) 6.2 M
6. What volume of 0.300 M KCl will contain 15.3 g KCl?
   (a) 1.46 L   (b) 684 mL   (c) 61.5 mL   (d) 4.60 L
7. What weight of $BaCl_2$ will be required to prepare 200 mL of 0.150 M solution?
   (a) 0.750 g   (b) 156 g   (c) 6.25 g   (d) 31.2 g
8. If 15 g of H_2SO_4 are dissolved in water to make 80 mL of solution, the normality is:
   (a) 0.38 N (b) 1.9 N (c) 6.1 N (d) 3.8 N

Problems 9–11 relate to the reaction $CaCO_3 + 2 HCl \longrightarrow CaCl_2 + H_2O + CO_2$

9. What volume of 6.0 M HCl will be needed to react with 0.350 mole of $CaCO_3$?
   (a) 42.0 mL   (b) 1.17 L   (c) 117 mL   (d) 583 mL
10. If 400 mL of 2.0 M HCl react with excess $CaCO_3$, the volume of CO_2 produced, measured at STP, is:
   (a) 18 L   (b) 5.6 L   (c) 9.0 L   (d) 56 L
11. If 5.3 g of $CaCl_2$ were produced in the reaction, what was the molarity of the HCl used if 25 mL of it reacted with excess $CaCO_3$?
   (a) 3.8 M (b) 0.19 M (c) 0.38 M (d) 0.42 M
12. In the reaction $Ca + 2 HCl \longrightarrow CaCl_2 + H_2$, the equivalent weight of Ca is:
   (a) 36.5 g   (b) 20.0 g   (c) 40.1 g   (d) 80.2 g

13. In the reaction $3\,HNO_3 + Cr(OH)_3 \longrightarrow Cr(NO_3)_3 + 3\,H_2O$, the equivalent weight of $Cr(OH)_3$ is:
    (a) 34.3 g   (b) 52 g   (c) 103 g   (d) 309 g

14. If 20.0 g of the nonelectrolyte urea, $CO(NH_2)_2$, are dissolved in 25.0 g of water, the freezing point of the solution will be:
    (a) $-2.48°C$ (b) $-1.40°C$ (c) $-24.8°C$ (d) $-3.72°C$

15. When 256 g of an unknown nonvolatile nonelectrolyte were dissolved in 500 g of H_2O, the freezing point was found to be $-2.79°C$. The molecular weight of the unknown solute is:
    (a) 357   (b) 62.0   (c) 768   (d) 341

16. How many millilitres of 6.0 M H_2SO_4 must you use to prepare 500 mL of 0.20 M sulfuric acid solution?
    (a) 60   (b) 17   (c) 1.6   (d) 5.9×10^3

17. How many millilitres of water must be added to 200 mL of 1.40 M HCl to make a solution that is 0.500 M HCl?
    (a) 360 mL   (b) 560 mL   (c) 140 mL   (d) 280 mL

18. Which procedure is most likely to increase the solubility of most solids in liquids?
    (a) Stirring   (c) Heating the solution
    (b) Pulverizing the solid   (d) Increasing the pressure

19. The addition of a crystal of $NaClO_3$ to a solution of $NaClO_3$ causes additional crystals to precipitate. The original solution was:
    (a) Unsaturated   (b) Dilute   (c) Saturated   (d) Supersaturated

20. Which of the following anions will not form a precipitate with silver ions, Ag^+?
    (a) Cl^- (b) NO_3^- (c) Br^- (d) CO_3^{2-}

21. Which of the following salts are considered to be soluble in water?
    (a) $BaSO_4$ (b) NH_4Cl (c) AgI (d) PbS

CHAPTER 15
IONIZATION.
ACIDS, BASES,
SALTS

True–False: Answer the following as either true or false.

1. Arrhenius defined a base as a hydroxide-containing substance that dissociates in water to produce hydroxide ions.

2. The Brønsted–Lowry theory defined an acid as a proton donor.

3. The Lewis theory defined an acid as an electron-pair donor.

4. In the reaction $HCl + NH_3 \longrightarrow NH_4^+ + Cl^-$, according to the Brønsted–Lowry theory, the conjugate base of HCl is NH_4^+.

5. In the reaction $HNO_3 + H_2O \longrightarrow H_3O^+ + NO_3^-$, the conjugate base of HNO_3 is NO_3^-.

6. When an acid reacts with a carbonate, the products are a salt, water, and carbon dioxide.

7. When electrolytes dissociate, they break into individual molecules.

8. The equation for the ionization of acetic acid is:

$$HC_2H_3O_2 + H_2O \rightleftharpoons H_3O^+ + C_2H_3O_2^-$$

9. In water, zinc nitrate dissociates as shown by the equation

$$Zn(NO_3)_2 \xrightarrow{H_2O} Zn^{2+} + 2\,NO_3^-$$

10. The equation for the dissociation of sodium carbonate in water is

$$Na_2CO_3 \xrightarrow{H_2O} Na_2^{2+} + CO_3^{2-}$$

11. When 1 mole of $CaCl_2$ dissolves in water, it will give 1 mole of Ca^{2+} and 2 moles of Cl^- ions in solution.

12. A solution with $[H^+] = 1.0 \times 10^{-9}$ M has a pH of -9.0.
13. A solution with $[H^+] = 7.8 \times 10^{-2}$ M has a pH between 1 and 2.
14. The reaction of an acid and a base to form a salt and water is known as neutralization.
15. In neutralizations between a strong acid and a strong base, the net ionic equation is

$$H^+ + OH^- \longrightarrow H_2O$$

16. For the aqueous reaction of NaOH with $FeCl_3$, the net ionic equation is

$$Fe^{3+} + 3\,OH^- \longrightarrow Fe(OH)_3(s)$$

17. A 1.0 M $HC_2H_3O_2$ solution will freeze at a lower temperature than a 1.0 M solution of KBr.
18. In writing net ionic equations, weak electrolytes are written in their un-ionized form.
19. Acids, bases, and salts are electrolytes.
20. Positive ions are called spectator ions.
21. Negative ions are called anions.
22. Titration is the process of measuring the volume of one reagent required to react with a measured weight or volume of another reagent.
23. A pH meter is an instrument used to measure the acidity of a solution.
24. As the acidity of a solution increases, the pH decreases.
25. When 50.0 mL of 0.20 M NaOH and 100 mL of 0.10 M HCl are mixed, the resulting solution will have a pH of 7.
26. The conjugate acid of HSO_4^- is SO_4^{2-}.
27. H_3O^+ is called a hydronium ion and NH_4^+ is called a nitronium ion.
28. A 0.1 M HCl and a 0.1 M $HC_2H_3O_2$ solution will have the same pH.

Multiple Choice: Choose the correct answer to each of the following.

1. When the reaction

$$Al + HCl \longrightarrow$$

is completed and balanced, a term appearing in the balanced equation is:
(a) 3 HCl   (b) $AlCl_2$ (c) $3\,H_2$ (d) 4 Al

2. When the reaction

$$CaO + HNO_3 \longrightarrow$$

is completed and balanced, a term appearing in the balanced equation is:
(a) H_2 (b) $2\,H_2$ (c) $2\,CaNO_3$ (d) H_2O

3. When the reaction

$$H_3PO_4 + KOH \longrightarrow$$

is completed and balanced, a term appearing in the balanced equation is:
(a) H_3PO_4 (b) $6\,H_2O$ (c) KPO_4 (d) $2\,H_3PO_4$

4. When the reaction

$$HCl + Cr_2(CO_3)_3 \longrightarrow$$

is completed and balanced, a term appearing in the balanced equation is:
(a) Cr_2Cl (b) 3 HCl   (c) $3\,CO_2$ (d) H_2O

5. When the reaction

$$KOH + BiCl_3 \longrightarrow$$

is completed and balanced, a term appearing in the balanced equation is:
(a) $6\,KOH$　　(b) $Bi(OH)_3$　　(c) $3\,Cl_2$　　(d) KCl_3

6. Which of the following is not a salt?
(a) $K_2Cr_2O_7$　　(b) $NaHCO_3$　　(c) $Ca(OH)_2$　　(d) $Na_2C_2O_4$

7. Which of the following is not an acid?
(a) H_3PO_4　　(b) H_2S　　(c) H_2SO_4　　(d) NH_3

8. Which of the following is a weak electrolyte?
(a) NH_4OH　　(b) $Ni(NO_3)_2$　　(c) K_3PO_4　　(d) $NaBr$

9. Which of the following is a nonelectrolyte?
(a) $HC_2H_3O_2$　　(b) $MgSO_4$　　(c) $KMnO_4$　　(d) CCl_4

10. Which of the following is a strong electrolyte?
(a) H_2CO_3　　(b) HNO_3　　(c) NH_4OH　　(d) H_3BO_3

11. Which of the following is a weak electrolyte?
(a) $NaOH$　　(b) $NaCl$　　(c) $HC_2H_3O_2$　　(d) H_2SO_4

12. A solution has a concentration of H^+ of $3.4 \times 10^{-5}\,M$. The pH is:
(a) 4.47　　(b) 5.53　　(c) 3.53　　(d) 5.47

13. A solution with a pH of 5.85 has an H^+ concentration of:
(a) $7.1 \times 10^{-5}\,M$　　(b) $7.1 \times 10^{-6}\,M$　　(c) $3.8 \times 10^{-4}\,M$　　(d) $1.4 \times 10^{-6}\,M$

14. 16.55 mL of $0.844\,M$ NaOH is required to titrate 10.00 mL of a hydrochloric acid solution. The molarity of the acid solution is:
(a) $0.700\,M$　　(b) $0.510\,M$　　(c) $1.40\,M$　　(d) $0.255\,M$

15. What volume of $0.462\,M$ NaOH is required to titrate 20.00 mL of $0.391\,M$ HNO_3?
(a) 23.6 mL　　(b) 16.9 mL　　(c) 9.03 mL　　(d) 11.8 mL

16. Dilute hydrochloric acid is a typical acid, as shown by its:
(a) Color　　(b) Odor　　(c) Solubility　　(d) Taste

17. What is the pH of a $0.00015\,M$ HCl solution?
(a) 4.0　　(b) 2.82　　(c) Between 3 and 4　　(d) No correct answer given

18. The chloride ion concentration in 300 mL of $0.10\,M$ $AlCl_3$ is
(a) $0.30\,M$　　(b) $0.10\,M$　　(c) $0.030\,M$　　(d) $0.90\,M$

19. The amount of $BaSO_4$ that will precipitate when 100 mL of $0.10\,M$ $BaCl_2$ and 100 mL of $0.10\,M$ Na_2SO_4 are mixed is:
(a) 0.010 mole　　(b) 0.10 mole　　(c) 23 g　　(d) No correct answer given

20. The freezing point of a 0.50 molal NaCl aqueous solution will be about:
(a) $-1.86°C$　　(b) $-0.93°C$　　(c) $-2.79°C$　　(d) No correct answer given

CHAPTER 16
CHEMICAL
EQUILIBRIUM

True–False:　Answer the following as either true or false.

1. A reversible reaction is one in which the products formed in a chemical reaction are reacting to produce the original reactants.

2. The study of reaction rates is known as chemical kinetics.

3. When the rate of the forward reaction is exactly equal to the rate of the reverse reaction, a condition of chemical equilibrium exists.

4. A statement of Le Chatelier's principle is that if a stress is applied to a system in equilibrium, the system will behave in such a way as to relieve that stress and restore equilibrium but under a new set of conditions.

5. A catalyst will shift the point of equilibrium of a reaction, but will not alter the reaction rates.

6. The reaction $CaCO_3 \rightleftharpoons CaO(s) + CO_2(g)$ will proceed to the right better in a closed container where the pressure of the CO_2 can build up than in an open container where the CO_2 can escape.

7. When heat is applied to a system in equilibrium, the reaction that absorbs the heat is favored.

8. The ionization constant expression for the ionization of the weak acid HCN would be $K_i = [H^+][CN^-]$.

9. If K_i for acetic acid is 1.8×10^{-5}, and K_i for nitrous acid is 4.5×10^{-4}, then at equal concentrations, acetic acid is a stronger acid.

10. If the K_{sp} for AgI is 1.6×10^{-16}, and the K_{sp} for CuS is 8×10^{-45}, then AgI is more soluble than CuS.

11. The amount of energy needed to form an activated complex is known as the activation energy of a reaction.

12. A catalyst can lower the activation energy, thus increasing the speed of a reaction.

13. If a reaction is exothermic, the speed of that reaction generally can be increased by increasing the temperature.

14. Generally, as the concentrations of the reactants increase in a chemical reaction, the speed of the reaction decreases.

15. When the temperature of an exothermic reaction is increased, the forward reaction is favored.

16. The ion product constant for water at 25°C is 1×10^{-14}.

17. A solution of pOH 12 has an H^+ concentration of 0.010 mole per litre.

18. A solution of pOH 3 will turn blue litmus red.

Multiple Choice: Choose the correct answer to each of the following.

1. The equation $HC_2H_3O_2 + H_2O \rightleftharpoons H_3O^+ + C_2H_3O_2^-$ implies that:
   (a) If you start with 1.0 mole of $HC_2H_3O_2$, 1.0 mole of H_3O^+ and 1.0 mole of $C_2H_3O_2^-$ will be produced.
   (b) An equilibrium exists between the forward reaction and the reverse reaction.
   (c) At equilibrium, equal molar amounts of all four substances will exist.
   (d) The reaction proceeds all the way to the products, then reverses, going all the way back to the reactants.

2. If the reaction $A + B \rightleftharpoons C + D$ is initially at equilibrium, and then more A is added, which of the following is not true?
   (a) More collisions of A and B will occur, thus the rate of the forward reaction will be increased.
   (b) The equilibrium will shift toward the right.
   (c) The moles of B will be increased.
   (d) The moles of D will be increased.

3. What will be the H^+ concentration in a 1.0 M HCN solution? ($K_i = 4.0 \times 10^{-10}$.)
   (a) $2.0 \times 10^{-5} M$ (b) $1.0 M$ (c) $4.0 \times 10^{-10} M$ (d) $2.0 \times 10^{-10} M$

4. If $[H^+] = 1 \times 10^{-5} M$, which of the following is not true?
   (a) pH = 5   (c) $[OH^-] = 1 \times 10^{-5} M$
   (b) pOH = 9   (d) The solution is acidic.

5. If $[H^+] = 2.0 \times 10^{-4} M$, then $[OH^-]$ will be:
   (a) $5.0 \times 10^{-9} M$ (b) 3.70   (c) $2.0 \times 10^{-4} M$ (d) $5.0 \times 10^{-11} M$

6. The solubility product of $BaCrO_4$ is 2×10^{-10}. The solubility of $BaCrO_4$ is:
   (a) $1.4 \times 10^{-5} M$ (b) $2 \times 10^{-10} M$ (c) $4 \times 10^{-20} M$ (d) $1.0 M$

7. The solubility of AgBr is $6.3 \times 10^{-7} M$. The value of the solubility product is:
   (a) 6.3×10^{-7} (b) 4.0×10^{-13} (c) 4.0×10^{-48} (d) 4.0×10^{-15}

8. Which of the following solutions would be the best buffer solution?
   (a) 0.10 M HC$_2$H$_3$O$_2$ + 0.10 M NaC$_2$H$_3$O$_2$
   (b) 0.10 M HCl
   (c) 0.10 M HCl + 0.10 M NaCl
   (d) Pure water

9. For the reaction H$_2$(g) + I$_2$(g) $\rightleftarrows$ 2 HI(g), at 700 K, K_{eq} = 56.6. If an equilibrium mixture at 700 K was found to contain 0.55 M HI and 0.21 M H$_2$, the I$_2$ concentration must be:
   (a) 0.046 M (b) 0.025 M (c) 22 M (d) 2.6 M

10. The equilibrium constant for the reaction 2A + B $\rightleftarrows$ 3C + D is

   (a) $\dfrac{[C]^3[D]}{[A]^2[B]}$ (b) $\dfrac{[2A][B]}{[3C][D]}$ (c) $\dfrac{[3C][D]}{[2A][B]}$ (d) $\dfrac{[A]^2[B]}{[C]^3[D]}$

11. In the equilibrium represented by N$_2$ + O$_2$ $\rightleftarrows$ 2 NO$_2$ (all gases), as the pressure is increased, the amount of NO$_2$ formed:
   (a) Increases   (c) Remains the same
   (b) Decreases   (d) Increases and decreases irregularly

12. Which factor will not increase the concentration of ammonia as represented by the following equation?

   $$3 \text{ H}_2(g) + \text{N}_2(g) \rightleftarrows 2 \text{ NH}_3(g) + 22 \text{ kcal}$$

   (a) Increasing the temperature
   (b) Increasing the concentration of N$_2$
   (c) Increasing the concentration of H$_2$
   (d) Increasing the pressure

13. If HCl(g) is added to a saturated solution of AgCl, the concentration of Ag$^+$ in solution:
   (a) Increases   (c) Remains the same
   (b) Decreases   (d) Increases and decreases irregularly

14. The solubility of CaCO$_3$ at 20°C is 0.013 g/L. What is the K_{sp} for CaCO$_3$?
   (a) 1.3 × 10^{-8} (b) 1.3 × 10^{-4} (c) 1.7 × 10^{-8} (d) 1.7 × 10^{-4}

15. The K_{sp} for BaCrO$_4$ is 8.5 × 10^{-11}. What is the solubility of BaCrO$_4$ in grams per litre?
   (a) 9.2 × 10^{-6} (b) 0.073   (c) 2.3 × 10^{-3} (d) 8.5 × 10^{-11}

16. Which would occur if a small amount of sodium acetate crystals, NaC$_2$H$_3$O$_2$, were added to 100 mL of 0.1 M HC$_2$H$_3$O$_2$ at constant temperature?
   (a) The number of acetate ions in the solution would decrease.
   (b) The number of acetic acid molecules would decrease.
   (c) The number of sodium ions in solution would decrease.
   (d) The H$^+$ concentration in the solution would decrease.

17

Oxidation – Reduction

After studying Chapter 17 you should be able to:

1. Understand the terms listed in Question A at the end of the chapter.
2. Assign oxidation numbers to all the atoms in a given compound or ion.
3. Determine what is being oxidized and what is being reduced in an oxidation–reduction reaction.
4. Identify the oxidizing agent and the reducing agent in an oxidation–reduction reaction.
5. Balance oxidation–reduction equations in molecular and in ionic forms.
6. Understand the general principles concerning the activity series of metals.
7. Use the activity series to determine whether a proposed single-replacement reaction will occur.
8. Distinguish between an electrolytic and a voltaic cell.
9. Draw a voltaic cell that will produce electric current from an oxidation–reduction reaction involving two metals and their salt solutions.
10. Identify the anode reaction and the cathode reaction in a given electrolytic or voltaic cell.
11. Write equations for the overall chemical reaction and for the oxidation and reduction reactions involved in the discharging or charging of a lead storage battery.
12. Explain how the charge condition of a lead storage battery can be estimated with the aid of a hydrometer.

17.1 Oxidation Number

The oxidation number of an atom (frequently called its *oxidation state*) can be considered to represent the number of electrons lost, gained, or unequally shared by the atom. Oxidation numbers can be zero, positive, or negative. When the oxidation number of an atom is zero, the atom has the same number of electrons assigned to it as there are in the free neutral atom. When the oxidation number is positive, the atom has fewer electrons assigned to it than there are in the neutral atom. When the oxidation number is negative, the atom has more electrons assigned to it than there are in the neutral atom.

The oxidation number of an atom that has lost or gained electrons to form a monatomic ion is the same as the plus or minus charge of the ion. In covalent compounds, where electrons are shared between two atoms, the atoms are assigned oxidation numbers by a somewhat arbitrary system. When two atoms share a pair of electrons, the atom with the higher electronegativity has a greater attraction for the electrons. Therefore, when a pair of electrons is shared unequally between two atoms, both electrons are assigned to the more electronegative element. Each element is then assigned an oxidation number based on the number of electrons it then has compared to the neutral atom.

In the ionic compound sodium chloride (NaCl), where one electron has completely transferred from a Na atom to a Cl atom, the oxidation number of Na is clearly established to be $+1$, and for Cl it is -1. The Na^+ ion has one electron less than the Na atom and hence it has a $+1$ charge or a $+1$ oxidation number. The Cl^- ion has one electron more than the Cl atom and has a -1 charge or a -1 oxidation number. In magnesium chloride ($MgCl_2$), two electrons have transferred from the Mg atom to the two Cl atoms; thus, the oxidation number of Mg is $+2$, and the oxidation number of each Cl is -1.

In symmetrical covalent molecules such as H_2 and Cl_2,

H:H $:\overset{..}{\underset{..}{C}}l:\overset{..}{\underset{..}{C}}l:$

electrons are shared equally between the two atoms. Neither atom is more positive or negative than the other; therefore, each is assigned an oxidation number of zero.

In compounds with covalent bonds, such as NH_3 and H_2O,

the pairs of electrons are unequally shared between the atoms and are attracted toward the more electronegative elements, N and O. This unequal sharing causes the N and O atoms to be relatively negative with respect to the H atoms. At the same time, it causes the H atoms to be relatively positive with respect to the N and O atoms. In H_2O, both pairs of shared electrons are assigned to the O atom, giving it two electrons more than the neutral O atom. At the same time, each H atom is assigned one electron less than the neutral H atom. Therefore, the O atom is assigned an oxidation number of -2, and each H atom is assigned an oxidation number of $+1$. In NH_3, the three pairs of shared electrons are assigned to the N atom, giving it three electrons more than the neutral N atom. At the same time, each H atom has one electron less than the neutral atom. Therefore, the N atom is assigned an oxidation number of -3, and each H atom is assigned an oxidation number of $+1$.

The assignment of correct oxidation numbers to elements is essential for balancing oxidation–reduction equations. Restudy Sections 7.11–7.14, regarding oxidation numbers, oxidation number tables, and the determination of oxidation numbers from formulas. Table 7.4 lists relative electronegativities of the elements. Rules for assigning oxidation numbers are given in Section 7.11 and are summarized in Table 17.1, given here. Examples showing oxidation numbers in compounds and ions are given in Table 17.2.

TABLE 17.1 Arbitrary rules for assigning oxidation numbers.

1. All elements in their free state (uncombined with other elements) have an oxidation number of zero (for example, Na, Cu, Mg, H_2, O_2, Cl_2, N_2).
2. H is $+1$, except in metal hydrides, where it is -1 (for example, NaH, CaH_2).
3. O is -2, except in peroxides, where it is -1, and in OF_2, where it is $+2$.
4. The metallic element in an ionic compound has a positive oxidation number.
5. In a covalent compound where a pair of electrons is unequally shared between two atoms, that pair is assigned to the more electronegative atom. This atom is then assigned a negative oxidation number for each electron it has in excess of the neutral atom. The other atom, since it has lost electrons, is assigned a positive oxidation number.
6. The algebraic sum of the oxidation numbers of the elements in a compound is zero.
7. The algebraic sum of the oxidation numbers of the elements in a polyatomic ion is equal to the charge of the ion.

TABLE 17.2 Oxidation numbers of atoms in selected compounds.

Ion or compound	Oxidation number
H_2O	H, $+1$; O, -2
SO_2	S, $+4$; O, -2
CH_4	C, -4; H, $+1$
CO_2	C, $+4$; O, -2
$KMnO_4$	K, $+1$; Mn, $+7$; O, -2
Na_3PO_4	Na, $+1$; P, $+5$; O, -2
$Al_2(SO_4)_3$	Al, $+3$; S, $+6$; O, -2
NO	N, $+2$; O, -2
BCl_3	B, $+3$; Cl, -1
SO_4^{2-}	S, $+6$; O, -2
NO_3^-	N, $+5$; O, -2
CO_3^{2-}	C, $+4$; O, -2

PROBLEM 17.1 Determine the oxidation number of each element in (a) KNO_3, and (b) SO_4^{2-}.

(a) K is a Group IA metal; therefore it has an oxidation number of $+1$. The oxidation number of each O atom is -2 (Table 17.1, Rule 3). Using these values and the fact that the sum of the oxidation numbers of all the atoms in a compound is zero, we can determine the oxidation number of N.

$$KNO_3$$
$$+1 + N + 3(-2) = 0$$
$$N = +6 - 1 = +5$$

The oxidation numbers are: K, $+1$; N, $+5$; O, -2.

(b) SO_4^{2-} is an ion; therefore, the sum of the oxidation numbers of the S and the O atoms must be -2, the charge of the ion. The oxidation number of each O atom is -2 (Table 17.1, Rule 3). Then,

$$SO_4^{2-}$$

$$S + 4(-2) = -2$$

$$S = -2 + 8 = +6$$

The oxidation numbers are: S, $+6$; O, -2.

■

17.2 Oxidation–Reduction

oxidation–
reduction
redox

oxidation
reduction

Oxidation–reduction, also known as **redox,** is a chemical process in which the oxidation number of an element is changed. The process may involve the complete transfer of electrons to form ionic bonds or only a partial transfer or shift of electrons to form covalent bonds.

Oxidation occurs whenever the oxidation number of an element increases as a result of losing electrons. Conversely, **reduction** occurs whenever the oxidation number of an element decreases as a result of gaining electrons. For example, a change in oxidation number from $+2$ to $+3$ or from -1 to 0 is oxidation; a change from $+5$ to $+2$ or from -2 to -4 is reduction (see Figure 17.1). Oxidation and reduction occur simultaneously in a chemical reaction; one cannot take place without the other.

Many combination and decomposition reactions, and all single-replacement reactions, involve oxidation–reduction. Let us examine the combustion of hydrogen and oxygen from this point of view:

$$2\,H_2 + O_2 \longrightarrow 2\,H_2O$$

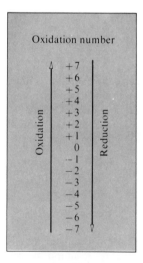

Figure 17.1 Oxidation and reduction: Oxidation results in an increase in the oxidation number, and reduction results in a decrease in the oxidation number.

Both reactants, hydrogen and oxygen, are elements in the free state and have an oxidation number of zero. In the product, water, hydrogen has been oxidized to $+1$ and oxygen reduced to -2. The substance that does the oxidizing is known as the **oxidizing agent.** The substance that does the reducing is the **reducing agent.** In this reaction, the oxidizing agent is free oxygen and the reducing agent is free hydrogen. In the reaction

oxidizing agent
reducing agent

$$Zn(s) + H_2SO_4(aq) \longrightarrow ZnSO_4(aq) + H_2\uparrow$$

metallic zinc is oxidized and hydrogen ions are reduced. Zinc is the reducing agent and the hydrogen ions are the oxidizing agent. Electrons are transferred from the zinc metal to the hydrogen ions. The reaction is better expressed as

$$Zn^0 + 2\,H^+ + SO_4^{2-} \longrightarrow Zn^{2+} + SO_4^{2-} + H_2^0\uparrow$$

Oxidation: **Increase in oxidation number**
Loss of electrons

Reduction: **Decrease in oxidation number**
Gain of electrons

The oxidizing agent is reduced and gains electrons. The reducing agent is oxidized and loses electrons. The loss and gain of electrons is a characteristic feature of all redox reactions.

17.3 Balancing Oxidation–Reduction Equations

Many simple redox equations may be easily balanced by inspection, or trial and error.

$$Na + Cl_2 \longrightarrow NaCl \quad \text{(Unbalanced)}$$
$$2\,Na + Cl_2 \longrightarrow 2\,NaCl \quad \text{(Balanced)}$$

Balancing this equation is certainly not complicated. But as we study more complex reactions and equations, such as

$$P + HNO_3 + H_2O \longrightarrow NO + H_3PO_4 \quad \text{(Unbalanced)}$$
$$3\,P + 5\,HNO_3 + 2\,H_2O \longrightarrow 5\,NO + 3\,H_3PO_4 \quad \text{(Balanced)}$$

the trial-and-error method of finding the proper numbers to balance the equation would take an unnecessarily long time.

One systematic method of balancing oxidation–reduction equations is based on the transfer of electrons between the oxidizing and reducing agents. Consider the first equation again.

$$Na^0 + Cl_2^0 \longrightarrow Na^+Cl^- \quad \text{(Unbalanced)}$$

In this reaction, sodium metal loses one electron per atom when it changes to a sodium ion. At the same time, chlorine gains one electron per atom. Because chlorine is diatomic, two electrons per molecule are needed to form a chloride ion from each atom. These electrons are furnished by two sodium atoms. Stepwise, the reaction may be written as two half-reactions, the oxidation half-reaction and the reduction half-reaction:

Oxidation half-reaction $2\,Na^0 \longrightarrow 2\,Na^+ + 2\,e^-$
Reduction half-reaction $\dfrac{Cl_2^0 + 2\,e^- \longrightarrow 2\,Cl^-}{2\,Na^0 + Cl_2^0 \longrightarrow 2\,Na^+Cl^-}$

When the two half-reactions, each containing the same number of electrons, are added together algebraically, the electrons cancel out. In this reaction there are no excess electrons; the two electrons lost by the two sodium atoms are utilized by chlorine. In all redox reactions, the loss of electrons by the reducing agent must equal the gain of electrons by the oxidizing agent. Sodium is oxidized; chlorine is reduced. Chlorine is the oxidizing agent; sodium is the reducing agent.

The following problems illustrate a systematic method of balancing redox equations.

PROBLEM 17.2   Balance the following equation:

$$Sn + HNO_3 \longrightarrow SnO_2 + NO_2 + H_2O \quad \text{(Unbalanced)}$$

Step 1. Assign oxidation numbers to each element, in order to identify the elements that are being oxidized and those that are being reduced. Write the oxidation numbers below each element in order to avoid confusing them with the charge on an ion or radical.

$$\underset{0}{Sn} + \underset{+1}{H}\ \underset{+5}{N}\ \underset{-2}{O_3} \longrightarrow \underset{+4}{Sn}\ \underset{-2}{O_2} + \underset{+4}{N}\ \underset{-2}{O_2} + \underset{+1}{H_2}\ \underset{-2}{O}$$

Note that the oxidation numbers of Sn and N have changed.

Step 2. Now write two new equations, using only the elements that change in oxidation number. Then add electrons to bring the equations into electrical balance. One equation represents the oxidation step; the other represents the reduction step. The oxidation step produces electrons; the reduction step uses electrons.

Oxidation $Sn^0 \longrightarrow Sn^{4+} + 4\,e^-$ (Sn^0 loses 4 electrons)
Reduction $N^{5+} + 1\,e^- \longrightarrow N^{4+}$ (N^{5+} gains 1 electron)

Step 3. Now multiply the two equations by the smallest integral numbers that will make the loss of electrons by the oxidation step equal to the number of electrons gained in the reduction step. In this reaction, the oxidation step is multiplied by 1 and the reduction step is multiplied by 4. For example, when multiplying the reduction half-reaction by 4, each component in the equation is multiplied by 4:

$$4 \times (N^{5+} + 1\,e^- \longrightarrow N^{4+}) = 4\,N^{5+} + 4\,e^- \longrightarrow 4\,N^{4+}$$

The equations become

$$\text{Oxidation} \qquad Sn^0 \longrightarrow Sn^{4+} + 4\,e^- \qquad (Sn^0 \text{ loses 4 electrons})$$
$$\text{Reduction} \qquad 4\,N^{5+} + 4\,e^- \longrightarrow 4\,N^{4+} \quad (4\,N^{5+} \text{ gain 4 electrons})$$

We have now established the ratio of the oxidizing to the reducing agent as being four atoms of N to one atom of Sn.

Step 4. Now transfer the coefficient that appears in front of each substance in the balanced oxidation–reduction equations to the corresponding substance in the original equation. We need to use 1 Sn, 1 SnO_2, 4 HNO_3, and 4 NO_2:

$$Sn + 4\,HNO_3 \longrightarrow SnO_2 + 4\,NO_2 + H_2O \quad \text{(Unbalanced)}$$

Step 5. In the usual manner, balance the remaining elements that are not oxidized or reduced to give the final balanced equation:

$$Sn + 4\,HNO_3 \longrightarrow SnO_2 + 4\,NO_2 + 2\,H_2O \quad \text{(Balanced)}$$

In balancing the final elements, we must not change the ratio of the elements that were oxidized and reduced. We should make a final check to ensure that both sides of the equation have the same number of atoms of each element. The final balanced equation contains 1 atom of Sn, 4 atoms of N, 4 atoms of H, and 12 atoms of O on each side.

■

Since each new equation may present a slightly different problem and since proficiency in balancing equations requires practice, we will work through a few more problems.

PROBLEM 17.3   Balance the following equation:

$$I_2 + Cl_2 + H_2O \longrightarrow HIO_3 + HCl \quad \text{(Unbalanced)}$$

Step 1. Assign oxidation numbers:

$$\underset{0}{I_2} + \underset{0}{Cl_2} + \underset{+1 \ -2}{H_2\ O} \longrightarrow \underset{+1+5-2}{H\ I\ O_3} + \underset{+1\ -1}{H\ Cl}$$

The oxidation numbers of I_2 and Cl_2 have changed. Iodine changes from 0 to $+5$. Chlorine changes from 0 to -1.

Step 2. Write the oxidation–reduction half-reactions, balancing with electrons:

$$\text{Oxidation} \qquad I_2 \longrightarrow 2\,I^{5+} + 10\,e^- \quad (I_2 \text{ loses 10 electrons})$$
$$\text{Reduction} \qquad Cl_2 + 2\,e^- \longrightarrow 2\,Cl^- \quad (Cl_2 \text{ gains 2 electrons})$$

Step 3. Adjust loss and gain of electrons so that they are equal. Multiply the oxidation step by 1 and the reduction step by 5:

$$\text{Oxidation} \qquad I_2 \longrightarrow 2\,I^{5+} + 10\,e^- \qquad (I_2 \text{ loses 10 electrons})$$
$$\text{Reduction} \qquad 5\,Cl_2 + 10\,e^- \longrightarrow 10\,Cl^- \quad (5\,Cl_2 \text{ gain 10 electrons})$$

Step 4. Transfer the coefficients from the balanced half-reactions into the original equation. We need to use 1 I_2, 2 HIO_3, 5 Cl_2, and 10 HCl.

$$I_2 + 5\,Cl_2 + H_2O \longrightarrow 2\,HIO_3 + 10\,HCl \quad \text{(Unbalanced)}$$

Step 5. Balance the remaining elements, H and O:

$$I_2 + 5 Cl_2 + 6 H_2O \longrightarrow 2 HIO_3 + 10 HCl \quad (Balanced)$$

Check: The final balanced equation contains 2 atoms of I, 10 atoms of Cl, 12 atoms of H, and 6 atoms of O on each side.

■

PROBLEM 17.4 Balance the following equation:

$$K_2Cr_2O_7 + FeCl_2 + HCl \longrightarrow CrCl_3 + KCl + FeCl_3 + H_2O$$
(Unbalanced)

Step 1. Assign oxidation numbers (Cr and Fe have changed):

$$\underset{+1}{K_2} \ \underset{+6}{Cr_2} \ \underset{-2}{O_7} + \underset{+2}{Fe} \ \underset{-1}{Cl_2} + \underset{+1}{H} \ \underset{-1}{Cl} \longrightarrow \underset{+3}{Cr} \ \underset{-1}{Cl_3} + \underset{+1}{K} \ \underset{-1}{Cl} + \underset{+3}{Fe} \ \underset{-1}{Cl_3} + \underset{+1}{H_2} \ \underset{-2}{O}$$

Step 2. Write the oxidation–reduction half-reactions, balancing with electrons:

Oxidation $Fe^{2+} \longrightarrow Fe^{3+} + 1 e^-$ (Fe^{2+} loses 1 electron)
Reduction $Cr^{6+} + 3 e^- \longrightarrow Cr^{3+}$

or

$$2 Cr^{6+} + 6 e^- \longrightarrow 2 Cr^{3+} \quad (2 Cr^{6+} \text{ gain 6 electrons})$$

Step 3. Balance the loss and gain of electrons. Multiply the oxidation step by 6 and the reduction step by 1 to equalize the transfer of electrons:

Oxidation $6 Fe^{2+} \longrightarrow 6 Fe^{3+} + 6 e^-$ (6 Fe^{2+} lose 6 electrons)
Reduction $2 Cr^{6+} + 6 e^- \longrightarrow 2 Cr^{3+}$ (2 Cr^{6+} gain 6 electrons)

Step 4. Transfer the coefficients from the balanced half-reactions into the original equation. (Note that one formula unit of $K_2Cr_2O_7$ contains two Cr atoms.) We need to use 1 $K_2Cr_2O_7$, 2 $CrCl_3$, 6 $FeCl_2$, and 6 $FeCl_3$.

$$K_2Cr_2O_7 + 6 FeCl_2 + HCl \longrightarrow 2 CrCl_3 + KCl + 6 FeCl_3 + H_2O$$
(Unbalanced)

Step 5. Balance the remaining elements in this order: K, Cl, H, O.

$$K_2Cr_2O_7 + 6 FeCl_2 + 14 HCl \longrightarrow 2 CrCl_3 + 2 KCl + 6 FeCl_3 + 7 H_2O$$
(Balanced)

Check: The final balanced equation contains 2 K atoms, 2 Cr atoms, 7 O atoms, 6 Fe atoms, 26 Cl atoms, and 14 H atoms on each side.

■

PROBLEM 17.5 Try the following equation, which has a little different twist to it.

$$Cu + HNO_3 \longrightarrow Cu(NO_3)_2 + NO + H_2O \quad (Unbalanced)$$

Step 1. Assign oxidation numbers [Cu and N (in NO) have changed]:

$$\underset{0}{Cu} + \underset{+5}{HNO_3} \longrightarrow \underset{+2 +5}{Cu(NO_3)_2} + \underset{+2}{NO} + H_2O$$

Step 2. Write the oxidation–reduction half-reactions, balancing with electrons:

Oxidation $Cu^0 \longrightarrow Cu^{2+} + 2\,e^-$ (Cu^0 loses 2 electrons)
Reduction $N^{5+} + 3\,e^- \longrightarrow N^{2+}$ (N^{5+} gains 3 electrons)

Step 3. Balance the loss and gain of electrons. Multiply the oxidation step by 3 and the reduction step by 2 to equalize the loss and gain of electrons:

Oxidation $3\,Cu^0 \longrightarrow 3\,Cu^{2+} + 6\,e^-$ (3 Cu^0 lose 6 electrons)
Reduction $2\,N^{5+} + 6\,e^- \longrightarrow 2\,N^{2+}$ (2 N^{5+} gain 6 electrons)

Step 4. Transfer the coefficients from the balanced half-reactions into the original equation:

$$3\,Cu + 2\,HNO_3 \longrightarrow 3\,Cu(NO_3)_2 + 2\,NO + H_2O \quad \text{(Unbalanced)}$$

Step 5. Balance the remaining elements. In doing this, we notice that there are 8 N atoms on the right side of the equation and 2 N atoms on the left. This imbalance indicates that 6 more HNO_3 molecules are needed on the left and also that 6 NO_3^- ions did not enter into the redox reaction. The use of 8 HNO_3 in the balanced equation does not destroy the ratio of 3 Cu/2 HNO_3 needed for oxidation–reduction. The balanced equation is

$$3\,Cu + 8\,HNO_3 \longrightarrow 3\,Cu(NO_3)_2 + 2\,NO + 4\,H_2O \quad \text{(Balanced)}$$

Check: The final balanced equation contains 3 Cu atoms, 8 H atoms, 8 N atoms, and 24 O atoms on each side.

17.4 Balancing Ionic Redox Equations

The main difference in balancing redox equations containing ions versus those not written in the ionic form is the handling of ions. In addition to having the same number of each kind of element, the net charges on both sides of the final equation must be equal to each other. In assigning oxidation numbers, we must be careful to consider the charge on the ions. In many respects, balancing ionic equations is much simpler than balancing molecular equations.

PROBLEM 17.6   Balance the following equation:

$$Fe^{2+} + Br_2 \longrightarrow Fe^{3+} + Br^- \quad \text{(Unbalanced)}$$

You might try to balance this equation simply by placing a 2 in front of the Br^-:

$$Fe^{2+} + Br_2 \longrightarrow Fe^{3+} + 2\,Br^- \quad \text{(Unbalanced)}$$

Although all the atoms in the equation are balanced, the equation is not balanced because the electrical charges on the left and the right sides of the equation are not equal. The left side has a charge of $+2$ and the right side has a charge of $+1\,(+3-2)$. The net charge on each side is determined by adding the charges of all the ions. The equation is correctly balanced by the half-reaction method:

Oxidation $Fe^{2+} \longrightarrow Fe^{3+} + 1\,e^-$ (Fe^{2+} loses 1 electron)
Reduction $Br_2 + 2\,e^- \longrightarrow 2\,Br^-$ (2 Br gain 2 electrons)

Equalize the loss and gain of electrons by multiplying the oxidation step by 2:

Oxidation $2\,Fe^{2+} \longrightarrow 2\,Fe^{3+} + 2\,e^-$ (2 Fe lose 2 electrons)
Reduction $Br_2 + 2\,e^- \longrightarrow 2\,Br^-$ (2 Br gain 2 electrons)

Finally, transfer the balanced half-reactions into the original equation:

$$2\,Fe^{2+} + Br_2 \longrightarrow 2\,Fe^{3+} + 2\,Br^- \text{ (Balanced)}$$
Net charge: $(+4)\ +\ (0) = +4$ $(+6)\ +\ (-2) = +4$

The balanced equation contains the same number of each kind of atom and the same electrical charge on each side of the equation. The charge on each side is $+4$.

■

PROBLEM 17.7   Try this more complex equation:

$$MnO_4^- + S^{2-} + H^+ \longrightarrow Mn^{2+} + S^0 + H_2O \text{ (Unbalanced)}$$

First assign oxidation numbers (Mn and S have changed):

$$MnO_4^- + S^{2-} + H^+ \longrightarrow Mn^{2+} + S^0 + H_2O$$
$$+7 \quad\ -2 \qquad\qquad\qquad\quad +2 \qquad 0$$

Write the oxidation–reduction half-reactions and balance with electrons:

Oxidation $S^{2-} \longrightarrow S^0 + 2\,e^-$ (S^{2-} loses 2 electrons)
Reduction $Mn^{7+} + 5\,e^- \longrightarrow Mn^{2+}$ (Mn^{7+} gains 5 electrons)

Multiply the oxidation step by 5 and the reduction step by 2 to balance the loss and gain of electrons:

Oxidation $5\,S^{2-} \longrightarrow 5\,S^0 + 10\,e^-$ ($5\,S^{2-}$ lose 10 electrons)
Reduction $2\,Mn^{7+} + 10\,e^- \longrightarrow 2\,Mn^{2+}$ ($2\,Mn^{7+}$ gain 10 electrons)

Transfer the coefficients from the balanced half-reactions into the original equation:

$$2\,MnO_4^- + 5\,S^{2-} + H^+ \longrightarrow 2\,Mn^{2+} + 5\,S^0 + H_2O \text{ (Unbalanced)}$$

At this point there remain to be balanced the electrical charge, the H atoms, and the O atoms. First the electrical charges are balanced by the use of additional H^+ ions. The H and O atoms are then balanced by the use of H_2O molecules as needed. We find that 16 H^+ ions are needed and 8 H_2O molecules are needed to bring the equation into balance.

$$2\,MnO_4^- + 5\,S^{2-} + 16\,H^+ \longrightarrow 2\,Mn^{2+} + 5\,S^0 + 8\,H_2O \text{ (Balanced)}$$
Net charge: $(-2)\ +\ (-10)\ +\ (+16) = +4$ $(+4)\ +\ (0)\ +\ (0) = +4$

Check: Both sides have a net charge of $+4$ and contain the same number of atoms of each element. The equation is balanced.

■

Another method for balancing ionic redox equations uses ionic charges and electrons instead of oxidation numbers. This method is as follows:

1. Write the two half-reactions containing the elements being oxidized and reduced. Use the full formulas containing the elements being oxidized and reduced.
2. Balance the elements other than oxygen and hydrogen.
3. Balance oxygen and hydrogen:
   (a) For reactions occurring in acid solution, use H_2O and H^+ to balance oxygen and hydrogen. For each oxygen needed, use one H_2O.
   (b) For reactions occurring in alkaline solutions, use H_2O and OH^- to balance oxygen and hydrogen. For each oxygen needed, use two OH^- ions.
4. Add electrons (e^-) to each half-reaction to bring them into electrical balance.
5. Since the loss and gain of electrons must be equal, multiply each half-reaction by appropriate numbers to make the number of electrons the same in each half-reaction.
6. Add the two half-reactions together, canceling electrons and any other identical substances that appear on opposite sides of the equation.

Let us apply this method to the following equation:

$$Ag + NO_3^- \longrightarrow Ag^+ + NO \quad \text{(Acid solution)}$$

Step 1. Write the two half-reactions:

Oxidation $\qquad Ag \longrightarrow Ag^+$

Reduction $\qquad NO_3^- \longrightarrow NO$

Step 2. Balance elements other than H and O. Both Ag and N are balanced.

$$Ag \longrightarrow Ag^+$$
$$NO_3^- \longrightarrow NO$$

Step 3. Balance H and O using H_2O and H^+ (acid solution). In the reduction step use 2 H_2O on the right to balance the O atoms. Then 4 H^+ will be needed on the left to balance the H atoms.

$$Ag \longrightarrow Ag^+$$
$$4 H^+ + NO_3^- \longrightarrow NO + 2 H_2O$$

Step 4. Balance electrically using e^- as needed. In the oxidation step, 1 e^- is needed on the right side of the equation. In the reduction step, 3 e^- are needed on the left side of the equation.

$$Ag \longrightarrow Ag^+ + e^- \qquad\qquad \text{(Net charge} = 0)$$
$$3 e^- + 4 H^+ + NO_3^- \longrightarrow NO + 2 H_2O \qquad \text{(Net charge} = 0)$$

Step 5. Equalize loss and gain of electrons. Multiply the oxidation half-reaction by 3 (in this case).

$$3 \, Ag \longrightarrow 3 \, Ag^+ + 3 \, e^-$$

$$3 \, e^- + 4 \, H^+ + NO_3^- \longrightarrow NO + 2 \, H_2O$$

Step 6. Add the two half-reactions together to give the balanced equation. The $3 \, e^-$ cancel.

$$3 \, Ag \longrightarrow 3 \, Ag^+ + \cancel{3 \, e^-}$$
$$\underline{\cancel{3 \, e^-} + 4 \, H^+ + NO_3^- \longrightarrow NO + 2 \, H_2O}$$
$$3 \, Ag + 4 \, H^+ + NO_3^- \longrightarrow 3 \, Ag^+ + NO + 2 \, H_2O$$

Use this method to balance the equation given in Problem 17.7 to see that it gives the same balanced equation as by the oxidation number method.

17.5 Activity Series of Metals

Knowledge of the relative chemical reactivity of the elements is useful for predicting the course of many chemical reactions.

Calcium reacts with cold water and magnesium reacts with steam to produce hydrogen in each case. Calcium, therefore, is considered to be a more reactive metal than magnesium.

$$Ca(s) + 2 \, H_2O(l) \longrightarrow Ca(OH)_2(aq) + H_2\uparrow$$

$$Mg(s) + \underset{\text{Steam}}{H_2O(g)} \longrightarrow MgO(s) + H_2\uparrow$$

The difference in their activity is attributed to the relative ease with which each loses its two valence electrons. It is apparent that calcium loses these electrons more easily than magnesium and is therefore more reactive and/or more readily oxidized than magnesium.

When a strip of copper is placed in a solution of silver nitrate ($AgNO_3$), free silver begins to plate out on the copper. After the reaction has continued for some time, we can observe a blue color in the solution, indicating the presence of copper(II) ions. If a strip of silver is placed in a solution of copper(II) nitrate [$Cu(NO_3)_2$], no reaction is visible. The equations are

$$Cu^0 + 2 \, AgNO_3(aq) \longrightarrow 2 \, Ag^0 + Cu(NO_3)_2(aq)$$

$$Cu^0 + 2 \, Ag^+ \longrightarrow 2 \, Ag^0 + Cu^{2+} \qquad \text{Net ionic equation}$$

$$Cu^0 \longrightarrow Cu^{2+} + 2 \, e^- \qquad \text{Oxidation of } Cu^0$$

$$Ag^+ + e^- \longrightarrow Ag^0 \qquad \text{Reduction of } Ag^+$$

$$Ag^0 + Cu(NO_3)_2(aq) \longrightarrow \text{No reaction}$$

In the reaction between Cu and $AgNO_3$, electrons are transferred from Cu^0 atoms to Ag^+ ions in solution. Since copper has a greater tendency than silver to lose electrons, an electrochemical force is exerted upon silver ions to accept electrons from copper atoms. When a Ag^+ ion adds an electron, it is reduced to a Ag^0 atom and is no longer soluble in solution. At the same time, Cu^0 is oxidized and goes into solution as Cu^{2+} ions. From this reaction we can conclude that copper is more reactive than silver.

Metals such as sodium, magnesium, zinc, and iron, which react with solutions of acids to liberate hydrogen, are more reactive than hydrogen. Metals such as copper, silver, and mercury, which do not react with solutions of acids to liberate hydrogen, are less reactive than hydrogen. By studying a series of reactions such as those just given, we may list metals according to their chemical activity, placing the most active at the top and the least active at the bottom. This list is called the **Activity Series of Metals.** Table 17.3 shows some of the common metals in the series. The arrangement corresponds to the ease with which the elements listed are oxidized or lose electrons. The most easily oxidizable element is listed first. More extensive tables are available in chemistry reference books.

Activity Series of Metals

TABLE 17.3 Activity Series of Metals. Relative ease of oxidation of selected metals and hydrogen.

Ease of oxidation

$$K \rightarrow K^+ + e^-$$
$$Ba \rightarrow Ba^{2+} + 2e^-$$
$$Ca \rightarrow Ca^{2+} + 2e^-$$
$$Na \rightarrow Na^+ + e^-$$
$$Mg \rightarrow Mg^{2+} + 2e^-$$
$$Al \rightarrow Al^{3+} + 3e^-$$
$$Zn \rightarrow Zn^{2+} + 2e^-$$
$$Cr \rightarrow Cr^{3+} + 3e^-$$
$$Fe \rightarrow Fe^{2+} + 2e^-$$
$$Ni \rightarrow Ni^{2+} + 2e^-$$
$$Sn \rightarrow Sn^{2+} + 2e^-$$
$$Pb \rightarrow Pb^{2+} + 2e^-$$
$$H_2 \rightarrow 2H^+ + 2e^-$$
$$Cu \rightarrow Cu^{2+} + 2e^-$$
$$As \rightarrow As^{3+} + 3e^-$$
$$Ag \rightarrow Ag^+ + e^-$$
$$Hg \rightarrow Hg^{2+} + 2e^-$$
$$Au \rightarrow Au^{3+} + 3e^-$$

The general principles governing the arrangement and use of the activity series are as follows:

1. The reactivity of the metals listed decreases from top to bottom.
2. A free metal can displace the ion of a second metal from solution provided that the free metal is above the second metal in the activity series.

3. Free metals above hydrogen react with nonoxidizing acids in solution to liberate hydrogen gas.
4. Free metals below hydrogen do not liberate hydrogen from acids.
5. Conditions such as temperature and concentration may affect the relative position of some of these elements.

Two examples of the application of the activity series are given in the following problems.

PROBLEM 17.8   Will zinc metal react with dilute sulfuric acid?

From Table 17.3, we see that zinc is above hydrogen; therefore, zinc atoms will lose electrons more readily than hydrogen atoms will. Hence, zinc atoms will reduce hydrogen ions from the acid to form hydrogen gas and zinc ions. In fact, these reagents are commonly used for the laboratory preparation of hydrogen. The equation is

$$Zn(s) + H_2SO_4(aq) \longrightarrow ZnSO_4(aq) + H_2\uparrow$$
$$Zn(s) + 2\,H^+ \longrightarrow Zn^{2+} + H_2\uparrow \qquad \text{Net ionic equation}$$

■

PROBLEM 17.9   Will a reaction occur when copper metal is placed in an iron(II) sulfate solution?

No, copper lies below iron in the series, loses electrons less easily than iron, and therefore will not replace iron(II) ions from solution. In fact, the reverse is true. When an iron nail is dipped into a copper(II) sulfate solution, It becomes coated with free copper. The equations are

$$Cu(s) + FeSO_4(aq) \longrightarrow \text{No reaction}$$
$$Fe(s) + CuSO_4(aq) \longrightarrow FeSO_4(aq) + Cu\downarrow$$

From Table 17.3 we may abstract the following pair in their relative position to each other.

$$Fe \longrightarrow Fe^{2+} + 2\,e^-$$
$$Cu \longrightarrow Cu^{2+} + 2\,e^-$$

According to the second principle given for the use of the activity series, we can predict that free iron will react with copper(II) ions in solution to form free copper metal and iron(II) ions in solution.

$$Fe(s) + Cu^{2+}(aq) \longrightarrow Fe^{2+}(aq) + Cu(s) \quad \text{Net ionic equation}$$

■

17.6 Electrolytic and Voltaic Cells

electrolysis
electrolytic cell

The process in which electrical energy is used to bring about a chemical change is known as **electrolysis**. An **electrolytic cell** uses electrical energy to produce a nonspontaneous chemical reaction. There are many applications of electrical energy in the chemical industry—for example, in the production of sodium, sodium hydroxide, chlorine, fluorine, magnesium, aluminum, and pure hydrogen and oxygen, and in the purification and electroplating of metals.

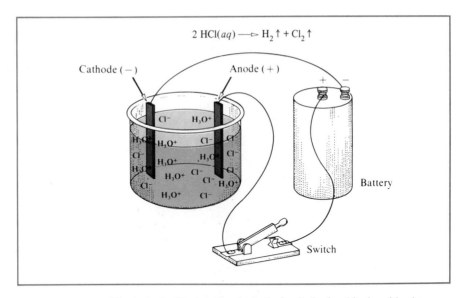

$$2\,HCl(aq) \longrightarrow H_2\uparrow + Cl_2\uparrow$$

Figure 17.2 Electrolysis: During the electrolysis of a hydrochloric solution, positive hydronium ions are attracted to the cathode, where they gain electrons and form hydrogen gas. Chloride ions migrate to the anode, where they lose electrons and form chlorine gas. The equation for this process is

$$2\,HCl(aq) \longrightarrow H_2\uparrow + Cl_2\uparrow$$

What happens when an electric current is passed through a solution? Let us consider a hydrochloric acid solution in a simple electrolysis cell, as shown in Figure 17.2. The cell consists of a source of direct current (a battery) connected to two electrodes that are immersed in a solution of hydrochloric acid. The cathode is attached to the negative pole of the battery and becomes the negative electrode. The anode is attached to the positive pole and becomes the positive electrode. The battery supplies electrons to the cathode.

When the switch is closed, the electric circuit is completed; positive hydronium ions (H_3O^+) migrate toward the cathode, where they pick up electrons and evolve hydrogen gas. At the same time, the negative chloride ions (Cl^-) migrate toward the anode, where they lose electrons, completing the cycle, and evolve chlorine gas.

Reaction at the cathode (Negative electrode)	$H_3O^+ + 1\,e^- \longrightarrow H^0 + H_2O$ $H^0 + H^0 \longrightarrow H_2\uparrow$
Reaction at the anode (Positive electrode)	$Cl^- \longrightarrow Cl^0 + 1\,e^-$ $Cl^0 + Cl^0 \longrightarrow Cl_2\uparrow$
Net reaction	$2\,HCl(aq) \xrightarrow{\text{Electrolysis}} H_2\uparrow + Cl_2\uparrow$

Note that oxidation–reduction has taken place. Chloride ions lose electrons (are oxidized) at the anode, and hydronium ions gain electrons (are reduced) at the cathode.

In electrolysis, oxidation always occurs at the anode and reduction always occurs at the cathode.

When sodium chloride brines are electrolyzed, the products are sodium hydroxide, hydrogen, and chlorine. The overall reaction is

$$2\,Na^+ + 2\,Cl^- + 2\,H_2O \xrightarrow{\text{Electrolysis}} 2\,Na^+ + 2\,OH^- + H_2\uparrow + Cl_2\uparrow$$

The net ionic equation is

$$2\,Cl^- + 2\,H_2O \longrightarrow 2\,OH^- + H_2\uparrow + Cl_2\uparrow$$

During the electrolysis, Na^+ ions move toward the cathode and Cl^- ions move toward the anode. The anode reaction is similar to that of hydrochloric acid; chlorine is liberated.

$$2\,Cl^- \longrightarrow Cl_2\uparrow + 2\,e^-$$

Even though Na^+ ions are attracted by the cathode, the facts show that hydrogen is liberated there. No evidence of metallic sodium is found, but the area around the cathode tests alkaline from accumulated OH^- ions. The reaction at the cathode is believed to be

$$2\,H_2O + 2\,e^- \longrightarrow H_2\uparrow + 2\,OH^-$$

If the electrolysis is allowed to continue until all the chloride is reacted, the solution remaining will contain only sodium hydroxide. Large tonnages of sodium hydroxide and chlorine are made by this process.

Sodium chloride melts at 800°C. When molten sodium chloride (without water) is subjected to electrolysis, metallic sodium and chlorine gas are formed:

$$2\,Na^+(l) + 2\,Cl^-(l) \xrightarrow{\text{Electrolysis}} 2\,Na(l) + Cl_2(g)$$

An important electrochemical application is the electroplating of metals. Electroplating is the art of covering a surface or an object with a thin adherent electrodeposited metal coating. Electroplating is done for the protection of the surface of the base metal or for a purely decorative effect. The layer deposited is surprisingly thin, varying from as little as 5×10^{-5} cm to 2×10^{-3} cm, depending on the metal and the intended use. The object to be plated is set up as the cathode and is immersed in a solution containing ions of the metal to be plated. When an electric current passes through the solution, metal ions migrating to the cathode are reduced, depositing on the object as the free metal. In most cases the metal deposited on the object is replaced in the solution by using an anode of the same metal. The following equations show the chemical changes in the electroplating of nickel:

Reaction at the cathode	$Ni^{2+} + 2\,e^- \longrightarrow Ni(s)$	Ni plated out on an object
Reaction at the anode	$Ni(s) \longrightarrow Ni^{2+} + 2\,e^-$	Ni replenished to solution

Metals most commonly used in commercial electroplating are copper, nickel, zinc, lead, cadmium, chromium, tin, gold, and silver.

In the electrolytic cell shown in Figure 17.2, an electric current flows through the circuit when the switch is closed. The driving force responsible for the current is supplied by the source of direct current. Electrons are moving through the wires and electrodes, and ions (H_3O^+ and Cl^-) are moving in the solution. As a result of the transfer of electrons and ions, hydrochloric acid is converted to hydrogen and chlorine. In electrolytic processes of this kind, electrical energy is used to bring about nonspontaneous redox reactions. The hydrogen and chlorine produced have more potential energy than was present in the hydrochloric acid before electrolysis.

Conversely, some spontaneous redox reactions can be made to supply useful amounts of electrical energy. When a piece of zinc is put in a copper(II) sulfate solution, the zinc quickly becomes coated with metallic copper. We expect this to happen because zinc is above copper in the activity series; copper(II) ions are therefore reduced by zinc atoms:

$$Zn^0(s) + Cu^{2+}(aq) \longrightarrow Zn^{2+}(aq) + Cu^0(s)$$

This reaction is clearly a spontaneous redox reaction. But simply dipping a zinc rod into a copper(II) sulfate solution will not produce useful electric current! However, when we carry out this reaction in the cell shown in Figure 17.3, an electric current is produced. The cell consists of a piece of zinc immersed in a zinc sulfate solution and connected by a wire through a voltmeter to a piece of copper immersed in copper(II) sulfate solution. The two solutions are connected by a salt bridge. Such a cell produces an electric current and a potential of about 1.1 volts when both solutions are 1.0 M in concentration. A cell that produces electric current from a spontaneous chemical reaction is called a **voltaic cell** is also known as a *galvanic cell.*

voltaic cell

Although the zinc–copper voltaic cell is no longer used commercially, it was used to energize the first transcontinental telegraph lines. Such cells are the direct ancestors of the many different kinds of "dry" cells (batteries) that operate portable radio and television sets, automatic cameras, tape recorders, and so on.

The driving force responsible for the electric current in the zinc–copper cell originates in the great tendency of zinc atoms to lose electrons relative to the tendency of copper(II) ions to gain electrons. In the cell shown in Figure 17.3, zinc atoms lose electrons and are converted to zinc ions at the zinc electrode; the electrons flow through the external circuit (the wire) to the copper electrode. Here, copper(II) ions pick up electrons and are reduced to copper atoms, which

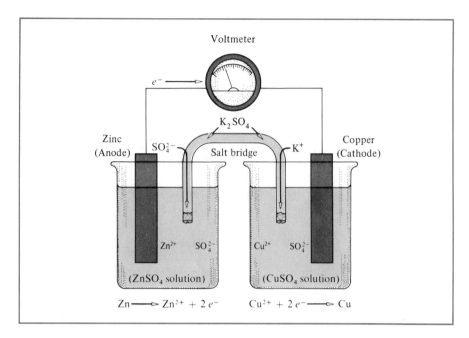

Figure 17.3 Zinc–copper voltaic cell. The cell has a potential of 1.1 volts when $ZnSO_4$ and $CuSO_4$ solutions are 1.0 M. The salt bridge provides electrical contact between the two half-cells.

plate out on the copper electrode. In order to keep the solutions electrically neutral, SO_4^{2-} ions diffuse out of the salt bridge into the $ZnSO_4$ solution as Zn^{2+} ions are formed; K^+ ions diffuse out of the salt bridge into the $CuSO_4$ solution as Cu^{2+} ions are plated out. The equations for the reactions of this cell are

Anode	$Zn^0 \longrightarrow Zn^{2+} + 2\,e^-$	(Oxidation)
Cathode	$Cu^{2+} + 2\,e^- \longrightarrow Cu^0$	(Reduction)
Net ionic	$Zn^0 + Cu^{2+} \longrightarrow Zn^{2+} + Cu^0$	
Overall	$Zn(s) + CuSO_4(aq) \longrightarrow Cu(s) + ZnSO_4(aq)$	

The redox reaction, the movement of electrons in the metallic or external part of the circuit, and the movement of ions in the solution or internal part of the circuit of the copper–zinc cell are very similar to the actions that occur in the electrolytic cell of Figure 17.2. The only important difference is that the reactions of the zinc–copper cell are spontaneous. This is the crucial difference between all voltaic and electrolytic cells.

Voltaic cells use chemical reactions to produce electrical energy.
Electrolytic cells use electrical energy to produce chemical reactions.

An ordinary automobile storage battery is an energy reservoir. The charged battery acts as a voltaic cell and through chemical reactions furnishes electrical energy to operate the starter, lights, radio, and so on. When the engine is running, a generator, or alternator, produces and forces an electric current through the battery and, by electrolytic chemical action, restores the battery to the charged condition.

The cell unit consists of a lead plate filled with spongy lead and a lead dioxide plate, both immersed in dilute sulfuric acid solution, which serves as the electrolyte (see Figure 17.4). When the cell is discharging, or acting as a voltaic cell, these reactions occur.

Pb plate (anode) $\quad$ $Pb^0 \longrightarrow Pb^{2+} + 2\,e^-$ $\hspace{2cm}$ (Oxidation)

PbO$_2$ plate (cathode) $\quad$ $PbO_2 + 4\,H^+ + 2e^- \longrightarrow Pb^{2+} + 2\,H_2O$ $\quad$ (Reduction)

Net ionic redox reaction $\quad$ $Pb^0 + PbO_2 + 4\,H^+ \longrightarrow 2\,Pb^{2+} + 2\,H_2O$

Precipitation reaction on plates $\quad$ $Pb^{2+} + SO_4^{2-} \longrightarrow PbSO_4(s)$

Since lead(II) sulfate is insoluble, the Pb^{2+} ions combine with SO_4^{2-} ions to form a coating of $PbSO_4$ on each plate. The overall chemical reaction of the cell is

$$Pb(s) + PbO_2(s) + 2\,H_2SO_4(aq) \xrightarrow[\text{cycle}]{\text{Discharge}} 2\,PbSO_4(s) + 2\,H_2O + Energy$$

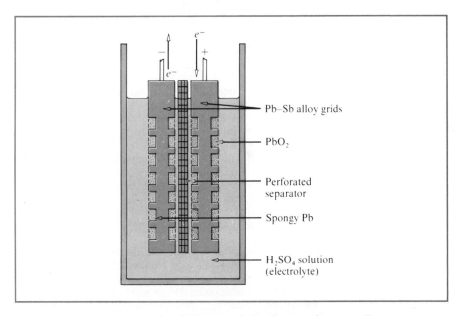

Figure 17.4 Cross-sectional diagram of a lead storage battery cell.

If the active material on both plates is converted to $PbSO_4$, the cell is discharged and no more electrical energy is to be had. The cell can be recharged by reversing the chemical reaction. This is accomplished by forcing an electric current through the cell in the opposite direction. Lead sulfate and water are reconverted to lead, lead dioxide, and sulfuric acid:

$$2\ PbSO_4(s) + 2\ H_2O \xrightarrow[\text{cycle}]{\text{Charge}} Pb(s) + PbO_2(s) + 2\ H_2SO_4(aq)$$

The electrolyte in a lead storage battery is a 38% by weight sulfuric acid solution having a density of 1.29 g/mL. As the battery is discharged, sulfuric acid is removed, thereby decreasing the density of the electrolyte solution. The state of charge or discharge of the battery can be estimated by measuring the density of the electrolyte solution with a hydrometer. When the density has dropped to about 1.05 g/mL, the battery needs recharging.

In an actual battery, each cell consists of a series of cell units of alternating lead–lead dioxide plates separated and supported by wood, asbestos, or glass wool. The energy storage capacity of a single cell is limited and its electrical potential is only about 2 volts. Therefore, a bank of six cells is connected in series to provide the 12 volt output of the usual automobile battery.

QUESTIONS

A. Review the Meanings of the New Terms Introduced in this Chapter

1. Oxidation–reduction	6. Reducing agent
2. Redox	7. Activity Series of Metals
3. Oxidation	8. Electrolysis
4. Reduction	9. Electrolytic cell
5. Oxidizing agent	10. Voltaic cell

B. Answers to the Following Questions Will Be Found in Tables and Figures

1. In the equation

   $$I_2 + 5\ Cl_2 + 6\ H_2O \longrightarrow 2\ HIO_3 + 10\ HCl$$

   (a) Has iodine been oxidized or has it been reduced?
   (b) Has chlorine been oxidized or has it been reduced?
   (See Figure 17.1.)
2. Based on Table 17.3, which element of each pair is more active?
   (a) Ag or Al   (b) Na or Ba   (c) Ni or Cu
3. Based on Table 17.3, will the following combinations react in aqueous solution?
   (a) $Zn + Cu^{2+}$ (c) $Sn + Ag^+$ (e) $Ba + FeCl_2$ (g) $Ni + Hg(NO_3)_2$
   (b) $Ag + H^+$ (d) $As + Mg^{2+}$ (f) $Pb + NaCl$ (h) $Al + CuSO_4$
4. The reaction between powdered aluminum and iron(III) oxide (in the Thermite process), producing molten iron, is very exothermic.
   (a) Write the equation for the chemical reaction that occurs.
   (b) Explain in terms of Table 17.3 why a reaction occurs.
   (c) Would you expect a reaction between powdered iron and aluminum oxide?
   (d) Would you expect a reaction between powdered aluminum and chromium(III) oxide?
5. Write equations for the chemical reaction of each of the following metals with dilute solutions of hydrochloric acid and sulfuric acid: aluminum, chromium, gold, iron, copper, magnesium, mercury, and zinc. If a reaction will not occur, write "no reaction" as the product. (See Table 17.3.)
6. A $NiCl_2$ solution is placed in the apparatus shown in Figure 17.2, instead of the HCl solution shown. Write equations for:
   (a) The anode reaction   (b) The cathode reaction
   (c) The net electrochemical reaction

7. What is the major distinction between the reactions occurring in Figure 17.2 versus Figure 17.3?
8. In the cell shown in Figure 17.3:
    (a) What would be the effect of removing the voltmeter and connecting the wires shown coming to the voltmeter?
    (b) What would be the effect of removing the salt bridge?

C. **Review Questions**
1. What is the oxidation number of the underlined element in each compound:

    (a) $\underline{Na}Cl$ (d) $Na\underline{N}O_3$ (g) $K\underline{Mn}O_4$ (j) $KC\underline{l}O_3$

    (b) $\underline{Fe}Cl_3$ (e) $H_2\underline{S}O_3$ (h) $\underline{I}_2$ (k) $K_2\underline{Cr}O_4$

    (c) $Pb\underline{O}_2$ (f) $\underline{N}H_4Cl$ (i) $\underline{N}H_3$ (l) $K_2\underline{Cr}_2O_7$

2. What is the oxidation number of the underlined elements?
    (a) $\underline{S}^{2-}$ (d) $\underline{Mn}O_4^-$ (f) $\underline{O}_2$ (h) $\underline{Fe}(OH)_3$

    (b) $\underline{N}O_2^-$ (e) $\underline{Bi}^{3+}$ (g) $\underline{As}O_4^{3-}$ (i) $\underline{I}O_3^-$

    (c) $Na_2\underline{O}_2$

3. In the following half-reactions, which element is changing oxidation state? Is the half-reaction an oxidation or a reduction? Supply the proper number of electrons to the proper side to balance each equation.
    (a) $Zn^{2+} \longrightarrow Zn$
    (b) $2\,Br^- \longrightarrow Br_2$
    (c) $MnO_4^- + 8\,H^+ \longrightarrow Mn^{2+} + 4\,H_2O$
    (d) $Ni \longrightarrow Ni^{2+}$
    (e) $SO_3^{2-} + H_2O \longrightarrow SO_4^{2-} + 2\,H^+$
    (f) $NO_3^- + 4\,H^+ \longrightarrow NO + 2\,H_2O$
    (g) $S_2O_4^{2-} + 2\,H_2O \longrightarrow 2\,SO_3^{2-} + 4\,H^+$
    (h) $Fe^{2+} \longrightarrow Fe^{3+}$

4. In the following unbalanced equations:
    (a) Identify the element oxidized and the element reduced.
    (b) Identify the oxidizing agent and the reducing agent.
        (1) $Cr + HCl \longrightarrow CrCl_3 + H_2$
        (2) $SO_4^{2-} + I^- + H^+ \longrightarrow H_2S + I_2 + H_2O$
        (3) $AsH_3 + Ag^+ + H_2O \longrightarrow H_3AsO_4 + Ag + H^+$
        (4) $Cl_2 + NaBr \longrightarrow NaCl + Br_2$

5. For the following equations:
    (a) Assign oxidation numbers to each element.
    (b) Write equations for the half-reactions of each oxidation and reduction, adding the correct number of electrons to the correct side.
    (c) Balance the original equation.
        (1) $Zn + S \longrightarrow ZnS$
        (2) $AgNO_3 + Pb \longrightarrow Pb(NO_3)_2 + Ag$
        (3) $Fe_2O_3 + CO \longrightarrow Fe + CO_2$
        (4) $H_2S + HNO_3 \longrightarrow S + NO + H_2O$
        (5) $MnO_2 + HBr \longrightarrow MnBr_2 + Br_2 + H_2O$
        (6) $PH_3 + I_2 + H_2O \longrightarrow H_3PO_2 + I^- + H^+$
        (7) $I^- + ClO^- \longrightarrow Cl^- + IO_3^-$
        (8) $Cr_2O_7^{2-} + H_3AsO_3 + H^+ \longrightarrow Cr^{3+} + H_3AsO_4 + H_2O$
        *(9) $Br_2 + OH^- \longrightarrow BrO_3^- + Br^- + H_2O$
        (10) $Cr_2O_7^{2-} + HNO_2 + H^+ \longrightarrow Cr^{3+} + NO_3^- + H_2O$

6. Balance the following redox equations:
   (a) $Ag + HNO_3 \longrightarrow AgNO_3 + NO + H_2O$
   (b) $CuO + NH_3 \longrightarrow N_2 + Cu + H_2O$
   (c) $PbO_2 + Sb + NaOH \longrightarrow PbO + NaSbO_2 + H_2O$
   (d) $H_2O_2 + KMnO_4 + H_2SO_4 \longrightarrow O_2 + MnSO_4 + K_2SO_4 + H_2O$
   *(e) $Cl_2 + KOH \longrightarrow KClO_3 + KCl + H_2O$
   (f) $Zn + Cr_2O_7^{2-} + H^+ \longrightarrow Zn^{2+} + Cr^{3+} + H_2O$
   (g) $I^- + NO_3^- + H^+ \longrightarrow NO + I_2 + H_2O$
   *(h) $AsH_3 + Cu^{2+} + H_2O \longrightarrow H_3AsO_4 + Cu + H^+$
   *(i) $CrO_4^{2-} + Fe(OH)_2 + H_2O \longrightarrow Cr(OH)_3 + Fe(OH)_3 + OH^-$
   *(j) $MnO_4^- + SO_3^{2-} + H_2O \longrightarrow MnO_2 + SO_4^{2-} + OH^-$

7. Balance the following ionic redox equations: [*Note:* Supply H^+ or OH^- ions and H_2O molecules if needed to balance.]
   (a) $Zn + NO_3^- + H^+ \longrightarrow Zn^{2+} + NH_4^+ + H_2O$
   (b) $NO_3^- + S + H^+ \longrightarrow NO_2 + SO_4^{2-} + H_2O$
   (c) $Zn + AsO_4^{3-} + H^+ \longrightarrow AsH_3 + Zn^{2+} + H_2O$
   *(d) $ClO_3^- + Cl^- + H^+ \longrightarrow Cl_2 + H_2O$
   (e) $Cr_2O_7^{2-} + Fe^{2+} + H^+ \longrightarrow Cr^{3+} + Fe^{3+} + H_2O$
   (f) $HNO_2 + MnO_4^- + H^+ \longrightarrow Mn^{2+} + NO_3^- + H_2O$
   (g) $SeO_3^{2-} + Cl_2 + OH^- \longrightarrow SeO_4^{2-} + Cl^- + H_2O$
   (h) $H_2C_2O_4 + MnO_4^- \longrightarrow Mn^{2+} + CO_2$ (Acidic solution)
   *(i) $CuS + NO_3^- \longrightarrow Cu^{2+} + S + NO$ (Acidic solution)
   *(j) $Se \longrightarrow Se^{2-} + SeO_3^{2-}$ (Basic solution)

8. Why are oxidation and reduction said to be complementary processes?
9. When molten $CaBr_2$ is electrolyzed, calcium metal and bromine are produced. Write equations for the two half-reactions that occur at the electrodes. Label the anode half-reaction and the cathode half-reaction.
10. Why is direct current used instead of alternating current in the electroplating of metals?
11. The chemical reactions taking place during discharge in a lead storage battery are

$$Pb + SO_4^{2-} \longrightarrow PbSO_4$$

$$PbO_2 + SO_4^{2-} + 4\,H^+ \longrightarrow PbSO_4 + 2\,H_2O$$

   (a) Complete each half-reaction by supplying electrons.
   (b) Which reaction is oxidation and which is reduction?
   (c) Which reaction occurs at the anode of the battery?
12. What property of lead dioxide and lead(II) sulfate makes it unnecessary to have salt bridges in the cells of a lead storage battery?
13. Explain why the density of the electrolyte in a lead storage battery decreases during the discharge cycle.
14. In one type of alkaline cell used to power devices such as portable radios, Hg^{2+} ions are reduced to metallic mercury when the cell is being discharged. Does this occur at the anode or the cathode? Explain.
15. Differentiate between an electrolysis cell and a voltaic cell.
16. Why is a porous barrier or a salt bridge necessary in some voltaic cells?
17. Which of the following statements are correct?
   (a) An atom of an element in the uncombined state has an oxidation number of zero.
   (b) The oxidation number of molybdenum in Na_2MoO_4 is $+4$.

(c) The oxidation number of an ion is the same as the electrical charge on the ion.

(d) The process in which an atom or an ion loses electrons is called reduction.

(e) The reaction $Fe^{3+} + e^- \longrightarrow Fe^{2+}$ is a reduction reaction.

(f) In the reaction $2\ Al + 3\ CuCl_2 \longrightarrow 2\ AlCl_3 + 3\ Cu$, aluminum is the oxidizing agent.

(g) In a redox reaction the oxidizing agent is reduced and the reducing agent is oxidized.

(h) $Cu^0 \longrightarrow Cu^{2+}$ is a balanced oxidation half-reaction.

(i) In the electrolysis of sodium chloride brine (solution), Cl_2 gas is formed at the cathode and hydroxide ions are formed at the anode.

(j) In any cell, electrolytic or voltaic, reduction takes place at the cathode and oxidation occurs at the anode.

(k) In the Zn–Cu voltaic cell, the reaction at the anode is $Zn \longrightarrow Zn^{2+} + 2\ e^-$.

The statements in (l) to (o) pertain to this activity series:

Ba   Mg   Zn   Fe   H   Cu   Ag

(l) The reaction $Zn + MgCl_2 \longrightarrow Mg + ZnCl_2$ is a spontaneous reaction.

(m) Barium is more active than copper.

(n) Silver metal will react with acids to liberate hydrogen gas.

(o) Iron is a better reducing agent than zinc.

(p) Oxidation and reduction occur simultaneously in a chemical reaction; one cannot take place without the other.

(q) A free metal can displace from solution the ions of a metal that lies below the free metal in the activity series.

(r) In electroplating, the piece to be electroplated with a metal is attached to the cathode.

D. Review Problems

1. How many moles of NO gas will be formed by the reaction of 20.0 g of silver with nitric acid? [See the equation given in Question C.6, part (a).]

2. What volume of chlorine gas, measured at STP, is required to react with excess KOH to form 0.200 mole $KClO_3$? [See the equation given in Question C.6, part (e).]

3. What weight of $KMnO_4$ would be needed to react with 200 mL of H_2O_2 solution ($d = 1.031$ g/mL, 9.0% H_2O_2 by weight)? [See the equation given in Question C.6, part (d).]

4. What volume of 0.200 M $K_2Cr_2O_7$ will be required to oxidize 6.00 g H_3AsO_3? [See the equation given in Question C.5, part (8).]

5. What volume of 0.200 M $K_2Cr_2O_7$ will be required to oxidize the Fe^{2+} ion in 50.0 mL of 0.200 M $FeSO_4$ solution? [See the equation given in Question C.7, part (e).]

6. A sample of crude potassium iodide was analyzed using the following reaction (not balanced):

$$KI + H_2SO_4 \longrightarrow I_2 + H_2S + K_2SO_4 + H_2O$$

If a 3.50 g sample of KI produced 2.26 g of iodine, what is the percent purity of the crude KI?

7. What weight of copper is formed when 25.0 L of ammonia gas, measured at STP, reacts with copper(II) oxide? [See the equation given in Question C.6, part (b).]

*8. What volume of NO gas, measured at 28°C and 744 mm Hg, will be formed by the reaction of 120 mL of 0.400 M KI solution reacting with excess nitric acid? [See the equation given in Question C.6, part (g).]

*9. How many moles of chlorine gas can be formed by the reaction of 100 mL of 0.100 M $KClO_3$ with 50.0 mL of 0.400 M KCl? [See the equation given in Question C.7, part (d).]

18

Radioactivity and Nuclear Chemistry

After studying Chapter 18 you should be able to:

1. Understand the terms listed in Question A at the end of the chapter.
2. Outline the historical development of nuclear chemistry, giving the major contributions of Antoine Henri Becquerel, Marie Curie, Ernest Rutherford, Paul Villard, E. O. Lawrence, Irene Joliot-Curie, Otto Hahn, Fritz Strassmann, and Edwin McMillan.
3. Write balanced nuclear chemical equations using isotopic notation.
4. Determine the amount of a radioactive isotope remaining after a given period of time when the starting amount and half-life are given.
5. List the characteristics that distinguish alpha particles, beta particles, and gamma rays from the standpoint of mass, charge, relative velocities, and penetrating power.
6. Describe the effect of a magnetic field on alpha particles, beta particles, and gamma rays.
7. Describe a radioactive disintegration series and predict what isotope would be formed by the loss of specified numbers of alpha and beta particles from a given radioactive element.
8. Distinguish between radioactive disintegration and nuclear fission reactions.
9. Tell how the fission of $^{235}_{92}U$ can lead to a chain reaction and why a critical mass is needed for the chain reaction.
10. Describe how the energy from nuclear fission is converted to electrical energy.
11. Tell what the essential difference is between the fission reactions of a nuclear reactor and those of an atomic bomb.
12. Explain what is meant by nuclear fusion and why large-scale efforts to develop controlled nuclear fusion are under way.
13. Understand the concept of mass defect and its relationship to the binding energy of the nucleus of an atom.
14. Explain how nuclear radiation can cause acute, long-term, and genetic damage to living organisms.
15. Describe the formation of carbon-14 in the atmosphere and tell how this isotope is used to determine the age of carbon-containing material.
16. Explain the use of radioactive tracers.

Up to this point, we have considered only atomic changes that involve the electron structures of atoms and molecules. In this chapter we introduce an entirely different aspect of the atom—reactions of the nucleus. Since the dramatic ending of World War II, nuclear chemistry has become a tremendously important branch of science. The immense amounts of energy available from nuclear reactions can be used for constructive or destructive purposes. The controlled release of nuclear energy has become an important source of power. Nuclear power plants are in operation or are being built in virtually every industrialized nation in the world. The use of radioactive tracers is a routine technique in industry and in biological and medical research and applications.

18.1 Discovery of Radioactivity

One of the most important events leading to the discovery of radioactivity occurred in 1895. Wilhelm Konrad Roentgen (1845–1923) discovered X rays when he observed that a vacuum discharge tube, enclosed in a thin, black cardboard box, caused a nearby piece of paper coated with barium platinocyanide to glow with a brilliant luminescence. From this and other experiments he concluded that certain rays, which he called X rays, were emitted from the discharge tube, penetrated the box, and caused the salt to glow. Roentgen also showed that X rays could penetrate other bodies and affect photographic plates. This observation led to the development of X-ray photography.

Shortly after this discovery, Antoine Henri Becquerel (1852–1908) attempted to show a relationship between X rays and the luminescence of uranium salts. In one of his experiments, he wrapped a photographic plate in black paper, placed a sample of uranium salt on it, and exposed it to sunlight. The developed photographic plate showed that rays emitted from the salt had penetrated the paper. Later, Becquerel prepared to repeat the experiment, but because the sunlight was intermittent, he placed the entire setup in a drawer. Several days later he decided to develop the photographic plate, expecting to find it only slightly affected. He was amazed to observe an intense image on the plate. He repeated the experiment in total darkness and obtained the same results, proving that the uranium salt emitted rays that affected the photographic plate without its being exposed to sunlight. In this way, Becquerel discovered radioactivity. The name *radioactivity* was given to this phenomenon 2 years later (in 1898) by Marie Curie. Becquerel later showed that the rays coming from uranium were able to ionize air and were also capable of penetrating thin sheets of metal.

In 1898, Marie Sklodowska Curie (1867–1934) and her husband, Pierre Curie (1859–1906), turned their research interests to radioactivity. In a short time, the Curies discovered two new elements, polonium and radium, both of which are radioactive. To confirm their work on radium, they processed about 8 tons of pitchblende, an ore rich in uranium, to obtain 0.1 g of pure radium chloride, which they used for atomic weight determinations.

In 1899, Ernest Rutherford began to investigate the nature of the rays emitted from uranium. He found two rays, which he called *alpha* and *beta rays.* Soon after, he realized that uranium, while emitting these rays, was changing into another element. By 1912, over 30 radioactive isotopes (radioisotopes) were known, and many more are known today. The *gamma ray,* a third ray emitted from radioactive materials and similar to an X ray, was discovered by Paul Villard in 1900. After the description of the nuclear atom by Rutherford, the phenomenon of radioactivity was attributed to reactions taking place in the nuclei of atoms.

The symbolism and notation described for isotopes in Chapter 5 are very useful in nuclear chemistry and are briefly reviewed here.

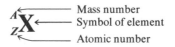

Mass number
Symbol of element
Atomic number

For example, $^{238}_{92}U$ represents a uranium isotope with an atomic number of 92 and a mass number of 238. This isotope is also designated as U-238 or uranium-238 and contains 92 protons and 146 neutrons. The protons and neutrons collectively are known as **nucleons.** The mass number is the total number of nucleons in the nucleus. Table 18.1 shows the isotopic notation for several particles associated with nuclear chemistry.

nucleons

TABLE 18.1 Symbols in isotopic notation for several particles (and small isotopes) associated with nuclear chemistry.

Particle	Symbol	Atomic number, Z	Atomic mass, A
Neutron	$^{1}_{0}n$	0	1
Proton	$^{1}_{1}H$	1	1
Beta particle (electron)	$^{0}_{-1}e$	-1	0
Positron (positive electron)	$^{0}_{+1}e$	1	0
Alpha particle (helium nucleus)	$^{4}_{2}He$	2	4
Deuteron (heavy hydrogen nucleus)	$^{2}_{1}H$	1	2

18.2 Natural Radioactivity

radioactivity

Radioactivity is the spontaneous emission of radiation from the nucleus of an atom. Elements having this property are said to be radioactive. Radioactive elements continually undergo **radioactive decay,** or disintegration, to form different elements. The chemical properties of an element are associated with its electronic structure, but radioactivity is a property of the nucleus. Therefore, neither ordinary changes of temperature and pressure nor the chemical or physical state of an element has any effect on its radioactivity.

radioactive decay

The principal emissions from the nuclei of radioisotopes are known as alpha rays (or particles), beta rays (or particles), and gamma rays. Upon losing an alpha or beta particle, the radioactive element changes into a different element. This process will be explained in detail later.

Each radioactive isotope disintegrates at a specific and constant rate, which is expressed in units of half-life. The **half-life** ($t_{1/2}$) is the time required for one-half of a specific amount of a radioactive element to disintegrate. The half-lives

half-life

of the elements vary from a fraction of a second to billions of years. For example, $^{238}_{92}U$ has a half-life of 4.5 x 10⁹ years, $^{226}_{88}Ra$ has a half-life of 1620 years, and $^{15}_{6}C$ has a half-life of 2.4 seconds. To illustrate, if we start today with 1.0 g of $^{226}_{88}Ra$, we will have 0.50 g of $^{226}_{88}Ra$ remaining at the end of 1620 years; at the end of another 1620 years, 0.25 g will remain, and so on.

$$1.0 \text{ g } ^{226}_{88}Ra \xrightarrow[\substack{1620 \text{ years}}]{t_{1/2}} 0.50 \text{ g } ^{226}_{88}Ra \xrightarrow[\substack{1620 \text{ years}}]{t_{1/2}} 0.25 \text{ g } ^{226}_{88}Ra$$

The half-lives of the various radioisotopes of the same element are different from each other. Those for radium are as follows:

Radium isotope	Half-life
Ra-223	11.2 days
Ra-224	3.64 days
Ra-226	1620 years
Ra-228	6.7 years

A radioactive decay curve is illustrated in Figure 18.1. Note that half of the quantity remaining decays in the half-life of the radioisotope.

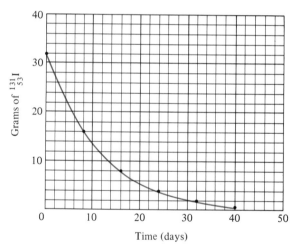

Figure 18.1   Radioactive decay curve for $^{131}_{53}I$, which has a half-life of 8 days.

$$^{131}_{53}I \longrightarrow ^{131}_{54}Xe + _{-1}^{0}e$$

Half of the $^{131}_{53}I$ remaining decays in 8 days. Starting with 32 g, there will be 1 g of $^{131}_{53}I$ left after 5 half-lives (40 days).

PROBLEM 18.1   In how many half-lives will 10.0 g of a radioactive element decay to less than 10% of its original value?

Ten percent of the original amount is 1.0 g. After the first half-life, half of the original material remains and half has decayed. After the second half-life, one-fourth of the original material remains, which is one-half of the starting amount, at the end of the first half-life. This progression continues, reducing the quantity remaining by half for each half-life that passes.

Half-lives	0	1	2	3	4
Percent remaining	100%	50%	25%	12.5%	6.25%
Amount remaining	10.0 g	5.00 g	2.50 g	1.25 g	0.625 g

Therefore, the amount remaining will be less than 10% sometime between the third and the fourth half-life.

∎

Isotopes are said to be either *stable* (nonradioactive) or *unstable* (radioactive). All isotopes of elements that have atomic numbers greater than 83 (bismuth) are naturally radioactive, although some of these isotopes have extremely long half-lives. Some of the naturally occurring isotopes of elements 81, 82, and 83 are radioactive and some are stable. Only a few naturally occurring isotopes of elements that have atomic numbers less than 81 are radioactive. However, no stable isotopes of element 43 (technetium) or of element 61 (promethium) are known.

Radioactivity is believed to be a result of an unstable ratio of neutrons to protons in the nucleus. Stable isotopes of elements up to about atomic number 20 generally have about a 1:1 neutron to proton ratio. In elements above number 20 the neutron to proton ratio in the stable isotopes gradually increases to about 1.5:1 in element number 83 (bismuth). When the neutron to proton ratio is too high or too low, alpha, beta, or other particles are emitted to achieve a more stable nucleus.

18.3 Properties of Alpha Particles, Beta Particles, and Gamma Rays

alpha particle

Alpha particle. An **alpha particle** consists of two protons and two neutrons, has a mass of about 4 amu, a charge of $+2$, and is identical to a doubly charged helium atom. It is usually given one of the following symbols: α, He^{2+}, or ^4_2He. When an alpha particle is emitted from the nucleus, a different element is formed. The atomic number of the new element is 2 less and the mass is 4 amu less than the starting element. For example, when $^{238}_{92}U$ loses an alpha particle, $^{234}_{90}Th$ is formed, because two neutrons and two protons are lost from the uranium nucleus. This may be written as an equation:

$$^{238}_{92}U \longrightarrow {}^{234}_{90}Th + \alpha \quad \text{or} \quad {}^{238}_{92}U \longrightarrow {}^{234}_{90}Th + {}^4_2He$$

146 n	144 n	2 n
92 p	90 p	2 p

Change in the nucleus due to loss of an α particle

For the loss of an alpha particle from $^{226}_{88}Ra$, the equation is

$$^{226}_{88}Ra \longrightarrow {}^{222}_{86}Rn + {}^4_2He \quad \text{or} \quad {}^{226}_{88}Ra \longrightarrow {}^{222}_{86}Rn + \alpha$$

A nuclear equation, like a chemical equation, consists of reactants and products and must be balanced. To have a balanced nuclear equation, the sum of the mass numbers (superscripts) on both sides of the equation must be equal, and the sum of the atomic numbers (subscripts) on both sides of the equation must be equal.

Sum of mass numbers equals 226

$$^{226}_{88}\text{Ra} \longrightarrow {}^{222}_{86}\text{Rn} + {}^{4}_{2}\text{He}$$

Sum of atomic numbers equals 88

What new element will be formed when $^{230}_{90}\text{Th}$ loses an alpha particle? This loss is equivalent to two protons and two neutrons. The new element will have a mass of 226 amu (230 − 4) and will contain 88 protons (90− 2) so that its atomic number is 88. Locate the corresponding element on the periodic chart. It is $^{226}_{88}\text{Ra}$ or radium-226.

beta particle

Beta particle.   The **beta particle** is identical in mass and charge to an electron; its charge is −1. Both a beta particle and a proton are produced by the decomposition of a neutron. The beta particle leaves and the proton remains in the nucleus. When an atom loses a beta particle from its nucleus, a new element is formed, having essentially the same mass but an atomic number that is 1 greater than the starting element. The beta particle is written as β or $_{-1}^{0}e$. Here are some examples of equations in which a beta particle is lost:

$$^{234}_{90}\text{Th} \longrightarrow {}^{234}_{91}\text{Pa} + \beta$$

$$^{234}_{91}\text{Pa} \longrightarrow {}^{234}_{92}\text{U} + {}_{-1}^{0}e$$

$$^{210}_{82}\text{Pb} \longrightarrow {}^{210}_{83}\text{Bi} + \beta$$

gamma ray

Gamma ray.   Gamma rays are photons of energy with properties similar to those of X rays. They have no electrical charge and no measurable mass. Gamma rays emanate from the nucleus in many radioactive changes along with either alpha or beta particles. The designation for a gamma ray is γ. Gamma radiation does not result in a change in atomic number or the mass of an element.

PROBLEM 18.2   (a) Write an equation for the loss of an alpha particle from the isotope $^{194}_{78}\text{Pt}$. (b) What isotope is formed when $^{228}_{88}\text{Ra}$ loses a beta particle from the nucleus?

(a) Loss of an alpha particle, $^{4}_{2}\text{He}$, means the loss of two neutrons and two protons. This results in a decrease of 4 in the mass and a decrease of 2 in the atomic number.

Mass of new isotope: $A − 4$ or $194 − 4 = 190$

Atomic number of new isotope: $Z − 2$ or $78 − 2 = 76$

Looking up element number 76 on the periodic table, we find it to be osmium, Os. The equation, then, is

$$^{194}_{78}Pt \longrightarrow ^{190}_{76}Os + ^4_2He$$

(b) The loss of a beta particle from a $^{228}_{88}Ra$ nucleus means a gain of 1 in the atomic number with no essential change in mass. This can result from

$$n \longrightarrow p + e^-(\beta)$$

The new isotope will have an atomic number of 89 $(Z + 1)$, which is actinium, Ac.
  The isotope formed is $^{228}_{89}Ac$.

■

PROBLEM 18.3   What isotope will be formed when $^{214}_{82}Pb$ successively emits β, β, and α particles from its nucleus? Write successive equations showing these changes.
  The changes brought about in the three steps outlined are as follows:

  β loss: Increase of 1 in the atomic number; no change in mass

  β loss: Increase of 1 in the atomic number; no change in mass

  α loss: Decrease of 2 in the atomic number; decrease of 4 in the mass

The equations are

$$^{214}_{82}Pb \xrightarrow{-\beta} ^{214}_{83}X \xrightarrow{-\beta} ^{214}_{84}X \xrightarrow{-\alpha} ^{210}_{82}X$$

where X stands for the new isotope formed. Looking up each of these elements by their atomic numbers, we rewrite the equations

$$^{214}_{82}Pb \xrightarrow{-\beta} ^{214}_{83}Bi \xrightarrow{-\beta} ^{214}_{84}Po \xrightarrow{-\alpha} ^{210}_{82}Pb$$

■

The ability of radioactive rays to pass through various objects is in proportion to the speed at which they leave the nucleus. Gamma rays travel at the velocity of light (186,000 miles per second) and are capable of penetrating several inches of lead. The velocities of beta particles are variable, the fastest being about nine-tenths the velocity of light. Alpha particles have velocities less than one-tenth the velocity of light. Figure 18.2 (page 454) illustrates the relative penetrating power of these rays. A sheet of paper will stop alpha particles; a thin sheet of aluminum will stop both alpha and beta particles; and a 5 cm block of lead will reduce, but not completely stop, gamma radiation. In fact, it is difficult to stop all gamma radiation. Table 18.2 summarizes the properties of alpha, beta, and gamma radiation.

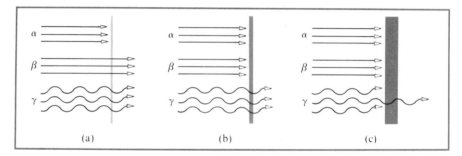

Figure 18.2 Relative penetrating ability of alpha, beta, and gamma radiation. (a) Thin sheet of paper; (b) thin sheet of aluminum; (c) 5 cm lead block.

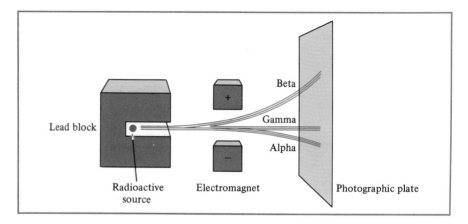

Figure 18.3 The effect of an electromagnetic field on the three radioactive rays. Lighter beta particles are deflected considerably more than alpha particles. Alpha and beta particles are deflected in opposite directions. Gamma radiation is not affected by the electromagnetic field.

TABLE 18.2. Characteristics of nuclear radiation.

Radiation	Symbol	Mass (amu)	Electrical charge	Velocity	Composition
Alpha	α ^4_2He	4	$+2$	Variable—less than 10% the speed of light	Identical to He^{2+}
Beta	β $^{\ 0}_{-1}e$	1/1837	-1	Variable—up to 90% the speed of light	Identical to an electron
Gamma	γ	None	0	Speed of light	Photons or electromagnetic waves of energy

The classical experiment proving that alpha and beta particles are oppositely charged was conducted by Marie Curie. A radioactive source was placed in a hole in a lead block, and the rays given off were allowed to pass between the poles of a strong electromagnet. The paths of the charged particles were deflected as they passed through the field of the electromagnet and were finally detected by striking a photographic plate (see Figure 18.3). The lighter beta particles were deflected considerably more than the alpha particles, and in the opposite direction; but the gamma rays were not affected by the electromagnet and struck the photographic plate in paths straight out of the lead block.

18.4 Radioactive Disintegration Series

The naturally occurring radioactive elements above lead (Pb) fall into three orderly disintegration series. Each series proceeds from one element to the next by the loss of either an alpha or a beta particle, finally ending in a nonradioactive stable isotope. The uranium series starts with $^{238}_{92}U$ and ends with $^{206}_{82}Pb$. The thorium series starts with $^{232}_{90}Th$ and ends with $^{208}_{82}Pb$. The actinium series starts with $^{235}_{92}U$ and ends with $^{207}_{82}Pb$. A fourth series, the neptunium series, starts with the synthetic element $^{241}_{94}Pu$ and ends with the stable bismuth isotope $^{209}_{83}Bi$. The uranium series is shown in Figure 18.4; gamma radiation, which accompanies alpha and beta radiation, is not shown in the figure.

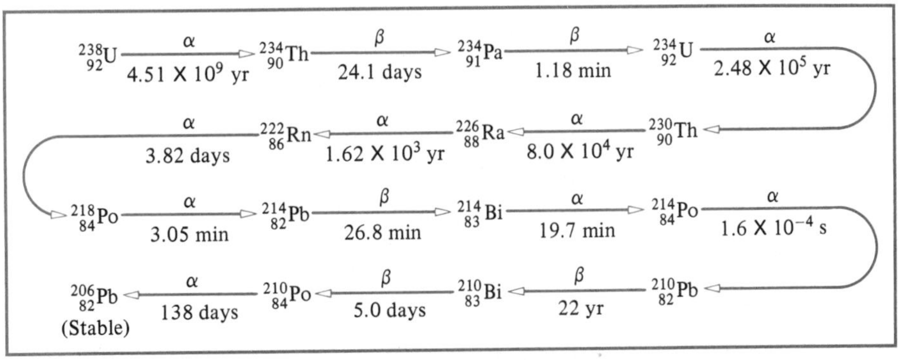

Figure 18.4   The uranium disintegration series. Uranium-238 decays by a series of alpha (α) and beta (β) emissions to the stable isotope lead-206. The type of decay and the corresponding half-life are shown with the arrow.

By using such a series and the half-lives of its members, scientists have been able to approximate the age of certain geologic deposits. This was done by comparing the amount of $^{238}_{92}U$ with the amount of $^{208}_{82}Pb$ and other isotopes in the series that are present in a particular geologic formation. Rocks found in Canada and Finland have been calculated to be about 3.0×10^9 (3 billion) years old. Some meteorites have been determined to be 4.5×10^9 years old.

18.5 Transmutation of Elements

transmutation

Transmutation is the conversion of one element into another by either natural or artificial means. Transmutation occurs spontaneously in natural radioactive disintegrations. Alchemists tried for centuries to convert lead and mercury into gold by artificial means. But transmutation by artificial means was not achieved until 1919, when Ernest Rutherford succeeded in bombarding the nuclei of nitrogen atoms with alpha particles and produced oxygen isotopes and protons. The nuclear equation for this transmutation can be written as

$$^{14}_{7}N + \alpha \longrightarrow {}^{17}_{8}O + {}^{1}_{1}H \quad \text{or} \quad {}^{14}_{7}N + {}^{4}_{2}He \longrightarrow {}^{17}_{8}O + {}^{1}_{1}H$$

It is believed that the alpha particle enters the nitrogen nucleus, forming the compound nucleus $^{18}_{9}F$, which then decomposes into the products.

Rutherford's experiments opened the door to nuclear transmutations of all kinds. Atoms were bombarded by alpha particles, neutrons, protons, deuterons ($^{2}_{1}H$), electrons, and so on. Massive instruments were developed for accelerating these particles to very high speeds and energies to aid their penetration of the nucleus. Some of these instruments are the famous cyclotron, developed by E. O. Lawrence at the University of California; the Van de Graaff electrostatic generator; the betatron; and the electron and proton synchrotron. With these instruments, many nuclear transmutations became possible. Equations for a few of these follow.

$$^{7}_{3}Li + {}^{1}_{1}H \longrightarrow 2\,{}^{4}_{2}He$$

$$^{40}_{18}Ar + {}^{1}_{1}H \longrightarrow {}^{40}_{19}K + {}^{1}_{0}n$$

$$^{23}_{11}Na + {}^{1}_{1}H \longrightarrow {}^{23}_{12}Mg + {}^{1}_{0}n$$

$$^{114}_{48}Cd + {}^{2}_{1}H \longrightarrow {}^{115}_{48}Cd + {}^{1}_{1}H$$

$$^{2}_{1}H + {}^{2}_{1}H \longrightarrow {}^{3}_{1}H + {}^{1}_{1}H$$

$$^{209}_{83}Bi + {}^{2}_{1}H \longrightarrow {}^{210}_{84}Po + {}^{1}_{0}n$$

$$^{16}_{8}O + {}^{1}_{0}n \longrightarrow {}^{13}_{6}C + {}^{4}_{2}He$$

$$^{27}_{13}Al + {}^{4}_{2}He \longrightarrow {}^{30}_{15}P + {}^{1}_{0}n$$

18.6 Artificial Radioactivity

Irene Joliot-Curie, a daughter of Pierre and Marie Curie, and her husband, Frederic Joliot-Curie, observed that when aluminum-27 was bombarded with alpha particles, neutrons and positrons (positive electrons) were emitted as part of the products. When the source of alpha particles was removed, neutrons ceased to be produced but positrons continued to be emitted. This suggested that the neutrons and positrons were coming from two separate reactions. It also indicated that one of the products of the first reaction was radioactive.

After further investigation they discovered that when aluminum-27 is bombarded with alpha particles, phosphorus-30 and neutrons are produced. Phosphorus-30 is radioactive, has a half-life of 2.5 minutes, and decays to silicon-30 with the emission of a positron. The equations for these reactions follow.

$$^{27}_{13}\text{Al} + ^{4}_{2}\text{He} \longrightarrow ^{30}_{15}\text{P} + ^{1}_{0}\text{n}$$

$$^{30}_{15}\text{P} \longrightarrow ^{30}_{14}\text{Si} + ^{0}_{+1}\text{e}$$

artificial radioactivity induced radioactivity

The radioactivity of isotopes produced in this manner is known as **artificial radioactivity** or **induced radioactivity.** Artificial radioisotopes behave like natural radioactive elements in that they disintegrate in a definite fashion and have a specific half-life for each isotope. The Joliot-Curies received the Nobel Prize in chemistry in 1935 for the discovery of artificial or induced radioactivity.

A list of some commonly used radioisotopes and their mode of decay is given in Table 18.3.

TABLE 18.3 Some commonly used radioisotopes.

Radioisotope	Radiation emitted	Half-life
$^{14}_{6}\text{C}$	β	5668 years
$^{32}_{15}\text{P}$	β	14.3 days
$^{35}_{16}\text{S}$	β	87 days
$^{60}_{27}\text{Co}$	β	5.3 years
$^{90}_{38}\text{Sr}$	β	28 years
$^{131}_{53}\text{I}$	β	8.1 days
$^{137}_{55}\text{Cs}$	β	30 years

18.7 Measurement of Radioactivity

Radiation from radioactive sources is so energetic that it is called *ionizing radiation.* When it strikes an atom or a molecule, one or more electrons are knocked off and an ion is created. One of the common instruments used to detect and measure radioactivity, the Geiger counter, depends on this fact. The instrument consists of a Geiger–Müller detecting tube and a counting device. The detector tube is a pair of oppositely charged electrodes in an argon gas–filled chamber fitted with a thin window. When radiation, such as a beta particle, passes through the window into the tube, some argon is ionized and a momentary pulse of current (discharge) flows between the electrodes. These current pulses are electronically amplified in the counter and appear as signals in the form of audible clicks, flashing lights, meter deflections, or numerical readouts. Figure 18.5 on page 458 illustrates a simple Geiger counter.

The curie is the unit used to express the amount of radioactivity produced by an element. **One curie** is defined as the quantity of radioactive material

1 curie

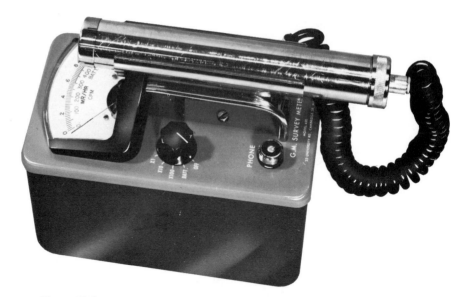

Figure 18.5   A Geiger counter, or battery-operated portable Geiger–Müller survey meter. Radioactivity is indicated in a headphone by clicks and by a graduated meter indicating the number of counts per minute or hour. The Geiger tube located on top is the detector; the rest of the instrument is the counting device. (*Courtesy Atomic Accessories, Inc.*)

TABLE 18.4   Radiation units.

Curie (Ci)	A unit of radioactivity indicating the rate of decay of a radioactive substance. 1 Ci = 3.7×10^{10} disintegrations per second
Roentgen (R)	A unit of exposure of gamma radiation based on the quantity of ionization produced in air
Rad	A unit of absorbed dose of radiation indicating the energy absorbed from any ionizing radiation. 1 rad = 0.01 joule of energy absorbed per 1 kg of matter
Rem	A unit of radiation dose equivalent. The rem takes into account that the same dose in rads from different sources of radiation does not produce the same degree of biological effect. 1 rem is equal to the dose in rads multiplied by a factor dependent on the particular type of radiation.

giving 3.7×10^{10} disintegrations per second. The basis for this figure is pure radium, which has an activity of 1 curie per gram. Because the curie is such a large quantity, the millicurie and microcurie, representing one-thousandth and one-millionth of a curie, respectively, are more practical and more commonly used. Common units in which radioactivity is expressed are given in Table 18.4.

18.8 Nuclear Fission

nuclear fission

In **nuclear fission,** a heavy atomic nucleus, when struck in the right way by a neutron, splits into two or more large fragments. The fragments are called *fission products.* As the atom splits, it releases energy and two or three neutrons, each of which can cause another nuclear fission. The first instance of nuclear fission was reported in January 1939 by the German scientists Otto Hahn and Fritz Strassmann. In experiments in which they were attempting to prepare elements with atomic numbers greater than 92 they bombarded uranium with neutrons. To their amazement, they detected isotopes of barium, krypton, cerium, and lanthanum among the products. This led Hahn and Strassmann to believe that the uranium nucleus had been split into smaller fragments.

Characteristics of nuclear fission are

1. Upon absorption of a neutron, a heavy nucleus splits into two or more smaller nuclei (fission products).
2. The mass of the nuclei formed range from about 70 to 160 amu.
3. Two or more neutrons are produced from the fission of each atom.
4. Large quantities of energy are produced as a result of the conversion of a small amount of mass into energy.
5. All nuclei formed are radioactive and give off beta and gamma radiation.

One suggested process by which this fission takes place is illustrated in Figure 18.6. When a heavy nucleus captures a neutron, the energy increase may be sufficient to cause deformation of the nucleus until the mass finally splits into two fragments, releasing radiant energy and usually two or more neutrons.

In a typical fission reaction, a $^{235}_{92}U$ nucleus captures a neutron and forms unstable $^{236}_{92}U$. This $^{236}_{92}U$ nucleus undergoes fission, quickly disintegrating into

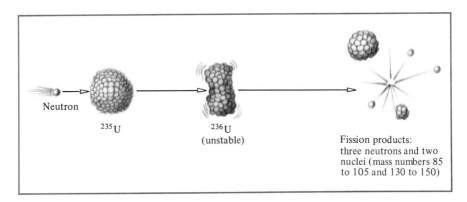

Neutron

^{235}U

^{236}U
(unstable)

Fission products:
three neutrons and two
nuclei (mass numbers 85
to 105 and 130 to 150)

Figure 18.6 The fission process: When a neutron is captured by a heavy nucleus, the nucleus becomes more unstable. The more energetic nucleus begins to deform, resulting in fission. Two nuclear fragments and three neutrons are produced by this fission process.

two fragments such as $^{139}_{56}\text{Ba}$ and $^{94}_{36}\text{Kr}$ and three neutrons. The three neutrons in turn may be captured by three other $^{235}_{92}\text{U}$ atoms, each of which undergoes fission, producing nine neutrons, and so on. A reaction of this kind, where the products cause the reaction to continue or magnify, is known as a **chain reaction.** For a nuclear chain reaction to continue, there must be enough fissionable material present so that each atomic fission causes, on the average, at least one additional fission reaction. The minimum quantity of fissionable material needed to support a self-sustaining chain reaction is called the **critical mass.** Since energy is released in each atomic fission, chain reactions constitute a possible source of a steady supply of energy. A chain reaction is illustrated in Figure 18.7. Two of the many possible ways U-235 may fission are shown by the following equations:

chain reaction

critical mass

$$^{235}_{92}\text{U} + ^{1}_{0}\text{n} \longrightarrow ^{139}_{56}\text{Ba} + ^{94}_{36}\text{Kr} + 3\,^{1}_{0}\text{n}$$

$$^{235}_{92}\text{U} + ^{1}_{0}\text{n} \longrightarrow ^{144}_{54}\text{Xe} + ^{90}_{38}\text{Sr} + 2\,^{1}_{0}\text{n}$$

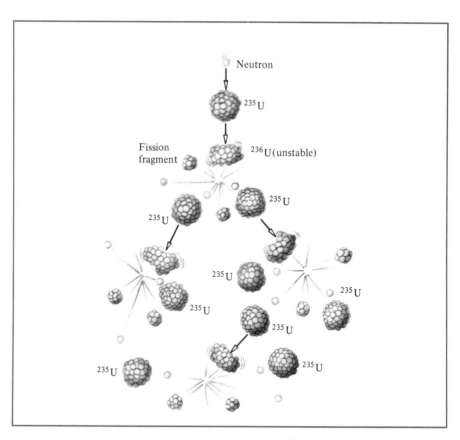

Figure 18.7 Fission and chain reaction of $^{235}_{92}\text{U}$. Each fission produces two major fission fragments and three neutrons, which may be captured by other $^{235}_{92}\text{U}$ nuclei, continuing the chain reaction.

18.9 Nuclear Power

Nearly all electricity is produced by machines consisting of a turbine linked by a drive shaft to an electrical generator. The energy required to run the turbine may be supplied from falling water, used in hydroelectric power plants, or by steam generated by heat from fuel, used in thermal power plants. Thermal power plants burn fossil fuel—coal, oil, or natural gas.

The world's demand for energy, largely from fossil fuels, has continued to grow at an ever-increasing rate for about 250 years. Even at present rates of consumption, the estimated world supply of fossil fuels is sufficient for only a few centuries. The United States has large coal and oil shale deposits, which are many times greater than our estimated petroleum deposits. However, the mining, transportation, and coal-burning plants required to handle the tremendous quantity of coal or oil shale needed to supplant the oil and natural gas we now use are not available. Consequently, to meet our energy needs, we are currently importing, at great expense, about 40% of our oil requirements. We clearly need to develop alternative energy sources. At present, uranium is by far the most productive alternative energy source; about 12% of the electrical energy used in the United States is generated in power plants using uranium fuel.

The major disadvantage of nuclear-fueled power plants is that they produce highly radioactive waste in the form of isotopes, some of which have half-lives of thousands of years. As yet, no general agreement has been reached on how to dispose of these dangerous wastes safely.

A nuclear power plant is a thermal power plant in which heat is produced by a nuclear reactor instead of by combustion of fossil fuel. The major components of a nuclear reactor are (1) an arrangement of nuclear fuel, called the reactor core; (2) a control system, which serves to regulate the rate of fission and thereby the rate of heat generation; and (3) a cooling system, which serves to remove the heat from the reactor and also to keep the core at the proper temperature. One type of reactor uses metal slugs containing uranium enriched from the normal 0.7% U-235 to about 3% U-235. These slugs are placed into a graphite block to form a nuclear pile. The self-sustaining fission reaction is moderated by the graphite and by adjustable control rods containing substances that slow down and capture some of the neutrons produced. Ordinary water, heavy water, and sodium are typical coolants used. Energy obtained from nuclear reactions in the form of heat is used in the production of steam to drive turbines for generating electricity.

Reactors used for commercial power production in the United States use uranium that is enriched with the relatively scarce fissionable U-235 isotope. Because the supply of U-235 is limited, a new type of reactor known as a *breeder reactor* has been developed. In a breeder reactor more fissionable material is produced at the same time that the fission reaction is occurring. In an ordinary reactor, excess neutrons are absorbed by the control rods. In a breeder reactor these excess neutrons are allowed to be captured by nonfissionable material, converting them to fissionable material. Thus, Th-232 can be converted by neutron bombardment to fissionable U-233; nonfissionable U-238 is converted to fissionable Pu-239. These transmutations have made it possible to greatly extend the supply of fuel for nuclear reactors. Unfortunately, the fissionable isotopes can also be used for nuclear weapons (see Section 18.10). No breeder reactors are commercially operating in the United States, but several of them are in operation in Europe and Great Britain. The equations for these nuclear transmutations are

$$^{238}_{92}\text{U} + ^{1}_{0}\text{n} \longrightarrow \, ^{239}_{92}\text{U} \xrightarrow{\,-\beta\,} \, ^{239}_{93}\text{Np} \xrightarrow{\,-\beta\,} \, ^{239}_{94}\text{Pu}$$

$$^{232}_{90}\text{Th} + ^{1}_{0}\text{n} \longrightarrow \, ^{233}_{90}\text{Th} \xrightarrow{\,-\beta\,} \, ^{233}_{91}\text{Pa} \xrightarrow{\,-\beta\,} \, ^{233}_{92}\text{U}$$

18.10 The Atomic Bomb

The atomic bomb is a fission bomb; it operates on the principle of a very fast chain reaction that releases a tremendous amount of energy. An atomic bomb and a nuclear reactor both depend on self-sustaining nuclear fission chain reactions. The essential difference is that in a bomb the fission is "wild," or uncontrolled, whereas in a nuclear reactor the fission is moderated and carefully controlled. A minimum critical mass of fissionable material is needed for a bomb, or a major explosion will not occur. When a quantity smaller than the critical mass is used, too many neutrons formed in the fission step escape without combining with another nucleus, and a chain reaction does not occur. Therefore, the fissionable material of an atomic bomb must be stored as two or more subcritical masses and brought together to form the critical mass at the desired time of explosion. The temperature developed in an atomic bomb is believed to be about 10 million °C.

On July 16, 1945, the first atomic bomb was successfully tested in the desert near Alamogordo, New Mexico. Three weeks later, on August 6, 1945, an atomic bomb was dropped on the city of Hiroshima, Japan, followed three days later by another bomb at Nagasaki, Japan. Figure 18.8 shows an atomic bomb of the "Little Boy" type, the nuclear weapon detonated over Hiroshima.

The isotopes used in atomic bombs are U-235 and Pu-239. Uranium deposits contain about 0.7% of the U-235 isotope, the remainder being U-238. Uranium-238 does not undergo fission, except with very high-energy neutrons. It was discovered, however, that U-238 captures a low-energy neutron without undergoing fission and that the product, U-239, changes to $^{239}_{94}\text{Pu}$ (plutonium) by a beta decay process. Plutonium-239 readily undergoes fission upon capture of a neutron and is therefore useful for nuclear weapons.

The hazards of an atomic bomb explosion include not only shock waves from the explosive pressure and tremendous heat, but also intense radiation in the form of alpha, beta, gamma, and ultraviolet rays. Gamma rays and X rays can penetrate deep into the body, causing burns, sterilization, and mutation of the genes, which may have an adverse effect on future generations. Both radioactive fission products and unfissioned material are present after the explosion. If the bomb explodes near the ground, many tons of dust are lifted into

Figure 18.8 Atomic bomb of the "Little Boy" type detonated over Hiroshima, Japan, in World War II. The bomb is 28 inches in diameter and 120 inches long. The first nuclear weapon ever detonated, it weighed about 9000 pounds and had a yield equivalent to approximately 20,000 tons of high explosives (TNT). (*Courtesy U.S. Department of Energy, Los Alamos Scientific Laboratory.*)

the air. Radioactive material adhering to this dust, known as *fallout,* is spread by air currents over wide areas of the land and constitutes a lingering source of radiation hazard.

The possibility of a nuclear war is certainly one of the most awesome threats facing civilization today. Only two rather primitive atomic bombs were used to destroy the Japanese cities of Hiroshima and Nagasaki and to bring World War II to an early end. Many thousands of nuclear weapons, some at least several hundred times more powerful than the Hiroshima and Nagasaki bombs, are now in existence. The threat of nuclear war is further complicated by the fact that the number of nations possessing nuclear weapons is steadily increasing.

18.11 Nuclear Fusion

nuclear fusion

The process of uniting the nuclei of two light elements to form one heavier nucleus is known as **nuclear fusion.** Such reactions are used for producing energy because the mass of the two individual nuclei that are fused into a single nucleus is greater than the mass of the nucleus formed by their fusion. The mass differential is liberated in the form of energy. Fusion reactions are responsible for the tremendous energy output of the sun. Thus, aside from relatively small amounts from nuclear fission and radioactivity, fusion reactions are the ultimate source of our energy, even that obtained from fossil fuels. They are also responsible for the devastating power of the thermonuclear, or hydrogen, bombs.

Fusion reactions require temperatures on the order of tens of millions of degrees for initiation. Such temperatures are present in the sun but have been produced only momentarily on earth. For example, the hydrogen, or fusion, bomb is triggered by the temperature of an exploding fission bomb.

Two typical fusion reactions are

$$\underset{\text{Tritium}}{_{1}^{3}\text{H}} + \underset{\text{Deuterium}}{_{1}^{2}\text{D}} \longrightarrow {_{2}^{4}\text{He}} + {_{0}^{1}\text{n}} + \text{Energy}$$

$$\underset{\substack{3.01495 \\ \text{amu}}}{_{1}^{3}\text{H}} + \underset{\substack{1.00782 \\ \text{amu}}}{_{1}^{1}\text{H}} \longrightarrow \underset{\substack{4.00260 \\ \text{amu}}}{_{2}^{4}\text{He}} + \text{Energy}$$

The total mass of the reactants in the second equation is 4.02277 amu, which is 0.02017 amu greater than the mass of the product. This difference in mass is manifested in the great amount of energy liberated.

During the past 35 to 40 years, a large amount of research on controlled nuclear fusion reactions has been done in the United States and in other countries, especially the Soviet Union. So far, the goal of controlled nuclear fusion has not been attained, although the required ignition temperature has been reached in several devices. Evidence to date leads us to believe that it is possible to develop a practical fusion power reactor. Fusion power, if it can be developed, will be far superior to fission power, for two reasons: (1) Virtually infinite amounts of energy are to be had from fusion. Uranium supplies for fission power are limited, but heavy hydrogen, or deuterium (the most likely fusion fuel), is relatively abundant. It is estimated that the deuterium in a cubic mile of seawater is an energy resource (as fusion fuel) greater than the petroleum reserves of the entire world. (2) From an environmental viewpoint, fusion power is much "cleaner" than fission power. This is because fusion reactions, in contrast to uranium and plutonium fission reactions, do not produce large amounts of long-lived and dangerously radioactive isotopes.

18.12 Mass–Energy Relationship in Nuclear Reactions

Since large amounts of energy are released in nuclear reactions, significant amounts of mass are converted to energy. It was stated earlier that the amount of mass converted to energy in chemical changes is considered insignificant. In fission reactions, about 0.1% of the mass is converted into energy. In fusion reactions, as much as 0.5% of the mass may be changed into energy. The Einstein equation, $E = mc^2$, may be used to calculate the energy liberated, or available, when the mass loss is known. For example, in the reaction

$$\underset{\substack{7.0160 \text{ g}}}{_{3}^{7}\text{Li}} + \underset{\substack{1.0078 \text{ g}}}{_{1}^{1}\text{H}} \longrightarrow \underset{\substack{4.0026 \text{ g}}}{_{2}^{4}\text{He}} + \underset{\substack{4.0026 \text{ g}}}{_{2}^{4}\text{He}} + \text{Energy}$$

the mass difference between the reactants and the products $(8.0238 - 8.0052)$ is 0.0186 g. The energy equivalent to this amount of mass is 4.0×10^{11} cal. By comparison, this is more than 4 million times greater than the 9.4×10^4 cal of energy obtained from the complete combustion of 12.0 g (1 mole) of carbon.

The mass of an atomic nucleus is actually less than the sum of the masses of the protons and neutrons that make up that nucleus. The difference between the mass of the protons and the neutrons in a nucleus and the mass of the nucleus is known as the *mass defect.* The energy equivalent to this difference in mass is known as the *binding energy* of the nucleus. This is the energy that would be required to break a nucleus up into its individual protons and neutrons. The higher the binding energy, the more stable the nucleus is. Elements of intermediate atomic masses have high binding energies. For example, iron (element number 26) has a very high binding energy and therefore has a very stable nucleus. Just as electrons attain less energetic and more stable arrangements through ordinary chemical reactions, neutrons and protons attain less energetic and more stable arrangements through nuclear fission or fusion reactions. Thus, when uranium undergoes fission, the products have less mass (and greater binding energy) than the original uranium. In like manner, when hydrogen and lithium fuse to form helium, the helium has less mass (and greater binding energy) than the hydrogen and lithium. It is this conversion of mass to energy that accounts for the very large amounts of energy associated with both nuclear fission and fusion reactions.

18.13 Transuranium Elements

transuranium
elements

The elements following uranium on the periodic chart and having atomic numbers greater than 92 are known as the **transuranium elements.** All of them are synthetic radioactive elements; none of them occur naturally.

The first transuranium element, number 93, was discovered in 1939 by Edwin M. McMillan (1907–   ) at the University of California while he was investigating the fission of uranium. He named it neptunium for the planet Neptune. In 1941, element 94, plutonium, was identified as a beta decay product of neptunium:

$$^{238}_{93}\text{Np} \longrightarrow {}^{238}_{94}\text{Pu} + {}^{0}_{-1}e$$

$$^{239}_{93}\text{Np} \longrightarrow {}^{239}_{94}\text{Pu} + {}^{0}_{-1}e$$

Plutonium is one of the most important fissionable elements known today.

Element 104 was first reported in 1964 by a Russian research group, but their work has not yet been independently confirmed. They have suggested the name kurchatovium (Ku) after the late Russian physicist Igor Kurchatov (1903–1960). In 1969, a research group at the Lawrence Radiation Laboratory of the University of California announced the synthesis and positive identification of two isotopes of element 104. They suggested the name rutherfordium (Rf) for this element. Nuclear reactions for the syntheses of element 104 follow.

$$^{242}_{94}\text{Pu} + ^{22}_{10}\text{Ne} \longrightarrow ^{260}_{104}\text{Ku} + 4\,^{1}_{0}\text{n}$$

$$^{249}_{98}\text{Cf} + ^{12}_{6}\text{C} \longrightarrow ^{257}_{104}\text{Rf} + 4\,^{1}_{0}\text{n}$$

$$^{249}_{98}\text{Cf} + ^{13}_{6}\text{C} \longrightarrow ^{259}_{104}\text{Rf} + 3\,^{1}_{0}\text{n}$$

In April 1970, the Lawrence Radiation group, headed by Albert Ghiorso, reported the discovery of element 105. They suggested that it be named Hahnium (Ha) after the German chemist Otto Hahn, who won the Nobel Prize for discovering nuclear fission. Element 105 was synthesized by bombarding Cf-249 with a stream of N-15 nuclei; it has a half-life of 1.6 seconds.

$$^{249}_{98}\text{Cf} + ^{15}_{7}\text{N} \longrightarrow ^{260}_{105}\text{Ha} + 4\,^{1}_{0}\text{n}$$

Both Russian and American nuclear scientists announced the synthesis of element 106 in 1974.

Table 18.5 lists all the presently known transuranium elements.

TABLE 18.5 Transuranium elements.

Element	Symbol	Atomic number	Discovery date
Neptunium	Np	93	1939
Plutonium	Pu	94	1941
Americium	Am	95	1944
Curium	Cm	96	1944
Berkelium	Bk	97	1949
Californium	Cf	98	1950
Einsteinium	Es	99	1953
Fermium	Fm	100	1953
Mendelevium	Md	101	1955
Nobelium	No	102	1957
Lawrencium	Lr	103	1961
Unnilquadium[a]	Unq[a]	104	1964
Unnilpentium[a]	Unp[a]	105	1970
Unnilhexium[a]	Unh[a]	106	1974

[a]These are names and symbols suggested by the International Union of Pure and Applied Chemistry.

18.14 Biological Effects of Radiation

As we have seen, radiation that has enough energy to dislocate bonding electrons and create ions when passing through matter is classified as *ionizing radiation*. Alpha, beta, and gamma rays, along with X rays, fall into this classification. Ionizing radiation can damage or kill living cells. This damage is particularly devastating when it occurs in the cell nucleus and affects molecules involved in cell reproduction. The overall effects of radiation on living organisms fall into these general categories: (1) acute, or short-term, effects; (2) long-term effects; and (3) genetic effects.

Acute radiation damage. High levels of radiation (100–200 rems), especially of gamma or X rays, produce nausea, vomiting, and diarrhea. The effect has been likened to a sunburn throughout the body. If the dosage is high enough (200–500 rems), death may occur in a few days. The damaging effects of radiation appear to be centered in the nuclei of the cells, and cells that are undergoing rapid cell division are most susceptible to damage. It is for this reason that cancers are often treated with gamma radiation from a cobalt-60 source. Cancerous cells are multiplying rapidly and are destroyed by a level of radiation that does not seriously damage normal cells.

Long-term radiation damage.   Protracted exposure to low levels of any form of ionizing radiation can weaken the organism and lead to the onset of malignant tumors, even after fairly long time delays. The largest exposure to man-made sources of radiation is from X rays. There is evidence that a number of early workers in radioactivity and X-ray technology may have had their lives shortened by long-term radiation damage.

A number of women who, in the early 1920s, had been employed to paint luminous numbers on watch dials died some years later from the effects of radiation. These women had ingested radium by using their lips to point the brushes used on the job. Radium was retained in their bodies and, as an alpha emitter with a half-life of about 1620 years, continued to inflict radiation damage.

Strontium-90 isotopes are present in the fallout from atmospheric testing of nuclear weapons. Strontium is in the same periodic table group as calcium, and its chemical behavior is similar to that of calcium. Hence, when foods contaminated with Sr-90 are eaten, Sr-90 ions are laid down in the bone tissue along with ordinary calcium ions. Strontium-90 is a beta emitter with a half-life of 28 years. Blood cells that are manufactured in bone marrow are affected by the radiation from Sr-90. Hence, there is concern that Sr-90 accumulation in the environment may cause an increase in the incidence of leukemia and bone cancers. Fortunately, the United States and the Soviet Union have agreed to stop atmospheric testing of nuclear weapons; however, some countries are still doing testing in the atmosphere.

Genetic effects.   All the information needed to create an individual of a particular species—be it a bacterial cell or a human being—is contained within the nucleus of a cell. This genetic information is encoded in the structure of DNA (deoxyribonucleic acid) molecules, which make up genes. The DNA molecules form precise duplicates when cells divide, thus passing genetic information from one generation to the next. Radiation can damage DNA molecules. If the damage is not severe enough to prevent reproduction, a mutation (a sudden heritable variation in the offspring) may result. Most mutation-induced traits are undesirable. Unfortunately, if the bearer of the altered genes survives to reproduce, these traits are passed along to succeeding generations. In other words, the genetic effects of increased radiation exposure are found in future generations, not in the present generation.

Because radioactive rays are hazardous to health and living tissue, special precautions must be taken in designing laboratories and nuclear reactors, in disposing of waste materials, and in monitoring the radiation exposure of people working in this field. For example, personnel working in areas of hazardous radiation wear film badges or pocket dosimeters to provide them with an accurate indication of cumulative radiation exposure.

18.15 Applications of Radioisotopes

To date, the largest uses of radioactive materials have been for making weapons and for the generation of electricity in nuclear power plants. Aside from these major uses, radioisotopes have innumerable applications. They are used extensively in chemical, physical, biological, and medical research. Radioisotopes now serve in a wide variety of almost routine technological applications in medicine and various branches of industry, including chemical, petroleum, and metallurgical processing. A few of these applications are briefly described here.

Cancer treatment. For many years, radium has been used in the treatment of cancer; cobalt-60 and cesium-137 are now extensively used for therapy in this field. The effectiveness of radiation therapy for cancer is dependent on the fact that rapidly growing or dividing malignant cells are more susceptible to radiation damage than are normal cells. Therefore, radiation therapy is effective when it can be applied at levels that will destroy cancer cells but not normal cells.

Isotopic tracers. Compounds containing a radioactive isotope are described as being *labeled* or *tagged.* These compounds undergo their normal chemical reactions. In addition, their location can be detected because of their radioactivity. When such compounds are given to a plant or to an animal, the movement of the radioisotope can be traced through the organism by the use of a Geiger counter or other detecting device. The following are examples of biological research in which such tracer techniques have been employed: the rate of phosphate intake by plants, using radiophosphorus; the utilization of carbon dioxide in photosynthesis, using radioactive carbon; the accumulation of iodine in the thyroid gland, using radioactive iodine; and the absorption of iron by the hemoglobin of the blood, using radioactive iron. In chemistry, the applications are unlimited: the study of reaction mechanisms, the measurement of the rates of chemical reactions, and the determination of physical constants are just a few of the areas of application.

Doctors can examine the heart's pumping performance and check for evidence of obstruction in the coronary arteries by *nuclear scanning.* The radioisotope thallium-201, when injected into the bloodstream, lodges in healthy heart muscle. T1-201 emits gamma radiation, which is detected by a special imaging device called a *scintillation camera.* The data obtained are simultaneously translated into pictures by a computer. With this technique doctors can observe if heart tissue has died after a heart attack and whether blood is flowing freely through the coronary passages.

Agriculture. Agricultural research scientists use gamma radiation from cobalt-60 or other sources to develop disease-resistant and highly productive grains and other crops. The seeds are exposed to gamma radiation to induce mutations. The most healthy and vigorous plants grown from the irradiated seed are then selected and propagated to obtain new and improved varieties for commercial use. Preservation of foodstuffs by radiation is another beneficial application of radioactivity (see Figure 18.9).

Insect control.   Radioactivity has been used to control and, in some areas, to eliminate the screw-worm fly. The larvae of this obnoxious insect pest burrow

(a)

(b)

Figure 18.9   Small doses of radiation have successfully inhibited the sprouting of onions and potatoes. Sprouting in onions can be inhibited by a dose of 2000–4000 rads of radiation; potatoes require about 7500 rads. There is no detectable change in the product at these dosage levels. The onions (a) and potatoes (b) illustrated have been stored over 5 months. In each case, sprouting was greatest in the control, which was not irradiated. The control basket of potatoes is the upper left one; those in each basket (left to right) have been treated with a larger dose of radiation. (*Courtesy Agriculture Canada.*)

into wounds in livestock. The female fly, like a queen bee, mates only once. When large numbers of gamma ray–sterilized male flies are released at the proper time in an area infested with screw-worm flies, the majority of the females mate with sterile males. As a consequence, the flies fail to reproduce sufficiently to maintain their numbers. This technique has also been used to eradicate the Mediterranean fruit fly in some areas.

Age dating.   An interesting outgrowth of the use of radioisotope techniques is *radiocarbon dating.* The method is based on the decay rate of carbon-14 and was devised by the American chemist W. F. Libby, who received the Nobel Prize in chemistry in 1960 for this work. The principle of radiocarbon dating is as follows: Carbon dioxide in the atmosphere contains a fixed ratio of radioactive carbon-14 to ordinary carbon-12. This is because carbon-14 is produced at a steady rate in the atmosphere by bombardment of nitrogen-14 by neutrons from cosmic ray sources.

$$^{14}_{7}N + ^{1}_{0}n \longrightarrow ^{14}_{6}C + ^{1}_{1}H$$

Plants that consume carbon dioxide during photosynthesis and animals that eat the plants contain the same proportion of C-14 to C-12 as long as they are alive. When an organism dies, the amount of C-12 remains fixed, but the C-14 content diminishes according to its half-life (5668 years). By comparing the ratio of C-14 to C-12 in an object to the same ratio in living plants, one can estimate the age of the object being evaluated. In 5668 years, one-half the radiocarbon initially present will have undergone decomposition. In 11,336 years, one-fourth of the original C-14 will be left. The age of fossil material, archaeological specimens, and old wood can be determined by this method. The age of specimens from ancient Egyptian tombs calculated by radiocarbon dating correlates closely with the chronological age established by Egyptologists. Charcoal samples obtained at Darrington Walls, a wood-henge in Great Britain, were determined to be about 4000 years old. Radiocarbon dating instruments currently in use enable researchers to date specimens back as far as 50,000 years.

Radioactive decay has been used to date samples other than those containing carbon. For example, the age of rock formations containing uranium has been approximated by determining the ratio of U-238 to Pb-206. Lead-206 is the last isotope formed in the U-238 disintegration series. Thus, a geologic deposit containing a 1:1 ratio of U-238 to Pb-206 would correspond to a time lapse of one half-life of U-238, which is 4.5×10^9 years.

QUESTIONS   A. **Review the Meanings of the New Terms Introduced in this Chapter**

1. Nucleons	9. Artificial radioactivity
2. Radioactivity	10. Induced radioactivity
3. Radioactive decay	11. 1 curie
4. Half-life	12. Nuclear fission
5. Alpha particle	13. Chain reaction
6. Beta particle	14. Critical mass
7. Gamma ray	15. Nuclear fusion
8. Transmutation	16. Transuranium elements

B. **Answers to the Following Questions Will Be Found in Tables and Figures**
1. To afford protection from radiation injury, which kind of radiation requires:
   (a) The most shielding?   (b) The least shielding?
2. Why is an alpha particle deflected less than a beta particle in passing through an electromagnetic field?
3. In the uranium disintegration series, Figure 18.4, the series stops with $^{206}_{82}Pb$, which is stable. Before reaching this isotope of lead, the series passed through $^{214}_{82}Pb$ and $^{210}_{82}Pb$. Suggest a possible reason that the series did not stop at one of these isotopes of lead.
4. List the first five isotopes in Table 18.3 in order of increasing half-life.
5. What is the half-life of strontium-90?
6. Name three pairs of isotopes that might be obtained by fissioning three U-235 atoms.
7. List the transuranium elements that are named for:
   (a) People   (b) Geographical place-names   (c) Planets

C. **Review Questions**
1. Identify the following people and their associations with the early history of radioactivity:
   (a) Antoine Henri Becquerel   (c) Wilhelm Roentgen   (e) Paul Villard
   (b) Marie and Pierre Curie   (d) Ernest Rutherford
2. Why is the radioactivity of an element unaffected by the usual factors that affect the rate of chemical reactions, such as ordinary changes of temperature and concentration?
3. Which of the following elements do not have stable isotopes?
   (a) Californium   (d) Palladium   (g) Thorium
   (b) Francium   (e) Platinum   (h) Vanadium
   (c) Lead   (f) Technetium   (i) Xenon
4. Explain the term *half-life*.
5. Tell how alpha, beta, and gamma radiation are distinguished from the standpoint of:
   (a) Charge   (c) Nature of particle or ray
   (b) Relative mass   (d) Relative penetrating power
6. How are the mass and the atomic number of a nucleus affected by the loss of the following:
   (a) An alpha particle   (b) A beta particle
7. Distinguish between natural and artificial radioactivity.
8. Write nuclear equations for the alpha decay of:
   (a) $^{218}_{85}At$ (b) $^{221}_{87}Fr$ (c) $^{192}_{78}Pt$ (d) $^{210}_{84}Po$
9. Write nuclear equations for the beta decay of:
   (a) $^{14}_{6}C$ (b) $^{137}_{55}Cs$ (c) $^{239}_{93}Np$ (d) $^{90}_{38}Sr$
10. Stable $^{208}_{82}Pb$ is formed from $^{232}_{90}Th$ in the thorium disintegration series by successive $\alpha, \beta, \beta, \alpha, \alpha, \alpha, \alpha, \beta, \beta, \alpha$ particle emissions. Write the symbol (including mass and atomic number) for each isotope formed in this series.
11. The isotope $^{237}_{93}Np$ loses a total of seven alpha particles and four beta particles. What isotope remains after these losses?
12. Bismuth-211 decays by alpha emission to give an isotope that in turn decays by beta emission to yield a stable isotope. Show these two steps with nuclear equations.
13. Write nuclear equations for the following:
   (a) Conversion of $^{13}_{6}C$ to $^{14}_{6}C$ (b) Conversion of $^{30}_{15}P$ to $^{30}_{14}Si$
14. Complete and balance the following nuclear equations:
   (a) $^{27}_{13}Al + ^{4}_{2}He \longrightarrow ^{30}_{15}P +$
   (b) $^{27}_{14}Si \longrightarrow ^{0}_{+1}e +$

(c) $\quad\quad + {}^2_1H \longrightarrow {}^{13}_7N + {}^1_0n$

(d) $\quad\quad \longrightarrow {}^{82}_{36}Kr + {}^0_{-1}e$

(e) ${}^{66}_{29}Cu \longrightarrow {}^{66}_{30}Zn +$

(f) ${}^0_{-1}e + \quad\quad \longrightarrow {}^7_3Li$

(g) ${}^{27}_{13}Al + {}^4_2He \longrightarrow {}^{30}_{14}Si +$

(h) ${}^{85}_{37}Rb + \quad\quad \longrightarrow {}^{82}_{35}Br + {}^4_2He$

(i) ${}^{214}_{83}Bi \longrightarrow {}^4_2He +$

15. In nuclear reactions brought about by bombarding nuclei, protons and alpha particles are often used, but they must be accelerated to high energies by devices like a cyclotron. However, low-energy neutrons are often used for nuclear bombardments, and are often more effective than higher-energy neutrons. How can you account for this difference?

16. What was the contribution to nuclear physics of Otto Hahn and Fritz Strassmann?

17. What is a breeder reactor? Explain how it accomplishes the "breeding."

18. What is the essential difference between the nuclear reactions in a nuclear reactor and those in an atomic bomb?

19. Why must a certain minimum amount of fissionable material be present before a self-supporting chain reaction can occur?

20. Are the terms *atomic bomb* and *hydrogen bomb* synonymous? If not, what is the major distinction between them?

21. Explain why radioactive rays are classified as ionizing radiation.

22. Give a brief description of the biological hazards associated with radioactivity.

23. Strontium-90 has been found to occur in radioactive fallout. Why is there so much concern about this radioisotope being found in cow's milk? (See Table 18.3.)

24. What is a radioactive tracer? How is it used?

25. Describe the radiocarbon method for dating archaeological artifacts.

26. Which of the following statements are correct?

(a) Radioactivity was discovered by Marie Curie.

(b) There are 59 neutrons in an atom of ${}^{59}_{28}Ni$.

(c) The loss of a beta particle by an atom of ${}^{75}_{33}As$ forms an atom of increased atomic number.

(d) The emission of an alpha particle from the nucleus of an atom lowers its atomic number by 4 and lowers its atomic mass by 2.

(e) Emission of gamma radiation from the nucleus of an atom leaves both the atomic number and atomic mass unchanged.

(f) There are relatively few naturally occurring radioactive isotopes with atomic numbers below 81.

(g) The longer the half-life of a radioisotope, the more slowly it decays.

(h) The beta ray has the greatest penetrating power of all the rays emitted from the nucleus of an atom.

(i) The symbol ${}^0_{+1}e$ is used to indicate a positron, which is a positively charged particle with the mass of an electron.

(j) The disintegration of ${}^{226}_{88}Ra$ into ${}^{214}_{83}Po$ involves the loss of three alpha particles and two beta particles.

(k) If 1.0 g of a radioisotope has a half-life of 7.2 days, the half-life of 0.50 g of that isotope is 3.6 days.

(l) If the mass of a radioisotope is reduced by radioactive decay from 12 g to 0.75 g in 22 hours, the half-life of the isotope is 5.5 hours.

(m) A very high temperature is required to initiate nuclear fusion reactions.

(n) Radiocarbon dating of archaeological artifacts is based on an increase in the ratio of carbon-14 to carbon-12 in the object.

(o) Cancers are often treated by radiation from cobalt-60, which destroys the rapidly growing cancer cells.

(p) High levels of radiation produce nausea, vomiting, and diarrhea.
(q) Carbon-14, used in radiocarbon dating, is produced in living matter by the beta decay of nitrogen-14.
(r) Radioactive tracers are small amounts of radioactive isotopes of selected elements whose progress in chemical or biological processes can be followed using a radiation detection instrument.

D. Review Problems

1. Strontium-90 has a half-life of 28 years. If a 1.00 mg sample of this isotope was stored for 84 years, what weight of this isotope would remain?
2. If radium costs $50,000 a gram, how much will 0.0200 g of $^{226}RaCl_2$ cost if the price is based only on the radium content?
3. Cesium-137 has a half-life of 30 years. If a sample was tested in 1980 and found to be emitting 220 counts/minute, in what year would the same sample be found to be emitting 27 counts/minute?
4. An archaeological specimen was analyzed and found to be emitting only 25% as much carbon-14 radiation per gram of carbon as newly cut wood. How old is this specimen?
5. Barium-141 is a beta emitter. What is the half-life if a 12.0 g sample of the isotope decays to 0.750 g in 72 minutes?
*6. Calculate (a) the mass defect and (b) the binding energy of $^{7}_{3}Li$. Mass data: $^{7}_{3}Li$ = 7.0160 g; n = 1.0087 g; p = 1.0073 g; e = 0.00055 g; 1.0 g = 2.2 × 10^{13} cal (from $E = mc^2$).
*7. For the fission reaction

$$^{235}_{92}U + ^{1}_{0}n \longrightarrow ^{94}_{38}Sr + ^{139}_{54}Xe + 3\,^{1}_{0}n + energy$$

Mass data: U-235 = 235.0439; Sr-94 = 93.9154; Xe-139 = 138.9179; n = 1.0087; 1.0 g = 2.2 × 10^{13} cal. Calculate:
(a) The energy released in calories for a single event (one uranium atom splitting)
(b) The energy released in calories per mole of uranium splitting
(c) The percentage of the mass that is lost in the reaction
*8. For the fusion reaction

$$^{1}_{1}H + ^{2}_{1}H \longrightarrow ^{3}_{2}He + energy$$

Mass data: $^{1}_{1}H$ = 1.00782; $^{2}_{1}H$ = 2.01410; $^{3}_{2}He$ = 3.01603; 1.0 g = 2.2 × 10^{13} cal. Calculate:
(a) The energy released in calories per mole of $^{3}_{2}He$ formed
(b) The percentage of the mass that is lost in the reaction

E. Review Exercises

1. The half-life of plutonium-244 is 76 million years. If the age of the earth is about 5 billion years, discuss the feasibility of this isotope's being found as a naturally occurring element.
2. Suggest why nuclear reactions using alpha particles are easier to carry out with lighter elements than with heavier ones.
3. Agricultural researchers have produced improved crop species by cross-breeding. Explain how the same purpose might be accomplished by exposing seeds to radioactive rays.
4. How might radioactivity be used to locate a leak in an underground pipe?
5. Anthropologists have found bones whose age suggests that the human line may have emerged in Africa as much as 4 million years ago. If wood or charcoal was found with such bones, would carbon-14 dating be useful in dating the bones? Explain.

19

Organic Chemistry

After studying Chapter 19 you should be able to:

1. Understand the terms listed in Question A at the end of the chapter.
2. Understand the tetrahedral nature of the carbon atom.
3. Identify the different types of bonding between carbon atoms.
4. Write the names and formulas for the first ten straight-chain alkanes.
5. Write the names and structural formulas for the common alkyl groups (C_nH_{2n+1}).
6. Write structural formulas and IUPAC names for the isomers of an alkane or a halogenated alkane.
7. Give the IUPAC name of a hydrocarbon or an alkyl halide when given the structural formula.
8. Write equations for the combustion, halogenation, and dehydrogenation of an alkane.
9. Write the names and the formulas for all the monohalogen substitution isomers of a given alkane.
10. Distinguish by structure an alkane, an alkene, and an alkyne.
11. Name and write structural formulas of alkenes and alkynes.
12. Write equations for addition reactions of alkenes and alkynes.
13. Explain Baeyer's test for unsaturation.
14. Understand the general makeup of polymers (macromolecules).
15. Write structural formulas for polymers derived from modified ethylene monomers when given the monomer, or vice versa.
16. Describe the nature of benzene and how its properties differ from open-chain unsaturated hydrocarbons.
17. Write formulas and names for the common monosubstituted derivatives of benzene.
18. Name and draw structural formulas for disubstituted and polysubstituted benzene compounds.
19. Recognize and write formulas for the common fused aromatic ring compounds.
20. Write equations for the following reactions of benzene: halogenation (chlorination or bromination), nitration, and alkylation (Friedel–Craft reaction).
21. Give the IUPAC and common names for alcohols, or write the structural formula when given the name.
22. Recognize and identify primary, secondary, and tertiary alcohols.
23. Write structural formulas for all the isomeric alcohols when given a molecular formula.
24. Write equations for the following reactions of alcohols: oxidation to form aldehydes or ketones, dehydration to an alkene or an ether, and the reaction with an active metal such as sodium.
25. Be familiar with the methods of preparing methanol and ethanol and the general properties and uses of these two alcohols.

26. Understand the differences in the properties of the hydroxyl group (—OH) when bonded to an aromatic ring (a phenol) and to an alkyl group (an alcohol).
27. Write structural formulas and names for common phenolic compounds.
28. Name and write formulas of ethers.
29. Write equations for the preparation of ethers by the sulfuric acid and the Williamson methods.
30. Give IUPAC and common names for aldehydes and ketones.
31. Write structural formulas for aldehydes and ketones when given their names.
32. Show that aldehydes and ketones are isomeric with each other.
33. Select the proper alcohol to prepare an aldehyde or a ketone by oxidation.
34. Discuss the Tollens', Fehling's, and Benedict's tests, including reagents used, evidence of a positive test, and the equations for the reactions that occur in positive tests.
35. Use Tollens' and Fehling's tests to distinguish between aldehydes and ketones.
36. Write equations for the reduction of aldehydes and ketones to form alcohols.
37. Give the common and IUPAC names for carboxylic acids, or write the structural formula when given the name.
38. Write equations for (a) the oxidation of an alcohol to a carboxylic acid, (b) the reaction of carboxylic acids with sodium metal; (c) the reaction of carboxylic acids with a base; and (d) the reduction of carboxylic acids to alcohols.
39. Write equations for the preparation of esters from carboxylic acids and alcohols.
40. Write common names, IUPAC names, and formulas for esters.
41. Identify the alcohol and the carboxylic acid that would be needed to prepare a given ester.

19.1 Organic Chemistry: History and Scope

Chemists during the late 18th and 19th centuries were baffled by the fact that compounds obtained from animal and vegetable sources defied the established rules for inorganic compounds (namely, that compound formation was due to the attraction between positive and negative charged elements). They observed that a group of four elements—carbon, hydrogen, oxygen, and nitrogen—gave rise to a large number of different compounds that often were remarkably stable. This was in contrast to only one, or at most a few, known compounds composed of any other group of only two or three elements.

Since no organic compounds had been synthesized from inorganic substances and since there was no other explanation for the complex nature of organic compounds, chemists were led to believe that organic compounds were formed by some "vital force." The **vital force theory** held that organic substances could originate only from some form of living material. In 1828, Friedrich Wöhler (1800–1882), a German chemist, did a simple experiment that eventually proved to be the death blow to this theory. In attempting to prepare ammonium cyanate (NH_4CNO) by heating cyanic acid (HCNO) and ammonia, Wöhler obtained a crystalline white substance, which he identified

vital force theory

as urea (H_2N—CO—NH_2). Urea is an authentic organic substance because it is a product of metabolism, and Wöhler knew that it had been isolated from urine. Although Wöhler's discovery was not immediately and generally recognized, the vital force theory was overthrown by this simple experiment since *one* organic compound had been made from nonliving materials.

After the work of Wöhler, it was apparent that no vital force, other than skill and knowledge, was needed to make organic chemicals in the laboratory and that inorganic as well as other organic substances could be used as raw materials. Today, **organic chemistry** simply designates the branch of chemistry that deals with carbon compounds but does not imply that these compounds must originate from some form of life. A few special kinds of carbon compounds—for example, carbon oxides, metal carbides, and metal carbonates—are excluded from the organic classification because their chemistry is more conveniently related to that of inorganic substances.

organic chemistry

The field of organic chemistry is vast, for it includes the composition not only of all living organisms but also of a great many other materials that we use daily. Examples of organic materials are foodstuffs (fats, proteins, carbohydrates); fuels of all kinds; fabrics (cotton, wool, rayon, nylon); wood and paper products; paints and varnishes; plastics; dyes; soaps and detergents; cosmetics; medicinals; rubber products; and explosives.

The sources of organic compounds are carbon-containing raw materials—petroleum and natural gas, coal, carbohydrates, fats, and oils. In the United States we produce about 250 billion pounds of organic chemicals per year from these sources. This amounts to more than 1100 pounds per year for every man, woman, and child in the United States. About 90% of this 250 billion pounds comes from petroleum and natural gas. Since there is a finite amount of petroleum and natural gas in the world, we will, sometime in the future, have to rely on the other sources to make the voluminous amount of organic chemicals that we use. Fortunately, we know how to do this, but at much greater expense than from petroleum and natural gas.

An immense and ever-growing number of organic chemical compounds have been prepared and are described in the chemical literature. The number is estimated to be over 4.5 million. The elements that make up most of the organic compounds are carbon, hydrogen, oxygen, nitrogen, the halogens (Cl, Br, I), and sulfur. There is no theoretical limit on the number of organic compounds that can exist. Several thousand different organic chemicals are available commercially.

19.2 The Need for Classification of Organic Compounds

It is physically impossible for anyone to study the properties of each of the hundreds of thousands of known organic compounds. Hence, organic compounds with similar structural features are grouped into series or classes. Even within a single class there is such a large number of possible compounds that it is virtually impossible to study each individual compound. The properties of a few members of a series are studied, and the information obtained is used to make predictions about the behavior of other members of the series. For example, ethyl alcohol, a two-carbon alcohol, can be oxidized to a two-carbon acid. On the basis of this information, it can be predicted and found to be true that hexyl alcohol, a six-carbon alcohol, can be oxidized to a six-carbon acid, because the oxidizable structures of the two alcohols are identical. Some of the classes of organic compounds we will study are hydrocarbons, alcohols, phenols, aldehydes, ketones, ethers, carboxylic acids, esters, carbohydrates, and proteins. Each of these classes of compounds is identified by certain characteristic structural features.

19.3 The Carbon Atom: Tetrahedral Structure

The carbon atom is central to all organic compounds. The atomic number of carbon is 6 and its electron structure is $1s^2 2s^2 2p^2$. Two stable isotopes of carbon exist, C-12 and C-13. In addition, there are several radioactive isotopes; C-14 is the most widely known of these because of its use in radiocarbon dating. Having four electrons in its outer shell, carbon has oxidation numbers ranging from $+4$ to -4 and forms predominantly covalent bonds. Carbon occurs as the free element in diamond, graphite, coal, coke, carbon black, charcoal, and lampblack.

A carbon atom generally forms four covalent bonds. The most common geometric arrangement of these bonds is tetrahedral (see Figure 19.1). In this tetrahedral structure, the four covalent bonds are not planar about the carbon atom but are directed from the center toward the corners of a regular tetrahedron. (A tetrahedron is a solid figure with four sides.) Figure 19.1 illustrates (a) a regular tetrahedron, (b) a carbon atom with its bonds in tetrahedral arrangement, (c) a carbon atom placed inside a regular tetrahedron, and (d) a model of a methane molecule with the carbon–hydrogen bonds tetrahedral. The angle between these bonds is 109.5°.

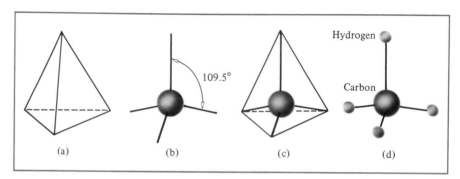

Figure 19.1   Tetrahedral structure of carbon: (a) a regular tetrahedron; (b) a carbon atom with tetrahedral bonds; (c) a carbon atom within a regular tetrahedron; (d) a methane molecule, CH_4.

19.4 Carbon–Carbon Bonds

With four outer-shell electrons, the carbon atom ($\cdot\overset{\cdot}{\underset{\cdot}{C}}\cdot$), following the octet rule, forms four single covalent bonds by sharing electrons with other atoms. The structures of methane and carbon tetrachloride illustrate this point:

$$
\begin{array}{c}
H \\
H\!:\!\overset{\cdot\cdot}{\underset{\cdot\cdot}{C}}\!:\!H \\
H
\end{array}
\qquad
\begin{array}{c}
H \\
| \\
H-C-H \\
| \\
H
\end{array}
\qquad
\begin{array}{c}
Cl \\
Cl\!:\!\overset{\cdot\cdot}{\underset{\cdot\cdot}{C}}\!:\!Cl \\
Cl
\end{array}
\qquad
\begin{array}{c}
Cl \\
| \\
Cl-C-Cl \\
| \\
Cl
\end{array}
$$

Methane   Carbon tetrachloride

Actually, the bond angles (109.5°) in these compounds are tetrahedral (see Figure 19.1), but for convenience in writing, the bonds are drawn at right angles. In methane, each bond is formed by the sharing of electrons between a carbon and a hydrogen atom. The ability of carbon to bond to itself is based on the sharing of electrons between carbon atoms. One, two, or three pairs of electrons can be shared between two carbon atoms, forming a single, double, or triple bond, respectively:

$$
\cdot\overset{\cdot}{C}\!:\!\overset{\cdot}{C}\cdot
\qquad
\cdot\overset{\cdot}{C}\!::\!\overset{\cdot}{C}\cdot
\qquad
\cdot C\!:::\!\overset{\cdot}{C}\cdot
$$

$$
\cdot\overset{\cdot}{\underset{\cdot}{C}}\!-\!\overset{\cdot}{\underset{\cdot}{C}}\cdot
\qquad
\cdot\overset{\cdot}{C}\!=\!\overset{\cdot}{C}\cdot
\qquad
\cdot C\!\equiv\!C\cdot
$$

Single bond   Double bond   Triple bond

Each dash represents a covalent bond. Carbon, more than any other element, has the ability to form short or very long chains of atoms covalently bonded together. This bonding ability is the main reason for the large number of organic compounds. Three examples follow. It is easy to see how, because of this bonding ability, a chain of carbon atoms can be formed, linking one carbon to another through covalent bonds.

$$
\cdot\overset{\cdot}{C}\!:\!\overset{\cdot}{C}\!:\!\overset{\cdot}{C}\cdot
\qquad
\cdot\overset{\cdot}{\underset{\cdot}{C}}\!-\!\overset{\cdot}{\underset{\cdot}{C}}\!-\!\overset{\cdot}{\underset{\cdot}{C}}\cdot
\qquad
\cdot\overset{\overset{\overset{\cdot}{C}}{\diagup\diagdown}}{C}\underset{109.5°}{}\overset{\cdot}{C}\cdot
$$

Three carbon atoms bonded by single bonds

Seven-carbon chain

Ten carbon atoms bonded together

19.5 Hydrocarbons

hydrocarbons

Hydrocarbons are compounds that are composed entirely of carbon and hydrogen atoms bonded to each other by covalent bonds. Several classes of hydrocarbons are known. These include the alkanes, alkenes, alkynes, and aromatic hydrocarbons (see Figure 19.2).

Fossil fuels—natural gas, petroleum, and coal—are the principal sources of hydrocarbons. Natural gas is primarily methane with small amounts of ethane, propane, and butane. Petroleum is a complex mixture of hydrocarbons. Gasoline, kerosene, fuel oil, lubricating oil, paraffin wax, and petrolatum—all of which are simply mixtures of hydrocarbons—are separated from petroleum. Coal tar, a volatile product driven off in the process of making coke from coal for use in the steel industry, is the source of many valuable chemicals including the aromatic hydrocarbons benzene, toluene, and naphthalene.

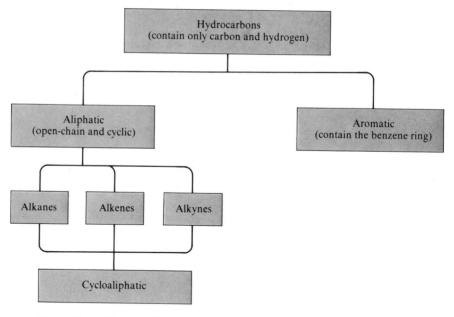

Figure 19.2   Classes of hydrocarbons.

19.6 Saturated Hydrocarbons: Alkanes

alkanes

The **alkanes,** also known as *paraffins* or *saturated hydrocarbons,* are straight- or branched-chain hydrocarbons with only single covalent bonds between the carbon atoms. We shall study the alkanes in some detail because many other classes of organic compounds may be considered derivatives of these substances. For example, it is necessary to learn the names of the first ten members of the alkane series because these names are used (with slight modifications) for corresponding compounds belonging to other classes.

Methane (CH_4) is the first member of the alkane series. Succeeding members having two, three, and four carbon atoms are ethane, propane, and butane, respectively. The names of the first four alkanes are of common or trivial origin and must simply be memorized, but the names beginning with the fifth member, pentane, are derived from Greek numbers and are relatively easy to recall. The names and formulas of the first ten members of the series are given in Table 19.1.

TABLE 19.1 Names, formulas, and physical properties of straight-chain alkanes.

Name	Formula, C_nH_{2n+2}	Boiling point (°C)	Melting point (°C)
Methane	CH_4	−161	−183
Ethane	C_2H_6	−88	−172
Propane	C_3H_8	−45	−187
Butane	C_4H_{10}	0	−138
Pentane	C_5H_{12}	36	−130
Hexane	C_6H_{14}	69	−95
Heptane	C_7H_{16}	98	−90
Octane	C_8H_{18}	125	−57
Nonane	C_9H_{20}	151	−54
Decane	$C_{10}H_{22}$	174	−30

Successive compounds in the alkane series differ in composition by one carbon and two hydrogen atoms. This difference between any two successive alkanes can be observed in Table 19.1. When each member of a series differs from the next higher member by a CH_2 group, as do the alkanes, the series is called a **homologous series**. The members of a homologous series are similar in structure but have a regular difference in formula. All common classes of organic compounds exist in homologous series. A homologous series can be represented by a general formula. For all open-chain alkanes the general formula is C_nH_{2n+2}, where n corresponds to the number of carbon atoms in the molecule. The formula of any specific alkane is easily determined from this general formula. Thus, for pentane, $n = 5$ and $2n + 2 = 12$, so the formula is C_5H_{12}. For hexadecane, a 16-carbon alkane, the formula is $C_{16}H_{34}$.

homologous series

19.7 Structural Formulas and Isomerism

The properties of an organic substance are dependent on its molecular structure. By structure we mean the way in which the atoms are bonded together within the molecule. The majority of organic compounds are made from relatively few elements—namely, carbon, hydrogen, oxygen, nitrogen, and the halogens. In these compounds, carbon is tetravalent, hydrogen is monovalent, oxygen is divalent, nitrogen is trivalent, and the halogens are monovalent. The valence bonds or points of attachment may be represented in structural formulas by a corresponding number of dashes attached to each atom:

$$-\overset{|}{\underset{|}{C}}- \quad H- \quad -O- \quad -\overset{}{\underset{|}{N}}- \quad Cl- \quad Br- \quad I- \quad F-$$

Thus, carbon will have four bonds to each atom, nitrogen three bonds, oxygen two bonds, and hydrogen and the halogens one bond to each atom.

In an alkane, each carbon atom is joined to four other atoms by single covalent bonds. These bonds are separated by angles of 109.5° (the angles correspond to those formed by lines drawn from the center of a regular tetrahedron to its corners). Alkane molecules contain only carbon–carbon and carbon–hydrogen bonds and are essentially nonpolar. Alkane molecules are nonpolar because (1) carbon–carbon bonds are nonpolar since they are between like atoms; (2) carbon–hydrogen bonds are only slightly polar since there is only a small difference in electronegativity between carbon and hydrogen atoms; and (3) the bonds in an alkane are symmetrically directed toward the corners of a tetrahedron. Because of this nonpolarity, alkane molecules have very little intermolecular attraction and therefore relatively low boiling points when compared with other organic compounds of similar molecular weight.

Without the use of models or perspective drawings, the three-dimensional character of atoms and molecules is difficult to portray accurately. However, concepts of structure can be conveyed to some extent by electron-dot (or Lewis-dot) diagrams of structural formulas. Methane and ethane are represented by electron-dot diagrams as

$$\begin{array}{cc} \text{H} & \text{H H} \\ \text{H:}\overset{..}{\underset{..}{\text{C}}}\text{:H} & \text{H:}\overset{..}{\underset{..}{\text{C}}}\text{:}\overset{..}{\underset{..}{\text{C}}}\text{:H} \\ \text{H} & \text{H H} \\ \text{Methane} & \text{Ethane} \end{array}$$

But it is more convenient to use conventional structural formulas representing electron pairs by single short lines:

$$\begin{array}{cc} \text{H} & \text{H H} \\ | & | \ | \\ \text{H}-\text{C}-\text{H} & \text{H}-\text{C}-\text{C}-\text{H} \\ | & | \ | \\ \text{H} & \text{H H} \end{array}$$

To write the correct structural formula for propane (C_3H_8), the next member of the alkane series, we must determine how to place each atom in the molecule. An alkane contains only single bonds, and carbon is tetravalent. Therefore, each carbon atom must be bonded to four other atoms by either C—C or C—H bonds. Hydrogen is univalent and therefore must be bonded to only one carbon atom by a C—H bond, since C—H—C bonds do not occur, and an H—H bond would simply represent a hydrogen molecule. Applying this information, we find that the only possible structure for propane is

$$
\begin{array}{ccccccc}
 & H & & H & & H & \\
 & | & & | & & | & \\
H- & C & - & C & - & C & -H \\
 & | & & | & & | & \\
 & H & & H & & H & \\
\end{array}
$$

Propane

However, it is possible to write two structural formulas corresponding to the molecular formula C_4H_{10}:

$$
\begin{array}{ccccccccc}
 & H & & H & & H & & H & \\
 & | & & | & & | & & | & \\
H- & C & - & C & - & C & - & C & -H \\
 & | & & | & & | & & | & \\
 & H & & H & & H & & H & \\
\end{array}
$$

Normal butane   and

$$
\begin{array}{ccccc}
 & & H & & \\
 & H & | & H & \\
 & \diagdown & C & \diagup & \\
 & H & | & H & \\
 & | & | & | & \\
H- & C - & C & - C & -H \\
 & | & | & | & \\
 & H & H & H & \\
\end{array}
$$

Isobutane

Two C_4H_{10} compounds with the structural formulas shown actually exist. The butane with the unbranched carbon chain is called *normal butane* (abbreviated *n*-butane); it boils at 0.5°C and melts at −138.3°C. The branched-chain butane is called *isobutane*; it boils at −11.7°C and melts at −159.5°C. These differences in physical properties are sufficient to establish that the two compounds, both with the same molecular formula, are actually different substances. Models illustrating the structural arrangement of the atoms in methane, ethane, propane, butane, and isobutane are shown in Figure 19.3 (page 484).

This phenomenon of two or more compounds having the same molecular formula but different structures is called **isomerism.** The various individual compounds are called **isomers.** For example, there are 2 isomers of butane, C_4H_{10}. Isomerism is very common among organic compounds and is another reason for the large number of known compounds. There are 3 isomers of pentane, 5 isomers of hexane, 9 isomers of heptane, 18 isomers of octane, 35 isomers of nonane, and 75 isomers of decane. It is calculated that there are 366,319 isomers of eicosane, $C_{20}H_{42}$ (but not all have been made). The phenomenon of isomerism is a very compelling reason for the use of structural formulas.

isomerism
isomers

Isomers are compounds that have the same molecular formula but different structural formulas.

To save time and space in writing, condensed structural formulas are often used. In the condensed structural formulas, atoms and groups attached to a carbon atom are generally written to the right of that carbon atom. Examine the examples in Table 19.2 on page 485.

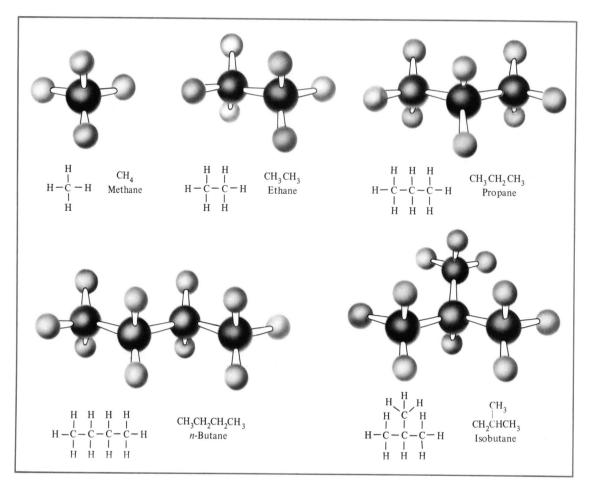

Figure 19.3 Ball-and-stick models illustrating structural formulas of methane, ethane, propane, butane, and isobutane.

Let us interpret the condensed structural formula for propane:

$$\overset{1}{CH_3} - \overset{2}{CH_2} - \overset{3}{CH_3}$$

Carbon number 1 has three hydrogen atoms attached to it and is bonded to carbon number 2, which has two hydrogen atoms on it and is bonded to carbon number 3, which has three hydrogen atoms bonded to it.

PROBLEM 19.1   There are three isomers of pentane, C_5H_{12}. Write structural formulas and condensed structural formulas for these isomers.

In a problem of this kind it is best to start by writing the carbon skeleton of the compound containing the longest continuous carbon chain. In this case it is five carbon atoms:

C — C — C — C — C

TABLE 19.2 Representation of molecular, structural, and condensed structural formulas of hydrocarbons.

Molecular formula	Structural formula	Condensed structural formula
CH_4 Methane	H | H—C—H | H	CH_4
C_2H_6 Ethane	H H | | H—C—C—H | | H H	$CH_3—CH_3$ or CH_3CH_3
C_3H_8 Propane	H H H | | | H—C—C—C—H | | | H H H	$CH_3—CH_2—CH_3$ or $CH_3CH_2CH_3$
C_4H_{10} Butane	H H H H | | | | H—C—C—C—C—H | | | | H H H H	$CH_3—CH_2—CH_2—CH_3$, $CH_3CH_2CH_2CH_3$, or $CH_3(CH_2)_2CH_3$
C_4H_{10} 2-Methyl propane (Isobutane)	H H＼ | ／H C H | H | | | H—C—C—C—H | | | H H H	CH_3 $\quad$ CH_3 | $\qquad$ | $CH_3—CH—CH_3$, CH_3CHCH_3, $CH_3CH(CH_3)CH_3$, $CH_3CH(CH_3)_2$ or $(CH_3)_3CH$

Now complete the structure by attaching hydrogen atoms around each carbon atom so that each carbon atom has four bonds attached to it. The carbon atoms at each end of the chain need three hydrogen atoms. The three inner carbon atoms each need two hydrogen atoms to give them four bonds.

```
    H   H   H   H   H
    |   |   |   |   |
H — C — C — C — C — C — H
    |   |   |   |   |
    H   H   H   H   H
```

For the next isomer, start by writing a four-carbon chain and attach the fifth carbon atom to either of the middle carbon atoms—do not use the end ones.

```
        C                 C
        |                 |
  C—C—C—C           C—C—C—C
```

Both of these structures represent the same compound.

Now add the 12 hydrogen atoms to complete the structure:

$$
\begin{array}{cccc}
 & & H & \\
 & H & | & H \\
 & H & \diagup C \diagdown & H \\
H & H & C & H \\
| & | & | & | \\
H-C-C-C-C-H \\
| & | & | & | \\
H & H & H & H
\end{array}
$$

For the third isomer, write a three-carbon chain, attach the other two carbon atoms to the central carbon atom, and complete the structure by adding the 12 hydrogen atoms:

$$
\begin{array}{ccc}
 & C & \\
 & | & \\
C & - C & - C \\
 & | & \\
 & C &
\end{array}
\qquad
\begin{array}{ccc}
 & H & \\
 H & | & H \\
 H & \diagup C \diagdown & H \\
 | & | & | \\
H-C-C-C-H \\
 | & | & | \\
 H & C & H \\
 H & \diagdown | \diagup & H \\
 & H &
\end{array}
$$

The condensed structural formulas are derived from the structural formulas by placing the hydrogen atoms after the carbon atom to which they are attached:

$$
\begin{array}{ccc}
 & CH_3 & CH_3 \\
 & | & | \\
CH_3CH_2CH_2CH_2CH_3 & CH_3CH_2CHCH_3 & CH_3CCH_3 \quad or \quad C(CH_3)_4 \\
 & & | \\
or \quad CH_3(CH_2)_3CH_3 & or \quad CH_3CH_2CH(CH_3)_2 & CH_3
\end{array}
$$

■

19.8 Naming Organic Compounds

In the early stages of the development of organic chemistry, each new compound that was recognized was simply given a name, usually by the person who had isolated or synthesized it. Names were not systematic but often did carry some information—usually about the origin of the substance. Wood alcohol (methanol), for example, was so named because it was obtained by destructive distillation or pyrolysis of wood. Methane is formed during the decomposition of vegetable matter underwater in marshes, and was originally called *marsh gas.* Often, a single compound came to be known by several names. For example, the active ingredient in alcoholic beverages has been called *alcohol, ethyl alcohol, methyl carbinol, grain alcohol, spirit,* and *ethanol.*

Beginning with a meeting in Geneva in 1892, an international system for naming compounds was developed. In its present form, the system recommended by the International Union of Pure and Applied Chemistry is systematic, generally unambiguous, and internationally accepted. It is called the

IUPAC System. Despite the existence of the official IUPAC System, a great many well-established common, or trivial, names and abbreviations (such as TNT and DDT) are used because of their brevity and/or convenience. So it is necessary to have a knowledge of both the IUPAC System and many common names.

alkyl group

In order to name organic compounds systematically, you must be able to recognize certain common alkyl groups. **Alkyl groups** have the general formula C_nH_{2n+1} (one less hydrogen atom than the corresponding alkane). The name of the group is formed from the name of the corresponding alkane by simply dropping -*ane* and substituting a -*yl* ending. The names and formulas of selected alkyl groups are given in Table 19.3. The letter "R" is often used in formulas to mean any of the many possible alkyl groups.

The following relatively simple rules are all that are needed to name a great many alkanes according to the IUPAC System. In later sections these rules will be extended to cover other classes of compounds, but advanced texts or references must be consulted for the complete system.

IUPAC rules for naming alkanes:

1. Select the longest continuous chain of carbon atoms as the parent compound, and consider all alkyl groups attached to it as branch chains that have replaced hydrogen atoms of the parent hydrocarbon. The name of the alkane consists of the name of the parent compound prefixed by the names of the branch-chain alkyl groups attached to it.

TABLE 19.3 **Names and formulas of selected alkyl groups.**

Formula	Name[a]	Formula	Name[a]
CH_3-	Methyl	CH_3CH- with CH_3 above	Isopropyl
CH_3CH_2-	Ethyl		
$CH_3CH_2CH_2-$	n-Propyl	CH_3CHCH_2- with CH_3 above	Isobutyl
$CH_3CH_2CH_2CH_2-$	n-Butyl		
$CH_3(CH_2)_3CH_2-$	n-Pentyl	CH_3CH_2CH- with CH_3 above	sec-Butyl (secondary butyl)
$CH_3(CH_2)_4CH_2-$	n-Hexyl		
$CH_3(CH_2)_5CH_2-$	n-Heptyl		
$CH_3(CH_2)_6CH_2-$	n-Octyl	CH_3C- with CH_3 above and CH_3 below	tert-Butyl (tertiary butyl)
$CH_3(CH_2)_7CH_2-$	n-Nonyl		
$CH_3(CH_2)_8CH_2-$	n-Decyl		

[a]The lowercase n in n-propyl, n-butyl, and so on, means that the hydrocarbon chains are *normal*; that is, the alkyl group represents an unbranched hydrocarbon with a hydrogen atom missing from carbon number 1.

2. Number the carbon atoms in the parent carbon chain starting from the end closest to the first carbon atom that has an alkyl group substituted for a hydrogen atom.
3. Name each branch-chain alkyl group and designate its position on the parent carbon chain by a number (for example, 2-methyl means a methyl group attached to carbon number 2).
4. When the same alkyl group branch chain occurs more than once, indicate this by a prefix, *di-*, *tri-*, *tetra-*, and so forth, written in front of the alkyl group name (for example, *dimethyl* indicates two methyl groups). The numbers indicating the positions of these alkyl groups are separated by a comma, followed by a hyphen, and are placed in front of the name (for example, 2,3-dimethyl).
5. When several different alkyl groups are attached to the parent compound, list them in alphabetical order; for example, ethyl before methyl in 3-ethyl-4-methyloctane.

The compound that follows is commonly called *isopentane*. Consider naming it by the IUPAC System:

$$\overset{4}{C}H_3 - \overset{3}{C}H_2 - \overset{2}{C}H - \overset{1}{C}H_3$$
$$| $$
$$CH_3$$

2-Methylbutane
(Isopentane)

The longest continuous chain contains four carbon atoms. Therefore, the parent compound is butane and the compound is named as a butane. The methyl group (CH_3-) attached to carbon number 2 is named as a prefix to butane, the "2-" indicating the point of attachment of the methyl group on the butane chain.

How would we write the structural formula for 2-methylpentane? An analysis of its name gives us this information:

1. The parent compound, pentane, contains five carbons. Write and number the five-carbon skeleton of pentane:

$$\overset{5}{C} - \overset{4}{C} - \overset{3}{C} - \overset{2}{C} - \overset{1}{C} \quad \text{or} \quad \overset{1}{C} - \overset{2}{C} - \overset{3}{C} - \overset{4}{C} - \overset{5}{C}$$

2. Put a methyl group on carbon number 2 (2-methyl in the name gives this information):

$$\overset{5}{C} - \overset{4}{C} - \overset{3}{C} - \overset{2}{C} - \overset{1}{C}$$
$$|$$
$$CH_3$$

3. Add hydrogens to give each carbon four bonds. The structural formula is

$$CH_3-CH_2-CH_2-CH-CH_3$$
$$\quad\quad\quad\quad\quad\quad\quad\quad |$$
$$\quad\quad\quad\quad\quad\quad\quad\quad CH_3$$
2-Methylpentane

Should this compound be called 4-methylpentane? No, since the IUPAC System specifically states that the parent carbon chain shall be numbered starting from the end nearest the branch chain.

It is very important to understand that it is the sequence of atoms and groups that determines the name of a compound, and not the way the sequence is written. Each of the following formulas represents 2-methylpentane:

$$\overset{1}{CH_3}-\overset{2}{CH}-\overset{3}{CH_2}-\overset{4}{CH_2}-\overset{5}{CH_3} \qquad \overset{5}{CH_3}-\overset{4}{CH_2}-\overset{3}{CH_2}-\overset{2}{CH}-\overset{1}{CH_3}$$
$$\quad\quad\quad |$$
$$\quad\quad CH_3$$

The following formulas and names demonstrate other aspects of the official nomenclature system:

$$\overset{4}{CH_3}-\overset{3}{CH}-\overset{2}{CH}-\overset{1}{CH_3}$$
$$\quad\quad\quad |\quad\quad |$$
$$\quad\quad CH_3\ \ CH_3$$
2,3-Dimethylbutane

The name of this compound is 2,3-dimethylbutane. The longest carbon atom chain is four, indicating butane; dimethyl indicates two methyl groups; "2,3-" means that one CH_3 is on carbon 2 and one is on carbon 3.

$$\quad\quad\quad\quad\quad CH_3$$
$$\quad\quad\quad\quad\quad |2$$
$$\overset{4}{CH_3}-\overset{3}{CH_2}-\overset{2}{C}-\overset{1}{CH_3}$$
$$\quad\quad\quad\quad\quad |$$
$$\quad\quad\quad\quad\quad CH_3$$
2,2-Dimethylbutane

(Both methyl groups are on the same carbon atom; both numbers are required.)

$$\quad\quad\quad\quad\quad\quad\quad CH_3$$
$$\quad\quad\quad\quad\quad\quad\quad |4$$
$$\overset{1}{CH_3}-\overset{2}{CH}-\overset{3}{CH_2}-\overset{4}{CH}-\overset{5}{CH_2}-\overset{6}{CH_3}$$
$$\quad\quad\quad |$$
$$\quad\quad CH_3$$
2,4-Dimethylhexane

(The molecule is numbered from left to right.)

$$CH_3 - \overset{3}{\underset{\underset{\underset{CH_3}{\overset{1|}{\vert}}}{\overset{2|}{\underset{CH_2}{\vert}}}}{CH}} - \overset{4}{CH_2} - \overset{5}{CH_2} - \overset{6}{CH_3}$$

3-Methylhexane

(There are six carbons in the longest continuous chain.)

$$\overset{8}{CH_3} - \overset{7}{CH_2} - \overset{6}{CH_2} - \overset{5}{CH_2} - \overset{\overset{\textstyle CH_2 - CH_3}{\overset{|4}{\vert}}}{\underset{\underset{CH_3}{\vert}}{C}} {\overset{3}{\underset{\underset{Cl}{\vert}}{CH}}} - {\overset{2}{\underset{\underset{CH_3}{\vert}}{CH}}} - \overset{1}{CH_3}$$

3-Chloro-4-ethyl-2,4-dimethyloctane

(The groups attached to the octane chain are listed in alphabetical order.)

PROBLEM 19.2   Write the formulas for (a) 3-ethylpentane, and (b) 2,2,4-trimethylpentane.

(a) The name *pentane* indicates a five-carbon chain. Write five connecting carbon atoms and number them:

$$\overset{1}{C} - \overset{2}{C} - \overset{3}{C} - \overset{4}{C} - \overset{5}{C}$$

An ethyl group is written as $CH_3CH_2 -$. Attach this group to carbon number 3:

$$\overset{1}{C} - \overset{2}{C} - \overset{3}{\underset{\underset{CH_2CH_3}{\vert}}{C}} - \overset{4}{C} - \overset{5}{C}$$

Now add hydrogen atoms to give each carbon atom four bonds. Carbons 1 and 5 each need three H atoms; carbons 2 and 4 each need two H atoms; and carbon 3 needs one H atom. The formula is complete.

$$CH_3CH_2\underset{\underset{CH_2CH_3}{\vert}}{CH}CH_2CH_3$$

(b) Pentane indicates a five-carbon chain. Write five connecting carbon atoms and number them:

$$\overset{1}{C} - \overset{2}{C} - \overset{3}{C} - \overset{4}{C} - \overset{5}{C}$$

There are three methyl groups ($CH_3 -$) in the compound (trimethyl), two attached to carbon 2 and one attached to carbon 4. Attach these three methyl groups to their respective carbon atoms:

$$\overset{1}{C} - \overset{2}{\underset{\underset{CH_3}{\vert}}{\overset{\overset{CH_3}{\vert}}{C}}} - \overset{3}{C} - \overset{4}{\overset{\overset{CH_3}{\vert}}{C}} - \overset{5}{C}$$

Now add H atoms to give each carbon atom four bonds. Carbons 1 and 5 each need three H atoms; carbon 2 does not need any H atoms; carbon 3 needs two H atoms; and carbon 4 needs one H atom. The formula is complete.

$$\begin{array}{c} \quad \text{CH}_3 \ \ \text{CH}_3 \\ \quad | \qquad | \\ \text{CH}_3\text{CCH}_2\text{CHCH}_3 \\ \quad | \\ \quad \text{CH}_3 \end{array}$$

■

PROBLEM 19.3   Name the following compounds:

$$\begin{array}{c} \qquad\qquad \text{CH}_3 \\ \qquad\qquad | \\ \text{(a)}\ \ \text{CH}_3\text{CH}_2\text{CH}_2\text{CH}_2\text{CHCH}_3 \end{array}$$

$$\begin{array}{c} \qquad\qquad \text{CH}_2\text{CH}_3 \\ \qquad\qquad | \\ \text{(b)}\ \ \text{CH}_3\text{CH}_2\text{CH}_2\text{CHCH}_2\text{CHCH}_3 \\ \qquad\qquad\qquad\qquad | \\ \qquad\qquad\qquad\quad \text{CH}_2\text{CH}_3 \end{array}$$

(a) The longest continuous carbon chain contains six carbon atoms. Thus, the parent name of the compound is hexane. Number the carbon chain from right to left so that the methyl group attached to carbon 2 is given the lowest possible number. With a methyl group on carbon 2 the name of the compound is 2-methylhexane.

(b) There are several sequences of carbon chains in this compound. The longest continuous carbon chain contains eight carbon atoms.

$$\begin{array}{c} \qquad\qquad \text{C}-\text{C} \\ \ 8 \quad 7 \quad 6 \quad |5 \quad 4 \quad 3 \\ \text{C}-\text{C}-\text{C}-\text{C}-\text{C}-\text{C}-\text{C} \\ \qquad\qquad\quad |2 \quad 1 \\ \qquad\qquad\quad \text{C}-\text{C} \end{array}$$

Thus, the parent name is octane. As the chain is numbered, there is a methyl group on carbon 3 and an ethyl group on carbon 5. Thus, the name of the compound is 5-ethyl-3-methyloctane. Note that ethyl is named before methyl (alphabetical order).

■

19.9 Reactions of Alkanes

One single type of reaction of alkanes has inspired men to explore equatorial jungles, endure the heat and sandstorms of the deserts of Africa and the Middle East, mush across the frozen arctic, and drill holes in the earth more than 30,000 feet deep! These strenuous and expensive activities have been undertaken because alkanes, as well as other hydrocarbons, undergo combustion with oxygen with the evolution of large amounts of heat energy. Methane, for example, reacts with oxygen:

$$\text{CH}_4(g) + 2\,\text{O}_2(g) \longrightarrow \text{CO}_2(g) + 2\,\text{H}_2\text{O}(g) + 191.8\ \text{kcal}$$

The thermal energy can be converted to mechanical and electrical energy. In order of economic importance, combustion reactions overshadow all other reactions of alkanes. But combustion reactions are not usually of great interest to organic chemists, since carbon dioxide and water are the only chemical products of complete combustion.

Aside from their combustibility, alkanes are sluggish and limited in their reactivity. But with proper activation, such as high temperature and/or catalysts, alkanes can be made to react in a variety of ways. Some industrially important noncombustion reactions of alkanes are the following:

1. **Halogenation** (substitution of halogens for hydrogen): Halogenation is a general term. When a specific halogen, such as chlorine, is used, the reaction is called chlorination.

$$RH + X_2 \longrightarrow RX + HX \quad (X = Cl \ \text{ or } \ Br)$$

$$CH_3CH_3 + Cl_2 \longrightarrow \underset{\text{Chloroethane}}{CH_3CH_2Cl} + HCl$$

This reaction yields alkyl halides (RX), which are useful as intermediates for the manufacture of other substances.

2. **Dehydrogenation** (removal of hydrogen):

$$C_nH_{2n+2} \xrightarrow{700-900°C} C_nH_{2n} + H_2$$

$$CH_3CH_2CH_3 \xrightarrow{\Delta} \underset{\text{Propene}}{CH_3CH=CH_2} + H_2$$

This reaction yields alkenes, which, like alkyl halides, are useful chemical intermediates. Hydrogen is a valuable by-product.

3. **Cracking** (breaking up large molecules to form smaller ones):

Example: $C_{16}H_{34} \xrightarrow{\Delta} \underset{\text{Alkane}}{C_8H_{18}} + \underset{\text{Alkene}}{C_8H_{16}}$
   $\underset{\text{Alkane}}{}$

4. **Isomerization** (rearrangement of molecular structures):

Example:
$$CH_3-CH_2-CH_2-CH_2-CH_3 \longrightarrow CH_3-CH_2-CH(CH_3)-CH_3$$

Halogenation is used extensively in the manufacture of petrochemicals (chemicals derived from petroleum and used for purposes other than fuels). The other three reactions—cracking, dehydrogenation, and isomerization—singly or in combination, are of great importance in the production of both motor fuels and petrochemicals.

A well-known reaction of methane and chlorine is shown by the equation

$$CH_4 + Cl_2 \xrightarrow[\text{light}]{uv} \underset{\substack{\text{Chloromethane} \\ \text{(Methyl chloride)}}}{CH_3Cl} + HCl$$

Methane can be chlorinated to yield mono-, di-, tri-, or tetrachloromethane. The formulas and names for these compounds are given in Table 19.4.

$$CH_4 \xrightarrow{Cl_2} CH_3Cl \xrightarrow{Cl_2} CH_2Cl_2 \xrightarrow{Cl_2} CHCl_3 \xrightarrow{Cl_2} CCl_4 + 4\,HCl$$

TABLE 19.4 **Chlorination products of methane.**

Formula	IUPAC name	Common name
CH_3Cl	Chloromethane	Methyl chloride
CH_2Cl_2	Dichloromethane	Methylene chloride
$CHCl_3$	Trichloromethane	Chloroform
CCl_4	Tetrachloromethane	Carbon tetrachloride

The term *monosubstitution product* refers to a compound in which one hydrogen atom from an organic molecule is substituted by one other atom or group of atoms. In hydrocarbons, for example, when we substitute one chlorine atom for a hydrogen atom, the new compound is a monosubstitution (mono-chloro substitution) product. In a like manner, we can have di-, tri-, tetra-, and so on, substitution products. Chlorination (or bromination) is a general reaction of alkanes. There are nine different chlorination products of ethane. See if you can write the structural formulas for all of them.

When propane is chlorinated, two isomeric monosubstitution products are obtained because a hydrogen atom may be replaced on either the first or second carbon.

$$CH_3CH_2CH_3 + Cl_2 \xrightarrow[25°]{\text{Light}} CH_3CH_2CH_2Cl + CH_3CHClCH_3 + HCl$$

1-Chloropropane   2-Chloropropane
(n-Propyl chloride)   (Isopropyl chloride)
45%   55%

These two compounds are isomers. They both have the same molecular formula, C_3H_7Cl, but they have different structural formulas.

The letter X is commonly used in organic compounds to indicate a halogen atom. The formula RX indicates a halogen atom attached to an alkyl group and **alkyl halide** represents the class of compounds known as the **alkyl halides.** When $R = CH_3$, CH_3X can be CH_3F, CH_3Cl, CH_3Br, or CH_3I.

Alkyl halides are named systematically in the same general way as alkanes. Halogen atoms are identified by the prefixes *fluoro-, chloro-, bromo-,* or *iodo-,* and are named as substituents like side-chain alkyl groups. Study these examples:

$$CH_3 - CHCl - CH_2 - CH_3 \qquad CH_2Cl - CHBr - CH_3$$

2-Chlorobutane   2-Bromo-1-chloropropane

$$CH_3 - CH_2 - CH - CHCl - CH_3$$
$$\qquad\qquad\quad |$$
$$\qquad\qquad\quad CH_3$$

2-Chloro-3-methylpentane

PROBLEM 19.4 How many monochlorosubstitution products can be obtained from pentane? First write the formula for pentane:

$$\overset{5}{C}H_3\overset{4}{C}H_2\overset{3}{C}H_2\overset{2}{C}H_2\overset{1}{C}H_3$$

Now rewrite the formula five times substituting a Cl atom for an H atom on each C atom:

I $CH_3CH_2CH_2CH_2\overset{1}{CH_2}Cl$ Cl on carbon 1

II $CH_3CH_2CH_2\overset{2}{CH}ClCH_3$ Cl on carbon 2

III $CH_3CH_2\overset{3}{CH}ClCH_2CH_3$ Cl on carbon 3

IV $CH_3\overset{4}{CH}ClCH_2CH_2CH_3$ Cl on carbon 4

V $\overset{5}{CH_2}ClCH_2CH_2CH_2CH_3$ Cl on carbon 5

Compound IV is identical to II and compound V is identical to I, because if we number the carbon chain from left to right, we find the Cl atoms substituted on carbons 1 and 2. By naming the compounds we find that both I and V are 1-chloropentane and that II and IV are both 2-chloropentane. Thus, there are three monochloro substitution products of pentane—compounds I, II, and III, 1-chloropentane, 2-chloropentane, and 3-chloropentane.

■

19.10 Functional Groups

functional group

Organic compounds were obtained originally from plants and animals; even today these are the direct sources of many important chemicals. As a case in point, millions of tons of sucrose (table sugar) are obtained from cane and beet juices each year. As chemical knowledge developed, many naturally occurring compounds were synthesized, often at costs far less than those of the natural products. Of even greater significance than the cheaper manufacture of existing substances was the synthesis of new substances totally unlike any natural product. The synthesis of new substances was aided greatly by the realization that organic chemicals can be divided into a relatively small number of classes and studied on the basis of similar chemical properties. The various classes of compounds are identified by the presence of certain characteristic groups called **functional groups.** For example, when a halogen atom (F, Cl, Br, I) is substituted for a hydrogen atom in an alkane molecule, the resulting compound is an alkyl halide. Thus, alkyl halides are a class of compounds in which the functional group is the halogen atom. Each class of compounds contains a different functional group, and the members comprise a homologous series. Furthermore, the members of a homologous series exhibit similar chemical properties as a result of their common functional group. Examples of the more common classes of compounds and their functional groups are shown in Table 19.5 on pages 496 and 497.

19.11 Alkenes and Alkynes

The bulk of the organic compounds manufactured in the United States are derived from seven starting materials: ethylene, propylene, benzene, butylene, toluene, xylene, and methane. Of the seven substances listed, all are unsaturated hydrocarbons except methane. Ethylene alone is the base for almost half the petrochemicals made in the country. About 29 billion pounds of ethylene are produced annually, making it the sixth ranking chemical in production volume in the United States.

alkene
alkyne

Both **alkenes** and **alkynes** are classified as unsaturated hydrocarbons. They are said to be unsaturated because, unlike alkanes, their molecules do not contain the maximum possible number of hydrogen atoms. Alkenes (also known as *olefins*) contain at least one double bond between adjacent carbon atoms. Alkynes (also known as *acetylenes*) contain at least one triple bond between adjacent carbon atoms.

Alkenes contain a carbon–carbon double bond.
Alkynes contain a carbon–carbon triple bond.

The simplest alkene is ethylene (or ethene), $CH_2\!=\!CH_2$, and the simplest alkyne is acetylene (or ethyne), $CH\!\equiv\!CH$ (Figure 19.4). Both ethylene and acetylene are the first members of homologous series in which the formulas of successive members differ by increments of $-CH_2-$ (for example, $CH_2\!=\!CH_2$, $CH_3CH\!=\!CH_2$, and $CH_3CH_2CH\!=\!CH_2$). Huge quantities of alkenes are made by cracking and dehydrogenating alkanes in processing crude oils. These alkenes are used to manufacture motor fuels, polymers, and petrochemicals. Alkene molecules, like those of alkanes, have very little polarity. Hence, the physical

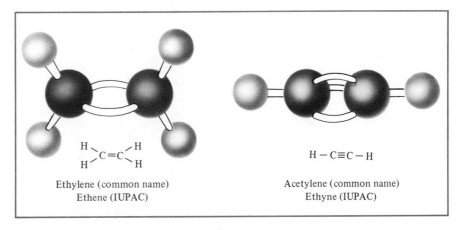

Ethylene (common name)
Ethene (IUPAC)

Acetylene (common name)
Ethyne (IUPAC)

Figure 19.4 Ball-and-stick models for ethylene and acetylene.

TABLE 19.5 Classes of organic compounds.

Class	General formula	Structure of functional group	Sample structural formula	Name Common	Name IUPAC
Alkane	RH	R—H	CH_4	Methane	Methane
			CH_3CH_3	Ethane	Ethane
Alkene	R—CH=CH$_2$	\C=C/	CH_2=CH_2	Ethylene	Ethene
			CH_3CH=CH_2	Propylene	Propene
Alkyne	R—C≡C—H	—C≡C—	CH≡CH	Acetylene	Ethyne
			CH_3C≡CH	Methyl acetylene	Propyne
Alkyl halide	RX	—X X=F, Cl, Br, I	CH_3Cl	Methyl chloride	Chloro-methane
			CH_3CH_2Cl	Ethyl chloride	Chloro-ethane
Alcohol	ROH	—OH	CH_3OH	Methyl alcohol	Methanol
			CH_3CH_2OH	Ethyl alcohol	Ethanol
Ether	R—O—R	R—O—R	CH_3—O—CH_3	Dimethyl ether	Methoxy-methane
			CH_3CH_2—O—CH_2CH_3	Diethyl ether	Ethoxy-ethane
Aldehyde	R—C=O with H	—C=O with H	H—C=O with H	Formalde-hyde	Methanal
			CH_3—C=O with H	Acetalde-hyde	Ethanal
Ketone	R—C—R ‖ O	R—C—R ‖ O	CH_3—C—CH_3 ‖ O	Acetone	Propanone
			CH_3—C—CH_2CH_3 ‖ O	Methyl ethyl ketone	2-Butanone
Carboxylic acid	R—C with =O and OH	—C with =O and OH	HCOOH	Formic acid	Methanoic acid
			CH_3COOH	Acetic acid	Ethanoic acid
Ester	R—C with =O and OR	—C with =O and OR	$HCOOCH_3$	Methyl formate	Methyl methanoate
			CH_3COOCH_3	Methyl acetate	Methyl ethanoate

Class	General formula	Structure of functional group	Sample structural formula	Name Common	Name IUPAC
Amide	$R-C$ $\overset{O}{\underset{NH_2}{\diagdown}}$	$-C$ $\overset{O}{\underset{NH_2}{\diagdown}}$	$\overset{O}{\overset{\|}{HC-NH_2}}$	Formamide	Methana-mide
			$\overset{O}{\overset{\|}{CH_3C-NH_2}}$	Acetamide	Ethanamide
Amine	$R-NH_2$	$-NH_2$	CH_3NH_2	Methyl-amine	Amino-methane
			$CH_3CH_2NH_2$	Ethylamine	Aminoethane
Nitrile	$R-C\equiv N$	$-C\equiv N$	$CH_3C\equiv N$	Acetonitrile	Ethanonitrile
			$CH_3CH_2C\equiv N$	Propiono-nitrile	Propano-nitrile
Amino acid	$R-\underset{NH_2}{\overset{\|}{CH}}-COOH$	$-COOH$ $-NH_2$	H_2NCH_2COOH $CH_3\underset{NH_2}{\overset{\|}{CH}}-COOH$	Glycine Alanine	2-Amino-ethanoic acid 2-Aminopro-panoic acid

properties of alkenes are very similar to those of the corresponding saturated hydrocarbons.

General formula for alkenes: C_nH_{2n}
General formula for alkynes: C_nH_{2n-2}

Table 19.6 gives the names and formulas for several alkenes and alkynes.

TABLE 19.6 Names and formulas for several alkenes and alkynes.

Formula	IUPAC name	Common name
$CH_2=CH_2$	Ethene	Ethylene
$CH_3CH=CH_2$	Propene	Propylene
$CH_3CH_2CH=CH_2$	1-Butene	Butylene
$CH_3CH=CHCH_3$	2-Butene	
$CH_3\underset{CH_3}{\overset{\|}{C}}=CH_2$	2-Methylpropene	Isobutylene
$CH\equiv CH$	Ethyne	Acetylene
$CH_3C\equiv CH$	Propyne	Methyl acetylene
$CH_3CH_2C\equiv CH$	1-Butyne	Ethyl acetylene
$CH_3C\equiv CCH_3$	2-Butyne	Dimethyl acetylene

19.12 Naming Alkenes and Alkynes

The names of alkenes and alkynes are derived from the names of the corresponding alkanes. To name an alkene (or alkyne) by the IUPAC System,

1. Select the longest carbon–carbon chain that contains the double or triple bond.
2. Name this parent compound as you would an alkane but change the -*ane* ending to -*ene* for an alkene or to -*yne* for an alkyne; for example, propane is changed to propene or propyne.
3. Number the carbon chain of the parent compound so that the double or triple bond carries the lowest possible numbers. Use the smaller of the two numbers on the double- or triple-bonded carbon atoms to indicate the position of the double or triple bond. Place this number in front of the alkene or alkyne name; for example, 2-butene means that the carbon–carbon double bond is between carbon numbers 2 and 3.
4. Side chains and other groups are treated as in naming alkanes, by numbering and assigning them to the carbon atom to which they are attached.

Study the following examples of named alkenes and alkynes:

$$\overset{4}{C}H_3\overset{3}{C}H_2\overset{2}{C}H=\overset{1}{C}H_2 \qquad \overset{1}{C}H_3\overset{2}{C}H=\overset{3}{C}H\overset{4}{C}H_3$$

1-Butene   2-Butene

$$\begin{array}{c} CH_3 \\ \overset{4}{C}H_3\overset{3}{C}H\overset{2}{C}H=\overset{1}{C}H_2 \end{array}$$

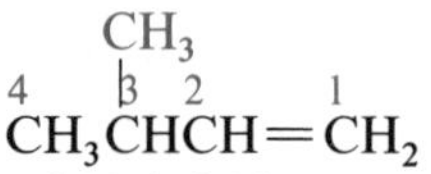

3-Methyl-1-butene

$$\overset{4}{C}H_3\overset{3}{C}H_2\overset{2}{C}\equiv\overset{1}{C}H \qquad \overset{1}{C}H_3-\overset{2}{C}\equiv\overset{3}{C}-\overset{4}{C}H-\overset{5}{C}H-\overset{6}{C}H_3$$

1-Butyne

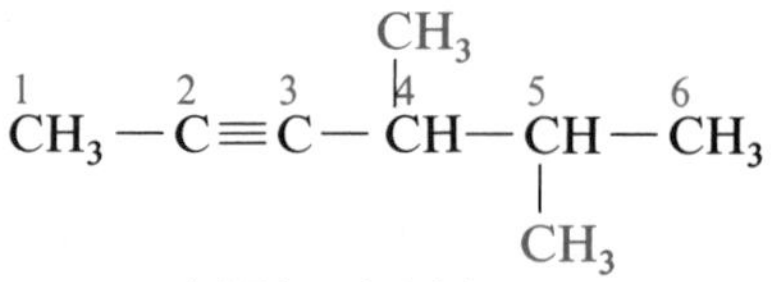

4,5-Dimethyl-2-hexyne

To write a structural formula from a systematic name, the naming process is, in effect, reversed. For example, how would we write the structural formula for 4-methyl-2-pentene? The name indicates:

1. Five carbons in the longest chain
2. A double bond between carbons 2 and 3
3. A methyl group on carbon 4

Write five carbon atoms in a row. Place a double bond between carbons 2 and 3, and place a methyl group on carbon 4:

$$\overset{1}{C}-\overset{2}{C}=\overset{3}{C}-\overset{4}{C}-\overset{5}{C}$$
$$\underset{CH_3}{\overset{\displaystyle|}{}}$$

Carbon skeleton

Now add hydrogen atoms to give each carbon atom four bonds. Carbons 1 and 5 each need three H atoms; carbons 2, 3, and 4 each need one H atom. The complete formula is

$$CH_3CH=CHCHCH_3$$
$$\underset{CH_3}{\overset{\displaystyle|}{}}$$

4-Methyl-2-pentene

PROBLEM 19.5

Write structural formulas for (a) 2-pentene, (b) 7-methyl-2-octene, and (c) 3-hexyne.

(a) The name pentene indicates a five-carbon chain with a carbon–carbon double bond; the number 2 locates the double bond between carbons 2 and 3. Write five carbon atoms in a row and place a double bond between carbons 2 and 3:

$$\overset{1}{C}-\overset{2}{C}=\overset{3}{C}-\overset{4}{C}-\overset{5}{C}$$

Add hydrogen atoms to give each carbon atom four bonds. Carbons 1 and 5 each need three H atoms; carbons 2 and 3 each need one H atom; carbon 4 needs two H atoms. The complete formula is

$$CH_3CH=CHCH_2CH_3$$
2-Pentene

(b) Octene, like octane, indicates an eight-carbon chain. The chain contains a double bond between carbons 2 and 3 and a methyl group on carbon 7. Write eight carbon atoms in a row, place a double bond between carbons 2 and 3, and place a methyl group on carbon 7:

$$\overset{1}{C}-\overset{2}{C}=\overset{3}{C}-\overset{4}{C}-\overset{5}{C}-\overset{6}{C}-\overset{7}{C}-\overset{8}{C}$$
$$\underset{CH_3}{\overset{\displaystyle|}{}}$$

Now add hydrogen atoms to give each carbon atom four bonds. The complete formula is

$$CH_3CH=CHCH_2CH_2CH_2CHCH_3$$
$$\underset{CH_3}{\overset{\displaystyle|}{}}$$

7-Methyl-2-octene

(c) The stem *hex-* indicates a six-carbon chain; the suffix *-yne* indicates a carbon–carbon triple bond; the number 3 locates the triple bond between carbons 3 and 4. Write six carbon atoms in a row and place a triple bond between carbons 3 and 4:

$$\overset{1}{C}-\overset{2}{C}-\overset{3}{C}\equiv\overset{4}{C}-\overset{5}{C}-\overset{6}{C}$$

Now add hydrogen atoms to give each carbon atom four bonds. Carbons 3 and 4 do not need any H atoms. The complete formula is

$$CH_3CH_2C\equiv CCH_2CH_3$$
3-Hexyne

■

19.13 Reactions of Alkenes

The alkenes are much more reactive than the corresponding alkanes. This greater reactivity is due to the carbon–carbon double bonds.

addition reaction

Addition. A reaction in which two substances join together to produce one compound is called an **addition reaction.** Addition at the carbon–carbon double bond is the most common reaction of alkenes. Hydrogen, halogens (Cl_2 or Br_2), hydrogen halides, sulfuric acid, and water are some of the reagents that can be added to unsaturated hydrocarbons. Ethylene (ethene), for example, reacts in this fashion:

$$CH_2{=}CH_2 + H_2 \xrightarrow[\text{1 atm}]{\text{Pt, 25°C}} CH_3{-}CH_3$$
Ethylene   Ethane

$$CH_2{=}CH_2 + Br{-}Br \longrightarrow CH_2Br{-}CH_2Br$$
1,2-Dibromoethane

Visible evidence of the Br_2 addition is the disappearance of the reddish brown color of bromine as it reacts.

$$CH_2{=}CH_2 + HCl \longrightarrow CH_3CH_2Cl$$
Chloroethane
(Ethyl chloride)

$$CH_2{=}CH_2 + HOSO_3H(conc.) \longrightarrow CH_3CH_2OSO_3H$$
Sulfuric acid   Ethyl hydrogen sulfate

$$CH_2{=}CH_2 + HOH \xrightarrow{H^+} CH_3CH_2OH$$ (The H^+ indicates that the
Ethanol   reaction is carried out under
(Ethyl alcohol)   acid conditions.)

Note that the double bond is broken and the unsaturated alkene molecules become saturated by an addition reaction.

The preceding examples dealt with ethylene, but reactions of this kind can be made to occur on almost any molecule that contains a carbon–carbon double bond.

Oxidation. Another typical reaction of alkenes is oxidation at the double bond. For example, when shaken with a cold, dilute solution of potassium permanganate ($KMnO_4$), an alkene is converted to a glycol (glycols are dihydroxy alcohols). Ethylene reacts in this manner.

$$CH_2{=}CH_2 + KMnO_4(aq) + H_2O \longrightarrow \underset{\substack{| \quad\ | \\ OH \quad OH}}{CH_2{-}CH_2} + MnO_2 + KOH$$

Ethylene   (Purple)   (Brown)

                                                                                Ethylene glycol

Baeyer's test makes use of this reaction to detect or confirm the presence of double (or triple) bonds in hydrocarbons. Evidence of reaction (positive Baeyer's test) is the disappearance of the purple color of permanganate ions. Baeyer's test is not specific for detecting unsaturation in hydrocarbons, since other classes of compounds may also give a positive Baeyer's test.

Carbon–carbon double bonds (ethylenic linkages) are found in many different kinds of molecules. Most of these substances will react with potassium permanganate and will undergo somewhat similar reactions with other oxidizing agents including oxygen in the air and, especially, with ozone. Such reactions are frequently troublesome. For example, premature aging and cracking of tires in smoggy atmosphere occurs because ozone attacks the ethylenic linkages in rubber molecules. Cooking oils and fats sometimes develop disagreeable odors and flavors because the oxygen of the air reacts with the double bonds present in these materials. Potato chips are especially subject to flavor damage caused by oxidation of the unsaturated cooking oils that they contain.

19.14 Acetylene: Preparation and Properties

Acetylene (ethyne), $HC{\equiv}CH$, is the simplest alkyne and is an industrial chemical of prime importance. It can be prepared from relatively inexpensive raw materials—lime and coke. These are heated to form calcium carbide, which is then reacted with water to form acetylene:

$$\underset{\substack{\text{Calcium oxide} \\ \text{(Lime)}}}{CaO} + \underset{\substack{\text{Carbon} \\ \text{(Coke)}}}{3\ C} \xrightarrow[\text{Electric furnace}]{2500°C} \underset{\text{Calcium carbide}}{CaC_2} + CO$$

$$CaC_2 + 2\ H_2O \longrightarrow \underset{\text{Acetylene}}{CH{\equiv}CH} + Ca(OH)_2$$

Until about 1940, nearly all acetylene was made from calcium carbide. Cheaper methods of production, using methane as the raw material, were devised in Germany during World War II. When methane is subjected to high temperatures for a very short time (0.01–0.1 second) and the products are cooled rapidly, a reasonably high yield of acetylene is obtained. The equation for the reaction, which is endothermic, is

$$\underset{\text{Methane}}{2\ CH_4} \xrightarrow{1500°C} \underset{\text{Acetylene}}{CH{\equiv}CH} + 3\ H_2 \quad -\ 95.5\ \text{kcal}$$

Special precautions must be taken when handling acetylene. Like all hydrocarbon gases, it forms explosive mixtures with air or oxygen. In addition to this hazard, acetylene is unstable at room temperature. When highly compressed or liquefied, acetylene may decompose violently (explode) either spontaneously or from a slight shock:

$$CH \equiv CH \longrightarrow H_2 + 2C + 54.3 \text{ kcal}$$

To eliminate the danger of explosion, acetylene is dissolved under pressure in acetone and is packed in cylinders that contain a porous inert material.

Acetylene is used mainly (1) as fuel for oxyacetylene cutting and welding torches, and (2) as an intermediate in the manufacture of other substances. Both uses are dependent upon the great reactivity of acetylene. Acetylene and oxygen mixtures produce flame temperatures of about 2800°C. Acetylene readily undergoes addition reactions rather similar to those of ethylene. It will react with chlorine and bromine and will decolorize a permanganate solution (Baeyer's test). Either one or two molecules of bromine or chlorine may be added:

$$CH \equiv CH + Br_2 \longrightarrow \underset{\text{1,2-Dibromoethene}}{CHBr = CHBr}$$

or

$$CH \equiv CH + 2 Br_2 \longrightarrow \underset{\text{1,1,2,2-Tetrabromoethane}}{CHBr_2 - CHBr_2}$$

Either unsaturated or saturated compounds may be obtained as addition products of acetylene. Often, unsaturated compounds capable of undergoing further reactions are made from acetylene. The monomer vinyl chloride, used to make the plastic polyvinyl chloride (PVC), can be made by simple addition of HCl to acetylene:

$$CH \equiv CH + HCl \longrightarrow \underset{\substack{\text{Vinyl chloride} \\ \text{(Chloroethene)}}}{CH_2 = CHCl}$$

[*Note:*   The common name for $CH_2 = CH-$ radical is *vinyl.*]

19.15 Polymers—Macromolecules

There exist in nature some very large molecules (macromolecules) containing tens of thousands of atoms. Some of these, such as starch, glycogen, cellulose, proteins, and DNA, have molecular weights in the millions and are central to many of our life processes. Man-made macromolecules touch every phase of modern living. It is hard to imagine a world today without polymers. Textiles for clothing, carpeting, and draperies; shoes; toys; automobile parts; construction materials; synthetic rubber; chemical equipment; medical supplies; cooking utensils; synthetic leather; recreational equipment—the list could go on and on. All these and a host of others that we consider to be essential in our daily life are wholly or partly man-made polymers. Most of these modern-day polymers were unknown 50 years ago. The vast majority of these polymeric materials are based on petroleum. Since petroleum is a nonreplaceable resource, our dependence on polymers is another good reason for not squandering our limited world supply of petroleum.

polymerization
polymer
monomer
copolymer

The process of forming very large, high-molecular-weight molecules from smaller units is called **polymerization.** The large molecule, or unit, is called the **polymer** and the small unit, the **monomer.** Polymers containing more than one kind of monomer are called **copolymers.** The term *polymer* is derived from the Greek word *polumerēs,* meaning "having many parts." Ethylene is a monomer and polyethylene is a polymer. Because of their large size, polymers are often called *macromolecules.* Another commonly used term is *plastics.* The word *plastic* means "capable of being molded, or pliable." Although all polymers are not pliable and capable of being remolded, the word *plastics* has gained general use and has come to mean any of a variety of polymeric substances.

Under proper conditions, ethylene reacts with itself to form a substance known as polyethylene (or polythene). Polyethylene is a long-chain hydrocarbon made from many ethylene units:

$$n\ CH_2{=}CH_2 \longrightarrow {+}CH_2{-}CH_2\overline{)}_n$$
Polyethylene

A typical polyethylene molecule is made up from about 2500–25,000 ethylene molecules joined in a continuous structure:

Polyethylene

$$\cdot = CH_2$$

Over 12 billion pounds of polyethylene are produced annually in the United States. Its uses are as varied as any single substance known and include chemical equipment, packaging material, electrical insulation, films, industrial protective clothing, and toys.

Ethylene derivatives, in which one or more hydrogen atoms have been replaced by other atoms or groups, can also be polymerized. Many of our commercial synthetic polymers are made from such modified ethylene monomers. The names, structures, and some uses for several of these polymers are given in Table 19.7 on page 504.

19.16 Aromatic Hydrocarbons: Structure

aromatic
compound

Benzene and all substances that have structures and chemical properties resembling benzene are classified as **aromatic compounds.** The word *aromatic* originally referred to the rather pleasant odor possessed by many of these substances, but this meaning has been dropped. Benzene, the parent substance of the aromatic hydrocarbons, was first isolated by Michael Faraday in 1825; its correct molecular formula, C_6H_6, was established a few years later. The establishment of a reasonable structural formula that would account for the properties of benzene was a very difficult problem for chemists in the mid-19th century.

TABLE 19.7 Polymers derived from modified ethylene monomers.

Monomer	Polymer	Uses
$CH_2 = CH_2$ Ethylene	$+CH_2 - CH_2\,\overset{}{\rightarrow}_n$ Polyethylene	Packaging material, molded articles, containers, toys
$CH_2 = CH$ $\quad\ \ \|$ $\quad\ \ CH_3$ Propylene	$\left(CH_2 - CH\right)$ $\qquad\quad \| \quad /n$ $\qquad\quad CH_3$ Polypropylene	Textile fibers, molded articles, lightweight ropes, autoclavable biological equipment
$CH_2 = C\big\langle{}^{CH_3}_{CH_3}$ Isobutylene	$\left(CH_2 - C\big\langle{}^{CH_3}_{CH_3}\right)_n$ Polyisobutylene	Pressure-sensitive adhesives, butyl rubber (contains some isoprene as copolymer)
$CH_2 = CH$ $\quad\ \ \|$ $\quad\ \ Cl$ Vinyl chloride	$\left(CH_2 - CH\right)$ $\qquad\quad \|\quad /n$ $\qquad\quad Cl$ Polyvinyl chloride (PVC)	Phonograph records, garden hoses, pipes, molded articles, floor tile, electrical insulation, vinyl leather
$CH_2 = CCl_2$ Vinylidene chloride	$+CH_2 - CCl_2\,\overset{}{\rightarrow}_n$ Saran[a]	Food packaging, textile fibers, pipes, tubing
$CH_2 = CH$ $\quad\ \ \|$ $\quad\ \ CN$ Acrylonitrile	$\left(CH_2 - CH\right)$ $\qquad\quad \|\quad /n$ $\qquad\quad CN$ Orlon, Acrilan	Textile fibers, tents
$CF_2 = CF_2$ Tetrafluoroethylene	$+CF_2 - CF_2\,\overset{}{\rightarrow}_n$ Teflon	Gaskets, valves, insulation, heat-resistant and chemically resistant coatings, linings for pots and pans
$CH_2 = CH$ ⬡ Styrene	$\left(CH_2 - CH\right)$ ⬡ $\qquad\qquad\quad /n$ Polystyrene	Molded articles, styrofoam, insulation
$CH_2 = CH$ $\quad\ \ \|$ $\quad\ \ OC - CH_3$ $\qquad | $ $\qquad O$ Vinyl acetate	$\left(CH_2 - CH\right)$ $\qquad\quad \|$ $\qquad\quad OC - CH_3$ $\qquad\qquad \|$ $\qquad\qquad O\ /n$ Polyvinyl acetate	Adhesives, paint and varnish, starting material for polyvinyl alcohol
$CH_2 = C - CH_3$ $\qquad\ \ \|$ $\qquad\ \ C - O - CH_3$ $\qquad\quad \|$ $\qquad\quad O$ Methylmethacrylate	$\left(CH_2 - C(CH_3)\right)$ $\qquad\qquad\quad\ \|$ $\qquad\qquad\quad\ C - OCH_3$ $\qquad\qquad\qquad \|$ $\qquad\qquad\qquad O\ /n$ Lucite, Plexiglas (acrylic resins)	Contact lenses, clear sheets for windows and optical uses, molded articles, automobile finishes

[a]Contains some vinyl chloride as copolymer.

Finally, in 1865, August Kekulé proposed that the carbon atoms in a benzene molecule are arranged in a six-membered ring with one hydrogen atom bonded to each carbon atom and with three carbon–carbon double bonds:

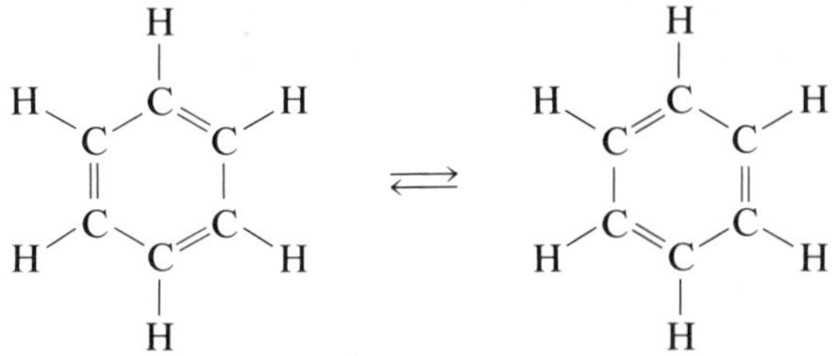

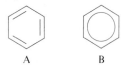

Kekulé's concepts are a landmark in the history of chemistry. They are the basis of the best representation of the benzene molecule devised in the 19th century, and they mark the beginning of our understanding of structure in aromatic compounds.

Kekulé's formulas do have one serious shortcoming in that they represent benzene and related substances as highly unsaturated compounds. Yet benzene does not react like a typical alkene (olefin); it does not decolorize bromine solutions rapidly, nor does it react with and destroy the purple color of permanganate ions (Baeyer's test). Instead, the chemical behavior of benzene resembles that of an alkane. Its typical reactions are the substitution type, wherein a hydrogen atom is replaced by some other group; for example,

$$C_6H_6 + Cl_2 \xrightarrow{\text{Fe}} C_6H_5Cl + HCl$$

This problem was not fully resolved until the technique of X-ray diffraction, which was developed in the years following 1912, permitted us to determine the actual distances between the nuclei of carbon atoms in molecules. The center-to-center carbon atom distances in different kinds of hydrocarbon molecules are

Ethane (single bond)	1.54 Å
Ethylene (double bond)	1.34 Å
Benzene	1.40 Å

Since only one carbon–carbon distance (bond length) is found in benzene, it is apparent that alternating single and double bonds do not exist in the benzene molecule. Modern theory suggests that the benzene molecule is a hybrid of the two Kekulé structures shown above.

For convenience, the present-day chemist usually writes the structure of benzene as one or the other of these abbreviated forms:

A   B

In both representations, it is understood that there is a carbon atom and a hydrogen atom at each corner of the hexagon. The classical Kekulé structure is represented by A; the modern molecular orbital structure is represented by B. Hexagons are used in representing the structural formulas of benzene derivatives—that is, substances in which one or more hydrogen atoms in the ring have been replaced by other atoms or groups. Chlorobenzene (C_6H_5Cl), for example, is written in this fashion:

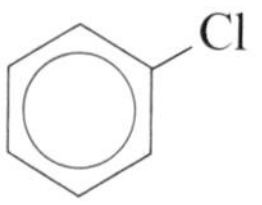

Chlorobenzene, C_6H_5Cl

This notation indicates that the chlorine atom has replaced a hydrogen atom and is bonded directly to a carbon atom in the ring. Thus, the correct formula for chlorobenzene is C_6H_5Cl, not C_6H_6Cl.

19.17 Naming Aromatic Compounds

A substituted benzene is derived by replacing one or more hydrogen atoms of benzene by another atom or group of atoms. Thus, a monosubstituted benzene has the formula C_6H_5G, where G is the group replacing a hydrogen atom.

Monosubstituted benzenes.   Some monosubstituted benzenes are named by adding the name of the substituent group as a prefix to the word *benzene.* The name is written as one word. Several examples follow:

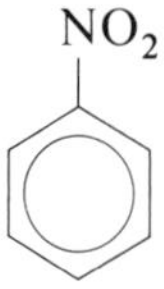

Nitrobenzene 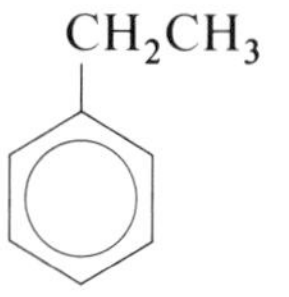   Ethylbenzene 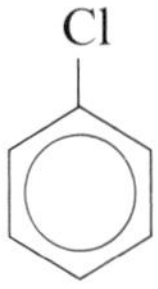   Chlorobenzene 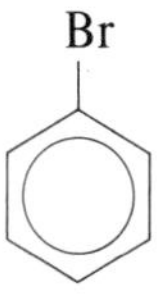   Bromobenzene

Certain monosubstituted benzenes have special names. These are used as parent names for further substituted compounds, so they should be learned.

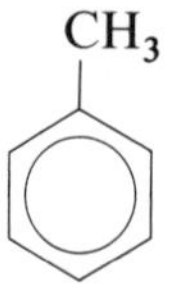

Toluene
(Methylbenzene)

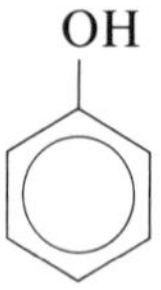

Phenol
(Hydroxybenzene)

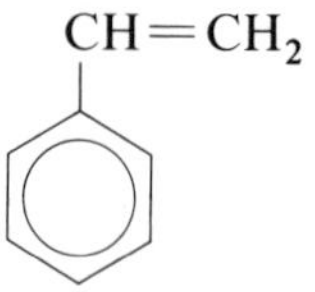

Styrene
(Vinylbenzene)

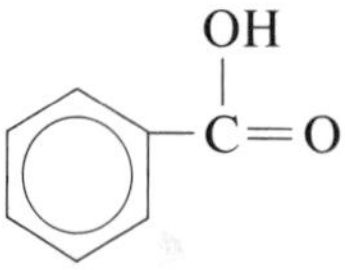

Benzoic acid
(Benzene carboxylic acid)

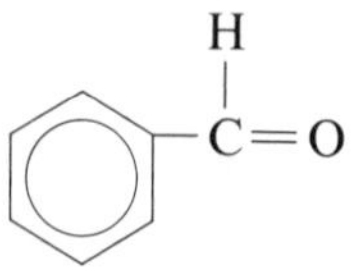

Benzaldehyde
(Benzene carboxaldehyde)

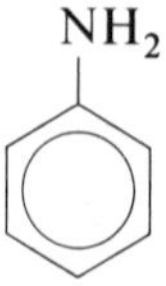

Aniline

All the hydrogen atoms in benzene are equivalent; therefore, it does not matter at which corner of the ring the substituted group is located. Each one of the following formulas represents chlorobenzene:

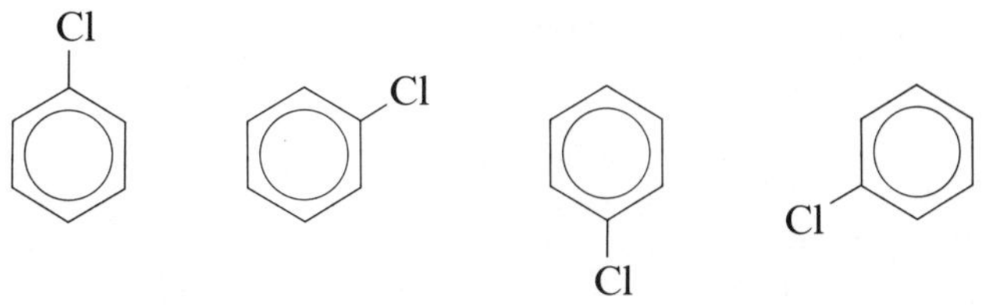

The C_6H_5 — group is known as the phenyl group (pronounced *fen-ill*), and the name *phenyl* is used for compounds that cannot easily be named as benzene derivatives. For example, the following compounds are named as derivatives of alkanes:

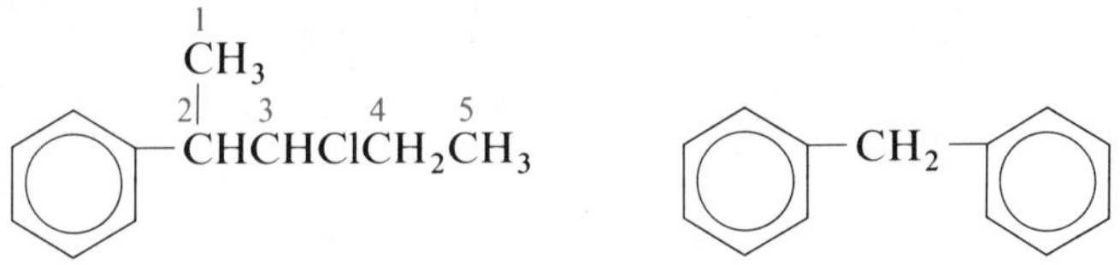

3-Chloro-2-phenylpentane   Diphenylmethane

Disubstituted benzenes.   When two substituent groups replace two hydrogen atoms in a benzene molecule, three different isomeric compounds are possible. The prefixes *ortho-*, *meta-*, and *para-* (abbreviated *o-*, *m-*, and *p-*) are used to name these disubstituted benzenes. In the ortho compound the substituents are located on adjacent carbon atoms. In the meta compound they are one carbon apart. And in the para compound the substituents are located on opposite sides of the ring:

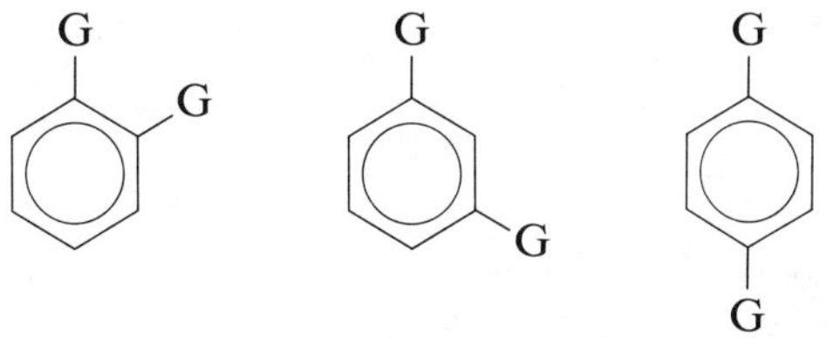

Ortho   Meta   Para

Let us examine the dichlorobenzenes, $C_6H_4Cl_2$, shown at the top of the next page. Note that the three isomers have different physical properties, indicating that they are truly different substances. Note that the para isomer is a solid and the two others are liquids at room temperature.

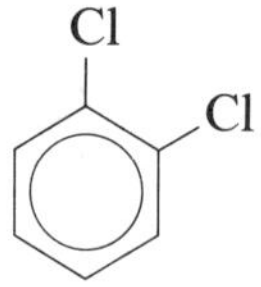

ortho-Dichlorobenzene
(1,2-Dichlorobenzene)
mp −17.2°C, bp 180.4°C

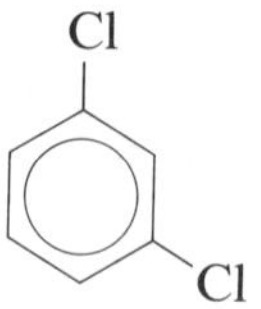

meta-Dichlorobenzene
(1,3-Dichlorobenzene)
mp −24.8°C, bp 172°C

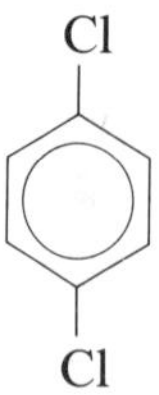

para-Dichlorobenzene
(1,4-Dichlorobenzene)
mp 53.1°C, bp 174.4°C

The dimethylbenzenes have the special name *xylene*:

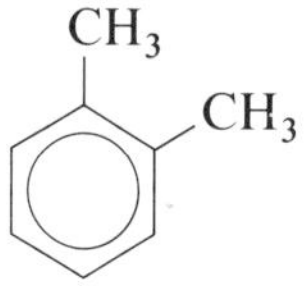

ortho-Xylene

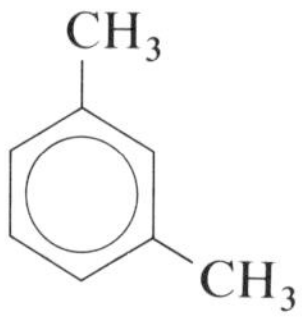

meta-Xylene

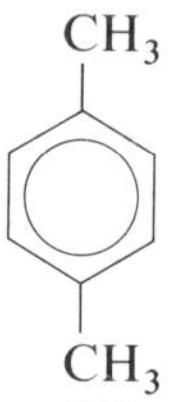

para-Xylene

When one of the substituents corresponds to a monosubstituted benzene that has a special name, the disubstituted compound is named as a derivative of that parent compound. In the following examples the parent compounds are phenol, aniline, and toluene:

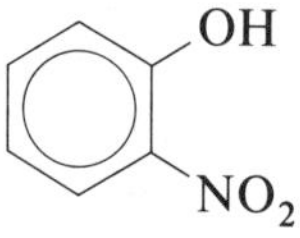

o-Nitrophenol

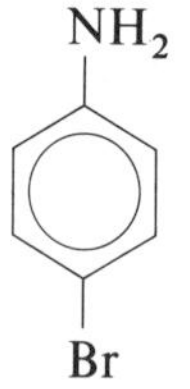

p-Bromoaniline

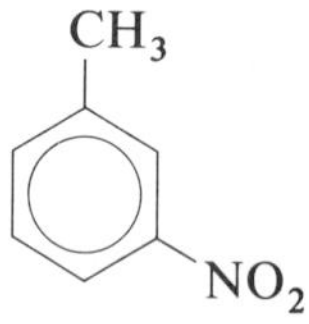

m-Nitrotoluene

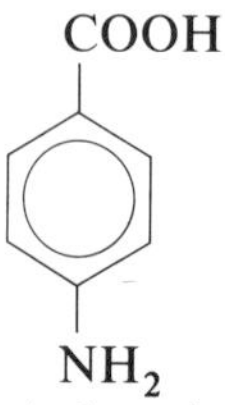

p-Aminobenzoic acid

Polysubstituted benzenes. When there are more than two substituents on a benzene ring the carbon atoms in the ring are numbered. Numbering starts at one of the substituted groups and may go either clockwise or counterclockwise, but must be done in the direction that gives the lowest possible numbers to the substituent groups. When the compound is named as a derivative of one of the special parent compounds, the substituent of the parent compound is considered to be on carbon 1 of the ring (the CH_3 group is on carbon 1 in 2,4,6-tribromotoluene). The following examples illustrate this system:

1,3,5-Trinitrobenzene　　1,2,4-Tribromobenzene　　2,4,6-Trinitrotoluene　　5-Bromo-2-chlorophenol
　　　　　　　　　　　　　　(not 1,4,6-)　　　　　　　　(TNT)

PROBLEM 19.6　Write formulas and names for all the possible isomers of (a) chloronitrobenzene [$C_6H_4Cl(NO_2)$] and (b) tribromobenzene ($C_6H_3Br_3$).

(a) The name and formula indicate a chloro group (Cl) and a nitro group (NO_2) attached to a benzene ring. There are six positions in which to place these two groups. They can be ortho, meta, or para to each other.

　　　o-Chloronitrobenzene　　　　m-Chloronitrobenzene　　　p-Chloronitrobenzene

(b) For tribromobenzene start by placing the three bromo groups in the 1, 2, and 3 positions; then the 1, 2, and 4 positions; and so on until all the possible isomers are formed. Name each isomer to check that no duplicate formulas have been written.

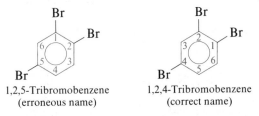

There are only three isomers of tribromobenzene. If one erroneously writes the 1,2,5 compound, a further check will show that by numbering the ring as indicated it is in reality the 1,2,4 isomer.

19.18 Fused Aromatic Rings

fused aromatic ring system

There are many other aromatic ring systems besides benzene. Their structures consist of two or more rings in which two carbon atoms are common to two rings. These compounds are known as **fused aromatic ring systems.** Three of the most common hydrocarbons in this category are naphthalene, anthracene, and phenanthrene. As you can see in the following structures, one hydrogen is attached to each carbon atom except at the carbons that are common to two rings.

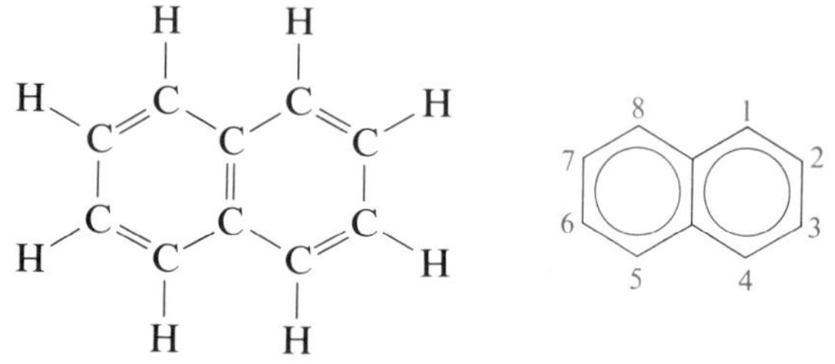

Naphthalene, $C_{10}H_8$

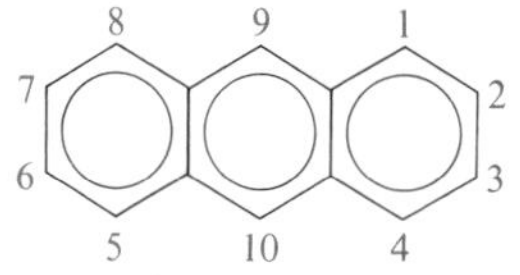

Anthracene, $C_{14}H_{10}$ 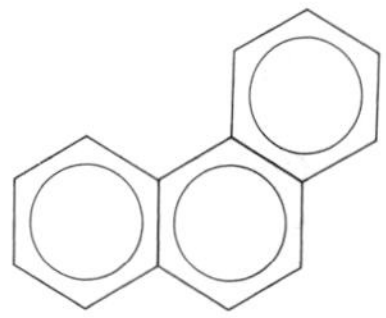   Phenanthrene, $C_{14}H_{10}$

All three of these substances may be obtained from coal tar. Naphthalene is known as mothballs and has been used as a moth repellent for many years.

19.19 Reactions of Aromatic Hydrocarbons

The most characteristic reactions of aromatic hydrocarbons involve the substitution of some group for a hydrogen on one of the ring carbons. The following are examples of typical aromatic substitution reactions. In each of these reactions a functional group is substituted for a hydrogen atom.

1. **Halogenation** (chlorination or bromination):

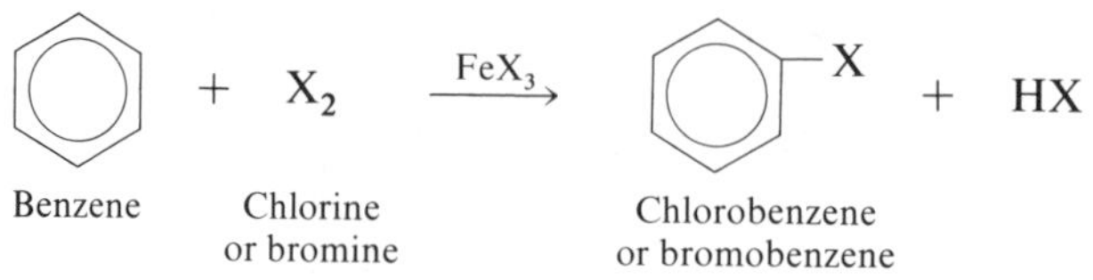

Benzene   Chlorine   Chlorobenzene
or bromine   or bromobenzene

2. Nitration:

Benzene + Nitric acid $\xrightarrow{H_2SO_4}$ Nitrobenzene + H_2O

3. Alkylation (Friedel–Craft reaction):

Benzene + Chloroethane $\xrightarrow{AlCl_3}$ Ethylbenzene + HCl

Aromatic substitution products of hydrocarbons are valuable in themselves and also serve as intermediates in the manufacture of other products. Thus, benzene can be nitrated to introduce a nitro group into the molecule. The product, nitrobenzene, is capable of a variety of reactions. Numerous drugs and dyes can be made by syntheses starting with nitrobenzene. These syntheses usually involve several steps. As an illustration, acetanilide, which is used in some pain- and fever-reducing drugs, can be made by the following sequence of reactions:

Nitrobenzene $\xrightarrow[HCl]{Sn}$ Aniline $\xrightarrow[\text{Acetic anhydride}]{(CH_3CO)_2O}$ Acetanilide

Organic compounds containing nitro groups are toxic and may also have explosive properties. Therefore, special safety precautions must be taken when handling nitro compounds. A number of nitro compounds are synthesized specifically for use as explosives. For example, the high explosive trinitrotoluene, or TNT, is made by nitrating toluene:

Toluene $\xrightarrow[\Delta]{HNO_3/H_2SO_4}$ 2,4,6-Trinitrotoluene (TNT)

19.20 Alcohols

The compounds derived from the alkane hydrocarbons by substituting an —OH group for a H atom are called *alcohols*. The functional group of the alcohols is —OH. The general formula for alcohols is ROH.

Alcohols differ from metal hydroxides in that they do not dissociate or ionize in water. The —OH group is attached to the carbon atom by a covalent bond and not by an ionic bond. The alcohols form a homologous series, with methanol, CH_3OH, the first member of the series. Models illustrating the structural arrangements of the atoms in methanol and ethanol are shown in Figure 19.5.

primary alcohol
secondary alcohol
tertiary alcohol

Alcohols are classified as **primary** (1°), **secondary** (2°), or **tertiary** (3°), depending on whether the carbon atom to which the —OH group is attached is bonded to one, two, or three other carbon atoms, respectively. Generalized formulas for 1°, 2°, and 3° alcohols are as follows:

H	R	R
|	|	|
R—C—OH	R—C—OH	R—C—OH
|	|	|
H	H	R
Primary alcohol	Secondary alcohol	Tertiary alcohol

Formulas of specific examples of these classes of alcohols are shown in Table 19.8. Methanol (CH_3OH) is grouped with the primary alcohols.

Molecular structures with more than one —OH group attached to a single carbon atom are generally not stable. But an alcohol molecule may contain two

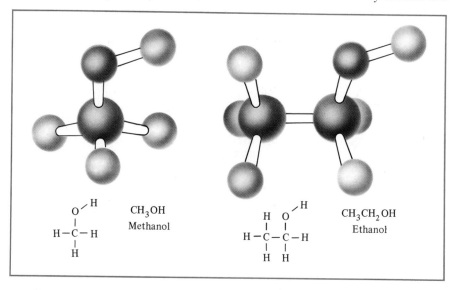

Figure 19.5   Ball-and-stick models illustrating structural formulas of methanol and ethanol.

TABLE 19.8 Naming and classification of alcohols.

Class	Formula	IUPAC name	Common name[a]	Boiling point, °C
Primary	CH_3OH	Methanol	Methyl alcohol	65.0
Primary	CH_3CH_2OH	Ethanol	Ethyl alcohol	78.5
Primary	$CH_3CH_2CH_2OH$	1-Propanol	n-Propyl alcohol	97.4
Primary	$CH_3CH_2CH_2CH_2OH$	1-Butanol	n-Butyl alcohol	118
Primary	$CH_3(CH_2)_3CH_2OH$	1-Pentanol	n-Amyl or n-pentyl alcohol	138
Primary	$CH_3(CH_2)_6CH_2OH$	1-Octanol	n-Octyl alcohol	195
Primary	CH_3CHCH_2OH | CH_3	2-Methyl-1-propanol	Isobutyl alcohol	108
Secondary	CH_3CHCH_3 | OH	2-Propanol	Isopropyl alcohol	82.5
Secondary	$CH_3CH_2CHCH_3$ | OH	2-Butanol	sec-Butyl alcohol	91.5
Tertiary	CH_3 | CH_3-C-OH | CH_3	2-Methyl-2-propanol	t-Butyl alcohol	82.9
Dihydroxy	$HOCH_2CH_2OH$	1,2-Ethanediol	Ethylene glycol	197
Trihydroxy	$HOCH_2CHCH_2OH$ | OH	1,2,3-Propanetriol	Glycerol or glycerine	290

[a]The abbreviations n, sec, and t stand for normal, secondary, and tertiary, respectively.

or more —OH groups if each —OH is attached to a different carbon atom. Accordingly, alcohols are further classified as monohydroxy, dihydroxy, trihydroxy, and so on, on the basis of the number of hydroxyl groups per molecule. **Polyhydroxy alcohols** and polyols are general terms for alcohols that have more than one —OH group per molecule.

polyhydroxy alcohol

Methanol. When wood is heated to a high temperature in an atmosphere lacking oxygen, methanol (or wood alcohol) and other products are formed and driven off. The process is called *destructive distillation*, and until about 1925, nearly all methanol was obtained in this way. In the early 1920s the synthesis of methanol by high-pressure catalytic hydrogenation of carbon monoxide was developed in Germany. The reaction is

$$\text{CO} + 2\,\text{H}_2 \xrightarrow[\text{300-400°C, 200 atm}]{\text{ZnO-Cr}_2\text{O}_3} \text{CH}_3\text{OH}$$

Nearly all methanol is now manufactured by this method.

Methanol is a volatile (bp 65°C), highly flammable liquid. It is poisonous and capable of causing blindness or death if taken internally. Exposure to methanol vapors for even short periods of time is dangerous. Despite these disadvantages, large quantities (over 6 billion pounds annually) are manufactured and used for

1. Conversion to formaldehyde (methanal) used primarily in the manufacture of polymers
2. Manufacture of other chemicals, especially various kinds of esters
3. Denaturing or rendering ethyl alcohol unfit for use as a beverage
4. An industrial solvent

The experimental use of 5 to 30% methanol in gasoline (gasahol) has shown promising results in reducing the amount of air pollutants emitted in automobile exhausts. Another important plus factor for using methanol in gasoline is that methanol can be made from nonpetroleum sources. The most economical non-petroleum source of carbon monoxide for methanol is coal. In addition to coal, burnable materials such as wood, agricultural wastes, and sewage sludge are potential sources of methanol.

Ethanol.   Ethanol is without doubt the oldest and most widely known alcohol. It is or has been known by a variety of names, such as ethyl alcohol, "alcohol," grain alcohol, methyl carbinol, spirit, and *aqua vitae*. The preparation of ethanol by fermentation is recorded in the Old Testament. Today, huge quantities of this substance are still prepared by fermentation. Starch and sugar are the raw materials. Starch must first be converted to sugar by enzyme- or acid-catalyzed hydrolysis. Conversion of simple sugars to ethanol is accomplished by the yeast enzyme zymase:

$$\underset{\text{Glucose}}{\text{C}_6\text{H}_{12}\text{O}_6} \xrightarrow{\text{Zymase}} \underset{\text{Ethanol}}{2\,\text{CH}_3\text{CH}_2\text{OH}} + 2\,\text{CO}_2$$

For legal use in beverages, ethanol must be made by fermentation; but a large part of the alcohol intended for industrial uses (about 1.3 billion pounds annually) is made from petroleum-derived ethylene. Some of the economically significant uses of ethanol are

1. Intermediate in the manufacture of other chemicals such as acetaldehyde, acetic acid, ethyl acetate, and diethyl ether
2. Solvent for many organic substances
3. Compounding ingredient for pharmaceuticals, perfumes, flavorings, and so on
4. Essential ingredient of alcoholic beverages

Physiologically, ethanol acts as a food, as a drug, and as a poison. It is a food in the limited sense that the body is able to metabolize small amounts of it to carbon dioxide and water with the production of energy. As a drug, ethanol is often mistakenly considered to be a stimulant, but it is in fact a depressant. In moderate quantities, ethanol causes drowsiness and depresses brain functions so that activities requiring skill and judgment (such as automobile driving) are impaired. In larger quantities, ethanol causes nausea, vomiting, impaired perception, and incoordination. If a very large amount is consumed, complete unconsciousness and, ultimately, death may occur.

Authorities maintain that the effects of ethanol on automobile drivers are a factor in about half of all fatal traffic accidents in the United States. The gravity of this problem can be grasped when you realize that traffic accidents are responsible for more than 50,000 deaths each year in the United States.

19.21 Naming Alcohols

If you know how to name alkanes, it is easy to name alcohols by the IUPAC System. Unfortunately, several of the alcohols are generally known by common or nonsystematic names, so it is often necessary to know more than one name for a given alcohol. The common name is usually formed from the name of the organic radical, which is attached to the $-OH$ group, followed by the word *alcohol*. See the examples that follow and Table 19.8. To name an alcohol by the IUPAC System,

1. Select the longest continuous chain of carbon atoms containing the hydroxyl group.
2. Number the carbon atoms in this chain so that the one bearing the $-OH$ group has the lowest possible number.
3. Form the alcohol name by dropping the final *-e* from the corresponding alkane name and adding *-ol*. Locate the $-OH$ group by putting the number (hyphenated) of the carbon atom to which it is attached immediately before the alcohol name.
4. Name each alkyl side chain (or other group) and designate its position by number.

Study the application of this naming system to these examples and to those shown in Table 19.8.

$$CH_3 - CH_2 - CH_2OH$$
1-Propanol
(*n*-Propyl alcohol)

$$CH_3 - CH - CH_3$$
$$|$$
$$OH$$
2-Propanol
(Isopropyl alcohol)

$$CH_3 - CH - CH_2 - CH_2OH$$
$$|$$
$$CH_3$$
3-Methyl-1-butanol
(Isoamyl alcohol or isopentyl alcohol)

$$HOCH_2 - CH_2OH$$
1,2-Ethanediol
(Ethylene glycol)

PROBLEM 19.7   Name this alcohol by the IUPAC method:

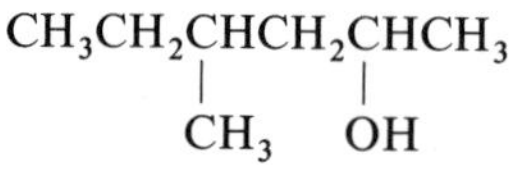

$$CH_3CH_2CHCH_2CHCH_3$$

with substituents CH_3 and OH below

Step 1. There are six carbon atoms in the longest continuous carbon chain containing the —OH group.

Step 2. This carbon chain is numbered from right to left so that the —OH group has the smallest possible number:

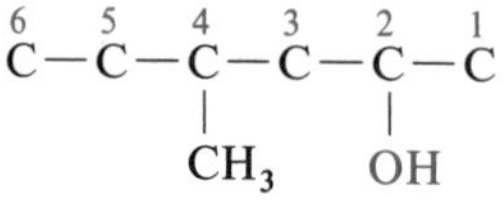

$$\overset{6}{C}-\overset{5}{C}-\overset{4}{C}-\overset{3}{C}-\overset{2}{C}-\overset{1}{C}$$

with CH_3 on carbon 4 and OH on carbon 2

In this case, the —OH is on carbon number 2.

Step 3. The alcohol name is derived from the six-carbon hydrocarbon hexane by dropping the -*e* and adding -*ol*. Thus, the alcohol name is 2-hexanol since the —OH group is on carbon 2.

Step 4. A methyl group (—CH_3) is located on carbon 4. Therefore, the full name of the compound is 4-methyl-2-hexanol.

■

PROBLEM 19.8   Write the structural formula of 3,3-dimethyl-2-hexanol.

Step 1. The 2-hexanol refers to a six-carbon chain with an —OH group on carbon number 2. Write the skeleton structure as follows:

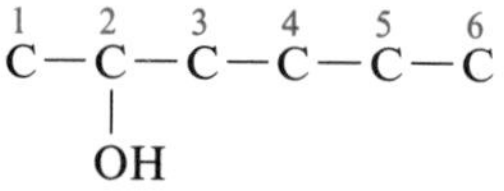

$$\overset{1}{C}-\overset{2}{C}-\overset{3}{C}-\overset{4}{C}-\overset{5}{C}-\overset{6}{C}$$

with OH on carbon 2

Step 2. Place the two methyl groups (3,3-dimethyl) on carbon number 3:

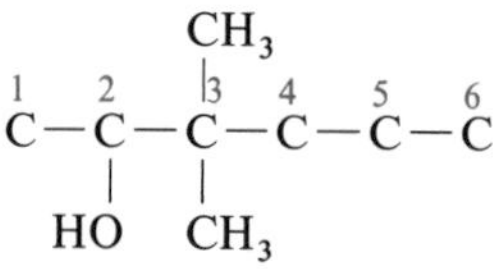

$$\overset{1}{C}-\overset{2}{C}-\overset{3}{C}-\overset{4}{C}-\overset{5}{C}-\overset{6}{C}$$

with CH_3 above carbon 3, HO below carbon 2, and CH_3 below carbon 3

Step 3. Finally, add H atoms to give each carbon atom four bonds:

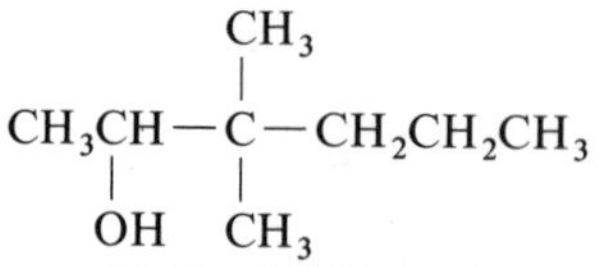

$$CH_3CH-\underset{|}{\overset{|}{C}}-CH_2CH_2CH_3$$

with CH_3 above the central carbon, OH below the first carbon, and CH_3 below the central carbon

3,3-Dimethyl-2-hexanol

■

19.22 Reactions of Alcohols

Oxidation. Alcohols may be oxidized to aldehydes, ketones, and carboxylic acids. Primary alcohols yield aldehydes and carboxylic acids; secondary alcohols are oxidized to ketones; tertiary alcohols resist oxidation.

Primary alcohol $\xrightarrow{\text{Oxidation}}$ Aldehyde $\xrightarrow{\text{Oxidation}}$ Carboxylic acid

Secondary alcohol $\xrightarrow{\text{Oxidation}}$ Ketone

Tertiary alcohol $\xrightarrow{\text{Oxidation}}$ No reaction

$$CH_3CH_2OH \xrightarrow[\text{or KMnO}_4]{\text{H}^+, \text{K}_2\text{Cr}_2\text{O}_7} CH_3\overset{\displaystyle H}{\underset{}{C}}=O \xrightarrow[\text{or KMnO}_4]{\text{H}^+, \text{K}_2\text{Cr}_2\text{O}_7} CH_3COOH$$

Ethanol   Ethanal   Acetic acid
               (an aldehyde)   (a carboxylic acid)

$$CH_3\underset{\underset{\text{OH}}{|}}{CH}CH_3 \xrightarrow[\text{or KMnO}_4]{\text{H}^+, \text{K}_2\text{Cr}_2\text{O}_7} CH_3\underset{\underset{\text{O}}{||}}{C}CH_3$$

2-Propanol   Propanone (Acetone)
                 (a ketone)

Dehydration. The term *dehydration* implies the elimination of water. Alcohols can be dehydrated to form alkenes or ethers. One of the more effective dehydrating reagents is sulfuric acid. Whether an ether or an alkene is formed depends on the ratio of alcohol to sulfuric acid, the reaction temperature, and the type of alcohol. With primary alcohols dehydration to an ether occurs with excess alcohol and a temperature lower than that required to dehydrate the alcohol to an alkene.

When ethanol and concentrated sulfuric acid are heated to 140°C, one molecule of water is eliminated per two molecules of alcohol to form diethyl ether. At 180°C the dehydration occurs within individual molecules and ethylene is formed. The equations are

$$\begin{matrix} CH_3CH_2O\boxed{H} \\ CH_3CH_2\boxed{OH} \end{matrix} \xrightarrow[140°C]{96\% \text{ H}_2\text{SO}_4} CH_3CH_2OCH_2CH_3 + H_2O$$

Diethyl ether

$$H-\underset{\underset{\boxed{\text{H}}}{|}}{\overset{\overset{\text{H}}{|}}{C}}-\underset{\underset{\boxed{\text{OH}}}{|}}{\overset{\overset{\text{H}}{|}}{C}}-H \xrightarrow[180°]{96\% \text{ H}_2\text{SO}_4} CH_2=CH_2 + H_2O$$

Ethylene

The dehydration to ethers is useful only for primary alcohols. Dehydration of secondary and tertiary alcohols yields predominantly the most highly substituted alkene. Thus, the dehydration of 2-butanol gives chiefly 2-butene and the dehydration of 2-methyl-2-butanol gives mainly 2-methyl-2-butene.

$$CH_3CH_2CHCH_3 \xrightarrow[100°]{60\% \ H_2SO_4} CH_3CH=CHCH_3 + CH_3CH_2CH=CH_2 + H_2O$$

$$\underset{OH}{|}$$

2-Butanol

2-Butene
(chief product)

1-Butene

$$CH_3CH_2\overset{\overset{\displaystyle CH_3}{|}}{C}CH_3 \xrightarrow[90°]{45\% \ H_2SO_4} CH_3CH=\overset{\overset{\displaystyle CH_3}{|}}{C}CH_3 + CH_3CH_2\overset{\overset{\displaystyle CH_3}{|}}{C}=CH_2 + H_2O$$

$$\underset{OH}{|}$$

2-Methyl-2-butanol

2-Methyl-2-butene
(chief product)

2-Methyl-1-butene

Reaction with active metals.   Although alcohols are less acidic than water, they readily react with active metals such as sodium and potassium to replace the hydrogen atom of the alcohol —OH group. The products formed are a salt and hydrogen gas. The salt formed is an ionic compound and is called an *alkoxide* (CH_3CH_2ONa is sodium ethoxide). The equation is

$$2 \, ROH + 2 \, Na \longrightarrow 2 \, RONa + H_2(g)$$
Alkoxide

$$2 \, CH_3CH_2OH + 2 \, Na \longrightarrow 2 \, CH_3CH_2ONa + H_2(g)$$
Ethanol   Sodium ethoxide

Methanol and ethanol react quite vigorously with sodium, but not as vigorously as sodium and water. Alcohols are less acidic than water. The reactivity of an alcohol decreases as its molecular weight increases. This is because the —OH group becomes a smaller and relatively less significant part of the molecule.

19.23 Phenols

phenol

The term **phenol** is used for the class of compounds that have a hydroxy group attached to an aromatic ring. The parent compound is called *phenol* (C_6H_5OH) and is also known as *carbolic acid.*

In the pure state phenol is a colorless crystalline solid with a melting point of about 41°C and a characteristic odor. Phenol is highly poisonous. Ingestion of even small amounts may cause nausea, vomiting, circulatory collapse, and death from respiratory failure.

Phenol is a weak acid—more acidic than alcohols and water, but less acidic than acetic and carbonic acids. The pH values are as follows: acetic acid (0.1 M), 2.87; water, 7.0; phenol (0.1 M), 5.5 Thus, phenol will react with sodium hydroxide solution to form a salt but will not react with sodium bicarbonate. The salt formed is called sodium phenoxide or sodium phenolate. Sodium hydroxide will not remove a hydrogen atom from an alcohol, since alcohols are weaker acids than water.

Phenol + NaOH ⟶ Sodium phenoxide (Sodium phenolate) + H₂O

In general, the phenols are toxic to microorganisms. They are widely used as antiseptics and disinfectants. Phenol was the first compound to be used extensively as an operating room disinfectant. Joseph Lister (1827–1912) first used phenol for this purpose in 1867.

Many phenols are named as derivatives of the parent compound using the general methods for naming aromatic compounds. For example:

Phenol   *m*-Bromophenol   *p*-Aminophenol   2,4,6-Trinitrophenol (Picric acid)

The ortho, meta, and para methylphenols are present in coal tar and are known as *cresols*. They are all useful disinfectants.

o-Cresol (*o*-Methylphenol)   *m*-Cresol (*m*-Methylphenol)   *p*-Cresol (*p*-Methylphenol)

Many well-known natural substances have phenolic groups in their structures. Several examples follow.

Vanillin   Eugenol   Thymol

Vanillin is the principal odorous component of the vanilla bean. It is one of the most widely used flavorings and is also used for masking undesirable odors in many products such as paints. Eugenol is the essence of oil of cloves. It is used as a dental analgesic and in the manufacture of synthetic vanillin. Thymol occurs in the oil of thyme. It has a fairly pleasant odor and flavor and is used as an antiseptic in many preparations such as mouthwashes. Thymol is the starting material for the synthesis of menthol, the main constituent of oil of peppermint. Thymol is a widely used flavoring and pharmaceutical.

The common acid–base indicator phenolphthalein is a phenol derivative. Phenolphthalein is also used as a medical cathartic. Epinephrine (adrenaline) is a substance that is secreted by the adrenal gland; it stimulates the conversion of glycogen to glucose in the body as a result of stress, fear, anger, or other heightened emotional states. It is also used medicinally as a stimulant in cardiac arrest.

Phenolphthalein

HO—CHCH$_2$NHCH$_3$
Epinephrine
(Adrenaline)

19.24 Ethers

ether

Ethers have the general formula ROR′. The radicals, R and R′, may be derived from saturated, unsaturated, or aromatic hydrocarbons, and for a given ether may be alike or different. Table 19.9 shows structural formulas and names for some of the different kinds of ethers.

Ethers have little chemical reactivity, but since a great many organic substances dissolve readily in ethers, they are often used as solvents in both laboratory and manufacturing operations.

Alcohols (ROH) and ethers (ROR′) are isomeric, having the same molecular formula but different structural formulas. For example, the molecular formula for ethanol and dimethyl ether is C$_2$H$_6$O but the structural formulas are

CH$_3$CH$_2$OH   CH$_3$—O—CH$_3$
Ethanol   Dimethyl ether

These two molecules are extremely different in both physical and chemical properties. Ethanol boils at 78.3°C, and dimethyl ether boils at −23.7°C. Ethanol has a greater solubility in water than dimethyl ether. Ethanol reacts with metallic sodium to give hydrogen and sodium ethylate and can also be oxidized by dichromate and permanganate solutions. Dimethyl ether is nonreactive with all these reagents.

TABLE 19.9 Names and structural formulas of ethers.

Name[a]	Formula	Boiling point (°C)
Dimethyl ether (Methoxymethane)	CH_3-O-CH_3	−24
Methyl ethyl ether (Methoxyethane)	$CH_3CH_2-O-CH_3$	8
Diethyl ether (Ethoxyethane)	$CH_3CH_2-O-CH_2CH_3$	35
Ethyl isopropyl ether (2-Ethoxypropane)	$CH_3CH_2-O-\underset{\underset{CH_3}{\vert}}{C}HCH_3$	54
Divinyl ether	$CH_2=CH-O-CH=CH_2$	39
Anisole (Methoxybenzene)	⬡—OCH_3	154

[a]The IUPAC name is in parentheses when given.

19.25 Naming Ethers

Individual ethers, like alcohols, may be known by several names. The ether having the formula $CH_3CH_2-O-CH_2CH_3$ and formerly widely used as an anesthetic is called diethyl ether, ethyl ether, ethoxyethane, or simply *ether*. Common names of ethers are formed from the names of the groups attached to the oxygen atom followed by the word *ether*.

$$CH_3 \longrightarrow \boxed{O} \longrightarrow CH_3 \qquad\qquad CH_3 \longrightarrow \boxed{O} \longrightarrow CH_2CH_3$$

↑	↑	↑	↑	↑	↑
Methyl	Ether	Methyl	Methyl	Ether	Ethyl

Dimethyl ether   Methyl ethyl ether

In the IUPAC System, ethers are named as alkoxy $(RO-)$ derivatives of the alkane corresponding to the longest carbon–carbon chain in the molecule. To name an ether by this system,

1. Select the longest carbon–carbon chain and label it with the name of the corresponding alkane.
2. Change the *-yl* ending of the other hydrocarbon radical to *-oxy* to obtain the alkoxy radical name. For example, CH_3O- is called *methoxy*.
3. Combine the two names from Steps 1 and 2, giving the alkoxy name first, to form the ether name, as diagrammed at the top of the next page.

$$CH_3 - O - CH_2CH_3$$

This is the longest C—C chain, so call it *ethane.*

This is the other radical; modify its name to *methoxy* and combine with ethane to obtain the name of the ether, *methoxyethane.*

Thus,

$CH_3CH_2 - O - CH_2CH_3$ is ethoxyethane;

$CH_3CH_2CH_2 - O - CH_2CH_2CH_2CH_3$ is *n*-propoxybutane.

19.26 Preparation of Ethers

We have seen that ethers can be made by intermolecular dehydration of alcohols by heating in the presence of an acid (see Section 19.20, under Dehydration). Ethers are also made from sodium alkoxides or sodium phenoxides and alkyl halides (Williamson synthesis).

$$\underset{\text{Sodium alkoxide}}{RONa} + \underset{\text{Alkyl halide}}{R'X} \longrightarrow \underset{\text{Ether}}{ROR'} + \underset{\text{Sodium halide}}{NaX}$$

The Williamson synthesis is especially useful in the preparation of mixed ethers (where $R \neq R'$) and aromatic ethers.

$$\underset{\substack{\text{Sodium}\\\text{ethoxide}}}{CH_3CH_2ONa} + \underset{\substack{\text{Bromo-}\\\text{methane}}}{CH_3Br} \longrightarrow \underset{\substack{\text{Methoxyethane}\\\text{(Methyl ethyl ether)}}}{CH_3CH_2 - O - CH_3} + NaBr$$

Sodium phenoxide + Bromomethane → Methyl phenyl ether (Anisole) + NaBr

19.27 Aldehydes and Ketones

carbonyl group

aldehyde

ketone

The aldehydes and ketones are closely related classes of compounds. The structure of both aldehydes and ketones contains the **carbonyl group,** $>C=O$, a carbon double-bonded oxygen. **Aldehydes** have at least one hydrogen atom bonded to the carbonyl group, whereas **ketones** have only alkyl or aryl (aromatic, Ar) groups bonded to the carbonyl group.

Aldehydes

$$\underset{}{R - \overset{\overset{\displaystyle H}{|}}{C}=O} \qquad Ar - \overset{\overset{\displaystyle H}{|}}{C}=O$$

Ketones

$$R - \overset{\displaystyle ||}{\underset{O}{C}} - R \qquad R - \overset{\displaystyle ||}{\underset{O}{C}} - Ar \qquad Ar - \overset{\displaystyle ||}{\underset{O}{C}} - Ar$$

The aldehyde group is often written as CHO or CH=O. In a linear expression of a ketone, the carbonyl group is written as CO; for example,

$$CH_3COCH_3 \quad \text{is equivalent to} \quad CH_3\underset{\underset{O}{\|}}{C}CH_3$$

Formaldehyde is the most widely used aldehyde. It is a poisonous, irritating gas, which is very soluble in water. It is marketed as a 40% aqueous solution called *formalin*. Since formaldehyde is a powerful germicide, it is used in embalming and to preserve biological specimens. Formaldehyde is also used for disinfecting dwellings, ships, and storage houses; for destroying flies; for tanning hides; and as a fungicide for plants and vegetables. But by far the largest use of this chemical is in the manufacture of polymers. About 2.4 billion pounds (1.1×10^9 kg) of formaldehyde are manufactured annually in the United States.

Formaldehyde vapors are intensely irritating to the mucous membranes. Ingestion may cause severe abdominal pains, leading to coma and death.

Acetone and methyl ethyl ketone are widely used organic solvents. Acetone, in particular, is used in very large quantities for this purpose. United States production of acetone is about 2.5 billion pounds annually. It is used as a solvent in the manufacture of drugs, chemicals, and explosives; for removal of paints, varnishes, and fingernail polish; and as a solvent in the plastics industry. Methyl ethyl ketone (MEK) is also widely used as a solvent, especially for lacquers.

19.28 Naming Aldehydes and Ketones

The IUPAC names of aliphatic aldehydes are obtained by dropping the final -*e* and adding -*al* to the name of the parent hydrocarbon (that is, the longest carbon–carbon chain carrying the —CHO group). The aldehyde carbon is always at the end of the carbon chain and is understood to be carbon number 1. The first member of the homologous series, $H_2C{=}O$, is methanal. The name *methanal* is derived from the hydrocarbon methane, which contains one carbon atom. The second member of the series is ethanal; the third member of the series is propanal; and so on.

The common name of aliphatic aldehydes is obtained by dropping -*ic acid* and adding -*aldehyde* to the common name of the corresponding carboxylic acid; for example, $H_2C{=}O$ is formaldehyde (from formic acid, HCOOH). It must be emphasized that the common names of aldehydes are derived from the common names of the corresponding acids, not from IUPAC names. See Table 19.10 for the common names of acids.

$\overset{\mathrm{H}}{\underset{\mathrm{HC}=O}{|}}$	$\overset{\mathrm{H}}{\underset{CH_3C=O}{|}}$	$\overset{\mathrm{H}}{\underset{CH_3CH_2C=O}{|}}$	$\overset{\mathrm{H}}{\underset{CH_3CH_2CH_3C=O}{|}}$	$\overset{CH_3\ \ \mathrm{H}}{\underset{CH_3CH-C=O}{|\ \ \ \ |}}$
Methanal	Ethanal	Propanal	Butanal	2-Methylpropanal
(Formaldehyde)	(Acetaldehyde)	(Propionaldehyde)	(Butyraldehyde)	(Isobutyraldehyde)

The IUPAC name of a ketone is derived from the name of the alkane corresponding to the longest carbon chain that contains the carbonyl group. The name is formed by changing the -e ending of the alkane to -one. If the chain is longer than four carbons, it is numbered so that the carbonyl carbon has the smallest number possible, and this number is prefixed to the name of the ketone. See the following examples:

$$CH_3-\overset{\underset{\displaystyle O}{\|}}{C}-CH_3$$

Propanone
(Acetone, dimethyl ketone)

$$\overset{5}{C}H_3\overset{4}{C}H_2\overset{3}{C}H_2-\overset{\underset{\displaystyle O}{\|}}{\overset{2}{C}}-\overset{1}{C}H_3$$

2-Pentanone
(Methyl-n-propyl ketone)

$$\overset{1}{C}H_3\overset{2}{C}H_2-\overset{\underset{\displaystyle O}{\|}}{\overset{3}{C}}-\overset{4}{C}H\overset{5}{C}H_2\overset{6}{C}H_3$$
$$\underset{\displaystyle CH_3}{|}$$

4-Methyl-3-hexanone
(Ethyl-s-butyl ketone)

Note that in 4-methyl-3-hexanone the carbon chain is numbered from left to right to give the ketone group the lowest possible number.

An alternative non-IUPAC method commonly used to name simple ketones is to list the names of the alkyl or aromatic groups attached to the carbonyl carbon together with the word *ketone*. Thus, butanone ($CH_3COCH_2CH_3$) is methyl ethyl ketone:

$$CH_3 \underset{\underset{\displaystyle \text{Methyl}}{\uparrow}}{} \overset{\displaystyle O}{\overset{\displaystyle \|}{\underset{\underset{\displaystyle \text{Ketone}}{\uparrow}}{C}}} \underset{\underset{\displaystyle \text{Ethyl}}{\uparrow}}{} CH_2CH_3$$

Two of the most widely used ketones have special common names: propanone is commonly called acetone, and butanone is known as methyl ethyl ketone, or MEK.

Aromatic aldehydes are commonly named after the corresponding acids; for example, benzaldehyde from benzoic acid and *p*-tolualdehyde from *p*-toluic acid. Aromatic ketones are named in a similar fashion to aliphatic ketones and often have special names as well.

Benzaldehyde

p-Tolualdehyde

Methyl phenyl ketone
(acetophenone, 1-phenylethanone)

PROBLEM 19.9 Name and write the formulas for the straight-chain five- and six-carbon aldehydes.

The systematic names are based on the five- and six-carbon alkanes. Drop the -e of the alkane name and add the suffix -al. Pentane (C_5) becomes pentanal and hexane (C_6) becomes hexanal. The common names are derived from valeric acid and caproic acid, respectively.

$$\begin{matrix} & H & & & & H \\ & | & & & & | \\ CH_3CH_2CH_2CH_2C=O & & & & CH_3CH_2CH_2CH_2CH_2C=O \end{matrix}$$

Pentanal (Valeraldehyde)   Hexanal (Caproaldehyde)

■

PROBLEM 19.10   Give two names for each of the following ketones:

$$\begin{matrix} & & CH_3 \\ & & | \\ (a)\ CH_3CH_2CCH_2CHCH_3 & & (b)\ CH_3CH_2CH_2C{-}\bigcirc \\ & \underset{O}{\overset{||}{}} & & \quad \underset{O}{\overset{||}{}} \end{matrix}$$

(a) Number the longest carbon chain that includes the carbonyl group to give the carbonyl group the smallest number. The longest carbon chain has six carbons. The ketone group is on carbon 3 and a methyl group is on carbon 5. Drop the -e from hexane and add -one to give hexanone. Prefix the name hexanone with 3- to locate the ketone group and with 5-methyl- to locate the methyl group. The name is 5-methyl-3-hexanone. The common name is ethyl isobutyl ketone since the C=O has an ethyl group and an isobutyl group attached to it.

(b) The longest aliphatic chain has four carbon atoms, making the parent ketone a butanone by dropping the -e of butane and adding -one. The butanone has a phenyl group attached to carbon 1. The name is therefore 1-phenyl-1-butanone. The other name for this compound is phenyl n-propyl ketone, since the C=O group has a phenyl and an n-propyl group attached to it.

■

19.29 Reactions of Aldehydes and Ketones

Oxidation.   We have seen (Section 19.20) that aldehydes are easily oxidized to carboxylic acids. Ketones resist oxidation except under drastic conditions where carbon–carbon bonds are broken and a variety of products are formed.

Since aldehydes are easily oxidized they can readily be distinguished from ketones. The Tollens' or silver-mirror test for aldehydes is based on the ability of silver ions to oxidize aldehydes. The Ag^+ ions are thereby reduced to metallic silver. In practice, a little of the suspected aldehyde is added to a solution of silver nitrate and ammonia in a clean test tube. The appearance of a silver mirror on the inner wall of the tube is a positive test for the aldehyde group. The abbreviated equation is

$$\begin{matrix} H & & & O \\ | & & & || \\ RC=O + 2\ Ag^+ & \xrightarrow[H_2O]{NH_3} & RC{-}O^- \ NH_4^+ + 2\ Ag\downarrow & (General\ reaction) \end{matrix}$$

$$\begin{matrix} H \\ | \\ CH_3C=O + 2\ Ag^+ \xrightarrow[H_2O]{NH_3} CH_3COO^- \ NH_4^+ + 2\ Ag\downarrow \end{matrix}$$

Fehling's and Benedict's solutions contain Cu^{2+} ions in an alkaline medium. In the Fehling's and Benedict's tests, the aldehyde group is oxidized to a carboxylic acid by Cu^{2+} ions. The blue Cu^{2+} ions are reduced and form brick-red copper(I) oxide (Cu_2O), which precipitates during the reaction. These tests can be used for detecting carbohydrates that have an available aldehyde group. The abbreviated equation is

$$\underset{Blue}{\overset{\overset{\displaystyle H}{\displaystyle |}}{RC}=O + Cu^{2+}} \xrightarrow[H_2O]{NaOH} RCOO^-\ Na^+ + \underset{Brick\text{-}red}{Cu_2O\downarrow} \quad \text{(General reaction)}$$

Ketones do not give a positive test with Tollens', Fehling's, or Benedict's solutions.

Reduction.   Aldehydes and ketones are easily reduced to alcohols, either by elemental hydrogen in the presence of a catalyst or by chemical reducing agents such as lithium aluminum hydride ($LiAlH_4$) or sodium borohydride ($NaBH_4$). Aldehydes yield primary alcohols; ketones yield secondary alcohols:

$$\underset{Aldehyde}{\overset{\overset{\displaystyle H}{\displaystyle |}}{R-C}=O} \xrightarrow[\Delta]{H_2/Ni} \underset{1° \text{ alcohol}}{RCH_2OH} \quad \text{(General reaction)}$$

$$\underset{Ketone}{\overset{\displaystyle R-C-R}{\underset{\displaystyle O}{\displaystyle \|}}} \xrightarrow[\Delta]{H_2/Ni} \underset{2° \text{ alcohol}}{\overset{\displaystyle R-CH-R}{\underset{\displaystyle OH}{\displaystyle |}}} \quad \text{(General reaction)}$$

The carbonyl group undergoes a great variety of additional reactions. Although there are differences, aldehydes and ketones undergo many similar reactions. However, ketones are generally less reactive than aldehydes.

19.30 Carboxylic Acids and Esters

carboxylic acid
carboxyl group

Organic acids, known as **carboxylic acids,** are characterized by the functional group called a **carboxyl group.** The carboxyl group is represented in the following ways:

$$\overset{\overset{\displaystyle O}{\displaystyle \|}}{-C}-OH \quad \text{or} \quad -COOH \quad \text{or} \quad -CO_2H$$

Aliphatic carboxylic acids form a homologous series. The carboxyl group is always at the end of a carbon chain, and the C atom in this group is understood to be carbon number 1 in naming the compound.

To name a carboxylic acid by the IUPAC System, first identify the longest carbon chain including the carboxyl group. Then form the acid name by dropping the *-e* from the corresponding parent hydrocarbon name and adding *-oic acid.* Thus, the names corresponding to the one-, two-, and three-carbon acids are methanoic acid, ethanoic acid, and propanoic acid. These names are, of course, derived from methane, ethane, and propane.

CH_4	Methane	HCOOH	Methanoic acid
CH_3CH_3	Ethane	CH_3COOH	Ethanoic acid
$CH_3CH_2CH_3$	Propane	CH_3CH_2COOH	Propanoic acid

The IUPAC method is neither the only nor the most generally used method of naming acids. Organic acids are ordinarily known by common names. Methanoic, ethanoic, and propanoic acids are commonly called formic, acetic, and propionic acids, respectively. These names usually refer to a natural source of the acid and are not really systematic. Formic acid was named from the Latin word *formica,* meaning "ant." This acid contributes to the stinging sensation of ant bites. Acetic acid is found in vinegar and is so named from the Latin word for vinegar. The name of butyric acid is derived from the Latin term for butter, since it is a constituent of butterfat. Many of the carboxylic acids, principally those having even numbers of carbon atoms ranging from 4 to about 20, exist in combined form in plant and animal fats. These acids are called *fatty acids.* Table 19.10 lists the common and IUPAC names of the more important saturated aliphatic acids.

The simplest aromatic acid is benzoic acid. *ortho*-Hydroxybenzoic acid is known as salicylic acid, the basis for many salicylate drugs such as aspirin. There are three methylbenzoic acids, known as *o*-, *m*-, and *p*-toluic acids.

Benzoic acid   Salicylic acid (*o*-Hydroxybenzoic acid)   Acetylsalicylic acid (Aspirin)   *p*-Toluic acid

TABLE 19.10 Formulas and names of saturated carboxylic acids.

Formula	IUPAC name	Common name
HCOOH	Methanoic acid	Formic acid
CH_3COOH	Ethanoic acid	Acetic acid
CH_3CH_2COOH	Propanoic acid	Propionic acid
$CH_3(CH_2)_2COOH$	Butanoic acid	Butyric acid
$CH_3(CH_2)_3COOH$	Pentanoic acid	Valeric acid
$CH_3(CH_2)_4COOH$	Hexanoic acid	Caproic acid
$CH_3(CH_2)_6COOH$	Octanoic acid	Caprylic acid
$CH_3(CH_2)_8COOH$	Decanoic acid	Capric acid
$CH_3(CH_2)_{10}COOH$	Dodecanoic acid	Lauric acid
$CH_3(CH_2)_{12}COOH$	Tetradecanoic acid	Myristic acid
$CH_3(CH_2)_{14}COOH$	Hexadecanoic acid	Palmitic acid
$CH_3(CH_2)_{16}COOH$	Octadecanoic acid	Stearic acid
$CH_3(CH_2)_{18}COOH$	Eicosanoic acid	Arachidic acid

One of the many known methods of preparing carboxylic acids is by the oxidation of primary alcohols. The following two equations illustrate this method:

$$CH_3CH_2CH_2CH_2OH \xrightarrow[H^+]{Cr_2O_7^{2-}} CH_3CH_2CH_2COOH$$

1-Butanol   Butyric acid (Butanoic acid)

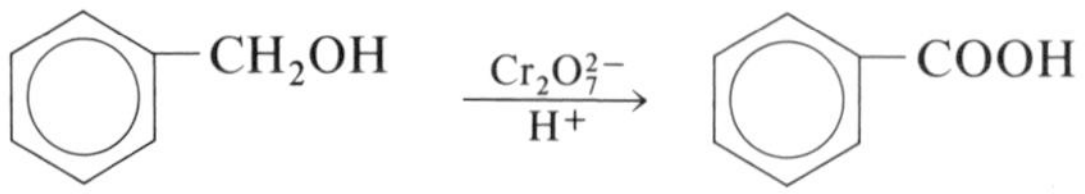

Benzyl alcohol   Benzoic acid

The carboxylic acids are weak acids, ionizing in water only to a small degree. They react with active metals to produce hydrogen and a salt, and with bases to form salts:

$$2\ CH_3COOH(aq) + 2\ Na(s) \longrightarrow 2\ CH_3COONa(aq) + H_2 \uparrow$$

Sodium acetate

$$CH_3COOH(aq) + NaOH(aq) \longrightarrow CH_3COONa(aq) + H_2O$$

Acids are reduced to the corresponding primary alcohol by reaction with lithium aluminum hydride, $LiAlH_4$. This is a source of long carbon chain alcohols.

$$CH_3(CH_2)_{14}COOH \xrightarrow{LiAlH_4} CH_3(CH_2)_{14}CH_2OH$$

Palmitic acid   1-Hexadecanol (Cetyl alcohol)

ester

Carboxylic acids react with alcohols in an acidic medium to form esters. **Esters** have the general formula RCOOR′ where R′ can be an aliphatic or an aromatic group. The functional group of the ester is —COOR′.

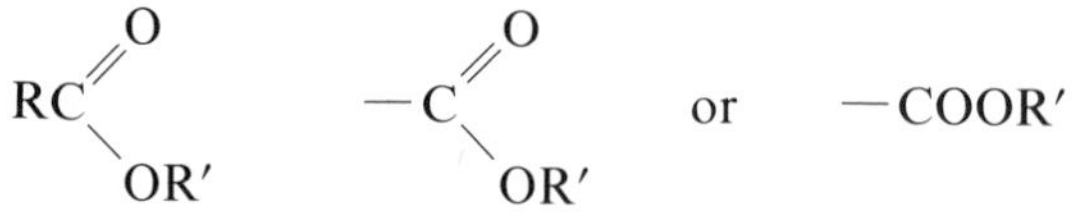

Ester   Functional group of an ester

The reaction of acetic acid and ethyl alcohol is shown as an example. In addition to the ester, a molecule of water is formed as a product. The method is called *esterification*.

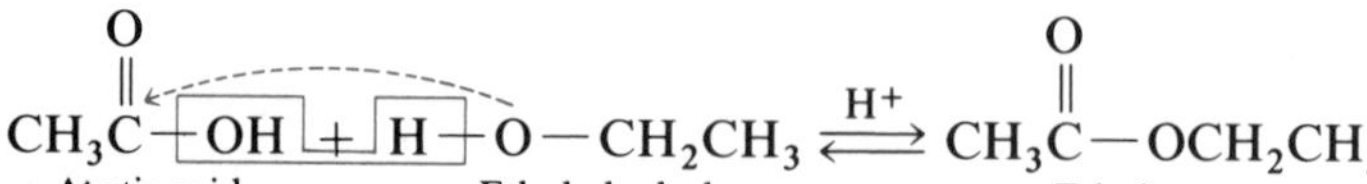

Acetic acid   Ethyl alcohol   Ethyl acetate
(Ethanoic acid)   (Ethanol)   (Ethyl ethanoate)

Esters are alcohol derivatives of carboxylic acids. They are named in much the same way as salts. The alcohol part is named first, followed by the name of the acid modified to end in *-ate*. The *-ic* ending of the organic acid name is replaced by the ending *-ate*. Thus, in the IUPAC System, ethano*ic acid* becomes ethano*ate*. In the common names, acet*ic acid* becomes acet*ate*. To name an ester it is necessary to recognize the portion of the ester molecule that comes from the acid and the portion that comes from the alcohol. In the general formula for an ester, the $RC=O$ comes from the acid and the $R'O$ comes from the alcohol:

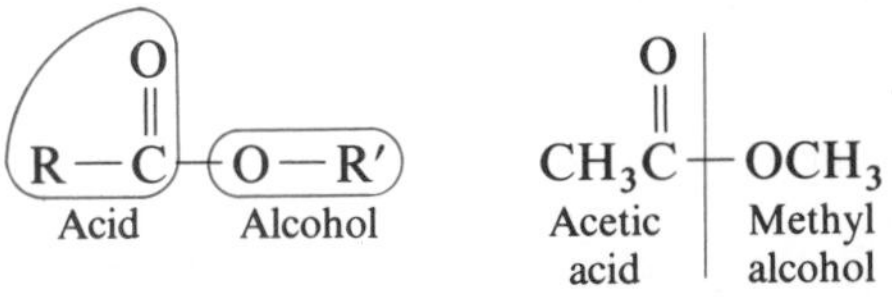

$$R-C\overset{\overset{\displaystyle O}{\|}}{}(O-R')$$
Acid   Alcohol

$$CH_3C\overset{\overset{\displaystyle O}{\|}}{}OCH_3$$
Acetic   Methyl
acid   alcohol

Methyl acetate
(Methyl ethanoate)

Esters occur naturally in many varieties of plant life. Many of them have pleasant, fragrant, fruity odors and are used as flavoring and scenting agents. Esters are insoluble in water, but soluble in alcohol. Table 19.11 gives a list of selected esters.

TABLE 19.11 Odors and flavors of selected esters.

Formula	IUPAC name	Common name	Odor or flavor
$CH_3COOCH_2CH_2CHCH_3$ $\quad\quad\quad\quad\quad\;\; CH_3$	Isopentyl ethanoate	Isoamyl acetate	Banana, pear
$CH_3CH_2CH_2COOCH_2CH_3$	Ethyl butanoate	Ethyl butyrate	Pineapple
$HCOOCH_2CHCH_3$ $\quad\quad\quad\quad CH_3$	Isobutyl methanoate	Isobutyl formate	Raspberry
$CH_3COOCH_2(CH_2)_6CH_3$	Octyl ethanoate	*n*-Octyl acetate	Orange
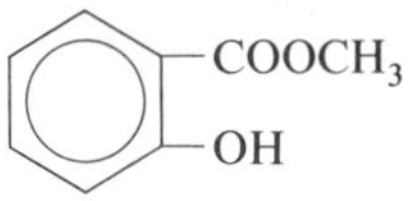 (benzene ring)—COOCH_3 —OH		Methyl salicylate	Wintergreen

QUESTIONS

A. Review the Meanings of the New Terms Introduced in this Chapter

1. Vital force theory
2. Organic chemistry
3. Hydrocarbons
4. Alkanes
5. Homologous series
6. Isomerism
7. Isomers
8. Alkyl group
9. Alkyl halide
10. Functional group
11. Alkene
12. Alkyne
13. Addition reaction
14. Polymerization
15. Polymer
16. Monomer
17. Copolymer
18. Aromatic compound
19. Fused aromatic ring system
20. Primary alcohol
21. Secondary alcohol
22. Tertiary alcohol
23. Polyhydroxy alcohol
24. Phenol
25. Ether
26. Carbonyl group
27. Aldehyde
28. Ketone
29. Carboxylic acid
30. Carboxyl group
31. Ester

B. Review Questions

1. What bonding characteristic of carbon is primarily responsible for the existence of so many organic compounds?
2. What is the most common geometric arrangement of covalent carbon bonds? Illustrate this structure.
3. How many electron pairs are associated with covalently bonded carbon atoms?
4. In addition to single bonds, what other types of bonds can carbon atoms form? Give examples.
5. Draw a Lewis structure for:
   (a) A single carbon atom
   (b) Molecules of methane, ethylene (ethene), and acetylene (ethyne)
6. Describe the geometry (shape) of a propane molecule. Are the three carbon atoms in the molecule linearly arranged? Do the three carbon atoms lie in a plane?
7. Write the names and draw the structural formulas for the first ten normal alkanes.
8. Write the names and draw the structural formulas for all possible alkyl groups (C_nH_{2n+1}) containing from one to four carbon atoms.
9. Name the following normal alkyl groups:
   (a) $C_5H_{11}-$ (b) $C_7H_{15}-$ (c) $C_8H_{17}-$ (d) $C_{10}H_{21}-$
10. What are the three principal sources of hydrocarbons?
11. How many hydrogen atoms must be removed from a hydrocarbon molecule to convert (a) a carbon–carbon single bond into a carbon–carbon double bond, and (b) a carbon–carbon double bond into a carbon–carbon triple bond?
12. Which one of the following compounds belongs to a different homologous series than the others?
   (a) C_2H_4 (b) CH_4 (c) C_6H_{14} (d) C_5H_{12}
13. Which word does not belong with the others?
   (a) Alkane   (c) Saturated   (e) Ethylene
   (b) Paraffin   (d) Ethane   (f) Pentane
14. Write the simplest (empirical) molecular and structural formulas for:
   (a) Ethane   (b) Benzene   (c) Propane
15. Draw structures for all the isomers of (a) butane, (b) hexane, and (c) pentene.
16. Name each of the compounds in Question 15.
17. Which compound is not an isomer of the others?
   (a) $CH_3(CH_2)_4CH_3$
   (b) $(CH_3)_2CH(CH_2)_2CH_3$
   (c) $(CH_3)_2CHCH(CH_3)_2$
   (d) $(CH_3)_2C{=}C(CH_3)_2$
18. Draw expanded structural formulas for each of the formulas in Question 17.
19. One name in each of the following pairs is incorrect. Draw structures corresponding to each name and indicate which name is incorrect.
   (a) 2-Methylbutane and 3-methylbutane
   (b) 2-Ethylbutane and 3-methylpentane
   (c) 2-Dimethylbutane and 2,2-dimethylbutane
   (d) 2,4-Dimethylhexane and 2-ethyl-4-methylpentane
20. What is the single most important reaction of alkanes?
21. Give IUPAC names and structural formulas of all the isomers having the following molecular formulas:
   (a) C_3H_5Br (b) C_4H_7Cl (c) $C_4H_8I_2$ (d) C_3H_6BrCl
22. Using alkanes and any other necessary inorganic reagents, write reactions showing the formation of:
   (a) CH_3Cl (d) $CH_3CHBrCH_3$
   (b) $CHCl_3$ (e) $CH_3CH{=}CHCH_3$
   (c) $CH_3CH_2CH_2Br$

23. Draw structures containing two carbon atoms for each of the following classes of compounds:

(a) Alkene   (e) Ether   (i) Amine
(b) Alkyne   (f) Aldehyde   (j) Nitrile
(c) Alkyl halide   (g) Carboxylic acid   (k) Amino acid
(d) Alcohol   (h) Amide   (l) Ester

24. Give the common name and the IUPAC name for each of the compounds in Question 23.

25. Which compounds in Question 23 are isomers?

26. Write formulas for the following:

(a) Chloromethane   (e) Iodoethyne
(b) Vinyl chloride   (f) 1-Bromo-4-methyl-2-hexene-5-yne
(c) Chloroform   (g) 1,1-Dibromoethene
(d) Hexachloroethane   (h) 1,2-Dibromoethene

27. Complete the following reactions and name the products:

(a) CH_2=$CHCH_3$ + $Br_2 \longrightarrow$
(b) CH_3CH=CH_2 + $KMnO_4(aq)$ + $H_2O \longrightarrow$
(c) CH_2=CH_2 + HBr $\longrightarrow$
(d) CH_3CH=$CHCH_3$ + $H_2 \xrightarrow[\text{1 atm}]{\text{Pt, 25°C}}$
(e) CH_2=CH_2 + $H_2O \xrightarrow{H^+}$
(f) CH≡CH + 2 $Br_2 \longrightarrow$

28. Write equations showing two methods for the preparation of acetylene.

29. Ethylene and its derivatives are the most common monomers for polymers. Write formulas for the following ethylene-based polymers:

(a) Polyethylene   (d) Saran
(b) Polyvinyl chloride   (e) Polyacrylonitrile (Orlon or Acrilan)
(c) Polystyrene

30. (a) Write a structural formula showing the polymer that can be formed from each of the following monomers (show four units): (a) propylene; (2) 1-butene; (3) 2-butene.

    (b) How many ethylene units are in a polyethylene molecule that has a molecular weight of 35,000?

31. Name the following aromatic compounds:

(a) 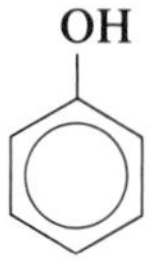 OH   (d) 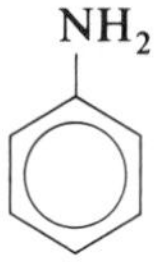 NH_2 (g) Cl   (j) 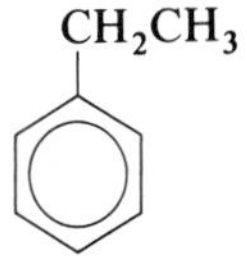 CH_2CH_3

(b) CH_3 (e) Cl   (h) 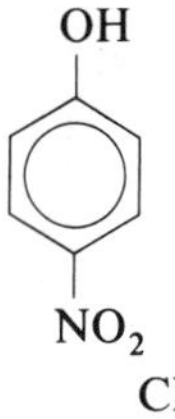 OH   (k) 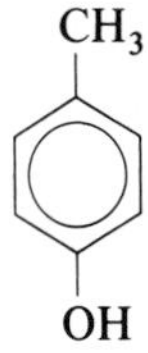 CH_3

(c) COOH   (f) Cl   (i) CH_3 (l) H

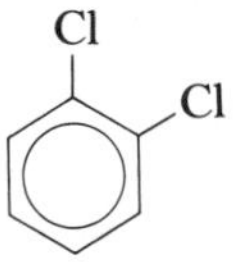

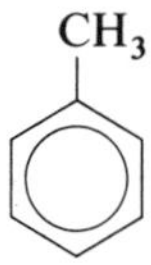

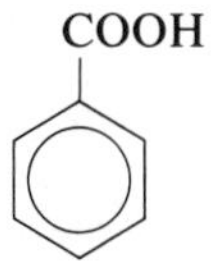

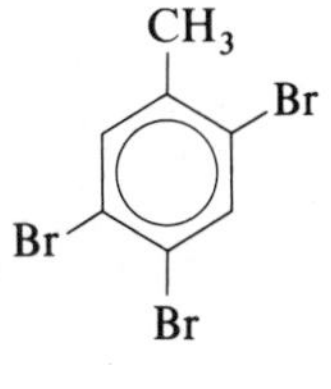

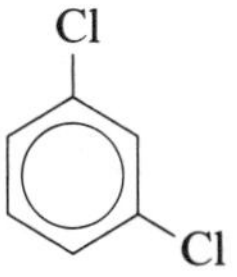

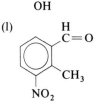

32. Write structural formulas for each of the following compounds:
    (a) *o*-Aminobenzoic acid
    (b) Styrene
    (c) Diphenylmethane
    (d) Diphenylether
    (e) *m*-Bromobenzaldehyde
    (f) 1,3,5-Trimethylbenzene
    (g) Salicylic acid
    (h) Phenanthrene
    (i) Phenylethanoic acid
    (j) *o*-Xylene

33. Using benzene, and other reagents as needed, write reactions showing the synthesis of the following:
    (a) Chlorobenzene
    (b) Nitrobenzene
    (c) Toluene
    (d) Aniline
    (e) 2,4,6-Trinitrotoluene

34. There are eight open-chain isomeric alcohols having the formula $C_5H_{11}OH$.
    (a) Write the structural formula and the IUPAC name for each of these alcohols.
    (b) Indicate which of these isomers are primary, secondary, and tertiary alcohols.

35. Write structural formulas for the following:
    (a) 2-Pentanol
    (b) Isopropyl alcohol
    (c) 2,2-Dimethyl-1-heptanol
    (d) 1,3-Propanediol
    (e) Glycerine

36. Alcohols can be made by the reaction of alkyl halides with aqueous sodium hydroxide:

$$RX + NaOH(aq) \longrightarrow ROH + NaX$$

What alkyl bromide should be used to prepare (a) isopropyl alcohol, (b) 3-methyl-1-butanol, and (c) 1,3-propanediol by this method? Write the equation for each reaction.

37. Complete the following equations and name the organic product formed in each reaction.
    (a) $CH_3CH_2CH_2OH \xrightarrow[\Delta]{H^+, K_2Cr_2O_7}$

    (b) $CH_3CHOHCH_3 \xrightarrow[\Delta]{H^+, K_2Cr_2O_7}$

    (c) $CH_3CH_2OH \xrightarrow[180°C]{96\% H_2SO_4}$

    (d) $CH_3OH \xrightarrow[140°C]{96\% H_2SO_4}$

    (e) $CH_3CH_2CH_2C(OH)(CH_3)_2 \xrightarrow[90°C]{45\% H_2SO_4}$

    (f) $CH_3CH_2CH_2OH + Na \longrightarrow$

38. Alcohols are considered to be toxic to the human body, with ethanol, the alcohol in alcoholic beverages the least toxic one. What are the hazards of ingesting (a) methanol; (b) ethanol?

39. What functional group is common to *o*-cresol, thymol, eugenol, and epinephrine?

40. Write the name and structure of the ether that is isomeric with (a) 1-propanol, (b) ethanol, and (c) isopropyl alcohol.

41. Give the names and structures of all ethers having the molecular formula $C_5H_{12}O$.

42. What possible combinations of RONa and RCl might be used in preparing each of the following ethers by the Williamson synthesis?
    (a) $CH_3CH_2OCH_3$ (b) $CH_3CH_2OCH_2CH_3$ (c) CH_3OCH_3

43. Write structural formulas for propanal and propanone. From these formulas, do you think that aldehydes and ketones are isomeric with each other? Show evidence and substantiate your answer by testing with a four-carbon aldehyde and ketone.

44. Name each of the following compounds:
   (a) $H_2C=O$
   (b) $CH_3(CH_2)_5CHO$
   (c) $CH_3CHBrCH_2CHO$
   (d) $(CH_3)_3CCHO$

   (e)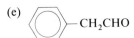

   (f)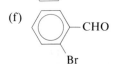

   (g) $CH_3\overset{\displaystyle O}{\underset{\|}{C}}CH_3$ (two names)

   (h) $CH_3CH_2\overset{\displaystyle O}{\underset{\|}{C}}CH_3$ (two names)

   (i) 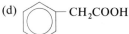 $-\overset{\displaystyle O}{\underset{\|}{C}}CH_2CH_2CH_3$

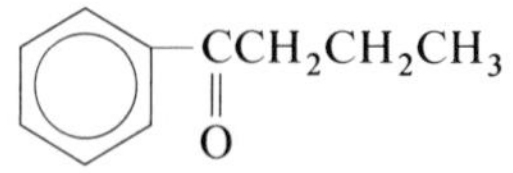

   (j) $CH_3\overset{\displaystyle O}{\underset{\|}{C}}CHClCH_3$

45. Ketones are prepared by oxidation of secondary alcohols. What alcohol should be used to prepare each of the following?
   (a) Dimethyl ketone
   (b) Methyl ethyl ketone
   (c) Ethyl isopropyl ketone
   (d) 1-Phenyl-3-hexanone

46. (a) What functional group is present in a compound that gives a positive Tollens' test?
   (b) What is the visible evidence for a positive Tollens' test?
   (c) Write an equation showing the reaction involved in a positive Tollens' test.

47. Give a chemical test or reaction by which it is possible to distinguish between the compounds in each pair:
   (a) $CH_3CH_2OCH_2CH_3$ and CH_3OH
   (b) CH_3CHO and $CH_3\overset{\displaystyle O}{\underset{\|}{C}}CH_3$

   (c) $CH_3(CH_2)_5CH=CH_2$ and $CH_3(CH_2)_6CH_3$
   (d) $CH_3(CH_2)_3CH_2OH$ and CH_3CH_2COOH
   (e) $CH_2=CHOCH=CH_2$ and $CH_3CH_2OCH_2CH_3$
       Divinyl ether   Diethyl ether

48. Give the common and the IUPAC names for the first five straight-chain carboxylic acids.

49. Write the formulas and IUPAC names for all the isomers of hexanoic acid, $CH_3(CH_2)_4COOH$.

50. Name each of the following acids:
   (a) $CH_3CHBrCOOH$
   (b) $CH_2=CHCH_2COOH$
   (c) $CH_3\underset{\underset{\displaystyle NH_2}{|}}{CH}COOH$

   (d) 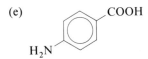 $-CH_2COOH$

   (e)
   (f) Cl
   (g) 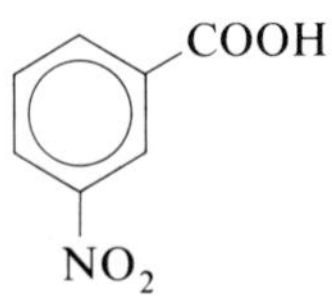
   (h)

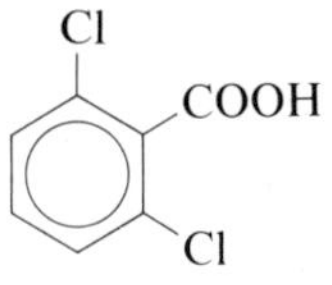

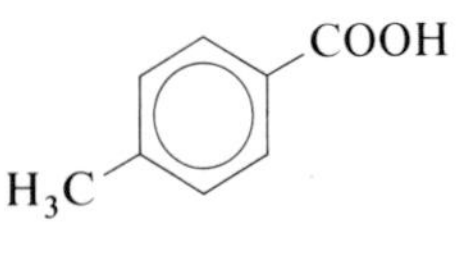

51. Write structural formulas for the following esters:
    (a) Ethyl formate
    (b) Methyl ethanoate
    (c) Isopropyl propanoate
    (d) *n*-Nonyl acetate
    (e) Ethyl benzoate
    (f) Methyl salicylate
    (g) Vinyl butanoate

52. Write IUPAC names for the following esters:
    (a) $CH_3COCH_2CH_3$
        $\overset{\|}{O}$
    (c) CH_3CO-⬡

    (b) $HCOCH(CH_3)_2$
        $\overset{\|}{O}$
    (d) 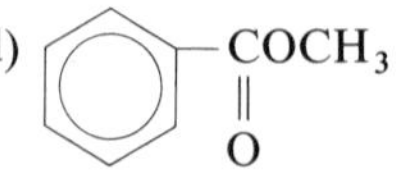 ⬡$-COCH_3$
        $\overset{\|}{O}$

53. Write equations for the preparation of the following esters:
    (a) Ethyl formate   (b) Methyl propanoate   (c) *n*-Propyl benzoate

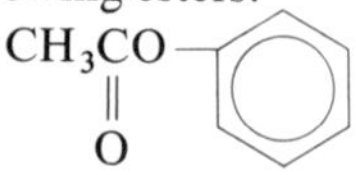

54. Complete the following equations:
    (a) $CH_3COOH + NaOH \longrightarrow$
    (b) $CH_3CHCOOH + NH_3 \longrightarrow$
        $\ \ \ \ |$
        $\ \ \ OH$
    (c) $CH_3COOH + Ca \longrightarrow$
    (d) 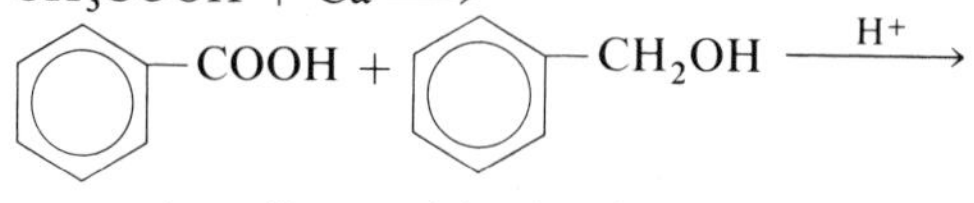 ⬡$-COOH +$ ⬡$-CH_2OH \xrightarrow{H^+}$

55. Using reactions discussed in the chapter, write equations showing the stepwise synthesis of the compounds that follow. Use only alkanes, benzene, and any necessary inorganic reagents as starting materials. Multisteps may be involved in some of these syntheses. A compound prepared in a previous synthesis may be used as a starting material in a subsequent synthesis.
    (a) CH_3CH_2Cl
    (b) $CH_2{=}CHCl$
    (c) $CH_2BrCHBr_2$
    (d) CH_3COOH
    (e) CH_2OHCH_2OH
    (f) $CH_3COCH_2CH_3$
        $\ \ \ \ \ \ \ \ \ \ \ \ \overset{\|}{O}$
    (g) $CH_3\overset{\bullet}{C}HO$
    (h) $CH_3CHOHCH_3$
    (i) CH_3CCH_3
        $\ \ \ \ \ \ \ \overset{\|}{O}$

    (j) $(CH_3)_2CHOCH_3$
    (k) $CH_3CH_2OCH_2CH_3$
    (l) 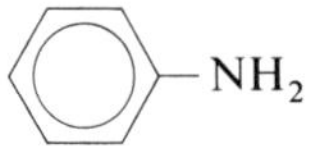 ⬡$-NH_2$
    (m) 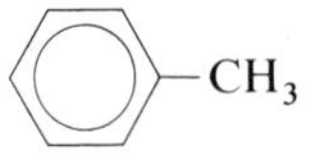 ⬡$-CH_3$
    (n)  ⬡$-CH_2Cl$
    (o) 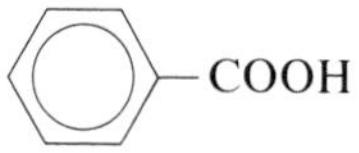 ⬡$-COOH$
    (p) 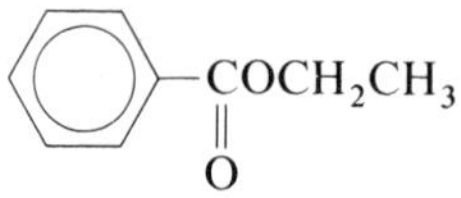 ⬡$-COCH_2CH_3$
        $\ \ \ \ \ \ \ \ \overset{\|}{O}$

    (q)  ⬡$-CH_2CH_3$
    (r) 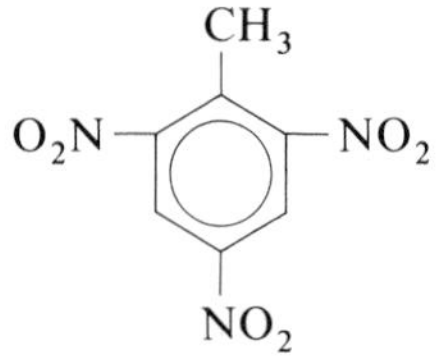 O_2N-⬡$-NO_2$ with CH_3 top and NO_2 bottom
    (s) 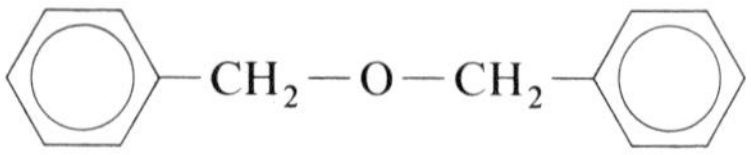 ⬡$-CH_2-O-CH_2-$⬡
    (t) 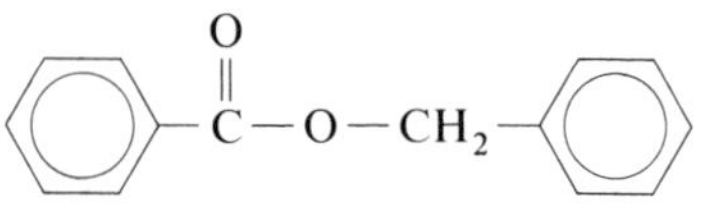 ⬡$-\overset{\overset{O}{\|}}{C}-O-CH_2-$⬡

56. Which of the following statements are correct?
   (a) In organic molecules eight bonding electrons are associated with each carbon atom except for those carbon atoms that are part of a double or triple bond.
   (b) Both primary and secondary carbon atoms are present in *n*-pentane.
   (c) In order for a molecule to contain a tertiary carbon atom, the molecule must contain a minimum of three carbon atoms.
   (d) Propane is an isomer of propene.
   (e) Isopropyl alcohol is an isomer of both methyl ethyl ether and methyl ethyl ketone.
   (f) Depending on which hydrogen atom is removed, it is possible to get two alkyl groups, *t*-butyl and isobutyl, from isobutane.
   (g) Isobutane, methylpropane, and 1,1-dimethylethane are all correct names for the same compound.
   (h) Compounds belonging to the same homologous series will generally exhibit similar chemical properties.
   (i) A single monosubstituted product will result from the chlorination of butane.
   (j) There are eight carbon atoms in a molecule of 3,3,4-trimethylpentane.
   (k) All isomers having the formula C_3H_8O will have similar chemical and physical properties.
   (l) The addition of HCl to ethylene and monochlorination of ethane yield the same product, chloroethane.
   (m) Dimethyl ketone, acetone, and propanone all have the same molecular formula but different structural formulas.
   (n) The simplest (empirical) formula for benzene is CH.
   (o) The number of isomers of dichlorobenzene, trichlorobenzene, and tetra-chlorobenzene is the same.
   (p) Baeyer's test is useful for distinguishing between alcohols and ketones.
   (q) Phenanthrene is isomeric with anthracene.
   (r) Polymerization is the process of forming very large high-molecular-weight molecules from smaller units.
   (s) *o*-Dichlorobenzene and 1,3-dichlorobenzene are different names for the same compound.
   (t) The monomer of Teflon is tetrafluoroethylene, $CF_2=CF_2$.
   (u) The chemical characteristics of CH_3OH and NaOH are similar.
   (v) Ethers, because of their relative unreactivity, are often used as solvents in organic reactions.
   (w) Both aldehydes and ketones are easily oxidized.
   (x) Esters are derivatives of alcohols and carboxylic acids.
   (y) Acetic acid is the most common and structurally the simplest carboxylic acid.
   (z) Methyl propanoate and propyl methanoate are isomers.

20

Introduction to Biochemistry

After studying Chapter 20 you should be able to:

1. Understand the terms listed in Question A at the end of the chapter.
2. Explain the factors involved in photosynthesis and the nature of the products formed.
3. Classify carbohydrates as mono-, di-, or polysaccharides.
4. Draw structural formulas in the open-chain and cyclic forms for glucose, fructose, galactose, and ribose.
5. Draw structural formulas for maltose, sucrose, and lactose.
6. State the properties of and the general occurrence of glucose, galactose, fructose, and ribose.
7. Understand the manner in which monosaccharides are linked together in maltose, lactose, and sucrose.
8. Write equations for the enzyme-catalyzed hydrolysis of maltose, lactose, and sucrose.
9. Give the monosaccharide composition of maltose, lactose, sucrose, starch, cellulose, and glycogen.
10. Discuss the similarities and differences between starch and cellulose.
11. Discuss, in simple terms, the metabolism of carbohydrates in the human body.
12. Rate the relative sweetness of the common mono- and disaccharides.
13. Give the general formula for fats and oils.
14. Write the names and formulas of the fatty acids that most commonly occur in fats and oils.
15. State which fatty acids are essential to human diets.
16. Write the structure of a triglyceride when given the fatty acid composition.
17. Write an equation for the saponification of a fat or an oil with caustic soda (NaOH).
18. Tell how a fat differs from an oil.
19. Explain the "hardening" of vegetable oils, and the purposes for this hardening.
20. Draw the structural formula for cholesterol and the structural feature that is common to all steroids.
21. Distinguish among these three lipids: fats, phospholipids, and glycolipids.
22. List five foods that are major sources of proteins.
23. Explain the meaning of α-amino acids and the significance of these compounds in naturally occurring protein material.
24. Show the structural formula of a di-, tri-, or polypeptide that will be formed by combining amino acids.
25. State the significance of the primary structure of a polypeptide or a protein.

26. Describe the functions of and the metabolic fate of amino acids and proteins.
27. Name and write the formulas of the six fundamental components of DNA.
28. Write the structure for a segment of a polynucleotide that contains up to four nucleotides.
29. Explain the three structural differences between DNA and RNA.
30. Describe the double-helix structure of DNA according to Watson and Crick.
31. Explain the concept of complementary bases and how it relates to DNA.
32. Discuss the role of DNA in genetics.
33. Distinguish between cell division in mitosis and meiosis.
34. Discuss the role of enzymes in the body and the theory of how they function.
35. Tell what is meant by the specificity of an enzyme.
36. Explain the function of vitamins in the body.
37. State, in a general way, the results of vitamin deficiencies.
38. Explain the function of hormones in the body.
39. State how vitamins and hormones differ in origin.

20.1 Introduction

The study of life has long fascinated people—and it is probably the most intriguing of all scientific studies, although the answer to the question, "What is life?" still eludes us.

The chemical substances present in all living organisms—from microbes to humans—range in complexity from water and simple salts to DNA (deoxyribonucleic acid) molecules containing tens of thousands of atoms. Four of the chemical elements, hydrogen, carbon, nitrogen, and oxygen, make up approximately 95% of the mass of living matter. Small amounts of sulfur, phosphorus, calcium, sodium, chlorine, magnesium, and iron, together with trace amounts of many other elements such as copper, manganese, zinc, cobalt, and iodine, are also found in living organisms. The human body consists of about 60% water with some tissues having a water content as high as 80%. An excellent article titled "The Chemical Elements of Life" by Earl Frieden (*Scientific American*, July 1972) discusses the elements essential for animal life.

biochemistry   **Biochemistry** is the branch of chemistry that is concerned with the chemical reactions occurring in living organisms. Its scope includes such processes as growth, respiration, digestion, metabolism, and reproduction.

The four major classes of biomolecules upon which all life depends are carbohydrates, lipids, proteins, and nucleic acids. Each kind of living organism has an amazing ability to select and synthesize a large portion of the many complicated molecules needed for its existence. In fact, the processes carried out in a living organism can be likened to those of a highly automated, smoothly running "chemical factory." But unlike a chemical factory, a living organism is able to expand (grow), repair damage (if not too severe), and, finally, reproduce itself.

Of necessity, we are limited here to brief consideration of only a few important aspects of biochemistry. These include photosynthesis, carbohydrates, lipids, proteins, nucleic acids, enzymes, vitamins, and hormones.

20.2 Photosynthesis

photosynthesis

Only simple inorganic substances are necessary to support green plants. These substances are water, carbon dioxide, nitrogen (or nitrogen compounds), and a variety of about 30 inorganic ions including Mg^{2+}, Ca^{2+}, Fe^{2+}, K^+, Na^+, PO_4^{3-}, SO_4^{2-}, NO_3^-, and Cl^-. These materials are absorbed directly by the plant from the environment. Energy is also required for all living things. Green plants absorb light energy and, through the process of photosynthesis, utilize it in making complex organic compounds. **Photosynthesis** is the process whereby plants utilize energy from the sun to synthesize carbohydrates from carbon dioxide and water. Nearly all nonphotosynthetic organisms, both plant and animal, are ultimately dependent on photosynthesis for their existence. This is because they obtain energy from organic nutrients or foods that are traceable, sometimes through a long food chain, back to photosynthetic organisms.

In most plants the sun's energy is absorbed by chlorophyll molecules in the plant cells and is used to transform carbon dioxide and water into carbohydrates and oxygen. The transformation goes through many complex intermediate reactions. The following equations represent the net overall process:

$$6\,CO_2 + 6\,H_2O + 673\ \text{kcal} \xrightarrow[\text{Chlorophyll}]{\text{Sunlight}} C_6H_{12}O_6 + 6\,O_2 \tag{1}$$

$$n\,CO_2 + n\,H_2O + \text{Energy} \xrightarrow[\text{Chlorophyll}]{\text{Sunlight}} (CH_2O)_n + n\,O_2 \tag{2}$$

Equation (1) indicates the formation of a simple sugar, glucose ($C_6H_{12}O_6$); equation (2) represents the formation of higher-molecular-weight polymeric substances such as starch and cellulose ($C_6H_{10}O_5)_n$, where n may have values running into the hundreds or even thousands. Photosynthesis is an endothermic series of reactions in which energy is absorbed and stored in the products until such time as they are utilized by plants in oxidative processes and by animals as food.

$$C_6H_{12}O_6 + 6\,O_2 \xrightarrow{\text{Enzymes}} 6\,CO_2 + 6\,H_2O + 673\ \text{kcal} \tag{3}$$

Equation (3) represents the oxidation of glucose and corresponds to the reversal of the overall photosynthesis reaction. Glucose actually can be burned in oxygen to produce carbon dioxide, water, and heat energy. But in the living cell, the oxidation of glucose does not proceed directly to carbon dioxide and water. Instead, like photosynthesis, the overall process proceeds by a series of enzyme-catalyzed intermediate reactions. These intermediate steps channel some of the liberated energy into uses other than heat production. Specifically, a portion of the energy is stored in the chemical bonds of adenosine triphosphate (ATP) for other uses when needed by the living organism.

The 1961 Nobel Prize in Chemistry was awarded to the American chemist Melvin Calvin (1911–) of the University of California at Berkeley for his work in photosynthesis. Calvin and his coworkers used radioactive carbon tracer techniques to establish the detailed and complicated sequence of chemical reactions that occur in the overall process of photosynthesis.

20.3 Carbohydrates

carbohydrates

Chemically, **carbohydrates** are polyhydroxy aldehydes or polyhydroxy ketones, or substances that will yield these compounds when hydrolyzed. The name *carbohydrate* was given to this class of compounds many years ago by French scientists who called them *hydrates de carbone* because their empirical formulas approximated $(C \cdot H_2O)_n$. However, the hydrogen and oxygen do not actually exist as water or in hydrate form as we have seen in compounds such as $BaCl_2 \cdot 2H_2O$. Empirical formulas used to represent carbohydrates are $C_x(H_2O)_y$ and $(CH_2O)_n$.

Carbohydrates occur naturally in plants and are one of the three principal classes of animal food. The other two classes of foods are lipids (fats) and proteins. We have seen that plants are able to synthesize carbohydrates by the photosynthetic process. Animals are incapable of this synthesis and are dependent on the plant kingdom for their source of carbohydrates. Animals are capable, however, of converting plant carbohydrate into glycogen (animal carbohydrate) and storing this glycogen throughout the body as reserve carbohydrate. The amount of energy available from carbohydrates is about 4 kcal/g.

Carbohydrates exist as sugars, starches, and cellulose. The simplest of these are the sugars, also called *saccharides*. The names of the sugars end in *-ose* (e.g., glucose, sucrose, maltose). Carbohydrates are classified as monosaccharides, disaccharides, oligosaccharides, and polysaccharides according to the number of monosaccharide units linked together to form a molecule.

MONOSACCHARIDES

monosaccharide

A **monosaccharide** is a carbohydrate that cannot be hydrolyzed to simpler carbohydrate units. They are often called simple sugars. The most common of these is glucose. Monosaccharides containing three, four, five, and six carbon atoms are called *trioses, tetroses, pentoses,* and *hexoses,* respectively. Mono-saccharides that contain an aldehyde ($-\overset{\overset{\displaystyle H}{|}}{C}=O$) group on one carbon atom and a hydroxy ($-OH$) group on each of the other carbon atoms are called *aldoses*. One important hexose, fructose, also known as levulose, contains a ketone group and is called a *ketohexose*. Fructose is the sweetest of the sugars, followed by sucrose, and then glucose. Most of the known sugars are not sweet tasting. Structural formulas of several monosaccharides are as follows:

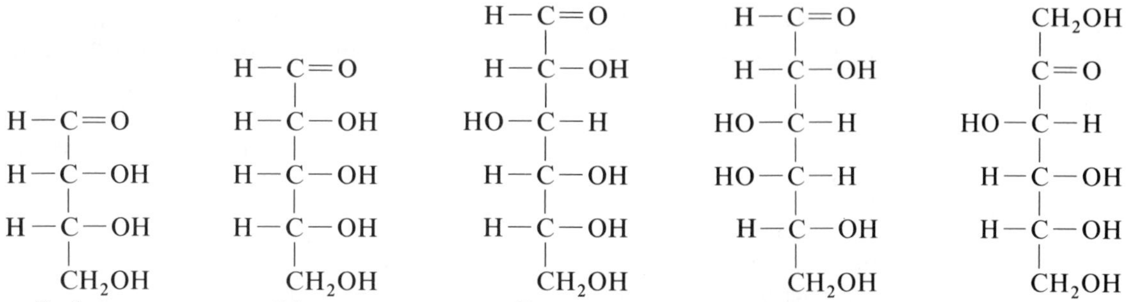

```
                             H—C=O         H—C=O          CH₂OH
                               |             |              |
               H—C=O         H—C—OH        H—C—OH         C=O
                 |             |             |              |
H—C=O          H—C—OH        HO—C—H        HO—C—H        HO—C—H
  |              |             |             |              |
H—C—OH         H—C—OH        H—C—OH        HO—C—H         H—C—OH
  |              |             |             |              |
H—C—OH         H—C—OH        H—C—OH        H—C—OH         H—C—OH
  |              |             |             |              |
 CH₂OH          CH₂OH         CH₂OH         CH₂OH          CH₂OH
Erythrose       Ribose        Glucose       Galactose      Fructose
(an aldotetrose) (an aldopentose) (an aldohexose) (an aldohexose) (a ketohexose)
```

Sixteen different isomeric aldohexoses of formula $C_6H_{12}O_6$ are known; glucose and galactose are the most important of these. Most sugars exist predominantly in a cyclic structure, forming a six-membered ring. For example, glucose exists as a ring in which carbon 1 is bonded through an oxygen atom to carbon 5.

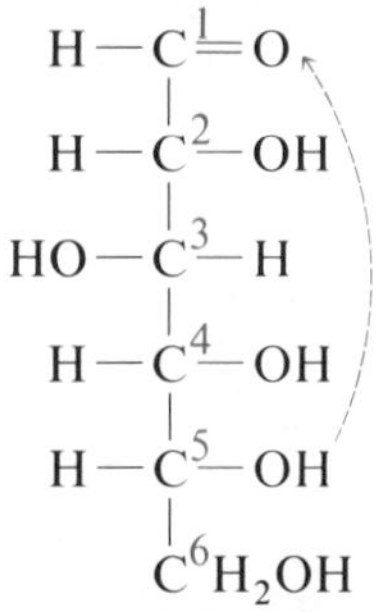

Open-chain form of glucose

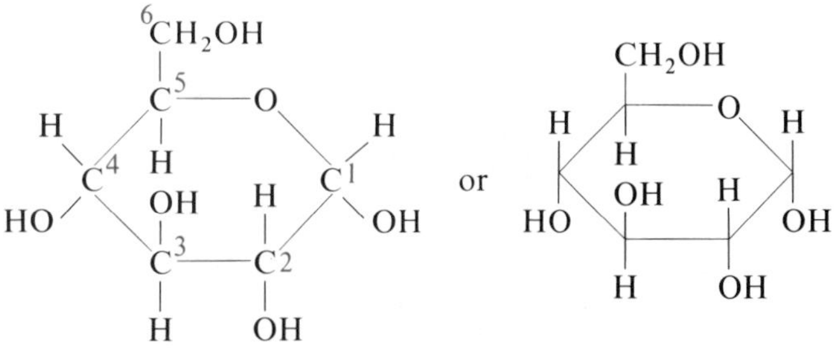

Cyclic form of glucose

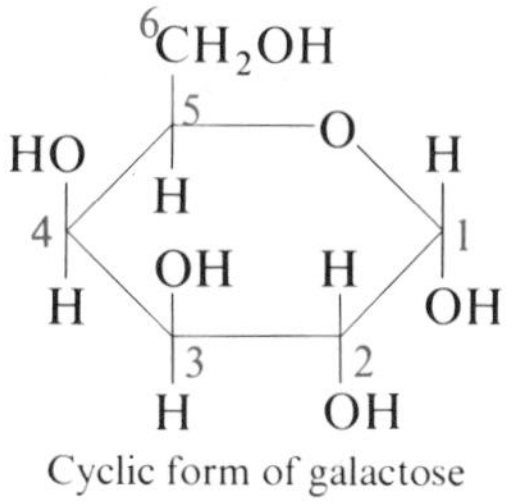

Cyclic form of galactose 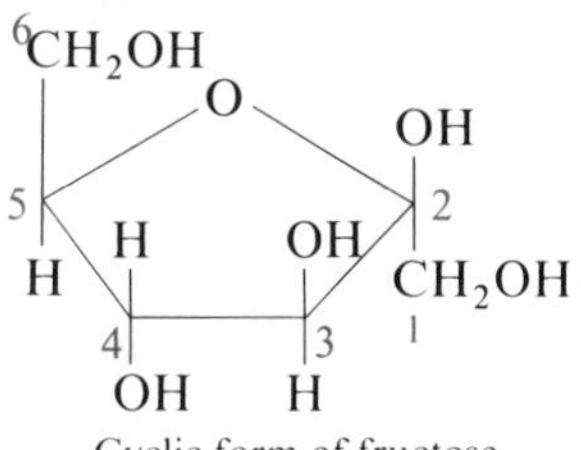   Cyclic form of fructose 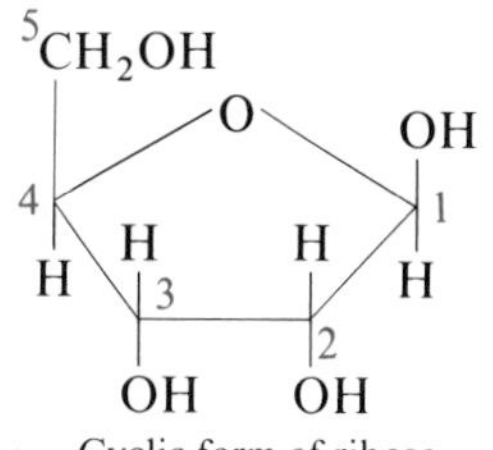   Cyclic form of ribose

The properties of four important monosaccharides follow.

Glucose. Glucose is the most important of the monosaccharides. It is an aldohexose and is found in the free state in plants and animal tissue. Glucose is commonly known as *dextrose* or *grape sugar*. It is a component of the disaccharides sucrose, maltose, and lactose, and is also the monomer of the polysaccharides starch, cellulose, and glycogen. Among the common sugars, glucose is of intermediate sweetness (See Table 20.1).

Glucose is the key sugar of the body and is carried by the bloodstream to all of its parts. The concentration of glucose in the blood is normally 80 to 100 mg/100 mL blood. Glucose requires no digestion; therefore, it may be given intravenously to patients who cannot take food by mouth. The body's heat is derived primarily from the oxidation of glucose. Glucose is found in the urine of those who have diabetes mellitus (sugar diabetes).

Galactose.   Galactose is also an aldohexose and occurs along with glucose in lactose and in many oligo- and polysaccharides such as pectin, gums, and mucilages. Galactose is an isomer of glucose, differing only in the spatial arrangement of the $-H$ and $-OH$ groups around carbon 4. Galactose is synthesized in the mammary glands to make the lactose of milk. Galactose is less than half as sweet as glucose.

A severe inherited disease, called galactosemia, is the inability of infants to metabolize galactose. The galactose concentration increases markedly in the blood and also appears in the urine. Galactosemia causes vomiting, diarrhea, enlargement of the liver, and often mental retardation. If not recognized within a few days after birth, it can lead to death. If diagnosis is made early and lactose is excluded from the diet, the symptoms disappear and normal growth may be resumed.

Fructose.   Fructose, also known as levulose, is a ketohexose and occurs in fruit juices, honey, and (along with glucose) as a constituent of sucrose. Fructose is the major constituent of the polysaccharide inulin, a starch-like substance present in many plants such as dahlia tubers, chicory roots, and Jerusalem artichokes. Fructose is the sweetest of all the sugars, being about twice as sweet as glucose. This accounts for the sweetness of honey; the enzyme invertase present in bees splits sucrose into glucose and fructose. Fructose is metabolized directly, but is also readily converted to glucose in the liver.

Ribose.   Ribose ($C_5H_{10}O_5$) is an aldopentose and is present in adenosine triphosphate (ATP), one of the chemical energy carriers in the body. Ribose and one of its derivatives, deoxyribose, are also important components of the nucleic acids DNA and RNA, the genetic information carriers in the body.

DISACCHARIDES

disaccharide

A **disaccharide** is a carbohydrate whose molecules can be hydrolyzed to yield two monosaccharides. The three disaccharides that are especially important from a biological viewpoint are sucrose, lactose, and maltose. Sucrose ($C_{12}H_{22}O_{11}$), which is commonly known as *table sugar*, is found in the free state throughout the plant kingdom. Sugar cane contains 15–20% sucrose, and sugar beets contain 10–17%. Maple syrup and sorghum are also good sources of sucrose.

Lactose ($C_{12}H_{22}O_{11}$), also known as *milk sugar*, is found free in nature mainly in the milk of mammals. Human milk contains about 6.7% lactose, and cow milk contains about 4.5% of this sugar.

TABLE 20.1 Relative sweetness of sugars.

Fructose	100	Galactose	19
Sucrose	58	Lactose	9.2
Glucose	43	Invert sugar	75
Maltose	19		

Maltose ($C_{12}H_{22}O_{11}$) is found in sprouting grain, but occurs much less commonly (in nature) than either sucrose or lactose. Maltose is prepared commercially by the partial hydrolysis of starch, catalyzed either by enzymes or by dilute acids.

Disaccharides are not utilized directly in the body, but are first hydrolyzed to monosaccharides. Upon hydrolysis, disaccharides yield two monosaccharide molecules. The hydrolysis is catalyzed by hydrogen ions (acids), usually at elevated temperatures, or by certain enzymes that act effectively at room or body temperatures. An enzyme is a protein that acts as a biochemical catalyst and is specific in its actions; that is, a particular enzyme catalyzes a specific biochemical reaction. Thus, a different enzyme is required for the hydrolysis of each of the three disaccharides:

$$\text{Sucrose} + \text{Water} \xrightarrow{\ H^+ \text{ or sucrase}\ } \text{Glucose} + \text{Fructose}$$

$$\text{Lactose} + \text{Water} \xrightarrow{\ H^+ \text{ or lactase}\ } \text{Galactose} + \text{Glucose}$$

$$\text{Maltose} + \text{Water} \xrightarrow{\ H^+ \text{ or maltase}\ } \text{Glucose} + \text{Glucose}$$

The structure of a disaccharide may be considered as being derived from two monosaccharide molecules by the elimination of a water molecule between them. In maltose, for example, the two monosaccharides are glucose. The water molecule is split out between the OH group on carbon 1 of one glucose unit and the OH group on carbon 4 of the other glucose unit. Thus the two glucose units are joined at carbon 1 and carbon 4. Sucrose consists of a glucose unit and a fructose unit linked together through an oxygen atom from carbon 1 on glucose to carbon 2 on fructose. In lactose, the linkage is from carbon 1 of galactose through an oxygen atom to carbon 4 of glucose. The structures of maltose and sucrose follow.

Maltose, a disaccharide

$$^6CH_2OH$$

Glucose unit   Fructose unit

Sucrose, a disaccharide

POLYSACCHARIDES

polysaccharide

Polysaccharides are complex carbohydrates that can be hydrolyzed to a large number of monosaccharide units. The molecular weights of polysaccharides range up to 1 million or more. Three of the most important polysaccharides are starch, glycogen, and cellulose.

Starch is a polymer of glucose. It is found mainly in the seeds, roots, and tubers of plants. Corn, wheat, potatoes, rice, and cassava are the chief sources of starch. The principal use of starch is for food.

Glycogen is the reserve carbohydrate of the animal kingdom. It is often called animal starch. Glycogen is formed in the body by polymerization of glucose and is stored especially in the liver and in muscle tissue. Glycogen also occurs in some insects and lower plants including fungi and yeasts.

Cellulose, like starch and glycogen, is also a polymer of glucose. It differs from starch and glycogen in the manner in which the cyclic glucose units are linked together to form chains. Cellulose is the most abundant organic substance found in nature. It is the chief structural component of plants and wood. Cotton fibers are almost pure cellulose, and wood, after removal of moisture, consists of about 50% cellulose. Cellulose is an important substance in the textile and paper industries. It is also used to make rayon fibers, photographic film, guncotton, celluloid, and cellophane. Humans cannot utilize cellulose as food because they lack the necessary enzymes to hydrolyze it to usable glucose.

The digestion or metabolism of carbohydrates is a very complex biochemical process. It starts in the mouth where the enzyme amylase in the saliva begins the hydrolysis of starch to maltose and temporarily stops in the stomach where the hydrochloric acid present deactivates the enzyme. Digestion continues again in the intestines where the hydrochloric acid is neutralized and pancreatic enzymes complete the hydrolysis to maltose. The enzyme maltase then catalyzes the digestion of maltose to glucose. Other specific enzymes in the intestines convert sucrose and lactose to monosaccharides.

$$\text{Starch} \xrightarrow{\text{Amylase}} \text{Dextrins} \xrightarrow{\text{Amylase}} \text{Maltose} \xrightarrow{\text{Maltase}} \text{Glucose}$$

Glucose is absorbed through the intestinal walls into the bloodstream and is rapidly removed by the liver and muscle tissue, where it is polymerized and stored as glycogen. As the body calls for it, glycogen is converted back to glucose, which is ultimately oxidized to carbon dioxide and water with the release of energy. This energy is used by the body for maintenance, growth, and other normal functions. The rate of carbohydrate metabolism and proper blood glucose level are controlled by three hormones: insulin, epinephrine (adrenaline), and glucagon. Insulin acts to reduce blood glucose levels by increasing the rate of glycogen formation. Epinephrine and glucagon increase the rate of glycogen breakdown and thereby increase blood glucose levels. The synthesis of glycogen from glucose is called **glycogenesis.** The hydrolysis or breakdown of glycogen to glucose is known as **glycogenolysis.**

glycogenesis
glycogenolysis

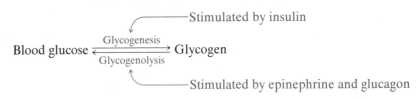

20.4 Lipids

lipids

Lipids are a group of organic substances found in living organisms that are water insoluble; soluble in fat solvents such as diethyl ether, benzene, chloroform, and carbon tetrachloride; and greasy to the touch.

The most abundant lipids are the fats and oils, which make up one of the three important classes of foods.

fat
oil
triglyceride

Fats and **oils** are esters of glycerol and predominantly long-chain fatty acids. Fats and oils are also called **triglycerides,** since each molecule is derived from one molecule of glycerol and three molecules of fatty acid:

Glycerol portion →

$$CH_2-O-\overset{\overset{\displaystyle O}{\|}}{C}-R$$
$$CH-O-\overset{\overset{\displaystyle O}{\|}}{C}-R'$$
$$CH_2-O-\overset{\overset{\displaystyle O}{\|}}{C}-R''$$

General formula
for a triglyceride

$$CH_2-O-\overset{\overset{\displaystyle O}{\|}}{C}-C_{17}H_{35}$$
$$CH-O-\overset{\overset{\displaystyle O}{\|}}{C}-C_{15}H_{31}$$
$$CH_2-O-\overset{\overset{\displaystyle O}{\|}}{C}-C_{11}H_{23}$$

Typical triglyceride
containing three different
fatty acids

The structural formulas of triglyceride molecules vary, for the following reasons:

1. The length of the fatty acid chain may vary from 4 to 20 carbons, but the number of carbon atoms in the chain is nearly always even.

2. Each fatty acid may be saturated, or it may be unsaturated and contain one, two, or three carbon–carbon double bonds.
3. An individual triglyceride may, and frequently does, contain three different fatty acids.

The most abundant saturated fatty acids in fats and oils are lauric, myristic, palmitic, and stearic acids (see Table 19.10). The most abundant unsaturated acids in fats and oils contain 18 carbon atoms and have one, two, or three carbon–carbon double bonds. Their formulas are

$$CH_3(CH_2)_7CH\!=\!CH(CH_2)_7COOH$$
Oleic acid

$$CH_3(CH_2)_4CH\!=\!CHCH_2CH\!=\!CH(CH_2)_7COOH$$
Linoleic acid

$$CH_3CH_2CH\!=\!CHCH_2CH\!=\!CHCH_2CH\!=\!CH(CH_2)_7COOH$$
Linolenic acid

The major physical difference between fats and oils is that fats are solid and oils are liquid at room temperature. Since the glycerol part of the structure is the same for a fat and an oil, the difference must be due to the fatty acid end of the molecule. Fats contain a higher proportion of saturated fatty acids, whereas oils contain higher amounts of unsaturated fatty acids. The term *polyunsaturated* has been popularized in recent years; this means that the molecules of a particular product each contain several double bonds.

Fats and oils are obtained from natural sources. In general, fats come from animal sources and oils from vegetable sources. Thus, lard is obtained from hogs and tallow from cattle and sheep. Olive, cottonseed, corn, soybean, linseed, and other oils are obtained from the fruit or seed of their respective vegetable sources. Table 20.2 shows the major constituents of several fats and oils.

TABLE 20.2 Fatty acid composition of selected fats and oils.

Fat or oil	Fatty acid (%)				
	Myristic acid	Palmitic acid	Stearic acid	Oleic acid	Linoleic acid
Animal fat					
Butter[a]	7–10	23–26	10–13	30–40	4–5
Lard	1–2	28–30	12–18	41–48	6–7
Tallow	3–6	24–32	14–32	35–48	2–4
Vegetable oil					
Olive	0–1	5–15	1–4	49–84	4–12
Peanut	—	6–9	2–6	50–70	13–26
Corn	0–2	7–11	3–4	43–49	34–42
Cottonseed	0–2	19–24	1–2	23–33	40–48
Soybean	0–2	6–10	2–4	21–29	50–59
Linseed[b]	—	4–7	2–5	9–38	3–43

[a]Butyric acid, 3–4%.
[b]Linolenic acid, 25–58%.

Solid fats are preferable to oils for the manufacture of soaps and for use as certain food products. Hardening of oils by hydrogenation to make them solid is carried out on a large commercial scale. In this process hydrogen, bubbled through hot oil containing a finely dispersed nickel catalyst, adds to the carbon–carbon double bonds of the oil to saturate the double bonds and form fats. In practice, only some of the double bonds are allowed to become saturated. The product that is marketed as solid "shortening" (Crisco, Spry, etc.) is used for cooking and baking. Oils and fats are also partially hydrogenated to improve their keeping qualities. Rancidity in fats and oils results from air oxidation at points of unsaturation, producing low-molecular-weight aldehydes and acids of disagreeable odor and flavor.

Fats are an important food source for humans and normally account for about 25–50% of caloric intake. When oxidized to carbon dioxide and water, fats supply about 9.4 kcal of energy per gram, which is more than twice the amount obtained from carbohydrates and proteins.

Fats are digested in the small intestine where they are first emulsified by the bile salts and then hydrolyzed to di- and monoglycerides, fatty acids, and glycerol. In this form, fats are able to pass through the intestinal walls for transport by the blood to various parts of the body where they are broken down in a series of enzyme-catalyzed reactions for the production of potential energy, stored in the form of ATP. Part of the hydrolyzed fat is converted back into body fat and stored for future use. Fats are the major constituent of adipose tissue, which is distributed throughout the body. In addition to being a source of reserve energy, fat deposits function to insulate the body against loss of heat and protect vital organs against mechanical injury.

Three unsaturated fatty acids—linoleic, linolenic, and arachidonic—are essential for animal nutrition and must be supplied in the diet. Diets lacking these fatty acids lead to impaired growth and reproduction, and skin disorders such as eczema and dermatitis. Fats are not required in our diet except as a source of these three fatty acids.

Soap is made by hydrolyzing fats or oils with caustic soda. This hydrolysis process is called *saponification* and requires 3 moles of NaOH per mole of fat:

$$
\begin{array}{l}
CH_2-O-\overset{\displaystyle O}{\overset{\|}{C}}-R \\[4pt]
CH-O-\overset{\displaystyle O}{\overset{\|}{C}}-R' \;+\; 3\,NaOH \longrightarrow \\[4pt]
CH_2-O-\overset{\displaystyle O}{\overset{\|}{C}}-R'' \\
\end{array}
\qquad
\begin{array}{l}
CH_2-OH \quad RCOONa \\[4pt]
CH-OH \;+\; R'COONa \\[4pt]
CH_2-OH \quad R''COONa \\
\end{array}
$$

A fat   Glycerol   Soap

The most common soaps are the sodium salts of long-chain fatty acids, such as sodium stearate, $C_{17}H_{35}COONa$; sodium palmitate, $C_{15}H_{31}COONa$; and sodium oleate, $C_{17}H_{33}COONa$.

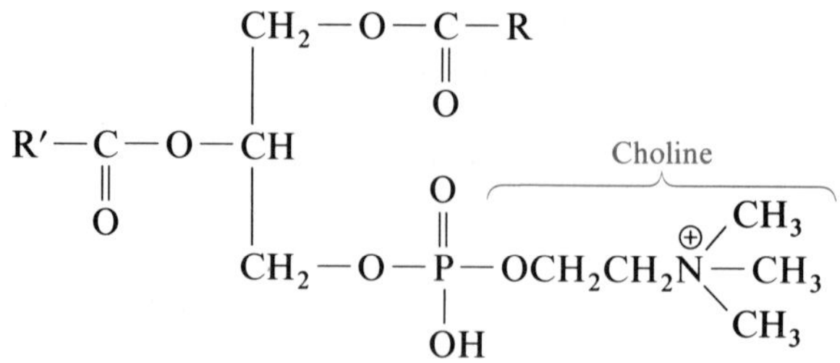

A lecithin (a phospholipid)

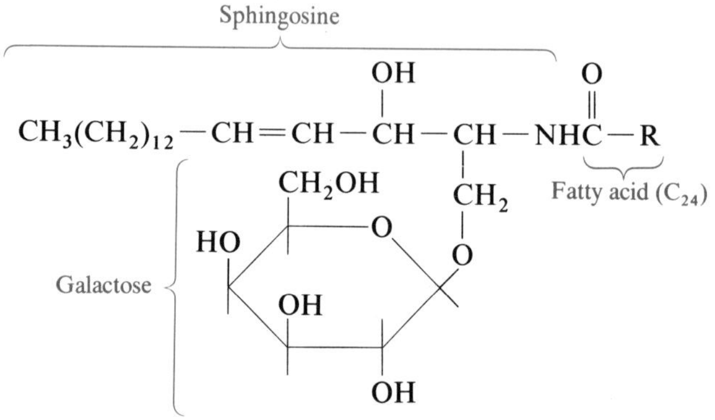

A cerebroside (a glycolipid)

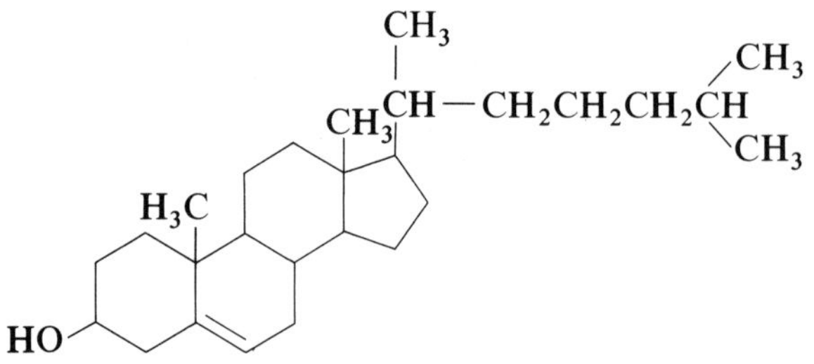

Cholesterol (a steroid)

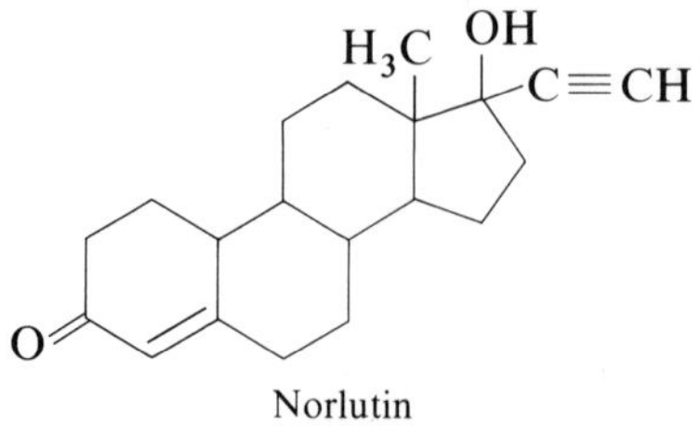

Norlutin
(a birth control pill)

Figure 20.1 Formulas for a phospholipid, a glycolipid, and steroids.

Other principal classes of lipids, besides fats and oils, are phospholipids, glycolipids, and steroids (see Figure 20.1). The phospholipids are found in all animal and vegetable cells and are abundant in the brain, the spinal cord, egg yolk, and liver. Glycolipids (cerebrosides) are not glycerol esters and contain a monosaccharide (usually galactose). Glycolipids are found in many different tissues, but, as the name *cerebroside* indicates, occur in large quantities in brain tissue.

Steroids all have a four-fused-carbocyclic-ring system (as in cholesterol) with various side groups attached to the rings. Cholesterol is the most abundant steroid in the body. It occurs in the brain, the spinal column, and nervous tissue, and it is the principal constituent of gallstones. The body synthesizes about 1 g of cholesterol per day, whereas about 0.3 g per day is ingested in the average diet. The major sources of cholesterol in the diet are meat, liver, and egg yolk. The cholesterol level in the blood generally rises with a person's age and body weight. In recent years, a high blood-level cholesterol has been associated with atherosclerosis (hardening of the arteries). This results in reduced flow of blood and high blood pressure. Cholesterol is needed by the body to synthesize other steroids, some of which regulate male and female sexual characteristics. Many of the synthetic birth control pills such as *norlutin* are modified steroids that interfere with the normal conception cycle in the female.

20.5 Amino Acids and Proteins

Proteins are the third important class of foodstuffs. Some common foods with high (over 10%) protein content are gelatin, fish, beans, nuts, cheese, eggs, poultry, and meat of all kinds. These are the kinds of foods that are most desired and needed and the least available to the undernourished people of the world. Proteins are present in all body tissue including hair, muscle, blood, and skin; they are also present in enzymes and some hormones. About 15% of the human body weight is protein. Chemically, proteins are polymers of amino acids with high molecular weights ranging up to more than 50 million.

amino acids

Amino acids are organic compounds containing two functional groups, an amino group ($-NH_2$) and a carboxyl group ($-COOH$), and another variable group, G. The G group represents any of the various groups that make up the specific amino acids. For example, when G is $H-$, the amino acid is glycine; when G is CH_3-, the amino acid is alanine; when G is $CH_3SCH_2CH_2-$, the amino acid is methionine.

Some amino acids have two amino groups and some contain two acid groups, but all naturally occurring ones have an amino group in the alpha (α) position to the carboxyl group. They are called α-amino acids. The alpha position is the carbon atom adjacent to the carboxyl group. The beta (β) position is the next adjacent carbon, the gamma (γ) position the next carbon, and so on.

$$\overset{\gamma}{C}H_3\overset{\beta}{C}H_2\overset{\alpha}{C}HCOOH \quad \alpha\text{-Aminobutyric acid}$$
$$\underset{NH_2}{|}$$

An alpha(α) amino acid

protein

peptide linkage

Proteins are polymeric substances that yield primarily amino acids on hydrolysis. The bond connecting the amino acids in a protein is commonly called a **peptide linkage** or peptide bond. If we combine two glycine molecules with the elimination of a water molecule between the amino group of one and the carboxyl group of the second glycine, we form a compound containing the amide structure and the peptide linkage. The compound containing the two amino acid groups is called a *dipeptide*.

Amide structure   Peptide linkage

$$CH_2C\overset{O}{\underset{NH_2}{\diagdown}}OH \;+\; \overset{CH_2COOH}{\underset{H-N-H}{|}} \longrightarrow CH_2C\overset{O}{\diagdown} CH_2COOH + H_2O$$

Glycine   Glycine   Glycylglycine (Gly-Gly)
(a dipeptide)

The product formed from two glycine molecules is called *glycylglycine* (abbreviated Gly-Gly). Note that there is still a free amino group at one end and a free carboxyl group at the other end of the molecule. The formation of glycylglycine may be considered to be the first step in the synthesis of a protein, since each end of the molecule is capable of joining to another amino acid. We can thus visualize the formation of a protein by joining a great many amino acids in this fashion. Another example, showing a *tripeptide* (three amino acids linked together), follows. This compound contains two peptide linkages.

$$HO\text{—}\langle \bigcirc \rangle\text{—}\overset{CH_2CHCOOH}{\underset{NH_2}{|}} \;+\; \overset{CH_3CHCOOH}{\underset{NH_2}{|}} \;+\; \overset{CH_2COOH}{\underset{NH_2}{|}}$$

Tyrosine   Alanine   Glycine

$$HO\text{—}\langle \bigcirc \rangle\text{—}\overset{CH_2CHC}{\underset{NH_2}{|}}\overset{O}{\diagdown}\overset{CH_3}{\underset{NH}{\overset{|}{CH}}}-C\overset{O}{\diagdown}CH_2COOH$$

Tyrosylalanylglycine (a tripeptide)   (Tyr-Ala-Gly)

There are five other tripeptide combinations of these three amino acids using only one unit of each amino acid. Peptides containing up to about 40–50 amino acid units in a chain are *polypeptides*. Still longer chains of amino acids are *proteins*.

The names, formulas, and abbreviations of the common amino acids are given in Table 20.3 on pages 552 and 553. Eight of these are considered essential amino acids, since the human body is not capable of synthesizing them. Therefore, they must be supplied in our diets if we are to enjoy normal health.

The number of peptides possible goes up very rapidly as the number of amino acid units increases. For example, there are $120 (1 \times 2 \times 3 \times 4 \times 5 = 120)$ different ways to combine five different amino acids to form a pentapeptide using each amino acid only once in each molecule. If the same constraints are applied to 15 different amino acids, the number of possible combinations is greater than 1 trillion (10^{12})! Since a protein molecule may contain several hundred amino acid units, with individual amino acids occurring several times, the number of possible combinations from 20 amino acids is simply beyond imagination.

There are a number of small, naturally occurring polypeptides with significant biochemical functions. The amino acid sequences of two of these, oxytocin and vasopressin, are shown in Figure 20.2. Oxytocin controls uterine contractions during labor in childbirth and also causes contraction of the smooth muscles of the mammary gland, resulting in milk excretion. Vasopressin in high concentration raises the blood pressure and has been used in surgical shock treatment for this purpose. Vasopressin is also an antidiuretic, regulating the excretion of fluid by the kidneys. The absence of vasopressin leads to diabetes insipidus. This condition is characterized by excretion of up to 30 litres of urine per day, but may be controlled by administration of vasopressin or its derivatives. Oxytocin and vasopressin are similar nonapeptides, differing only at positions 3 and 8.

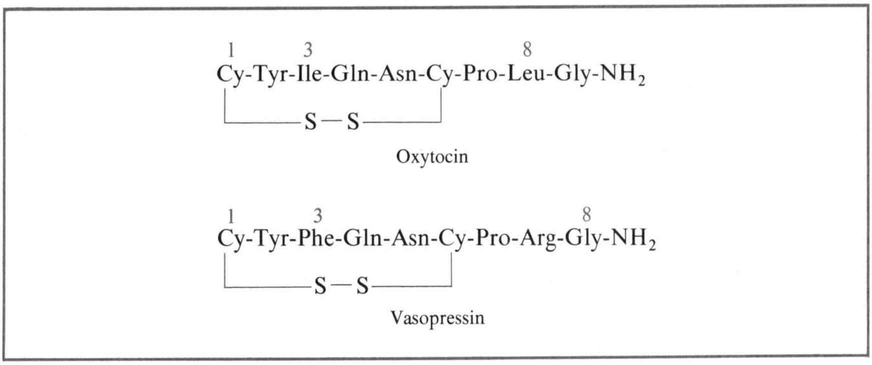

$$\underset{1}{Cy}-Tyr-\underset{3}{Ile}-Gln-Asn-Cy-Pro-Leu-\underset{8}{Gly}-NH_2$$

Oxytocin

$$\underset{1}{Cy}-Tyr-\underset{3}{Phe}-Gln-Asn-Cy-Pro-Arg-\underset{8}{Gly}-NH_2$$

Vasopressin

Figure 20.2 Amino acid sequences of oxytocin and vasopressin. The difference in only two amino acids in these two compounds results in very different physiological activity. The C-terminal amino acid has an amide structure instead of the free COOH (indicated as -Gly-NH$_2$).

TABLE 20.3 Common amino acids derived from proteins.

Name	Abbreviation	Formula
Alanine	Ala	$\underset{\underset{\text{NH}_2}{\mid}}{\text{CH}_3\text{CHCOOH}}$
Arginine	Arg	$\text{NH}_2\underset{\underset{\text{NH}}{\parallel}}{\text{C}}\text{NH}-\text{CH}_2\text{CH}_2\text{CH}_2\underset{\underset{\text{NH}_2}{\mid}}{\text{CHCOOH}}$
Asparagine	Asn	$\text{NH}_2\underset{\underset{\text{O}}{\parallel}}{\text{C}}-\text{CH}_2\underset{\underset{\text{NH}_2}{\mid}}{\text{CHCOOH}}$
Aspartic acid	Asp	$\text{HOOCCH}_2\underset{\underset{\text{NH}_2}{\mid}}{\text{CHCOOH}}$
Cysteine	Cys	$\text{HSCH}_2\underset{\underset{\text{NH}_2}{\mid}}{\text{CHCOOH}}$
Glutamic acid	Glu	$\text{HOOCCH}_2\text{CH}_2\underset{\underset{\text{NH}_2}{\mid}}{\text{CHCOOH}}$
Glutamine	Gln	$\text{NH}_2\underset{\underset{\text{O}}{\parallel}}{\text{C}}\text{CH}_2\text{CH}_2\underset{\underset{\text{NH}_2}{\mid}}{\text{CHCOOH}}$
Glycine	Gly	HCHCOOH with NH_2 below
Histidine	His	(imidazole ring structure) $\text{C}-\text{CH}_2\underset{\underset{\text{NH}_2}{\mid}}{\text{CHCOOH}}$
Isoleucine[a]	Ile	$\text{CH}_3\text{CH}_2\underset{\underset{\text{CH}_3}{\mid}}{\text{CH}}-\underset{\underset{\text{NH}_2}{\mid}}{\text{CHCOOH}}$
Leucine[a]	Leu	$(\text{CH}_3)_2\text{CHCH}_2\underset{\underset{\text{NH}_2}{\mid}}{\text{CHCOOH}}$
Lysine[a]	Lys	$\text{NH}_2\text{CH}_2\text{CH}_2\text{CH}_2\text{CH}_2\underset{\underset{\text{NH}_2}{\mid}}{\text{CHCOOH}}$
Methionine[a]	Met	$\text{CH}_3\text{SCH}_2\text{CH}_2\underset{\underset{\text{NH}_2}{\mid}}{\text{CHCOOH}}$

TABLE 20.3   *Continued*

Name	Abbreviation	Formula
Phenylalanine[a]	Phe	$-CH_2CHCOOH$ with NH_2
Proline	Pro	$-COOH$ (pyrrolidine ring, N-H)
Serine	Ser	$HOCH_2CHCOOH$ with NH_2
Threonine[a]	Thr	$CH_3CH-CHCOOH$ with OH NH_2
Tryptophan[a]	Trp	$C-CH_2CHCOOH$ (indole ring, CH, N-H) with NH_2
Tyrosine	Tyr	$HO-\bigcirc-CH_2CHCOOH$ with NH_2
Valine[a]	Val	$(CH_3)_2CHCHCOOH$ with NH_2

[a]Amino acids essential in human nutrition.

The isolation and synthesis of oxytocin and vasopressin was accomplished by Vincent du Vigneaud (1901–1978) and coworkers at Cornell University. Du Vigneaud was awarded the Nobel Prize in chemistry in 1955 for this work. Synthetic oxytocin is indistinguishable from the natural material.

The primary structure of a protein is established by the number, kind, and sequence of amino acid units comprising the polypeptide chain or chains making up the molecule.

In 1902, Emil Fischer proposed that proteins consisted of amino acid units joined by amide bonds in this way:

$$\sim\sim N-\underset{\underset{H}{|}}{\overset{\overset{H}{|}}{C}}-\underset{}{\overset{\overset{R}{|}}{C}}-N-\underset{\underset{R}{|}}{\overset{\overset{H}{|}}{C}}-\underset{\underset{O}{||}}{C}-N-\underset{\underset{H}{|}}{\overset{\overset{H}{|}}{C}}-\underset{}{\overset{\overset{R}{|}}{C}}-N-\underset{\underset{R}{|}}{\overset{\overset{H}{|}}{C}}-\underset{\underset{O}{||}}{C}\sim\sim$$

Determining the *sequence* of the amino acids in even one protein molecule was a formidable task. The amino acid sequence of beef insulin was announced in 1955 by the British biochemist Frederick Sanger (1918–   ). This structure determination required several years of effort by a team under Sanger's direction. He was awarded the 1958 Nobel Prize in chemistry for this work. Beef insulin consists of 51 amino acid units in two polypeptide chains. The two chains are connected by disulfide linkages ($-S-S-$) of two cysteine residues at two different sites. The structure is shown in Figure 20.3. Insulins from other animals, including humans, differ slightly by one, two, or three amino acid residues.

Protein digestion takes place in the stomach and the small intestine. Here digestive enzymes hydrolyze proteins to smaller peptides and amino acids, which pass through the walls of the intestines, are absorbed by the blood, and transported to the liver and other tissues of the body. The body does not store free (unbonded) amino acids. They are utilized in many ways: (1) to replace and repair body tissue, (2) to synthesize new proteins, (3) to synthesize other nitrogen-containing substances such as enzymes, certain hormones, and heme, (4) to synthesize nucleic acids, and (5) to synthesize other necessary foods such as carbohydrates and fats. Proteins are catabolized (degraded) to carbon dioxide, water, and urea. Urea, containing the protein nitrogen, is eliminated from the body in the urine.

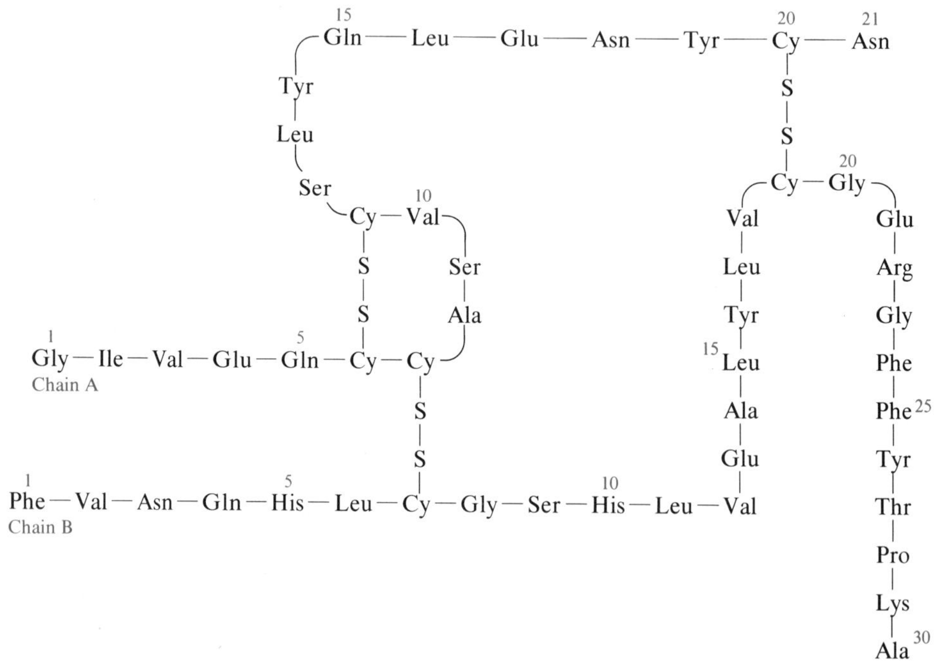

Figure 20.3   Amino acid sequence of beef insulin.

Carbohydrates and fats are used primarily to supply heat and energy to the body. Proteins, on the other hand, are used mainly to repair and replace worn-out tissue. Tissue proteins are continually being broken down and resynthesized. Therefore, protein must be continually supplied to the body in the diet. It is nothing short of amazing how the organism picks out the desired amino acids from the bloodstream and puts them together in proper order to synthesize a needed protein. The synthesis of proteins is controlled by nucleic acids.

20.6 Nucleic Acids

Explaining how hereditary material duplicates itself was one of the most baffling problems of biology. For many years biologists attempted in vain to solve this problem, and also to find an answer to the question, "Why are the offspring of a given species undeniably of that species?" Many thought the chemical basis for heredity lay in the structure of the proteins. But no one was able to provide sufficient evidence as to how protein could reproduce itself. The answer to the heredity problem was finally found in the structure of the nucleic acids.

The unit structure of all living things is the cell. Suspended in the nucleus of cells are chromosomes, which consist largely of proteins and nucleic acids. The nucleic acids and the proteins are intimately associated into complexes called nucleoproteins. There are two types of nucleic acids, those that contain the sugar deoxyribose, and those that contain the sugar ribose. Accordingly, they are called deoxyribonucleic acid (DNA) and ribonucleic acid (RNA). DNA was discovered in 1869 by Swiss physiologist Friedrich Miescher (1844 – 1895), who extracted it from the nuclei of cells.

DNA

DNA is a polymeric substance made up of thousands of units called *nucleotides.* The fundamental components of the nucleotides in DNA are phosphoric acid, deoxyribose (a pentose sugar), and the four nitrogen-containing bases, adenine, thymine, guanine, and cytosine (abbreviated as A, T, G, and C). Phosphoric acid is obtained from minerals in the diet; deoxyribose is synthesized in the body from glucose; and the four nitrogen bases are made in the body from amino acids. The formulas for these compounds are given in Figure 20.4 (page 556).

nucleotide

A **nucleotide** in DNA consists of one of the four bases linked to a deoxyribose sugar which in turn is linked to a phosphate group. Each nucleotide has the following sequence:

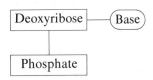

The structures for a single nucleotide and a segment of a polynucleotide (DNA) are shown in Figure 20.5.

Deoxyribonucleic acid (DNA) is a polymeric substance containing thousands of nucleotides. The order in which the four nucleotides (Figure 20.5) occur

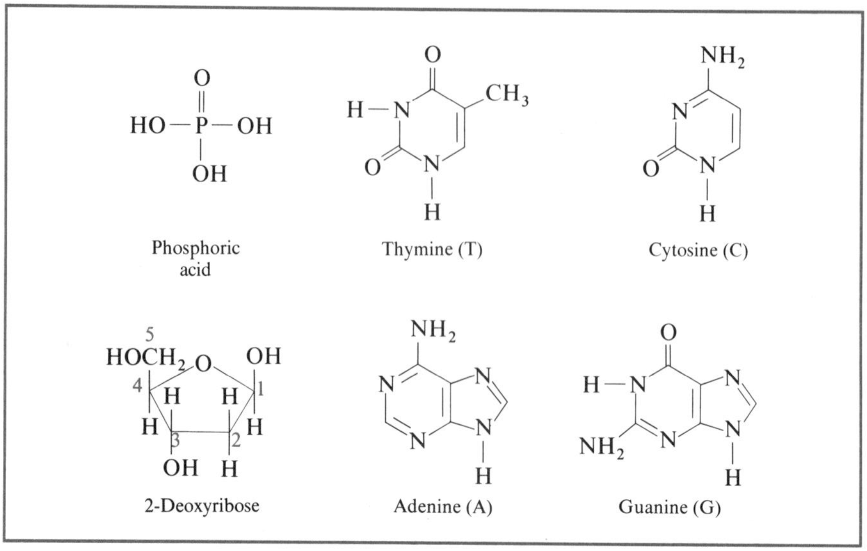

Figure 20.4 Fundamental components of nucleotides.

differs in different DNA molecules, and it is this order that determines the specificity of each DNA molecule.

In 1953, the American biologist James D. Watson (1928–   ) and the British physicist Francis H. C. Crick (1916–   ), announced their now famous double-stranded helix structure for DNA. This was a milestone in the history of biology, and in 1962, Watson and Crick, along with Maurice H. F. Wilkin (1916–   ), who did the brilliant X-ray diffraction studies on DNA, were awarded the Nobel Prize in medicine and physiology.

The structure of DNA, according to Watson and Crick, consists of two polymeric strands of nucleotides in the form of a double helix, with both nucleotide strands coiled around the same axis (see Figure 20.6). Along each strand are alternate phosphate and deoxyribose units with one of the four bases adenine, guanine, cytosine, or thymine attached to deoxyribose as a side group. The double helix is held together by hydrogen bonds extending from the base on one strand of the double helix to a complementary base on the other strand. Furthermore, Watson and Crick ascertained that adenine was always hydrogen bonded to thymine, and guanine was always hydrogen bonded to cytosine. Previous analytical work by others, substantiating this concept of complementary bases, showed that the molar ratio of adenine to thymine in DNA was approximately 1 to 1 and the molar ratio of guanine to cytosine was also approximately 1 to 1.

The structure of DNA has been likened to a ladder that has been twisted into a double helix, with the rungs of the ladder kept perpendicular to the twisted railings. The phosphate and deoxyribose units alternate along the two railings of the ladder and two nitrogen bases form each rung of the ladder. The

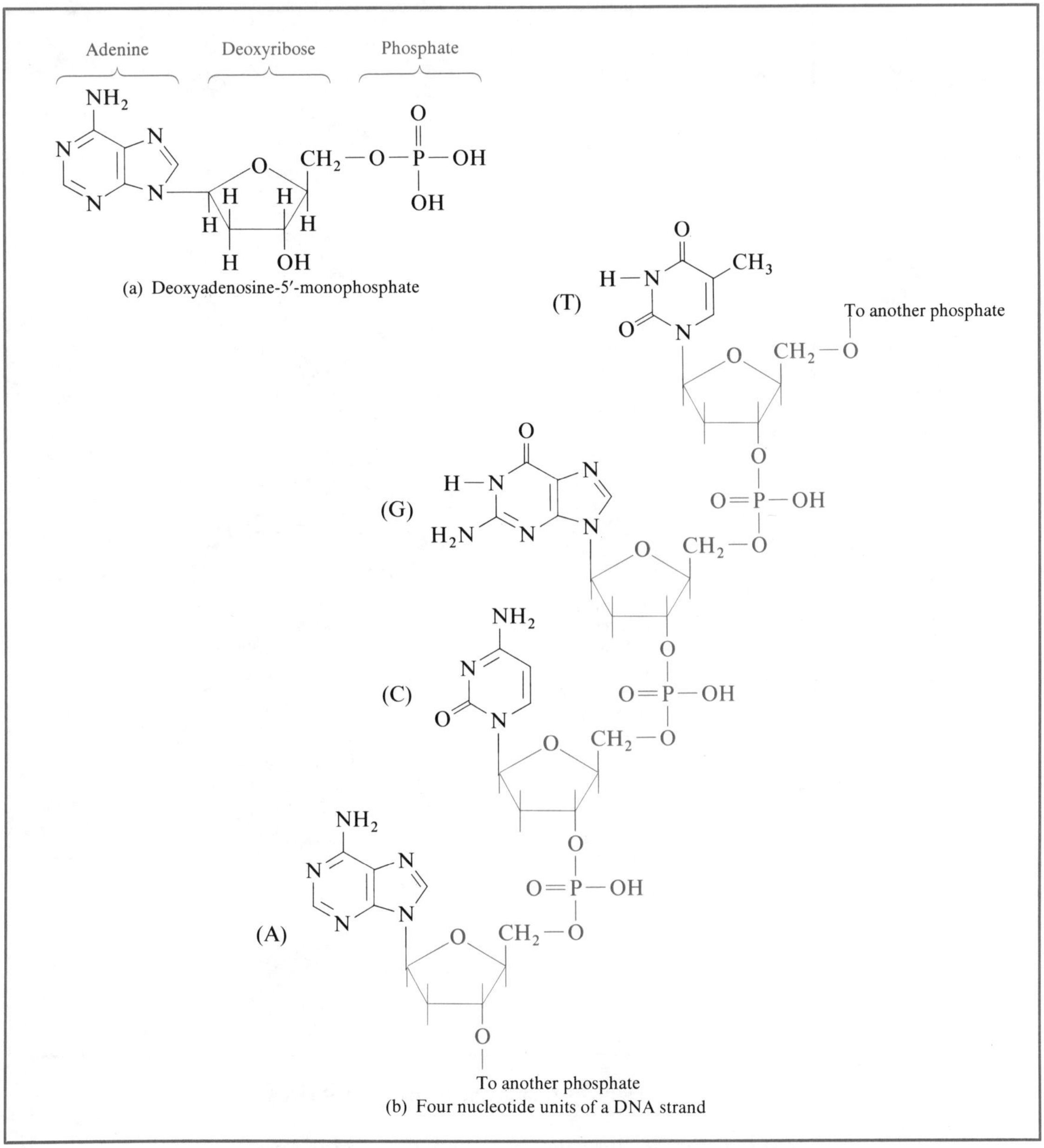

(a) Deoxyadenosine-5'-monophosphate

(b) Four nucleotide units of a DNA strand

Figure 20.5 (a) A single nucleotide, adenine deoxyribonucleotide (deoxy-adenosine-5'-monophosphate). (b) A segment of one strand of deoxyribonucleic acid (DNA) showing four nucleotides, including those of adenine (A), cytosine (C), guanine (G), and thymine (T). The names of the last three nucleotides, are, respectively: cytosine deoxyribonucleic acid (deoxycytidine-5'-monophosphate); guanine deoxyribonucleic acid (deoxyguanosine-5'-monophosphate); and thymine deoxyribonucleic acid (deoxythymidine-5'-monophosphate).

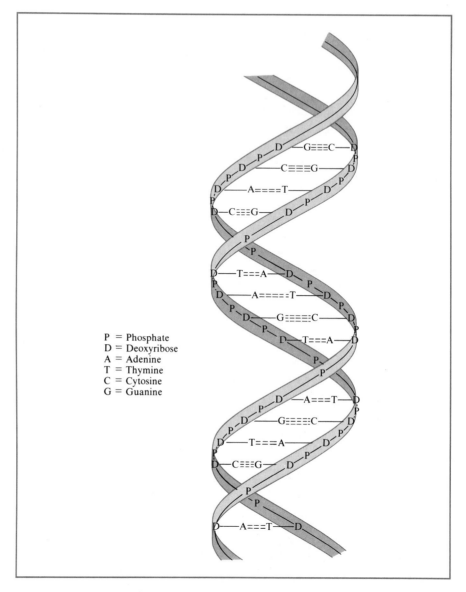

P = Phosphate
D = Deoxyribose
A = Adenine
T = Thymine
C = Cytosine
G = Guanine

Figure 20.6 Double-stranded helix structure of DNA.

DNA structure is illustrated in Figure 20.6, and the detailed pairing of complementary bases is shown in Figure 20.7.

For any individual of any species, the sequence of base combinations and the length of the nucleotide chains in DNA molecules contain the coded messages that determine all of the characteristics of that individual. The sequence of bases also contains the code for reproduction of that species. In this sense the DNA molecule is like a template or a computer that stores information for recall as needed. DNA contains the genetic code of life, which is passed on from one generation to another.

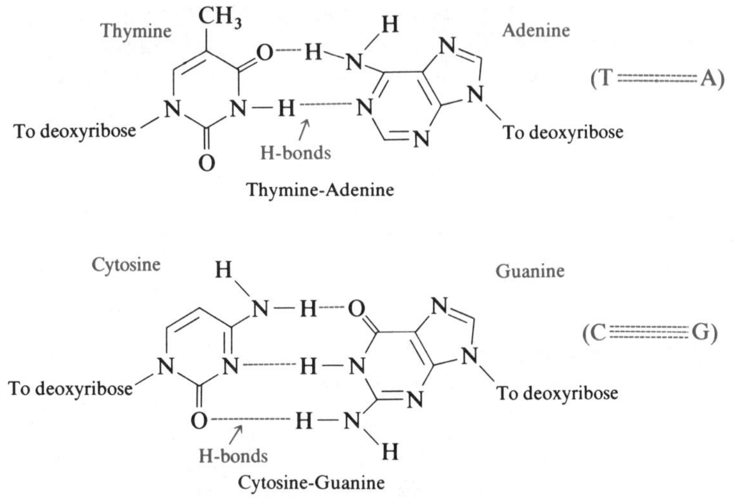

Figure 20.7 Hydrogen bonding between the complementary bases thymine and adenine (T═══A) and cytosine and guanine (C═══G). Note that one pair of bases has two hydrogen bonds and the other pair has three hydrogen bonds between them.

RNA

One important function of DNA is in the formation of RNA (ribonucleic acid). **RNA** is a polymer of nucleotides but differs from DNA in that (1) it exists in the form of a single-stranded helix, (2) it contains the pentose sugar ribose instead of deoxyribose, and (3) it contains uracil instead of thymine as one of its four nitrogen bases.

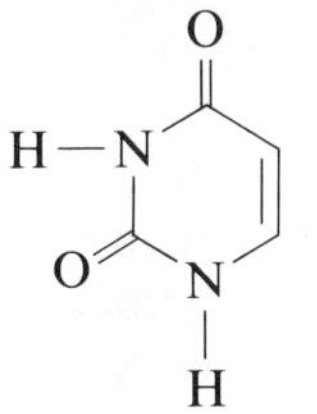

The making of RNA from DNA is called transcription. The nucleotide sequence of only one strand of DNA is transcribed into a single strand of RNA. This transcription occurs in a complementary fashion. Where there is a guanine base in DNA, a cytosine base will occur in RNA. Cytosine is transcribed to guanine, thymine to adenine, and adenine to uracil.

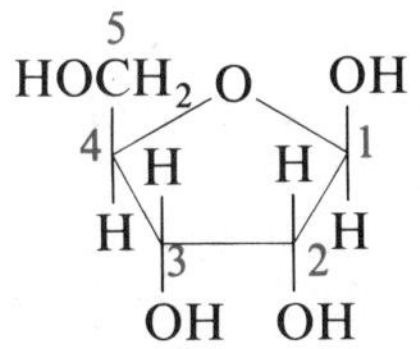

The main function of RNA is to direct the synthesis of proteins. RNA is produced in the cell nucleus but performs its function outside of the nucleus. Three kinds of RNA are produced directly from DNA: messenger RNA (*m*RNA), transfer RNA (*t*RNA), and ribosomal RNA (*r*RNA). Messenger RNA contains bases in the exact order prescribed by a strip of the master code from

DNA. The base sequence on the *m*RNA in turn establishes the sequence of amino acids that are put together to make a specific protein. The function of the relatively small transfer RNA molecules is to bring specific amino acids to the site of protein synthesis. There is at least one different *t*RNA for each amino acid. The actual site of protein synthesis is a *ribosome,* which is composed of *r*RNA and protein. The function of the *r*RNA is not completely understood. However, the ribosome is believed to move along the *m*RNA chain and to aid in the polymerization of amino acids in the order prescribed by the base sequence of the *m*RNA chain. The flow of genetic information is in one direction, from DNA to RNA to proteins.

20.7 DNA and Genetics

Heredity is the process by which the physical and mental characteristics of parents are transferred to their offspring. In order for this to occur it is necessary for the material responsible for genetic transfer to be able to make exact copies of itself. The design for replication is built into the DNA structure of Watson and Crick, first by the nature of its double helical structure and second by the complementary nature of its nitrogen bases where adenine will only bond to thymine and guanine to cytosine. The DNA double helix unwinds, or simply "unzips," into two separate helices at the hydrogen bonds between the bases. Each helix then serves as a template combining only with the proper free nucleotides to produce two identical replicas of itself. This replication of DNA occurs in the cell just before the cell divides, thereby giving each daughter cell the full genetic code of the cell that it came from. This process is illustrated in Figure 20.8.

DNA, as we have indicated before, is an integral part of the chromosomes. Each species carries a specific number of chromosomes in the nucleus of each of its cells. The number of chromosomes varies with different species. Humans have 23 pairs, or 46 chromosomes. Each chromosome contains strings of *genes,* or DNA molecules.

In ordinary cell division, known as *mitosis,* each DNA molecule forms a duplicate by uncoiling to single strands. Each strand then assembles the *complementary* portion from available free nucleotides to form a duplicate of the original DNA molecule. After cell division is completed, each daughter cell contains DNA molecules (genes) that correspond exactly to those that were present in the original cell before division.

However, in almost all higher forms of life reproduction takes place by union of the male sperm with the female egg. Cell splitting to form the sperm cell and the egg cell occurs by a different and more complicated process called *meiosis.* In meiosis the sperm cell carries only half of the chromosomes from its original cell and the egg cell also carries only half the chromosomes from its original cell. Between them they form a new cell that once again contains the correct number of chromosomes and all of the hereditary characteristics of the species. Thus, the offspring derives half of its genetic characteristics from the father and half from the mother.

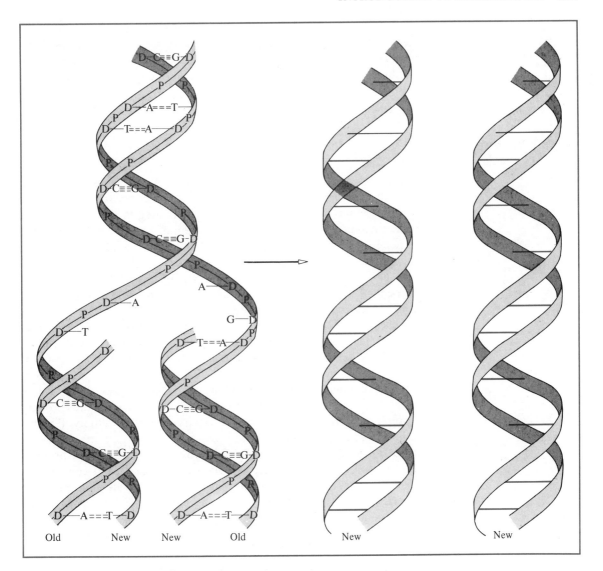

Figure 20.8   Method of replication of DNA. The two helices unwind, separating at the point of the hydrogen bonds. Each strand then serves as a template, recombining with the proper nucleotides to duplicate itself as a double-stranded helix.

Nature is not 100% perfect. Occasionally, DNA replication is not perfect or a section of the DNA molecule is damaged by X rays, radioactive rays, or drugs, and a mutant organism is produced. In the disease of sickle-cell anemia, a large proportion of the red blood cells form into sickle shapes instead of the usual globular shape. This limits the ability of the blood to carry oxygen and causes the person to be weak and unable to fight infection, leading to early death. Sickle-cell anemia is due to one misplaced amino acid in the structure of hemoglobin.

The sickle-cell-producing hemoglobin has a valine molecule where a glutamic acid molecule should be located. Sickle-cell anemia is an inherited disease indicating a fault in the DNA coding that is transmitted from parent to child. Many biological disorders and ailments have been traced directly to a deficiency in the genetic information of DNA.

20.8 Enzymes

enzymes

Enzymes are the catalysts of biochemical reactions. All enzymes are proteins and they catalyze nearly all of the myriad reactions that occur in living cells. Uncatalyzed reactions that may require hours of boiling in the presence of a strong acid or a strong base may occur in a fraction of a second in the presence of the proper enzyme at room temperature and nearly neutral pH. This is all the more remarkable when we realize that enzymes do not actually cause chemical reactions. They act as catalysts by greatly lowering the activation energy of specific biochemical reactions. The lowered activation energy permits these reactions to proceed at high speed at body temperature.

Louis Pasteur (1822–1895) was one of the first scientists to study enzyme-catalyzed reactions. He believed that living yeasts or bacteria were required for these reactions, which he called *fermentations*—for example, the conversion of glucose to alcohol by yeasts. In 1897, Eduard Büchner (1860–1917) made a cell-free filtrate that contained enzymes prepared by grinding yeast cells with very fine sand. The enzymes in this filtrate converted glucose to alcohol, thus proving that the presence of living cells was not required for enzyme activity. For this work Büchner received the Nobel Prize in chemistry in 1907.

Each organism contains thousands of enzymes. Some are simple proteins consisting only of amino acid units. Others are conjugated and consist of a protein part, or *apoenzyme*, and a nonprotein part, or *coenzyme*. Both parts are essential, and a functioning enzyme consisting of both the protein and non-protein parts is called a *holoenzyme*.

Apoenzyme + Coenzyme = Holoenzyme

Often the coenzyme is a vitamin, and the same coenzyme may be associated with many different enzymes.

For some enzymes an inorganic component such as a metal ion—for example, Ca^{2+}, Mg^{2+}, or Zn^{2+}—is required. This inorganic component is an *activator*. From the standpoint of function, an activator is analogous to a coenzyme, but inorganic components are not called coenzymes.

Another remarkable property of enzymes is their specificity of reaction; that is, a certain enzyme will catalyze the reaction of a specific type of substance. For example, the enzyme maltase catalyzes the reaction of maltose and water to form glucose. Maltase has no effect on the other two common disaccharides, sucrose and lactose. Each of these sugars requires a specific enzyme; sucrase to hydrolyze sucrose; lactase to hydrolyze lactose. (See hydrolysis equations in Section 20.3.)

The substance acted on by an enzyme is called the *substrate.* Sucrose is the substrate of the enzyme sucrase. Enzymes have been named by adding the suffix *-ase* to the root of the substrate name. Note the derivations of maltase, sucrase, and lactase from maltose, sucrose, and lactose. Many enzymes, especially digestive enzymes, have trivial names such as pepsin, rennin, trypsin, and so on. These names have no systematic significance.

Enzymes act according to the following general sequence. Enzyme (E) and substrate (S) combine to form an enzyme–substrate intermediate (E–S). This intermediate decomposes to give the product (P) and regenerate the enzyme:

$$E + S \rightleftharpoons E\text{–}S \longrightarrow E + P$$

For the hydrolysis of maltose the sequence is

$$\underset{\text{E}}{\text{Maltase}} + \underset{\text{S}}{\text{Maltose}} \longrightarrow \underset{\text{E–S}}{\text{Maltase–Maltose}}$$

$$\underset{\text{E–S}}{\text{Maltase–Maltose}} + H_2O \longrightarrow \underset{\text{E}}{\text{Maltase}} + \underset{\text{P}}{\text{Glucose}}$$

Enzyme specificity is believed to be due to the particular shape of a small part of the enzyme, which exactly fits a complementary-shaped part of the substrate (see Figure 20.9). It is analogous to a lock and key; the substrate is the lock and the enzyme the key. Just as a key will open only the lock it fits, the enzyme will act only on a molecule that fits its particular shape. When the substrate and the enzyme come together, they form a substrate–enzyme complex unit. The substrate, activated by the enzyme in the complex, reacts to form the products, regenerating the enzyme.

A more recently suggested model of the enzyme–substrate catalytic site is known as the "induced fit" model. In this model, the enzyme site of attachment

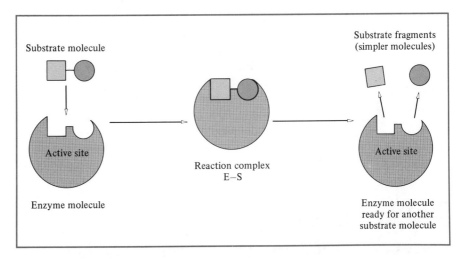

Figure 20.9 Enzyme–substrate interaction illustrating specificity of an enzyme by the lock-and-key analogy.

to the substrate is flexible, with the substrate inducing a change in the enzyme shape to fit the shape of the substrate. This theory allows for the possibility that in some cases the enzyme might wrap itself around the substrate and so form the correct shape of lock and key. Thus, the enzyme does not need to have an exact preformed catalytic site to match the substrate.

20.9 Vitamins

vitamins

Vitamins are a group of unrelated naturally occurring organic compounds that are essential for normal growth, nutrition, and maintenance of life. Animals maintained on a diet consisting of only proteins, fats, carbohydrates, and the necessary minerals and water are not able to sustain life. Although vitamins are required in only minute amounts, normal nutrition, growth, and development are not possible without them. The minimum daily adult requirement may be as much as 100 milligrams for some vitamins and as little as 0.1 microgram for others. The human body is unable to synthesize these substances and is dependent on vitamins being supplied in the diet. A substance that functions as a vitamin for one species does not necessarily function as a vitamin for another species.

A prolonged lack of vitamins in the diet leads to vitamin deficiency diseases such as beriberi, pellagra, pernicious anemia, rickets, and scurvy. Some of these vitamin deficiency diseases may be corrected by feeding supplementary amounts of vitamins. However, it is especially important for young children to have sufficient vitamins for proper growth and development. For example, it is difficult to correct distorted bone structures that have developed because of a lack of vitamin D. Most vitamins are manufactured synthetically and are available as dietary supplements, although a balanced diet should supply all the necessary vitamins.

The chemical composition of vitamins varies greatly; some are relatively simple substances, but others are extremely complex. On the basis of solubility characteristics, vitamins are generally divided into two groups: fat-soluble and water-soluble. The fat-soluble vitamins include A, D, E, and K. Those classified as water-soluble are vitamin C and the B complex, which includes about a dozen different compounds. One of the B vitamins, pantothenic acid, is a component of coenzyme A and therefore takes part in a great many metabolic reactions. Table 20.4 lists some of the important vitamins, their main food sources, and deficiency symptoms.

Most of the vitamins are known to function as coenzymes. For example, the B complex vitamins riboflavin, niacin, and pantothenic acid have coenzyme functions in glucose metabolism.

20.10 Hormones

hormones

Hormones are chemical substances that act as control or regulatory agents in the body. They help to regulate overall physiological processes such as

TABLE 20.4 Some of the most important vitamins. Alternative names are given in parentheses.

Vitamin	Important dietary sources	Some deficiency symptoms
Vitamin A (retinol)	Green and yellow vegetables, butter, eggs, nuts, cheese, fish liver oil	Poor teeth and gums, night blindness
Vitamin B_1 (thiamin)	Meat, whole-grain cereals, liver, yeast, nuts	Beriberi (nervous system disorders, heart disease, fatigue)
Vitamin B_2 (riboflavin)	Meat, cheese, eggs, fish, meat products, liver	Sores on the tongue and lips, bloodshot eyes, anemia
Vitamin B_6 (pyridoxine)	Cereals, liver, meat, fresh vegetables	Skin disorders (dermatitis)
Vitamin B_{12} (cyanocobalamin)	Meat, eggs, liver, milk	Pernicious anemia
Vitamin C (ascorbic acid)	Citrus fruits, tomatoes, green vegetables	Scurvy (bleeding gums, loose teeth, swollen joints, slow healing of wounds, weight loss)
Vitamin D (calciferol)	Egg yolk, milk, fish liver oils; formed from provitamin in the skin when exposed to sunlight	Rickets (low blood calcium level, soft bones, distorted skeletal structure)
Vitamin E (tocopherol)	Widely distributed in foods, meat, egg yolk, wheat germ oil, green vegetables	Not definitely known in humans
Vitamin K (phylloquinone)	Eggs, liver, green vegetables; produced in the intestines by bacterial reactions	Blood is slow to clot (antihemorrhagic vitamin)
Niacin (nicotinic acid and amide)	Meat, yeast, whole wheat	Pellagra (dermatitis, diarrhea, mental disorders)
Biotin (vitamin H)	Liver, yeast, egg yolk	Skin disorders (dermatitis)
Folic acid	Liver extract, wheat germ, yeast, green leaves	Macrocytic anemia, gastrointestinal disorders

digestion, metabolism, growth, and reproduction. For example, the concentration of glucose in the blood is maintained within definite limits by the action of hormones. Hormones are secreted by the endocrine, or ductless, glands directly into the bloodstream prior to use and are transported to various parts of the body to exert specific control functions. The endocrine glands include the pituitary, thyroid, parathyroid, pancreas, adrenal, ovaries, testes, placenta, and certain portions of the gastrointestinal tract. A hormone produced by one species is usually active in some other species. For example, the insulin used to treat diabetes mellitus in humans is obtained from the pancreas of animals slaughtered in meat packing plants. Hormones are often referred to as the chemical messengers of the body. They do not fit into any single chemical structural classification. Many are proteins or polypeptides, some are steroids, some are phenol or amino acid derivatives. Since a lack of any hormone may produce

TABLE 20.5 Selected hormones and their functions.

Hormone	Principal Functions
Insulin	Controls blood sugar level and storage of glycogen
Oxytocin	Stimulates contraction of the uterine muscles and secretion of milk by the mammary glands
Vasopressin	Controls water excretion by the kidneys; stimulates constriction of the blood vessels
Growth hormone	Stimulates and controls growth in all tissues
Prolactin	Stimulates milk production by mammary glands shortly after the birth of a baby
Epinephrine (adrenaline)	Stimulates rise in blood pressure, acceleration of heartbeat, decreased secretion of insulin, and increased blood sugar
Cortisone	Helps control carbohydrate metabolism, salt and water balance, formation and storage of glycogen
Thyroxine and triiodothyroxine	Increase the metabolic rate of carbohydrates and proteins; control the metabolic rate of most body cells
Calcitonin	Prevents the rise of calcium in the blood above the required level
Gastrin	Stimulates secretion of gastric juices
Estrogens	Stimulate development and maintenance of female sexual characteristics
Progesterone	Stimulates female sexual characteristics and prepares the uterine lining for the fertilized egg
Testosterone	Stimulates development and maintenance of male sexual characteristics

serious physiological disorders, many of them are produced synthetically or are extracted from their natural sources and made available for medical use.

Like the vitamins, hormones are generally needed in only minute amounts. Concentrations range from 10^{-6} to 10^{-12} M. Unlike vitamins, which must be supplied in the diet, the necessary hormones are produced in the body of a healthy person. A number of hormones and their functions are listed in Table 20.5

QUESTIONS A. **Review the Meanings of the New Terms Introduced in this Chapter**

1. Biochemistry
2. Photosynthesis
3. Carbohydrates
4. Monosaccharides
5. Disaccharides
6. Polysaccharides
7. Glycogenesis
8. Glycogenolysis
9. Lipids
10. Fats
11. Oils
12. Triglycerides
13. Amino acids
14. Proteins
15. Peptide linkage
16. DNA
17. Nucleotide
18. RNA
19. Enzymes
20. Vitamins
21. Hormones

B. Answers to the Following Questions Will Be Found in Tables and Figures

1. Of the sugars tabulated in Table 20.1, which is the sweetest disaccharide? Which is the sweetest monosaccharide? (Invert sugar is a mixture of glucose and fructose so it should not be considered.)
2. According to Table 20.2, are the fatty acids in vegetable oils more saturated or unsaturated than those in animal fats? Explain your answer.
3. Which of the common amino acids in Table 20.3 have more than one carboxyl group? Which have more than one amino group?
4. How many disulfide linkages are there in each molecule of beef insulin? (See Figure 20.3.)
5. In the four nucleotide units of DNA shown in Figure 20.5, which of these components are part of the backbone chain, and which are off to the side: the nitrogen bases, the deoxyribose, and the phosphoric acid?
6. In the double-stranded helix structure of DNA (Figure 20.6), which nitrogen bases are always hydrogen bonded to each of the following: cytosine, thymine, adenine, and guanine?
7. Give the dietary sources and deficiency symptoms for these vitamins: A, B_1, B_2, B_{12}, C, D, and niacin. (See Table 20.4.)
8. What are the principal functions of these hormones: insulin, vasopressin, epinephrine, cortisone, gastrin, estrogens, progesterone, and testosterone? (See Table 20.5.)

C. Review Questions

1. Life is dependent on four major classes of biomolecules. What are they?
2. What substances are responsible for the synthesis of carbohydrates in photosynthesis? Write an equation to illustrate this process.
3. What is an aldose, an aldotetrose, a ketose, a ketohexose? Give an example of each.
4. Classify each of the following as a monosaccharide, disaccharide, or polysaccharide: glucose, sucrose, maltose, fructose, cellulose, lactose, glycogen, galactose, starch, and ribose.
5. Draw structural formulas in the open-chain form for ribose, glucose, fructose, and galactose.
6. Draw structural formulas in the cyclic form for ribose, glucose, fructose, and galactose.
7. State the properties and the sources of ribose, glucose, fructose, and galactose.
8. The molecular formula for lactic acid is $C_3H_6O_3$, and its structural formula is $CH_3CH(OH)COOH$. Is this compound a carbohydrate? Explain.
9. What is the monosaccharide composition of (a) sucrose, (b) maltose, (c) lactose, (d) starch, (c) cellulose, and (f) glycogen?
10. Draw structural formulas in the cyclic form for sucrose and maltose.
11. If the most common monosaccharides have the formula $C_6H_{12}O_6$, why do the resulting disaccharides have the formula $C_{12}H_{22}O_{11}$, rather than $C_{12}H_{24}O_{12}$?
12. Write equations, using structural formulas in the cyclic form, for the hydrolysis of (a) sucrose, and (b) maltose. (c) What enzymes catalyze these reactions?
13. Discuss the similarities and differences between starch and cellulose.
14. In what form is carbohydrate stored in the body?
15. Discuss, in simple terms, the metabolism of carbohydrates in the human body.
16. State the natural sources of sucrose, maltose, lactose, and starch.
17. Invert sugar, obtained by the hydrolysis of sucrose to an equal-molar mixture of fructose and glucose, is commonly used as a sweetener in commercial food preparations. Why is invert sugar sweeter than the original sucrose?
18. What properties of molecules cause them to be classified as lipids?
19. Write structural formulas for glycerol, stearic acid, palmitic acid, oleic acid, and linoleic acid.

20. Distinguish both chemically and physically between a fat and a vegetable oil.
21. What is a triglyceride? Give an example.
22. Write the structure for tristearin, a fat in which all the fatty acid units are stearic acid.
23. Write the structure of a triglyceride that contains one unit each of linoleic, stearic, and oleic acids. How many other formulas are possible in which the triglyceride contains one unit each of these acids?
24. Write equations for the saponification of (a) tripalmitin, and (b) the triglyceride of Question 23. Name the product(s) that is a soap.
25. How can vegetable oils be "hardened"? What is the advantage of "hardening" these oils?
26. What functions do fats have in the human body?
27. Which fatty acids are essential to human diets?
28. Draw the structural formula of cholesterol.
29. Draw the ring structure that is common to all steroids.
30. List six foods that are major sources of proteins.
31. What functional groups are present in amino acids?
32. Why are the amino acids of proteins called α-amino acids?
33. Write out the full structure for the two possible dipeptides containing glycine and phenylalanine.
34. Write structures for (a) glycylglycine, (b) glycylglycylalanine, and (c) leucyl-methionylglycylserine.
35. Using amino acid abbreviations, write all the possible tripeptides containing one unit each of glycine, phenylalanine, and leucine.
36. What are essential amino acids? Write the names of the amino acids that are essential to humans.
37. What is meant by the primary structure of a protein? What is its significance?
38. When proteins are eaten by a human, what are the metabolic fates of the protein material?
39. Why should protein be continually included in a balanced diet?
40. Write structural formulas for the compounds that make up DNA.
41. (a) What are the three units that make up a nucleotide?
    (b) List the components of the four types of nucleotides found in DNA.
    (c) Write the structure and name of one of these nucleotides.
42. Briefly describe the structure of DNA as proposed by Watson and Crick.
43. What is the role of hydrogen bonding in the structure of DNA?
44. Explain the concept of complementary bases and how it relates to DNA.
45. A segment of a DNA strand has a base sequence of C-G-A-T-T-G-C-A. What is the base sequence of the other complementary strand of the double helix?
46. Explain the replication process of DNA.
47. Briefly discuss the relationship of DNA to genetics.
48. What are the three differences between DNA and RNA in terms of structure.
49. Distinguish between cell division in mitosis and in meiosis.
50. What are enzymes and what is their role in the body?
51. What is meant by specificity of an enzyme?
52. What are vitamins and what is their function in the body?
53. State, in a general way, the consequences of vitamin deficiencies.
54. What are hormones, and what is their function in the body?
55. How do vitamins and hormones differ in origin?
56. Which of the following statements are correct?
    (a) Biochemistry is the branch of chemistry that is concerned with the chemical reactions occurring in living organisms.
    (b) Photosynthesis is the name of the overall process by which green plants use sunlight to help in the digestion of carbohydrates.

(c) Carbohydrates are polyhydroxy aldehydes or polyhydroxy ketones, or compounds that will yield them when hydrolyzed.

(d) The body stores glucose as glycogen until it is needed.

(e) The ultimate use of carbohydrates in the body is oxidation to carbon dioxide and water and the utilization of the energy released.

(f) A disaccharide molecule consists of two monosaccharide units linked together, minus a water molecule.

(g) The most common lipids are fats and oils.

(h) Oils have a higher percentage of saturated fatty acids than fats.

(i) Fats are digested in the stomach and converted to monosaccharides.

(j) Fats are burned in the body or stored as fats for future use.

(k) Proteins are high-molecular-weight polymers of amino acids.

(l) The bond connecting the amino acids in a protein is commonly called a peptide linkage.

(m) The double helix of DNA is tied together by hydrogen bonds between the deoxyribose and the phosphoric acid units.

(n) When a cell divides, the DNA double helix unwinds and each helix serves as a template combining only with the proper free nucleotides to produce two identical replicas of itself.

(o) Examples of polysaccharides are starch, olive oil, and cellulose.

(p) A major difference between RNA and DNA is that RNA contains the pentose ribose and DNA contains deoxyribose.

(q) A nucleotide is made up of components called nucleic acids.

(r) The backbone of each strand of DNA is alternating molecules of deoxyribose and nitrogen bases.

(s) The disaccharide found in mammalian milk is galactose.

(t) Enzymes are proteins.

(u) The substrate acted on by an enzyme is called a coenzyme.

(v) DNA is a polymer made from nucleotides.

(w) The molar ratio of adenine to thymine and guanine to cytosine is about 1:1 in DNA.

(x) The main function of RNA is to see that each cell contains 46 chromosomes.

(y) The process of synthesizing glycogen from glucose is called glycogenesis.

(z) Vitamins must be supplied in the diet since they are not synthesized in the body.

Review Exercises for Chapters 17–20

CHAPTER 17
OXIDATION–
REDUCTION

True–False: Answer the following as either true or false.

1. In oxidation, the oxidation number of an element increases in a positive direction as a result of gaining electrons.
2. The oxidation number of chlorine in Cl_2 is -1.
3. Oxidation and reduction occur simultaneously in a chemical reaction; one cannot take place without the other.
4. In the electrolysis of water, the anode reaction is:

$$2\,H_2O \longrightarrow O_2 + 4\,H^+ + 4\,e^-$$

5. The cathode is the electrode at which oxidation takes place.
6. The change in the oxidation number of an element from -2 to 0 is reduction.
7. A free metal can displace from solution the ions of a metal that lies below the free metal in the activity series.
8. Metallic zinc will react with hydrochloric acid.
9. In electroplating, the piece to be electroplated with a metal is attached to the cathode.
10. In a lead storage battery, $PbSO_4$ is produced at both electrodes in the discharging cycle.
11. As a lead storage battery discharges, the electrolyte becomes less dense.
12. The algebraic sum of the oxidation numbers of all the atoms in $K_2Cr_2O_7$ is zero.
13. A reducing agent will always decrease in oxidation number.
14. Potassium is a better reducing agent than sodium.

Multiple Choice: Choose the correct answer to each of the following.

1. In K_2SO_4, the oxidation number of sulfur is:
   (a) $+2$ (b) $+4$ (c) $+6$ (d) -2
2. In $Ba(NO_3)_2$, the oxidation number of N is:
   (a) $+5$ (b) -3 (c) $+4$ (d) -1
3. In the reaction $H_2S + 4\,Br_2 + 4\,H_2O \longrightarrow H_2SO_4 + 8\,HBr$, the oxidizing agent is:
   (a) H_2S (b) Br_2 (c) H_2O (d) H_2SO_4
4. In the reaction $VO_3^- + Fe^{2+} + 4\,H^+ \longrightarrow VO^{2+} + Fe^{3+} + 2\,H_2O$, the element reduced is:
   (a) V   (b) Fe   (c) O   (d) H

Questions 5, 6, and 7 pertain to the activity series
K Ca Mg Al Zn Fe H Cu Ag

5. Which of the following pairs will not react in water solution?
   (a) $Zn, CuSO_4$ (b) $Cu, Al_2(SO_4)_3$ (c) $Fe, AgNO_3$ (d) $Ca, Al_2(SO_4)_3$
6. Which element is the most easily oxidized?
   (a) K   (b) Mg   (c) Zn   (d) Cu
7. Which element will reduce Cu^{2+} to Cu but will not reduce Zn^{2+} to Zn?
   (a) Fe   (b) Ca   (c) Ag   (d) Mg
8. In the electrolysis of fused (molten) $CaCl_2$, the product at the negative electrode is:
   (a) Ca^{2+} (b) Cl^- (c) Cl_2 (d) Ca
9. In its reactions, a free element from Group IIA in the periodic table is most likely to:
   (a) be oxidized   (b) be reduced   (c) be unreactive   (d) gain electrons
10. In the partially balanced redox equation

$$3\,Cu + HNO_3 \longrightarrow 3\,Cu(NO_3)_2 + 2\,NO + \quad H_2O$$

the coefficient needed to balance H_2O is:
   (a) 8   (b) 6   (c) 4   (d) 2

11. Which reaction does not involve oxidation–reduction?
    (a) Burning sodium in chlorine
    (b) Chemical union of Fe and S
    (c) Decomposition of $KClO_3$
    (d) Neutralization of NaOH with H_2SO_4

12. How many moles of Fe^{2+} can be oxidized to Fe^{3+} by 2.50 moles of Cl_2 according to the following equation?

$$Fe^{2+} + Cl_2 \longrightarrow Fe^{3+} + Cl^-$$

    (a) 2.50 moles   (b) 5.00 moles   (c) 1.00 mole   (d) 22.4 moles

13. How many grams of sulfur can be produced from 100 mL of 6.00 M HNO_3?

$$HNO_3 + H_2S \longrightarrow S + NO + H_2O$$

    (a) 28.9 g   (b) 19.3 g   (c) 32.1 g   (d) 289 g

Balancing Oxidation–Reduction Equations: Balance each of the following equations.
1. $P + HNO_3 \longrightarrow HPO_3 + NO + H_2O$
2. $MnSO_4 + PbO_2 + H_2SO_4 \longrightarrow HMnO_4 + PbSO_4 + H_2O$
3. $Cr_2O_7^{2-} + H^+ + Cl^- \longrightarrow Cr^{3+} + H_2O + Cl_2$
4. $MnO_4^- + AsO_3^{3-} + H^+ \longrightarrow Mn^{2+} + AsO_4^{3-} + H_2O$
5. $As_2O_3 + Cl_2 + H_2O \longrightarrow H_3AsO_4 + HCl$

CHAPTER 18
RADIOACTIVITY
AND NUCLEAR
CHEMISTRY

True–False: Answer the following as either true or false.
1. Radioactivity was first discovered by Antoine Henri Becquerel.
2. A gamma ray has more penetrating ability than either an alpha or beta particle.
3. A thin sheet of paper will normally block beta radiation.
4. In an electromagnetic field, an alpha particle undergoes greater deflection than does a beta particle.
5. The half-life is the time required for one-half of a specified amount of a radioactive element to disintegrate.
6. When a nucleus emits an alpha particle or a beta particle, it always changes into an isotope of a different element.
7. A beta particle consists of two protons and two neutrons.
8. When an atom loses a beta particle from its nucleus, a new element is formed, having essentially the same mass and an atomic number one greater than the starting element.
9. A radioactive disintegration series shows the succession of alpha and beta emissions by which naturally occurring radioactive elements decay to reach stability.
10. The fission of U-235 can become a chain reaction because of all the energy liberated.
11. The energy from nuclear fission can be harnessed to produce steam, which can drive turbines and produce electricity.
12. The process of uniting the nuclei of two light elements to form one heavier nucleus is known as nuclear fusion.
13. The high energy released from nuclear fusion gives other nuclei enough kinetic energy to sustain a chain reaction.
14. In radiocarbon dating, the ratio of C-14 to C-12 gives data relative to the age of the object being dated.
15. Cancers are often treated by gamma radiation from cobalt-60, which destroys the rapidly growing cancer cells.
16. Protracted exposure to low levels of any form of ionizing radiation can weaken the body and lead to the onset of malignant tumors.
17. There are 337 nucleons in an atom of $^{235}_{92}U$.

18. A positron has the mass of an electron and the electrical charge of a proton.
19. Transmutation is the changing of an isotope of one element to an isotope of another element.
20. *Rem* is the unit of radiation that relates to the biological effect of absorbed radioactive radiation.

Multiple Choice: Choose the correct answer to each of the following.

1. If $^{238}_{92}U$ loses an alpha particle, the resulting isotope is:
   (a) $^{237}_{92}U$ (b) $^{234}_{90}Th$ (c) $^{238}_{93}Np$ (d) $^{210}_{83}Bi$

2. If $^{210}_{82}Pb$ loses a beta particle, the resulting isotope is:
   (a) $^{209}_{83}Bi$ (b) $^{210}_{81}Ti$ (c) $^{206}_{80}Hg$ (d) $^{210}_{83}Bi$

3. In the equation $^{209}_{83}Bi + \underline{\qquad} \longrightarrow {}^{210}_{84}Po + {}^{1}_{0}n$, the missing bombarding particle would be:
   (a) $^{2}_{1}H$ (b) $^{1}_{0}n$ (c) $^{4}_{2}He$ (d) $_{-1}^{0}e$

4. Which of the following is not a characteristic of nuclear fission?
   (a) Upon absorption of a proton, a heavy nucleus splits into two or more smaller nuclei.
   (b) Two or more neutrons are produced from the fission of each atom.
   (c) Large quantities of energy are produced.
   (d) All nuclei formed are radioactive, giving off beta and gamma radiation.

5. The half-life of Sn-121 is 10 days. If you started with 40 g of this isotope, how much would you have left 30 days later?
   (a) 10 g   (b) None   (c) 15 g   (d) 5 g

6. $^{241}_{94}Pu$ successively emits $\beta, \alpha, \alpha, \beta, \alpha, \alpha$. At that point, the isotope has become:
   (a) $^{225}_{94}Pu$ (b) $^{225}_{88}Ra$ (c) $^{207}_{84}Po$ (d) $^{219}_{84}Po$

7. Calculate the binding energy of $^{56}_{26}Fe$. Mass data: $^{56}_{26}Fe = 55.9349$ g/mole; n = 1.0087 g/mole; p = 1.0073 g/mole; $e^{-} = 0.00055$ g/mole; 1.0 g = 2.2×10^{13} cal
   (a) 1.2×10^{13} cal/mole   (c) 0.5302 g/mole
   (b) 56.4651 g/mole   (d) 1.2×10^{15} cal/mole

8. The radioactive ray with the greatest penetrating ability is:
   (a) Alpha   (b) Beta   (c) Gamma   (d) Proton

9. In a nuclear reaction:
   (a) Mass is lost   (c) Mass is converted into energy
   (b) Mass is gained   (d) Energy is converted into mass

10. As the temperature of a radioisotope increases, its half-life:
    (a) Increases   (b) Decreases   (c) Remains the same   (d) Fluctuates

11. The isotope that has the longest half-life is:
    (a) $^{238}_{92}U$ (b) $^{210}_{82}Pb$ (c) $^{234}_{90}Th$ (d) $^{222}_{88}Ra$

12. Which of the following is not a unit of radiation?
    (a) Curie   (b) Roentgen   (c) Rod   (d) Rem

13. When $^{235}_{92}U$ is bombarded by a neutron, the atom can fission into:
    (a) $^{124}_{53}I + {}^{109}_{47}Ag + 2\,{}^{1}_{0}n$ (c) $^{134}_{56}Ba + {}^{128}_{36}Xe + 2\,{}^{1}_{0}n$
    (b) $^{123}_{50}Sn + {}^{110}_{42}Mo + 2\,{}^{1}_{0}n$ (d) $^{90}_{38}Sr + {}^{143}_{58}Ce + 2\,{}^{1}_{0}n$

14. In the nuclear equation

    $$^{45}_{21}Sc + {}^{1}_{0}n \longrightarrow X + {}^{1}_{1}H$$

    the isotope X that is formed is:
    (a) $^{45}_{22}Ti$ (b) $^{45}_{20}Ca$ (c) $^{46}_{22}Ti$ (d) $^{45}_{20}K$

CHAPTER 19
ORGANIC
CHEMISTRY

True–False: Answer the following as either true or false.
1. When a glucose solution ferments, carbon dioxide and ethyl alcohol are the products.
2. Organic chemistry is the chemistry of the organs of the body.

3. Hydrocarbons are compounds composed entirely of carbon and hydrogen atoms bonded to each other with covalent bonds.
4. Alkanes, alkenes, and alkynes are all hydrocarbons.
5. The IUPAC name for $CH_3-CH-CH-CH_2-CH_3$ is 3,4-dimethylpentane.

                                     CH_3 CH_3

6. If Cl_2 is added to $CH_3CH=CH_2$, the product is $CH_3CH_2CH_2Cl$.
7. Ethylene can be formed from ethanol by dehydration with sulfuric acid.
8. When each member of a series of compounds differs from the next higher member by a CH_2 group, the series is called a homologous series.
9. The IUPAC name for $CH_3CHClCH_3$ is chloropropane.

                            O
                            ||

10. The carbonyl group is $-C-OH$.

                            O
                            ||

11. Formaldehyde is $H-C-H$.
12. A positive Tollens' test is the appearance of a silver mirror on the inner walls of a test tube.
13. 2-Methylpentane and 2,3-dimethylbutane are isomers.
14. Dimethyl ketone and propanal have the same molecular weight.
15. A compound of formula $C_6H_{12}O$ cannot be a carboxylic acid.
16. Acetylene is the common name for ethyne.
17. Although ethyl alcohol is used in beverages, it is classified physiologically as a depressant and a poison.
18. Methyl alcohol is a poisonous substance that can lead to blindness if ingested.
19. Ethanal may be distinguished from propanal by reaction with Tollens reagent.
20. The oxidizing agent in Benedict's reagent is Cu^+.

Multiple Choice: Choose the correct answer to each of the following.
1. Which of the following is not a correct name for the alkane shown with it?
    (a) C_2H_6, ethane   (c) C_7H_{16}, heptane
    (b) C_5H_{12}, propane   (d) $C_{10}H_{22}$, decane
2. If CH_3CH_2OH is reacted with PBr_3, the product will be:
    (a) CH_3CH_2Br (c) $CH_3CH_2-O-CH_2CH_3$
    (b) CH_3CBr_3 (d) CH_3CH_2P
3. The structural formula of *o*-xylene is:

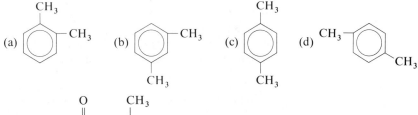

4. The ester $CH_3C-O-CHCH_3$ can be made from which alcohol and carboxylic acid?

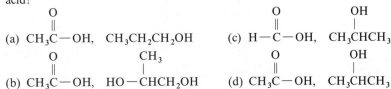

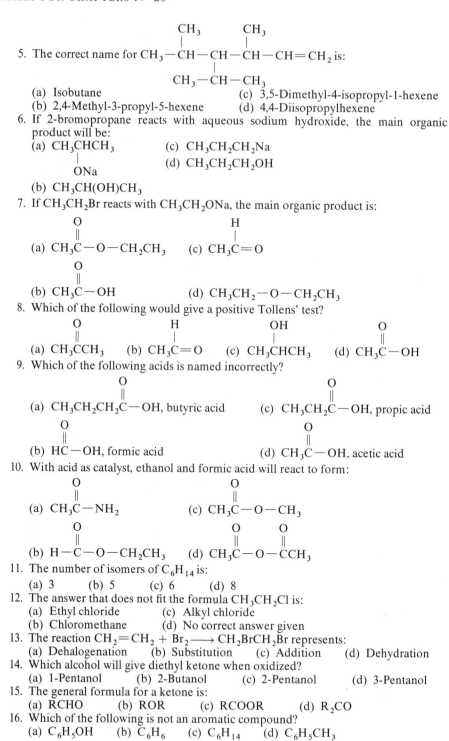

5. The correct name for $CH_3-CH-CH-CH-CH=CH_2$ is:

   (a) Isobutane   (c) 3,5-Dimethyl-4-isopropyl-1-hexene
   (b) 2,4-Methyl-3-propyl-5-hexene   (d) 4,4-Diisopropylhexene

6. If 2-bromopropane reacts with aqueous sodium hydroxide, the main organic product will be:

   (a) CH_3CHCH_3 (c) $CH_3CH_2CH_2Na$
         |
        ONa   (d) $CH_3CH_2CH_2OH$

   (b) $CH_3CH(OH)CH_3$

7. If CH_3CH_2Br reacts with CH_3CH_2ONa, the main organic product is:

   (a) $CH_3C-O-CH_2CH_3$ (c) $CH_3C=O$

   (b) CH_3C-OH (d) $CH_3CH_2-O-CH_2CH_3$

8. Which of the following would give a positive Tollens' test?

   (a) CH_3CCH_3 (b) $CH_3C=O$ (c) CH_3CHCH_3 (d) CH_3C-OH

9. Which of the following acids is named incorrectly?

   (a) $CH_3CH_2CH_2C-OH$, butyric acid   (c) CH_3CH_2C-OH, propic acid

   (b) $HC-OH$, formic acid   (d) CH_3C-OH, acetic acid

10. With acid as catalyst, ethanol and formic acid will react to form:

    (a) CH_3C-NH_2 (c) $CH_3C-O-CH_3$

    (b) $H-C-O-CH_2CH_3$ (d) $CH_3C-O-CCH_3$

11. The number of isomers of C_6H_{14} is:

    (a) 3   (b) 5   (c) 6   (d) 8

12. The answer that does not fit the formula CH_3CH_2Cl is:

    (a) Ethyl chloride   (c) Alkyl chloride
    (b) Chloromethane   (d) No correct answer given

13. The reaction $CH_2=CH_2 + Br_2 \longrightarrow CH_2BrCH_2Br$ represents:

    (a) Dehalogenation   (b) Substitution   (c) Addition   (d) Dehydration

14. Which alcohol will give diethyl ketone when oxidized?

    (a) 1-Pentanol   (b) 2-Butanol   (c) 2-Pentanol   (d) 3-Pentanol

15. The general formula for a ketone is:

    (a) RCHO   (b) ROR   (c) RCOOR   (d) R_2CO

16. Which of the following is not an aromatic compound?

    (a) C_6H_5OH (b) C_6H_6 (c) C_6H_{14} (d) $C_6H_5CH_3$

17. The 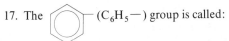 (C_6H_5-) group is called:

    (a) Phenol   (b) Hexyl   (c) Phenyl   (d) Benzyl

18. What is the correct name for the following?

    (a) *m*-Dinitrophenol   (c) 3,5-Dinitrophenol
    (b) 2,4-Dinitrophenol   (d) 1,3-Dinitrophenol

19. Which compound is the weakest acid of those listed?

    (a) H_2O (b) CH_3CH_2OH (c) C_6H_5OH (d) CH_3COOH

20. Which of the following can be oxidized to a carboxylic acid by $K_2Cr_2O_7 + H^+$?

    (a) C_3H_8 (b) CH_3CCH_3 (with $\parallel$ O below)   (c) CH_3CCH_3 (with OH below)   (d)

CHAPTER 20
INTRODUCTION TO BIOCHEMISTRY

True–False: Answer the following as either true or false.

1. Photosynthesis is the name of the overall process by which green plants use sunlight to help in the digestion of carbohydrates.
2. Photosynthesis consumes oxygen and produces carbon dioxide.
3. Carbohydrates are polyhydroxy aldehydes or polyhydroxy ketones, or compounds that yield them when hydrolyzed.
4. The body stores glucose as glycogen until it is needed.
5. The ultimate use of carbohydrates in the body is oxidation to carbon dioxide and the utilization of the energy released.
6. The structure of a disaccharide molecule consists of two monosaccharide units linked together minus a water molecule.
7. Polysaccharides are macromolecules made up of many monosaccharide units linked together.
8. Maltose is a disaccharide composed of two glucose units.
9. Starch and cellulose are both polysaccharides, but cellulose consists only of glucose units whereas starch is alternating glucose and fructose units.
10. The most common lipids are fats and oils.
11. Fats and oils are esters of glycerol and the higher-molecular-weight fatty acids.
12. Oils have a higher percentage of saturated fatty acids than fats.
13. Fats are digested in the stomach and converted to monosaccharides.
14. Fats are burned in the body or stored as fats for future use.
15. Palmitic acid contains one carbon–carbon double bond.
16. Vegetable oils are "hardened" by reacting them with sodium hydroxide.
17. When a fat is saponified, the products are glycerol and soap molecules.
18. Proteins are high-molecular-weight polymers of amino acids.
19. Amino acids are organic compounds containing two functional groups: an amino group and a carboxyl group.
20. $H_2N-\overset{\overset{O}{\parallel}}{C}-OH$ is an alpha amino acid.
21. The bond connecting the amino acids in a protein is commonly called a peptide linkage.

22. The double helix of DNA is tied together by hydrogen bonds between the deoxyribose and the phosphoric acid units.
23. For any individual of any species, the sequence of base combinations and the length of the nucleotide chains in DNA molecules contain the coded messages that determine all the characteristics of the individual.
24. The flow of genetic information is in one direction, from DNA to RNA to proteins.
25. Enzymes are protein molecules that act as catalysts by greatly lowering the activation energy of specific biochemical reactions.

Multiple Choice: Choose the correct answer to each of the following.
1. Sugars are members of a group of compounds with the general name:
   (a) Carbohydrates   (b) Lipids   (c) Proteins   (d) Steroids
2. The products formed when maltose is hydrolyzed are:
   (a) Glucose and fructose   (c) Glucose and glucose
   (b) Glucose and galactose   (d) Galactose and fructose
3. Which is not true about starch?
   (a) It is a polysaccharide.
   (b) It is hydrolyzed to maltose.
   (c) It is composed of glucose units.
   (d) It is not digestible by humans.
4. Lactose is:
   (a) A monosaccharide
   (b) A disaccharide composed of galactose and glucose
   (c) A disaccharide composed of two glucose units
   (d) A decomposition product of starch
5. Which is not true about glucose?
   (a) It is a monosaccharide.
   (b) It is a component of sucrose, maltose, lactose, starch, glycogen, and cellulose.
   (c) It is a ketohexose.
   (d) It is the main source of energy for the body.
6. The sweetest of the common sugars is:
   (a) Fructose   (b) Sucrose   (c) Glucose   (d) Maltose
7. Which of the following is not true?
   (a) Lactose is a disaccharide known as milk sugar.
   (b) Glycogen is known as animal starch.
   (c) Cellulose is the most abundant organic substance in nature.
   (d) The synthesis of glycogen from glucose is called glycogenolysis.
8. Which is not formed in the saponification of a fat?
   (a) Glycerol   (c) Soap
   (b) Amino acids   (d) A metal salt of a long chain fatty acid
9. Which is not an essential fatty acid?
   (a) Oleic   (b) Linoleic   (c) Linolenic   (d) Arachidonic
10. Which of the following lipids does not contain a glycerol unit as part of its structure?
   (a) A fat   (b) A phospholipid   (c) A glycolipid   (d) An oil
11. An alpha amino acid always contains:
   (a) An amino group on the carbon atom adjacent to the carboxyl group
   (b) A carboxyl group at each end of the molecule
   (c) Two amino groups
   (d) Alternating amino and carboxyl groups
12. Which of the following amino acids contains sulfur?
   (a) Alanine   (b) Histidine   (c) Cysteine   (d) Glycine
13. A compound containing ten amino acid molecules linked together is called a:
   (a) Protein   (b) Polypeptide   (c) Deca-amino acid   (d) Nucleotide
14. The major end-product(s) of protein nitrogen metabolism in humans is(are):
   (a) Amino acids   (b) Ammonium salts   (c) Urea   (d) Dipeptides

15. Which of the following is not a correct statement about DNA and RNA?
    (a) DNA contains deoxyribose, while RNA contains ribose.
    (b) Both DNA and RNA are polymers made up of nucleotides.
    (c) DNA directs the synthesis of proteins and RNA contains the genetic code of life.
    (d) DNA exists as a double helix, while RNA exists as a single helix.
16. Which of the following bases is found in RNA, but not in DNA?
    (a) Thymine   (b) Adenine   (c) Guanine   (d) Uracil
17. Complementary base pairs in DNA are linked through the formation of:
    (a) Phosphate ester bonds   (c) Hydrogen bonds
    (b) Peptide linkages   (d) Ionic bonds
18. In a DNA double helix, hydrogen bonding occurs between:
    (a) Adenine and thymine   (c) Adenine and uracil
    (b) Thymine and guanine   (d) Cytosine and thymine
19. Which of the following scientists did not receive the Nobel Prize for the structure of DNA?
    (a) Crick   (b) Watson   (c) Sanger   (d) Wilkins
20. The substance acted on by an enzyme is called a(an):
    (a) Catalyst   (b) Apoenzyme   (c) Coenzyme   (d) Substrate

Appendix I
Mathematical Review

1. Multiplication. Multiplication is a process of adding any given number or quantity a certain number of times. Thus, 4 times 2 means 4 added two times, or 2 added together four times, to give the product 8. Various ways of expressing multiplication are

$$ab \qquad a \times b \qquad a \cdot b \qquad a(b) \qquad (a)(b)$$

All mean a times b, or a multiplied by b, or b times a.

When $a = 16$ and $b = 24$, we have $16 \times 24 = 384$.

The expression $°F = (1.8 \times °C) + 32$ means that we are to multiply 1.8 times $°C$ and add 32 to the product. When $°C$ equal 50,

$$°F = (1.8 \times 50) + 32 = 90 + 32 = 122°F$$

The result of multiplying two or more numbers together is known as the *product*.

2. Division. The word *division* has several meanings. As a mathematical expression, it is the process of finding how many times one number or quantity is contained in another. Various ways of expressing division are

$$a \div b \qquad \frac{a}{b} \qquad a/b$$

All mean a divided by b.

When $a = 15$ and $b = 3$, $\dfrac{15}{3} = 5$.

The number above the line is called the *numerator*; the number below the line is the *denominator*. Both the horizontal and the slanted (/) division signs also mean "per." For example, in the expression for density, the mass per unit volume:

$$\text{Density} = \text{Mass}/\text{Volume} = \frac{\text{Mass}}{\text{Volume}} = \text{g}/\text{mL}$$

The diagonal line still refers to a division of grams by the number of millilitres occupied by that weight.

The result of dividing one number into another is called the *quotient*.

3. Fractions and decimals.   A fraction is an expression of division, showing that the numerator is divided by the denominator. A *proper fraction* is one in which the numerator is smaller than the denominator. In an *improper fraction,* the numerator is the larger number. A decimal or a decimal fraction is a proper fraction in which the denominator is some power of 10. The decimal fraction is determined by carrying out the division of the proper fraction. Examples of proper fractions and their decimal fraction equivalents are shown in the following table.

Proper fraction		*Decimal fraction*		*Proper fraction*
$\dfrac{1}{8}$	$=$	0.125	$=$	$\dfrac{125}{1000}$
$\dfrac{1}{10}$	$=$	0.1	$=$	$\dfrac{1}{10}$
$\dfrac{3}{4}$	$=$	0.75	$=$	$\dfrac{75}{100}$
$\dfrac{1}{100}$	$=$	0.01	$=$	$\dfrac{1}{100}$
$\dfrac{1}{4}$	$=$	0.25	$=$	$\dfrac{25}{100}$

4. Addition of numbers with decimals.   To add numbers with decimals, we use the same procedure as that used when adding whole numbers, but always line up the decimal points in the same column. For example, add 8.21 + 143.1 + 0.325

```
      8.21-
 +  143.1--
 +    0.325
    151.635
```

When adding numbers expressing units of measurement, always be certain that the numbers added together represent the same units. For example, what is the total length of these three pieces of glass tubing: 10.0 cm, 125 mm, 8.4 cm? If we add these directly, we obtain a value of 143.4, but we are not certain what the unit of measurement is. To add these lengths correctly, first change 125 mm to 12.5 cm. Now all the lengths are expressed in the same units and can be added.

10.0 cm
12.5 cm
 8.4 cm
30.9 cm

5. Subtraction of numbers with decimals. To subtract numbers containing decimals, we use the same procedure as for subtracting whole numbers, but always line up the decimal points in the same column. For example, subtract 20.60 from 182.49.

$$
\begin{array}{r}
182.49 \\
- \ \ 20.60 \\
\hline
161.89
\end{array}
$$

6. Multiplication of numbers with decimals. To multiply two or more numbers together that contain decimals, we first multiply as if they were whole numbers. To locate the decimal point in the product, we add together the number of digits to the right of the decimal in all the numbers multiplied together. The product should contain this total number of digits to the right of the decimal point.

Multiply 2.05×2.05 (total of four digits to the right of the decimal):

$$
\begin{array}{r}
2.05 \\
\times \ 2.05 \\
\hline
1025 \\
4100 \\
\hline
\end{array}
$$
4.2025   (Four digits to the right of the decimal)

Here are more examples:

$14.25 \times 6.01 \times 0.75 = 64.231875$ (Six digits to the right of the decimal)

$39.26 \times 60 = 2355.60$ (Two digits to the right of the decimal)

[*Note:*   When at least one of the numbers that is multiplied is a measurement, the answer must be adjusted to contain the correct number of significant figures. (See Section 2.2 on significant figures.)]

7. Division of numbers with decimals. To divide numbers containing decimals, we first relocate the decimal points of the numerator and denominator by moving them to the right as many places as needed to make the denominator a whole number. (Move the decimal of both the numerator and the denominator the same amount and in the same direction.) For example,

$$
\frac{136.94}{4.1} = \frac{1369.4}{41}
$$

The decimal point adjustment in this example is equivalent to multiplying both numerator and denominator by 10. Now we carry out the division normally, locating the decimal point immediately above its position in the dividend.

$$
\begin{array}{r}
33.4 \\
41\overline{)1369.4} \\
123 \\
\hline
139 \\
123 \\
\hline
164 \\
164 \\
\hline
\end{array}
\qquad
\frac{0.441}{26.25} = \frac{44.1}{2625} =
\begin{array}{r}
0.0168 \\
2625\overline{)44.1000} \\
2625 \\
\hline
17850 \\
15750 \\
\hline
21000 \\
21000 \\
\hline
\end{array}
$$

[*Note:* When at least one of the numbers in the division is a measurement, the answer must be adjusted to contain the correct number of significant figures. (See Section 2.2 on significant figures.)]

The foregoing examples are merely guides to the principles used in performing the various mathematical operations illustrated. There are, no doubt, shortcuts and other methods, and the student will discover these with experience. Every student of chemistry should use either an electronic calculator or a slide rule for solving problems. The use of these devices will save many hours of time that would otherwise be spent in doing tedious longhand calculations. After solving a problem, the student should check for errors and evaluate the answer to see if it is logical and consistent with the data given.

8. Algebraic equations.   Many mathematical problems that are first encountered in chemistry fall into the following algebraic forms. Solutions to these problems are simplified by first isolating the desired term on one side of the equation. This is accomplished by treating both sides of the equation in an identical manner (so as not to destroy the equality) until the desired term is isolated.

(a) $a = \dfrac{b}{c}$

To solve for a, simply divide b by c.
To solve for b, multiply both sides of the equation by c.

$$a \times c = \frac{b}{\cancel{c}} \times \cancel{c}$$

$$b = a \times c$$

To solve for c, multiply both sides of the equation by $\dfrac{c}{a}$.

$$\cancel{a} \times \frac{c}{\cancel{a}} = \frac{b}{\cancel{c}} \times \frac{\cancel{c}}{a}$$

$$c = \frac{b}{a}$$

(b) $\dfrac{a}{b} = \dfrac{c}{d}$

To solve for a, multiply both sides of the equation by b.

$$\frac{a}{\cancel{b}} \times \cancel{b} = \frac{c}{d} \times b$$

$$a = \frac{c \times b}{d}$$

To solve for b, multiply both sides of the equation by $\dfrac{b \times d}{c}$.

$$\frac{a}{\cancel{b}} \times \frac{\cancel{b} \times d}{c} = \frac{\cancel{c}}{\cancel{d}} \times \frac{b \times \cancel{d}}{\cancel{c}}$$

$$b = \frac{a \times d}{c}$$

(c) $a \times b = c \times d$

To solve for a, divide both sides of the equation by b.

$$\frac{a \times \cancel{b}}{\cancel{b}} = \frac{c \times d}{b}$$

$$a = \frac{c \times d}{b}$$

(d) $\dfrac{(b - c)}{a} = d$

To solve for b, first multiply both sides of the equation by a:

$$\frac{\cancel{a}(b - c)}{\cancel{a}} = d \times a$$

$$b - c = d \times a$$

Then add c to both sides of the equation:

$$b - \cancel{c} + \cancel{c} = d \times a + c$$

$$b = (d \times a) + c$$

When $a = 1.8$, $c = 32$, and $d = 35$,

$$b = (35 \times 1.8) + 32 = 63 + 32 = 95$$

9. Exponents; powers of 10; expression of large and small numbers.   In scientific measurements and calculations, we often encounter very large and very small numbers; for example, 0.00000384 and 602,000,000,000,000,000,000,000.

These numbers are troublesome to write and awkward to work with, especially in calculations. A convenient method of expressing these large and small numbers in a simplified form is by means of exponents or powers of 10. This method of expressing numbers is known as **scientific** or **exponential notation.**

An *exponent* is a number written as a superscript following another number; it is also called a *power* of that number, and it indicates how many times the number is used as a factor. In the number 10^2, 2 is the exponent and the number means 10 squared, or 10 to the second power, or $10 \times 10 = 100$. Three other examples are

$$3^2 = 3 \times 3 = 9$$

$$3^4 = 3 \times 3 \times 3 \times 3 = 81$$

$$10^3 = 10 \times 10 \times 10 = 1000$$

For ease of handling, large and small numbers are expressed in powers of 10. Powers of 10 are used because multiplying or dividing by 10 coincides with moving the decimal point in a number by one place. Thus, a number multiplied by 10^1 would move the decimal point one place to the right; 10^2, two places to the right; 10^{-2}, two places to the left. To express a number in powers of 10, we move the decimal point in the original number to a new position, placing it so that the number is a value between 1 and 10. This new decimal number is multiplied by 10 raised to the proper power. For example, to write the number 42,389 in exponential form (powers of 10), the decimal point is placed between the 4 and the 2 (4.2389) and the number is multiplied by 10^4; thus, the number is 4.2389×10^4.

$$42,389 = 4.2389 \times 10^4$$
$$\overset{\frown\frown\frown}{4\ 3\ 2\ 1}$$

The power of 10 (4) tells us the number of places that the decimal point must be moved to restore it to its original position. The exponent of 10 is determined by counting the number of places that the decimal point is moved from its original position. If the decimal point is moved to the left, the exponent is a positive number; if it is moved to the right, the exponent is a negative number. To express the number 0.00248 in exponential notation (as a power of 10), the decimal point is moved three places to the right; the exponent of 10 is -3, and the number is 2.48×10^{-3}.

$$0.00248 = 2.48 \times 10^{-3}$$
$$\overset{\smile\smile\smile}{1\ 2\ 3}$$

Study the following examples.

$$1237 = 1.237 \times 10^3$$

$$988 = 9.88 \times 10^2$$

$$147.2 = 1.472 \times 10^2$$

$$2,200,000 = 2.2 \times 10^6$$

$$0.0123 \quad = 1.23 \times 10^{-2}$$

$$0.00005 \quad = 5 \times 10^{-5}$$

$$0.000368 = 3.68 \times 10^{-4}$$

Exponents in multiplication and division.   The use of powers of 10 in multiplication and division greatly simplifies locating the decimal point in the answer. In multiplication, first change all numbers to powers of 10, then multiply the numerical portion in the usual manner, and finally add the exponents of 10 algebraically, expressing them as a power of 10 in the product. In multiplication, the exponents (powers of 10) are added algebraically.

$$10^2 \times 10^3 = 10^{(2+3)} = 10^5$$

$$10^2 \times 10^2 \times 10^{-1} = 10^{(2+2-1)} = 10^3$$

Multiply: $\qquad\qquad$ $40,000 \times 4200$

Change to powers of 10: $4 \times 10^4 \times 4.2 \times 10^3$

Rearrange: $\qquad\qquad$ $4 \times 4.2 \times 10^4 \times 10^3$

$\qquad\qquad\qquad\qquad$ $16.8 \times 10^{(4+3)}$

$\qquad\qquad\qquad\qquad$ 16.8×10^7 or 1.68×10^8 (Answer)

Multiply: 380×0.00020

$\qquad\qquad$ $3.80 \times 10^2 \times 2.0 \times 10^{-4}$

$\qquad\qquad$ $3.80 \times 2.0 \times 10^2 \times 10^{-4}$

$\qquad\qquad$ $7.6 \times 10^{(2-4)}$

$\qquad\qquad$ 7.6×10^{-2} or 0.076 (Answer)

Multiply: $125 \times 284 \times 0.150$

$\qquad\qquad$ $1.25 \times 10^2 \times 2.84 \times 10^2 \times 1.50 \times 10^{-1}$

$\qquad\qquad$ $1.25 \times 2.84 \times 1.50 \times 10^2 \times 10^2 \times 10^{-1}$

$\qquad\qquad$ $5.325 \times 10^{(2+2-1)}$

$\qquad\qquad$ 5.32×10^3 (Answer)

In division, after changing the numbers to powers of 10, move the 10 and its exponent from the denominator to the numerator, changing the sign of the exponent. Carry out the division in the usual manner and evaluate the power of

10. The following is a proof of the equality of moving the power of 10 from the denominator to the numerator.

$$1 \times 10^{-2} = 0.01 = \frac{1}{100} = \frac{1}{10^2} = 1 \times 10^{-2}$$

In division, change the sign(s) of the exponent(s) of 10 in the denominator and move the 10 and its exponent(s) to the numerator. Then add all the exponents of 10 together. For example,

$$\frac{10^5}{10^3} = 10^5 \times 10^{-3} = 10^{(5-3)} = 10^2$$

$$\frac{10^3 \times 10^4}{10^{-2}} = 10^3 \times 10^4 \times 10^2 = 10^{(3+4+2)} = 10^9$$

Divide: $\dfrac{2871}{0.0165}$

Change to powers of 10: $\dfrac{2.871 \times 10^3}{1.65 \times 10^{-2}}$

Move 10^{-2} to the numerator, changing the sign of the exponent. This is mathematically equivalent to multiplying both numerator and denominator by 10^2.

$$\frac{2.87 \times 10^3 \times 10^2}{1.65}$$

$$\frac{2.87 \times 10^{(3+2)}}{1.65} = 1.74 \times 10^5 \quad \text{(Answer)}$$

Divide: $\dfrac{0.000585}{0.00300}$

$$\frac{5.85 \times 10^{-4}}{3.00 \times 10^{-3}}$$

$$\frac{5.85 \times 10^{-4} \times 10^3}{3.00} = \frac{5.85 \times 10^{(-4+3)}}{3.00}$$

$$1.95 \times 10^{-1} \quad \text{or} \quad 0.195 \quad \text{(Answer)}$$

Calculate: $\dfrac{760 \times 300 \times 40.0}{700 \times 273}$

$$\frac{7.60 \times 10^2 \times 3.00 \times 10^2 \times 4.00 \times 10^1}{7.00 \times 10^2 \times 2.73 \times 10^2}$$

$$\frac{7.60 \times 3.00 \times 4.00 \times 10^2 \times 10^2 \times 10^1}{7.00 \times 2.73 \times 10^2 \times 10^2}$$

$$4.77 \times 10^1 \quad \text{or} \quad 47.7 \quad \text{(Answer)}$$

10. Significant figures in calculations. The result of a calculation based on experimental measurements cannot be more precise than the measurement that has the greatest uncertainty. (See Section 2.2 for additional discussion.)

Addition and subtraction. The result of an addition or subtraction should contain no more digits to the right of the decimal point than are contained in the quantity that has the least number of digits to the right of the decimal point.

Perform the operation indicated and then round off the number to the proper number of significant figures.

$$
\begin{array}{r}
142.8 \\
18.843 \\
\underline{36.42} \\
198.063
\end{array}
\qquad
\begin{array}{r}
93.45 \\
-\ \underline{18.0} \\
75.45
\end{array}
$$

198.1 (Answer)   75.4 (Answer)

Multiplication and division. In calculations involving multiplication or division, the answer should contain the same number of significant figures as the measurement that has the least number of significant figures. In multiplication or division the position of the decimal point has nothing to do with the number of significant figures in the answer. Study the following examples:

	Round off to
$2.05 \times 2.05 = 4.2025$	4.20
$18.48 \times 5.2 = 96.096$	96
$0.0126 \times 0.020 = 0.000252$ or	
$1.26 \times 10^{-2} \times 2.0 \times 10^{-2} = 2.52 \times 10^{-4}$	2.5×10^{-4}
$\dfrac{1369.4}{41} = 33.4$	33
$\dfrac{2268}{4.20} = 540$	540

11. Dimensional analysis. Many problems of chemistry can be solved readily by dimensional analysis using the factor-label or conversion factor method. Dimensional analysis involves the use of proper units of dimension for all factors that are multiplied, divided, added, or subtracted in setting up and solving a problem. Dimensions are physical quantities such as length, mass, and time, which are expressed in such units as centimetres, grams, and seconds, respectively. In solving a problem, these units are treated mathematically just as though they were numbers, giving us an answer that contains the correct dimensional units.

A measurement or quantity given in one kind of unit can be converted to any other kind of unit having the same dimension. To convert from one kind of unit to another, the original quantity or measurement is multiplied or divided by a conversion factor. The key to success lies in choosing the correct conversion factor. This general method of calculation is illustrated in the following examples.

Suppose we want to change 24 ft to inches. We need to multiply 24 ft by a conversion factor containing feet and inches. Two such conversion factors can be written relating inches to feet.

$$\frac{12 \text{ in.}}{1 \text{ ft}} \quad \text{or} \quad \frac{1 \text{ ft}}{12 \text{ in.}}$$

We choose the factor that will mathematically cancel feet and leave the answer in inches. Note that the units are treated in the same way we treat numbers, multiplying or dividing as required. Two possibilities then arise to change 24 ft to inches:

$$24 \text{ ft} \times \frac{12 \text{ in.}}{1 \text{ ft}} \quad \text{or} \quad 24 \text{ ft} \times \frac{1 \text{ ft}}{12 \text{ in.}}$$

In the first case (the correct method), feet in the numerator and the denominator cancel, giving us an answer of 288 in. In the second case, the units of the answer are $ft^2/in.$, the answer being 2.0 $ft^2/in.$ In the first case, the answer is reasonable since it is expressed in units having the proper dimensions. That is, the dimension of length expressed in feet has been converted to length in inches according to the mathematical expression

$$\text{ft} \times \frac{\text{in.}}{\text{ft}} = \text{in.}$$

In the second case, the answer is not reasonable since the units ($ft^2/in.$) do not correspond to units of length. The answer is therefore incorrect. The units are the guiding factor for the proper conversion.

The reason we can multiply 24 ft times 12 in./ft and not change the value of the measurement is because the conversion factor is derived from two equivalent quantities. Therefore, the conversion factor 12 in./ft is equal to unity. And when you multiply any factor by 1, it does not change the value.

$$12 \text{ in.} = 1 \text{ ft} \quad \text{and} \quad \frac{12 \text{ in.}}{1 \text{ ft}} = 1$$

Convert 16 kg to milligrams. In this problem it is best to proceed in this fashion:

$$\text{kg} \longrightarrow \text{g} \longrightarrow \text{mg}$$

The possible conversion factors are

$$\frac{1000 \text{ g}}{1 \text{ kg}} \quad \text{or} \quad \frac{1 \text{ kg}}{1000 \text{ g}} \qquad \frac{1000 \text{ mg}}{1 \text{ g}} \quad \text{or} \quad \frac{1 \text{ g}}{1000 \text{ mg}}$$

We use the conversion factor that leaves the proper unit at each step for the next conversion. The calculation is

$$16 \text{ kg} \times \frac{1000 \text{ g}}{1 \text{ kg}} \times \frac{1000 \text{ mg}}{1 \text{ g}} = 1.6 \times 10^7 \text{ mg}$$

Many problems may be solved by a sequence of steps involving unit conversion factors. This sound, basic approach to problem solving, together with neat and orderly setting up of data, will lead to correct answers having the right units, fewer errors, and considerable saving of time.

12. Graphical representation of data.   A graph is often the most convenient way to present or display a set of data. Various kinds of graphs have been devised, but the most common type uses a set of horizontal and vertical coordinates to show the relationship of two variables. It is called an $x-y$ graph because the data of one variable are represented on the horizontal or x axis (abscissa) and the data of the other variable are represented on the vertical or y axis (ordinate). (See Figure I.1.)

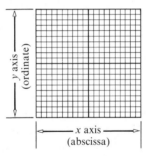

Figure I.1

As a specific example of a simple graph, let us graph the relationship between Celsius and Fahrenheit temperature scales. Assume that initially we have only the information in the following table.

°C	°F
0	32
50	122
100	212

On a set of horizontal and vertical coordinates (graph paper), scale off at least 100 Celsius degrees on the x axis and at least 212 Fahrenheit degrees on the y axis. Locate and mark the three points corresponding to the three temperatures given and draw a line connecting these points (see Figure I.2, page A-12).

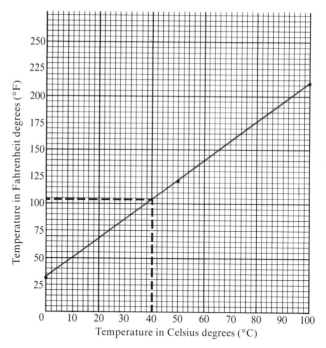

Figure I.2

Here is how a point is located on the graph: Using the 50°C–122°F data, trace a vertical line up from 50°C on the *x* axis and a horizontal line across from 122°F on the *y* axis and mark the point where the two lines intersect. This process is called *plotting*. The other two points are plotted on the graph in the same way. [*Note:* The number of degrees per scale division was chosen to give a graph of convenient size. In this case there are 5 Fahrenheit degrees per scale division and 2 Celsius degrees per scale division.]

The graph in Figure I.2 shows that the relationship between Celsius and Fahrenheit temperature is that of a straight line. The Fahrenheit temperature corresponding to any given Celsius temperature between 0 and 100° can be determined from the graph. For example, to find the Fahrenheit temperature corresponding to 40°C, trace a perpendicular line from 40°C on the *x* axis to the line plotted on the graph. Now trace a horizontal line from this point on the plotted line to the *y* axis and read the corresponding Fahrenheit temperature (104°F). See the dotted lines on Figure I.2. In turn, the Celsius temperature corresponding to any Fahrenheit temperature between 32 and 212° can be determined from the graph. This is accomplished by tracing a horizontal line from the Fahrenheit temperature to the plotted line and reading the corresponding temperature on the Celsius scale directly below the point of intersection.

The mathematical relationship of Fahrenheit and Celsius temperatures is expressed by the equation °F = 1.8°C + 32. Figure I.2 is a graph of this equation. Since the graph is a straight line, it can be extended indefinitely at either end. Any desired Celsius temperature can be plotted against the corresponding Fahrenheit temperature by extending the scales along both axes as necessary. Negative, as well as positive, values can be plotted on the graph (see Figure I.3).

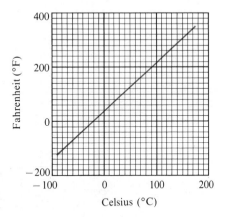

Figure I.3

Figure I.4 (page A-14) is a graph showing the solubility of potassium chlorate in water at various temperatures. The solubility curve on this graph was plotted from the data in the following table.

Temperature (°C)	Solubility (g KClO₃/100 g water)
10	5.0
20	7.4
30	10.5
50	19.3
60	24.5
80	38.5

In contrast to the Celsius–Fahrenheit temperature relationship, there is no known mathematical.equation that describes the exact relationship between temperature and the solubility of potassium chlorate. The graph in Figure I.4 was constructed from experimentally determined solubilities at the six temperatures shown. These experimentally determined solubilities are all located on the smooth curve traced by the unbroken line portion of the graph. We are therefore confident that the unbroken line represents a very good approximation of the solubility data for potassium chlorate covering the temperature range from 10 to 80°C. All points on the plotted curve represent the composition of saturated solutions. Any point below the curve represents an unsaturated solution.

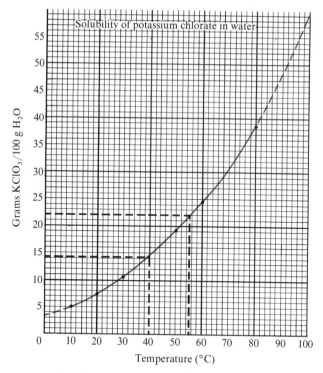

Figure I.4

The dotted-line portions of the curve are *extrapolations*; that is, they extend the curve above and below the temperature range actually covered by the plotted solubility data. Curves such as this are often extrapolated a short distance beyond the range of the known data, although the extrapolated portions may not be highly accurate. Extrapolation is justified only in the absence of more reliable information.

The graph in Figure I.4 can be used with confidence to obtain the solubility of $KClO_3$ at any temperature between 10 and 80°C, but the solubilities between 0 and 10°C and between 80 and 100°C are less reliable. For example, what is the solubility of $KClO_3$ at 55°C, at 40°C, and at 100°C?

First draw a perpendicular line from each temperature to the plotted solubility curve. Now trace a horizontal line to the solubility axis from each point on the curve and read the corresponding solubilities. The values that we read from the graph are

    55°   22.0 g $KClO_3$/100 g water

    40°   14.2 g $KClO_3$/100 g water

  100°   59 g $KClO_3$/100 g water

Of these solubilities, the one at 55°C is probably the most reliable because experimental points are plotted at 50°C and at 60°C. The 40°C solubility value is a bit less reliable because the nearest plotted points are at 30°C and 50°C. The 100°C solubility is the least reliable of the three values because it was taken from the extrapolated part of the curve, and the nearest plotted point is at 80°C. Actual handbook solubility values are 14.0 and 57.0 g of $KClO_3$/100 g of water at 40°C and 100°C, respectively.

The graph in Figure I.4 can also be used to determine whether a solution is saturated or unsaturated. For example, a solution contains 15 g of $KClO_3$/100 g of water and is at a temperature of 55°C. Is the solution saturated or unsaturated? *Answer:*   The solution is unsaturated because the point corresponding to 15 g and 55°C on the graph is below the solubility curve—and all points below the curve represent unsaturated solutions.

Appendix II
Vapor Pressure of Water at Various Temperatures

Temperature (°C)	Vapor pressure (mm Hg)	Temperature (°C)	Vapor pressure (mm Hg)
0	4.6	26	25.2
5	6.5	27	26.7
10	9.2	28	28.3
15	12.8	29	30.0
16	13.6	30	31.8
17	14.5	40	55.3
18	15.5	50	92.5
19	16.5	60	149.4
20	17.5	70	233.7
21	18.6	80	355.1
22	19.8	90	525.8
23	21.2	100	760.0
24	22.4	110	1074.6
25	23.8		

Appendix III
Units of Measurement

Prefixes and numerical values for SI units

Prefix	Symbol	Numerical value	Power of 10 equivalent
exa	E	1,000,000,000,000,000,000	10^{18}
peta	P	1,000,000,000,000,000	10^{15}
tera	T	1,000,000,000,000	10^{12}
giga	G	1,000,000,000	10^{9}
mega	M	1,000,000	10^{6}
kilo	k	1,000	10^{3}
hecto	h	100	10^{2}
deka	da	10	10^{1}
deci	d	0.1	10^{-1}
centi	c	0.01	10^{-2}
milli	m	0.001	10^{-3}
micro	μ	0.000001	10^{-6}
nano	n	0.000000001	10^{-9}
pico	p	0.000000000001	10^{-12}
femto	f	0.000000000000001	10^{-15}
atto	a	0.000000000000000001	10^{-18}

Length

1 in. = 2.54 cm
10 mm = 1 cm
100 cm = 1 m
1000 mm = 1 m
1000 m = 1 km
1 mile = 1.61 km
1 Å = 10^{-8} cm

Mass

1 lb = 453.6 g
1000 mg = 1 g
1000 g = 1 kg
1 ounce = 28.3 g
2.20 lb = 1 kg

Volume

1 mL = 1 cm^3
1000 mL = 1 litre
1 fluid ounce = 29.6 mL
1 qt = 0.946 litre
1 gal = 3.785 litres

Temperature

$°F = 1.8°C + 32$

$°C = \dfrac{(°F - 32)}{1.8}$

$K = °C + 273$

$°F\ 1.8(°C + 40) - 40$

Absolute zero = $-273.18°C$ or $-459.72°F$

Appendix IV
Solubility Table

	F^-	Cl^-	Br^-	I^-	O^{2-}	S^{2-}	OH^-	NO_3^-	CO_3^{2-}	SO_4^{2-}	$C_2H_3O_2^-$
Na^+	S	S	S	S	S	S	S	S	S	S	S
K^+	S	S	S	S	S	S	S	S	S	S	S
NH_4^+	S	S	S	S	—	S	S	S	S	S	S
Ag^+	S	I	I	I	I	I	—	S	I	I	I
Mg^{2+}	I	S	S	S	I	d	I	S	I	S	S
Ca^{2+}	I	S	S	S	I	d	I	S	I	I	S
Ba^{2+}	I	S	S	S	s	d	s	S	I	I	S
Fe^{2+}	s	S	S	S	I	I	I	S	s	S	S
Fe^{3+}	I	S	S	—	I	I	I	S	I	S	I
Co^{2+}	S	S	S	S	I	I	I	S	I	S	S
Ni^{2+}	s	S	S	S	I	I	I	S	I	S	S
Cu^{2+}	s	S	S	—	I	I	I	S	I	S	S
Zn^{2+}	s	S	S	S	I	I	I	S	I	S	S
Hg^{2+}	d	S	I	I	I	I	I	S	I	d	S
Cd^{2+}	s	S	S	S	I	I	I	S	I	S	S
Sn^{2+}	S	S	S	s	I	I	I	S	I	S	S
Pb^{2+}	I	I	I	I	I	I	I	S	I	I	S
Mn^{2+}	s	S	S	S	I	I	I	S	I	S	S
Al^{3+}	I	S	S	S	I	d	I	S	—	S	S

Key: S = soluble in water
s = slightly soluble in water
I = insoluble in water (less than 1 g/100 g H_2O)
d = decomposes in water

Appendix V
Glossary

absolute zero   See Kelvin scale.

acid   (1) A substance that produces H^+ (H_3O^+) when dissolved in water. (2) A proton donor. (3) An electron-pair acceptor. A substance that bonds to an electron pair.

acid anhydride   A nonmetal oxide that reacts with water to form an acid.

activated complex   The intermediate high-energy species formed when reactants collide. The complex can decompose to form either reactants or products.

activation energy   The amount of energy needed to form the activated complex.

Activity Series of Metals   A listing of metallic elements in descending order or reactivity.

alchemists   Practitioners of chemistry during the Middle Ages whose aims were to change the baser metals into gold and to discover the "philosopher's stone," which was thought to bring eternal youth.

alcohol   An organic compound consisting of an $-OH$ group bonded to a carbon atom in a nonaromatic hydrocarbon group.

aldehyde   An organic compound that contains the $-CHO$ group. The general formula is RCHO.

alkali metal   An element (except H) from Group IA of the periodic table.

alkaline earth metal   An element from Group IIA of the periodic table.

alkane or paraffin hydrocarbon   A straight- or branched-chain compound composed of carbon and hydrogen and having only single bonds between the atoms.

alkene   An unsaturated hydrocarbon whose molecules have at least one carbon–carbon double bond.

alkyl group or radical   An organic group derived from an alkane by removal of one H atom. The general formula is C_nH_{2n+1} (for example, CH_3- methyl). Alkyl groups are generally indicated by the letter R.

alkyl halide   A halogen atom bonded to an alkyl group. The general formula is RX (for example, CH_3Cl).

alkyne   An unsaturated hydrocarbon whose molecules have at least one carbon–carbon triple bond.

allotropy   A phenomenon in which an element exists in two or more molecular or crystalline forms. Graphite and diamond are two allotropic forms of carbon.

alpha particle   A particle emitted from a nucleus during radioactive decay; consists of a helium-4 nucleus with a mass of 4 and a charge of $+2$.

amide   An organic compound containing the $-\overset{\displaystyle O}{\overset{\displaystyle \|}{C}}-N\diagdown^{\diagup}$ group.

amino acid   An organic compound containing two functional groups—an amino group (NH_2) and a carboxyl group (COOH). Amino acids are the building blocks for proteins.

amorphous   A solid without definite crystalline form.

amphoteric substance   A substance that can react as either an acid or a base.

anion   A negatively charged ion.

anode   The electrode where oxidation occurs in an electrochemical reaction.

aromatic compound   An organic compound whose molecules contain a benzene ring or a structure resembling benzene.

artificial or induced radioactivity   Radioactivity produced during some types of transmutations. Artificial radioactive isotopes behave like natural radioactive elements in that they disintegrate in a definite fashion with a specific half-life.

atmospheric pressure   The pressure experienced by objects on the earth as a result of the layer of air surrounding our planet. A pressure of 1 atmosphere (1 atm) is the pressure that will support a column of mercury 760 mm high at 0°C.

atom   The smallest particle of an element that can enter into a chemical reaction.

atomic mass unit (amu)   A unit of mass equal to one-twelfth the mass of a carbon-12 atom.

atomic number   The number of protons in the nucleus of an atom.

atomic theory   The theory that substances are composed of atoms, and that chemical reactions are explained by the properties and the interactions of these atoms.

atomic weight   An average relative mass of the isotopes of an element referred to the atomic mass of carbon-12 as exactly 12 amu.

Avogadro's hypothesis Equal volumes of different gases at the same temperature and pressure contain equal numbers of molecules.

Avogadro's number 6.02×10^{23}; the number of formula units in 1 mole of whatever is indicated by the formula.

balanced equation A chemical equation having the same number and kind of atoms and the same electrical charge on each side of the equation.

barometer A device used to measure pressure.

base (1) A substance that produces OH^- when dissolved in water. (2) A proton acceptor. (3) An electron-pair donor.

basic anhydride A metal oxide that reacts with water to form a base.

beta particle A particle identical in charge and mass to an electron.

binary compound A compound composed of two different elements.

binding enery Energy equivalent to the mass difference between the theoretical sum of the masses of protons and neutrons in a nucleus and the actual mass of the nucleus.

biochemistry The branch of chemistry concerned with the chemical reactions occurring in living organisms.

boiling point The temperature at which the vapor pressure of a liquid is equal to the pressure above the liquid.

bond dissociation energy The energy required to break a covalent bond.

bond length The distance between two nuclei that are joined by a chemical bond.

Boyle's law At constant temperature, the volume of a given mass of gas is inversely proportional to the pressure (PV = constant).

buffer solution A solution that resists changes in pH when diluted or when small amounts of a strong acid or strong base are added.

calorie (cal) One calorie is a quantity of heat energy that will raise the temperature of 1 gram of water 1°C (from 14.5 to 15.5°C).

carbohydrate A polyhydroxy aldehyde or polyhydroxy ketone, or a compound that upon hydrolysis yields a polyhydroxy aldehyde or ketone. Sugars, starch, and cellulose are examples.

carbonyl group The structure $\diagdown\!C\!=\!O$

carboxyl group The functional group of carboxylic acids;

$$-C\overset{\displaystyle O}{\diagdown_{OH}}$$

carboxylic acid An organic compound having a carboxyl group.

catalyst A substance that influences the rate of a reaction and can be recovered in its original form at the end of the reaction.

cathode The electrode where reduction occurs in an electrochemical reaction.

cation A positively charged ion.

Celsius scale (°C) The temperature scale on which water freezes at 0°C and boils at 100°C at 1 atm pressure.

chain reaction A self-sustaining reaction where the prod-

ucts cause the reaction to continue or to increase in magnitude.

Charles' law At constant pressure, the volume of a gas is directly proportional to the absolute (K) temperature (V/T = constant).

chemical bond The attractive force that holds atoms together in a compound.

chemical change A change producing products that differ in composition from the original substances.

chemical equation An expression showing the reactants and the products of a chemical change (for example, $2 H_2 + O_2 \rightarrow 2 H_2O$).

chemical equilibrium The state in which the rate of the forward reaction equals the rate of the reverse reaction for a chemical change.

chemical formula A shorthand method for showing the composition of a compound using symbols of the elements.

chemical kinetics The study of reaction rates or the speed of a particular reaction.

chemical properties Properties of a substance related to its chemical changes.

chemistry The science dealing with the composition of substances and the transformations they undergo.

colligative properties Properties of a solution that depend on the number of solute particles in solution and not on the nature of the solute (for example, vapor pressure lowering, freezing point lowering, boiling point elevation).

combination reaction A direct union or combination of *two* substances to produce *one* new substance.

combustion In general, the process of burning or uniting a substance with oxygen, which is accompanied by the evolution of light and heat.

common ion effect A decrease in the solubility or the ionization of a solute in solution caused by the addition of an ion common to the solute in solution.

compound A substance composed of two or more elements combined in a definite proportion by weight.

concentrated solution A solution containing a relatively large amount of solute.

concentration of a solution A quantitative expression of the amount of dissolved solute in a certain quantity of solvent or solution.

conjugate acid–base Two molecules or ions whose formulas differ by one H^+. (The acid is the species with the H^+, and the base is the species without the H^+.)

coordinate covalent bond A covalent bond in which the shared pair of electrons is furnished by only one of the bonded atoms.

copolymer A polymer containing two different kinds of monomer units.

covalent bond A chemical bond formed between two atoms by sharing a pair of electrons.

critical mass The mass of a fissionable material required to sustain a nuclear chain reaction.

curie (Ci) A unit of radioactivity indicating the rate of decay of a radioactive substance: $1 \text{ Ci} = 3.7 \times 10^{10}$ disintegrations per second.

Dalton's atomic theory The first modern atomic theory to state that elements are composed of tiny, individual particles called atoms.

Dalton's law of partial pressures The total pressure in a mixture of gases is equal to the sum of the partial pressures of each gas in the mixture.

decomposition reaction A breaking down of one substance into two or more different substances.

deliquescence The absorption of water by a compound beyond the hydrate stage to form a solution.

density The mass of an object divided by its volume.

diffusion The process by which gases and liquids mix spontaneously because of the random motion of their particles.

dilute solution A solution containing a relatively small amount of solute.

dipeptide Two α-amino acids joined by a peptide linkage.

dipole A molecule with separation of charge causing it to be positive at one end and negative at the other end.

disaccharide A carbohydrate that yields two monosaccharide units when hydrolyzed.

disintegration series The spontaneous decay of a certain radioactive isotope by emission of alpha and beta particles from the nucleus, finally stopping at a stable isotope of lead or bismuth.

dissociation The process by which a salt separates into individual ions when dissolved in water.

DNA Deoxyribonucleic acid; a high-molecular-weight polymer of nucleotides, present in all living matter, that contains the genetic code and transmits hereditary characteristics.

double bond A covalent bond in which two pairs of electrons are shared.

double-replacement reaction A reaction of two compounds to produce two different compounds by exchanging the components of the reacting compounds.

efflorescence The spontaneous loss of water of hydration by a compound when exposed to air.

effusion The passage of gas through a tiny orifice from a region of high pressure to a region of lower pressure.

Einstein's mass–energy equation $E = mc^2$; the relationship between mass and energy.

electrolysis The process whereby electrical energy is used to bring about a chemical change.

electrolyte A substance whose aqueous solution conducts electricity.

electrolytic cell An electrolysis apparatus in which electrical energy from an outside source is used to produce a chemical change.

electron A subatomic particle that exists outside the nucleus and has an assigned electrical charge of -1.

electron affinity The energy released when an electron is added to an atom or an ion.

electron-dot structure (Lewis structure) A method of indicating the covalent bonds between atoms in a molecule or an ion where a pair of electrons (:) represents the valence electrons forming the covalent bond.

electronegativity The relative attraction that an atom has for the electrons in a covalent bond.

electron shell *See* energy levels of electrons.

electrovalent bond A chemical bond between a positively charged ion and a negatively charged ion.

element A basic building block of matter that cannot be broken down into simpler substances by ordinary chemical changes.

empirical formula A chemical formula that gives the smallest whole-number ratio of atoms in a compound.

endothermic reaction A chemical reaction that absorbs energy from the surroundings as it proceeds from reactants to products.

energy The capacity or ability of matter to do work.

energy levels of electrons Areas in which electrons are located at various distances from the nucleus.

energy sublevels The *s, p, d,* and *f* orbitals within a principal energy level occupied by electrons in an atom.

enzyme A protein that catalyzes a biochemical reaction.

equilibrium A dynamic state in which two or more opposing processes are taking place at the same time and at the same rate.

equilibrium constant, K_{eq} A value representing the equilibrium state of a chemical reaction involving the concentrations of the reactants and the products.

equivalent weight That weight of a substance that will react with, combine with, contain, replace, or in any other way be equivalent to 1 atomic weight of hydrogen. In oxidation–reduction reactions, an amount of substance that will lose or gain 1 mole of electrons.

essential amino acid An amino acid that is not synthesized by the body and therefore must be supplied in the diet.

ester An organic compound derived from a carboxylic acid and an alcohol. The general formula is $R-C\overset{O}{\underset{OR'}{\big<}}$

ether An organic compound having two hydrocarbon groups attached to an oxygen atom. The general formula is $R-O-R'$.

evaporation The escape of molecules from the liquid state to the gas or vapor state.

exothermic reaction A chemical reaction in which heat is released as a product.

Fahrenheit scale (°F) The temperature scale on which water freezes at $32°F$ and boils at $212°F$ at 1 atm pressure.

fats and oils Esters of fatty acids and glycerol.

fatty acids Long-chain carboxylic acids present in lipids (fats and oils).

formula weight The sum of the atomic weights of all the atoms in a chemical formula.

free radical A species having one or more unpaired electrons.

freezing or melting point The temperature at which the solid and liquid states of a substance are in equilibrium.

functional group An atom or group of atoms that characterizes a class of organic compounds. For example, $-COOH$ is the functional group of carboxylic acids.

gamma ray High-energy photons emitted by radioactive nuclei.

gas The state of matter that is the least compact of the three physical states; a gas has no shape or definite volume, and completely fills its container.

Gay-Lussac's Law of Combining Volumes of Gases At constant temperature and pressure, the ratios of the volumes of reacting gases are small whole numbers.

glycogenesis   The synthesis of glycogen from glucose in the body.

glycogenolysis   The hydrolysis, or breakdown, of glycogen to glucose in the body.

Graham's law of effusion   The rates of effusion of different gases are inversely proportional to the square root of their molecular weights or densities.

gram-atomic weight   The atomic weight of an element expressed in grams.

gram-formula weight   The formula weight of a substance expressed in grams (1 mole of a substance).

gram-molecular weight   *See* gram-formula weight.

groups or families of elements   Vertical groups of elements in the periodic table (IA, IIA, etc.). Families of elements have similar outer orbital electron structures.

half-life ($t_{1/2}$)   The time required for one-half of a specific amount of a radioactive element to disintegrate. For example, the half-life for $^{32}_{15}P$ is 14 days; the half-life for $^{14}_{6}C$ is over 5000 years.

halogen family   Group VIIA of the periodic table; consists of the elements fluorine, chlorine, bromine, iodine, and astatine.

heat   A form of energy associated with the motion of small particles of matter.

heat capacity   The quantity of heat required to change the temperature of 1 gram of any substance by 1°C.

heat of fusion   The amount of heat required to change 1 gram of a solid into a liquid at its melting point.

heat of reaction   The quantity of heat produced by a chemical reaction.

heat of vaporization   The amount of heat required to change 1 gram of a liquid to a vapor at its normal boiling point.

heterogeneous   Matter without uniform composition; two or more phases present.

homogeneous   Matter having uniform properties throughout.

homologous series   A series of compounds in which the members differ from one another by a regular increment. For example, each member of the alkane series of hydrocarbons differs by a CH_2 group.

hormone   A chemical substance that acts as a control or regulatory agent in the body. Hormones are often called *chemical messengers.*

hydrate   A substance that contains water molecules as a part of its crystalline structure. $CuSO_4 \cdot 5H_2O$ is an example.

hydrocarbon   A compound composed entirely of carbon and hydrogen.

hydrogen bond   A chemical bond between polar molecules that contain hydrogen covalently bonded to the highly electronegative atoms F, O, or N:

hydrometer   An instrument used to measure the specific gravity of a liquid. It consists of a weighted bulb at the end of a sealed calibrated tube.

hydronium ion H_3O^+; formed when an H^+ ion reacts with a water molecule.

hygroscopic substance   A substance that readily absorbs and retains water vapor.

hypothesis   A tentative explanation of the results of data to provide a basis for further experimentation.

ideal gas   A gas that obeys the gas laws and the Kinetic-Molecular Theory exactly.

ideal gas equation $PV = nRT$; a single equation relating the four variables—P, V, T, and n—used in the gas laws. R is a proportionality constant known as the ideal or universal gas constant.

immiscible   Incapable of mixing. Immiscible liquids do not form solutions with each other.

induced radioactivity   *See* artificial radioactivity.

inorganic chemistry   The chemistry of the elements and their compounds other than the carbon compounds.

ion   An electrically charged atom or group of atoms. A positively charged ($+$) ion is called a *cation* and a negatively charged ($-$) ion is called an *anion.*

ionic bond   *See* electrovalent bond.

ionization   The formation of ions.

ionization constant, K_i   The equilibrium constant for the ionization of a weak electrolyte in water.

ionization energy   The energy required to remove an electron from an atom, an ion, or a molecule.

ion product constant for water, K_w $K_w = [H^+][OH^-] = 1 \times 10^{-14}$ at 25°C.

isomerism   The phenomenon of two or more compounds having the same molecular formula but different structures.

isomers   Compounds having identical molecular formulas but different structural formulas.

isotopes   Atoms of an element having the same atomic number but different atomic masses. Since the atomic numbers are identical, isotopes vary only in the number of neutrons in the nucleus.

joule, J   The SI unit of energy; $4.184 J = 1$ cal.

Kelvin (absolute) scale (K)   Absolute temperature scale starting at absolute zero, the lowest temperature possible. Freezing and boiling points of water on this scale are 273 K and 373 K, respectively, at 1 atm pressure.

kernel of an atom   The nucleus of the atom plus all the electrons except the outermost shell of electrons.

ketone   An organic compound that contains a carbonyl group between two other carbon atoms. The general formula is $R_2C{=}O$.

kilocalorie, kcal   1000 cal; the kilocalorie is also known as the nutritional or large Calorie, used for measuring the energy produced by food.

kilogram, kg   The standard unit of mass in the metric system.

kinetic energy   Energy of motion; $KE = \frac{1}{2}mv^2$.

Kinetic-Molecular Theory   A group of assumptions used to explain the behavior and properties of ideal gas molecules.

law   A statement of the occurrence of natural phenomena that occur with unvarying uniformity under the same conditions.

Law of Conservation of Energy   Energy cannot be created or destroyed, but it may be transformed from one form to another.

Law of Conservation of Mass   There is no detectable change in the total mass of the substances in a chemical reaction; the mass of the products equals the mass of the reactants.

Law of Definite Composition   A compound always contains the same elements in a definite proportion by weight.

Le Chatelier's principle   If the conditions of an equilibrium system are altered, the system will shift to establish a new equilibrium system under the new set of conditions.

Lewis structure   *See* electron-dot structure.

limiting reagent   A reactant in a chemical reaction that limits the amount of product formed. The limitation is imposed because an insufficient quantity of the reagent, compared to amounts of other reactants, was used in the reaction.

lipids   Organic compounds found in living organisms that are water insoluble, but soluble in such fat solvents as diethyl ether, benzene, and carbon tetrachloride. Examples are fats, oils, and steroids.

liquid   One of the three physical states of matter. The particles in a liquid move about freely while the liquid still retains a definite volume. Thus, liquids flow and take the shape of their containers.

litre, L   A unit of volume commonly used in chemistry; 1 L = 1000 mL; the volume of a kilogram of water at 4°C.

macromolecule   *See* polymer.

mass   The quantity or amount of matter that an object possesses.

mass defect   The difference between the actual mass of an atom of an isotope and the calculated mass of the protons and neutrons in the nucleus of that isotope.

mass number (A)   The sum of the number of protons and neutrons in the nucleus of a given isotope of an atom.

matter   Anything that has mass and occupies space.

mechanism of a reaction   The route or steps by which a reaction takes place. The mechanism describes the manner in which atoms or molecules are transformed from reactants into products.

metal   An element that is lustrous, ductile, malleable, and a good conductor of heat and electricity. Metals tend to lose their valence electrons and become positive ions.

metalloid   An element having properties that are intermediate between those of metals and nonmetals.

metre, m   The standard unit of length in the SI and metric systems.

metric system   A decimal system of measurements.

miscible   Capable of mixing and forming a solution.

mixture   Matter containing two or more substances that can be present in variable amounts.

molality, m   The number of moles of solute dissolved in 1000 grams of solvent.

molarity, M   The number of moles of solute per litre of solution.

molar solution   A solution containing 1 mole of solute per litre of solution.

molar volume of a gas   The volume of 1 mole of a gas at STP, 22.4 L/mole.

mole   The amount of a substance containing the same number of formula units (6.02×10^{23}) as there are in exactly 12 grams of carbon-12. One mole is equal to the gram-formula weight of any substance.

molecular formula   The true formula representing the total number of atoms of each element present in one molecule of a compound.

molecular weight   The total mass of all the atoms in a molecule of a particular compound.

molecule   A small, uncharged individual unit of a compound formed by the union of two or more atoms.

mole ratio   A ratio of the number of moles of any two species in a balanced chemical equation. The mole ratio can be used as a conversion factor in stoichiometric calculations.

monomer   The small unit or units that undergo polymerization to form a polymer.

monosaccharide   A carbohydrate that cannot be hydrolyzed to simpler carbohydrate units; for example, simple sugars like glucose or fructose.

net ionic equation   A chemical equation that includes only those molecules and ions that have changed in the chemical reaction.

neutralization   The reaction of an acid and a base to form water plus a salt.

neutron   A subatomic particle that is electrically neutral and has an assigned mass of 1 amu.

noble gases   A family of elements in the periodic table—helium, neon, argon, krypton, and xenon—that contain a particularly stable electron structure.

nonelectrolyte   A substance whose aqueous solutions do not conduct electricity.

nonmetal   Any of a number of elements that do not have the characteristics of metals. They are located mainly in the upper right-hand corner of the periodic table.

nonpolar covalent bond   A covalent bond between two atoms with the same electronegativity value. Thus, the electrons are shared equally between the two atoms.

normal boiling point   The temperature at which the vapor pressure of a liquid equals 1 atm or 760 mm Hg pressure.

normality, N   The number of equivalent weights (equivalents) of solute per litre of solution.

nuclear fission   The splitting of an atom with a heavy nucleus into two or more large fragments when the nucleus is struck by a neutron. As fission occurs, energy and two or three more neutrons are released.

nuclear fusion   The uniting of two light nuclei to form one heavier nucleus, accompanied by the release of energy.

nuclear power   The energy derived from a nuclear fission or fusion reaction.

nucleons   A general term for the neutrons and protons in the nucleus of an atom.

nucleotide   The building-block unit for nucleic acids. A phosphate group, a sugar residue, and a nitrogenous organic base are bonded together to form a nucleotide.

nucleus   The central part of an atom where all the protons and neutrons of the atom are located. The nucleus is very dense and has a positive electrical charge.

octet rule (rule of eight) An atom tends to lose or gain electrons until it has eight electrons in its outer shell.

oil *See* fats and oils.

orbital A cloudlike region around the nucleus where electrons are located. Orbitals are considered to be energy sublevels within the principal levels and are labeled *s, p, d,* and *f.*

organic chemistry The branch of chemistry that deals with carbon compounds.

oxidation An increase in the oxidation number of an atom as a result of losing electrons.

oxidation number A small number representing the state of oxidation of an atom. For an ion, it is the positive or negative charge on the ion; for covalently bonded atoms, it is a positive or negative number assigned to the more electronegative atom; in free elements, it is zero.

oxidation–reduction A chemical reaction wherein electrons are transferred from one element to another.

oxidizing agent The substance that oxidizes another substance. The oxidizing agent is reduced during the course of the reaction.

peptide linkage The amide bond in a protein molecule; bonds one amino acid to another.

percentage composition of a compound The elemental composition of a chemical compound, expressed on a weight-percent basis; the weight percent of each element in a compound.

percent yield $\dfrac{\text{Actual yield}}{\text{Theoretical yield}} \times 100\%$.

periodic law The properties of the elements are a periodic function of their atomic numbers.

periodic table An arrangement of the elements according to their atomic numbers, illustrating the periodic law. The table consists of horizontal rows or periods and vertical columns or families of elements. Each period ends with a noble gas.

periods of elements The horizontal groupings of elements in the periodic table.

pH A method of expressing the H^+ concentration (acidity) of a solution. $pH = -\log[H^+]$; $pH = 7$ is a neutral solution, less than 7 is acidic and greater than 7 is basic.

phase A homogeneous part of a system separated from other parts by a physical boundary.

phenol An organic compound in which an OH group is attached to a benzene ring.

photosynthesis The process by which green plants utilize light energy to synthesize carbohydrates.

physical change A change in form (such as size, shape, physical state) without a change in composition.

physical properties Characteristics associated with the existence of a particular substance. Inherent characteristics such as color, taste, density, and melting point are physical properties of various substances.

physical states of matter Solids, liquids, and gases.

polar covalent bond A covalent bond between two atoms with differing electronegativity values resulting in unequal sharing of bonding electrons.

polyatomic ion An ion composed of more than one atom.

polyhydroxy alcohol An alcohol that has more than one OH group.

polymer (macromolecule) A natural or synthetic giant molecule formed from smaller molecules (monomers).

polymerization The process of forming large, high-molecular-weight molecules from smaller units.

polysaccharide A carbohydrate that can be hydrolyzed to many monosaccharide units; cellulose, starch, and glycogen are examples.

positron A particle with a +1 charge having the mass of an electron (a positive electron).

potential energy Stored energy or the energy an object has because of its relative position.

pressure Force per unit area; expressed in many units, such as mm Hg, atm, in./cm^2, torr.

primary alcohol An alcohol in which the carbon atom bonded to the —OH group is bonded to only one other carbon atom.

product A chemical substance produced from reactants by a chemical change.

properties The characteristics, or traits, of substances. Properties are classified as physical or chemical.

protein A polymer consisting mainly of α-amino acids linked together; occurs in all animal and vegetable matter.

proton A subatomic particle found in the nucleus of all atoms; has a charge of +1 and a mass of about 1 amu. An H^+ ion is a proton.

quantum mechanics or wave mechanics The modern theory of atomic structure based on the wave properties of matter.

rad A unit of absorbed radiation indicating the energy absorbed from any ionizing radiation. 1 rad = 0.01 joule of energy absorbed per kilogram of matter.

radioactive decay The process by which an unstable nucleus emits particles or rays and is transformed into an atom of another element.

radioactivity The spontaneous emission of radiation from the nucleus of an atom.

rate of reaction The rate at which the reactants of a chemical reaction disappear and the products form.

reactant A chemical substance entering into a reaction.

redox An abbreviation for *oxidation–reduction.*

reducing agent The substance that reduces another substance. The reducing agent is oxidized during the course of a reaction.

reduction A decrease in the oxidation number of an element as a result of gaining electrons.

rem A unit of radiation dose equivalent taking into account that the energy absorbed from different sources does not produce the same degree of biological effect.

representative element An element in one of the A groups in the periodic table.

reversible reaction A chemical reaction in which products can react to form the original reactants. A double arrow is used to indicate that a reaction is reversible.

RNA Ribonucleic acid; a high-molecular-weight polymer of nucleotides present in all living matter. Its main function is to direct the synthesis of proteins.

roentgen   A unit of exposure of gamma radiation based on the quantity of ionization produced in air.

salt   A chemical compound formed by replacing a hydrogen in an acid by a metal ion or other electropositive group.

saturated solution   A solution containing dissolved solute in equilibrium with undissolved solute.

scientific method   A method of solving problems by observation; recording and evaluating data of an experiment; formulating hypotheses and theories to explain the behavior of nature; and devising additional experiments to test the hypothesis and theories to see if they are correct.

scientific notation   A number between 1 and 10 (the decimal point after the first nonzero digit) multiplied by 10 raised to a power; for example, 6.02×10^{23}.

secondary alcohol   An alcohol in which the carbon atom bonded to the $-OH$ group is bonded to two other carbon atoms.

significant figures   The number of digits that are known plus one that is uncertain are considered significant in a measured quantity.

single bond   A covalent bond in which one pair of electrons is shared between two atoms.

single-replacement reaction   A reaction of an element and a compound to produce a different element and a different compound.

soap   A salt of a long-carbon-chain fatty acid.

solid   One of the three physical states of matter; matter in the solid state has a definite shape and a definite volume.

solubility   An amount of solute that will dissolve in a specific amount of solvent.

solubility product constant, K_{sp}   The equilibrium constant for the solubility of a slightly soluble salt.

solute   The substance that is dissolved in a solvent to form a solution.

solution   A homogeneous mixture of two or more substances.

solvent   The substance present to the largest extent in a solution. The solvent dissolves the solute.

specific gravity   The ratio of the density of one substance to the density of another substance taken as a standard. Water is usually the standard for liquids and solids; air, for gases.

spectator ion   An ion in solution that does not undergo chemical change during a chemical reaction.

standard boiling point   *See* normal boiling point.

standard conditions   *See* STP.

stoichiometry   The area of chemistry that deals with the quantitative relationships among reactants and products in a chemical reaction.

STP, standard temperature and pressure   0°C (273 K) and 1 atm (760 mm Hg).

strong electrolyte   An electrolyte that is essentially 100% ionized in aqueous solution.

subatomic particles   Mainly protons, neutrons, and electrons.

sublimation   The process of going directly from the solid state to the vapor state without becoming a liquid.

substance   Matter that is homogeneous and has a definite, fixed composition. Substances occur in two forms—as elements and as compounds.

supersaturated solution   A solution containing more solute than a saturated solution at a particular temperature. Supersaturated solutions tend to be unstable; jarring the container or dropping in a "seed" crystal will cause crystallization of the excess solute.

symbol   In chemistry, an abbreviation for the name of an element.

temperature   A measure of the intensity of heat or how hot or cold a system is.

tertiary alcohol   An alcohol in which the carbon atom bonded to the $-OH$ group is bonded to three other carbon atoms.

theoretical yield   The maximum amount of product that can be produced according to a balanced equation.

theory   A well-tested hypothesis or explanation of the results of many experiments.

titration   The process of measuring the volume of one reagent required to react with a measured weight or volume of another reagent.

total ionic equation   An equation that shows compounds in the form in which they actually exist. Strong electrolytes are written as ions in solution, whereas nonelectrolytes, weak electrolytes, precipitates, and gases are written in the un-ionized form.

transition elements   The metallic elements characterized by increasing numbers of d and f electrons in an inner shell. These elements are located in Groups IB through VIIB and in Group VIII of the periodic table.

transmutation   The conversion of one isotope into another.

transuranium element   An element having an atomic number higher than that of uranium (> 92).

triglyceride   An ester of glycerol and three fatty acids.

triple bond   A covalent bond in which three pairs of electrons are shared between two atoms.

un-ionized equation (molecular equation)   A chemical equation in which all the reactants and products are written in their molecular or normal formula expression.

unsaturated hydrocarbon   A hydrocarbon whose molecules contain one or more double or triple covalent bonds.

unsaturated solution   A solution containing less solute per unit volume than its corresponding saturated solution.

valence electrons   Electrons in the outer energy level of an atom. These electrons are primarily involved in chemical reactions.

vapor pressure   The pressure exerted by a vapor in equilibrium with its liquid.

vapor pressure curve   A graph generated by plotting the temperature of a liquid on the x axis and its vapor pressure on the y axis. Any point on the curve represents an equilibrium between the vapor and liquid.

vital force theory   A theory that held that organic substances could originate only from some form of living material. The theory was overthrown early in the 19th century.

vitamin   An organic compound essential in trace amounts in the diet for normal growth and development. Many vitamins function as coenzymes.

volatile substance   A substance that evaporates readily; a liquid with a high vapor pressure and a low boiling point.

voltaic cell   An electrochemical cell that produces an electric current from a spontaneous chemical reaction.

volume percent solution   The volume of solute in 100 mL of solution.

water of crystallization or hydration   Water molecules that are part of a crystalline structure, as in a hydrate.

weak electrolyte   A substance that is ionized to a small extent in aqueous solution.

weight   An extraneous property that an object possesses. The weight of an object depends on the gravitational attraction of the earth for that object. Therefore, an object's weight depends on its location in relation to the earth.

weight percent solution   The grams of solute in 100 g of a solution.

yield   The amount of product obtained from a chemical reaction.

Appendix VI
Answers to Problems

CHAPTER 1   **A.3** The following statements are correct: a, b, d, f, g.

CHAPTER 2   **C.6** The following statements are correct: b, c, f, g, h, i, k, m, n, o, q, u, w.
D.1 (a) 2   (b) 3   (c) 3   (d) 2   (e) 3   (f) 6   (g) 3   (h) 4
D.2 (a) 82.2   (b) 3.88   (c) 0.0385   (d) 62.2   (e) 8.94   (f) 217
(g) 25.5   (h) 1.84×10^6
D.3 (a) 3.4×10^6 (b) 2.73×10^{-2} (c) 8.8×10^{-1} (d) 3.0662×10^3
(e) 3.70×10^{-3} (f) 2.040×10^1 (g) 6.2×10^7 (h) 7.7×10^{-5}
D.4 (a) 20.8   (b) 112   (c) 2.22×10^8 (d) 23   (e) 121   (f) 349.5
(g) 0.435   (h) 2.59×10^{-7} (i) 1.69×10^{-3} (j) 3.03×10^{-4}
D.5 (a) 0.625   (b) 0.286   (c) 0.667   (d) 0.500
D.6 (a) 3.1   (b) 139   (c) 1.8   (d) 36.6   (e) 7.47   (f) 0.038
D.7 (a) 71   (b) 63   (c) 298   (d) 3.4 mL
D.8 (a) 0.160 m   (b) 0.268 km   (c) 85.2 mm   (d) 1.25×10^{-4} km
(e) 5.25×10^{-6} km   (f) 2.5×10^8 Å   (g) 6.5×10^5 Å
(h) 8.07×10^2 cm   (i) 4.0×10^3 m   (j) 29.5 cm   (k) 4.2×10^7 mm
(l) 2.4×10^{-6} cm   (m) 5.25×10^4 cm   (n) 0.4662 nm   (o) 85.3 cm
(p) 4.4×10^4 in.   (q) 652 km   (r) 6.45 cm^2 (s) 152 ft
(t) 7.76 miles   (u) 1.3×10^2 mm
D.9 (a) 4.68×10^3 mg   (b) 8.6×10^{-2} kg   (c) 4.54×10^{-3} kg   (d) 84.6 lb
(e) 0.078 g   (f) 7.25×10^5 mg   (g) 3.2×10^3 g   (h) 5.68×10^4 g
D.10 (a) 1.23×10^{-2} L   (b) 2.25×10^3 mL   (c) 5.1×10^3 mL
(d) 8.5×10^4 m^3 (e) 862 mL   (f) 10.7 gal   (g) 0.015 mL
(h) 21.2 L
D.11 (a) 89 km/hr   (b) 81 ft/second
D.12 1.86×10^5 miles/second
D.13 290 g
D.14 (a) 4.7×10^4 miles/hr   (b) 7.5×10^4 km/hr
D.15 70.4 kg
D.16 0.32 g
D.17 5.0×10^2 seconds
D.18 3×10^4 mg

D.19 1.1×10^5 tons/day; 1.0×10^8 kg/day

D.20 $\$3.57 \times 10^3$

D.21 3.0×10^3 times heavier

D.22 $23

D.23 39 miles/gal

D.24 2.2×10^2 L

D.25 $20

D.26 2.7×10^4 cm^3

D.27 Less; 750 mL vs. 757 mL

D.28 74 L; 20 gal

D.29 2×10^5 m^2

D.30 (a) 65°C   (b) −18°C   (c) 255 K   (d) 10°F   (e) 72°F
(f) −27°C   (g) 546 K   (h) −73°C

D.31 37.0°C

D.32 −100°C; −100°C = −148°F

D.33 (a) −40°C = −40°F   (b) 11.4°F = −11.4°C

D.34 4.0×10^3 cal

D.35 4.5×10^2 cal

D.36 0.092 cal/g°C

D.37 18°C

D.38 46.8°C

D.39 1.170 g/mL

D.40 3.12 g/mL

D.41 7.1 g/mL

D.42 1.19×10^3 g

D.43 0.879 g/mL

D.44 680 g

D.45 71.9 g

D.46 A = magnesium, B = aluminum, C = silver

D.47 2.10×10^3 g

D.48 0.965 g/mL

D.49 (a) 3.20 g/mL   (b) 3.20

D.50 Less; 0.716 vs. 1.00

D.51 Ethyl alcohol; 127 mL vs. 100 mL

D.52 No. Density of bar = 17.2 g/cm^3, vs. 19.3 g/cm^3 for pure gold

D.53 4.83×10^3 cm^3; 4.83 L

D.54 Alcohol = 0.790 g/mL; metal slug = 2.7 g/mL

D.55 (a) 1.26   (b) 144 mL

CHAPTER 3   **C.18** The following statements are correct: c, g, h, i, k, m, o, p, q.

D.1 (a) −30.3°F   (b) 238 K

D.2 58.9°C

D.3 78.7 g Hg

D.4 62% Fe

D.5 (a) 6.9 g O   (b) 60.3% Mg

D.6 (a) 1.2×10^{14} cal   (b) 4.0×10^8 gal

CHAPTER 4   **C.11** (a) 2   (b) 5   (c) 2   (d) 9   (e) 11   (f) 8   (g) 5   (h) 17
(i) 16
C.12 2
C.13 (a) 2 atoms H, 1 atom O   (b) 2 atoms H, 2 atoms O
C.15 (a) 4   (b) 12   (c) 2   (d) 9   (e) 9
C.16 (a) 10   (b) 12   (c) 24   (d) 8
C.29 The following statements are correct: c, e, g, j, k, m, n, p, r, u, v, w, y, aa.
D.1 21.2 g Cl
D.2 69% Cu; 31% Zn
D.3 13.4 g BaO
D.4 2.72 g/mL
D.5 (a) 44.4% S   (b) Ca   (c) 24.0 g S
D.6 18 carat
D.7 75% C
D.8 7.6×10^3 g Au
D.9 1.5×10^3 kg alloy

CHAPTER 5   **C.6** The following statements are correct: a, b, c, f, g, i.
C.20 The following statements are correct: a, c, d, e, f, h, i, j, k, l, n, p, q.
C.27 The following statements are correct: b, c, f, g, h, k, m.
C.38 (a) 6.02×10^{23} (b) 6.02×10^{23} (c) 1.20×10^{24} (d) 16.0
(e) 32.0
C.39 The following statements are correct: a, b, c, d.
D.1 (a) 5.1×10^4 (b) 2.74×10^{-3} (c) 1×10^{-6} (d) 3.4×10^2
D.2 $K = 2; L = 8; M = 18; N = 32; O = 50; P = 72$
D.3 (a) 16   (b) 18   (c) 14   (d) 28   (e) 118   (f) 142
D.4 (a)   5   11   B   5   6
(b) 42   96   Mo   42   54
(c) 50   118   Sn   50   68
D.5 (a) 0.23   (b) 0.0347   (c) 8.35   (d) 6.5×10^{-4}
D.6 (a) 1.16×10^{23} atoms Cd   (b) 5.8×10^{21} atoms Sr
(c) 7.71×10^{24} atoms Fe   (d) 1.9×10^{20} atoms Bi
(e) 2.5×10^{27} atoms Be   (f) 9.6×10^{20} atoms Ti
D.7 (a) 3.332×10^{-22} g   (b) 3.82×10^{-23} g   (c) 1.33×10^{-22} g
(d) 3.442×10^{-22} g
D.8 (a) 15   P   15   31.0   31.0 g 5.15×10^{-23} g
(b) 29   Cu   29   63.5   63.5 g 1.05×10^{-22} g
(c) 56   Ba   56   137.3   137.3 g 2.28×10^{-22} g
(d) 19   K   19   39.1   39.1 g 6.49×10^{-23} g
D.9 (a) 0.332 g Ag   (b) 1.6×10^{-10} g Mn   (c) 395 g Se   (d) 5.20×10^3 g W
D.10 (a) 1.20×10^3 g I   (b) 9.3 kg Pb   (c) 1.51 moles S   (d) 1.8×10^5 g Cd
(e) 5.31×10^{23} atoms V   (f) 2.41 moles Hg
D.11 (a) 6.4×10^{-2} mole Zn   (b) 9.26 g C   (c) 60 g Cl_2 (d) 20.1 moles Rb
(e) 6.4×10^{23} atoms Mg   (f) 7.91 moles Br_2
D.12 (a) 6.02×10^{23} molecules CO_2 (b) 6.02×10^{23} atoms C
(c) 1.20×10^{24} atoms O

D.13 4.8×10^{23} atoms P

D.14 4.71 g Na

D.15 190 g

D.16 5.42×10^{19} m

D.17 1.5×10^{14} dollars /person

D.18 (a) 8.3×10^{16} drops/mile^3 (b) 7.3×10^6 miles^3

D.19 7.2×10^{18} km; 4.5×10^{18} miles

D.20 (a) 7.12 cm^3 (b) 1.92 cm

E.6 B is the best bargain, at \$12.86/g, or \$400/troy oz

CHAPTER 6   **C.43** The following statements are correct: a, d, e, g, h, i, k, n, o, q, s, t, u.

CHAPTER 7   **C.34** The following statements are correct: a, b, e, h, i, j, l, o, q, r, s, t, u, v, w, x, z, cc, dd, ff, gg, kk, ll, nn.

CHAPTER 8   **24.** The following statements are correct: a, c, d, e, g, i.

CHAPTER 9   **B.6** The following statements are correct: a, b, e, g, h, k, l, n.

C.1 (a) 119.0   (b) 142.1   (c) 331.2   (d) 46.0   (e) 60.0   (f) 231.4
(g) 342.0   (h) 342.3   (i) 132.0

C.2 (a) 40.0 g   (b) 275.8 g   (c) 152.0 g   (d) 96.0 g   (e) 146.3 g
(f) 122.0 g   (g) 180.0 g   (h) 368.2 g   (i) 244.3 g

C.3 (a) 0.446 mole KOH   (b) 0.254 mole Cl_2 (c) 5.27×10^{-3} mole $CaCl_2$
(d) 0.272 mole C_2H_5OH (e) 0.0208 mole $NaNO_3$ (f) 5.2 moles FeI_3

C.4 (a) 0.32 mole Al atoms   (b) 0.121 mole I atoms   (c) 0.042 mole Cu atoms
(d) 12 moles N atoms

C.5 (a) 221 g H_2O (b) 44.1 g $CaCl_2$ (c) 828 g H_2SO_4 (d) 1.04 g $CuSO_4$
(e) 640 g O_2 (f) 4.9 g CH_3OH

C.6 (a) 1.8×10^{24} molecules O_2; 3.6×10^{24} atoms
(b) 1.3×10^{23} molecules CO_2; 4.0×10^{23} atoms
(c) 1.9×10^{23} molecules CH_4; 9.4×10^{23} atoms
(d) 3.30×10^{23} molecules HCl; 6.60×10^{23} atoms

C.7 (a) 3.27×10^{-22} g Au   (b) 2.99×10^{-23} g H_2O (c) 3.36×10^{-23} g Ne
(d) 2.82×10^{-23} g NH_3

C.8 (a) 1.661×10^{-21} mole CO_2 (b) 3.3×10^{-16} mole Mn
(c) 1.661×10^{-21} mole C_6H_6 (d) 1.4×10^{-3} mole Cl_2
(e) 6.64×10^{-24} mole C   (f) 25 moles H_2O

C.9 (a) 4.5×10^{23} atoms O   (b) 2.4×10^{23} atoms O   (c) 2.4×10^{23} atoms O
(d) 6.0×10^{24} atoms O   (e) 4.51×10^{24} atoms O
(f) 5.0×10^{16} atoms O

C.10 (a) 18.8 g Ag   (b) 1.70 g Cl   (c) 358 g N -   (d) 7.6 g O   (e) 4.7 g H

C.11 0.188 mole NH_4NO_3; 13.9 moles H_2O

C.12 6.19 moles H_2SO_4

C.13 1.41 moles HNO_3

C.14 (a) 32.9% K; 67.1% Br   (b) 27.3% C; 72.7% O
(c) 58.8% Ba; 13.8% S; 27.4% O   (d) 27.1% Na; 16.5% N; 56.5% O
(e) 34.6% Al; 61.5% O; 3.8% H   (f) 43.4% Na; 11.3% C; 45.3% O

C.15 (a) 20.5% Zn; 79.5% I   (b) 18.2% N; 9.1% H; 31.1% C; 41.5% O
(c) 30.2% Cr; 28.0% S; 41.8% O   (d) 29.5% Cr; 15.9% N; 54.5% O
(e) 26.6% K; 35.4% Cr; 38.1% O   (f) 28.2% N; 8.1% H; 20.8% P; 43.0% O

C.16 (a) K = +1   (b) C = +4   (c) Ba = +2   (d) Na = +1
(e) Al = +3   (f) Na = +1

C.17 (a) 77.7% Fe   (b) 69.9% Fe   (c) 72.3% Fe   (d) 17.0% Fe

C.18 Highest percentage Cl, $CHCl_3$; lowest, $BaCl_2$

C.19 63.2% Mn; 36.8% O

C.20 42.80% Mo; 57.20% S

C.21 (a) 86.99% Hg   (b) 64.7% O   (c) 35.0% N   (d) 6.7% H

C.22 (a) H_2O (b) N_2O_3 (c) Equal   (d) KCl   (e) $KHSO_4$
(f) Na_2CrO_4

C.23 (a) N_2O (b) NO   (c) NO_2 (d) Na_2CO_3 (e) $NaClO_4$
(f) K_2SO_3

C.24 (a) CuBr   (b) $CuBr_2$ (c) Cr_2S_3 (d) K_3PO_4 (e) $BaCr_2O_7$ (f) ZnN_2O_4

C.25 SnO_2

C.26 V_2O_3

C.27 Yes—ratio of Zn to S is 0.234 mole to 0.299 mole; need ratio of 1 : 1

C.28 No—ratio of Na to O is 3.04 moles to 1.88 moles; need ratio of 2 : 1

C.29 $C_6H_6O_2$

C.30 $C_6H_{12}O_6$

C.31 $C_9H_8O_4$

C.32 $MnI_2 \cdot 4H_2O$

C.33 (a) CH_4; CH_4 (b) CH_3; C_2H_6 (c) CH_2; C_3H_6 (d) C_5H_4; $C_{10}H_8$

CHAPTER 10   **C.18** The following statements are correct: a, d, e, f, h, i, j, l, n, o, p.

CHAPTER 11   **B.1** (a) 0.148 mole KNO_3 (b) 0.0294 mole $Ca(NO_3)_2$
(c) 44 moles $(NH_4)_2C_2O_4$ (d) 14.3 moles $NaHCO_3$
(e) 2.82×10^{-3} mole $ZnCl_2$ (f) 0.035 mole NaOH   (g) 14 moles CO_2
(h) 6.86 moles C_2H_5OH (i) 0.493 mole H_2SO_4

B.2 (a) 521 g $Fe(OH)_3$ (b) 13.1 g $NiSO_4$ (c) 3.88×10^5 g $CaCO_3$
(d) 2.40 g $HC_2H_3O_2$ (e) 430 g NH_3 (f) 470 g Bi_2S_3 (g) 1.0 g HCl
(h) 0.81 g $C_6H_{12}O_6$ (i) 1.87×10^3 g Br_2 (j) 19 g K_2CrO_4

B.3 (a) 5.0 g H_2O (b) 20.0 g HCl

B.4 (a) $\dfrac{9 \text{ moles } O_2}{2 \text{ moles } C_3H_7OH}$ (b) $\dfrac{2 \text{ moles } C_3H_7OH}{9 \text{ moles } O_2}$ (c) $\dfrac{9 \text{ moles } O_2}{6 \text{ moles } CO_2}$

(d) $\dfrac{8 \text{ moles } H_2O}{2 \text{ moles } C_3H_7OH}$ (e) $\dfrac{6 \text{ moles } CO_2}{8 \text{ moles } H_2O}$

B.5 (a) 3.8 moles O_2 (b) 0.663 mole H_2O (c) 28.0 g N_2 (d) 486 g NH_3

B.6 (a) 0.500 mole Fe_2O_3 (b) 9.1 moles O_2 (c) 8.44 moles SO_2
(d) 79.7 g SO_2 (e) 0.543 mole O_2 (f) 182 g FeS_2

B.7 6.08 moles HCl

B.8 (a) 2.52 moles H_2 (b) 184 g HCl

B.9 (a) 656 g HCl   (b) 38 g H_3PO_4

B.10 175 kg Fe

B.11 109 g steam; 253 g iron

B.12 (a) 3.6×10^2 g ethyl alcohol   (b) 9.9×10^2 g glucose

B.13 (a) 35.0 moles O_2 (b) 9.78 g CO_2 (c) 249 g CO_2

B.14 $MgCl_2$ (has greater weight percent of Cl than $CaCl_2$)

B.15 9.1×10^2 g C

B.16 73.2% CaC_2

B.17 (a) KOH limiting, HNO_3 in excess   (b) NaOH limiting, H_2SO_4 in excess
(c) $Bi(NO_3)_3$ limiting, H_2S in excess   (d) Fe limiting, H_2O in excess

B.18 (a) 3.0 moles CO_2 (b) 9.0 moles CO_2 (c) 1.8 moles CO_2
(d) 4.0 moles O_2; 6.0 moles CO_2; 8.0 moles H_2O (e) 16.5 g CO_2
(f) 60.0 g CO_2 (g) 60.0 g CO_2

B.19 (a) 8.9×10^2 g $AlBr_3$ (b) 111 g $AlBr_3$ (c) 39.8% yield

B.20 45.7 g CH_3OH

B.21 90.6% yield

B.22 3.72×10^2 kg Li_2O

B.23 14 kg of concentrated H_2SO_4

B.24 1.4×10^3 g air

B.25 54.6% $KClO_3$

B.26 The following statements are correct: a, c, d, f.

B.27 The following statements are correct: a, c, e.

CHAPTER 12
C.13 The following statements are correct: b, d, f (1, 4), h, i, n, o, p, q.

D.1 (a) 0.954 atm   (b) 28.5 in. Hg   (c) 14.0 lb/in.^{2} (d) 725 torr
(e) 966 millibar

D.2 (a) 0.024 atm   (b) 92 atm   (c) 0.980 atm   (d) 1.00 atm

D.3 (a) 214 mL   (b) 500 mL   (c) 197 mL

D.4 (a) 400 mm Hg   (b) 914 mm Hg

D.5 83 atm

D.6 (a) 5.50 L   (b) 5.14 L   (c) 4.03 L   (d) 6.55 L

D.7 590 K or 317°C

D.8 1.48×10^3 mm Hg

D.9 (a) 186 mL   (b) 337 mL

D.10 2.65 L

D.11 3.88 atm or 2.95×10^3 mm Hg

D.12 34 L

D.13 2×10^5 L

D.14 32 L

D.15 101 L

D.16 45 moles H_2; 1.6×10^2 g H_2

D.17 0.22 g H_2

D.18 29.1 g/mole

D.19 (a) 9.91 g/L   (b) 0.179 g/L   (c) 2.86 g/L   (d) 0.714 g/L

D.20 (a) 3.17 g/L   (b) 2.88 g/L

D.21 $-78°C$

D.22 48.8 g/mole

D.23 (a) 22.4 L   (b) 11.2 L   (c) 44.8 L

D.24 (a) 1100 mm Hg   (b) 0.058 mole

D.25 4.93×10^{21} molecules; 1.48×10^{22} atoms; 0.184 L; density greater at STP

D.26 722 mm Hg

D.27 0.630 g

D.28 (a) 2.1×10^2 L H_2 (b) 9.4 moles H_2SO_4

D.29 (a) 7.0 moles NH_3 (b) 5.6 moles NH_3 (c) 3.0×10^2 g NO
      (d) 0.400 L NO   (e) 1.6 L NO   (f) 89 g O_2 (g) 6.1 g NH_3

D.30 (a) 257 L O_2 (b) 187 L SO_2

D.31 (a) 0 moles CO; 2.0 moles O_2; 12 moles CO_2
      (b) 31 atm or 2.4×10^4 mm Hg

D.32 3 ft^3 of H_2; 1 ft^3 of CO

D.33 (a) 3.2 L CH_4 (b) The volume remains the same.

D.34 73.0% $KClO_3$

D.35 (a) 9.0 L   (b) 6.84 g CH_4 (c) 4.08 g/L   (d) 50.6 g/mole

E.4 (a) $He/CH_4 = 2:1$ (b) 67 cm from He end

E.5 C_2H_6

E.6 2.28×10^7 L SO_2

CHAPTER 13

C.41 The following statements are correct: a, b, c, f, h, l, m, o, p, s, t, u, w, y.

D.1 (a) 0.401 mole $CuSO_4 \cdot 5H_2O$ (b) 0.262 mole $FeI_2 \cdot 4H_2O$

D.2 (a) 1.17 moles H_2O (b) 1.84 moles H_2O

D.3 51.1% H_2O

D.4 1.36×10^5 cal

D.5 8.8×10^3 cal

D.6 9.0×10^4 cal

D.7 (a) 9.72 kcal/mole   (b) 5.49×10^4 cal

D.8 Not sufficient ice

D.9 Yes, sufficient steam

D.10 48.2 g H_2O

D.11 18.0 mL vs. 22.4 L

D.12 (a) 4.0 moles O_2 (b) 1.1×10^2 L

D.13 7.9 L O_2

D.14 $MgCO_3 \cdot 3H_2O$

D.15 88.1 g $NaC_2H_3O_2 \cdot 3H_2O$

D.16 (a) 18.0 g H_2O (b) 36.0 g H_2O (c) 0.460 g H_2O (d) 18.0 g H_2O
      (e) 0.321 g H_2O (f) 0.167 g H_2O

D.17 6.97×10^{18} molecules/second

D.18 54 g acid

D.19 (a) Yes; O_2 remains   (b) 15.0 mL of O_2

D.20 $ZnSO_4 \cdot 7H_2O$

CHAPTER 14   **C.28** The following statements are correct: a, b, f, h, j, k, n, p, s.

D.1 42 g NaBr

D.2 29.0 g $KClO_3$

D.3 (a) 200 g solution   (b) 453 g solution

D.4 (a) 20.0% NaCl   (b) 36% $HC_2H_3O_2$ (c) 25% $C_{12}H_{22}O_{11}$

D.5 (a) 3.8 g KCl   (b) 22.5 g NaCl   (c) 7.5 g $NaHCO_3$

D.6 (a) 2.7 g NaCl   (b) 270 g H_2O

D.7 About 8 g H_2O

D.8 40 g NaOH solution

D.9 193 g NaOH solution

D.10 (a) 2.4×10^2 g sugar   (b) 0.70 M (c) 0.825 m

D.11 (a) 441 g H_2SO_4 (b) 1.13 L

D.12 (a) 0.50 M (b) 4.6 M NaCl   (c) 1.5 M HCl   (d) 0.42 M $BaCl_2$

D.13 (a) 0.500 M Na_2CO_3 (b) 2.78 M $C_6H_{12}O_6$

(c) 1.46×10^{-3} M $Al_2(SO_4)_3$ (d) 0.104 M $Ca(NO_3)_2$

D.14 (a) 25 moles LiCl   (b) 0.15 mole H_2SO_4 (c) 2.5×10^{-4} mole NaOH

(d) 16 moles $AgNO_3$

D.15 (a) 5.9×10^3 g NaCl   (b) 9.1 g HCl   (c) 2.8×10^2 g H_2SO_4

(d) 6.57 g $Na_2C_2O_4$

D.16 (a) 1.16×10^3 mL   (b) 2.70×10^4 mL   (c) 723 mL

(d) 5.39×10^3 mL

D.17 4.50 M H_2SO_4

D.18 (a) 6.0 M HCl   (b) 0.074 M $ZnSO_4$ (c) 1.5 M HCl

D.19 (a) 250 mL   (b) 16 mL   (c) 6.2 mL   (d) 97 mL

D.20 5.1 L

D.21 (a) 0.42 M (b) 0.73 M (c) 1.2 M

D.22 360 mL

D.23 (a) 0.33 mole KCl   (b) 0.67 mole $CrCl_3$ (c) 0.24 mole $FeCl_2$

(d) 69 mL 0.060 M $K_2Cr_2O_7$ (e) 31 mL 6.0 M HCl

D.24 (a) 1.2 moles Cl_2 (b) 16 moles HCl   (c) 2.7×10^2 mL 6.0 M HCl

(d) 1.8 L Cl_2

D.25 (a) 7.60 g $BaCrO_4$ (b) 15 mL 1.0 M $BaCl_2$

D.26 (a) 0.200 mole H_2 (b) 5.02 L H_2

D.27 1.72 M HCl

D.28 $Al(OH)_3$ will neutralize more acid.

D.29 (a) 36.5 g HCl, 40.0 g NaOH   (b) 36.5 g HCl, 85.6 g $Ba(OH)_2$

(c) 49.0 g H_2SO_4, 37.0 g $Ca(OH)_2$ (d) 98.1 g H_2SO_4, 56.1 g KOH

(e) 49.0 g H_3PO_4, 23.9 g LiOH

D.30 (a) 39.1 g K   (b) 12.2 g Mg   (c) 18.6 g Fe   (d) 32.7 g Zn

(e) 63.5 g Cu   (f) 23.2 g Ga

D.31 (a) 0.769 m (b) $-1.43°C$

D.32 (a) 101.2°C   (b) 2.4 m

D.33 (a) $-10.0°C$ (b) 102.8°C   (c) 5.38 m

D.34 (a) 0.713 m (b) 1.9°C   (c) 81.9°C

D.35 258 g/mole

D.36 $C_8H_4N_2$

D.37 1000 g H_2O

D.38 (a) 6.00×10^3 g $C_2H_6O_2$ (b) 5.41 L $C_2H_6O_2$ (c) $+5.00°F$

CHAPTER 15   **C.29** The following statements are correct: a, c, g, i, j, k, l, p, r, t.

D.1 (a) 0.010 M Na^+; 0.010 M Cl^- (b) 10.0 M Na^+; 10.0 M K^+; 10.0 M SO_4^{2-}
(c) 0.55 M Zn^{2+}; 1.1 M Br^- (d) 2.50 M Al^{3+}; 3.75 M SO_4^{2-}
(e) 0.10 M Ca^{2+}; 0.20 M Cl^- (f) 0.217 M K^+; 0.217 M I^-
(g) 0.758 M NH_4^+; 0.379 M SO_4^{2-} (h) 0.0523 M Mg^{2+}; 0.105 M ClO_3^-

D.2 (a) 0.023 g Na^+; 0.035 g Cl^- (b) 23.0 g Na^+; 39.1 g K^+; 96.1 g SO_4^{2-}
(c) 3.6 g Zn^{2+}; 8.8 g Br^- (d) 6.75 g Al^{3+}; 36.0 g SO_4^{2-}
(e) 0.40 g Ca^{2+}; 0.71 g Cl^- (f) 0.848 g K^+; 2.75 g I^-
(g) 1.36 g NH_4^+; 3.64 g SO_4^{2-} (h) 0.127 g Mg^{2+}; 0.877 g ClO_3^-

D.3 0.360 M Ca^{2+}

D.4 (a) 1.0 M Na^+; 1.0 M Cl^- (b) 0.50 M Na^+; 0.50 M Cl^-
(c) 1.0 M K^+; 0.50 M Ca^{2+}; 2.0 M Cl^-
(d) 0.15 M K^+; 0.30 M Cl^-; 0.15 M H^+ (e) No ions
(f) 0.67 M Na^+; 0.67 M NO_3^-

D.5 1.0×10^4 mL

D.6 (a) 0.821 M HCl   (b) 0.716 M HCl   (c) 0.513 M HCl
(d) 0.169 M NaOH   (e) 1.03 M NaOH   (f) 0.414 M NaOH

D.7 0.362 M $Ba(OH)_2$

D.8 (a) 33.3 mL 0.300 M HCl   (b) 1.28×10^3 mL 0.300 M HCl

D.9 0.187 M HCl

D.10 1.79 moles solute; 4.33 moles solvent

D.11 (a) 159 mL benzene; 418 mL carbon tetrachloride   (b) 1.40 g/mL

D.12 (a) 0.602 L H_2 (b) 1.20 L H_2

D.13 (a) 2   (b) 0.00   (c) 8.19   (d) 7   (e) 0.30   (f) 4.0

D.14 (a) 3.68   (b) 2.82   (c) 4.21   (d) 10.47

D.15 (a) 316 mL   (b) 4.0 M $HC_2H_3O_2$ (c) 5.0 m

D.16 (a) 20.0% $BaCl_2$ (b) 1.15 M $BaCl_2$ (c) 1.20 m $BaCl_2$

D.17 11 L 18.0 M H_2SO_4

D.18 (a) 2.23 g AgI   (b) 2.35 g AgI

D.19 Basic; 0.050 mole NaOH + 0.025 mole HCl

CHAPTER 16   **C.21** The following statements are correct: c, d, g, h, j, k, l, n, q, s, t, u, v.

D.1 4.20 moles HI

D.2 (a) 3.16 moles HI   (b) 0.30 mole H_2; 0.57 mole I_2; 3.40 moles HI
(c) $K_{eq} = 56.6$

D.3 2.75 moles H_2; 0.538 mole I_2; 0.500 mole HI

D.4 0.15 mole I_2

D.5 128 times faster

D.6 HOCl, $K_i = 3.5 \times 10^{-8}$; $HC_3H_5O_2$, $K_i = 1.3 \times 10^{-5}$; HCN, $K_i = 3.9 \times 10^{-10}$

D.7 (a) $[H^+] = 1.9 \times 10^{-3}$ M (b) pH = 2.72   (c) 0.95%

D.8 $K_i = 3.7 \times 10^{-5}$

D.9 $K_i = 4 \times 10^{-10}$

D.10 (a) 0.42%; pH = 2.4   (b) 1.3%; pH = 2.9   (c) 4.2%; pH = 3.4

D.11 (a) 1.5×10^{-9} (b) 1.9×10^{-12} (c) 1.2×10^{-23} (d) 2.6×10^{-13}
(e) 1.8×10^{-10} (f) 2.4×10^{-5} (g) 5.13×10^{-17} (h) 1.81×10^{-18}

D.12 (a) $6 \times 10^{-10}\,M$ (b) $3.7 \times 10^{-4}\,M$ (c) $1.1 \times 10^{-17}\,M$
(d) $5.8 \times 10^{-6}\,M$

D.13 (a) $5 \times 10^{-9}\,g/100\,mL$ (b) $4.4 \times 10^{-3}\,g/100\,mL$
(c) $2.7 \times 10^{-16}\,g/100\,mL$ (d) $5.2 \times 10^{-5}\,g/100\,mL$

D.14 $[Ca^{2+}] = 2.1 \times 10^{-4}\,M; [F^-] = 4.2 \times 10^{-4}\,M$
(b) $1.6 \times 10^{-3}\,g\,CaF_2/100\,mL\,H_2O$

D.15 (a) Precipitate   (b) Precipitate   (c) No precipitate

D.16 (a) $[H^+] = 4.5 \times 10^{-5}\,M$; pH = 4.3   (b) $[H^+] = 1.8 \times 10^{-5}\,M$; pH = 4.7

D.17 (a) Initial pH = 7.00; change of pH = 5.30
(b) Initial pH = 4.74; change of pH = 0.02

D.18 (a) pH = 4; pOH = 10   (b) pH = 12; pOH = 2
(c) pH = 11.4; pOH = 2.6   (d) pH = 4.2; pOH = 9.8
(e) pH = 8.8; pOH = 5.2

CHAPTER 17   **C.17** The following statements are correct: a, c, e, g, j, k, m, p, q, r.
D.1 0.0618 mole NO
D.2 13.4 L Cl_2
D.3 34 g $KMnO_4$
D.4 79.4 mL 0.200 M $K_2Cr_2O_7$
D.5 8.33 mL 0.200 M $K_2Cr_2O_7$
D.6 84.6% pure KI
D.7 106 g Cu
D.8 0.404 L NO
D.9 0.012 mole Cl_2

CHAPTER 18   **C.26** The following statements are correct; c, e, f, g, i, l, m, o, p, r.
D.1 0.12 mg
D.2 $761
D.3 In the year A.D. 2070
D.4 Approximately 11,000 years old
D.5 18 min
D.6 (a) 0.0423 g/mole   (b) $9.3 \times 10^{11}\,cal/mole$
D.7 (a) $7.1 \times 10^{-12}\,cal/$one atom uranium
(b) $4.3 \times 10^{12}\,cal/mole$ uranium fissioning   (c) 0.08185% mass loss
D.8 (a) $1.3 \times 10^{11}\,cal/mole$ of ^3_2He formed   (b) 0.195% mass loss

CHAPTER 19   **B.56** The following statements are correct: b, f, h, j, l, n, o, q, r, t, v, x, z.

CHAPTER 20   **C.56** The following statements are correct: a, c, d, e, f, g, j, k, l, n, p, t, v, w, y, z.

Index

NAMES AND FORMULAS OF COMMON IONS

Positive Ions	
Ammonium	NH_4^+
Copper(I)	Cu^+
(Cuprous)	
Hydrogen	H^+
Potassium	K^+
Silver	Ag^+
Sodium	Na^+
Barium	Ba^{2+}
Cadmium	Cd^{2+}
Calcium	Ca^{2+}
Cobalt(II)	Co^{2+}
Copper(II)	Cu^{2+}
(Cupric)	
Iron(II)	Fe^{2+}
(Ferrous)	
Lead(II)	Pb^{2+}
Magnesium	Mg^{2+}
Manganese(II)	Mn^{2+}
Mercury(II)	Hg^{2+}
(Mercuric)	
Nickel(II)	Ni^{2+}
Tin(II)	Sn^{2+}
(Stannous)	
Zinc	Zn^{2+}
Aluminum	Al^{3+}
Antimony(III)	Sb^{3+}
Arsenic(III)	As^{3+}
Bismuth(III)	Bi^{3+}
Chromium(III)	Cr^{3+}
Iron(III)	Fe^{3+}
(Ferric)	
Titanium(III)	Ti^{3+}
(Titanous)	
Manganese(IV)	Mn^{4+}
Tin(IV)	Sn^{4+}
(Stannic)	
Titanium(IV)	Ti^{4+}
(Titanic)	
Antimony(V)	Sb^{5+}
Arsenic(V)	As^{5+}

Negative Ions	
Acetate	$C_2H_3O_2^-$
Bromate	BrO_3^-
Bromide	Br^-
Chlorate	ClO_3^-
Chloride	Cl^-
Chlorite	ClO_2^-
Cyanide	CN^-
Fluoride	F^-
Hydride	H^-
Bicarbonate	HCO_3^-
(Hydrogen carbonate)	
Bisulfate	HSO_4^-
(Hydrogen sulfate)	
Bisulfite	HSO_3^-
(Hydrogen sulfite)	
Hydroxide	OH^-
Hypochlorite	ClO^-
Iodate	IO_3^-
Iodide	I^-
Nitrate	NO_3^-
Nitrite	NO_2^-
Perchlorate	ClO_4^-
Permanganate	MnO_4^-
Thiocyanate	SCN^-
Carbonate	CO_3^{2-}
Chromate	CrO_4^{2-}
Dichromate	$Cr_2O_7^{2-}$
Oxalate	$C_2O_4^{2-}$
Oxide	O^{2-}
Peroxide	O_2^{2-}
Silicate	SiO_2^{2-}
Sulfate	SO_4^{2-}
Sulfide	S^{2-}
Sulfite	SO_3^{2-}
Arsenate	AsO_4^{3-}
Borate	BO_3^{3-}
Phosphate	PO_4^{3-}
Phosphite	PO_3^{3-}